STRUCTURAL ANALYSIS AND DESIGN

STRUCTURAL ANALYSIS AND DESIGN

Robert L. Ketter
Professor of Engineering and Applied Sciences
State University of New York at Buffalo

George C. Lee
Professor of Engineering and Applied Sciences
State University of New York at Buffalo

Sherwood P. Prawel, Jr.
Associate Professor of Engineering and Applied Sciences
State University of New York at Buffalo

McGraw-Hill Book Company

New York   St. Louis   San Francisco   Auckland   Bogotá   Düsseldorf
Johannesburg   London   Madrid   Mexico   Montreal   New Delhi
Panama   Paris   São Paulo   Singapore   Sydney   Tokyo   Toronto

STRUCTURAL ANALYSIS AND DESIGN

Copyright © 1979 by McGraw-Hill, Inc. All rights
reserved. Printed in the United States of America. No
part of this publication may be reproduced, stored in a
retrieval system, or transmitted, in any form or by any
means, electronic, mechanical, photocopying, recording, or
otherwise, without the prior written permission of the
publisher.

1234567890 FGRFGR 7832109

This book was set in Times Roman.
The editors were Frank J. Cerra and Frances A. Neal;
the cover was designed by Antonia Goldmark;
the production supervisor was Leroy A. Young.
The drawings were done by J & R Services, Inc.
Fairfield Graphics was printer and binder.

Library of Congress Cataloging in Publication Data

Ketter, Robert L date
    Structural analysis and design.

    Includes index.
    1. Structures, Theory of. 2. Structural engineering.
I. Lee, George C., joint author. II. Prawel, Sherwood P., date
joint author. III. Title.
TA645.K47   624′.17   78-14942
ISBN 0-07-034291-1

CONTENTS

PREFACE

There is no real consensus on the approach that should be used to give a true appreciation of the significant factors in structural engineering to the beginning student. In the past, most texts have presumed that it is first necessary to master most of the concepts of analysis, and only then to proceed to a consideration of the design problem. It is our belief that fundamentals are more easily understood—and therefore best learned—when *analysis* and *design* are interwoven from the outset. The early and continuous introduction of design serves to hold the attention of the student, to reinforce the concepts being discussed and, in general, to make the work more understandable and meaningful. This approach has shaped the organization of this book.

In the teaching of design, it has been past practice to sequentially restrict attention to single structural materials. Whole courses have been devoted to concrete design, steel design, timber design, etc. Each of these has been presented in terms of the nomenclature and restrictive domains of current practice as defined by applicable design codes and specifications, with little attention being given to consistency of presentation or to the use of standard symbols among the various courses. Again, this book deviates from that practice. Examples using steel, aluminum, concrete, etc., are found in almost all the chapters. More than that, general design provisions are developed which are applicable to different materials. This text, then, is essentially *code-independent*, although instructors can, if they so desire, use any codes available.

Most of the developed design requirements are written in nondimensional form, or in forms that easily can be made nondimensional. Their use in the metric system, therefore, is appropriate. A table of metric equivalents is provided, however, to facilitate the development of specific dimensional equations, should this be desired. (The current ACI Code contains many of the alternative forms for particular concrete details.)

It also is to be noted that this text differs from others in the amount of material included which has to do with structural stability. Chapter 6 is quite complete—some may say, too much so. It is our feeling that for too long this topic has been inadequately treated at the undergraduate level, and students are not appreciative of the limitations associated with many design provisions. For those who wish less coverage, there has been provided as an appendix to chapter 6 a summary of the resulting basic equations. The concepts of stability covered in the first parts of the various sections of the chapter could be introduced, and the student then referred to the appendix, if this is preferred.

This text is intended to be used in a three-semester sequence of courses. Students should have had statics, strength of materials, calculus and (if they have not taken the course previously) should be enrolled simultaneously in ordinary differential equations.

Much of the material contained in chapters 1 to 3 is a review of particular aspects of statics and strength of materials. For the most part chapters 4 and 5 contain new material for the student. It is assumed that chapters 1 to 5 will be covered in the first course of the sequence. Chapters 6 and 7 constitute the second course. (It is the opinion of the authors that all students in civil engineering, whether they wish to specialize in structures or not, should complete these first seven chapters as part of their required program of study.) For students who wish to specialize, chapters 8 and 9 are required. With appropriately selected special supplemental design problems, these nine chapters should provide a sound basis of beginning structural engineering.

The presentation of this material in this particular sequence has been class-tested over the past several years. Students have provided many valuable suggestions, and for these the authors are most grateful.

Finally, the authors wish to express their appreciation to all who have assisted them in making this work possible: to those who taught and those who learned from the various chapters; to Julio Espinoza and Tzu-Li Hsu, who did the drawings and checked a number of the calculations; and to Edna Thill, Jane Golebiewski, and to the others who typed the manuscript in its various stages of development. To all we acknowledge our gratitude.

Robert L. Ketter
George C. Lee
Sherwood P. Prawel, Jr.

METRIC EQUIVALENTS

U.S. CUSTOMARY SYSTEM

LENGTH
1 in
0.394 in
12 in = 1 ft
36 in = 3 ft = 1 yd
39.37 in

AREA
1 in^2
0.155 in^2
1000 ft^2
10.76 ft^2

VOLUME
1 in^3
0.0610 in^3
1728 in^3 = 1 ft^3
61.025 in^3

MOMENT OF INERTIA
1 in^4
0.02403 in^4

METRIC SYSTEM (mks)

LENGTH
2.54 cm
1 cm
30.5 cm = 0.3048 m
0.915 m
1 m

AREA
6.45 cm^2
1 cm^2
92.9 m^2
1 m^2

VOLUME
16.39 cm^3
1 cm^3
28,322 cm^3 = 0.02832 m^3
1 liter

MOMENT OF INERTIA
41.62 cm^4
1 cm^4

WEIGHT

1 lb	0.454 kg
2.2046 lb	1 kg
1000 lb = 1 kip	453.60 kg
2204.62 lb = 2.205 kips	1000 kg = 1 metric ton
1 lb/ft^3	16 kg m^3
100 lb/ft^3	1602 kg/m^3
62.42 lb/ft^3	1000 kg/m^3

STRESS

1 lb/in^2	0.07 kg/cm^2
1000 lb/in^2 = 1 ksi	70.3 kg/cm^2
14.225 lb/in^2	1 kg/cm^2

TEMPERATURE

0°F	−18°C
32°F	0°C
40°F	4°C
60°F	16°C
80°F	27°C
212°F	100°C

STRUCTURAL ANALYSIS AND DESIGN

INTRODUCTION

Fundamentally, engineering is an end-product-oriented discipline that is innovative, cost-conscious, and mindful of human factors. It is concerned with the creation of new entities, devices, or methods of solution: a new process, a new material, an improved power source, a more efficient arrangement of tasks to accomplish a desired goal, or a new structure. Engineering is also more often than not concerned with obtaining economical solutions. And, finally, human safety is always a key consideration.

Engineering is concerned with the use of abstract scientific ways of thinking and of defining real world problems. The use of idealizations and the development of procedures for establishing bounds within which behavior can be ascertained are part of that process.

Many problems, by their very nature, cannot be fully described—even after the fact, much less at the outset. Yet acceptable engineering solutions to these problems must be found which satisfy the defined needs. Engineering, then, frequently concerns the determination of possible solutions within a context of limited data. Intuition or judgment is a key factor in establishing possible alternative strategies, processes, or solutions. And this, too, is all a part of engineering.

Because of the variety and kinds of input that must be taken into account at each and every stage in the solution process, engineering problems are most often approached in what can be referred to as a hierarchically structured sequential decision process. Certain things must be known, and the influence of these ascertained before subsequent steps in the solution process can be approached.

Finally, emphasizing one of the concepts alluded to above, engineering is most often concerned with the development of permissible alternatives—not necessarily the identification of a unique solution. True, economics, aesthetics, or other considerations may eventually dictate a particular, singular choice from among several or even many possibilities, but those acceptable ones will have been established prior to that final stage, using normal engineering methods.†

1.1 STRUCTURAL ENGINEERING

More often than not, structural engineering encompasses the following five major areas of activity:

1. Development of a desired form and necessary associated dimensions to accomplish the intended functions of the structure in question.
2. Determination of the nature, magnitude, and distribution of forces that can be expected to exist during the anticipated life of the structure
3. Resolution of the manner in which these externally applied forces, and those other ones caused by fabrication, temperature changes, accelerations, etc., are transmitted among the various components which make up the structure (a process known as *structural analysis*)
4. Establishment of the desired construction or fabrication material from which the structure is to be made, and based on the results of the analysis, and having established the necessary safety requirements and economic considerations, determination of the necessary geometric proportions of the resulting structural components (a process known as *structural design*)
5. Construction (which includes fabrication and erection)

Although these five areas of activity appear to be independent and more or less self-contained, they are, in fact, closely related. For example, looking at items (1), (4), and (5), a particular structural form could be excluded at the outset because of fabrication problems or material availability.

In terms of influence on success, perhaps the most important stage in any design is that most often referred to as *planning*. *Givens*, such as design objectives, imposed loadings, desired margins of safety, etc., would be here noted. Alternative design possibilities also would be specified and the criteria to be used in arriving at a final solution would be defined. If constraints on design time are to be realized, then it is here that they should be clearly stated—and admitted. The "best" solution, two years after the scheduled completion date, is hardly a solution at all.

More often than not structural design is concerned with a systematic study of alternatives. A preliminary design is usually suggested by previous knowledge, intuition, or what have you. An analysis is carried out, and based on the results obtained from that or possibly several trials, an evaluation is made to determine how well the

---

† A first-time reading of the above statements concerning engineering may seem to the beginning student to be abstract, obtuse, or even confusing. For this as well as other reasons the student, while proceeding through the material contained herein, should periodically turn back and reread this brief overview to ascertain where in the total process the information being discussed lies.

design objectives have been met. This provides new knowledge and, therefore, the possibility of new trials for parts of, or for the entire structure in question. At each stage, factors such as safety, strength, and cost are examined and a decision is made as to whether or not additional trials (redesigns) are required. If one of the design objectives to be realized is, for example, "least cost of total structure," and if at the same time the member sizes available for incorporation in the structure do not vary continuously but rather exist only at discrete points on the continuum, then a number of trials may be required to establish the true minimum. Fortunately, *relative* rather than *absolute* minima problems are more often than not required.

Safety is a design objective that is almost without exception specified. Within the scope of this text, structural safety can be presumed when it can be demonstrated using acceptable engineering procedures and principles that the structure in question (*a*) will perform its intended function over its expected life, (*b*) will retain an appearance throughout that life that does not cause the public to become concerned as to its safety, and (*c*) will give visible warning signs sufficiently in advance of any danger. Moreover, it is to be understood that these danger signs referred to in (*c*) should not be evident under working-load conditions.

Cost (minimum, or within certain acceptable limits) is also an objective that is often specified or, at least, presumed. Cost can be examined in a number of ways: initial cost, and here one would be concerned with the total value or the value of one of the component parts; maintenance cost; or, in certain cases, total dollar investment over a specified period of time, including insurance premiums, depreciation, etc. Depending on the objective specified, the factors to be considered will vary. In many cases costs are directly related to (or at least strongly influenced by) the weight of the structure, and for these cases minimum-weight solutions are usually required. In certain other special situations, the cost of erection is the critical consideration and a minimization of this quantity is required. The point being made is that minimum cost, in and of itself, is not a sufficient design objective. The definition of cost must be more explicit.

Other objectives also may be prescribed. For example, if two machine parts are to operate effectively together with a minimum of distortion and wear, an imposed condition of limited deformation at their points of attachment or interaction could well be the defined design criteria. Another example: if a given structure is subjected to induced vibrations, a prohibition against the realization of resonance frequency may be appropriate.

Design constraints or conditions are also defined at the planning stage. Not only is the space to be enclosed specified, but also, to ensure harmony with surrounding architectural or natural elements, acceptable construction materials are prescribed. In like manner, to ensure a particular aesthetic effect, maximum values of certain cross-sectional dimensions of key design elements are defined or emphasized.

In closing this section, we acknowledge that the bringing into being of a truly significant structure requires more than just the knowledge and practice of structural engineering as a technology, although this is surely a necessary ingredient. Aesthetic considerations play a major part. Engineers in this creative role must assume the attitude of the philosopher, must work within the limits prescribed by society, and, at the same time, help to bring about positive social change.

1.2 LOADS

Structures are subjected to a great variety of loads. These may be due to gravity (or acceleration), wind, earthquake, water pressure, gas pressure, temperature change, support movement, friction, shrinkage, creep, vibration, impact, blast, etc.

Loads and loading conditions are most often classified as to whether they are static or dynamic. They are further subdivided according to whether they are time-dependent, time-independent, or random in nature. For example, the weight of a beam, which is the result of gravitational attraction, is considered to be static and time-independent. The influence of shrinkage in a prestressed concrete beam, on the other hand, is normally thought of as being static but time-dependent. The aerodynamic forces acting on a spacecraft are dynamic, as are the forces induced by blasts. In this latter case, however, the time duration of the loading may be exceedingly short. Earthquakes are generally irregular and therefore random in nature.

Frequently loads are subdivided into the following two groupings: dead loads and live loads. Dead loads are presumed to remain essentially constant over the intended life of the structure. (The weights of a structure and its component parts are examples of "dead loads.") Live loads, on the other hand, usually vary markedly. (Occupancy loads, wind loads, snow loads, the loads associated with movement of vehicles, are all examples of live loads, as are earthquake loads, loads due to blasts, etc.)

Dead loads The value of the dead load acting on a structure is the easiest to establish and can be determined with the greatest degree of precision. It is to be recognized, however, that such definition requires a knowledge of the materials in question and the sizes of the resulting members—something that is usually not known until the design has been completed. An experienced engineer will have a feel for the relative, if not absolute, sizes of the various members that compose the structure being designed, and the initial assumptions often will be quite close to the resulting values. In any event it is good procedure to compare the final weights with those initially assumed—and if necessary to make appropriate adjustments.

Table 1.1, a compilation of unit weights of construction and other materials, is a resource for the calculation of dead loads and affords a comparison of effective density of various construction materials.

Occupancy loads Within the dead-load versus live-load categorizations, it is normally assumed, by definition, that all loads other than the weight of the structure are live loads. Some of these, such as the weights associated with large pieces of fixed equipment, are essentially permanent and constant if the structure continues to be used for its original purpose. Other loads, as indicated earlier in this section, are random and almost unpredictable. Therefore, to ensure an adequate margin for safety most national, state, and local building codes specify for various types of occupancy minimum design values of distributed live loads. The values listed in table 1.2 are from the 1955 edition of the National Building Code of the National Board of Fire Underwriters.

Table 1.1 Unit weights of construction and other materials, lb/ft^3

Metals			Various other solids	
Aluminum	165		Brick	
Iron			pressed	140
cast	450		common	120
wrought	485		soft	100
Lead	710			
Manganese	475		Coal	
Nickel	565		anthracite	47–58
Steel	490		bituminous	40–54
Zinc	440		coke	23–32
Concrete				
Lightweight	70–110		Glass	
Semilightweight	120–130		common	156
Normal	150		plate	161
Bituminous substances			Various liquids	
Asphalt	80		Alcohol	49
Pitch	69		Vegetable oil	58
Tar	75		Water	
			maximum density	62
Minerals			ice	56
Granite, ashlar	165		100°C	60
Gypsum, alabaster	159		seawater	64
Limestone, marble	165		snow, fresh fallen	8
Sandstone, bluestone	147			
Shale, slate	175			

	Condition			
Wood	Green	Air-dry†	Earth, soils	
Ash	48	41	Clay	
Cedar, eastern	37	33	dry	63
Elm, American	54	35	in water	80
Fir, commercial white	46	27	damp, plastic	110
Fir, Douglas	38	34	Clay and gravel	
Hemlock, eastern	50	28	dry	100
Hemlock, western	41	29	in water	65
Maple, red	50	38	Sand, gravel	
Maple, sugar	56	44	dry, loose	90–105
Oak, red	64	44	dry, packed	100–120
Oak, white	63	47	wet	118–120
Pine, eastern white	36	25	in water	60
Pine, southern yellow	53	36		
Pine, western white	35	27		
Spruce, eastern	34	28		
Spruce, Sitka	33	28		
Walnut, black	58	38		

† 12 percent moisture content.

Table 1.2 Minimum uniform live loads

Occupancy	Live load, lb/ft²	Occupancy	Live load, lb/ft²
Assembly		*Institutional*	
Armories	150	Hospital, asylums,	
Assembly halls, churches,		infirmaries	
auditoriums		operating rooms	60
fixed seats	60	private rooms	40
movable seats	100	wards	40
Dance halls, dining rooms,		x-ray rooms	100
reviewing stands	100	Penal institutions	
Theaters		cell blocks	40
aisles, corridors,		corridors	100
lobbies, projection		Mercantile	
rooms	100	retail	
dressing rooms	40	first-floor rooms	100
balconies	60	upper floors	75
stage floors	150	wholesale	125
Business		*Residential*	
Offices	80	Dormitories	
Card file rooms	125	partitioned	40
		nonpartitioned	80
Educational		Dwellings	
Libraries		first floor	40
corridors	100	upper floors and	
reading rooms	60	habitable attics	30
stacks	150	uninhabitable attics	20
Schools and colleges		Hotels, motels	
classrooms	40	corridors serving	
corridors	100	public rooms	100
		guest rooms	40
Industrial		public corridors	60
Bakeries	150	public rooms	100
Factories and manu-		private corridors	40
facturing plants	125	Multifamily houses	
Foundries	600	corridors	60
Laboratories, scientific	100	private apartments	40
Printing plant			
composing room	100	*Storage*	
press	150	Parking garages	
paper storage, 50 lb/ft		passenger cars	100
of clear		trucks, 3 to 10 tons	150
story height		trucks, with load,	
		more than 10 tons	200
		warehouse, light	125
		warehouse, heavy	250

Source: Adapted from the National Board of Fire Underwriters, National Building Code, 1955.

For purposes of design, live loads should be positioned to give maximum effect on each of the elements being examined. This may require separate consideration of partial loadings, alternate span loadings, various combinations of different systems of loads, etc.

In many cases it is reasonable to assume for design purposes that all spaces are simultaneously loaded maximally—although the probability of this occurring is almost zero. When assumption is made, many design codes provide for a reduction in applied live loading provided certain conditions are met. For example, in the National Building Code referred to above, for other than roof live loads or live loads associated with assembly halls or parking garages, and providing that the design live load is less than 100 lb/ft^2 and that the member in question supports a floor area of 150 ft^2 or more, the live load can be reduced by the least of the following percentages:

1. By 0.08 times the square feet of area supported by the member

2. By $R = 100 \dfrac{D + L}{4.33L}$

   where R = reduction in percent

   D = dead load per square foot of area supported by the member

   L = design live load per square foot of area supported by the member
3. By 60 percent

Using these provisions, it has been observed that the greatest reductions in loading will be realized in the design of columns for tall buildings.

Bridge live loads Standard highway bridge vehicle live loads in the United States are prescribed by the American Association of State Highway and Transportation Officials (AASHTO). Two general approaches are defined: one where a standard truck load is given and the other where a "lane load" is used to approximate a series of trucks. For the first of these, two separate systems of loading, designated as H and HS loadings, depending on whether two or three axles per truck are presumed, are defined. For an H-20 and HS-20 loading, the requirements would be those shown in fig. 1.1. For an H-15 or HS-15 loading 75 percent of that shown would be used. Similarly, for H-10 and HS-10, 50 percent would be the appropriate value.

Railroad bridge loadings have been prescribed in a similar fashion by the American Railway Engineering Association (AREA). Because of the decreasing number of new railway bridge designs, examples have not been included.

For both highway and railway bridge design it is understood that the prescribed loadings should be so placed that they will produce the maximum design effects for each of the conditions to be examined. That is, maximum bending may occur for one placement of the loads, maximum shear for another, and for bearing still a third case may be required. When a continuous structure is to be designed, combinations of various span loadings may have to be examined with the possibility that a zero live loading may contribute in a positive fashion.

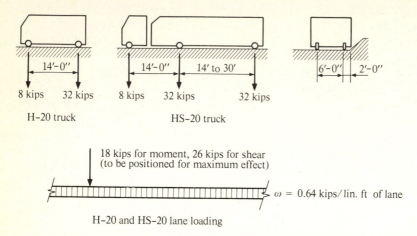

Figure 1.1 AASHTO-1969 highway H-20 and HS-20 loadings.

Snow loads Depending primarily on relative moisture content, snow has a specific gravity which varies from 0.1 for newly fallen snow to 0.2 for packed snow. This is from 6 to 12 lb/ft³. It must be recognized, however, that depths of snow accumulations on roofs and roof live loads are not readily predictable. These loads depend on the depth of snow on the ground, the velocity of the wind, the geographic relationship of the structure to other structures or natural barriers in the vicinity, the shape of the roof, the type of insulation or heating adjacent to the roof, and in the extreme case the possibility of rain or ice accumulations on top of the snow, owing to frozen drains, ice bridges, etc. Drifting or sliding of snow also occurs and, to allow for this, reductions are permitted for steeper slopes. Valleys accumulate snow, and this added factor must be taken into account. The melting of snow on one side of a roof can result in an antisymmetrical load that can be critical.

To facilitate the consideration of snow loads on structures, design maps have been prepared which give minimum acceptable snow-load values. Figure 1.2 indicates the 50-year mean recurrence intervals for snow loads on the ground in pounds per square foot of horizontal area as determined by the United States Weather Bureau. Generally, snow loads on roofs are somewhat less than those on the ground although no reliable equations have thus far been advanced relating these two quantities. Local building codes frequently require the use of this larger value in design.

Wind loads For weights of structures, occupancy loads, highway and railway live loads, and snow loads, it has been presumed that the distributed and concentrated forces act in a vertical direction. Moreover, for design purpose, these can be considered to be static loads. When taking into account the influence of wind, such assumptions may not be valid.

Wind forces on structures and structural elements result from the differential pressure that such objects cause as they inhibit the free flow of the wind. Moving air has both mass and velocity, and therefore kinetic energy. If an object is placed in its

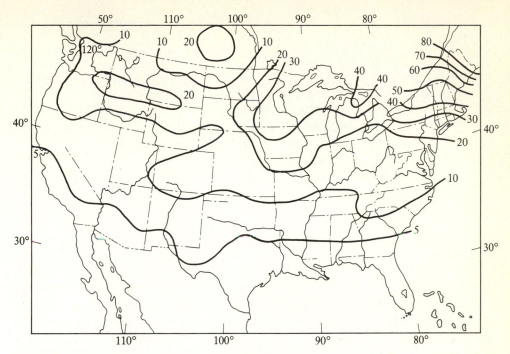

Figure 1.2 Minimum snow loads on the ground in pounds per square foot of horizontal area. *(United States Weather Bureau Map 12158. From White, Gergely, and Sexmith, "Structural Engineering," Wiley, New York, 1976.)*

path, and if the air is stopped or deflected, then all or part of that kinetic energy will be transferred into potential energy of pressure, that is, there will result wind forces on the object. The magnitude and distribution of such wind forces depend on a number of factors including the velocity and density of the air, the shape, and in some special cases the size, of the object in question, the coefficient of friction, and the angle of incidence of the wind.

It is accepted practice to treat wind forces on nonmovable structures as static live loads, providing the time variation of the wind and the natural frequencies of the structure are significantly different, and further, providing the shape and size of the structure are not such as to produce self-excitation due to vortex shedding. Transmission lines, tall stacks, hangers on bridges, and certain types of suspension bridges fall into this latter category and require special attention.

The increase in static pressure on an object (that is, the wind force, or wind load) can be determined from eq. (1.1).

$$p = C_s q = C_s (\tfrac{1}{2}\rho v^2) \tag{1.1}$$

where p = the equivalent static pressure
q = the dynamic (or velocity) pressure
ρ = the mass density of the air
v = the velocity of the air
C_s = a coefficient to account for the shape of the object, the angle of incidence of the wind, the effects of nearby topographic features, etc.

Using appropriate values for air at sea level and at a temperature of 15°C, this becomes

$$p = 0.00256 C_s V^2 \qquad (1.2)$$

where p is in pounds per square foot, V is in miles per hour, and C_s is a nondimensional shape coefficient obtained from appropriate charts or tables.

If, for example, wind is directed at 90° against the face of a rectangular, box-type, airtight building, the shape factor for the pressure on the windward face is $+0.8$. In addition, suction occurs on the leeward face, and its shape factor is -0.5. The combined effects of these are additive and, therefore, the combined influence of the wind would be an equivalent static pressure in the direction of the wind of

$$p = (0.00256)(0.8 + 0.5)V^2 = 0.00332 V^2 \qquad (1.3)$$

Using this equation a pressure of 20 lb/ft², the design value specified in many building codes in this country, would be equivalent to a wind velocity of approximately 78 mph.

Shape factors for a large number of structural forms and wind directions have been determined in wind-tunnel tests. These are summarized in the Final Report of the Task Committee on Wind Force that is published in the 1961 edition of *Transactions* of the American Society of Civil Engineers.† Certain of these cases also are shown in some more recent structural specifications and building codes.

In considering wind loads, it is important to recognize that openings in buildings such as doors, windows, vents, etc., give rise to internal pressures when winds are applied to the external surfaces. When these openings are primarily on the windward face, the internal pressures are positive and shape factors up to $+0.25$ can be realized. When the openings are essentially on the leeward face, suction occurs throughout the entire interior and its magnitude may reach a value as high as that associated with a shape factor of -0.35. These additional forces also must be taken into account in any design.

Wind velocities 30 ft above the ground and associated with a 50-year mean recurrence interval are most often referenced in design specifications. The information given in fig. 1.3 is based on statistical analysis of records of weather station data and corresponds to *open-level-country* velocities where friction is essentially uniform. For urban areas, for areas near large bodies of water, and for locations where winds are "channeled," modifications of these values are required.

Variation of wind velocity with height above the ground must be considered for tall structures. Again for open-level-country it has been observed that up to a height of approximately 1000 ft a variation according to the $\frac{1}{7}$ power law is reasonable.

$$V_z = V_{30}\left(\frac{h_z}{h_{30}}\right)^{1/7} \qquad (1.4)$$

For rough wooded country or for the outskirts of cities a 0.25 to 0.30 power variation

† Task Committee on Wind Force "Wind Forces on Structures," *Trans. Am. Soc. Civ. Eng.*, vol. 126, pt. 2, p. 1124, 1961.

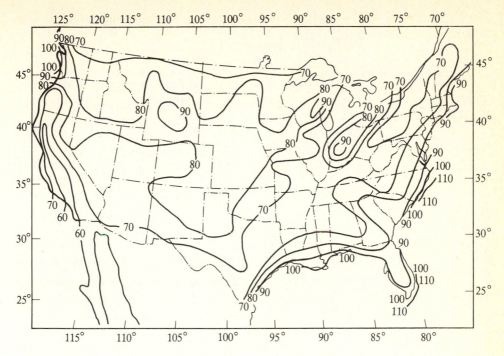

Figure 1.3 Annual extreme fastest wind speed in miles per hour, 30 ft above the ground, 50-year mean recurrence interval. *(From H. C. S. Thom, J. Struct. Div. ASCE, July 1968.)*

would be more realistic. For the center of large cities the value of the exponent may reach 0.40.

Throughout this discussion the influence of *gusts* (that is, fluctuations in velocity) have been neglected. This is reasonable for comparatively large structures, or for gusts of relatively short duration. There are times, however, when gusts become important factors; for example, in the design of highway signs. In such cases gusts are taken into account by multiplying the design velocity by a gust factor and handling the resulting velocity as if it were static. For this approach to be reasonable, however, the velocity and time duration of the gust should be such that the structure is enveloped in the mass of moving air. For a 60 mph basic wind and a 1-s gust, a 90 ft downstream length could be covered, and a gust factor of 1.3 is required. For the same 60 mph basic wind and a 10-s gust, a 900-ft downwind length would be enveloped and a factor of 1.1 should be used.

Another factor that has not been specifically noted, but one that accounts for a large number of problems is the fact that wind can exert considerable uplift forces, and these must be taken into account in design. Values as high as 40 lb/ft^2 have been observed. Again, these can be successfully predicted using appropriate shape factors.

Loads due to fluid and soil pressure Fluids, at rest, exert pressures normal to the surface in question. The magnitude of these pressures depends on the value of the unit weight of the fluid, γ, and the vertical distance from the point in question to the free surface of the fluid, h. That is

$$p = \gamma h \tag{1.5}$$

If a structure or structural element is subjected to fluid forces, and if deflection results, these fluid forces will normally continue to follow the element. Such may not be the case when structures are subjected to soil pressures.

Cohesion and friction are factors that must be taken into account when determining soil pressures on structures. The vertical pressure will be equal to the unit weight multiplied by the depth, but the lateral pressure may be significantly less than this. For example, for sand the lateral pressure may be only 60 percent of the vertical value. For cohesive clay, it may be as low as 20 percent. Moreover, it is to be recognized that the relative amount of water in the soil may be a most significant factor, and this must be known before any predictions can be made.

The significance of soil pressure is demonstrated most graphically by the fuel oil tanks that are buried, empty, and which pop out of the ground after several days or weeks of installation.

Earthquake loads Unlike the preceding cases, earthquake loading is a direct inertial response of a structure to an imposed ground motion, and not the reverse. Forces, shears, and moments develop, but these are the consequence of—and not the cause for—the deformations. Conceptually, this is illustrated in fig. 1.4.

Earthquakes have been recorded in almost all areas of the world. It has been observed, however, that they occur more frequently and with greater intensity in certain regions than they do in others. For example, one major earthquake belt encircles almost the entire Pacific Ocean. Another stretches from the Mediterranean through Southern Asia. While they may not be as severe or as destructive as ones

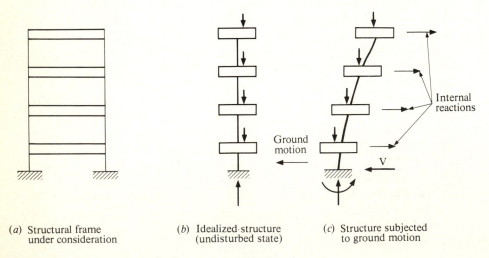

(a) Structural frame under consideration

(b) Idealized structure (undisturbed state)

(c) Structure subjected to ground motion

Figure 1.4 Response of an idealized structure to ground motions.

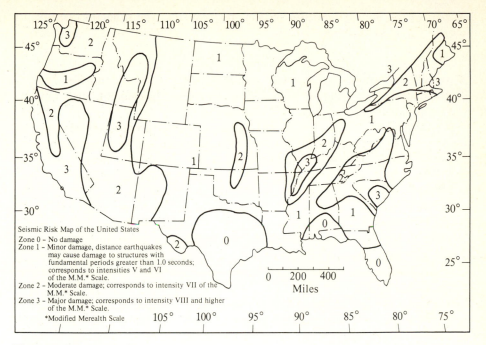

Figure 1.5 Seismic risk map.

that occur in these belts, quakes of major significance do occur in most areas of the United States. This is indicated in the seismic risk map of fig. 1.5.

Earthquakes may be of a variety of types and kinds. The most destructive travel over the surface of the earth in a more or less wave fashion. The major motion may last from several seconds to several minutes, and there may be related a number of aftershocks. Periods usually vary from 0.1 to 0.2 s, and accelerations may exceed that of gravitation. Amplitudes, that is, ground displacements, for certain extreme cases have been in excess of 9 to 10 in.

In the truest sense, earthquake analysis requires a consideration of the dynamic response of the structure in question. Factors such as the characteristics of the motion, the distribution of mass and stiffness of the structure in the directions of the imposed displacements, damping, subsoil conditions must all be taken into account, to mention only the more obvious factors. Such an analysis is most complex, but may be required for special structures and for certain high-risk areas. More often, approximate, semiempirical procedures have been developed to facilitate design. The group that has been most active in the development of these procedures is the Structural Engineers' Association of California (SEAOC).

The Uniform Building Code requirements for earthquake design, which are based on the recommendations of SEAOC, prescribe that the total seismic shear forces at the bases of the columns, acting in the directions of the major axes of the building, shall be at least equal to that specified by eq. (1.6).

$$V = ZIKCSW \qquad (1.6)$$

where Z = zone factor, as indicated in fig. 1.5 (1.0 for the areas of highest seismicity)

$\quad I$ = occupancy importance coefficient ($I \geq 1.0$. $I = 1.5$ for essential buildings)

$\quad K$ = coefficient which represents the relative ductility of the structure (for most structures varies from 0.67 to 2.5—with the smaller values associated with the more ductile structures)

$\quad C$ = the seismic coefficient (which need not exceed 0.12) = $1/15\sqrt{T}$

$\quad W$ = weight of the structure

$\quad S$ = numerical coefficient for the site-structure resonance ($S \geq 1.0$. $S = 1.5$ unless determined by more precise methods)

$\quad T$ = natural (or fundamental) period of the structure

If the natural period T is not readily known, the approximation defined in eq. (1.7) may be used.

$$T = \frac{0.05H}{\sqrt{D}} \tag{1.7}$$

where H = height of building in feet

$\quad D$ = dimension of the building in feet in the direction parallel to the applied forces

Loads due to temperature change When structures or structural elements are subjected to differential changes in temperature, relative movements take place. If these cannot freely occur, that is, if they are somehow constrained, there will develop within the various elements internal forces, shears, and moments.

The *unit elongation* of a structural member due to temperature change is governed by the relationship

$$\varepsilon = \alpha \, \Delta T \tag{1.8}$$

where ε = strain

$\quad \alpha$ = linear coefficient of thermal expansion

$\quad \Delta T$ = change in temperature

(It should be noted that the cross-sectional area of the member is not a factor.) Thermal coefficients for a variety of structural materials are given in table 1.3.

Table 1.3 Coefficients of thermal expansion α ($\times 10^{-6}/°F$)

Metals		*Structural clay masonry*	
Aluminum	12.8	Brick	4.0
Copper	9.3	Tile	3.3
Iron, wrought	6.7		
Lead	15.9	*Minerals*	
Manganese	12.0	Granite	4.4
Nickel	7.0	Limestone	4.2
Steel, mild	6.5	Sandstone	5.4
Steel, stainless	9.9	Shale	4.4
Zinc	17.3		
		Wood (parallel to the grain)	
		Fir	2.1
Concrete	5.5	Maple	3.6
Common rock aggregates	(0.5–9.0)	Oak	2.7
Hydrated Portland cement	(5.9–9.0)	Pine	3.0

Loads due to support settlement, construction misfits, etc. For statically determinate structures, support settlements and construction misfits do not result in the introduction of additional forces, shears, and moments in the member. When these occur in indeterminate structures, however, this is not the case. In fact, for certain types of structures, where less than adequate attention has been given to design details, or where the supervision of construction has been somewhat inadequate, it has been calculated that the additional shears, forces, and moments due to these nonplanned variations may exceed those caused by the design loading.

Impact factors In all the discussions above it has been presumed that the applied loads increase from zero to their maximum design value in a continuous, steady fashion. Impact loading was not considered, in spite of the fact that such loading can and does occur.

Except for dead load, almost all loading situations can have a dynamic effect on the structure, and therefore—at least conceptually—all should be adjusted to include the influence of impact. Unfortunately, dynamic analyses of other than the most simple structures are exceedingly complicated, and for this reason empirical formulas and factors to account for this effect are usually used.

The 1969 AASHTO highway bridge specification requires that all live loads be increased to account for this factor. The factor to be used is that given by eq. (1.9).

$$I = \frac{50}{L + 125} \le 30\% \tag{1.9}$$

where I = the impact factor

L = the lengths of the span which is loaded to give maximum stress in the member in question, expressed in feet

In a similar fashion and to facilitate design, the American Institute of Steel Construction (AISC) requires the following impact factors:

For supports for elevators	100%
For traveling crane support girders and their connections	25%
For supports of reciprocating machinery or power-drive units	50%
For hangers supporting floors and balconies	33%

(It should be noted that these are only a few of the conditions and factors specified.)

1.3 CLASSIFICATION AND IDEALIZATION OF STRUCTURES

Real world structures, by the very nature of the spaces that they enclose or the loading to which they are subjected, are three-dimensional. From a structural mechanics point of view, however, certain of these structures, or the individual subassemblages or components of which they are made, can frequently be examined as if they were two-dimensional, or even one-dimensional.

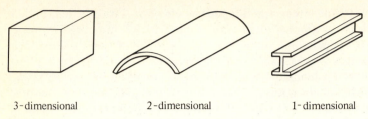

3-dimensional   2-dimensional   1-dimensional

Figure 1.6 Examples of structural components.

In general, a three-dimensional structural element has all these spatial dimensions of comparable magnitude. Examples would be concrete building blocks, spherical bearings, a gear in an automobile transmission, etc. Studies of the stress in and deformations of these bodies due to the application of external or body forces belong to the fields of *elasticity* and *plasticity*. This text will not be concerned with these types of problems.

A two-dimensional structural component or element has two of its three mutually perpendicular dimensions of comparable magnitude. By definition, the third

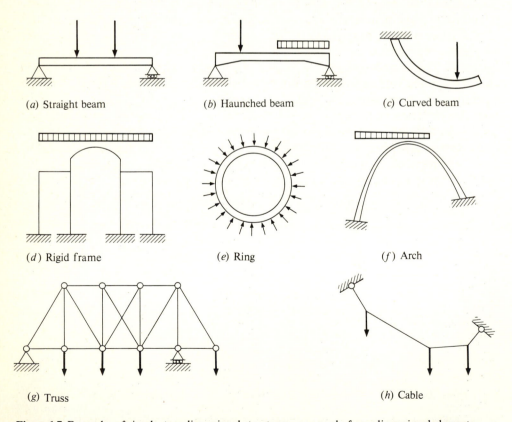

(*a*) Straight beam   (*b*) Haunched beam   (*c*) Curved beam

(*d*) Rigid frame   (*e*) Ring   (*f*) Arch

(*g*) Truss   (*h*) Cable

Figure 1.7 Examples of simple, two-dimensional structures composed of one-dimensional elements.

dimension is small by comparison. The floor of a building, the roof of a dome structure, the wing of an airplane, the bottom of a ship, the shell containment structure of a rocket are all examples of two-dimensional elements. The ones used as examples are normally studied in courses of the *plates and shells* type.

This volume is primarily concerned with the behavior of structures composed of interconnected one-dimensional members: those for which two of the dimensions are small when compared with the third. (It should be noted that in such members the two smaller dimensions may or may not be of comparable magnitudes.) Examples of one-dimensional structural components are the following: I beams, piston rods, cables, box-type columns, etc.

It must be clearly recognized and understood at the outset that, although attention will be focused on the determination of the behavior of one-dimensional structural elements, the structures which are made up of these elements are three- or two-dimensional. Furthermore, the one-dimensional structural elements (or structural components, or structural members) may be straight or curved, prismatic or nonprismatic, and may be composed of more than one material. Figures 1.7 and 1.8 contain idealized examples of two- and three-dimensional structures made up of one-dimensional elements—fig. 1.7 for two-dimensional cases, fig. 1.8 for three-dimensional ones.

It should be evident, but also should be here noted for emphasis, that both the structures and the structural elements illustrated in figs. 1.7 and 1.8 are idealizations of real situations. For example, with the exceptions of fig. 1.7a, b, c, e, and f, members have been represented as lines (with no depth or thickness), all having the same thickness. This is not to suggest a particular assumption as to relative sizes among and between the various members, but rather to indicate the general geometry of the centerline of the presumed member. In like fashion, in fig. 1.7a, b, c, e, and f the relative ratios of thickness to length shown are not important, other than to emphasize that they are small by comparison. In fig. 1.7b, it is shown to emphasize the types of haunches. In 1.7f, it is to force attention to the fact that the cross-sectional properties at the points of connection to the ground are significantly different from those at the uppermost point. There are no hard and fast rules as to when

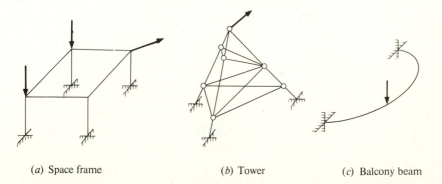

(*a*) Space frame   (*b*) Tower   (*c*) Balcony beam

Figure 1.8 Examples of three-dimensional structures composed of one-dimensional elements.

one particular representation should be used. It is at the discretion of the individual designer. The designer should remember, however, that others may—most often will—be required to check the design, and anything that will clarify the assumptions that were made is desirable.

Idealizations of the type shown also presume representation in the *undeformed state*, that is, in the original position. While under design loading, deformations are small; this is a fundamental assumption. Among other things the principle requires that loads applied be less than those which would produce buckling (or resonance frequencies) of the entire structure or any of its component parts.

Assumptions with regard to internal and external supports also are made. The most often used ones will be discussed in detail later in this section.

The loading to which a structure will be subjected is also idealized into concentrated or distributed forces. For beams, rigid frames, arches, etc., both these situations are found. For trusses, however, it is assumed—by definition—that body forces are zero, and applied loads can occur only at the points of connection of the various members, that is, at the pin joints. Such applied loads must necessarily be concentrated loads, and under these assumptions, the forces in the various members must necessarily be axial forces.

Concentrated loads are presumed to act at points, and not over distributed areas—although it is known and appreciated that in reality this could never be the case. This is possible because it has been demonstrated in a large number of laboratory tests that such an assumption is quite adequate for predicting the overall behavior of structures. (Correspondingly, it should be understood that the regions of high stress intensity and high relative deformation that occur in the immediate vicinity of the concentrated load are not considered in the material contained in this text. Books having to do with elasticity or advanced mechanics of materials deal with these types of problems.)

For two-dimensional structures, a particular assumption is normally made that may not be immediately obvious: the cross sections of the various elements must have at least one axis of symmetry, which axis also is presumed to lie in the plane of the applied loads and the support reactions. Such an assumption is required to ensure that the resulting deformations remain in the same plane.

A special assumption has been made in the rigid frame cases of figs. 1.7*d* and 1.8*a*. Here the joints are presumed *rigid*; that is, the original joint angles between the various members remain constant during loading and deformation. (This does not require that they remain in an original horizontal-vertical direction, however.)

While the balcony beam in fig. 1.8*c* is, in itself, two-dimensional owing to the fact that the applied loading is out of the plane of the structure, the system must be considered as three-dimensional. For such cases, it is necessary that the supports be capable of resisting twisting as well as bending moments.

In general, it is desirable to "reduce" or idealize three-dimensional structures or structural components to two-dimensional ones. The two-dimensional circular ring shown in fig. 1.7*e*, for example, could well be the idealization of a long cylindrical shell subjected to external pressure. If the shell is sufficiently long, this two-dimensional presentation represents an intermediate section along the shell, with negligible error. It should be evident, however, that not all three-dimensional struc-

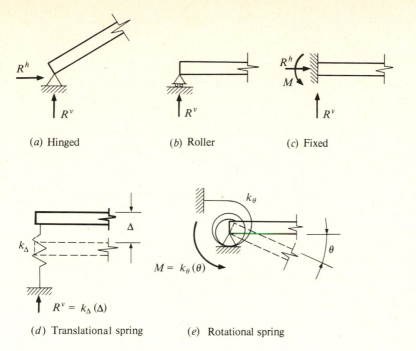

(a) Hinged   (b) Roller   (c) Fixed

(d) Translational spring   (e) Rotational spring

Figure 1.9 Two-dimensional support symbols.

tures or loadings can be so conveniently idealized. The balcony beam, the tower, and the space-frame structures, for example cannot be readily reduced to two-dimensional cases.

To facilitate analysis, idealizations also are made with regard to systems of structural supports. The more common two-dimensional support symbols and their respective assumptions are illustrated in fig. 1.9. The roller support illustrated in fig. 1.9b can resist force in only one direction, perpendicular to the indicated cross-hatched line. A hinged (or pin) support can have any two mutually perpendicular reactive force components, but no bending resistance. A built-in or fixed support, on the other hand, resists components of force in any two mutually perpendicular directions, as well as bending moment.

A support that is not perfectly rigid, or absolutely free, is usually referred to as an elastic, or spring, support. For the translational spring of fig. 1.9d, the vertical reaction (in the direction of the spring) equals $k_\Delta(\Delta)$, the product of the spring constant and the total displacement Δ. The units of the spring constant would be pounds per unit of deformation. Similarly, the magnitude of the bending moment resisting the rotational deformation shown in fig. 1.9e would be $M = k_\theta(\theta)$. That the end-support idealizations may be extended to three-dimensional cases will be demonstrated in later chapters.

It is also frequently necessary to make certain assumptions with regard to support conditions that exist in the interior of a structure. The one most often encountered is the frictionless hinge (pin). It should be noted, however, that while such

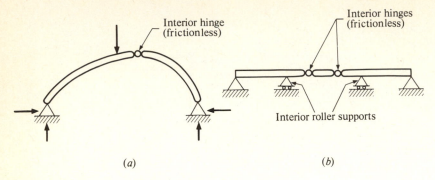

Figure 1.10 Two-dimensional structures with interior hinges.

an assumption may for certain situations be only an approximation to the actual condition that exists in the real structure, there are cases where it very closely approaches reality. Two cases of internal hinges are illustrated diagrammatically in fig. 1.10. The basic assumptions associated with interior frictionless hinges are: (*a*) that they can transmit fully both shear and axial thrust (or any other two, mutually perpendicular components of reaction), but (*b*) that they provide no resistance, whatsoever, to bending.

Individual members in a structure also can be interconnected at various points in a semirigid fashion. For these, shear and thrust are transmitted as is moment, but the magnitudes of these are dependent upon the relative deformations of the adjoining members. Since conceptually these conditions approach those illustrated in fig. 1.9*d* and *e*, no additional sketch will be here shown.

1.4 EQUILIBRIUM

Of all of the concepts that constitute the theoretical base for structural mechanics, the most important one, by any measure, is that of *equilibrium*. It is an extremely broad, all-encompassing idea that allows for consideration of both static and dynamic conditions of loading. Furthermore, it provides the conditions for determining the relationships that must necessarily exist between applied loads and resulting stresses, between stresses and stress resultants, among stress resultants, and even between internal and external energies. It is assumed that the student has had previous course work (in physics, statics, etc.) where the concept has been defined, developed, and proven. This section, then, will be only a selective review the purpose of which will be to highlight certain key points. A detailed discussion of equilibrium of stress resultants in statically determinate systems will be given in chap. 3.

Consider the two-dimensional structure and loading shown in fig. 1.11. If this structure is in a state of static equilibrium under the action of the applied forces and the presumed reactions, the resultant of all the stress resultants (both forces and

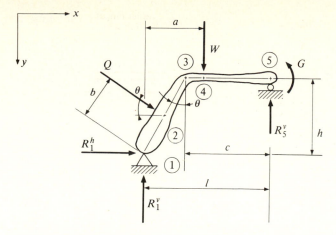

Figure 1.11 Generalized two-dimensional structure.

moments—including reactions) must vanish. In mathematical form this requirement is simply

$$\sum R = 0$$
$$\sum M_i = 0 \tag{1.10}$$

Expressing the resultant force in component form, the more normal set of equations would be

$$\sum F_x = 0$$
$$\sum F_y = 0 \tag{1.11}$$
$$\sum M_i = 0$$

where i represents any general axis perpendicular to the plane of the structure, about which the sum of the moment of the forces must vanish. Assuming W, Q, G, a, b, c, l, h, and θ are known, and R_1^v, R_1^h, and R_5^v are to be determined, the three equilibrium equations corresponding to eqs. (1.11) would be

$$\sum F_x = 0 \qquad R_1^h + Q \cos\theta = 0$$
$$\sum F_y = 0 \qquad W + Q \sin\theta - R_1^v - R_5^v = 0$$
$$\sum M_1 = 0 \qquad Qb + Wa - R_5^v(l) - G = 0$$

Since there are three unknown components of reaction and three independent equations of equilibrium, the system is determinate. (Quite arbitrarily, plus signs for components of forces were presumed when they were in the direction of the indicated coordinate. Similarly, moments were presumed positive when acting clockwise about the reference point 1.) Simultaneous solution yields the magnitudes of the three reaction components, R_1^h, R_1^v, and R_5^v in terms of the applied forces.

Although there are only three independent equilibrium equations to satisfy for any two-dimensional structure, eqs. (1.11) are not the only (nor necessarily the best) form that can be used to ensure equilibrium. For certain structures and loading, it may be more convenient to use

$$\sum F_x = 0$$
$$\sum M_A = 0 \qquad\qquad (1.12)$$
$$\sum M_B = 0$$

or

$$\sum M_A = 0$$
$$\sum M_B = 0 \qquad\qquad (1.13)$$
$$\sum M_C = 0$$

where subscripts A, B, and C represent three different moment centers in the plane of the structure. For any given situation the actual selection of the more desirable form depends on the problem being solved and the preference of the user. The only requirement is that the three selected equations be independent.

Suppose for the structure shown in fig. 1.11 that it is desired to determine the value of the bending moment, the thrust, and the shearing force just to the right of location 3. The free body for 3-4-5 shown in fig. 1.12, is selected for consideration. The three equations of static equilibrium defined in (1.11) will again be used.

$$\sum F_x = 0 \qquad P_3 = 0$$
$$\sum F_y = 0 \qquad W - V_3 - R_5^v = 0$$
$$\sum M_5 = 0 \qquad M_3 + V_3(c) - W(l - a) - G = 0$$

Again, three independent equations and three unknowns—the system is determinate.

In general, in two-dimensional structures, any *cut* can introduce no more than three unknown components of force and moment. Therefore, it is always possible to determine the internal stress resultants at a cut section by applying the three equilibrium conditions on a free body which encompasses at one extreme an end of the member or one of its supports, and further provided that the free body has only one cut.

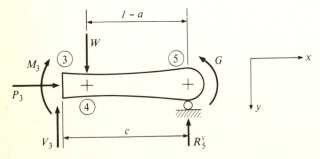

Figure 1.12 Free body diagram of beam segment 3-4-5.

The above discussion presumes certain important conditions. First, the deformations of the structure (and also the selected free bodies) are negligibly small. There are many problems for which this is a valid assumption, but it is equally true that there are situations where this is not reasonable and the changes in geometry must be taken into account. (When deformations are not considered, the resulting solutions are said to be the result of *first-order* considerations. When deflections are taken into account in the initial formulation—*second-order* considerations.) A comparatively detailed discussion and clarification of small versus large displacements and linear versus nonlinear strains will be given later. For the purposes of this preliminary discussion, however, it may be presumed that a structure or free body is in a state of static equilibrium—using the initially prescribed geometrical relationships defining the locations of the various members—if the constraints (reactions, stress resultants at a cut) permit no rigid body motion under the application of external loads.

It also has been assumed that the actual peak stresses in the immediate vicinity of concentrated applied loads or reactions are not under examination, but rather that overall effects away from the loads are significant. This idealization is often referred to as *St. Venant's principle*, and can be stated as follows:

> *A system of stress resultants may be replaced by any statically equivalent system if the point under consideration is sufficiently far removed from the points of application of the systems in question.*

(It should be recognized that this principle does not eliminate the actual local problem of stress concentrations.)

For a general three-dimensional structure defined in rectangular cartesian coordinates, where applied and reactive forces are given in their component forms, there are six independent conditions of static equilibrium to be satisfied—three forces and three moments, each with respect to the three mutually perpendicular axes.

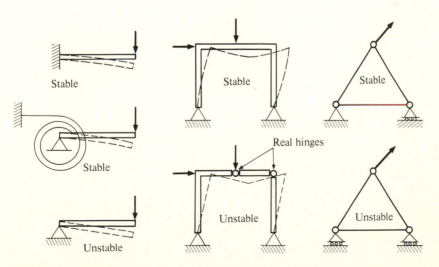

Figure 1.13 Examples of statically stable and unstable two-dimensional structures.

From the previous discussion it is to be recognized that a necessary condition for a structure (or part of a structure) to be in a state of static equilibrium is for it to be properly constrained by external supports (or internally by unknown stress resultants at a cut). A structure that satisfies this requirement is said to be in a *statically stable* equilibrium position. On the other hand, if there is an insufficient number of external or internal constraints, the structure will undergo rigid body motion under the application of loads, and is said to be statically unstable. Several examples of statically stable and unstable structures are given in fig. 1.13.

Not only geometry and supports but also the condition of the loading will influence whether or not a structure is in a state of statically stable equilibrium. At times, and under certain loads, a given structure will behave in a statically stable manner. For some other loading conditions, however, the same structure will become unstable. Figure 1.14 illustrates a beam resting on two roller supports. When the applied load Q acts at an angle as is shown in (a), the beam will tend to roll in the x direction. This system is unstable! If, on the other hand, the load Q is applied in the vertical direction as in (b), and no imperfections are presumed, there will be no tendency toward motion. This situation is usually referred to as one of *conditional stability* or *neutral equilibrium*. Figure 1.14c represents a statically stable situation under all directions of the applied load.

Stability, conditional stability (or neutral equilibrium), and instability can best be ascertained by examining the response of the system in question to an assumed virtual disturbance in the direction of the presumed problem. This concept is illustrated in fig. 1.15 where there is shown a spherical ball resting on three different types of frictionless surfaces. In (a) the ball is on a horizontal plane, with equilibrium satisfied in the vertical direction (weight of ball = reaction). In the horizontal direction, however, if the ball is gently pushed, a displacement will occur. Removing the horizontal disturbance, a new location of the ball—one different from that originally presumed—will be observed. The ball is, therefore, said to be in a state of neutral equilibrium: it can choose any one of a number of equally possible equilibrium positions. Using the same line of reasoning, the ball illustrated in fig. 1.15b is in a state of stable equilibrium—it will return to its original equilibrium position—and the one shown in (c) is statically unstable.

While it is necessary for students to be able to readily distinguish between statically stable structures and unstable ones, it is to be understood that this text will be concerned primarily with statically stable situations. Dynamic responses of structures are considered in later chapters.

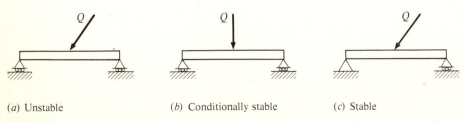

(*a*) Unstable   (*b*) Conditionally stable   (*c*) Stable

Figure 1.14 Influence of support conditions and direction of loading.

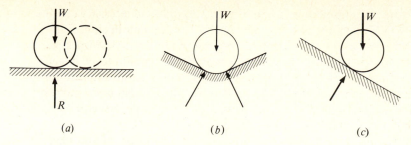

Figure 1.15 Indifference, stability, and instability.

Statically stable structures may be categorized as to whether they are *determinate* or *indeterminate*. A statically determinate structure is one for which all the unknown stress resultants, both internally and externally, can be determined by application of the static equilibrium equations above. If there are more unknowns than the number of these equations, the structure is said to be statically indeterminate. The number of unknowns in excess of the available independent static equilibrium conditions is the degree of indeterminacy or redundancy of the indeterminate structure. Figure 1.16 gives several examples of externally indeterminate structures.

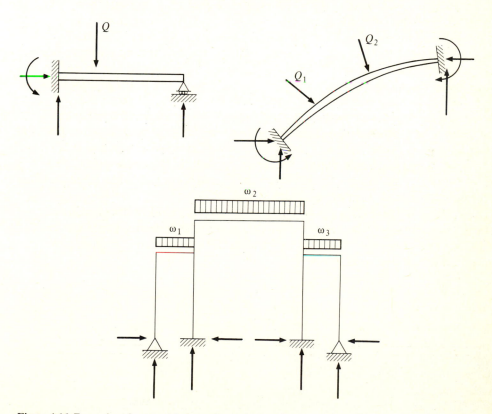

Figure 1.16 Examples of external indeterminacy (redundancy).

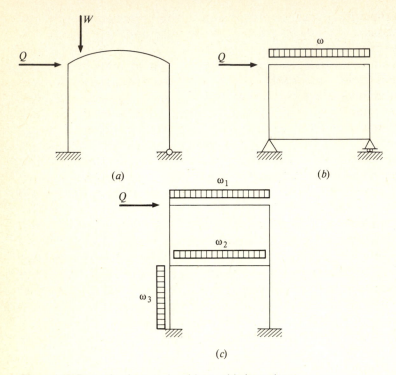

Figure 1.17 Example of external and internal indeterminacy.

When the redundant stress resultant appears in a free body of only a part of the structure, the structure is said to be *internally indeterminate*. It is to be understood that a structure may be both internally and externally indeterminate. Consider the examples shown in fig. 1.17.

A relatively simple procedure for determining the degree of redundancy of a given structure is to continuously reduce the indeterminacy by removing reactions, cutting members, etc., until finally the structure is reduced to one that is both determinate and statically stable. The number of components of reaction and internal stress resultants so removed is the degree of redundancy of the original structure.

1.5 ENGINEERING MATERIALS

The unit behavior of engineering materials is a topic that is much too broad to be extensively developed in a text of this type—one primarily concerned with the overall response of structures to imposed loading. Just to describe the basic properties would require a consideration of (at least) the mechanical, physical, electrical, chemical, magnetic, and thermal properties. For the purposes of this presentation, the mechanical properties are of primary interest. It is to be understood, however, that these properties can vary with time and temperature, and for particular problem situations, these additional factors may need to be taken into account.

Mechanical properties are concerned primarily with descriptions of the unit deformational responses (strains) of a material to imposed unit forces (stresses). A knowledge of these properties allows a prediction of the response to imposed loading of whole members (and structures) made from these materials. Strength (resistance to rupture) and stiffness (resistance to deformation) are two basic mechanical properties that are significant in many if not most design situations. Elasticity (recoverability of deformations upon removal of loads), resilience (energy absorption in the elastic range), and toughness (total energy absorption capacity) are other properties frequently tabulated. All these and other properties of this type vary with the type and kind of stress imposed—tension, compression, shear, multiaxial stress, etc. They also vary depending on the rate of application of the loading and the type and size of test specimens being examined, to mention but a few of the more obvious influencing factors.

Because of the difficulty in applying and controlling biaxial and triaxial states of stress and in measuring multidimensional deformational responses, and further, because it is possible within certain limits to predict these responses from a knowledge of uniaxial behavior, the behavior of materials subjected to simple tension or compression is most often referenced for design purposes. Moreover, *nominal* (as opposed to *true*) behavior is that condition most often described. By definition, nominal stress-strain curves are obtained by dividing the imposed loading by the original cross-sectional area, and the total elongation (or shortening) by the original gauge length.

Figure 1.18 is a nominal stress-strain curve for a material that has markedly different properties in tension and compression. Consider the tension portion of the

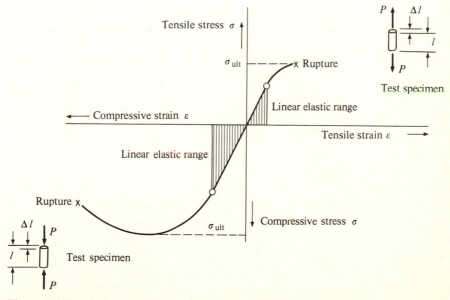

Figure 1.18 Nominal stress-strain curve for material having markedly different properties in tension and compression.

diagram and observe first a linear elastic range (shown as the shaded portion) wherein strains are proportional to stresses. They are related by Young's modulus of elasticity, E, according to the relationship

$$E = \frac{\Delta\sigma}{\Delta\varepsilon} = \frac{d\sigma}{d\varepsilon} \qquad (1.14)$$

where $\sigma = \dfrac{P}{A}$

$\varepsilon = \dfrac{\Delta l}{l}$

A = original cross-sectional area of the test specimen

(It should be noted that within this elastic range, if applied stresses are removed, strains return to zero.) When stresses are increased beyond the elastic stress limit, strains increase at an ever-increasing rate. Ultimately a particular value of stress at which rupture occurs is reached.

Beyond the elastic limit, if stresses are removed, strains decrease—but not along the path of original loading. This is illustrated in fig. 1.19. Unloading, as is noted in the diagram, is along a path parallel to the original elastic behavior. Upon complete removal of the applied stress there will be observed a permanent set. Reloading from this new position is again linear, but this time it continues almost to the point of initial unloading; that is, we see an increase in the apparent value of the limiting linear elastic stress. (It should be apparent from this discussion that the shape of the stress-strain curve can be altered by preloading the test specimens into the inelastic range prior to carrying out the actual stress-strain diagram test. Inelastic preloading will increase the range of elastic behavior—*providing* all subsequent loads are in the same direction and are of the same sense.)

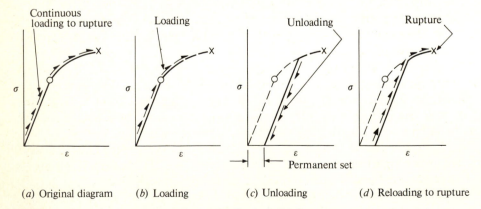

(*a*) Original diagram   (*b*) Loading   (*c*) Unloading   (*d*) Reloading to rupture

Figure 1.19 Influence of loading and unloading in the inelastic range.

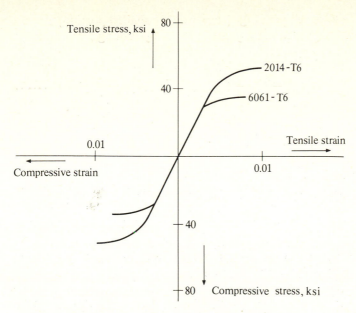

Figure 1.20 Stress-strain curves for structural aluminum alloys.

The behavior in compression of the material described in fig. 1.18 is different from that in tension in several ways. There is a significantly larger elastic range (as shown by the shaded areas). Moreover, unlike its tension counterpart, a maximum stress is reached, but this stress does not correspond to rupture. Beyond this point increasing values of strain correspond to decreasing values of stress—until rupture occurs. It is observed (*a*) that the moduli of elasticity in tension and compression are equal—and this is typical of most engineering materials; (*b*) that the resilience of the material is greater in compression than in tension—and is directly related to the increased linear range; and (*c*) that the toughness in compression is appreciably greater than it is in tension.

As examples of the stress-strain curve types shown in fig. 1.18, consider the behavior of the two structural grade aluminum alloys (2014-T6 and 6061-T6) shown in fig. 1.20. Note that the behavior in tension is for all practical purposes the same as it is in compression.

The two types of behavior illustrated in fig. 1.18 are not the only ones that engineering materials display. If the material were rubber, an elastic behavior of the type shown in fig. 1.21 would more likely be found. (Rubber is composed of inter-tangled long-chain molecules. The comparatively large range of nonlinear elastic behavior in tension is due to the straightening of these chains.)

Another quite different type of stress-strain curve is observed for mild structural steel. The tension part of the diagram is shown in fig. 1.22, and again, linear elastic behavior is observed first. A point is reached, however, and for this material it is quite sudden and pronounced, where this ceases to be the case. For only a slight increase in the load (and possibly even holding the load constant) large strains occur. This

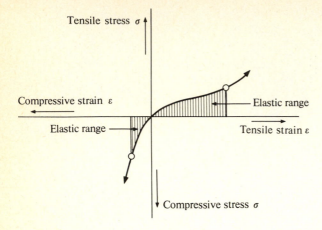

Figure 1.21 Elastic portion of stress-strain curve for rubber.

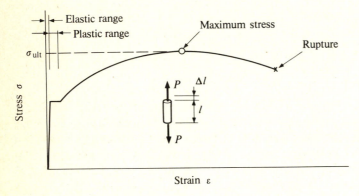

Figure 1.22 Stress-strain curve for mild structural steel.

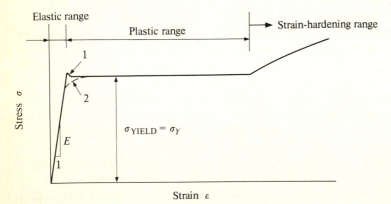

Figure 1.23 Enlarged diagram of elastic and plastic portion of stress-strain curves for mild structural steel.

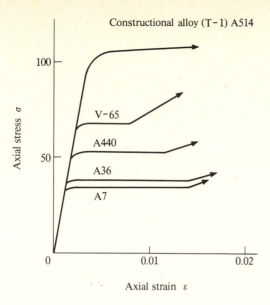

Figure 1.24 Elastic and plastic portions of uniaxial stress-strain curves for various structural steels.

plastic range may be 8 to 20 times the elastic range. It ends when *strain hardening* occurs. Thereafter both stress and strain increase, but in a nonlinear fashion, until the maximum stress is reached. Beyond this point, at one or more locations along the test specimens, necking down begins to occur, and increasing strains have associated with them decreasing values of nominal stress. Finally, rupture occurs. Because the initial portions of this diagram are the more important ones for structural design, fig. 1.23 is an enlarged diagram of the elastic and plastic ranges. It should be noted that an alternate transition curve (curve 2) has been shown connecting the elastic and plastic ranges. If the loads on the tension test specimen are not absolutely centered, or if the test were run in compression rather than in tension, the behavior indicated in curve 2 would be the one observed.

Figure 1.24 summarizes the stress-strain curves for a series of structural steels. Since the compression curves are essentially the same as the tension ones, only one set has been shown. It should be noted that the modulus of elasticity is the same for all of these.

Time of testing is significant for certain materials, for example, concrete. In fig. 1.25 is shown a series of compressive stress-strain curves for concrete. The same mix and specimen size was used in all of the tests. The only variable is the number of days from the original mix to testing. Because of this extreme variation, and because concrete develops most of its strength by that time, stress-strain properties for concrete are most often listed for tests carried out 28 days after pouring.

Concrete is a material that is also markedly influenced by the rate of loading. Using as a reference (for the purpose of nondimensionalizing the ordinate) the maximum compressive strength σ'_c, fig. 1.26 illustrates the variations that may be expected.

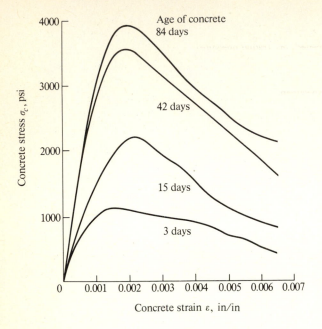

Figure 1.25 Concrete stress-strain curves: compression tests. (*From Bureau of Reclamation Report Sp-12, 1947.*)

Soils display load-deformational response properties not unlike other engineering materials. The loading, however, most often produces shear failures in the soil. Figure 1.27*a* and *b* is illustrative of cohesionless soils (sand) and cohesive soils (clay).

From figs. 1.18 to 1.27 it should be evident that there exists a large number of different types and kinds of stress-strain curves. Therefore, to facilitate tabulation of

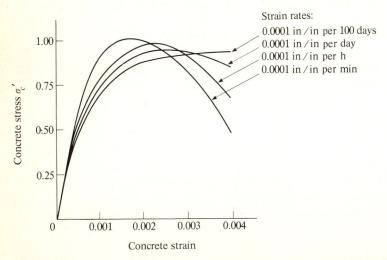

Figure 1.26 Variations in concrete stress-strain curves due to various strain rates of testing. (*Based on work by H. Rusch, J. ACI, 1970.*)

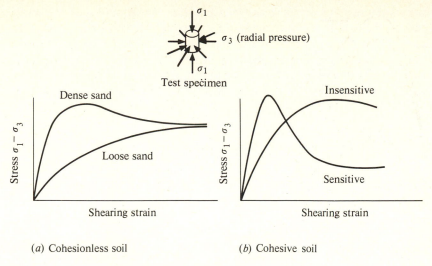

Test specimen

(*a*) Cohesionless soil   (*b*) Cohesive soil

Figure 1.27 Stress-strain curves for undisturbed soils.

critical values, a certain set of definitions has become more or less standard:

Proportional limit. The greatest stress which can be applied without a deviation from the laws of proportionality between stress and strain (Hooke's law).

Elastic limit. The greatest stress which can be applied without a permanent deformation remaining after complete removal of the stress.

Yield strength. The stress at which a material exhibits a specified limiting deviation from the proportionality of stress to strain.

Yield point. The stress for which a marked increase in strain occurs without a corresponding increase in stress.

Ultimate strength. The stress obtained by dividing the maximum load by the original cross-sectional area.

Percent elongation. The change in gauge length (after fracture) divided by the original gauge length, multiplied by 100.

While thus far it has not been specifically noted, it should be understood that when a specimen is loaded in one direction there will be associated with that loading not only a strain in the same direction, but also strains of opposite signs in perpendicular directions. The absolute value of the ratio of these lateral strains to the axial strain is known as *Poisson's ratio.* It is usually designated by the symbol μ and is a basic mechanical property of the material.

Table 1.4 lists typical mechanical properties for various construction materials at normal temperatures. Where different values are obtained for compression loading, as opposed to tensile loading, the values have been noted with (C) and (T).

It has been observed that under constant loads less than those which would be associated with rupture certain materials show a gradual change in dimensions. This flow under constant stress is referred to as *creep*. It will occur for almost all materials at elevated temperatures. More importantly, when stresses for certain materials are above the elastic limit of the material, it will even take place at room temperatures. Concrete is particularly susceptible to this type of behavior, and in some cases it is important to consider it in design. For these and other reasons, it will be discussed in detail in a later chapter.

Table 1.4 Typical mechanical properties for various materials at normal temperatures

Material	Young's modulus of elasticity E, $\times 10^6$ psi	Poisson's ratio μ	Ultimate stress σ_{ult}, $\times 10^3$ psi	Elongation at rupture, 2-in gauge length, %	Yield stress σ_y or σ_{pl},† $\times 10^3$ psi
Aluminum, 6061-T6	10.0	0.33	38	8–10	35‡
2014-T6	10.6	0.33	60	3–4	53‡
Copper, annealed	17.0	0.33	32	45	10
Lead, rolled	2.0	0.43	2.5	50	2
Nickel, annealed	30	...	70	40	20
Steel, mild, 0.2% C, hot-rolled	29–30	0.27	70	35	40
A7	29–30	0.29	60	28	33
A36	29–30	0.28	60	24	36
A440	29–30	0.27	63–70§	24	42–50§
T-1	29–30	0.26	105–135§	18	90–100§
Stainless, type 302, cold-rolled	29	0.30	140	30	100
Polycrystalline glass	12.5	0.25	2.0(B)		
Fused silica glass	10.5	0.17	4.0–1000.0		
Brick	2.0	...	4.0(C) 0.2(T)		
Granite	7.0	...	12.0(C) 1.2(T)		
Limestone	7.0	...	8.0(C) 0.8(T)		
Sandstone	3.0	...	5.0(C) 0.15(T)		
Rubber, natural	0.3	0.5	3.0	800	
Concrete, low-strength	2	0.10–0.20	2(C)		
medium-strength	3–4	0.10–0.20	3.4(C)		
high-strength	5	0.10–0.20	5(C)		
Fir	1.5	0.55	5.0		
Maple	1.6	0.39	7.5		
Oak	1.5	0.59	4.7		
Pine	1.6	0.41	6.5		

† σ_{pl} is the stress corresponding to the proportional limit.
‡ Proportional limit defined as stress at 0.2 percent of permanent-set strain.
§ Stress values vary with thickness of material.
 (B) Value determined from bending test.
 (C) Value in compression.
 (T) Value in tension.

When the loading to which a material is subjected varies in a cyclical fashion between certain prescribed minimum and maximum values, the maximums must necessarily be less than those which could be sustained under a condition of static loading. The actual value of the loading that can be sustained will depend on the type of loading (bending, axial, shear, etc.), the anticipated number of cycles of loading to be sustained, the ratio of the maximum to minimum stresses that are to be imposed, as well as the material itself. For example, if the material is a ferrous alloy, it has been observed that there can be established an *endurance limit* for the material. For stresses below this value an infinite number of cycles can be imposed. Such is not the case for nonferrous materials. For these, fatigue strengths at various stress levels must be ascertained, and the smaller the stress the larger the number of cycles that can be imposed. Actually, many factors affect the data in a fatigue test, not the least of which is the surface finish of the test specimen: fatigue cracks usually start at the surface. The atmosphere also is significant, and a corrosive environment will generally lower fatigue strength. If constant tensile stress is applied, in addition to the presumed cyclic variation, the fatigue life is lowered. On the other hand, if cyclic variation about an imposed compressive stress is a loading condition, the fatigue life will increase.

1.6 STRUCTURAL FAILURE CRITERIA AND STRUCTURAL SAFETY

A structure is said to perform adequately when it sustains the imposed loading over the structure's expected life without appreciable maintenance and retains an appearance of structural soundness. Another necessary feature of a good design is that the structure is constructed so that visible warning signs are given sufficiently in advance of collapse.

Failure, however, need not be just the sudden separation of one element from another, or the continuing increase in deformation under constant load. For example, if a structure were to be designed to house individuals, and if that structure were so flexible that under normal conditions it experienced significant vibrations which, in turn, subjected the occupants to discomfort motion sickness, then insofar as the usability of the structure is concerned, failure has occurred although the structure may remain structurally sound. Another example: if a structure were designed as an integral part of a machine, and if a necessary condition for the successful operation of that part were that it return to its original position on removal of the imposed loading, then a load which would subject that part to inelastic deformations would preclude the continued usefulness of the structure, and the structure would have failed.

To the structural engineer, failure of an entire structure or even a single component of that structure is a matter of definition. It need not correspond necessarily to fracture. Other conditions could equally well limit the usefulness of the member(s), and could, therefore, constitute a design limitation. The major failure criteria that are most often considered by structural engineers are:

1. *Strength*. This may be elastic strength, maximum strength, energy absorption, etc.

2. *Deflection.* Either recoverability of deformations, limiting deformations—which may exist in either the elastic or inelastic range, etc.
3. *Stiffness.* This governs such phenomena as stability, natural frequencies of vibration, etc.
4. *Fatigue*
5. *Brittle fracture*

Because of its relatively greater importance in the design of structures composed of slender members and because the procedures used could equally well apply to the other cases, only the first of these criteria will be examined in this section—and then only in an illustrative way. It is important to recognize that at this stage attention is directed to properties of the structure or structural elements of which it is composed, and not to the combined structure-load "real" system. True, the behavior of the structure may and probably will be described in terms of presumed limiting loads applied at given presupposed points, but the fact remains that the probability of the assumed loads occurring at those places, or the importance of one particular set of loads as opposed to another, is not under consideration. That a particular structure may be used subsequently for other than its intended design use also, at this stage, will not be considered.

The strength, or stiffness, or deformational response of a structure is dependent, first and foremost, on the properties of the materials of which it is made. Workmanship and inspection, however, are important also, as are such other obvious factors as the dimensioning of elements prior to fabrication or construction. In terms of analytically describing the limiting loads to which a structure may be subjected, it also is to be recognized that the accuracy of the method of analysis used is significant, as is the tolerance of the numerical computations carried out. Assumptions having to do with boundary conditions, initial internal stresses, and stress concentrations are important and where appropriate must be taken into account.

But as noted above, the most important variable in describing structural behavior is material behavior. Unfortunately, no two specimens of the same nominal material respond in the same fashion. Even for metals such as steel where quality is carefully controlled, an occasional relatively large deviation from average behavior is observed. For concrete, where control is less stringent, the variation is greater. This is also true for timber, soils, etc. Material properties, therefore, are statistical quantities, and this must be recognized. The values shown in Table 1.4 are, as noted, typical values and are not to be construed as absolute, or even necessarily guaranteed minimums. (For these and other reasons most specifications prescribe minimum values for design purposes. These values, over the years, have been demonstrated to be adequate for that purpose, although most would consider them to be unduly conservative, considering the large number of other design factors that are even less controlled.)

For the purposes of illustration it will be assumed that 100 structures of identical design are built, and these are then subjected to identical conditions of loading. Maximum load carrying capacity has been defined as the criteria for examination, and load tests are carried out. Because of the reasons given above, the results vary. If these strength values are grouped in relatively small increments, a diagram can be

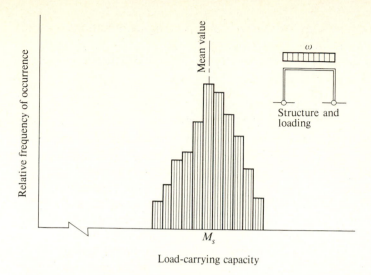

Figure 1.28 Typical histogram of load-carrying capacity of given structural type.

plotted where the observed maximum loads are shown as the abscissa and the frequency of observations of that particular interval of strength is the ordinate. One such typical *histogram* is shown in fig. 1.28. While this histogram is a true representation of observed values of strength, to be able to work analytically with the description, it is necessary that an algebraic expression be developed that represents the data within the desired limits.

Although several methods are available for determining this frequency distribution curve, all are more or less based on the following procedure. First, it is assumed that the mean value of the algebraic expression will correspond to the mean value of the observed data (in terms of the bar graph values of the histogram). Second, it is required that the first moments of the frequencies about the mean value be equal. Next, the second moments, etc., until a curve of the desired accuracy is obtained. One such analytic curve for the data given in fig. 1.28 is illustrated in fig. 1.29. W_{S_M} is the value of the mean carrying capacity for the structural tests in question.

Once the algebraic function has been established, it can be integrated to obtain limiting values above which or below which any certain percentage of observations could be exceeded. For example, in fig. 1.29 a load W_{S_2} will be exceeded by 3 percent of all the observed strengths. This conclusion or observation however, points up one of the serious drawbacks of using such statistical procedures for evaluating load capacity. That is, the load W_{S_2} is in a region of either no observations or a limited number of observations—and is, therefore, questionable.

For these same 100 structures, let it now be presumed that it is possible to determine all the loads in the same sense, same direction, and of the same distribution assumed in fig. 1.28 that might possibly be applied during the useful life of the structures. Further, let a histogram be developed for those loads, and let it be represented by a suitably chosen algebraic function. Again, W_{L_M} is the mean value. Figure

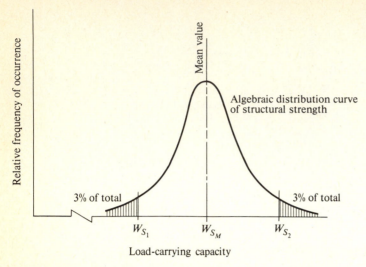

Figure 1.29 Analytical representation of histogram of carrying capacity.

1.31 is a composite plot of the algebraic functions of strength and applied loading on the same diagram.

For the hypothetical case just considered, what is the margin of safety? Is it the ratio of the two mean values, W_{S_M}/W_{L_M}? Is it a function of the ratio of the area beyond the point of intersection, F, of the two curves to the total under the curve? Is it the ratio of a conservative low strength W_{S_1} to a conservative high load W_{L_1}? Engineers' specifications will differ on which of these and other approaches should be used, depending on the type of structure, location, construction control, etc. But on one point they all seem to agree: there is no such thing as absolute safety! Failure of a structure cannot be prevented with certainty, but only with a high degree of proba-

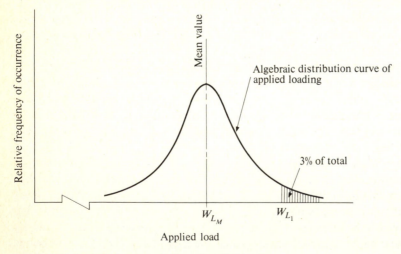

Figure 1.30 Analytical representation of applied load.

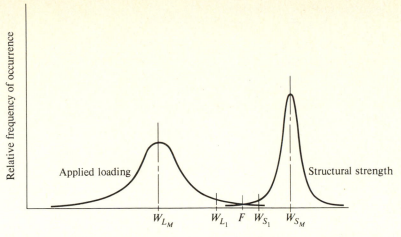

Figure 1.31 Strength versus loading.

bility! The concept of *factor of safety* is, therefore, a questionable one. More realistically, the appropriate notion would be the *probability of failure*.

If one were to analytically establish the probability curves described above from basic information rather than by designing, building, and observing a large number of such structures, it would be necessary to first obtain probability relations for each of the various variables: strength of the material, workmanship, dimensions, etc. In fact, there would have to be available at least one such relationship for each of the variables entering into the solution of the problem. A composite distribution would then be obtained, the magnitude and pattern of which would be governed by the statistical laws of superposition. Because the questions raised above regarding the use of single variable probabilities apply even more when the probability of combinations of influences are involved, it is preferable not to go into further detail on this method but rather to discuss other general methods of ensuring margins of safety.

In the past, using what has come to be known as *conventional elastic strength design* or *elastic design*, a margin of safety has been provided by two means: by using an allowable stress—which is a fraction of the breaking strength or yield strength of the material—and a design load which is set at a reasonably high level above the average load, but not too high to be unrealistic, and specifying the ratio between these two quantities.

For the design loads, one might select the highest load that is likely to occur in the lifetime of the structure. But this is difficult to do. If, for example, the loads were due to snow, there is a reasonable chance that previous records will be broken. Wind loads are subject to the same caveat. For these and other reasons, design loads that have been observed to be adequate for a large number of structures over a relatively long period of time have been incorporated into specifications. These were described in sec. 1.2.

Table 1.5 Coefficients for load-factor design

A factors		B factors	
Dead loading: These can be estimated fairly well, and need not be further substantially increased.	1.10	Material properties: Where minimum values have been specified, no need to add additional.	1.00
Live loading: These depend on the use of the structure; for example,	1.33	Dimensions $\pm\,5\%$ Analysis $\pm15\%$ Workmanship $\pm25\%$ Corrosion $\pm\,5\%$	1.50

Water, grain, etc. ±0
    Office building or school $\pm20\%$
    Factory building, warehouse $\pm33\%$			

Snow and wind loads: These have probably been specified too high.	1.00		

More recently, the concept of *load-factor design* has been applied to a variety of structural situations. Under these procedures the anticipated working load is first multiplied by the desired margin of safety, and this larger value is then used in the selection of member sizes. (It is to be understood that theoretically the structure is designed to collapse under these prorated loads.) The concept of load-factor design has been extended to include variable load factors. For example, although it is generally accepted that dead loads can be predicted with a reasonable degree of certainty, floor loadings in an office building or school are not so well prescribed. It would seem to be reasonable therefore to multiply these two types of loadings by different values. Material properties, workmanship, dimensions, etc., also vary, but not necessarily with the type or kind of loading. Therefore, a second factor must be included to take into account these variations. The appropriate load factor would then be determined by use of the following equation:

$$\text{Load factor} = (A \text{ factor})(B \text{ factor}) \tag{1.15}$$

where A and B are described in table 1.5. Using the values shown in this table to indicate a relative amount of uncertainty, the following load factors might be used for ultimate strength designs:

Dead loads	$(1.10)(1.50) = 1.65$	
Live loads	$(1.33)(1.50) = 2.00$	(1.16)
Snow or wind	$(1.00)(1.50) = 1.50$	

An even more sophisticated method for assigning variable load factors was published in *The Structural Engineer*.[†] This report, which is the result of a special committee of the Institution of Structural Engineers, has been based on the philosophy that a competent designer can estimate within reasonable limits the accuracy of the quantities that have been used in a given design. The major factors that should

† Institution of Structural Engineers, Report on Structural Safety, *Struct. Eng.*, vol. 34, no. 5, pp. 141–149, May 1955.

be taken into account on the question of failure probability are:

A   Material, workmanship, inspection, and maintenance
B   Accuracy of loading assumption, control of use of structure
C   Accuracy of analysis

For each of these the designer should be able to state whether it is vg (very good), g (good), f (fair), or p (poor). The designer should also be able to ascertain the seriousness of a failure should it occur. Again, the appropriate subfactors would be:

D   Loss of life due to collapse
E   Property damage due to collapse

For these cases the qualifying terms are ns (not serious), s (serious), or vs (very serious). The Institution of Structural Engineers has recommended the values shown in table 1.6. For any given problem the ultimate load factor would be determined as in eq. 1.17.

$$\text{Ultimate load factor} = (X \text{ factor})(Y \text{ factor}) \qquad (1.17)$$

Table 1.6a Values of X factors

			B =			
Characteristics			vg	g	f	p
A = vg	C =	vg	1.1	1.3	1.5	1.7
		g	1.2	1.45	1.7	1.95
		f	1.3	1.6	1.9	2.2
		p	1.4	1.75	2.1	2.45
A = g	C =	vg	1.3	1.55	1.8	2.05
		g	1.45	1.75	2.05	2.35
		f	1.6	1.95	2.3	2.65
		p	1.75	2.15	2.55	2.95
A = f	C =	vg	1.5	1.8	2.1	2.4
		g	1.7	2.05	2.4	2.75
		f	1.9	2.3	2.7	3.1
		p	2.1	2.55	3.0	3.45
A = p	C =	vg	1.7	2.15	2.4	2.75
		g	1.95	2.35	2.75	3.15
		f	2.2	2.65	3.1	3.55
		p	2.45	2.95	3.45	3.95

Table 1.6b Values of Y factors

		D =	
Characteristics	ns	s	vs
E = { ns	1.0	1.2	1.4
s	1.1	1.3	1.5
vs	1.2	1.4	1.6

Several examples of how these values would be used are as follows:

1. Reinforced concrete water tank designed for a water supply in a country district:
   A = g, B = vg, C = g, D = ns, E = s;
   therefore

$$X = 1.45, Y = 1.1$$

Ultimate load factor = (1.45)(1.1) = 1.60

2. Gallery for a theater in structural steel: A = g, B = vg, C = g, D = vs, E = s;

   therefore $\qquad X = 1.45, Y = 1.5$

Ultimate load factor = (1.45)(1.5) = 2.18

3. Load bearing wall in brick masonry built in a small community: A = p, B = g,
   C = p, D = vs, E = s;

   therefore $\qquad X = 2.95, Y = 1.5$

Ultimate load factor = (2.95)(1.5) = 4.4

It should be recognized, of course, that the procedures for determining appropriate factors of safety that have been suggested in this section—particularly the numerical values listed in tables 1.5 and 1.6—were advanced as illustrative of what might be appropriate in a given design situation. The actual procedures and values that must be used are set forth in Codes controlling the type of construction or fabrication in question.

1.7 DESIGN CODES

Various national organizations, governmental bodies, industries, etc. provide digests of knowledge, judgment, and experience required for structural design. These legal documents are referred to as *Codes, Construction Manuals, Building Codes,* etc. They provide officials (and the public) with an enforceable document to which the designer must adhere.

The requirements contained in the various codes are far from being uniform. Moreover, most are frequently revised. For this reason, no particular set of requirements, much less editions, have been selected for illustration. Rather, the reader either will be given certain (realistic) hypothetical code provisions or will be requested to consult the latest edition of the applicable document.

The following design codes and specifications are most commonly encountered in structural engineering practices. Their publication dates are purposely not given.

Aluminium Construction Manual, The Aluminium Association, 420 Lexington Avenue, New York, NY
   10017.
Building Code Requirements for Reinforced Concrete, American Concrete Institute, P.O. Box 4754,
   Redford Station, Detroit, MI 48219.

Light Gage Cold-Formed Steel Design Manual, American Iron and Steel Construction, 150 East 42d Street, New York, NY 10017.

National Building Code, The National Board of Fire Underwriters, 85 John Street, New York, NY 10038.

National Design Specifications for Wood Construction, National Forest Products Association, 1619 Massachusetts Avenue, NW, Washington, DC 20036.

Recommended Lateral Force Requirements and Commentary, Seismology Committee of the Structural Engineers Association of California, 171 Second Street, San Francisco, CA 94105.

Specifications of the American Railroad Engineering Association, American Railroad Engineering Association, 59 East Van Buren Street, Chicago, IL 60605.

Standard Specifications for Highway Bridges, American Association of State Highway and Transportation Officials, 341 National Press Building, Washington, DC 20045.

Steel Construction Manual, American Institute of Steel Construction, 1221 Avenue of the Americas, New York, NY 10020.

Uniform Building Code, International Conference of Building Officials, 5360 South Workman Mill Road, Whittier, CA 90601.

TWO

STRESSES IN STRUCTURAL MEMBERS

2.1 INTRODUCTION

Summarizing a major portion of chap. 1, adequately designed structures possess sufficient strength for the job intended, are serviceable, and are economical. It is the primary purpose of this chapter to consider the first of these criteria—sufficient strength—and to restrict examination to determinate structural systems. Moreover, it will be presumed that sufficient strength is assured if the maximum stresses can be adequately predicted, since these can then be related to allowable stress values which take into account appropriate factors of safety. In more specific terms, the problems with which chap. 2 deals are as follows:

1. For analysis: given a determinate structure and its loading, calculate the maximum stresses that exist in each of the members of the structure in question.
2. For design: given a determinate structure of prescribed geometry, an imposed loading, and, moreover, having specified a set of allowable maximum stresses, determine the sizes of the various members that will result in stresses that will approach—but not exceed—the stated allowable stress values.

Partly because structural members must support their own weight in addition to externally imposed forces, but also because of the basic iterative nature of the process itself, *design* consists of repetitive cycles of *analysis*. Therefore stress analysis of determinate structural members and frames will be the major focus of this chapter.

The process of computing stresses in given structural members consists of two distinct and separate operations:

1. Reduction of the externally imposed loading to sets of stress resultants, that is, forces and moments, that act on each of the separate members
2. At critical sections along these members, reduction of these stress resultants to unit stress distributions from which maximum stresses can be calculated

For statically determinate systems, that is, for those systems where the equations of statics are sufficient to establish the distribution of forces and moments, the first of these two operations is straightforward and relatively easy to accomplish. For indeterminate systems, however, additional conditions of continuity must be taken into account, and these require the introduction of deformational considerations. For all but the most elementary cases, the second operational step referred to above, that is, the determination of the actual stress distribution, is undertaken only after the first step has been completed. The methods and procedures that are used are those covered in courses in mechanics of materials (or strength of materials) and in elasticity. In this chapter the general ideas established in those courses will be applied to structural systems that are statically determinate. The structural materials that will be presumed will be those that respond to loads in a linear or approximately linear fashion (nonlinear response will be considered in a later chapter). Furthermore, since determinate structures are nearly always fabricated or constructed from one-dimensional (slender) members, discussion will be limited to those types of elements.

2.2 LINEAR ELASTICITY

The following assumptions of material behavior will be made:

1. The stress-strain relationship in tension is the same as it is in compression
2. The unit material response to imposed loading is linearly elastic
3. The material is both isotropic and homogeneous

A material so presumed is known as a *hookean material*—one that obeys Hooke's laws.

For such materials, it is possible to define in a rectangular cartesian coordinate system the interrelationships that must exist at a point among and between the various components of stress and strain. Defining σ_x, σ_y, and σ_z as the normal stress components in the x, y, and z directions, respectively as shown in fig. 2.1; ε_x, ε_y, and ε_z as the corresponding normal strain components; $\tau_{xy} = \tau_{yx}$, $\tau_{yz} = \tau_{zy}$, and $\tau_{zx} = \tau_{xz}$ as components of shearing stress, and γ_{xy}, γ_{yz}, and γ_{zx} as shearing strains; and

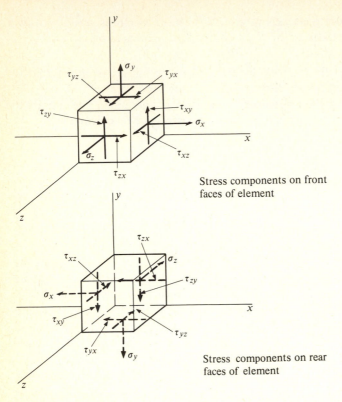

Stress components on front faces of element

Stress components on rear faces of element

Figure 2.1 Components of stress on a unit element.

assuming that there are no body forces acting, the governing stress-strain equations are:†

$$\sigma_x = \frac{E}{(1 + \mu)(1 - 2\mu)} [(1 - \mu)\varepsilon_x + \mu(\varepsilon_y + \varepsilon_z)]$$

$$\sigma_y = \frac{E}{(1 + \mu)(1 - 2\mu)} [(1 - \mu)\varepsilon_y + \mu(\varepsilon_x + \varepsilon_z)]$$

$$\sigma_z = \frac{E}{(1 + \mu)(1 - 2\mu)} [(1 - \mu)\varepsilon_z + \mu(\varepsilon_x + \varepsilon_y)]$$

$$\tau_{xy} = \tau_{yx} = \frac{E}{2(1 + \mu)} \gamma_{xy} = G\gamma_{xy}$$

$$\tau_{yz} = \tau_{zy} = \frac{E}{2(1 + \mu)} \gamma_{yz} = G\gamma_{yz}$$

$$\tau_{zx} = \tau_{xz} = \frac{E}{2(1 + \mu)} \gamma_{zx} = G\gamma_{zx}$$

(2.1)

† For a detailed development of these equations, see S. P. Timoshenko and J. N. Goodier, *Theory of Elasticity*, McGraw-Hill Book Company, 1951, or S. H. Crandall, N. C. Dahl, and T. J. Lardner, *An Introduction to the Mechanics of Solids*, 2d ed., McGraw-Hill Book Company, 1972.

Here, E is Young's modulus of elasticity, G is the shearing modulus, and μ is Poisson's ratio. (It is to be noted that all the stress components shown in fig. 2.1 are in the assumed positive directions.)

Equations (2.1) define the stress components at a point in terms of the corresponding strains. Frequently, it is more convenient to define the components of strain as functions of stress, that is, to invert the relationship. The resulting equations are given in eq. (2.2).

$$\varepsilon_x = \frac{1}{E}[\sigma_x - \mu(\sigma_y + \sigma_z)]$$

$$\varepsilon_y = \frac{1}{E}[\sigma_y - \mu(\sigma_x + \sigma_z)]$$

$$\varepsilon_z = \frac{1}{E}[\sigma_z - \mu(\sigma_y + \sigma_x)]$$

$$\gamma_{xy} = \gamma_{yx} = \frac{1}{G}\tau_{xy}$$

$$\gamma_{yz} = \gamma_{zy} = \frac{1}{G}\tau_{yz}$$

$$\gamma_{zx} = \gamma_{xz} = \frac{1}{G}\tau_{zx}$$

(2.2)

As noted in chap. 1, structural components are normally considered to be functionally one-, two-, or three-dimensional, depending upon the relative comparisons of their various overall dimensions. In some two- and one-dimensional cases, particularly those subjected to certain restricted types of loading, stress or strain components in one of the directions will be quite small and can be presumed equal to zero with negligible resulting error. Two such special cases are of particular importance: *plane stress* and *plane strain*.

For the case of plane stress shown in fig. 2.2, it is presumed that the stress components in the z direction equal zero; that is, $\sigma_z = \tau_{xz} = \tau_{yz} = 0$. The normal strain in that direction, however, may have a value; that is, $\varepsilon_z \neq 0$. Substituting these values into eq. (2.2) yields

$$\varepsilon_x = \frac{1}{E}(\sigma_x - \mu\sigma_y)$$

$$\varepsilon_y = \frac{1}{E}(\sigma_y - \mu\sigma_x)$$

$$\varepsilon_z = -\frac{1}{E}[\mu(\sigma_x + \sigma_y)]$$

(2.3)

$$\gamma_{xy} = \frac{1}{G}\tau_{xy}$$

$$\gamma_{yz} = \gamma_{xz} = 0$$

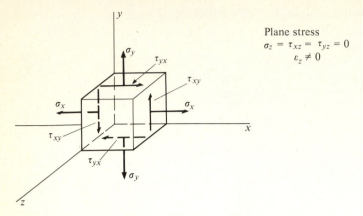

Plane stress
$$\sigma_z = \tau_{xz} = \tau_{yz} = 0$$
$$\varepsilon_z \neq 0$$

Figure 2.2 Plane-stress components on a unit element.

(It is to be recognized that a condition of plane stress does not rule out the possibility that, under certain combinations of applied stress, strains in the z direction may equal zero.)

Plane strain presumes $\varepsilon_z = \gamma_{xz} = \gamma_{yz} = 0$, with normal stress in the z direction allowed. The corresponding stress-strain relationships are those of eq. (2.4).

$$\varepsilon_x = \frac{1}{E}[\sigma_x - \mu(\sigma_y + \sigma_z)]$$

$$\varepsilon_y = \frac{1}{E}[\sigma_y - \mu(\sigma_x + \sigma_z)]$$

$$\varepsilon_z = 0 \qquad\qquad (2.4)$$

$$\gamma_{xy} = \gamma_{yx} = \frac{1}{G}\tau_{xy}$$

$$\gamma_{yz} = \gamma_{zx} = 0$$

(Note here that it is possible under these conditions for normal stress in the z direction to exist.)

For a truly one-dimensional member, only normal stress in that singular direction could be considered. Assuming this to be the z direction, the resulting stress-strain relationship becomes

$$\varepsilon_z = \frac{1}{E}\sigma_z$$

$$\qquad\qquad (2.5)$$

$$\varepsilon_x = \varepsilon_y = -\frac{\mu}{E}\sigma_z$$

But slender structural members are not one-dimensional members—in the strictest sense of the definition. They are frequently called upon to resist shear. For such cases, the stress-strain relationships will be those shown in eq. (2.6).

$$\varepsilon_z = \frac{1}{E}\sigma_z$$

$$\varepsilon_x = \varepsilon_y = -\frac{\mu}{E}\sigma_z$$

$$\gamma_{xy} = \frac{1}{G}\tau_{xy}$$

$$\gamma_{xz} = \frac{1}{G}\tau_{xz}$$

(2.6)

2.3 STRESSES IN SLENDER STRUCTURAL MEMBERS

Slender structural members are those having two of their three dimensions small when compared to the third dimension. Thus, a slender member is one whose cross-sectional dimensions are small when compared to its length. Examples of these are rods, columns, strings, cables, beams, arches, etc. It is to be noted that a slender member, when defined in this fashion, need not be straight or even prismatic.

In a rectangular cartesian coordinate system, there are six stress resultant components that produce stresses in the member. If the coordinate directions are taken as x, y, and z as shown in fig. 2.3, one possible (and convenient) set of components consists of the three forces and three moments shown. V_x and V_y are shearing forces that are assumed to act in the plane of the cross section in the x and y directions, respectively. P_z is an axial force (or thrust) assumed to act at the centroid of the cross section. The bending moments M_x and M_y occur about the x and y axes, respectively. M_z is a twisting moment, or torque, about the longitudinal axis of the member at the section in question.

While at any point within the member six separate components of stress can exist, only three are usually considered in the case of slender members. These are

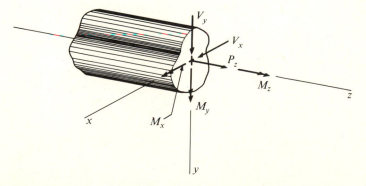

Figure 2.3 Stress resultant components in a cartesian coordinate system.

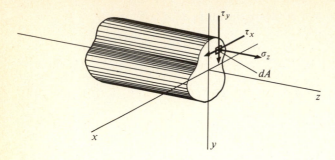

Figure 2.4 Typical components of stress in slender members.

shown in fig. 2.4 as two shearing stresses τ_x and τ_y and the normal stress σ_z. In most real cases, using St. Venant's principle, it can be demonstrated that the remaining stresses can be ignored.

Any one of the stress components shown in fig. 2.4 can be produced by one or several of the stress resultants illustrated in fig. 2.3. The interaction between stresses and the stress resultants mentioned previously is indicated in fig. 2.5. For linear problems these three stress components—σ_z, τ_x, and τ_y—can be considered separately. For nonlinear cases, however, the effects must be examined simultaneously. (To satisfy the condition of linearity, it is necessary that the material be hookean and that the effects of deformation on the stress resultants be small enough to be ignored. Since small deformations are presumed, formulation is in the undeformed state.)

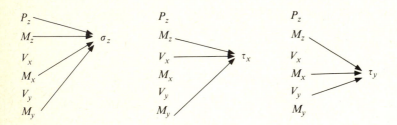

Figure 2.5

2.4 CROSS-SECTIONAL PROPERTIES

Slender structural members are classified according to the shape of their cross section. A convenient system is the following one:

1. Members with solid cross section: a solid circular bar, solid rectangular shaft, etc.
2. Members with thin-walled, closed cross section: a pipe, tube, or box beam
3. Members with thin-walled, open cross section: an angle, I shape, channel, etc.

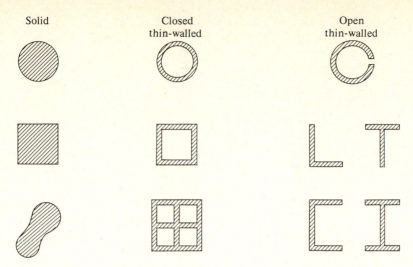

Figure 2.6 Typical cross-sectional shapes.

Typical examples of each of these are shown in fig. 2.6. The advantage of such a grouping lies in the similarity of the torsional response.

In the most general case a cross section has no axis of symmetry, for example, a structural angle having unequal legs. In many, if not most cases, however, members having at least one axis of symmetry are used. Doubly symmetric cross sections are used whenever possible—because of ease of erection, less complex analysis, etc.

A number of geometrical properties are important in structural design. To define these, consider the cross section shown in fig. 2.7. (For reference a rectangular xy coordinate system has been shown. It is presumed that the origin of that system coincides with the centroid of the cross section.) The following definitions are to be noted:

1. *Centroid*. The centroid of the cross section shown in fig. 2.7 is designated as 0. Its location corresponds to that particular point where the first moments of the cross-sectional area about any two mutually perpendicular (orthogonal) axes equal zero.

2. *Centroidal axes*. Any pair of orthogonal axes that pass through the centroid are known as centroidal axes. The x and y axes shown in fig. 2.7, for example, are centroidal axes, as are the ξ and η axes.

3. *First moment of area*. The integral of the product of the area times the distance from the area to the axis about which the first moment is to be taken is defined as

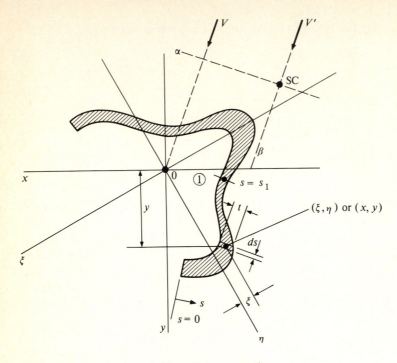

Figure 2.7 Generalized open, thin-walled cross section.

the first moment of the area. Thus, if Q denotes this first moment

$$Q_x = \int_0^{s_1} yt \, ds$$

and
$$\text{(2.7)}$$

$$Q_\eta = \int_0^{s_1} \xi t \, ds$$

4. *Moment of inertia* (or second moment of area). Denoted by the symbol I, the governing equations for determining the moments of inertia are

$$I_x = \int_s ty^2 \, ds$$

and
$$\text{(2.8)}$$

$$I_\xi = \int_s t\eta^2 \, ds$$

where I_x is the moment of inertia of the cross-sectional area about the x axis, and I_ξ is the moment of inertia of the cross-sectional area about the ξ axis.

5. *Product of inertia.* By definition, the product of inertia is defined by eq. (2.9).

$$I_{xy} = \int_s txy \, ds$$

or $\hspace{8cm}$ (2.9)

$$I_{\xi\eta} = \int_s t\xi\eta \, ds$$

(It is to be recognized that, when computing the moments and product of inertia, the entire area of the cross section is considered. For the first moment of the area, however, only a part may be selected for examination.)

6. *Principal axes.* Principal axes are a pair of centroidal axes about which the product of inertia is zero. Thus, if in fig. 2.7 the axes ξ and η are principal axes, $I_{\xi\eta} = 0$. (In general, the moment of inertia about one of the principal axes is a maximum for the cross section. About the other, it is a minimum. The principal axes having the largest moment of inertia is termed the major principal axes. Correspondingly, the one at 90° to that axis is the minor principal axis. The moments of inertia about other centroidal axes have values between these extremes.) It is to be recognized that an axis of symmetry is always a principal axis.

7. *Shear-center axis.* An axis of loading for which no torsional stresses result is defined as a shear-center axis. (An axis of symmetry is always a shear-center axis.)

8. *Shear center.* The shear center of a cross section is defined as the intersection of two shear-center axes. For a symmetrical section, the shear center must lie on the axis of symmetry. In fig. 2.7 the shear center is located at point SC, and the orthogonal axes α and β are shear axes. (A load V acting through the centroid of the cross section will produce torsional stresses, whereas one applied through SC will not.)

Example 2.1 The cross section shown in fig. 2.8*a* is built up of two C 8 × 11.5 channel sections. They are rigidly attached to form the L shape. The problem is

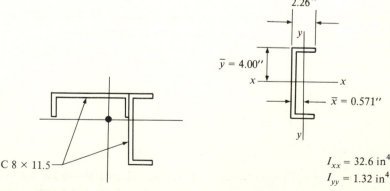

$\bar{y} = 4.00''$

$\bar{x} = 0.571''$

$I_{xx} = 32.6 \text{ in}^4$

$I_{yy} = 1.32 \text{ in}^4$

C 8 × 11.5

Figure 2.8a

Figure 2.8b Geometric properties of the C 8 × 11.5.

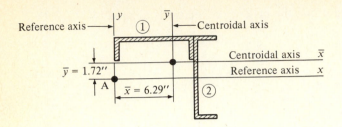

Figure 2.8c Location of the centroid.

to establish the major and minor principal axes of the total configuration. The cross-sectional properties of a single-channel C 8 × 11.5 are indicated in fig. 2.8*b*.

The location of the centroid of the composite section can be obtained by considering the first moments of the areas about the reference axes x and y drawn through point A as shown in fig. 2.8*c*.

$$\int_A dA = \sum A = 6.76 \text{ in}^2$$

$$\int_A x\, dA = \sum xA = 42.49 \text{ in}^3$$

$$\int_A y\, dA = \sum yA = 11.59 \text{ in}^3$$

From which

$$\bar{x} = \frac{\int_A x\, dA}{\int_A dA} = \frac{42.49}{6.76} = 6.29 \text{ in}$$

and

$$\bar{y} = \frac{\int_A y\, dA}{\int_A dA} = \frac{11.59}{6.76} = 1.72 \text{ in}$$

The moments and products of inertia about the centroidal axes x and y are

$$I_{\bar{x}} = I_{\bar{x}_1} + I_{\bar{x}_2} = 11.25 + 42.54 = 53.79 \text{ in}^4$$
$$I_{\bar{y}} = I_{\bar{y}_1} + I_{\bar{y}_2} = 50.26 + 18.97 = 69.23 \text{ in}^4$$

and

$$I_{\bar{x}\bar{y}} = \sum abA = 3.38[(1.714)(-2.286) + (-1.715)(2.285)] = -26.49 \text{ in}^4$$

From statics it is known that the moment of inertia about any axis ξ located at an angle θ from the x axis is given by the equation

$$I_\xi = I_x \cos^2 \theta - I_{xy}(2 \sin \theta \cos \theta) + I_y \sin^2 \theta$$

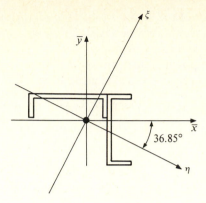

Figure 2.8d Principal axis of the built-up cross section.

The direction of the principal axes can be ascertained by equating to zero the first derivative with respect to θ of the moment of inertia about the ξ axis

$$\frac{dI_\xi}{d\theta} = -2I_x \cos\theta \sin\theta - 2I_{xy}(\cos^2\theta - \sin^2\theta) + 2I_y \sin\theta \cos\theta = 0$$

or

$$-I_x \sin 2\theta - 2I_{xy} \cos 2\theta + I_y \sin 2\theta = 0$$

From this equation

$$\tan 2\theta = \frac{2I_{xy}}{I_y - I_x}$$

or

$$\theta = -36.89°$$

(Since at this value of θ the second derivative of I_ξ with respect to θ is positive, this will be the direction of the minor principal axis.) Substituting this angle (and $-36.89° + 90°$) into the general expression for I_ξ yields for maximum and minimum values of the moment of inertia.

$$I_{max} = 89.10 \text{ in}^4 = I_\xi$$

$$I_{min} = 33.92 \text{ in}^4 = I_\eta$$

2.5 STRESSES AND STRESS RESULTANTS DUE TO BENDING

For structures that are in equilibrium, the resultant of all the various stress resultants (both forces and moments—and for that matter any of their components) must vanish. This is true for entire structures, for individual members, or for free bodies of segments of individual members. Therefore, to facilitate the determination of internal forces and moments, and from these the values of the stresses that exist at various locations across the section, it is necessary that free bodies be selected in such a way that the desired resultants act external to the body in question.

It also is necessary that there be assumed a possible distribution of strains at the cut section. For members subjected to bending, experimental evidence suggests that a linear variation of bending strains across the section is not unreasonable. Using such a distribution, selecting the appropriate stress-strain relationship from eqs. (2.2), (2.3), (2.4), or (2.6), and making use of the integral equilibrium equations relating stresses to forces and moments at the cut section, equations can be obtained which define internal stresses in terms of external loads.

As an example of the use of this procedure, consider the development of the relationship between the normal stress σ_z, the bending moment M_x, and the externally applied uniformly distributed loading q, shown in fig. 2.9a. The cross-sectional shape is a rectangle and the distributed lateral load is applied in direction y. It is further assumed that the beam is made from a homogeneous, isotropic, hookean material and that the resulting deformations are small.

The free body diagram shown in fig. 2.9b is selected for examination. At the cut section have been shown the coordinate axes and the resisting bending moment and

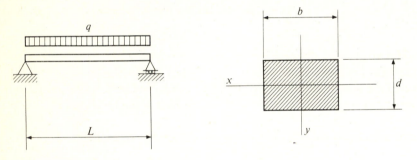

Figure 2.9a Simple beam subjected to a uniformly distributed lateral load.

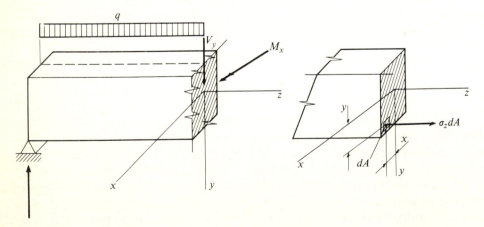

Figure 2.9b Free body diagram of the left-hand portion of the beam.

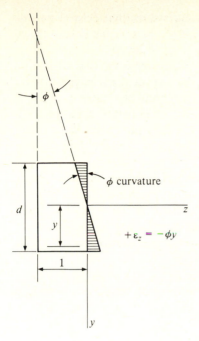

Figure 2.9c Unit length of a beam subjected to a curvature ϕ.

shearing force due to the influence of the remainder of the beam M_x and V_y. (It is to be remembered that for each of the elements of the cross-sectional area of the cut face, again due to the action of the remainder of the beam, there exists the possibility of the development of stresses. This is shown in the second of the two sketches of this figure.)

For equilibrium at the cut face, the following conditions must be met:

$$\int_A \sigma_z \, dA = 0$$

$$\int_A \sigma_z y \, dA = M_x \qquad (2.10)$$

$$\int_A \sigma_z x \, dA = 0$$

A linear strain distribution across the section is next presumed. This is indicated in fig. 2.9c, where positive strains ε_z are chosen in tension. Since normal strains ε_z are measured in inches per inch, and since curvature ϕ is the change in slope per unit of length along the member in question, only a unit length of the beam has been shown in the illustration. In formulating the relationship between assumed positive (tensile) strains and curvature in fig. 2.9c, we see that curvatures which produce increasing values of the slope with increasing values of z should be presumed positive. (For the case illustrated, there would result a decrease in slope with increasing z.) Moreover, for small deformations the value of the tangent of the curvature approaches the value of the angle itself. Therefore,

$$\varepsilon_z = -\phi y \qquad (2.11)$$

Using the first of the stress-strain equations (2.6), we find that these strains can be related to the normal stress by the relationship

$$\sigma_z = E\varepsilon_z \qquad (2.12)$$

Combining these two expressions yields

$$\sigma_z = -E\phi y$$

From the first of the equilibrium equations (2.10)

$$\int_A \sigma_z \, dA = \int_A -E\phi y \, dA = -E\phi \int_A y \, dA = 0$$

Since E or ϕ cannot necessarily be presumed to equal zero, it must be concluded that

$$\int_A y \, dA = 0$$

This requires that the x axis, as originally selected in fig. 2.9a and b, be the centroidal axis of the cross section. Further, since y is zero along this line, σ_z must also there be zero. This line of zero normal stress is defined as the *neutral axis* (NA) and will coincide with a centroidal axis when there are no axial forces acting on the section in question. When axial forces are present, the neutral axis shifts from this location.

Considering the third of the equilibrium equations (2.10), it is to be recognized that there will be no variation in normal stress σ_z in the x direction. Finally, from the second of these equations

$$\int_A \sigma_z y \, dA = -\int_A E\phi y^2 \, dA = -E\phi \int_A y^2 \, dA = M_x$$

But by definition $\int_A y^2 \, dA = I_x$, the moment of inertia of the cross section about the x axis. Therefore

$$E\phi I_x = -M_x$$

or

$$\phi = -\frac{M_x}{EI_x} \qquad (2.13)$$

Since $\sigma_z = -E\phi y$,

$$\sigma_z = \frac{M_x}{EI_x}(Ey) = \frac{M_x y}{I_x} \qquad (2.14)$$

This is the desired relationship between normal stress and bending moment at the section in question.

Since for a given structure, loading and cross section, M_x and I_x are constant, the maximum normal stress $\sigma_{z,\,max}$ will occur at the point on the cross section that is farthest from the neutral axis.

$$\sigma_{z,\,max} = \frac{M_x y_{max}}{I_x} = \frac{M_x(d/2)}{I_x} = \frac{M_x}{S_x} \tag{2.15}$$

where $S_x = I_x/(d/2)$. S_x is referred to as the *section modulus*.

The sign convention that was used in the preceding illustration is the one that will be used throughout most of the remainder of this text. In summary, it can be described as follows:

1. *Forces and moments.* A vector that is in the direction of a positive axis and acts on a positive face is defined to be a positive stress resultant. A positive face is the one that is farther in the direction of the positive coordinate axis. (All the quantities in fig. 2.10 are positive as shown.)
2. *Stresses.* Normal stresses are positive if in tension. A shearing stress is positive if it acts in the direction of a positive axis on a positive face.
3. *Loads.* Loads are positive if they act in the direction of a positive axis.
4. *Displacements.* Displacements are positive if they occur in the direction of a positive axis. Slopes and curvature are positive as indicated in the second sketch of fig. 2.10.

It is desirable to determine the conditions for equilibrium of the forces and moments acting on the structural element illustrated in fig. 2.10. For example, considering only the yz plane, equilibrium of forces in the y direction would require that

$$q_y\, dz + V_y + dV_y = V_y$$

or

$$\frac{dV_y}{dz} = -q_y \tag{2.16}$$

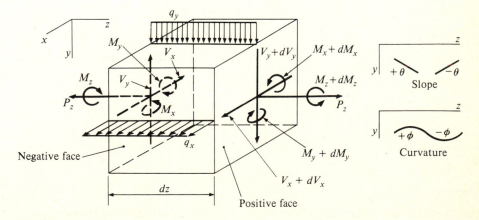

Figure 2.10 Positive sign convention.

That is, the rate of change in the z direction of the resultant shear V_y is equal to the negative of the unit value of the distributed load q. (This fact will be of help in the construction of shear diagrams.) Next, taking moments about an axis parallel to the x axis which passes through the center of the element

$$M_x + V_y \frac{dz}{2} + V_y \frac{dz}{2} + \frac{dV\ dz}{2} = M_x + dM_x$$

Collecting terms and omitting quantities of higher order,

$$\frac{dM_x}{dz} = V_y \tag{2.17}$$

(Again, this relationship will be of considerable help in the development of moment diagrams.) Differentiating eq. (2.17) with respect to z, and substituting the value shown in eq. (2.16) yields

$$\frac{d^2M_x}{dz^2} = \frac{dV_y}{dz} = -q_y$$

But, from eq. (2.13) $\phi = -M_x/EI_x$, or

$$EI_x \frac{d^2\phi}{dz^2} = -q_y$$

Since curvature ϕ is the second derivative with respect to z of the displacement of the beam, that is, since

$$\phi = \frac{d^2y}{dz^2}$$

the differential equation which describes the beam's deformation in terms of the applied loading will be given by eq. (2.18).

$$\frac{d^4y}{dz^4} = \frac{q_y}{EI_x} \tag{2.18}$$

The expressions just developed are readily adapted to the design of determinate beams subjected to uniformly distributed lateral loads.

Example 2.2 Consider the 15-ft-long member of rectangular cross section shown in fig. 2.11a. A uniformly distributed lateral load of 100 lb/ft is prescribed, as is

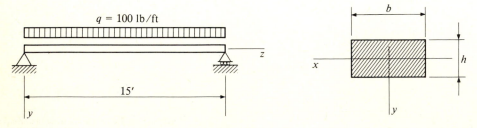

$q = 100$ lb/ft

15'

Figure 2.11a

an allowable normal bending stress of 10,000 lb/in². Cross-sectional dimensions *b* and *h* are desired. From statics it is known that the maximum bending moment occurs at the center of the span and has a magnitude of

$$M_{max} = \frac{ql^2}{8} = \frac{(100)(15)^2}{8} = 2812.5 \text{ ft} \cdot \text{lb}$$

From eq. (2.15)

$$\sigma_{max} = \frac{M_{max}}{S}$$

or

$$S_{rqd} = \frac{M_{max}}{\sigma_{max}} = \frac{(2812.5)(12)}{10,000} = 3.375 \text{ in}^3$$

Since for a rectangular cross section

$$S = \frac{I}{h/2} = \frac{1}{6}bh^2$$

the required value of bh^2 is

$$bh^2_{rqd} = (6)(3.375) = 20.25 \text{ in}^3$$

It should be recognized that insofar as particular values of *b* and *h* are desired, no unique solution to this equation exists. For a chosen *b*, a minimum allowable value of *h* can be computed. But from the data supplied, it is impossible to say that that particular set is superior to any other. For example, if it is assumed that the beam will be 6 in wide (that is, $b = 6$ in), then *h* must be at least 1.837 in. If, however, *b* were selected as 5 in, then *h* must be at least 2.01 in. This is typical of most design situations. There are a number of possibilities. The permissible design range is indicated in fig. 2.11*b*.

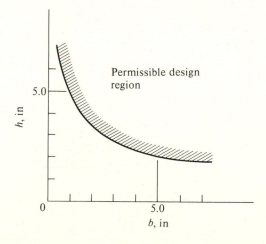

Figure 2.11b Permissible design region for example 2.2.

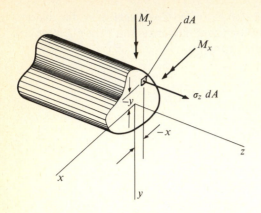

Figure 2.12 A slender structural member subjected to bending-moment components.

A variety of simple beam bending situations can be examined in the manner described above. For example, consider the development of the relationship between normal stress and bending moments for a beam subjected to bending about two orthogonal cross-sectional axes. As in the preceding case, the assumptions of a hookean material and a linear variation in bending strains will be made.

For the generalized member shown in fig. 2.12, the equilibrium equations to be satisfied are

$$\int_A \sigma_z \, dA = 0$$

$$\int_A \sigma_z y \, dA = M_x \tag{2.19}$$

$$\int_A \sigma_z x \, dA = M_y$$

Bending strains can be described by the relationship

$$\varepsilon_z = -(\phi_x y + \phi_y x)$$

where ϕ_x and ϕ_y are curvatures about the x and y axes, respectively. Since $\sigma_z = E\varepsilon_z$,

$$\sigma_z = -E(\phi_x y + \phi_y x)$$

From the first of the equilibrium equations (2.19)

$$\sigma_z \, dA = 0 = -\int_A E(\phi_x y + \phi_y x) \, dA$$

or

$$\phi_x E \int_A y \, dA + \phi_y E \int_A x \, dA = 0$$

Since ϕ_x, ϕ_y, and E cannot be assumed to be necessarily equal to zero,

$$\int_A y \, dA = \int_A x \, dA = 0$$

That is, the x and y axes must be centroidal axes.

From the second and third of the equilibrium equations

$$M_x = -\int_A E(\phi_x y + \phi_y x) y \, dA$$

$$M_y = -\int_A E(\phi_x y + \phi_y x) x \, dA$$

or

$$-M_x = EI_x \phi_x + EI_{xy} \phi_y$$

$$-M_y = EI_{xy} \phi_x + EI_y \phi_y$$

Solving for the curvatures ϕ_x and ϕ_y,

$$\phi_x = -\frac{1}{E}\left[\frac{-M_x I_y + M_y I_{xy}}{I_{xy}^2 - I_x I_y}\right]$$

$$\phi_y = -\frac{1}{E}\left[\frac{M_x I_{xy} - M_y I_x}{I_{xy}^2 - I_x I_y}\right]$$

Introducing these into the equation for σ_z defined previously, the general expression for normal stress in terms of cross-sectional properties and applied bending moments becomes

$$\sigma_z = \frac{(-M_x I_y + M_y I_{xy})y + (M_x I_{xy} - M_y I_x)x}{I_{xy}^2 - I_x I_y} \tag{2.20}$$

If the x and y axes are selected as principal axes, that is, $I_{xy} = 0$, eq. (2.20) reduces to

$$\sigma_z = \frac{M_x y}{I_x} + \frac{M_y x}{I_v} \tag{2.21}$$

These equations are used in exactly the same way as those developed earlier for single-axis bending. Consider the following three examples. All will presume the same cross section, but the selected axes or loading will vary. The cross section is shown in fig. 2.13a. The normal stress at point A on the cross section is desired.

Assume for the first case that both a shear load and a bending moment act through the centroid of the cross section as shown in fig. 2.13b. Further, assume that the axes of reference x and y are parallel to the sides of the member and are directed

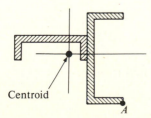

Centroid

A

Figure 2.13a A built-up cross section consisting of two channels.

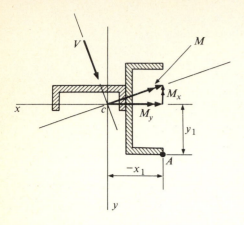

Figure 2.13b Shear and bending moment acting through the centroid of cross section.

as indicated. Since it cannot be assumed that the product of inertia equals zero, the most general expression [eq. (2.20)] must be used when computing σ_z. Moreover, for determining σ_{z_A}, $y = +y_1$ and $x = -x_1$. (The components of the applied moment M would be vectorally determined as indicated. Both would be negative in terms of the selected positive coordinate directions.)

For the case illustrated in fig. 2.13c, it is assumed that the axes $\bar{x}$ and $\bar{y}$ are principal axes. Therefore, $I_{\bar{x}\bar{y}} = 0$. Moreover, for the loading shown, $M_{\bar{y}} = 0$. Under these conditions, the normal stress at point A would be

$$\sigma_{z_A} = \frac{M\bar{y}_1}{I_{\bar{x}}}$$

In fig. 2.13d the $\bar{x}$ and $\bar{y}$ axes are again taken to be principal axes; however, M_x is assumed to act about the original x axis. Since $M_{\bar{y}} \neq 0$, the bending normal stress at point A on the cross section is given by eq. (2.22).

$$\sigma_z = \frac{M_x\bar{y}_1}{I_{\bar{x}}} + \frac{M_y(-\bar{x}_1)}{I_{\bar{y}}} \tag{2.22}$$

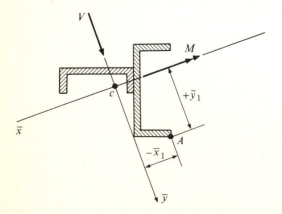

Figure 2.13c Shear and bending moment acting along the principal axes of cross section.

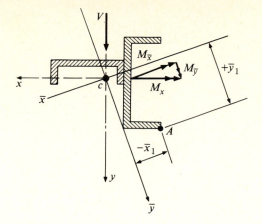

Figure 2.13d Shear and bending moment acting through the centroid parallel to sides of the cross section.

Example 2.3 As a numerical example, assume that the cross section defined in fig. 2.8a is to be used as a simple beam of 12-ft length as shown in fig. 2.14. Further, presuppose that the beam is subjected to a uniformly distributed lateral load of $q = 500$ lb/ft in the direction and through the point shown. The cross-sectional dimensions and principal axes are indicated in fig. 2.15a. (Coordinate dimensions to each of the reference locations A through F also have been shown.)

First, the applied load is resolved into its component parts—in the directions of the principal axes.

$$q_\xi = -400 \text{ lb/ft}$$

$$q_\eta = 300 \text{ lb/ft}$$

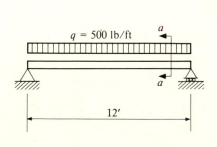

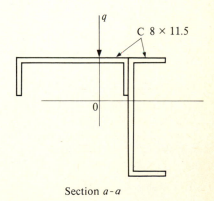

Section a-a

Figure 2.14

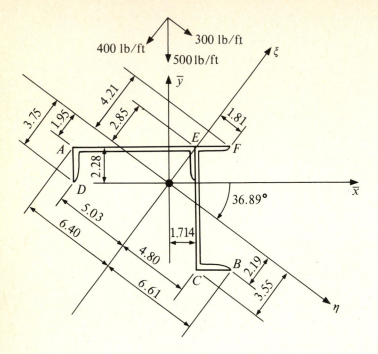

Figure 2.15a Dimensions of the cross section in the directions of the principal axes.

Therefore, the maximum bending moments (about the principal axes) at the center of the beam are

$$M_{\eta,\,max} = \frac{(-400)(12)^2}{8} = -7200 \text{ ft} \cdot \text{lb} = -86{,}400 \text{ in} \cdot \text{lb}$$

$$M_{\xi,\,max} = \frac{(300)(12)^2}{8} = 5400 \text{ ft} \cdot \text{lb} = 64{,}800 \text{ in} \cdot \text{lb}$$

The normal stress σ_z, at any general point on the cross section, is given by

$$\sigma_z = \frac{M_\xi \eta}{I_\xi} + \frac{M_\eta \xi}{I_\eta}$$

or

$$\sigma_z = \frac{64{,}800}{89.10} \eta + \frac{-86{,}400}{33.92} \xi = 727.27\eta - 2547.17\xi$$

For point A

$$\sigma_{z_A} = (727.27)(-6.40) - (2547.17)(-1.95) = 312.45 \text{ psi} = 0.31 \text{ ksi}$$

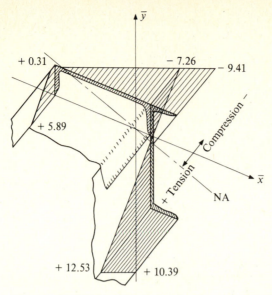

Figure 2.15b Distribution of normal stresses for example 2.3.

For the other reference points:

$$\sigma_{z_B} = (727.27)(6.61) - (2547.17)(-2.19) = 10{,}386 \text{ psi} = 10.39 \text{ ksi}$$

$$\sigma_{z_C} = 12{,}533 \text{ lb/in}^2 = 12.53 \text{ ksi}$$

$$\sigma_{z_D} = 5{,}893.7 \text{ lb/in}^2 = 5.89 \text{ ksi}$$

$$\sigma_{z_E} = -7{,}259.4 \text{ lb/in}^2 = -7.26 \text{ ksi}$$

$$\sigma_{z_F} = -9{,}407.2 \text{ lb/in}^2 = -9.41 \text{ ksi}$$

A plot of the resulting normal stress distribution is shown in fig. 2.15b.

2.6 STRESSES AND STRESS RESULTANTS DUE TO SHEAR

The relationship between bending shear stress τ_y and shear force V_y is established in a manner similar to that developed in the preceding section for beam bending, that is, by applying the equations of static equilibrium to an appropriately chosen free body diagram. For example, for the beams defined in fig. 2.9a the resultant shearing forces and moments acting on a differential element of length would be those shown in fig. 2.16a. But from the preceding discussion, it is evident that the bending moments acting on the faces of the element can be replaced by their equivalent normal stress distributions. This is shown in fig. 2.16b. Selecting for examination a portion of the element bounded by the bottom of the beam and by an xz plane located y distance

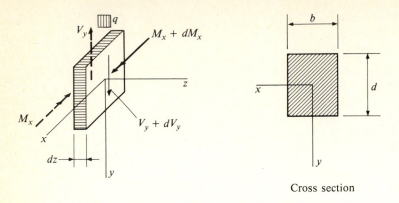

Cross section

Figure 2.16a Stress resultants acting on a unit length of a beam.

from the origin, the situation shown in fig. 2.16c is obtained. τ_z, the shearing stress on the horizontal cut face, is assumed to be uniform across the width of the section, since the normal stress distribution in that direction does not vary. Equilibrium of forces in the z direction requires that

$$\tau_z b \, dz = \frac{1}{2}\left[\frac{dM_x(d/2)}{I_x} + \frac{dM_x y}{I_x}\right]\left(\frac{d}{2} - y\right)b$$

or

$$\tau_z = \frac{dM_x}{dz}\left[\frac{1}{I_x}\left(\frac{d^2}{4} - y^2\right)\right] = \frac{V_x}{I_x}\left[\frac{1}{2}\left(\frac{d^2}{4} - y^2\right)\right]$$

But $(b/2)(d^2/4 - y^2) = Q_x$, the first moment of the area about the x axis of that part of

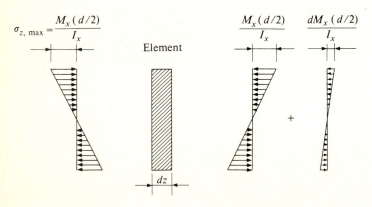

$$\sigma_{z,\,max} = \frac{M_x(d/2)}{I_x}$$

Element

$$\frac{M_x(d/2)}{I_x} \qquad \frac{dM_x(d/2)}{I_x}$$

$+$

Normal stress distribution
on rear face

Normal stress distribution
on front face

Figure 2.16b Distribution of longitudinal normal stresses due to bending moment.

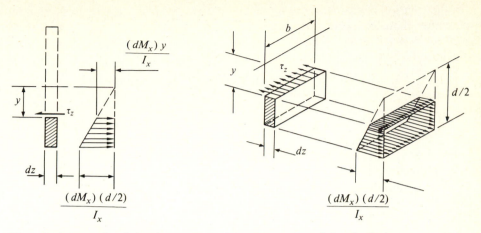

Figure 2.16c Normal and shearing stresses acting on a part of a unit length of a beam.

the face of the cross section between y and the bottom of the beam. Therefore,

$$\tau_z = \frac{V_y}{I_x} \frac{Q_x}{b} \tag{2.23}$$

(The first moment about the x axis, Q_x, refers only to the area outside of the location where the shearing stress is to be determined.)

Shearing stresses acting on the horizontal cut face shown in fig. 2.16c can be related to the vertical shearing stresses on the xy face of the element. In summation, they will equal the total applied vertical shear, V_y. Consider the element shown in fig. 2.16d. Taking moments about an axis parallel to the x axis through the center of the element yields

$$\tau_{yz} = \tau_{zy} = \tau_y$$

These shearing stresses—which must be equal—are often designated as τ_y.

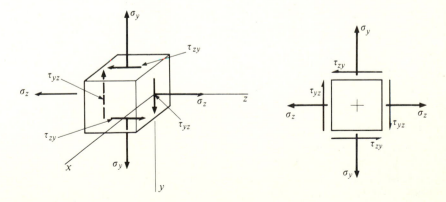

Figure 2.16d Stress components on a unit element.

In the solution of problems it is frequently more convenient to refer to shear-flow forces rather than to the shearing stresses. The shearing stress is a force per unit of area. *Shear flow*, on the other hand, is a shearing force per unit of length. For the case considered above

$$\text{Shear flow} = \tau_y b = \frac{V_y Q_x}{I_x} \tag{2.24}$$

(It is to be noted that for a particular location along a beam, the resultant shearing force and the moment of inertia will be known—and will be constant. The shear flow, therefore, will vary directly as the value of the first moment of the area varies.)

To illustrate the procedure that will be used in constructing shear-flow diagrams, consider the doubly symmetric I-shape cross section shown in fig. 2.17a. It is assumed that a vertical shearing force V is applied through the centroid of the section, in the direction of the web. Since V and I_x are constant, the only variable is Q_x. Further, since the section is symmetrical only one-half need be considered. Assuming that the thicknesses of the flanges and the web are small, behavior can be defined along the centerlines of the elements with reasonable accuracy. For the flanges, the first moment—and therefore the shear flow—will vary linearly, starting from zero at the outside edge. The maximum value would correspond to the use in the first-moments computation of the cross-sectional area shown in case 1 of

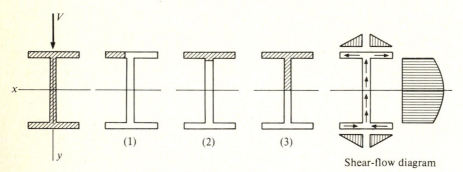

Figure 2.17a Shear flow in a wide-flange cross section.

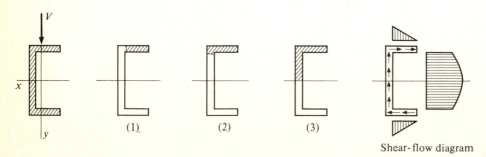

Figure 2.17b Shear flow in a channel cross section.

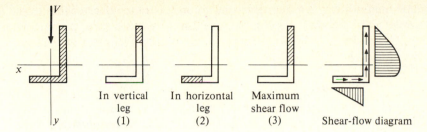

| In vertical leg (1) | In horizontal leg (2) | Maximum shear flow (3) | Shear-flow diagram |

Figure 2.17c Shear flow in an angle cross section.

fig. 2.17*a*. In the web, where it joins the flanges, the value of Q_x would be determined using the area indicated as case 2. Case 3 corresponds to the maximum value of the first moment of the area, and occurs at the midheight. (The reader should verify the direction of the shear flow shown in the final sketch.)

The shear flow in other shapes would be investigated in the same fashion. Figure 2.17*b* is for an idealized channel; 2.17*c* is for an angle.

Example 2.4 To illustrate these concepts, consider the simply supported prismatic beam of rectangular cross section shown in fig. 2.18*a*. The length is specified as 15 ft, and the member is composed of eight 2-in-thick layers of 8-in-wide lumber, laminated to provide an overall depth of 16 in. A load of $P = 5$ kips is applied at the midpoint of the beam.

From statics, the end reactions equal $1/2P$. The vertical shearing force is constant over each half of the span and equals this same value. The maximum bending moment is equal to $PL/4$ and occurs at the midpoint of the beam, directly under the applied load.

1. From the earlier discussion it should be evident that the maximum shearing stress occurs at the middepth of the section. Using eq. (2.23), the value would be

$$\tau_{max} = \frac{VQ}{Ib} = \frac{(P/2)(bd^2/8)}{\frac{1}{12}bd^3(b)} = \frac{3}{4}\frac{P}{bd} = \frac{3}{4}\frac{5000}{(8)(16)} = 29.3 \text{ psi}$$

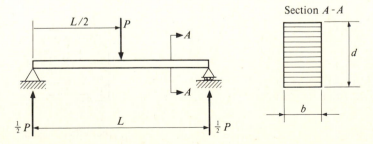

Section A-A

Figure 2.18a Laminated wood simple beam subjected to a concentrated lateral load.

The maximum normal bending stress occurs on the outer fiber of the beam. Therefore

$$\sigma_{max} = \frac{M(y_{max})}{I} = \frac{(PL/4)(d/2)}{\frac{1}{12}bd^3} = \frac{3}{2}\frac{P}{bd^2} = \frac{(3)(5000)(15)(12)}{(2)(8)(16)^2} = 659.2 \text{ psi}$$

2. For a second problem, assume that the maximum allowable shearing stress is $\tau_{all} = 100$ psi, and the maximum allowable normal bending stress is $\sigma_{all} = 1700$ psi. The problem is to determine the maximum load P that can be supported by the beam. Since for shear

$$\tau_{max} = \frac{3}{4}\frac{P}{bd}$$

$$P_{max, shear} = \frac{4}{3}\tau_{all}bd = \frac{(4)(100)(8)(16)}{(3)(1000)} = 17.07 \text{ kips}$$

For bending

$$\sigma_{max} = \frac{3}{2}\frac{PL}{bd^2}$$

or $$P_{max, bend} = \frac{2}{3}\frac{\sigma_{all}bd^2}{L} = \frac{(2)(1700)(8)(16)^2}{(3)(15)(12)(1000)} = 12.90 \text{ kips}$$

The lesser of these two must control. Therefore, the maximum load the beam can sustain is $P_{max} = 12.90$ kips, and the condition of failure is the attainment of the maximum allowable bending normal stress on the outside fiber of the midpoint section.

3. An obvious third problem for consideration is the determination of more suitable cross-sectional dimensions for the applied 5-kip load, assuming the allowable stress values given in problem 2. Since bending will probably govern

$$\sigma_{max} = \frac{3}{2}\frac{PL}{bd^2}$$

or

$$bd^2 = \frac{3}{2}\frac{PL}{\sigma_{all}} = \frac{3}{2}\frac{(5000)(15)(12)}{1700} = 794.12 \text{ in}^3$$

Assuming a width of member of 6 in, $d^2 = 794.12/6 = 132.35$, or $d = 11.5$ in. But since 2-in-thick strips are to be used, d will be chosen as 12 in.

Having selected a particular cross section on the basis of bending considerations, it now is necessary to check to make sure that the maximum shearing stress is not exceeded.

$$\tau_{max} = \frac{3}{4}\frac{P}{bd} = \frac{3\,(5000)}{4\,(6)(12)} = 52.08 \text{ psi}$$

Since this is less than the 100 psi allowed, the design is adequate. (It is to be recognized that the glue used to fasten the laminates together must have a long-term shearing strength of no less than τ_{max}.)

4. Is it possible, for the rectangular shape, to select a particular section which will simultaneously reach the allowable bending and shearing stresses specified in problem 2? Since

$$\sigma_{all} = \frac{3}{2}\frac{PL}{bd^2}$$

and

$$\tau_{all} = \frac{3}{4}\frac{P}{bd}$$

P/bd can be eliminated between the two, resulting in the expression

$$\frac{d}{L} = 2\,\frac{\tau_{all}}{\sigma_{all}}$$

or $d = 21.18$ in. Correspondingly, b would have to equal 1.77 in.

It must be here stated in the most emphatic terms that the design solution just obtained, while correct for the conditions specified, may not (and in all probability will not) correspond to an adequate solution. A 15-ft-long, 21.18-in-deep, 1.77-in-wide member is an exceedingly flexible member. Unless it is substantially braced in the lateral direction (that is, in the x direction), it may not even be able to support its own weight, much less the applied 5-kip force. (This is an obvious third design consideration that must be taken into account. The methods for handling this type of problem will be covered later in this text.) For completeness, the range of permissible cross-sectional dimensions for the two specified criteria are shown in fig. 2.18b.

Consideration has been given thus far to normal stresses or shearing stresses acting separately. In many cases, such an approach is valid, has been recognized in established procedures and codes, and leads to satisfactory design solutions. In others, however, particularly when using materials that show a pronounced weakness in tension, the combined effects of bending and shear must be taken into account.

The *unreinforced* concrete beam shown in fig. 2.19 represents such a case. Since failure (actual separation) may well depend on the tensile capacity of the material, it

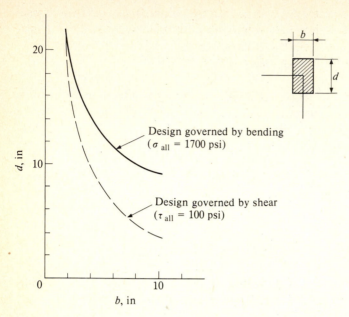

Figure 2.18b Permissible dimensions of the beam, based upon bending and shear.

is desirable to determine the magnitude and direction of the major and minor principal stresses at various points in the member.

Example 2.5 For illustration, the magnitudes and directions of the principal stresses at the points marked 1, 2, 3, and 4 in fig. 2.19 will be calculated. The length of the beam is given as 15 ft, as are the cross-sectional dimensions $b = 8$ in and $d = 12$ in. For convenience, an assumed uniformly distributed lateral load

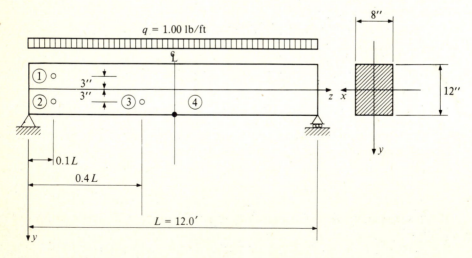

Figure 2.19 Simple beam subjected to a uniformly distributed lateral load.

$q = 1.0$ lb/ft has been selected for examination. (The stress produced by another load would simply be that new load magnitude multiplied by the stress produced by this unit case.)

The equations governing the stress components are

$$\sigma_z = \frac{My}{I_x} \qquad \text{and} \qquad \tau = \frac{V}{I_x b} Q_x$$

where

$$M = \frac{qx}{2}(L - x) \qquad \text{and} \qquad V = q\left(\frac{L}{2} - x\right)$$

For location 1, expressing all length quantities in feet,

$$\sigma_{z_1} = \frac{[(1)(1.2)/2](12 - 1.2)}{(\frac{1}{12})(\frac{2}{3})(1)^3}(-0.25) = -29.16 \text{ lb/ft}^2 \text{ (compression)}$$

$$\tau_1 = \frac{(1)(6 - 1.2)}{(\frac{1}{12})(\frac{2}{3})(1)^3(\frac{2}{3})}\left[\left(\frac{1}{4}\right)\left(\frac{2}{3}\right)\left(\frac{3}{8}\right)\right] = +8.11 \text{ lb/ft}^2$$

Table 2.1 summarizes the corresponding values for the location in question. These stress values are illustrated on unit elements in fig. 2.20a to e.

Table 2.1 Stresses at four locations in the beam

Point	Location, ft		Normal stress, lb/ft^2	Shearing stress, lb/ft^2
	z	y		
1	$+1.20$	-0.25	$\sigma_{z_1} = -29.16$ (compression)	$\tau_1 = +8.11$
2	$+1.20$	$+0.25$	$\sigma_{z_2} = +29.16$ (tension)	$\tau_2 = +8.11$
3	$+4.80$	$+0.25$	$\sigma_{z_3} = +77.76$ (tension)	$\tau_3 = +2.03$
4	$+6.00$	$+0.50$	$\sigma_{z_4} = +162.00$ (tension)	$\tau_4 = 0$

For combined normal and shearing stress the major and minor principal stresses can be computed from the relationship

$$\sigma_{\text{max, min}} = \frac{\sigma_z + \sigma_y}{2} \pm \sqrt{\left(\frac{\sigma_z - \sigma_y}{2}\right)^2 + \tau^2}$$

With reference to the z axis, the directions of the principal stresses are given by

$$\tan 2\theta = \frac{2\tau}{\sigma_z - \sigma_y}$$

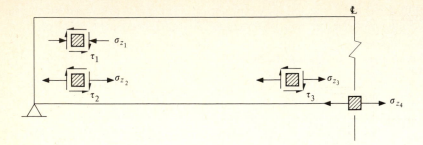

Figure 2.20a Normal and shearing stresses at four locations in the beam.

Consider each of the locations referred to above. In the first of the sketches is shown on an element the stress components in the z and y directions. The second sketch indicates the directions and magnitudes of the principal stresses at that same location.

Location 1

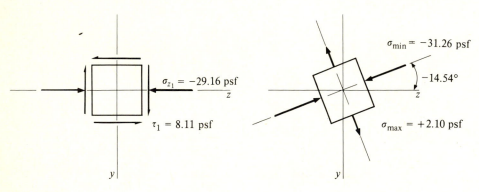

Figure 2.20b

$$\sigma_{max, min} = -\frac{29.16}{2} \pm \sqrt{\left(\frac{-29.16}{2}\right)^2 + (8.11)^2}$$

or

$$\sigma_{max, 1} = +2.10 \text{ lb/ft}^2$$

$$\sigma_{min, 1} = -31.26 \text{ lb/ft}^2$$

and

$$\theta_1 = -14.54°$$

Location 2

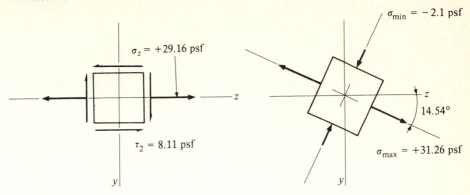

Figure 2.20c

$$\sigma_{\text{max, min}} = +\frac{29.16}{2} \pm \sqrt{\left(\frac{29.16}{2}\right)^2 + (8.11)^2}$$

$$\sigma_{\text{max, 2}} = +31.26 \text{ lb/ft}^2$$

$$\sigma_{\text{min, 2}} = -2.1 \text{ lb/ft}^2$$

and

$$\theta = +14.54°$$

Location 3

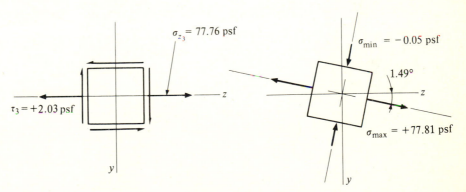

Figure 2.20d

$$\sigma_{\text{max, min}} = +\frac{77.76}{2} \pm \sqrt{\left(\frac{77.76}{2}\right)^2 + (2.03)^2}$$

$$\sigma_{\text{max, 3}} = +77.81 \text{ lb/ft}^2$$

$$\sigma_{\text{min, 3}} = -0.05 \text{ lb/ft}^2$$

and

$$\theta = +1.49°$$

Location 4

Since $\tau_4 = 0$, the computed σ_{z_4} is the principal stress.

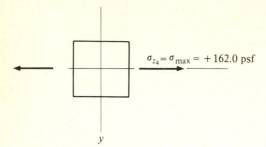

$\sigma_{z_4} = \sigma_{max} = +162.0 \text{ psf}$

Figure 2.20e

The following should be evident from these examples:

1. At the midpoint of the beam, where the bending moment is maximum and the shearing force is zero, the direction of the maximum normal stress is horizontal— throughout the entire depth of the beam.
2. At the end of the beam, where the bending moment is zero and the shearing force is at its maximum value, the direction of the maximum normal stress is inclined at 45° to the z axis—throughout the entire depth of the beam.
3. Along the entire length of the beam, at $y = +d/2$, the bending normal stress is maximum (and in tension), the shearing stress is zero, and the direction of the maximum tensile stress is horizontal.
4. Along the entire length of the beam, at $y = 0$, the bending normal stress is zero, the shearing stress is maximum, and the maximum tensile stress is inclined at 45° to the z axis.
5. At other locations along and through the depth of the beams, the magnitude and direction of the maximum tensile stresses vary and depend on the values of the normal and shearing stresses.

Consider now the determination of the location of the shear center (SC) of a channel type cross section. (It is to be recalled that the shear center is that particular

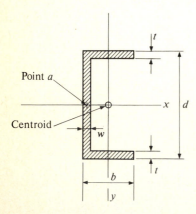

Figure 2.21a Dimensions of a channel cross section.

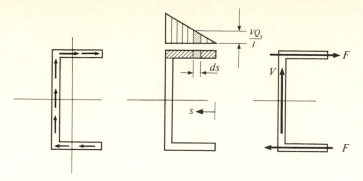

Figure 2.21b Shear flow in a channel cross section.

point through which applied shear loads will produce zero twisting moment, and for which the shearing stresses will be distributed in accordance with elementary bending theory of beams.) The dimensions of the cross section are shown in fig. 2.21a. It is assumed that the applied shear force is in the y direction.

The shear flow in the section would be directed as indicated in the first of the sketches of fig. 2.21b. The shear flow in the upper flange varies linearly as indicated. The integrated total force in that flange due to the shear flows, therefore, would be

$$F = \int_0^b \frac{VQ_s}{I_x}\, ds = \frac{V}{I_x}\left[\frac{A_f(d/2 - t/2)b}{2}\right]$$

where V = the applied vertical shearing force acting in the y direction

I_x = the moment of inertia of the entire cross section about the x axis

A_f = the area of one of the flanges, bt

In order to have zero torsional moment due to the developed shear flow forces in the flanges, the applied vertical shear force must act through the shear center as shown in fig. 2.21c. Taking moments about point a:

$$Ve = F(d - t)$$

or

$$e = \frac{A_f(d/2 - t/2)}{2I_x}\, bd \tag{2.25}$$

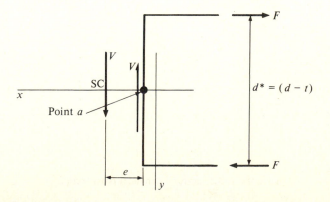

Figure 2.21c Equilibrium of shear-flow components.

Example 2.6 Locate the shear center for a C 15 × 50† channel cross section. The necessary geometrical properties for this are (see fig. 2.21d):

$A_f = 2.18$ in^2

$I_x = 404$ in^4

$b = 3.716$ in

$d = 15.00$ in

$t_f = 0.650$ in

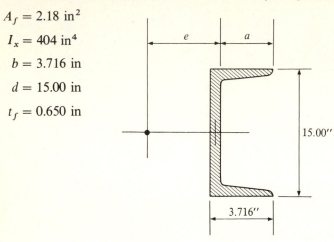

15.00″

3.716″

Figure 2.21d

(A_f is taken as the distance a times the average flange thickness.)

$$e = \frac{(2.18)(1.5/2 - 0.65/2)(3.716)(15.0)}{(2)(404.0)} = 1.079 \text{ in}$$

2.7 STRESSES AND STRESS RESULTANTS DUE TO TORSION

When slender members are subjected to moments about their longitudinal axis, torsional stresses result. The development of the relationships between these stresses and the applied twisting moment M_z is the purpose of this section.

Consider for illustration of the basic concepts the circular, tube-type cross section shown in fig. 2.22. The inside radius of the shaft is r_i, and the outside radius is r_o. τ is the torsional shearing stress at the general point (ψ, r).

For equilibrium between these torsional shearing stresses and the applied torsional moment

$$M_z = \int_{r_i}^{r_o} r\tau \, dA \tag{2.26}$$

where

$$dA = r \, d\psi \, dr$$

$$M_z = 2 \int_0^\pi \int_{r_i}^{r_o} \tau r^2 \, dr \, d\psi = 2\pi \int_{r_i}^{r_o} \tau r^2 \, dr \tag{2.27}$$

† C 15 × 50 is the standard AISC (American Institute of Steel Construction) designation for a rolled steel channel (C) 15 in deep and weighing 50 lb/ft of length. The properties of this and all other rolled steel shapes are contained in the AISC *Manual of Steel Construction*.

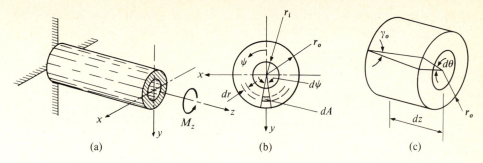

Figure 2.22 Circular tube-type cross section subjected to torsional moments.

It is presumed that the shearing strains γ produced by the twisting moment vary linearly with r to a maximum value at the outer edge of the shaft γ_o. Since for a hookean material $\tau = \gamma G$, where G is the shear modulus of elasticity, the stress also must vary linearly with r to a maximum τ_o on the outer surface.

$$\tau = \tau_o \frac{r}{r_o}$$

The torsional moment M_z therefore, is given by

$$M_z = 2\pi \int_{r_i}^{r_o} \frac{\tau_o}{r_o} r^3 \, dr = \frac{\tau_o}{r_o} \kappa_T \tag{2.28}$$

where κ_T is the St. Venant torsion constant. For this particular shape, κ_T equals the polar moment of inertia of the cross section. The maximum shearing stress is

$$\tau_{max} = \tau_o = \frac{M_z r_o}{\kappa_T} \tag{2.29}$$

For the differential length of member shown in fig. 2.22c, assuming that the angles in question are small

$$\gamma_o \, dz = r_o \, d\theta$$

$d\theta$ is the angle of twist of one face with respect to the other that occurs over the differential length in question. Therefore

$$\frac{d\theta}{dz} = \frac{\gamma_o}{r_o} = \frac{\tau_o}{r_o G} = \frac{M_z}{G\kappa_T} \tag{2.30}$$

which defines the angle of twist per unit of length along the shaft. The term $G\kappa_T$ is referred to as the *St. Venant torsional rigidity*. The total angle of twist, θ, at location z (assuming that $\theta = 0$ at $z = 0$) is given by

$$\theta = \int_0^z \frac{M_z}{G\kappa_T} \, dz \tag{2.31}$$

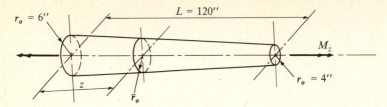

Figure 2.23 Solid, linearly tapered circular shaft subjected to torsional moments.

Example 2.7 To illustrate the use of the relationships just developed, consider the solid, linearly tapered, circular shaft shown in fig. 2.23. The member is 120 in long, has an outside radius at $z = 0$ of 6 in and at $z = 120$ in of 4 in. It is subjected to a twisting moment of M_z applied at each end as shown. (Since the shaft is solid, $r_i = 0$.)

The torsional constant κ_T is defined by the relationship

$$\kappa_T = 2\pi \int_0^{\bar{r}_o} r^3 \, dr = \frac{2\pi \bar{r}_o^4}{4} = \frac{\pi \bar{r}_o^4}{2}$$

where

$$\bar{r}_o = \text{outside radius at } z = 6\left(1 - \frac{z}{3L}\right) = 6\left(1 - \frac{z}{360}\right)$$

Therefore

$$\kappa_T = \frac{\pi}{2}\left[6\left(1 - \frac{z}{360}\right)\right]^4$$

The maximum shearing stress at section z due to the applied torsional moment M_z occurs at the surface and equals

$$\tau_{o_z} = \frac{M_z \bar{r}_o}{\kappa_T} = \frac{M_z}{(\pi/2)[6(1 - z/360)]^3}$$

Since M_z is constant along the length $\tau_{o,\,\text{max}}$ will occur where z has the largest value, and $\bar{r}_o$ has its smallest value.

$$\tau_{\text{max}} = \frac{M_z}{(\pi/2)[6(1 - \frac{1}{3})]^3} = \frac{M_z}{32\pi}$$

The total angle of twist of one end of the member with respect to the other is

$$\theta = \int_0^{120} \frac{M_z}{G\kappa_T} \, dz = \frac{2M_z}{G\pi} \int_0^{120} \left[6\left(1 - \frac{z}{360}\right)\right]^{-4} dz$$

Torsion of thin-walled closed cross sections For a thin-walled closed cross section the magnitude of the shearing force per unit of peripheral length around the cross section (that is, the shear flow) is constant regardless of whether or not the wall thickness

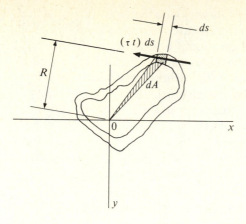

Figure 2.24 Shear flow in thin-walled closed cross sections due to torsion.

varies. This can be demonstrated by considering equilibrium of a differential element of dimensions $dz \times ds \times$ thickness along the edge in question.

The moment of the differential shearing force about a z axis through point 0 (see fig. 2.24) is

$$dT = R(\tau t)\, ds$$

where R is the perpendicular distance from the shearing force to point 0. From the sketch it is evident that $dA = \frac{1}{2}R\, ds$. Therefore

$$T_z = \int_S (\tau t)R\, ds = 2 \int_A (\tau t)\, dA = 2(\tau t)A$$

or

$$\tau t = \frac{T_z}{2A}$$

A is the total area enclosed by the centerline of the cross section of the thin-walled member, and t is the thickness of that particular location where the shearing stress is desired.

The total angle of twist from one end of the member to the other is

$$\theta = \frac{L}{2AG}\, \tau t \int_S \frac{ds}{t}$$

Torsion of noncircular and open cross sections When a circular-cross-section shaft is twisted and the deformations are relatively small, the cross section remains in a plane and the shearing stresses at every point in that plane act in a direction perpendicular to a radius vector. Such is not true when the shaft is other than circular in form. For such other cases, except for a few special locations on the cross section, the shearing stress has components both perpendicular to the radius vector and in the direction of the radius vector. This extra shearing force results in a shearing strain both within the plane of the cross section and normal to it. Since the shearing-force components vary from location to location, the cross section cannot remain flat. This out-of-plane distortion is called *warping*. It will exist for all but circular-cross-section shafts subject to twisting.

The magnitudes of the warping displacements vary with the type of cross section. Those which more nearly approximate the circular case warp less than those which deviate considerably from that type of configuration. For example, a circular shaft with a keyway cut into it would not warp as much as a shaft of square cross section. In turn, a square shaft would warp less than a shaft whose cross section is a thin-walled open cross section such as an I shape or a channel.

It should here be noted that warping displacements, in and of themselves, are not usually of major significance. In fact, seldom are they calculated. However, if these warping displacements are resisted, very high normal stresses—called *warping normal stresses*—will develop, and these can be important. (This is much like the problem of thermal stresses in that no stresses result if the member is allowed to expand and contract at will. If these changes in size and shape are resisted in any way, however, stresses are developed whose magnitudes are dependent upon the amount of thermal displacement that could not occur.) If a square bar is rigidly attached to two flat plates on its ends and then twisted, normal stresses of the type shown in fig. 2.25a develop. It is to be understood that the warping normal stresses σ_ω are self-equilibrating, that is, their integrated effect over the cross section is zero.

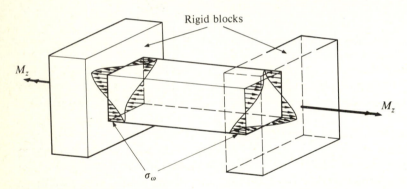

Figure 2.25a Warping normal stresses.

For bars of noncircular cross section subjected to twisting two types of phenomena are observed. The stresses associated with the first of these is referred to as *St. Venant* or *uniform* torsional shearing stresses and those associated with the second as *nonuniform* or *warping* stresses. (It should be evident that this problem is highly indeterminant and can only be resolved using procedures of a more sophisticated nature than those that have thus far been developed. These involve a rigorous continuum mechanics approach—frequently in conjunction with model studies.)

The St. Venant shearing stress in a rectangular bar normally is presented in the same form as that used for the circular shaft. That is

$$M_{z,\,\mathrm{sv}} = G\kappa_T \, \frac{d\theta}{dz} \qquad (2.32)$$

where $M_{z,\,\mathrm{sv}}$ = the torsional moment
κ_T = the St. Venant torsion constant
θ = the total angle of twist

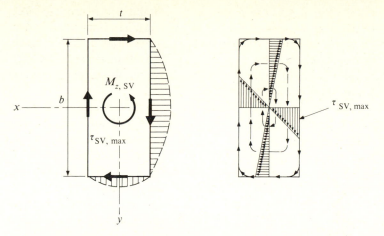

Figure 2.25b St. Venant shearing stresses in a rectangular cross section subjected to torsional moments.

The expression for the maximum shearing stress is

$$\tau_{SV,\,max} = \frac{M_{z,\,SV}\,t}{\kappa_T} = Gt\,\frac{d\theta}{dz} \tag{2.33}$$

Here t is the thickness of the rectangular bar, as indicated in fig. 2.25b. It is to be recognized from the typical stress drawing of fig. 2.25b that in the x and y coordinate directions the shearing stresses are parallel to the outside surface of the member and increase from zero at the center of the member to a maximum at the surface. $\tau_{SV,\,max}$ occurs at the center of the longer sides as indicated. The shearing stresses due to St. Venant type torsion at the four corners of the rectangular cross section equal zero.

One rectangular-cross-section case is of particular interest to the structural engineer—that corresponding to the case where the aspect ratio b/t becomes large. For this situation the torsion constant κ_T can be approximated with reasonable accuracy by the expression

$$\kappa_T \simeq \tfrac{1}{3}bt^3 \tag{2.34}$$

Moreover, it has been shown that when a cross section is of the open type and consists of several thin plate elements rigidly attached one to the other to form the "thin-walled" shape in question, κ_T can be taken as the sum of the torsion constants of each of the parts, that is,

$$\kappa_T = \sum_{i=1}^{n} \kappa_{Ti} = \frac{1}{3}\sum_{i=1}^{n} b_i t_i^3 \tag{2.35}$$

where i refers to any one of the n connected plate elements.

Example 2.8 To illustrate the procedure, consider the I shape shown in fig. 2.26. The material is steel, and has a shear modulus of $G = 12 \times 10^6$ lb/in². The length of the member is 12 ft. It is assumed that there is no resistance to warping

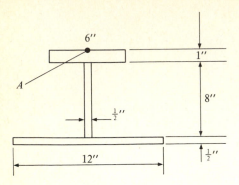

Figure 2.26

deformation (that is, $\sigma_\omega = 0$). The problem is to determine the maximum shearing stress and the angle of twist that results when the ends of the member are subjected to equal and opposite twisting moments of $M_z = 50,000$ in·lb.

The St. Venant torsion constant for the cross section in question is

$$\kappa_T = \frac{1}{3} \sum_{i=1}^{3} b_i t_i^3 = \frac{(6)(1)^3 + (8)(0.5)^3 + (12)(0.5)^3}{3} = 2.833 \text{ in}^4$$

Since all plate elements rotate through the same angle, the largest shearing stress will occur at the midpoint of the sides—with the maximum at the midpoint of the thickest plate. The maximum shearing stress in this case, therefore, occurs at point A in the figure and has a value of

$$\tau_{SV,\,max} = \frac{M_{z,\,SV}\,t}{\kappa_T} = \frac{(50,000)(1)}{2.833} = 17.649 \text{ ksi}$$

Since the bar is of prismatic cross section, $d\theta/dz$ has a constant value per unit of length. The total angle of twist θ is, therefore, given by

$$\theta_L = \frac{M_{z,\,SV}\,L}{G\kappa_T} = \frac{(50,000)(12)(12)}{(12,000,000)(2.833)} = 0.212 \text{ rad}$$

The presence or absence of warping normal stresses in a real, thin-walled, open cross-section member depends upon how the member is supported, and how it is loaded. To illustrate this, consider the three I type shapes loaded as shown in fig. 2.27.

For the first member shown, case a, twisting moments are applied to the ends of the member. These ends are supported in such a way that rotation about the z axis can occur, no vertical deformations are allowed, and there is no resistance to warping displacements. The twisted shape is illustrated. Since there are no warping stresses present, this is a case of uniform torsion, and the flanges of the member remain straight. The only torsional stresses induced are St. Venant stresses. (This case arises in practice when a simply supported member is twisted at its ends by another member.)

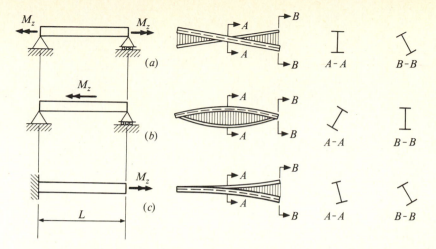

Figure 2.27 Rotations of wide-flange members due to torsional moments.

In case b of fig. 2.27, the member is also presumed to be simply supported—in terms of allowed rotation in the yz plane. However, the ends of the member are not assumed to be free to rotate about the z axis—although both are free to warp. The member is subjected to a concentrated torque M_z applied at the midpoint of the beam. (A load of this type would be produced, for example, by an eccentrically placed, concentrated, lateral load.) From the indicated deformed shape, it should be evident that the flanges do not remain straight—there will be normal stresses. Because of symmetry, all warping displacements at the midpoint of the beam are eliminated. At the ends of the member, however, warping is free to occur—and will. Thus, it can be concluded that for the case in question, warping normal stresses will exist in the member, and will vary from a maximum value at the center of the beam to zero at both ends.

The ability of a member such as the one shown in case b to resist torsion is clearly derived from both its St. Venant (or uniform torsional strength) and from its warping resistance. In this case the contribution of the uniform torsional strength is maximum at the ends of the member, and decreases toward the center. The warping strength is maximum at the center, and decreases toward the ends.

The structure and loading indicated in case c are interesting, not just because they appear different from the other cases, but because a critical comparison will indicate that c represents one-half the beam shown in case b.

In the most general form, it can be concluded that applied torques are resisted by the sum of the uniform (St. Venant) torsional resistance and the warping resistance. That is,

$$M_z = M_{z,\mathrm{sv}} + M_{z_\omega} \tag{2.36}$$

Since the St. Venant case already has been discussed, attention will be directed to an examination of the warping of I and W shapes.

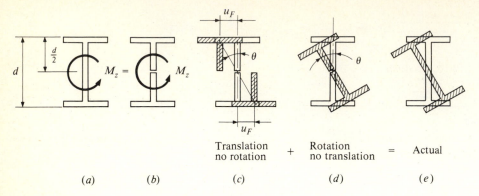

Translation   +   Rotation   =   Actual
no rotation   no translation

(a)   (b)   (c)   (d)   (e)

Figure 2.28 Warping of I and W shapes due to torsional moments.

If it is assumed that the angle of twist is small, and that the relative geometry of the cross section does not change as the member rotates, the warping effects can be described by considering lateral bending of the member as twisting occurs. Imagine that the deformation takes place in two stages. First, the member is presumed " cut " at the midheight as shown in fig. 2.28. Next, (in sketch c), translation—but no rotation—of the flanges is assumed. This is followed by rotation—but no translation. In simple terms, for this type of cross section, the translation of the flanges accounts for the primary warping effect. Rotation corresponds to St. Venant effects.

Consider the translational deformation indicated in sketch c. This is reproduced (for the top half) in fig. 2.29. If M_F is the moment that is induced in the flange due to the lateral deformation u_F, then

$$M_F = -EI_F \frac{d^2 u_F}{dz^2} \tag{2.37}$$

where I_F is the moment of inertia of the T section about a vertical axis through the web. For small angles

$$u_F = \frac{d}{2} \theta$$

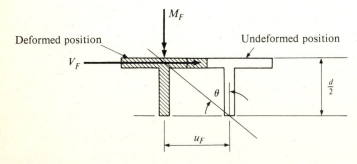

Figure 2.29 Translational deformation of the flange.

Therefore

$$M_F = -(EI_F)\frac{d}{2}\frac{d^2\theta}{dz^2} \tag{2.38}$$

The induced shear force in the flange is

$$V_F = -(EI_F)\frac{d}{2}\frac{d^3\theta}{dz^3} \tag{2.39}$$

The warping resistance of the cross section is then the result of the couple formed by these two equal and opposite flange shear forces, that is,

$$M_{z_\omega} = V_F d = -(EI_F)\frac{d^2}{2}\frac{d^3\theta}{dz^3} \tag{2.40}$$

For the I shape, I_F approximates the value $\frac{1}{2}I_y$ for the total cross section. Therefore

$$M_{z_\omega} = -(EI_y)\frac{d^2}{4}\frac{d^3\theta}{dz^3} = -EI_\omega\frac{d^3\theta}{dz^3} \tag{2.41}$$

where I_ω is referred to as the warping constant of the shape.

While this equation was developed specifically for the I or W shape, it can be shown that the form is valid for most thin-walled, open cross sections. The only difference is in the value of the warping constant I_ω. For this reason I_ω is normally included as a tabulated cross-sectional property in most design handbooks. The same is true for the St. Venant torsion constant κ_T.

It was noted previously that the torsional resistance of a member is the sum of that due to warping and to uniform rotational resistance. Twisting equilibrium is, therefore, defined by the relationship

$$M_z = M_{z,\text{sv}} + M_{z_\omega}$$

$I_W = d^2/4\,(I_y)$

or

$$M_z = G\kappa_T\frac{d\theta}{dz} - EI_\omega\frac{d^3\theta}{dz^3} \tag{2.42}$$

Defining a distributed torque as

$$m_z = \frac{dM_z}{dz}$$

the more normal fourth-order equation is obtained

$$m_z = G\kappa_T\frac{d^2\theta}{dz^2} - EI_\omega\frac{d^4\theta}{dz^4} \tag{2.43}$$

In its standard form

$$\frac{d^4\theta}{dz^4} - \frac{G\kappa_T}{EI_\omega}\frac{d^2\theta}{dz^2} = -\frac{m_z}{EI_\omega} \tag{2.44}$$

In summary, three types of stresses are induced in a thin-walled, open-type cross section subjected to twisting:

1. St. Venant shearing stresses τ_{SV}: The maximum value of these stresses is given by the equation

$$\tau_{SV, \, max} = Gt \frac{d\theta}{dz}$$

and occurs on the surface at the midpoint of the thickest part of the section.

2. Warping normal stress σ_{z_ω}: Considering the flange to be a rectangular beam subjected to bending,

$$M_F = -EI_F \frac{d^2 u_F}{dz^2} = -\frac{EI_y}{2} \frac{d}{2} \frac{d^2\theta}{dz^2}$$

Defining ζ as the horizontal distance from the center of web to the longitudinal fiber in question in the flange,

$$\sigma_\omega = \frac{M_F}{I_F} \zeta$$

The warping normal stress is therefore

$$\sigma_{\omega, \, max} = \frac{M_F b}{2(I_y/2)} = \frac{M_F b}{I_y} = -\frac{Edb}{4} \frac{d^2\theta}{dz^2} \tag{2.45}$$

where b is the flange width.

3. Warping shearing stress τ_ω: Since the induced shearing force in the flange is

$$V_F = -\frac{EI_y}{4} d \frac{d^3\theta}{dz^3}$$

the warping shearing stress is given by eq. (2.46)

$$\tau_\omega = \frac{V_F Q}{I_F t_F} = \frac{V_F Q}{(I_y/2)t_F} = -\frac{EdQ}{2t_F} \frac{d^3\theta}{dz^3} \tag{2.46}$$

The maximum value occurs at the center of the flange where Q is the greatest. For this location

$$Q = \frac{b^2 t_F}{8}$$

and

$$\tau_{\omega, \, max} = -\frac{Edb^2}{16} \frac{d^3\theta}{dz^3} \tag{2.47}$$

The distribution of these stresses for a typical I shape is shown in fig. 2.30.

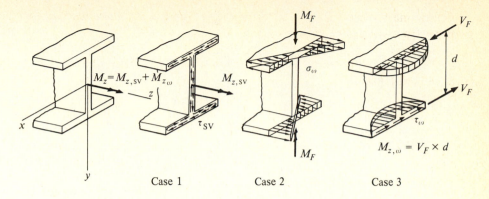

Figure 2.30 Normal and shearing stresses in a wide-flange shape due to torsional moments.

Example 2.9 The beam shown in fig. 2.31 is subjected to a concentrated vertical load of 10 kips, 7.5 ft from its end, and eccentric 5 in from the plane of the web. The member is a **W 12 × 50** and is simply supported at its ends as shown in the figure. It is desired to determine the normal stress distribution across the section at the center of the beam.

The maximum bending normal stress at the centerline section is given by

$$\sigma_b = \frac{M_b}{S} = \frac{PL/4}{S} = \frac{(10)(15)(12)/4}{64.7} = 6.95 \text{ ksi}$$

The warping normal stress is obtained using eq. (2.45)

$$\sigma_{\omega,\text{max}} = \frac{Edb}{4}\frac{d^2\theta}{dz^2}$$

where $d^2\theta/dz^2$ is defined by eq. (2.42). (For these conditions of support, $d^2\theta/dz^2$

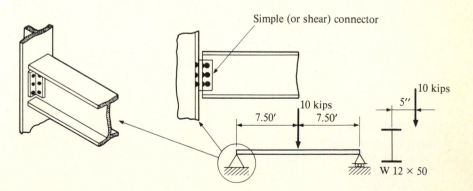

Figure 2.31

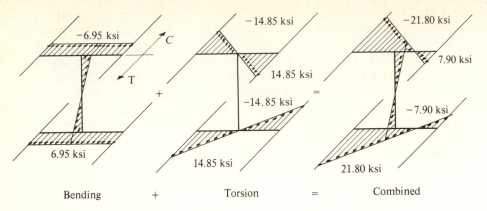

Figure 2.32 Normal stresses due to bending and torsion—example 2.9.

at the centerline section of the member is found to be -0.0219×10^{-3}.)[†] The maximum normal stress is, therefore,

$$\sigma_{\omega,\,max} = (-29.0 \times 10^{3})(23.3)(-0.0219 \times 10^{-3}) = 14.85 \text{ ksi}$$

The resulting combined bending and torsion normal stress distribution is shown in fig. 2.32. The very large increase in normal stress that is caused by twisting should be noted.

2.8 STRESSES AND STRESS RESULTANTS DUE TO AXIAL THRUST

An important consideration in structural analysis and design is the influence of axial thrust. When this is the only external force acting on the member, and when the applied loads are compressive, the members are referred to as *struts*, *columns*, or *stanchions*. When the loads are tensile, often they are called *tie bars*.

In most cases (for those where the applied axial thrust acts through the centroid of the cross section) the relationship between the stress resultant and the resulting normal stresses σ_z is obtained from a consideration of the first of the equilibrium equations defined by eq. (2.10). A uniform strain distribution will be presumed, and the governing relationship will be

$$\sigma_z = \frac{P_z}{A} \tag{2.48}$$

Here, A is the area of the cross section of the member. (It should be emphasized that this equation holds only so long as it is reasonable to presume that the strains remain uniform across the section. Good results are to be expected when investigating the

[†] See chap. 4 for a detailed discussion of the solutions associated with torsional deformations.

stresses at the center of a long tie bar. Near the connections at the ends of the bar, such will not be the case and other assumptions or analyses must be made.)

When the axial force acting on a member is compressive, the possibility of another phenomena occurring also must be considered—*buckling*. Assuming that the member in question is straight and that the applied axial thrust acts along the axis of the centroids of the cross section, and that it increases continuously from zero, the initially observed deformations are only axial shortening. These vary directly with the magnitude of the applied force. There can be reached, however, a particular load for which the member displays an equal affinity for a laterally bent configuration—as opposed to its initially straight form. That *critical load* is referred to as the *buckling load* and corresponds to bifurcation of the load-deformational relationship. As the thrust is increased in magnitude beyond this value—even by the smallest amount— lateral bending occurs.

In many cases safety against buckling or against instability is more critical to adequate design than, for example, safety against insufficiency of yield strength. For this reason, this topic will be treated in considerable detail in a later chapter of this text.

A large number of problems that are considered in structural analysis and design involve members subjected to combined axial thrust and bending. The bending can be either about one or both of the principal axes. For first-order type problems, that is, those that can be formulated in the undeformed state, the normal stresses that develop due to the various stress resultants can be separately considered and super-imposed one on the other. But many beam-column type problems do not fall in this category, and a more rigorous solution, one that includes the interaction between resultant bending moments along the member and the applied axial thrust times the lateral deflection, must be developed. These are often referred to as *second-order type problems*—where equilibrium expressions must be formulated in the *deformed state*.

For first-order type problems, and for the determination of axial normal stresses at sections for which the resultant thrust and bending moments are known,

$$\sigma_z = \sigma_{z,\,\text{axial force}} \pm \sigma_{z,\,\text{bending}}$$

Combining eqs. (2.14) and (2.48)

$$\sigma_x = \frac{P}{A} \pm \frac{M_x y}{I_x} \tag{2.49}$$

Bending about both the x and the y axes would yield

$$\sigma_z = \frac{P}{A} \pm \frac{M_x y}{I_x} \pm \frac{M_y x}{I_y} \tag{2.50}$$

(The sign convention associated with eqs. (2.48), (2.49), and (2.50) is not consistent with that previously defined. For these latter three equations compressive stresses have been presumed to be positive—as is normal convention in most discussions of columns and in most texts on stability. Since these stresses are to be combined with bending moments whose directions are either known or readily determined, there should result little, if any, confusion due to this inconsistency.)

Example 2.10 The rectangular-cross-section beam shown in fig. 2.33a is subjected to an axial compressive force of $P = 100$ kips and a bending moment about the x axis of $M_x = 1600$ in·kips. It is desired to establish the stress distribution across the section. From eq. (2.49)

$$\sigma_z = \frac{P}{A} \pm \frac{M_x y}{I_x} = \frac{(100)}{(6)(12)} \pm \frac{1600y}{(\frac{1}{12})(6)(12)^3} = 1.39 \pm 1.85y$$

This is shown diagrammatically in fig. 2.33b. Note that the neutral axis for the stress distribution shown is an axis parallel to x axis which passes through the point of zero normal stress.

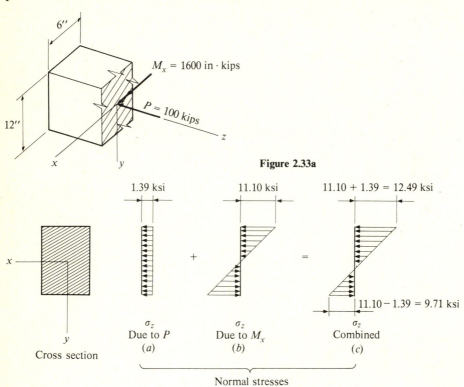

Figure 2.33a

Figure 2.33b Normal stresses due to axial thrust and bending moment.

Example 2.11 For the cross section shown in fig. 2.33a assuming $M_x = 1600$ in·kips in the direction shown, determine the axial force necessary to produce zero tensile stresses in the bottom fibers of the beam. (That is, the neutral axis should coincide with the bottom edge.) Since

$$\sigma_z = \frac{P}{A} \pm \frac{M_x y}{I_x}$$

with $\sigma_z = 0$ at $y = d/2$, $\quad 0 = \frac{P}{72} - \frac{(1600)(6)}{(864)}$

or $\qquad\qquad\qquad P = 800$ kips (compression)

Example 2.12 The 14×68 lb/ft wide-flange steel shape shown in fig. 2.34a is subjected to axial compression plus bending about both of its principal axes.

$$\sigma_{z,\,\text{ext}} = \frac{P}{A} \pm \frac{M_x}{S_x} \pm \frac{M_y}{S_y} = \frac{80}{20.00} \pm \frac{600}{103.0} \pm \frac{50}{24.1}$$

or

$$\sigma_{z,\,\text{ext}} = 4.00 \pm 5.83 \pm 2.08$$

The resulting stress diagram is shown in fig. 2.34b

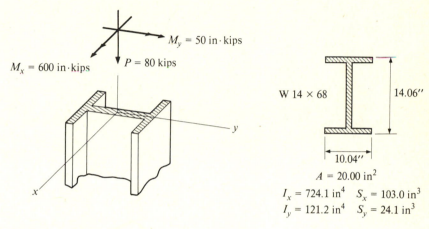

$M_y = 50 \text{ in} \cdot \text{kips}$

$M_x = 600 \text{ in} \cdot \text{kips}$ $P = 80 \text{ kips}$

y

x

W 14×68 $14.06''$

$10.04''$

$A = 20.00 \text{ in}^2$

$I_x = 724.1 \text{ in}^4$ $S_x = 103.0 \text{ in}^3$
$I_y = 121.2 \text{ in}^4$ $S_y = 24.1 \text{ in}^3$

Figure 2.34a

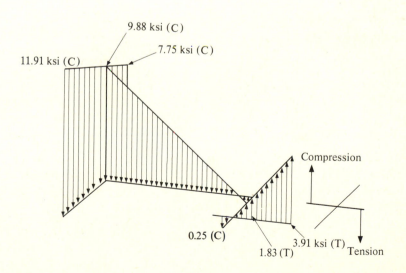

9.88 ksi (C)

7.75 ksi (C)

11.91 ksi (C)

Compression

0.25 (C)

3.91 ksi (T)

1.83 (T)   Tension

Figure 2.34b Normal stresses—example 2.12.

2.9 NONHOMOGENEOUS MEMBERS

Thus far in this chapter consideration has been given only to members that are homogeneous and isotropic. A single stress-strain relationship was sufficient to describe the load-deformational properties of such members. In many real cases, however, several materials may be joined together to form the structural system in question. For these nonhomogeneous members, the stress-strain relationships of each of the component parts must be considered in the analysis.

Actually, there are no substantial differences in the methods used to define stresses in nonhomogeneous members from those previously developed. As before, there is first considered the equilibrium of appropriately selected free bodies. Next, reasonable strain distributions are presumed. Finally, the appropriate stress-strain equations are introduced.

The assumption that is most often made regarding strains is that the various component parts from which the section is made are so attached at their point of junction that they do not move, one with respect to the other. The strains at the interface—regardless of which of the two parts is selected for examination—can be presumed to be the same. Another way to say the same thing: strain continuity exists. In certain cases, only partial continuity can be presumed. For these cases, further assumptions would have to be made relative to the amount of slippage that would correspond to a given strain distribution.

Probably the most commonly encountered nonhomogeneous members are those composed of *reinforced concrete*. Within the general category—steel reinforced concrete is most often observed. Other common examples of nonhomogeneous members are rolled structural steel beams bonded (or otherwise attached) to reinforced concrete flanges, members fabricated of various different grades of steel, "sandwich" members having metallic outer surfaces and substantially different cores, and metal-reinforced timber members.

To illustrate the general procedure, consider the rectangular-cross-section sandwich beam shown in fig. 2.35a. It consists of two flanges bonded to a core of another material. If it is assumed that the flanges are sufficiently thin to be ineffective in resisting shear and, further, if it is assumed that the core material is strong in shear but weak in normal stress (that is, bending), for all purposes the flanges will supply the entire bending resistance and the web will resist the entire shear. An approximation to the actual situation can, therefore, be obtained as follows: Considering only the flanges, and defining the forces developed in them as F_U and F_L, a summation of these forces requires that $F_U = F_L$. In turn, summing moments about the x axis yields

$$F_U d^* = M_x$$

where d^* is the distance between the lines of action of the flange forces. If a linear strain distribution is presumed as indicated in sketch a of fig. 2.35b, the stress distribution shown in b will follow. Therefore

In the flanges: $\qquad\qquad \varepsilon_z = \dfrac{\sigma_z}{E} \qquad$ or $\qquad \sigma_z = E\varepsilon_z$

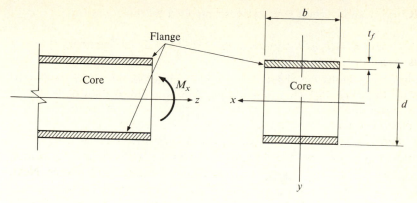

Figure 2.35a Sandwich-type bending member.

In the core: $\qquad\qquad\sigma_z = 0$

From the figures it also should be evident that

$$F_U = \tfrac{1}{2}(\sigma_{z_T} + \sigma_{z_L})bt_f = \sigma_{z_f} A_f$$

and

$$M_x = F_U d^* = \sigma_{z_f} A_f d^*$$

where $d^* = d - t_f$. But for the flanges $\sigma = E\varepsilon$ and $\varepsilon = -(d/2)\phi$. Recognizing that by definition $d^* \simeq d$, it follows that

$$M_x = -\frac{d}{2}\phi E A_f d = -E\frac{A_f d^2}{2}\phi$$

Since I_x for the bending resistant portion of the sandwich beam is given by

$$I_x = 2A_f\left(\frac{d}{2}\right)^2 = \frac{1}{2}A_f d^2$$

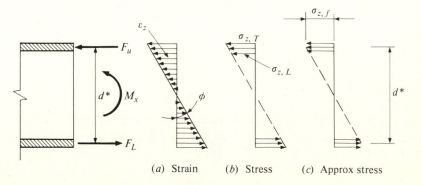

(*a*) Strain　　(*b*) Stress　　(*c*) Approx stress

Figure 2.35b Normal stresses in sandwich beam.

the bending moment and normal stresses are controlled by the equations

$$M_x = -EI_x \phi \quad \text{and} \quad \sigma_{z_f} = E\frac{M_x}{EI_x}\frac{d}{2} = \frac{M_x}{I_x}\frac{d}{2}$$

The use of nonhomogeneous members of this type is quite common in a variety of structural applications, particularly where a high bending-strength-to-weight ratio is needed.

2.9.1 Reinforced Concrete Members

Many materials are very strong in one sense, and weak in another. By appropriately combining the useful properties of such materials, desirable structural elements can be fabricated (or constructed). One very common combination of materials is concrete containing slender steel rods.

To illustrate how these two totally different materials work together, consider the compression block shown in fig. 2.36. The member consists of several steel bars embedded in the concrete such that their longitudinal axes are in the direction of the applied load P. Collectively, the bars have a cross-sectional area of A_s, and their modulus of elasticity is E_s. A_c represents that part of the cross-sectional area occupied by the concrete. The modulus of elasticity of the concrete is E_c. It is assumed that no slippage occurs between the steel bars and the concrete as deformations occur—that is, it is assumed that they are bonded together. Moreover, if the applied force P acts through the centroid of both the concrete and the steel portions of the cross section (and because of the assumed rigid caps) the strains will be uniform over the entire cross section. Equilibrium in the longitudinal direction then requires that

$$P = \int_A \sigma \, dA = \int_{A_c} \sigma_c \, dA_c + \int_{A_s} \sigma_s \, dA_s \qquad (2.51)$$

where σ_c and σ_s are the normal stresses in the concrete and steel, respectively.

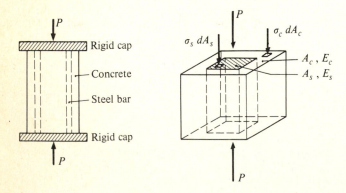

Figure 2.36 Reinforced concrete member subjected to axial compression.

Both materials are assumed to respond linearly to imposed loads according to the stress-strain relationships

$$\sigma_c = E_c \varepsilon_c \quad \text{and} \quad \sigma_s = E_s \varepsilon_s$$

Substituting these values in eq. (2.51) yields

$$P = E_c \varepsilon_c \int_{A_c} dA_c + E_s \varepsilon_s \int_{A_s} dA_s \tag{2.52}$$

or

$$P = E_c \varepsilon_c A_c + E_s \varepsilon_s A_s \tag{2.53}$$

But no slippage is permitted. Therefore, $\varepsilon = \varepsilon_c = \varepsilon_s$ or

$$P = \varepsilon(E_c A_c + E_s A_s)$$

This allows the strain to be expressed as a function of the load and the component cross-sectional areas.

$$\varepsilon = \frac{P}{E_c A_c + E_s A_s} \tag{2.54}$$

Introducing the *modular ratio*

$$n = \frac{E_s}{E_c} \tag{2.55}$$

the appropriate relationships would be

$$\varepsilon = \frac{P}{E_c(A_c + nA_s)} \tag{2.56}$$

$$\sigma_c = \frac{P}{A_c + nA_s} \tag{2.57}$$

and

$$\sigma_s = n \frac{P}{A_c + nA_s} = n\sigma_c \tag{2.58}$$

Note that the quantity nA_s has the same dimensional units as area. In concept it can be thought of as a equivalent area of concrete that has the same strength or load-carrying capacity as the "real" steel area. It is often referred to as the *transformed* area of the steel. The sum of $A_c + nA_s$ is called the transformed section and is denoted A_T. That is

$$\sigma_c = \frac{P}{A_T} \quad \text{and} \quad \sigma_s = n\sigma_c$$

Using the concept of the transformed section, consider the cross section shown in fig. 2.37. The original case is shown in sketch a; the transformed section is shown in sketch b. The area of the transformed section is

$$A_T = A_c + nA_s \tag{2.59}$$

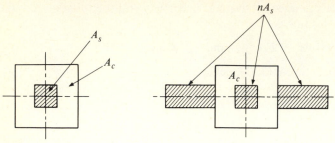

(a) Given cross section   (b) Transformed cross section

2.37 Properties of the transformed cross section.

Defining A_g as the original gross area, where $A_g = A_s + A_c$

$$A_T = A_g - A_s + nA_s$$

or

$$A_T = A_g + (n - 1)A_s \qquad (2.60)$$

Example 2.13 Consider the "short" reinforced compression member shown in fig. 2.38. The outside dimensions of the cross section are 16×16 in, and there are symmetrically placed within the member four no. 9 size steel reinforcing bars.† The axial force to which the cross section is subjected is 200 kips in compression. The given modulus of elasticity for the steel is 30,000 ksi, and for the concrete 3000 ksi. The modular ratio n is therefore

$$n = \frac{E_s}{E_c} = \frac{30,000}{3000} = 10$$

Since the gross area of the original section is

$$A_g = 16 \times 16 = 256 \text{ in}^2$$

and

$$(n - 1)A_s = 9(4.0) = 36 \text{ in}^2$$

The transformed area is

$$A_T = (256) + (36) = 292 \text{ in}^2$$

From eqs. (2.57) and (2.58) the stresses become

$$\sigma_c = \frac{P}{A_T} = \frac{200}{292} = 0.685 \text{ ksi}$$

$$\sigma_s = n\sigma_c = (10)(0.685) = 6.85 \text{ ksi}$$

---

† Steel reinforcing bars are normally designated by a number. This number is standard and uniquely defines a bar having a particular geometry. The more commonly used circular-cross-section bars are listed in table 2.2.

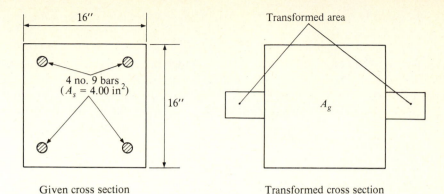

Figure 2.38 Reinforced concrete column cross section.

Next, consider the load-carrying capacity of such members. For convenience, the equations will be redefined as follows:

$$P = \sigma_c A_c + \sigma_s A_s$$

$$P = \sigma_c(A_g - A_s) + \sigma_s A_s$$

or

$$P = A_g\left[\sigma_c\left(1 - \frac{A_s}{A_g}\right) + \sigma_s\left(\frac{A_s}{A_g}\right)\right]$$

If the ratio of the area of the steel to the overall area of the member is defined by the quantity p, that is,

$$p = \frac{A_s}{A_g}$$

then

$$P = A_g[\sigma_c(1 - p) + \sigma_s p]$$

Table 2.2 Areas and perimeters of round reinforcing bars

Bar number designation	Diameter, in	Perimeter, in	Area, in^2
2	0.250	0.79	0.05
3	0.375	1.18	0.11
4	0.500	1.57	0.20
5	0.625	1.96	0.31
6	0.750	2.36	0.44
7	0.875	2.75	0.60
8	1.000	3.14	0.79
9	1.128	3.54	1.00
10	1.270	3.99	1.27
11	1.410	4.43	1.56

But p is usually quite small when compared to the number 1. Therefore

$$P = A_g(\sigma_c + \sigma_s p)$$

Defining σ'_c as the maximum useful stress to which the concrete can be subjected, that is, $\sigma'_c = \text{FS}_c \sigma_c$, where FS_c is the appropriate factor of safety for the concrete; and defining σ_y as the yield stress for the steel reinforcing bars such that $\sigma_y = \text{FS}_s \sigma_s$; the allowable maximum load to which the member can be subjected is

$$P_{\text{all}} = A_g \left[\frac{\sigma'_c}{\text{FS}_c} + \frac{\sigma_y p}{\text{FS}_s} \right]$$

Selecting $\text{FS}_c = 4.00$, $\text{FS}_s = 2.50$, $\sigma'_c = 3000 \text{ lb/in}^2$ and $\sigma_y = 60{,}000 \text{ lb/in}^2$

$$P_{\text{all}} = A_g(0.25\sigma'_c + 0.40\sigma_y p)$$

For the member defined in fig. 2.38, $A_g = 16 \times 16 = 256 \text{ in}^2$ and $p = 4/256 = 0.0156$. Therefore,

$$P_{\text{all}} = 256[(0.25)(3) + (0.40)(0.0156)] = 287.85 \text{ kips}$$

Concrete columns The design of reinforced concrete columns is essentially the inverse of the method of analysis just described. For design, there is given a set of material properties, the value of the applied loading, and in many cases, the overall proportions of the geometry of the desired section. The problem is to establish permissible dimensions of the cross-sectional areas of the concrete and steel so that the stipulated allowable stress values are not exceeded.

Before proceeding with an actual example of column design, a more general discussion of reinforced concrete columns is in order. Most concrete columns can be categorized as to one of two general types: *spiral columns* and *tied columns*. The two types differ in the manner in which the longitudinal steel reinforcing bars are positively constrained against bending (buckling) outward when the member is subjected to an external thrust. These are illustrated in fig. 2.39. Due to the efficiency of the continuous spiral in restraining the longitudinal steel, a spiral column will usually sustain more compressive load (per given cross-sectional area) than a tied column.

In addition to classifying reinforced concrete columns as to whether they contain spirals or lateral ties, such members are also described as being *long* or *short*. A short column is one whose response is not dependent upon overall member-buckling considerations. A long column, on the other hand, is one whose response is governed almost exclusively by this phenomena. For the examples in this chapter, the columns will be considered to be short. Long columns will be discussed in a later chapter.

As stated earlier, the design of a column requires the selection of a concrete cross-sectional area and an area of longitudinal steel reinforcing bars.† It also in-

---

† The examples that will be initially illustrated presume a design procedure for a pure compression member. Because of lack of control of workmanship—for example, misplacement of reinforcing bars—some structural design codes do not permit the use of pure compression members. The Code of the American Concrete Institute, for example, requires the assumption of a minimum eccentricity of the load not less than 1 in or $0.5t$ for a spiral column, and $0.10t$ for a tied column, where t is the overall thickness of the member.

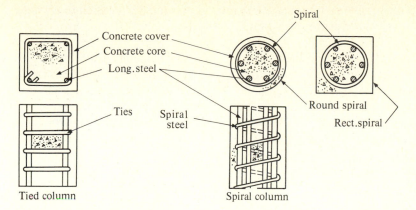

Figure 2.39 Spiral and tied concrete columns.

cludes specification of spirals or ties, the spacing of the bars, and other such details. At this stage attention will be restricted to primary design considerations—the selection of the concrete cross section and the associated steel reinforcing.

Example 2.14 Determine the required dimensions for a round spiral column subjected to an axial thrust of 500 kips. It is presumed that buckling is not a problem. ASTM type A432 steel reinforcing bars ($\sigma_y = 60{,}000$ lb/in^2) are to be used, as is a concrete whose ultimate compressive strength is $\sigma'_c = 3500$ lb/in^2. It is also given as desirable to keep the percentage of steel in the member as close to 2 percent as possible.

Assuming the same factors of safety as in the previous example; that is,

$$\text{FS}_c = 4.00 \qquad \text{and} \qquad \text{FS}_s = 2.50$$

the allowable column load can be expressed as

$$P_{\text{all}} = A_g(0.25\sigma'_c + 0.40\sigma_y p)$$

For the conditions specified

$$P(\text{kips}) = A_g[(0.25)(3.5) + (0.40)(60)(0.02)] = 1.355A_g$$

The required gross area of the column is therefore

$$A_g = \frac{P}{1.355} = \frac{500}{1.355} = 369.0 \text{ in}^2$$

Since

$$A_g = \frac{\pi t^2}{4}$$

where t is the diameter of the round column, there is obtained

$$t_{\text{rqd}} = \sqrt{\frac{(4)(369.0)}{\pi}} = 21.675 \text{ in}$$

For ease of forming, it is desirable to keep the diameter of the column to a whole inch. In this case, one of two possibilities exists:

$$t = 21 \text{ in} \qquad p > 0.02$$

or

$$t = 22 \text{ in} \qquad p < 0.02$$

The 22-in size will be selected.

The area of steel required is determined by separating the portion of the load carried by the concrete from that carried by the steel. For a diameter of 22 in

$$P_{conc} = 0.25\sigma'_c A_g = (0.25)(3.5)\left[\frac{\pi(22)^2}{4}\right] = 332.62 \text{ kips}$$

Therefore

$$P_{steel} = P - P_{conc} = 500 - 332.62 = 167.38 \text{ kips}$$

But

$$P_{steel} = 0.40pA_g = 0.40\sigma_y A_s$$

Therefore

$$A_s = \frac{167.38}{(0.40)(60)} = 6.974 \text{ in}^2$$

In order for the circular column to have no weak directions, a minimum number of six bars is usually specified. These bars are spaced uniformly within the inner circle formed by the constraining spirals. From the bar area chart of table 2.2, the following would be obtained:

six no. 9 bars $\qquad A_s = 6.00 \text{ in}^2$

six no. 10 bars $\qquad A_s = 7.62 \text{ in}^2$

eight no. 8 bars $\qquad A_s = 6.32 \text{ in}^2$

eight no. 9 bars $\qquad A_s = 8.00 \text{ in}^2$

Using six no. 10 bars, and providing $1\frac{1}{2}$ in of concrete cover† outside of the spiral steel to provide protection against fire and other dangers; and further presuming a $\frac{1}{2}$-in-diameter bar for the spiral, the spacing‡ for the six bars is

$$\text{Spacing} = \frac{[22 - (2)(1.5) - (2)(0.5) - 1.27]\pi}{6} = 8.76 \text{ in}$$

---

† The concrete cover is required to protect the steel bars from corrosion, fire, etc. The amount of cover needed is dependent upon the use to which the member is subjected. For example, a member inside of a building normally would require less cover than one resting directly on the ground, or one that is submerged in saltwater. Most design specifications define the minimum amount necessary.

‡ Spacing of the longitudinal bars is of importance for several reasons. If the spacing is too small, there will be difficulty in pouring the concrete into the forms. The cement paste might also separate from the aggregate if too tight a situation exists. Too great a spacing, on the other hand, is also undesirable because, in effect, large portions of the member may be effectively unreinforced.

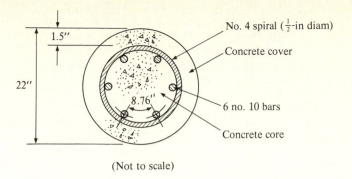

(Not to scale)

Figure 2.40 Dimension of reinforced concrete column—example 2.14.

As a check:

$$P_{all} = \frac{\pi(22)^2}{4}\left[\frac{3.5}{4} + 24\left(\frac{(7.62)(4)}{\pi(22)^2}\right)\right] = 515.08 \text{ kips} > 500.0 \text{ kips}$$

$$p = \frac{7.62}{380.13} = 0.020$$

The final cross section is shown in fig. 2.40.

Concrete beams The bending of a reinforced concrete beam can be approached in exactly the same manner. The only additional assumption that would be made (and this results in a conservative design solution) is that the tensile strength of the concrete is so small that it can be ignored.

Figure 2.41 shows a rectangular, reinforced concrete beam of width b and total depth h. The reinforcing bars are located only in the bottom of the beam, and for purposes of this analysis can be presumed concentrated at a depth d from the upper surface. The loading to which the member is subjected is bending. As shown, since the concrete is presumed to fail in tension, cracks develop from the bottom surface up

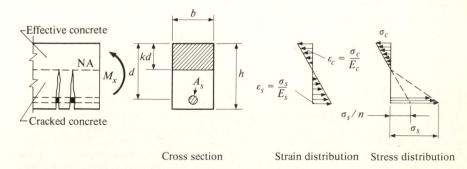

Figure 2.41 Reinforced concrete beam subjected to bending.

to the neutral axis. The beam is then effectively composed of a compressive zone of kd depth and a developed tensile force in the reinforcing bar. (Since the thickness of the bar is normally small when compared to the depth of the beam, an "average" bar stress will be used when computing the tensile force.) Referring to the stress-distribution diagram, and noting that only a bending moment is applied to the section in question,

$$\sum F_z = 0 \qquad \text{or} \qquad b \int_{NA}^{top} \sigma_c \, dy + \sigma_s A_s = 0$$

and

$$\sum M_{NA} = 0 \qquad \text{or} \qquad b \int_{NA}^{top} \sigma_c y \, dy + \sigma_s A_s(d - kd) = M_x$$

where kd is the depth of that portion of the concrete beam that is in compression and d is the effective depth of the member.

From the presumed strain distribution

$$\varepsilon_c = -\phi y \qquad \text{and} \qquad \varepsilon_s = \phi(d - kd)$$

or

$$\sigma_c = -E_c \phi y \qquad \text{and} \qquad \sigma_s = E_s \phi(d - kd)$$

Using the first of the equilibrium equations

$$-b \int_0^{-kd} E_c \phi y \, dy + E_s \phi(d - kd)A_s = 0$$

or

$$-\tfrac{1}{2} b E_c \phi(kd)^2 + E_s \phi(d - kd)A_s = 0 \tag{2.61}$$

which leads to

$$k^2 + 2npk - 2np = 0$$

or

$$k = \sqrt{2pn + (pn)^2} - pn \tag{2.62}$$

Using the second equilibrium equation, the curvature will be

$$\phi = \frac{M_x}{E_c \left[\dfrac{b(kd)^3}{3} + nA_s(d - kd)^2 \right]} \tag{2.63}$$

Substituting this into the expressions

$$\sigma_c = -E_c \phi y \qquad \text{and} \qquad \sigma_s = E_s \phi(d - kd) \tag{2.64}$$

yields the stresses in question.

The concept of the "transformed section" also can be used to analyze reinforced concrete beams. Consider the beam whose "real" and "effective"† cross sections are shown in fig. 2.42. The transformed section is shown in the third of the sketches.

---

† The area of the concrete in tension is not included in the effective section.

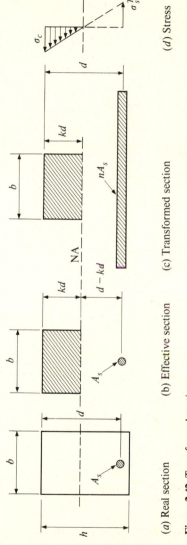

σ_c

σ_s^T

kd

b

nA_s

d

(c) Transformed section

(d) Stress

NA

kd

b

$d - kd$

A_s

(b) Effective section

b

d

h

A_s

(a) Real section

Figure 2.42 Transformed section.

It is to be recognized that in the transformed section, the effective steel area nA_s is located at the same depth d. Since that area is thought of as an equivalent area of concrete, the beam can now be considered as homogeneous, and the normal beam-bending equations developed earlier can be used.

The neutral axis passes through the centroid of the transformed section. By taking moments of the areas about that axis

$$-bkd\frac{kd}{2} + nA_s(d - kd) = 0$$

and substituting the value $p = A_s/A_g = A_s/bd$, the quadratic equation

$$k^2 + 2npk - 2np = 0$$

developed previously is obtained. The solution to this equation is

$$k = \sqrt{2pn + (pn)^2} - pn$$

The other required equations are

$$\sigma_c = \frac{M_x}{I_T}kd \qquad \sigma_s^T = \frac{M_x}{I_T}(d - kd) \qquad \text{or} \qquad \sigma_s = n\sigma_s^T$$

In some cases a better feel can be obtained using the stress block for developing these equations. The stress block for the example just considered is shown in fig. 2.43. The resultant compressive force C passes through the centroid of the compressive stress block. This is located $\frac{1}{3}kd$ down from the top of the beam. Its magnitude is equal to the volume of the block.

$$C = \sigma_c b\frac{kd}{2} \tag{2.65}$$

In similar fashion

$$T = \sigma_s A_s$$

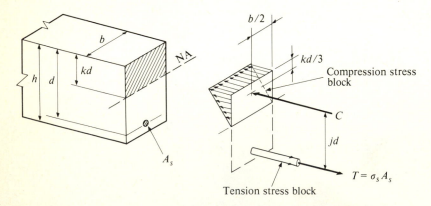

Figure 2.43 Normal stresses in a reinforced concrete beam.

Presuming the strain distribution of fig. 2.41

$$\varepsilon_s = \varepsilon_c \frac{d - kd}{kd} \tag{2.66}$$

or

$$\sigma_s = n\sigma_c \frac{d - kd}{kd} \tag{2.67}$$

and

$$T = nA_s\sigma_c \frac{d - kd}{kd} \tag{2.68}$$

For equilibrium $C = T$. Thus,

$$\sigma_c b \frac{kd}{2} = nA_s\sigma_c \frac{d - kd}{kd} \tag{2.69}$$

Solving for k, the following quadratic is again obtained:

$$k^2 + 2npk - 2np = 0$$

The bending moment is therefore

$$M_x = Cjd \qquad \text{or (in terms of the tensile force)} \qquad M_x = Tjd \tag{2.70}$$

where $jd = d - \frac{1}{3}kd$. It then follows that

$$M_x = \sigma_c b \frac{kd}{2}\left(d - \frac{kd}{3}\right)$$

or

$$\sigma_c = \frac{M_x}{(bd^2/2)k(1 - k/3)} \qquad \text{and} \qquad \sigma_s = \frac{M_x}{A_s d(1 - k/3)} \tag{2.71}$$

Example 2.15 The beam shown in fig. 2.41 has the dimensions $b = 12$ in, $d = 18$ in, and $A_s = 3$ in^2. For $M_x = 150$ ft·kips and $n = 10$, determine σ_c and σ_s.

Since

$$p = \frac{3}{(12)(18)} = 0.0139$$

$$k = \sqrt{(2)(0.0139)(10) + [(0.0139)(10)]^2} - (0.0139)(10) = 0.406$$

Therefore

$$\sigma_c = \frac{(150)(12)}{[(12)(18)^2/2](0.406)(1 - 0.406/2)} = 2.861 \text{ ksi}$$

and

$$\sigma_s = \frac{(150)(12)}{(3)(18)(1 - 0.406(2)} = \underset{38.56 \text{ ksc}}{41.824 \text{ ksi}}$$

SHOULD BE 3

If it were specified that the maximum allowable stress in the concrete is 1.35 ksi and in the steel is 20 ksi, what would be the allowable bending moment M_x?

$$M_x = Cjd = Tjd$$

$$C_{\text{max all}} = \frac{\sigma_{c,\,\text{max}}bkd}{2} = \frac{(1.35)(12)(18)(0.406)}{2} = 59.195 \text{ kips}$$

$$T_{\text{max all}} = \sigma_{s,\,\text{max}}A_s = (20)(3) = 60 \text{ kips}$$

(It should be recognized that when using this procedure the maximum allowable values of C and T need not and probably will not be equal.)

$$jd = d - \frac{kd}{3} = 18 - \frac{(0.406)(8)}{3} = 15.564 \text{ in}$$

Therefore, based on the stress in the concrete

$$M_{x,\,\text{max}_c} = 59.195\,\frac{15.564}{12} = 76.78 \text{ ft} \cdot \text{kips}$$

Based on the stress in the steel

$$M_{x,\,\text{max}_s} = 60\,\frac{15.564}{12} = 77.82 \text{ ft} \cdot \text{kips}$$

The maximum allowable moment must be the lesser of the two (why?), or 76.78 ft·kips.

From the above example, it should be evident that either the steel or the concrete could govern the maximum moment that a beam can sustain—depending on which reaches its maximum allowable stress first. If the moment is limited by the concrete, the beam is said to be overreinforced, or the percentage of steel in the beam is too high. If, on the other hand, the steel governs, the beam is underreinforced, and the percentage of the steel is possibly too small. A beam for which the allowable stresses in concrete and steel are simultaneously realized is referred to as a balanced design beam.

Laboratory tests of over- and underreinforced members have shown that over-reinforced beams fail in a brittle fashion with little or no warning. An underreinforced beam, on the other hand, fails ductilely and gives visible signs of distress long before actual collapse occurs. Since a ductile failure is to be preferred to a brittle one, most building codes prescribe that members be underreinforced. This can be accomplished by limiting the percentage of steel to something less than the balanced percentage. A value of 75 percent of the balanced case is not unusual.

Ultimate strength of concrete beams An alternative to the allowable-stress method of design, and one that is much more widely used in beam design practice, is similarly based upon an assumed linear distribution of bending strains across the member, but the actual nonlinear stress-strain relation for concrete—shown in fig. 2.44—is used rather than the linear one. If it is assumed that the concrete strain in the outermost

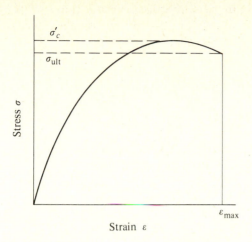

Compression
loading of material

Figure 2.44 Stress-strain properties of concrete in compression.

compressive fiber of a beam subjected to bending is the maximum strain ε_{max}, then the stress distribution across the section would be that shown in fig. 2.45b.

To facilitate design, a statically equivalent approximation to the actual stress distribution is commonly used. One such approximation is shown in fig. 2.45c. (The constants are determined in such a way that the approximate compressive stress block has the same volume as the actual one, and both stress blocks have the same moment about the neutral axis. Suggested values for these constants are contained in ACI publication 318-71.† Some of these are: $\sigma_{ult} = 0.85\sigma'_c$, $a = k_1 c$, where k_1 is 0.85 for values of σ'_c up to 4000 lb/in², and $\varepsilon_{max} = 0.003$.)

It is to be recognized that when using this type of procedure, where limiting values of stress or strain in the steel and concrete are presumed, appropriate load factors (another means for assuring an adequate margin of safety) must be introduced. Under these prorated loads ε_{max} is presumed to equal 0.003 and $\sigma_s = \sigma_y$.

† American Concrete Institute, *Building Code Requirements for Reinforced Concrete*, ACI 318-71.

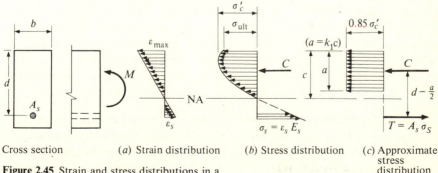

Cross section   (*a*) Strain distribution   (*b*) Stress distribution   (*c*) Approximate stress distribution

Figure 2.45 Strain and stress distributions in a beam subjected to bending moment.

Having assumed the strain distribution in fig. 2.45c, equilibrium must now be examined.

$$\sum F_z = 0 \quad \text{or} \quad C = T$$

Substituting the appropriate values from the stress blocks

$$0.85\sigma'_c ba = A_s \sigma_y = p_b bd\sigma_y \tag{2.72}$$

and

$$p_b = \frac{0.85\sigma'_c ba_b}{bd\sigma_y} = \frac{0.85\sigma'_c k_1 c_b}{d\sigma_y} \tag{2.73}$$

(The subscript b denotes a balanced design.)
From the presumed strain distribution

$$c_b = \frac{0.003}{0.003 + \sigma_y/E_s} d \tag{2.74}$$

Assuming a Young's modulus of elasticity for the steel of $E_s = 29,000,000$ lb/in²

$$c_b = \frac{87,000}{87,000 + \sigma_y} d \tag{2.75}$$

Thus,

$$p_b = \frac{0.85 k_1 \sigma'_c}{\sigma_y} \frac{87,000}{87,000 + \sigma_y} \tag{2.76}$$

To ensure ductile fracture of the member

$$p_{max} = 0.75 p_b$$

Example 2.16 If $\sigma'_c = 3000$ lb/in², $\sigma_y = 60,000$ lb/in², $b = 10$ in, and $d = 16$ in, what is the maximum desirable area of steel, A_s?
For $\sigma'_c = 3000$ lb/in², $k_1 = 0.85$. Therefore

$$p_b = \frac{(0.85)^2(3000)(87,000)}{(60,000)(60,000 + 87,000)} = 0.022$$

or

$$p_{max} = 0.75 p_b = 0.016$$

Thus

$$A_{s,max} = (0.016)(10)(16) = 2.560 \text{ in}^2$$

For a beam of this cross section and having this area of steel, what is the allowable moment that can be sustained?

$$M_{max} = A_s \sigma_y \left(d - \frac{a}{2} \right) = ab(0.85\sigma'_c)\left(d - \frac{a}{2} \right) \tag{2.77}$$

Noting that $A_s \sigma_y - (ab)(0.85\sigma'_c) = 0$

$$a = \frac{A_s}{b} \frac{\sigma_y}{0.85\sigma'_c} \qquad (2.78)$$

Substituting this into eq. (2.77) yields

$$M_{max} = \frac{A_s \sigma_y}{(0.85\sigma'_c)b} b(0.85\sigma'_c) \left[d - \frac{1}{2} \frac{A_s \sigma_y}{(0.85\sigma'_c)b} \right]$$

or

$$M_{max} = bd^2 \left[\sigma'_c p \frac{\sigma_y}{\sigma'_c} \left(1 - 0.59p \frac{\sigma_y}{\sigma'_c} \right) \right] \qquad (2.79)$$

Defining $q = p\sigma_y/\sigma'_c$, this becomes

$$M_{max} = bd^2 \sigma'_c q(1 - 0.59q) = A_s \sigma_y \left(d - \frac{a}{2} \right) \qquad (2.80)$$

For this case

$$q = (0.016) \left(\frac{60{,}000}{3000} \right) = 0.320$$

$$bd^2 = (10)(16)^2 = 2560 \text{ in}^3$$

Thus,

$$M_{max} = (2560)(3000)(0.32)(0.811) = 1993.11 \text{ in} \cdot \text{kips}$$

But as was noted earlier in this section—in discussing the basic philosophy of the ultimate strength design of concrete—it is necessary that there be introduced a load factor to ensure adequate safety. Defining

$$M_{max} = M_{all}\text{LF}$$

where LF is the load factor,†

$$M_{all} = \frac{M_{max}}{\text{LF}} \qquad (2.81)$$

If the ratio of the dead load of the beam to the applied live load is 0.5, an appropriate load factor LF would be

$$\text{LF} = 1.7 + (1.4)(0.5) = 2.4$$

Therefore

$$M_{all} = \frac{M_{max}}{\text{LF}} = \frac{1993.11}{2.4} = 830.5 \text{ in} \cdot \text{kips}$$

---

† One form for the load factor LF is

$$\text{LF} = 1.7 + 1.4 \frac{\text{dead load of the beam}}{\text{live load acting on the beam}}$$

It is to be noted, however, that some structural design codes prescribe the use of an additional factor (a multiplier) to take into account the type of structural element being considered. For flexural members, the suggested factor is 0.9. The current ACI Specification equation for load factors is

$$\text{LF} = 1.7(\text{live load}) + 1.4(\text{dead load})$$

Example 2.17 The equations developed also are applicable for design. For example, using the maximum allowable percentage of steel (that is, 75 percent of that required for a balanced design) what size beam and what area of steel are needed to carry a moment of 150.0 ft·kips, if $\sigma'_c = 3000$ lb/in² and $\sigma_y = 60,000$ lb/in²? The design (or factored) moment is

$$M_{max} = \text{LF } M_{all} = [1.7 + (1.4)(0.5)](150)(12) = 4320 \text{ in·kips}$$

The maximum permissible steel percentage, from eq. (2.76) is

$$p_{max} = 0.75p_b = 0.75\left[\frac{(0.85)^2(3000)(87,000)}{(60,000)(60,000 + 87,000)}\right] = 0.016$$

Therefore

$$q_{max} = p_{max}\frac{\sigma_y}{\sigma'_c} = 0.320$$

Since

$$M_{max} = bd^2\sigma'_c q(1 - 0.59q)$$

$$bd^2 = \frac{M_{max}}{\sigma'_c q(1 - 0.59q)} = \frac{4,320,000}{(3000)(0.320)(0.811)} = 5548.7 \text{ in}^3$$

Many combinations of b and d can be found that will satisfy this relationship. Some are given in Table 2.3.

The required area of the steel is determined from the expression

$$A_s = \frac{M_{max}}{\sigma_y\left(d - \dfrac{a}{2}\right)}$$

or

$$A_s = pbd$$

Table 2.3 Various possible design solutions

b, in	d^2	d, in
7.5	739.83	27.20
8.0	693.59	26.34
8.5	652.79	25.55
9.0	616.52	24.83
9.5	584.07	24.17
10.0	554.87	23.56
10.5	528.45	22.99
11.0	504.43	22.46
11.5	482.50	21.97
12.0	462.30	21.50

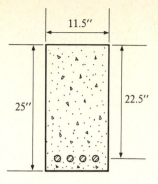

Figure 2.46 Final cross section design dimensions—example 2.17.

Selecting for further consideration the 11.5 in $\times$ 22 in member listed in table 2.3

$$A_s = (0.016)(11.5)(22.0) = 4.05 \text{ in}^2$$

Referring to the list of bar sizes in table 2.2, it is found that four no. 9 bars provide a steel area of 4.0 in². Since the selected value of d is slightly greater than necessary, the small deficiency in the steel area will probably not adversely affect the moment capacity of the beam. To be sure, however, a check will be made using the actual selected dimensions: $b = 11.5$ in, $d = 22.0$ in, and $A_s = 4.0$ in². From eq. (2.79)

$$M_{\text{all}} = \frac{(11.5)(22)^2}{(2.4)} \frac{(3)(4.0)(60)}{(11.5)(22)(3)} \left[1.0 - \frac{(0.59)(4)(60)}{(11.5)(22)(3)} \right]$$

$$= 1790.17 \text{ in} \cdot \text{kips}$$

$$M_{\text{rqd}} = \frac{M_{\text{max}}}{\text{LF}} = \frac{4320}{(2.4)} = 1800 \text{ in} \cdot \text{kips}$$

Since $M_{\text{all}} < M_{\text{rqd}}$ the section is (slightly) inadequate. If, however, d were increased to 22.5 in, $M_{\text{all}} = 1839.69$ in·kips. Specifying a minimum of 1.5 in of concrete cover over the steel bars, and allowing for the need to provide room for any additional secondary reinforcement,[†] an overall depth of the beam of 25 in is selected. The final section is shown in fig. 2.46.

Doubly reinforced concrete beams In many instances the depth of the cross section is limited by architectural or other considerations, and there is insufficient area of the concrete in compression to develop the required moment capacity. When such is the case, additional steel bars may be added to the compression zone to assist the concrete. A concrete beam containing compression reinforcement is referred to as a *doubly reinforced beam*.

The analysis of stresses in a doubly reinforced concrete beam is carried out in the same fashion as that for the more often observed singly reinforced case. The primary difference lies in the inclusion of compressive steel forces which may correspond to

---

† In most cases reinforcement for diagonal tension caused by shear or torsion is required. This will be discussed in detail in a later chapter.

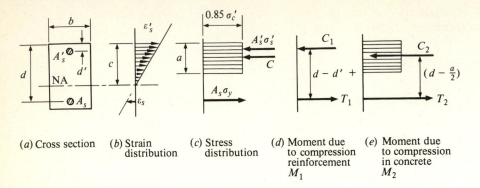

(a) Cross section   (b) Strain distribution   (c) Stress distribution   (d) Moment due to compression reinforcement M_1 (e) Moment due to compression in concrete M_2

Figure 2.47 Doubly reinforced concrete beam subjected to bending.

strains substantially below their values in tension. As with singly reinforced beams, it will be presumed that the maximum moment corresponds to the attainment of an ultimate compressive strain in the concrete of 0.003, and to a tensile steel stress equal to the yield value, σ_y.

As shown in fig. 2.47, the moment capacity of the doubly reinforced beam can be thought of as composed of two parts: that which is due to the compressive force in the concrete and part of the tensile steel force, and an additional couple that results from the compression steel force and the remaining portion of the tension reinforcement. The moment capacity is given in eq. (2.82)

$$M = M_1 + M_2 = T_1(d - d') + T_2(d - \tfrac{1}{2}a) \tag{2.82}$$

where d' is the depth from the compression face to the compression steel as shown. For equilibrium

$$T_1 + T_2 = A_s\sigma_y$$

and

$$C_1 + C_2 = A_s\sigma_y = A'_s\sigma'_s + 0.85\sigma'_c(ba - A'_s)$$

From the strain distribution

$$\varepsilon'_s = 0.003\frac{c - d'}{c} = 0.003\frac{a - d'k_1}{a} \tag{2.83}$$

where a is the depth of the compression stress block, and

$$k_1 = \frac{a}{c}$$

as previously defined. For this case, the depth a is given by eq. (2.84)

$$a = \frac{A_s\sigma_y - A'_s(\sigma'_s - 0.85\sigma'_c)}{0.85\sigma'_c b} \tag{2.84}$$

If it is neglected that the compressive steel occupies a small part of the concrete stress block, eq. (2.84) becomes

$$a = \frac{A_s\sigma_y - A_s'\sigma_s'}{0.85\sigma_c'b} \tag{2.85}$$

Substituting this into eq. (2.83) yields for the compressive steel strain

$$\varepsilon_s' = 0.003\frac{A_s\sigma_y - A_s'\sigma_s' - d'k_1(0.85\sigma_c'b)}{A_s\sigma_y - A_s'\sigma_s'} \tag{2.86}$$

Equation (2.86) is valid up to the point where $\sigma_s' = \sigma_y$. For cases where $\sigma_s' < \sigma_y$, the compressive stress in the steel can be found from the relationship

$$\sigma_s' = E_s\varepsilon_s'$$

If $\sigma_s' = \sigma_y$, the equation corresponding to eq. (2.85) would be

$$a = \frac{(A_s - A_s')\sigma_y}{0.85\sigma_c'b} \tag{2.87}$$

and eq. (2.86) would become

$$\varepsilon_s' = \varepsilon_y = \frac{\sigma_y}{E_s} = 0.003\frac{(A_s - A_s')\sigma_y - d'k_1(0.85\sigma_c'b)}{(A_s - A_s')\sigma_y} \tag{2.88}$$

Thus if

$$A_s - A_s' \geq 0.85k_1 d'b\frac{\sigma_c'/\sigma_y}{1 - \sigma_y/87{,}000} \tag{2.89}$$

(where σ_y and σ_c' are in lb/in^2)

$$\varepsilon_s' \geq \varepsilon_y$$

If $A_s - A_s'$ is less than the value given by eq. (2.88)

$$\varepsilon_s' < \varepsilon_y$$

Example 2.18 Consider the doubly reinforced cross section shown in fig. 2.48. It is given that $\sigma_c' = 4000$ lb/in^2 and $\sigma_y = 40{,}000$ lb/in^2. The moment capacity of the section about a horizontal axis is desired.

The level of stress in the compressive steel is first determined, using eq. (2.89). The right-hand side of the equation is

$$(0.85)(0.85)(2.5)(18)\frac{4000/40{,}000}{1 - 40{,}000/87{,}000} = 6.02 \text{ in}^2$$

while the left-hand side is

$$A_s - A_s' = (1.56)(4) - (1.00)(2) = 4.24 \text{ in}^2$$

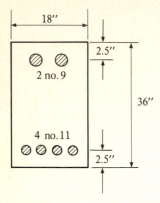

Figure 2.48

Since $4.24 < 6.02$ the value of the stress in the compression reinforcing is less than the yield value and is determined from eq. (2.86). Noting that $\sigma'_s = E_s \varepsilon'_s$

$$\sigma'_s = 0.003 E_s \frac{A_s \sigma_y - A'_s \sigma'_s - d' k_1 (0.85 \sigma'_c b)}{A_s \sigma_y - A'_s \sigma'_s}$$

Solving for σ'_s, assuming that $E_s = 29,000,000$ lb/in^2

$$\sigma'_s = 87,000 \frac{119,550 - 2\sigma'_s}{249,600 - 2\sigma'_s}$$

or

$$\sigma'_s = 28,347 \text{ lb/in}^2$$

a is therefore

$$a = \frac{A_s \sigma_y - A'_s \sigma'_s}{0.85 \sigma'_c b} = \frac{(6.24)(40) - (2.0)(28.347)}{(0.85)(4)(18)} = 3.15 \text{ in}$$

The moment capacity is

$$M = M_1 + M_2 = A'_s \sigma'_s (d - d') + 0.85 \sigma'_c ab(d - \tfrac{1}{2}a) = 7913 \text{ in·kips}$$

T beams and one-way slabs When constructing a concrete floor system, it is desirable (and not uncommon practice) that the floor slab and its supporting beams be poured at the same time. When such is the case, the slab and the beams are fully continuous, and no joint exists between the two. The effective result is a series of T beams joined together as shown in fig. 2.49a. In these members the supporting beam becomes the stem, and a portion of the slab contributes to the compressive capacity of the cross section. The effective portion of the slab is defined in various codes and specifications. For example, one such building code allows an effective flange width for a symmetrical T beam to be the smaller of:

1. One-fourth the span of the beam
2. The overhang on either side may not exceed eight times the slab thickness
3. The overhang may be no greater than one-half the distance to the face of the next T beam.

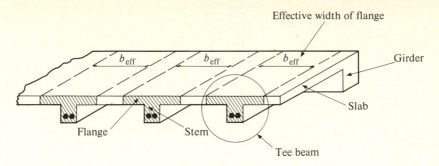

Figure 2.49a T-beam construction.

One of the major differences between the analysis of a T beam and a rectangular cross section is the uncertainty as to the location of the neutral axis. Depending on the geometry of the T and the arrangement and magnitude of the reinforcing steel, the neutral axis could be either in the flanges or in the stem. This is illustrated in fig. 2.49*b*.

The first case—neutral axis in the flange—would correspond to a relatively small steel-reinforcement ratio. The ratio in the second case would be higher. No special modification of the general methods of analysis is required for either of these two possibilities. (For the first case, it is to be recognized that the area of concrete effective in compression is a rectangle—as was true for the earlier considered conventional beams. Therefore, the relationships developed for that case can be directly applied.)

When the neutral axis is located in the stem, the area of concrete in compression has a T shape. This must be taken into account in the stress analysis. Consider the section shown in fig. 2.50.

The concrete strain in the outer compression fiber, as before, is presumed to be 0.003. The compressive stress, again as in the rectangular case, is retained at $0.85\sigma'_c$. (While this is not absolutely correct for the T shape, the associated error has been determined to be relatively small.) In addition, the same constant k_1 for the rectangular cross section is used.

For purposes of analysis and design, it has been found convenient to treat the T-shaped area of compression concrete as if it consisted of two parts. The first of these is directly associated with the stem and has a width b_w and a depth a. The remainder of the area is made up of two parts whose combined width is $b - b_w$ and whose depth is h_f. Correspondingly, the steel area is considered in two parts A_{s_w} and

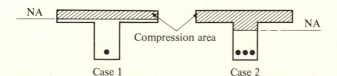

Figure 2.49b Location of the neutral axis in T beams.

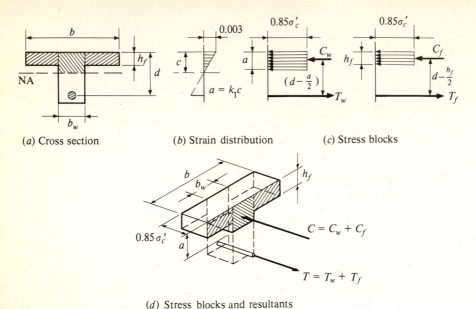

(*a*) Cross section   (*b*) Strain distribution   (*c*) Stress blocks

(*d*) Stress blocks and resultants

Figure 2.50 Normal stress distribution in T beams subjected to bending moments.

A_{s_f}. A_{s_w} is that part of the total reinforcement needed to balance the stem compression force. A_{s_f} is to counterbalance the flange compression. Thus

$$A_{s_w} = A_s - A_{s_f} \tag{2.90}$$

and the resultants of the compressive stress blocks are

$$C_w = A_{s_w}\sigma_y = 0.85\sigma'_c ab_w$$

and

$$C_f = A_{s_f}\sigma_y = 0.85\sigma'_c h_f(b - b_w) \tag{2.91}$$

The location of the neutral axis is given by eq. (2.92).

$$c = \frac{a}{k_1} = \frac{\sigma_y(A_s - A_{s_f})}{0.85\sigma'_c k_1 b_w}$$

where

$$A_{s_f} = \frac{0.85\sigma'_c}{\sigma_y} = h_f(b - b_w) \tag{2.92}$$

In like manner, the bending moment equation can be developed.

$$M_w = T_w\left(d - \frac{a}{2}\right)$$

and

$$M_f = T_f\left(d - \frac{h_f}{2}\right)$$

The total value is given by the relationship

$$M = M_w + M_f = (A_s - A_{s_f})\left(d - \frac{a}{2}\right)\sigma_y + A_{s_f}\left(d - \frac{h_f}{2}\right)\sigma_y \qquad (2.93)$$

As in the case of the rectangular beam, it is highly desirable for the T beam to be underreinforced—so that failure is ductile in nature. To accomplish this, one could again limit the steel ratio to 75 percent of that associated with a balanced design. For this type of cross section, however, the balanced ratio must be redefined to account for the different cross-sectional form.

Equilibrium of stress resultants requires that

$$A_s\sigma_y = 0.85\sigma'_c k_1 cb_w + 0.85\sigma'_c(b - b_w)h_f$$

or

$$A_s\sigma_y = 0.85\sigma'_c k_1 cb_w + A_{s_f}\sigma_y \qquad (2.94)$$

For this type of member, it is convenient to define the steel ratio, $\bar{p}$, in terms of the ratio of the total steel area, A_s, to only the concrete stem area, $b_w d$, that is,

$$\bar{p} = \frac{A_s}{b_w d}$$

Similarly, the flange steel ratio would be

$$p_f = \frac{A_{s_f}}{b_w d}$$

Substituting these into eq. (2.94) gives for the T beam the steel ratio

$$\bar{p} = \frac{0.85\sigma'_c k_1 cb_w}{\sigma_y b_w d} + \frac{A_{s_f}\sigma_y}{\sigma_y b_w d} = \frac{0.85\sigma'_c k_1 c}{\sigma_y d} + p_f \qquad (2.95)$$

The location of the neutral axis for balanced designs—that is, with $\varepsilon_{c,\,max} = 0.003$, and $\varepsilon_s = \varepsilon_y = \sigma_y/E_s$—is obtained from a consideration of the strain distribution diagram.

$$\frac{c}{d} = \frac{0.003}{0.003 + \varepsilon_y} = \frac{0.003}{0.003 + \sigma_y/E_s}$$

Assuming $E_s = 29{,}000{,}000$ lb/in^2, this gives

$$\bar{p} = \frac{0.85k_1\sigma'_c}{\sigma_y}\frac{87{,}000}{87{,}000 + \sigma_y} + p_f \qquad (2.96)$$

where σ_y and σ'_c are in lb/in^2. (It is to be noted that the first term of eq. (2.96) is identically the same as for the balanced design of a rectangular beam of b width and d depth.)

$$\bar{p}_{max} = 0.75\bar{p} \qquad (2.97)$$

would again be chosen to ensure a ductile failure.

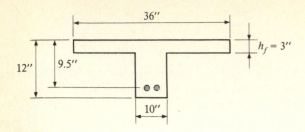

Figure 2.51

Example 2.19 For the section shown in fig. 2.51, the concrete has an ultimate strength of $\sigma_c' = 4000$ lb/in², and the steel $\sigma_y = 40,000$ lb/in². $A_s = 2.0$ in². Assuming that the neutral axis lies within the flange, the corresponding value of a (determined from the condition that $C = T$) would be

$$a = \frac{A_s \sigma_y}{0.85 \sigma_c' b} = \frac{(2.0)(40)}{(0.85)(4)(36)} = 0.653 \text{ in} < 3.00 \text{ in}$$

Since $a < h_f$, this assumption of the general location of the neutral axis is correct. The moment capacity is, therefore, given by the equations for a rectangular-cross-sectional beam.

$$M = A_s \sigma_y \left(d - \frac{a}{2} \right)$$

or

$$M = (2.0)(40)\left(9.5 - \frac{0.653}{2} \right) = 733.9 \text{ in} \cdot \text{kips}$$

As a second example, assume the same overall dimensions, but presume a reduction in the effective flange width to 28 in. Furthermore, the steel reinforcement is given as four no. 11 bars, and has a yield stress of 60 ksi. Again, compute a as if the neutral axis were located in the flange.

$$a = \frac{(6.24)(60)}{(0.85)(4)(28)} = 3.93 \text{ in} > h_f$$

Obviously, this assumption is incorrect! The neutral axis will therefore be located in the stem, and the modified equations must be used.

The steel area required to sustain the flange compressive force is given by the equation

$$A_{sf} = \frac{0.85 \sigma_c'}{\sigma_y} h_f (b - b_w) = \frac{(0.85)(4)}{(60)} (3)(28 - 10) = 3.06 \text{ in}^2$$

The depth of the compressive stress block in the stem is

$$a = \frac{(A_s - A_{sf})\sigma_y}{0.85 \sigma_c' b_w} = \frac{(6.24 - 3.06)(60)}{(0.85)(4)(10)} = 5.61 \text{ in}$$

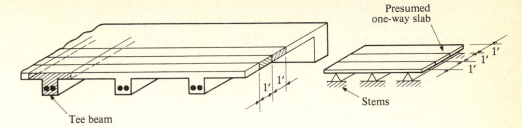

Floor system using tee beams construction   Idealization for slab design

Figure 2.52 Transverse one-way slab idealization for floor systems using T-beam construction.

The moment capacity is therefore given by eq. (2.93)

$$M = (6.24 - 3.06)\left(9.5 - \frac{5.61}{2}\right)(60) + (3.06)\left(9.5 - \frac{3.00}{2}\right)(60)$$

$$= 3,912.3 \text{ in} \cdot \text{kips}$$

Before ending this section, it should be pointed out that for floor systems where these types of members are used, the top slab is continuous over several stems. More than that: the top slab contains reinforcing bars that allow it to sustain bending moments in a direction perpendicular to the longitudinal axis of the stems. As an approximation—and a good one when the T beams in question are relatively long—the top continuous slab is frequently thought of as a series of 1-ft-wide one-way reinforced beams. This is shown in the idealization sketch of fig. 2.52.

2.9.2 Composite Beams

Where concrete one-way slabs are supported on steel members and both act together as one integral cross section, the beam that results is referred to as a *composite beam*. In general, to ensure interaction the two elements are attached at their interface by shear connectors. These are necessary components of this type of design. Without them, slippage along the interface would occur, and the bending resistance would consist of nothing more than the two parts acting independently. A typical composite beam cross section is shown in fig. 2.53.

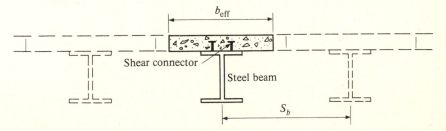

Figure 2.53 Use of shear connectors in composite beams.

The transformed section method of analysis is normally used in solving these types of problems. As illustrated earlier, this method involves transforming one of the elements into an equivalent area of the other using the modular ratio n.

$$n = \frac{E_s}{E_c}$$

E_s = Young's modulus of elasticity of the steel

E_c = Young's modulus of elasticity of the concrete

Since the transformed section is a homogeneous one, all the basic beam bending relationships can be employed.

For composite sections of this type, it is more convenient to transform the concrete into an equivalent area of steel rather than vice versa. Figure 2.54 illustrates the procedure.

Once the transformed section is defined, the analysis for stress is straightforward. The steel stress is calculated from the relationship

$$\sigma_s = \frac{My}{I_T} \tag{2.98}$$

where M = the moment applied to the cross section in question
y = is measured from the neutral axis
I_T = the moment of inertia of the transformed section

Similarly, shearing stresses are found from the equation

$$\tau = \frac{VQ}{I_T t} \tag{2.99}$$

Of particular interest is the shear flow force that exists at the interface between the steel and the concrete, that is

$$\text{Shear flow} = \tau t = \frac{VQ_c}{I_T} \tag{2.100}$$

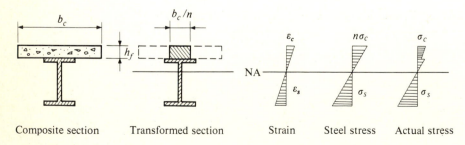

Composite section   Transformed section   Strain   Steel stress   Actual stress

Figure 2.54 Stress and strain distributions in composite beams.

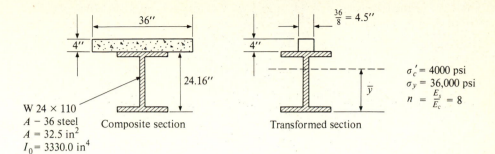

W 24 × 110
A − 36 steel
A = 32.5 in²
I_0 = 3330.0 in⁴

Composite section   Transformed section

σ_c' = 4000 psi
σ_y = 36,000 psi
$n = \dfrac{E_s}{E_c} = 8$

Figure 2.55

The shear flow is used to determine the size and spacing of the shear connectors. (It is to be recognized that the dimensional units associated with shear flow are force per unit of length, for example, pounds per foot of length along the beam.) Q_c in eq. (2.100) is the first moment of the transformed concrete area about the neutral axis.

Example 2.20 For the composite section shown in fig. 2.55, determine the allowable bending moment and resulting stress distribution across the section.

First, the location of the neutral axis must be established. This is most conveniently accomplished by setting up a table of the type 2.4.
Measuring $\bar{y}$ from the bottom of the beam

$$\bar{y} = \frac{863.48}{50.50} = 17.10 \text{ in}$$

The moment of inertia of the transformed section is therefore

$$I_T = \sum I_0 + \sum Ad^2 = 3354 + 2296 = 5650 \text{ in}^4$$

The section modulus associated with the bottom of the beam is

$$S_{bot} = \frac{I_T}{\bar{y}} = \frac{5650}{17.10} = 330.41 \text{ in}^3$$

Table 2.4 Properties of composite section: Example 2.20

Section	I_0, in⁴	A, in²	y_0, distance from bottom to centroid, in	$A \times y_0$	d, distance to neutral axis	$A \times d^2$
Beam	3330	32.50	12.08	392.60	− 5.02	818.69
Slab	24	18.00	26.16	470.88	9.06	1477.50
$\sum$	3354	50.50		863.48		2296.19

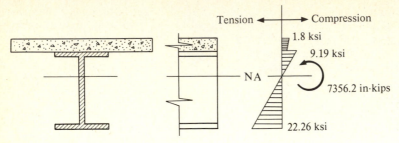

Figure 2.56 Stress distribution—example 2.20.

At the top of the concrete slab

$$S_{top} = \frac{I_T}{24.16 + 4.00 - 17.10} = \frac{5650}{11.06} = 510.85 \text{ in}^3$$

At the top of the steel beam

$$S_{steel} = \frac{I_T}{24.16 - 17.10} = \frac{5650}{7.06} = 800.28 \text{ in}^3$$

If the allowable stress for the steel is $\sigma_{s, all} = 24$ ksi, and for the concrete $\sigma_{c, all} = 1.8$ ksi, then the moment capacity of the beam, as determined by assuming each of the locations referred to above at its allowable stress value, would be:

At the bottom of the steel:

$$M = (S_{bot})(24.0) = (330.41)(24.0) = 7929.8 \text{ in} \cdot \text{kips}$$

At the top of the steel:

$$M = (S_{steel})(24.0) = (800.28)(24.0) = 19,206.7 \text{ in} \cdot \text{kips}$$

At the top of the concrete slab:

$$M = (S_{top})(n\sigma_{c, all}) = (510.85)(8)(1.8) = 7356.2 \text{ in} \cdot \text{kips}$$

The beam can sustain a moment only up to the smallest of these. Therefore, the moment capacity is governed by the stress at the top of the concrete slab. The stress distribution corresponding to that particular moment is shown in fig. 2.56.

2.10 PROBLEMS

2.1 For plane stress and plane strain develop expressions for normal stress as a function of strain.

2.2 An 8-in-long 2-in-diameter cylindrical aluminum block is placed between two rigid bearings at each end of the longitudinal axis. It is then heated to 200°F. Determine $\sigma_x, \sigma_y, \sigma_z, \varepsilon_x, \varepsilon_y$, and ε_z if the coefficient of thermal expansion for aluminum is $\alpha_A = 12.8 \times 10^{-6}/°F$, $\mu = 0.33$, and $E = 10.0 \times 10^6 \text{ lb/in}^2$.

2.3 An important case in two-dimensional linear elasticity is that of *pure shear*. For such a case, what will be the magnitudes, directions, and senses of the corresponding principal stresses? Using eq. (2.2), and letting $\sigma_z = 0$, $\sigma_\xi = +\tau_{xy}$, and $\sigma_\eta = -\tau_{xy}$, show that

$$G = \frac{E}{2(1 + \mu)}$$

2.4 A cantilever beam of rectangular cross section carries a concentrated load at its free end. Sketch three sets of stress contours, one for tension, one for compression, and one for maximum shearing stress.

2.5 It is given that the allowable vertical compressive stress for a wooden post in which the grain has a slope of 1 horizontal to 6 vertical shall not exceed 50 percent of the allowable stress in compression parallel to the grain. What ratio of shearing strength along the grain to compressive strength does this imply?

2.6 The *cubic dilatation* of an isotropic elastic material is defined as the change in volume per unit volume, that is

$$\text{Dilatation} = \frac{V - V_0}{V_0} = \frac{(1 + \varepsilon_x)(1 + \varepsilon_y)(1 + \varepsilon_z)\, dx\, dy\, dz - dx\, dy\, dz}{dx\, dy\, dz}$$

$$= \varepsilon_x + \varepsilon_y + \varepsilon_z$$

when the products of the strains are considered to be negligibly small. Show that the *bulk modulus*, that is, the average normal stress divided by the cubic dilatation, is given by

$$K = \frac{\frac{1}{3}(\sigma_x + \sigma_y + \sigma_z)}{\text{dilatation}} = \frac{E}{3(1 - 2\mu)}$$

2.7 For the T cross section shown, determine the following:
    (*a*) The location of the centroid
    (*b*) Q_x (about the centroidal axis) of the top flange
    (*c*) I_x (about the centroidal axis)
    (*d*) I_y (about the centroidal axis)

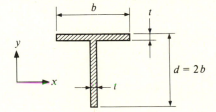

Figure P2.7

2.8 For the cross section shown, determine the following:
    (*a*) The location of the centroid
    (*b*) I_x (about the centroidal axis)
    (*c*) I_y (about the centroidal axis)

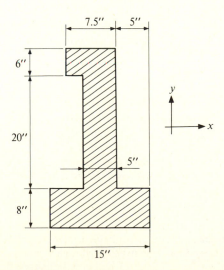

Figure P2.8

(d) I_{xy} (about the centroidal axes)
(e) Direction of the principal axes ξ and η
(f) I_η (about the centroidal axis)
(g) I_ξ (about the centroidal axis)
(h) Q_ξ (about the centroidal axis) of the bottom flange

2.9 For the solid, linearly tapered beam of constant width shown, express I_x as a function of z.

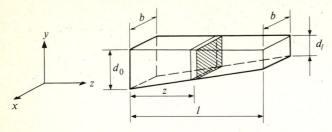

Figure P2.9

2.10 A built-up cross section consists of two channels C 12×30 laced together as shown. Ignoring the effects of the lacing, how long should be the distance d so that $I_x = I_y$?

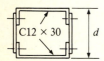 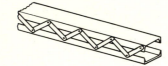

Figure P2.10

2.11 The T beam shown is subjected to bending about the x axis, and $P_z = 0$. Using (a) the equation of equilibrium and (b) stress blocks, compute M_x.

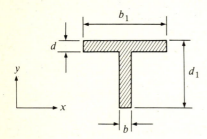

Figure P2.11

2.12 For the cross section shown, in terms of the extreme fiber stress and for the axis indicated, develop expressions for
(a) M_x (b) P_z (c) $\sigma_z = k_1 y$

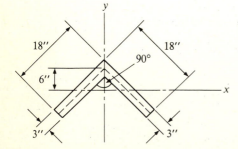

Figure P2.12

2.13 Determine M_x and M_y for the cross section shown if $\sigma_z = k_1 x^2 y$.

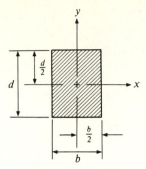

Figure P2.13

2.14 Assume for each of the cross sections shown that the loading is a bending moment M_x, that bending strains vary linearly across the section, and that normal stresses σ_z are related to normal strains ε_z according to the relationship $\varepsilon = \sigma/E$. Develop equations for the normal stresses as a function of the cross-sectional dimensions.

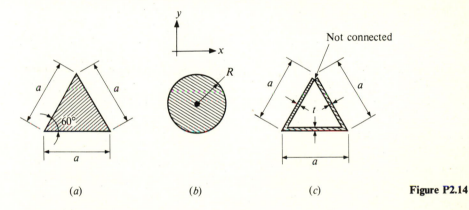

(a)   (b)   (c)   **Figure P2.14**

2.15 A solid square cross section is subjected to a twisting moment about its longitudinal axis. If it is given

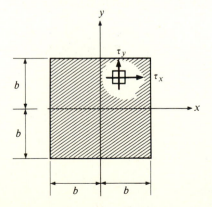

Figure P2.15

that the resulting shearing stresses are

$$\tau_x = \alpha\left(\frac{x}{b}\right)^2 - \alpha y$$

$$\tau_y = \alpha x - \left(\frac{y}{b}\right)^2$$

determine the value of M_z.

2.16 For the parabolically curved stress-strain relationship shown, determine the values of k_1 and k_2 such that the parabolic and rectangular areas have the same area and also have the same first moment of the areas about the stress axis.

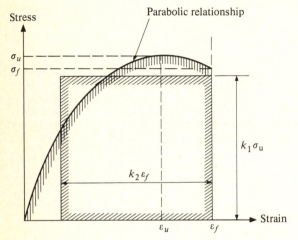

Figure P2.16

2.17 In terms of the applied load Q and the cross-sectional dimensions, derive expressions for the maximum normal stress in tension and compression for the T section shown.

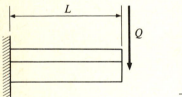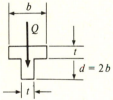

Figure P2.17

2.18 Determine the normal stress distribution for the T beam shown in prob. 2.17, if the member is subjected to a compressive force P applied eccentrically to the member at a distance a. What should be the value of a if it is desired to have zero normal stress at the bottom of the T?

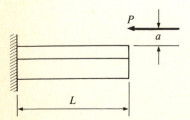

Figure P2.18

2.19 Calculate the maximum normal stress for the cross section defined in prob. 2.8, assuming a bending moment of 100 in·kips, applied about the x axis. If the maximum allowable stress is 1500 lb/in², what is the corresponding maximum M_x?

2.20 A circular shaft of 4 in in diameter is subjected to an axial end thrust of 10 kips, a bending moment of 8000 ft·lb, and a twisting moment about the longitudinal axis of 4000 ft·lb. Determine the numerical values of the maximum normal and shearing stresses in the shaft.

2.21 Determine the diameter of a circular-cross section steel shaft if the applied tensile load is $P = 5000$ lb and the torsional moment is $M_z = 8000$ in·lb. (Use a factor of safety of 2.0, $E = 30 \times 10^6$ lb/in², $\mu = 0.25$, and $\sigma_y = 30,000$ lb/in².) (*Note*: Solve by trial and error.)

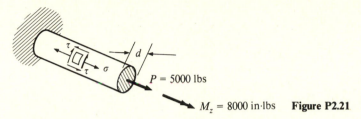

$P = 5000$ lbs

$M_z = 8000$ in·lbs   **Figure P2.21**

2.22 For the double T shown, find the maximum normal stress at the top and bottom of the section if the applied moment M_x is 100 ft·kips. If the maximum allowable stress is 1500 lb/in², what is the maximum allowable moment? Comment on the adequacy of the section to carry a moment about the x axis of 100 ft·kips.

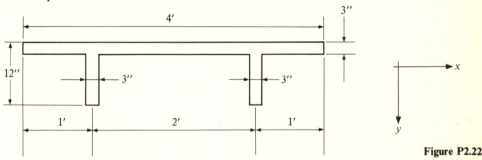

Figure P2.22

2.23 Select an appropriate rolled steel section of A36 steel if a safety factor of 1.66 is prescribed and
(a) $M_x = 650$ in·kips
(b) $M_x = 400$ in·kips, $M_y = 200$ in·kips
(c) $M_x = 500$ in·kips, $M_y = 100$ in·kips, $P_z = 100$ kips
Assume that the member is braced so that it cannot buckle laterally.

2.24 A horizontal, prismatic member whose weight is w lb/ft is to be picked up at two symmetrically placed points as shown in the sketch. How far from the ends of the member should those points be located if the bending stress is to be a minimum? If $w = 40$ lb/ft and $\sigma_{z,\,\text{all}} = 4000$ lb/in² and the cross section is 6 in × 6 in, what length of member can be handled in this manner?

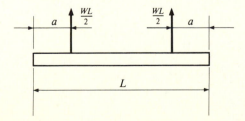

Figure P2.24

2.25 Determine the shearing stress distribution along the vertical axis for the member shown in prob. 2.12. What is the value of the maximum shearing stress?

2.26 The geometry of the floor framing for a typical interior panel of an apartment building is shown. In addition to applied live loads (see table 1.2), the members carry a 6-in-thick reinforced concrete slab. Neglecting the weights of the members themselves, determine the maximum moment and shearing force in each of the members—including the column loading. If the members are to be rolled steel shapes made from A36 steel, select an approximate shape for each member. (Include the weights of the members in these considerations.)

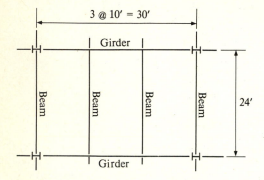

Figure P2.26

2.27 The plate girder section shown is fabricated by welding together a 44 in by $\frac{5}{16}$ in web plate and two 16 in by 1 in flange plates, all of A441 steel. If the maximum applied moment is 2000 ft·kips and the maximum applied shear is 150 kips, determine

  (a) The actual margin of safety in bending
  (b) Maximum total force per foot of length to be developed by the weld

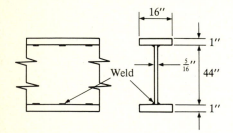

Figure P2.27

2.28 For the plate girder section and loading given in prob. 2.27, calculate the shear flow at 4 in from the upper flange tip.

2.29 Locate the shear centers for the cross sections shown.

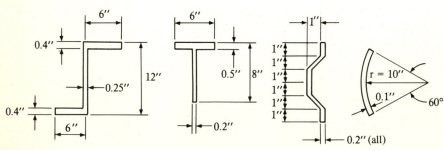

Figure P2.29

2.30 Locate the shear center for a thin-walled circular tube which contains a longitudinal slit along its entire length.

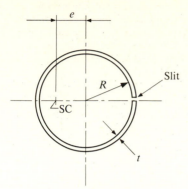

Figure P2.30

2.31 For the cross section shown, determine the torsion constant κ_T.

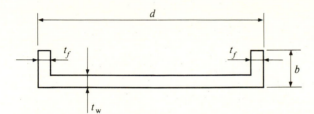

Figure P2.31

2.32 If in prob. 2.31, $b = 3$ in, $d = 12$ in, $t_w = 0.2$ in, and $t_f = 0.5$ in, what is the maximum St. Venant torsion stress if $M_z = 10,000$ in·lb? What is the total angle of rotation of one end of the member with respect to the other if the member is prismatic and 10 ft long?

2.33 Derive expressions for $\sigma_{\omega, \max}$ (maximum warping normal stress) and $\tau_{\omega, \max}$ (maximum warping shear stress) for the section shown in prob. 2.31.

2.34 The simply supported I beam shown is loaded by a uniformly distributed load of 1.0 kip/ft at a lateral eccentricity of 6 in. If the material is A36 steel, the shape W 8 × 31, and the length 15 ft, determine the maximum normal stress and the maximum shear stress. At location z from the end of the member, the angle of twist is given by:

$$\theta = \frac{qe}{G\kappa_T}\left(\frac{\cosh \lambda l - 1}{\lambda^2 \sinh \lambda l}\sinh \lambda z - \frac{1}{\lambda^2}\cosh \lambda z - \frac{l}{2}z + \frac{1}{\lambda^2} + \frac{1}{2}z^2\right)$$

$$\lambda = \frac{G\kappa_T}{EI_\omega}$$

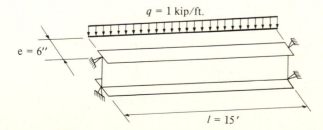

Figure P2.34

2.35 The "short" concrete column whose cross section is shown is subjected to a concentric load P. If $\sigma'_c = 4000$ lb/in^2 and $\sigma_y = 40,000$ lb/in^2, what is the maximum value of P that the column can sustain?

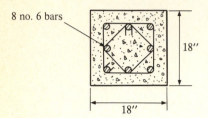

8 no. 6 bars

18"

18"

Figure P2.35

2.36 The rectangular cross section beam shown consists of two materials joined together. The elastic moduli are E_1 and E_2. Develop a relationship between the applied bending moment and curvature for this section.

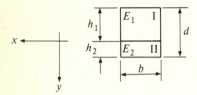

h_1 E_1 I

x

h_2 E_2 II

d

b

y

Figure P2.36

2.37 For the rectangular cross section shown, the modulus of elasticity E varies linearly from $1.0E$ at the top to $3.0E$ at the bottom of the section. Develop expressions for the moment-curvature relationship and the normal stress taking into account this variation.

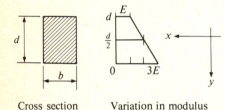

d

b

d

$\dfrac{d}{2}$

E

x

0 $3E$

y

Cross section   Variation in modulus of elasticity   **Figure P2.37**

2.38 A cross section consists of a 6 in × 8 in timber to which is attached at the bottom a steel strap. If the

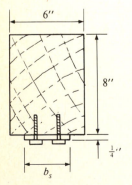

6"

8"

$\frac{1}{4}''$

b_s

Figure P2.38

applied moment is 13 ft·kips, how wide must be the $\frac{1}{4}$-in-thick steel strap?

$$E_{wd} = 1.8 \times 10^3 \text{ ksi}$$

$$E_s = 30 \times 10^3 \text{ ksi}$$

$$\sigma_{\text{all, } wd} = 1.7 \text{ ksi}$$

$$\sigma_{\text{all, } s} = 24 \text{ ksi}$$

2.39 If the beam in prob. 2.38 is 10 ft long, simply supported, and carries a uniformly distributed load of 1.2 kips/ft, what is the minimum spacing along the member of the connecting screws? (Each screw can develop 1000 lb of shearing force, and they are used in pairs as shown.)

2.40 The singly reinforced concrete beam section is constructed of concrete having an ultimate stress of $\sigma'_c = 3000$ lb/in² and steel whose yield stress is $\sigma_y = 40,000$ lb/in². According to the latest ACI code provisions, what is the moment capacity of this section?

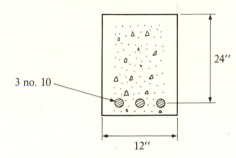

24″

3 no. 10

12″

Figure P2.40

2.41 The one-way slab shown has a total depth of 7 in. No. 6 bars are spaced uniformly at 4 in on center. The distance from the bottom of the slabs to the center of the steel is 2 in. If $\sigma'_c = 3000$ lb/in² and $\sigma_y = 40,000$ lb/in², what is the moment capacity of the slab per foot of width?

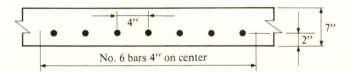

4″

7″

2″

No. 6 bars 4″ on center

Figure P2.41

2.42 A singly reinforced beam is designed to carry a moment of 3,000 in·kips. If its width is 12 in and $\sigma'_c = 4000$ lb/in² and $\sigma_y = 60,000$ lb/in², then, according to the latest ACI code, how many bars of what size would be required if the

    (a) Minimum steel ratio were used?

    (b) Maximum steel ratio were used?

What would be the depth of the beam in each case?

2.43 Select an appropriate rectangular cross section and the reinforcement for the mid- and quarter-points for the simply supported beam shown. The live load is uniformly distributed and equals 1.0 kip/ft. (The dead load depends upon the size of the beam selected.) Use as a load factor, LF = 1.40 + 1.7 (DL/LL), $\sigma'_c = 4000$ lb/in² and $\sigma_y = 40,000$ lb/in².

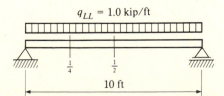

$q_{LL} = 1.0$ kip/ft

$\frac{1}{4}$ $\frac{1}{2}$

10 ft

Figure P2.43

2.44 Determine the moment capacity of the doubly reinforced beam shown.

$$\sigma'_c = 4000 \text{ lb/in}^2$$

$$\sigma_y = 40{,}000 \text{ lb/in}^2$$

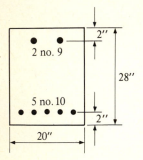

Figure P2.44

2.45 What is the moment capacity of the section shown? Comment upon the design based on the computed stresses. It is to be assumed that $\sigma'_c = 3000 \text{ lb/in}^2$, and $\sigma_y = 50{,}000 \text{ lb/in}^2$.

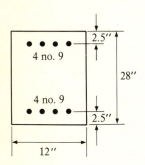

Figure P2.45

2.46 Determine the moment capacity of the T beam shown if
    (a) $A_s = $ three no. 9   (b) $A_s = $ six no. 9

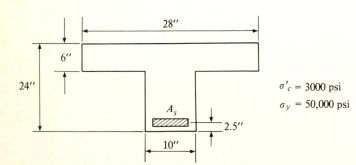

$\sigma'_c = 3000$ psi

$\sigma_y = 50{,}000$ psi

Figure P2.46

2.47 The concrete box beam shown contains four no. 10 bars in the top and bottom. Determine the bending capacity of the section about the x and y axes.

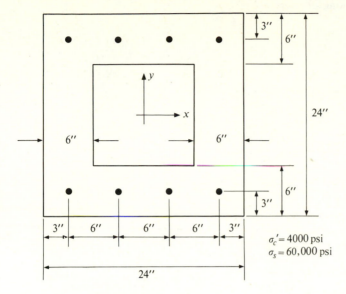

$$\sigma_c' = 4000 \text{ psi}$$
$$\sigma_s = 60,000 \text{ psi}$$

Figure P2.47

2.48 A composite section consists of a 6-in deep slab that is 48 in wide and is joined to a steel W 27 × 102 beam as shown in the sketch. In addition, the steel beam has welded to its lower flange a $\frac{1}{2}$ in by 9 in plate. What is the allowable moment for this section? ($\sigma_{s,\text{all}} = 24,000 \text{ lb/in}^2$, $\sigma_{c,\text{all}} = 1350 \text{ lb/in}^2$.)

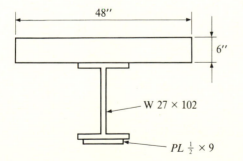

48″

6″

W 27 × 102

PL $\frac{1}{2}$ × 9   **Figure P2.48**

2.49 The concrete slab in prob. 2.48 is 5 in thick and has a 48-in width. Select an appropriate steel shape assuming the moment to be carried is 350 ft·kips and that $\sigma_{c,\text{all}} = 1350 \text{ lb/in}^2$ and $\sigma_{s,\text{all}} = 24,000 \text{ lb/in}^2$.

2.50 A concrete wall 2 ft 0 in in width exerts a load on a footing of 16,000 lb/linear foot along the wall. If the soil on which the foundation rests has a bearing capacity of 4000 lb/ft^2, determine the width of the symmetrical footing. Assuming d is 12 in and a 3-in cover over the steel is required, select an appropriate amount of reinforcing. The following are given:

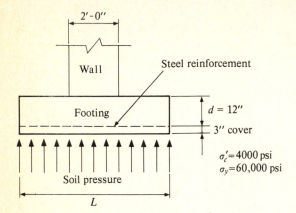

Footing

$d = 12''$

3'' cover

$\sigma_c' = 4000$ psi
$\sigma_y = 60,000$ psi

Soil pressure

L

Figure P2.50

THREE

EQUILIBRIUM OF STRESS RESULTANTS IN STATICALLY DETERMINATE STRUCTURES

3.1 INTRODUCTION

Chapter 2 dealt with the interrelationships that must necessarily exist between the forces and moments at a given cross section in a structure and the resulting normal and shearing stresses. It was presumed throughout the discussion that the resultants were either prescribed or could be readily determined from elementary procedures. In this chapter attention will be directed to the determination of these resultant forces and moments for a number of different types of structures subjected to a variety of applied forces. More specifically, chap. 3 deals with statically determinate systems. A later, counterpart chapter will deal with statically indeterminate ones.

As defined previously, statically determinate structures are those for which the equations of statics are sufficient to establish within the structure the distribution of forces and moments. The forces to which the structures will be subjected will be either the weights of the members themselves (or similarly induced forces due to mass times acceleration), applied concentrated or distributed forces, or bending or twisting moments. These include the resulting forces or moments due to induced reactions.

It generally will be assumed, in addition, that deformations of the structures are small and can be neglected in formulating the equilibrium equations. The expressions resulting will therefore correspond to first-order formulations.

For any structure in a state of static equilibrium, it is known that the overall resultant of all of the individual stress resultants (both forces and moments) must vanish. That is,

$$\sum R = 0$$
$$\sum M_i = 0$$

In component form, and presuming a three-dimensional situation, these two equations can be written in their more normal form:

$$\sum F_x = 0 \qquad \sum F_y = 0 \qquad \sum F_z = 0$$

$$\sum M_{xx} = 0 \qquad \sum M_{yy} = 0 \qquad \sum M_{zz} = 0 \tag{3.1}$$

Here the term $\sum F_i$ represents *all* the forces or components of force that exist in the i direction—including the reactions. $\sum M_{jj}$ is the sum of *all* the moments that exist about the jj axis (or an axis parallel to it).

For a two-dimensional case, only two of the three axes are involved. If these are the y and z axes, then one possible set of necessary equilibrium equations is

$$\sum F_y = 0 \qquad \sum F_z = 0 \qquad \sum M_x = 0 \tag{3.2}$$

But these are not the only ones (nor necessarily are they the best ones for a particular problem). As indicated in eqs. (1.12) and (1.13), other forms are available.

Internal stress resultants are obtained by cutting the structure in question at the point where the resultants are desired, and using this cut section as one end of a free body diagram. The reactive (internal) forces at the cut section must satisfy equilibrium—and this will be the general condition required for their determination.

Structural members are frequently classified according to the manner in which they resist internal stress resultants.

1. *Axial force members.* A member that is subjected to either an axial tensile or compressive force is referred to as an axial force member. For example, the cable shown in fig. 3.1*a* is an axially loaded tension member. The right-hand column of the portal frame (fig. 3.1*b*) is an axially loaded compression member. (A cut along either one of these members will result in the introduction of only one unknown stress resultant.)
2. *Bending and shear-resisting members.* Except in the very special, limiting case of constant moment, members subjected to bending also are normally presumed to be subjected to shearing forces. A cut of a beam, therefore, will introduce two stress resultants—M and V (fig. 3.1*c*).
3. *Bending, shear, and axial force resisting members.* The arch shown in fig. 3.1*d* represents the general two-dimensional case where all three components exist.

In all the instances described above, it should be understood that the cuts were made perpendicular to the longitudinal axis of the member at the point in question. This is not a necessary condition for the use of the equilibrium equations. The resulting values, however, correspond directly to the presumptions made in chap. 2 for stress analysis.

When structures are three-dimensional, a fourth general type of member is observed—a torsional member.

4. *Torsion-resisting member.* A member subjected to pure torsion has only one resultant. This is indicated for a circular bar in fig. 3.1*e*.

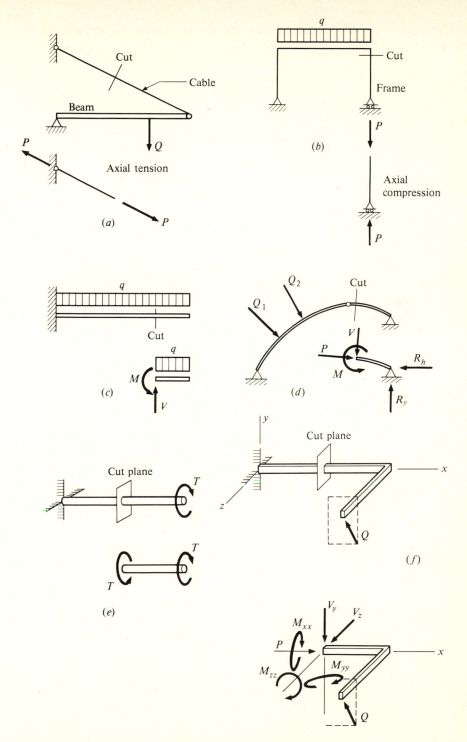

Figure 3.1 Internal stress resultants in various structural members.

When a three-dimensional member (or a two-dimensional member in a three-dimensional structure) is cut, a maximum of six unknown stress resultants can exist (fig. 3.1*f*).

3.2 BEAMS

Beams are used primarily to resist bending and shear. In this section two-dimensional cases are considered—although the procedures can readily be extended to include three-dimensional members and loadings. Since a two-dimensional problem is specified, there are three independent equations of static equilibrium available for the determination of reaction components. At cut sections, three components of resulting forces and moments are possible.

To facilitate erection, and in some special cases to reduce bending normal stresses, a statically determinate beam may take the form of a *compound beam*. A compound beam consists of two or more simple beams interconnected by frictionless pins or hinges. Such a beam requires more than three external reaction components to ensure stability. They often appear at a casual glance to be indeterminate. The interior hinges, however, provide sufficient additional independent equilibrium equations to allow the determination of these interior reactive forces. The most convenient procedure for solving members of this type is to take each interior simple beam and consider it as a separate free body. Use would be made of the fact that a frictionless pin transmits only shear and axial forces, and has no bending resistance. Several, typical determinate beams are shown in fig. 3.2.

In the preceding chapter it was demonstrated that a particular relationship must exist between (*a*) the rate of change of vertical shear and the lateral loading on the element, and (*b*) the rate of change of bending moment and the magnitude of the shearing force at the section in question:

Assuming the loading to be in the yz plane

$$\frac{dV_y}{dz} = -q_y$$

and
$$\frac{dM_x}{dz} = V_y \tag{3.3}$$

Therefore

$$\frac{d^2M_x}{dz^2} = \frac{dV_y}{dz} = -q_y \tag{3.4}$$

Given that the mathematical expression for the curvature of the curve of the deformed center line of the member is $d^2y/dz^2 = \phi = -M_x/EI_x$, the following fourth-order equation will result:

$$\frac{d^4y}{dz^4} = +\frac{q_y}{EI_x} \tag{3.5}$$

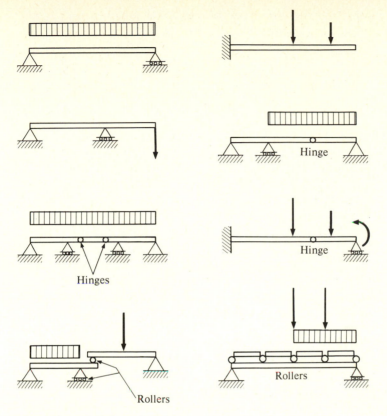

Figure 3.2 Typical determinate beam structures.

Equations (3.3), (3.4), and (3.5) are useful in determining the magnitudes and directions of the shear and bending-moment resultants along the length of the beam. Consider, for illustration of several of the concepts, the beam and loading shown in fig. 3.3. The distributed vertical load $q(z)$ is presumed to be a function of z. It is desired to determine the shear and bending moment as functions of z. Since in fig. 3.3 there are no horizontal components of applied force, it can be concluded from $\sum F_z = 0$ that there will be no horizontal component of reaction at end a. The vertical reactions can therefore be computed from the equations

$$\sum F_y = 0 \qquad \text{and} \qquad \sum M_a = 0$$

From eq. (3.3)

$$dV_y = -q(z)\, dz \qquad \text{and} \qquad dM = V_y\, dz$$

or

$$V = \int_0^z -q(z)\, dz + C_V \qquad M = \int_0^z V_y\, dz + C_M$$

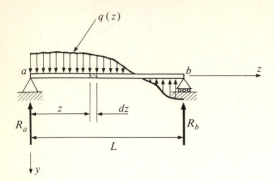

Figure 3.3 Simple beam subjected to a distributed lateral load.

C_V and C_M are constants and represent the values of shear and moment at $z = 0$. Therefore,

$$V(z) = V_a - \int_0^z q(z)\,dz$$

and

$$M(z) = M_a + \int_0^z V(z)\,dz$$

(Since the beam is simply supported, $M_a = M_b = 0$.)

A number of interesting observations can be made concerning these equations:

1. If $q(z) = 0$, shear will remain constant with z. (If a concentrated load is applied, and $q(z) = 0$, the shears on either side of the load will be constants. Moreover, the change in the value of the shear at the load point will equal the magnitude of the applied concentrated force at that point.)
2. If $V = 0$, the moment value must be either a maximum or a minimum—since $dM/dz = V$.
3. If $q(z)$ varies continuously over the range of integration, then V will also vary over the same range, but not in the same manner.

Example 3.1 Consider the cantilever beam shown in fig. 3.4, which is presumed to be carrying a distributed vertical load which varies as a parabola. The total length of the beam is L, and it is desired to determine the shear and bending-moment stress resultants at point $z = L/2$. (Note that for ease of mathematical expression the origin of the coordinate system coincides with the right-hand end of the beam. z is directed positively to the left.) For this case the shear at the center of the beam is given by the equation

$$V_{L/2} = V_0 - \int_0^{L/2} q(z)\,dz = 0 - \int_0^{L/2} q_L \left(\frac{z}{L}\right)^2 dz = -\left. \frac{q_L}{L^2} \frac{(z)^3}{3} \right|_0^{L/2}$$

or

$$V_{L/2} = -\frac{q_L(L)}{24}$$

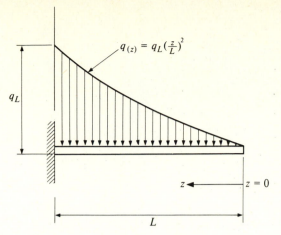

$q_{(z)} = q_L \left(\frac{z}{L}\right)^2$

q_L

$z \longleftarrow \quad z = 0$

L

Figure 3.4

The moment at the same section is

$$M_{L/2} = M_0 + \int_0^{L/2} V \, dz = 0 + \int_0^{L/2} \frac{-q_L}{L^3} \frac{(z)^3}{3} \, dz = \frac{q_L}{L^2} \frac{z^4}{12} \Big|_0^{L/2}$$

or $\qquad M_{L/2} = -\dfrac{q_L L^2}{192}$

In some cases it is convenient to express these problems in a nondimensional form prior to integrating. For example, selecting a nondimensional parameter

$$\zeta = \frac{z}{L} \qquad \text{or} \qquad dz = L \, d\zeta \tag{3.6}$$

the equations become

$$q(\zeta) = q_L \left(\frac{\zeta L}{L}\right)^2 = q_L \zeta^2 \tag{3.7}$$

and

$$V = V_0 - \int_0^\zeta q(\zeta) L \, d\zeta \tag{3.8}$$

The values of shear and moment at the midpoint of the beam are then

$$V_{L/2} = 0 - \int_0^{1/2} q_L \zeta^2 L \, d\zeta = -\frac{q_L L}{24}$$

and

$$M_{L/2} = 0 + \int_0^{1/2} -\frac{q_L L}{3} \zeta^3 \, d\zeta = -\frac{q_L L^2}{192}$$

which correspond identically to the values obtained above.

Example 3.2 For the simply supported beam shown in fig. 3.5, it is assumed that the distributed lateral load varies linearly with the coordinate axis. A nondimensional system $\zeta = z/L$ has been presumed.

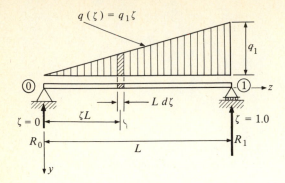

Figure 3.5

This example is different from that just considered. The end shear, $V_0 = R_0$, is not known at the outset. Therefore, this value must first be determined. Summing moments about end 0:

$$\sum M_0 = 0 = \int_0^1 q(\zeta)(\zeta L)\, d\zeta - R_1(1.0) = \int_0^1 q_1\, L\zeta^2\, d\zeta - R_1$$

or $\qquad R_1 = \tfrac{1}{3}q_1 L \qquad$ (assuming the direction shown!)

Considering equilibrium of forces in the vertical direction

$$\sum F_y = 0 = \int_0^1 q(\zeta)L\, d\zeta - R_0 - R_1 = \frac{q_1 L}{2} - (R_0 + R_1)$$

Thus $\qquad R_0 = \tfrac{1}{6}q_1 L \qquad$ (assuming the direction shown!)

Since the end shears are now known, integration can proceed as earlier indicated:

$$V = V_0 - \int_0^z q(z)\, dz = V_0 - \int_0^\zeta q(\zeta)L\zeta$$

or $\qquad V_\zeta = \frac{q_1 L}{6} - L\int_0^\zeta q_1 \zeta\, d\zeta = q_1 L\left(\frac{1}{6} - \frac{\zeta^2}{2}\right)$

(At $z = L$, $\zeta = 1.0$. Substituting this value into the equation yields

$$V_{(1.0)} = R_1 = q_1 L\left(\frac{1}{6} - \frac{1}{2}\right) = -\frac{q_1 L}{3}$$

The minus sign indicates that the load acts in the direction opposite to that of the coordinate axis.)

The bending moment at any general location ζ is given by the equation

$$M_\zeta = M_0 + \int_0^\zeta V_\zeta L\, d\zeta = 0 + \int_0^\zeta q_1 L^2\left(\frac{1}{6} - \frac{\zeta^2}{2}\right) d\zeta$$

or $\qquad M_\zeta = \frac{q_1 L^2}{6}(\zeta - \zeta^3)$

Its maximum value can be determined from the relationship $dM_\zeta/d\zeta = V_\zeta = 0$. That is,

$$q_1 L\left(\frac{1}{6} - \frac{\zeta^2}{2}\right) = 0$$

which gives as the location of the maximum moment

$$\zeta = \sqrt{\frac{1}{3}}$$

The corresponding moment value is

$$M_{max} = \frac{q_1 L^2}{6}\left(\sqrt{\frac{1}{3}} - \frac{1}{3}\sqrt{\frac{1}{3}}\right) = \frac{q_1 L^2}{9\sqrt{3}}$$

The shear and bending-moment diagrams for this beam and loading are shown in fig. 3.6.

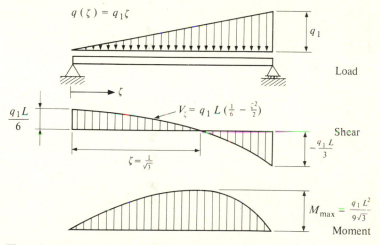

Figure 3.6 Shear and bending-moment diagrams—example 3.2.

When a beam is subjected to concentrated rather than distributed loads, the derived differential equations are not valid over the entire length of the beam. To directly include such possibilities, it is necessary to introduce into the development *singularity functions*.† There are, however, other methods for solving such problems—by dividing the beam into segments over which functions are continuous, and taking into account the boundary conditions at the ends of each of the segments.

Consider, for example, the beam shown in fig. 3.7. It is desired to construct the shear and moment diagrams for the loading specified. (Since the beam essentially consists of two separate parts joined by a frictionless hinge at point 3, this is a

---

† See S. H. Crandall, N. C. Dahl, and T. J. Lardner, *An Introduction to the Mechanics of Solids*, 2d ed., McGraw-Hill Book Company, 1972.

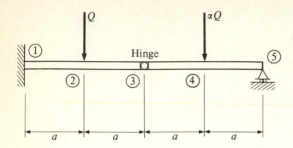

Figure 3.7

compound beam.) To facilitate examination, the two beams and the hinge connection will be considered separately. (It is to be understood that even though there are only three unknown external reactions, and this is a two-dimensional problem, the three unknowns cannot be determined from a direct application of the equilibrium equations to the total structure. There are no applied components of horizontal force. Nonetheless, one of the three independent equilibrium equations has to be used to rule out the possibility of a horizontal component at support 1.) Beam 3-5, however, involves only two unknown reactions, R_3 and R_5. Since two independent equations of equilibrium are available, these can be determined.

$$\sum M_3 = 0 \qquad (\alpha Q)(a) - (R_5)(2a) = 0$$

or
$$R_5 = +\frac{\alpha Q}{2}$$

From
$$\sum F_y = 0 \qquad R_3 + R_5 = \alpha Q$$

Therefore

$$R_3 = +\frac{\alpha Q}{2}$$

(The plus signs signify that the reactions act in the directions presumed.)

Having independently determined R_3, equilibrium of beam 1-3 now can be considered.

$$\sum M_1 = 0 \qquad (Q)(a) + (R_3)(2a) - M_1 = 0$$

or
$$M_1 = +Qa(1 + \alpha)$$

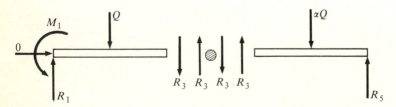

Figure 3.8

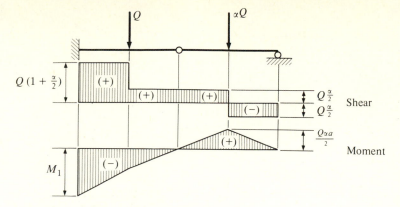

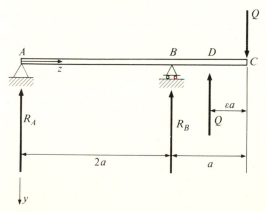

Figure 3.9 Shear and bending-moment diagram.

Summing forces in the vertical direction

$$\sum F_y = 0 \qquad R_1 - Q - R_3 = 0$$

or

$$R_1 = Q\left(1 + \frac{\alpha}{2}\right)$$

(Sign conventions previously defined were not rigorously followed in the solution of this problem. In fact, an assumed coordinate system was not even shown. For more complicated cases, however, where distributed as well as concentrated lateral forces are applied, as also might be concentrated or distributed bending moments, it is desirable to methodically adhere to a defined positive system.)

With the above determined reactions, and taking into account the known differential relationships [eq. (3.3)], the shear and bending-moment diagrams of fig. 3.9 are constructed.

Example 3.3 As a next example, consider the beam and loading shown in fig. 3.10. The beam is simply supported and of length $3a$, but it has an overhang of length a. At the end of this overhang there is applied a pair of forces equal in

Figure 3.10

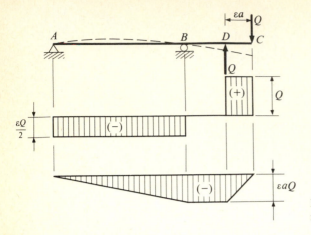

Figure 3.11 Shear and bending-moment diagram—example 3.3

magnitude but opposite in sense. They are separated by a distance εa. It is desired to construct the corresponding shear and bending-moment diagrams.

Beginning at the free end C, it is recognized that the shear between C and D has a constant magnitude of $+Q$. (It is plus because the stress resultant acts in the $+y$ direction on the most positive face, and constant because $dV/dz = q = 0$.) Between the points B and D the shear is zero. (*Why?*) Between A and B, it is constant and equal to the magnitude of the reaction R_A. (*Why?*) Summing moments about B (and assuming that clockwise moments are positive),

$$\sum M_B = 0 \qquad -R_A(2a) + Q(a) - Qa(1-\varepsilon) = 0$$

or

$$R_A = +\frac{Q\varepsilon}{2}$$

R_A acts in the direction of the coordinate axis y.

Since the shear is constant between A and B, the moment in this region will vary linearly. The value of the moment between points B and D is constant. It will change linearly between D and C. Thus, the moment diagram over the entire length of the member will be of the same sign, and will be negative. (*Why?*)

Based on this information and knowing that the bending-moment diagram has zero values at both points A and C, it is possible to construct the desired shear and moment diagrams. These are given in fig. 3.11. (It should be noted that the same general qualitative conclusions regarding the sign of the moment diagram could have been reached by sketching the deflected shape of the beam. This shape has been shown dashed in fig. 3.11.)

If for this same problem Q is increased substantially, while ε decreased, there is reached in the limit the case of an applied pure moment at point C. The corresponding moment and shear diagrams are given in fig. 3.12.

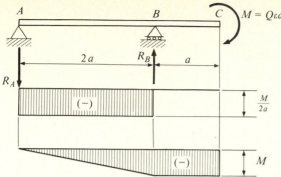

Figure 3.12 Shear and bending-moment diagram for limiting case.

Example 3.4 As a final example, consider the compound-beam type of arrangement shown in fig. 3.13. The particular case illustrated is typical of certain types of girder bridge designs. The main structural members are two outside girders that span the entire length $7a$. The floor beams are attached to these girders at equal intervals along the length—in a fashion that allows the assumption of a simple support. Stringers span between these floor beams—again in a simple fashion. The load per unit of deck area q is carried by the stringers, except for a strip along each girder. It is required to construct shear and bending-moment diagrams for a typical stringer, a typical floor beam, and a girder.

Consideration will first be given to the stringer. The distributed load per unit of length is q. The shear and bending-moment diagrams are given in sketch b.

The floor beam is next analyzed. The loading consists of four stringer reactions (one stringer reaction on each side of the floor beam). These are applied at the one-third points of the span. The shear and bending-moment diagrams are shown in fig. 3.13c.

The girders are treated last. The loads consist of the floor beam reactions plus the extra strip of distributed load indicated in sketch a. The method of analysis is illustrated graphically in part d. It is convenient to separate the effects of the concentrated and distributed forces in the drawing of the shear and moment diagrams, and this has been done. The two parts can then be superimposed to produce the final values as indicated.

For the sake of completeness, it is desirable to carry through to the actual selection of member sizes in this design example. Consider one of the floor beams of the bridge. It will be assumed that a rolled wide-flange shape is desired for these beams, and the material will be an A36 type steel.† Moreover, it will be presumed that the beam is fully supported laterally along its compression flange. q is given as 100 lb/ft, b is 10 ft, and a is 15 ft. From the moment diagram for live load, the maximum live-load moment is equal to

$$M_{LL, \max} = qb^2a \quad \text{or} \quad M_{LL, \max} = \frac{(100)(10)^2(15)}{(1000)} = 150 \text{ ft} \cdot \text{kips}$$

---

† See the AISC *Manual of Steel Construction* for a listing of the appropriate allowable stress values, member sizes, etc.

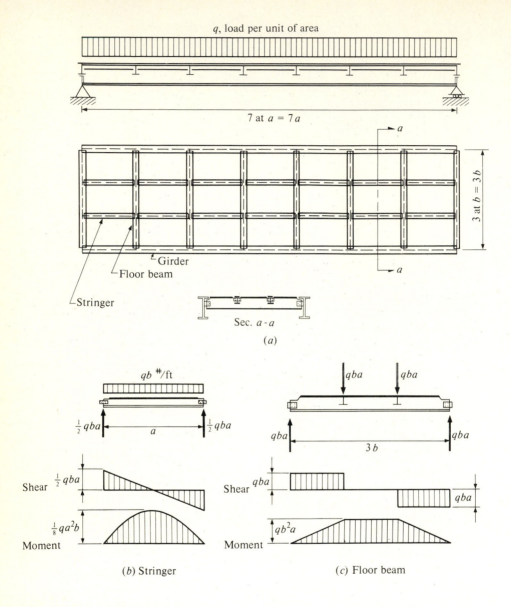

Figure 3.13a–c Girder bridge design.

To this must be added the dead-load moment due to the weight of the beam. Since DL is uniformly distributed, the moment is given by

$$M_{DL,\,max} = \frac{1}{8} w(3b)^2 = \frac{1}{8}(w)\frac{(30)^2}{1000} = \frac{9}{80}w \qquad \text{ft·kips}$$

where w is the weight per foot of the member selected. The total moment is therefore

$$M_{total} = M_{LL,\,max} + M_{DL,\,max} = 150 + \frac{9}{80}w \qquad \text{ft·kips}$$

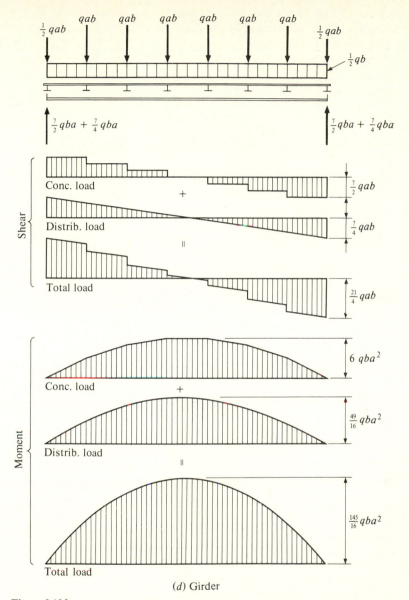

(d) Girder

Figure 3.13d

However, the member size is not known. A guess therefore must be made of the value of w. Estimating w to be 50 lb/ft, the total moment is

$$M_{total} = 150 + \left(\frac{9}{80}\right)(50) = 150 + 5.625 = 155.63 \text{ ft} \cdot \text{kips}$$

Using $\sigma_{all} = 24$ ksi, the required section modulus would be

$$S_{rqd} = \frac{M}{\sigma_{all}} = \left(\frac{155.63}{24}\right)(12) = 77.81 \text{ in}^3$$

The lightest wide-flange shape that will supply this S_{rqd} value is a W 18 × 45.[†] Since the weight per foot of member is less than our original guess, a reexamination with a lighter load is desirable:

$$M_{total} = 150 + \left(\frac{9}{80}\right)(45) = 155.06 \text{ ft} \cdot \text{kip}$$

and

$$S_{rqd} = \left(\frac{155.06}{24}\right)(12) = 77.53 \text{ in}^3$$

Again, from the section modulus table,[†] the lightest rolled wide-flange shape that will supply this section modulus is the W 18 × 45. This section therefore will be used for the floor beams.

3.3 PLANE FRAME STRUCTURES

Planar structures (as opposed to individual members) are normally one of two major types: trusses and rigid frames. Attention will be directed in this section to a consideration of rigid-frame type structures.

An important assumption that is usually made when analyzing rigid-frame structures is that the angles between the various members, at their points of connection, remain constant one with respect to the other as the frame responds to imposed loading. Such a stipulation requires among other things that the stress resultants be directly transmitted across the joint with no change in magnitude. (It is, of course, understood that what is an end-shear resultant in a horizontal beam connected to a vertical column will manifest itself as an applied vertical force in the column.)

In two-dimensional rigid frames the individual members are subjected to various combinations of bending moments, shearing forces, and axial thrusts. The relative proportions of these depend on the shape of the frame and the imposed loading.

Several simple examples of statically determinate rigid frames are shown in fig. 3.14.

Rigid frames can be indeterminate internally, or externally, or both. The degree of external indeterminacy is determined by subtracting from the number of unknown reaction components the number of available (independent) equations of static equilibrium for the structure as a whole. The degree of internal indeterminacy can be established by the successive and continuous cutting of members of the structure until finally there is obtained a structure that is both determinate and statically stable. The number of redundants (the degree of indeterminacy) will equal the number of such cuts times three—the number of possible component resultants at each cut section. Figure 3.15 illustrates several simple indeterminate rigid frames.

Possibly the easiest (and quickest) way to develop an understanding of the methods and procedures for analyzing determinate rigid frames is to consider in a

† See the current AISC *Manual of Steel Construction.*

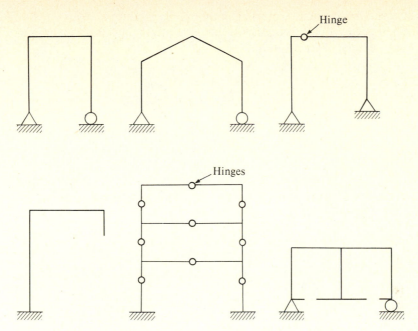

Figure 3.14 Typical statically determinate rigid frames.

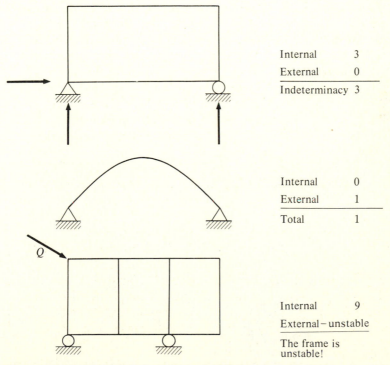

Figure 3.15 Several simple indeterminate rigid frames.

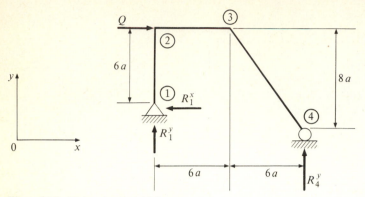

Figure 3.16a Determinate rigid frames.

step-by-step fashion the actual solution of a problem, for example, the one illustrated in fig. 3.16a. The rigid frame in question is composed of three members, and the external loading is a force Q applied horizontally at point 2. It is desired to construct the shear and bending-moment diagrams for each of the members.

First of all the structure is statically determinate. There are three components of reaction, and three independent equations of equilibrium available. From a summation of forces in the x direction

$$\sum F_x = 0 \qquad \text{or} \qquad R_1^x = Q \quad \leftarrow$$

Summing moments about location 1,

$$\sum M_1 = 0 \qquad 6aQ - 12aR_4^y = 0 \qquad \text{or} \qquad R_4^y = \frac{Q}{2} \quad \uparrow$$

From a consideration of the forces in the y direction

$$\sum F_y = 0 \qquad \text{or} \qquad R_1^y = \frac{Q}{2} \quad \downarrow$$

(Arrows have been shown to emphasize the directions of the resulting forces.)

Next, consider free bodies of each of the individual members and joints. This is shown in fig. 3.16b. It is to be noted that also indicated on these sketches are assumed positive coordinate axes for each of the members.

For the free bodies selected, three static conditions can be used to determine the unknowns at the cut sections. Starting with the left-hand column, it is known from the above reaction calculations that the shear and normal forces at point 1, that is, at $x = 0$, are Q and $\frac{1}{2}Q$, respectively. At the opposite end of the member there are three unknown stress resultants: a thrust, a bending moment, and a shear. Summing forces in the y direction the end shear is found to be equal to Q—since no lateral loads are applied along the column. Summing forces in the x direction, a tensile force of $\frac{1}{2}Q$ will exist at location 2. Finally, summing moments about point 2, $M_{21} = 6aQ$. (The directions of the various resultants are as shown.)

Consider next equilibrium of joint 2. Given are the applied horizontal force Q, the vertical tensile force from column 1-2, $\frac{1}{2}Q$, the horizontal shear from the column

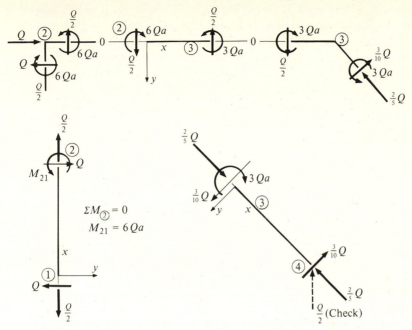

Figure 3.16b End forces and moments on the various components of the frame.

(directed to the left), Q, and the bending moment $6aQ$. Three unknowns again exist: the thrust, the shear, and the moment in the horizontal stem of the connection. Equilibrium requires that they be 0, $\frac{1}{2}Q$, and $6aQ$, respectively.

Similar considerations and procedures would be used for the other members and joints. The various values for the end resultants so obtained are shown. (Proceeding clockwise considering in turn the various members, and using the computed end values of each as the starting values for the next free body, a check on the numerical accuracy of the computations is obtained at the final step. There, the calculated values must equal the reaction originally defined from overall equilibrium considerations.)

Shear, bending-moment, and axial-force diagrams for the structure are shown in fig. 3.16c. Also indicated in the first of the sketches is an exaggerated illustration of the deflected shape.

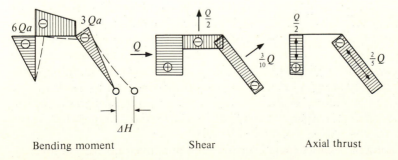

Bending moment   Shear   Axial thrust

Figure 3.16c Bending-moment, shear, and axial-thrust diagrams.

Let it be specified that $a = 3$ ft and $Q = 10$ kips. Further, let it be required that there be selected a timber beam column for member 1-2. (It will be presumed that deflections are sufficiently small that the vertical reaction at 1 times the resulting deflections will not materially alter the indicated bending moment values. It also will be assumed that the member is braced adequately in the lateral direction so that buckling out of the plane of the frame is prevented, and that joints can be designed which will transmit the necessary end forces and moments.) Design the member!

For beam columns, normal stresses are given by the relationship

$$\sigma = \frac{P}{A} \pm \frac{My}{I} \tag{3.9}$$

The axial force in member 1-2 is tensile. Therefore, the stress distribution at any section along the member is that indicated in fig. 3.16d. It is to be noted that the maximum (tensile) normal stress occurs on the intersurface of the member in question, and has a value

$$\sigma_{\max} = \frac{P}{A} + \frac{My_{\max}}{I} = \frac{P}{A} + \frac{M}{S}$$

Since S for a rectangle is $\frac{1}{6}bd^2$,

$$\sigma_{\max} = \frac{P}{bd} + \frac{6M}{bd^2}$$

For the structural dimensions and loading specified, the bending moment in member 1-2 is a maximum at 2. It is equal to $M_2 = 6aQ = (6)(3.0)(10) = 180$ ft·kips. The

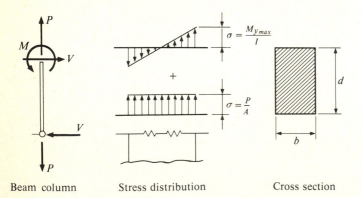

Beam column   Stress distribution   Cross section

Figure 3.16d End loading and normal stress distribution in the vertical column.

axial thrust is $P = \frac{1}{2}Q = 5$ kips. Therefore

$$\sigma_{max} = \frac{5}{bd} + \frac{(180)(12)}{bd^2} \quad \text{ksi}$$

The maximum shearing stress for a rectangular member occurs at the middepth of the cross section, and equals

$$\tau_{max} = \frac{VQ^*}{Ib} = \frac{3}{2}\frac{Q}{bd} = \frac{15}{bd} \quad \text{ksi}$$

where Q^* is the first moment of one-half the area of the cross section about the centroidal axis. If it is assumed† that

$$\sigma_{all} = 1700 \text{ lb/in}^2$$

$$\tau_{all} = 100 \text{ lb/in}^2$$

then the beam column must meet the conditions

$$\frac{5}{bd} + \frac{2160}{bd^2} \leq \frac{1700}{1000}$$

and

$$\frac{15}{bd} \leq \frac{100}{1000}$$

A 10-in × 15-in member will be selected for examination. It is observed that the shear criterion is just met, but the normal stresses are excessive.

$$\frac{5}{(10)(15)} + \frac{2160}{(10)(15)^2} = 5.793 > 1.7 \text{ ksi}$$

Bending therefore governs the design. Selecting a 10-in × 20-in member

$$\frac{5}{(10)(20)} + \frac{2160}{(10)(20)^2} = 3.265 > 1.7$$

and the section is still too small. A 12-in × 25-in member yields the following

$$\frac{5}{(12)(25)} + \frac{2160}{(12)(25)^2} = 1.745 > 1.7$$

which is quite close to the desired value. A final selection of a 12.5-in × 25-in member will satisfy the requirement and could be used.

Example 3.5 Consider the analysis of the hydraulic gate shown in fig. 3.17a. The structure is determinate, and it is desired to ascertain the value of the maximum bending moment.

---

† Design specifications for timber structures are contained in National Design Specifications for Wood Construction published by the National Forest Products Association. It is assumed that a laminated or Glulam section would be employed.

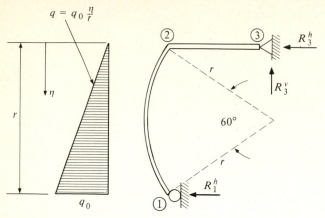

Figure 3.17a

The reaction components are determined as follows:

$$\sum M_3 = 0 \qquad R_1^h r = \tfrac{2}{3}r(\tfrac{1}{2}q_0 r) = \tfrac{1}{3}q_0 r$$

$$\sum F_h = 0 \qquad R_3^h = \tfrac{1}{6}q_0 r$$

$$\sum F_v = 0 \qquad R_3^v = 0$$

(It is to be recognized that if $R_3^v = 0$, then member 2-3 can have no bending moment. Thus, bending need be examined only for member 1-2.)

The bending moment in member 1-2, at a section η vertically down from point 2, is given by the expression

$$M_\eta = \frac{1}{6} q_0 r \eta - \frac{1}{2} q_\eta \eta \frac{\eta}{3}$$

or

$$M_\eta = \frac{q_0}{6}\left(r\eta - \frac{\eta^3}{r}\right)$$

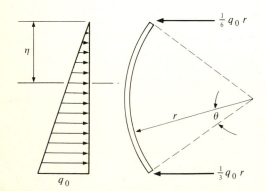

Figure 3.17b Presumed loading on hydraulic gate.

This holds for the entire range

$$0 \le \eta \le r$$

or

$$0 \le \frac{\eta}{r} \le 1.0$$

Introducing the dimensionless quantity λ, where

$$\lambda = \frac{\eta}{r}$$

there is obtained

$$M_\lambda = \tfrac{1}{6} q_0 r^2 (\lambda - \lambda^3)$$

The maximum value of the moment occurs where the derivative with respect to the variable λ is zero, that is,

$$\frac{dM_\lambda}{d\lambda} = 0 = \frac{1}{6} q_0 r^2 (1 - 3\lambda^2)$$

or

$$\lambda_{\text{cr}} = \frac{1}{\sqrt{3}}$$

This corresponds to

$$\eta_{\text{cr}} = \frac{r}{\sqrt{3}}$$

Substituting this value into the general expression for moment yields

$$M_{\text{max}} = \frac{2\sqrt{3}}{9} r^2$$

(It is to be recognized that the first derivative of the moment with respect to the coordinate direction is equal to an effective shear Q_λ in the direction perpendicular to the coordinate direction. But for the problem under consideration, only at one point will this equal the shear in the member—as defined in chap. 2—that is, in a direction normal to the longitudinal axis at the cut section. At all other locations, these values would be defined by the vector relationships of fig. 3.17c.)

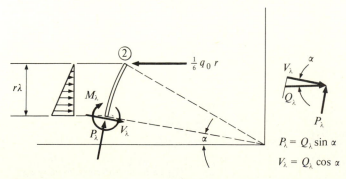

Figure 3.17c

3.4 THREE-HINGED ARCH

An arch is a type of beam that is curved in the plane of the applied loading and sustains most of that loading through compressive forces (as opposed to bending) that develop within the member. In addition, it is usually assumed when dealing with arches that the radius of curvature of the arch is large when compared to the depth of the member—so that the effect of curvature may be neglected when locating the neutral axis. (When the radius of curvature is not large, for example, in a crane hook, curved-beam theory must be used.)

An arch also must be supported at its ends so that both vertical and horizontal components of reaction can occur. If, for example, end supports are such that horizontal reactions will not develop, then for all purposes the member is a simple curved beam and the applied loads are primarily resisted by internally developed bending moments. It also should be noted that arches occur in the form of trusses, that is, pin-connected axial-force members. The trussed arch will be treated later.

Arches may be of the three-hinged type, the two-hinged type, or they may be hingeless (that is, fixed) as indicated in fig. 3.18. Of these three types, only the three-hinged arch is statically determinate.

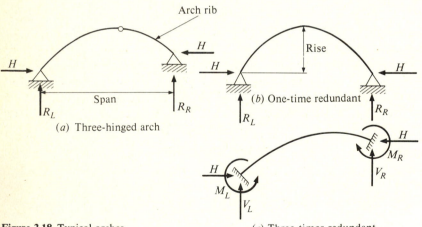

Figure 3.18 Typical arches.

(a) Three-hinged arch

(b) One-time redundant

(c) Three-times redundant

Consider the three-hinged circular arch shown in fig. 3.19a. As shown, there are four unknown reaction components: V_1, H_1, V_3, and H_3. The assumed plus directions for these components of reaction are those indicated. For the entire arch, it is determined from equilibrium considerations that

$$\sum F_z = 0 \qquad H_1 + H_3 = Q$$

$$\sum F_y = 0 \qquad V_1 + V_3 = Q$$

$$\sum M_3 = 0 \qquad V_1(r + r\sin 30°) + H_1(r\cos 30°) - Q(2r\sin 30°)$$

$$- Q(r\cos 30° - r\sin 30°) = 0$$

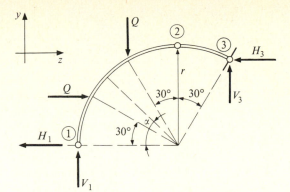

Figure 3.19a Three-hinged circular arch subjected to concentrated loads.

Considering just segment 2-3 as a free body and summing moments about end 2 yields

$$\sum M_2 = 0 \qquad V_3 r \sin 30° - H_3 r (1 - \cos 30°) = 0$$

These four expressions provide a sufficient number of independent equations to allow solution for the four reaction components.

$$H_1 = +6.68Q \qquad V_1 = -3.80Q$$

$$H_3 = -5.68Q \qquad V_3 = +4.80Q$$

(The negative signs signify that the reaction components must exist in a direction opposite to that indicated in fig. 3.19a.)

Once the reactions have been defined, the internal stress resultants can be determined. Consider, for example, that part of segment 1-2 that lies between the horizontal and vertical forces Q. An appropriate free body diagram is shown in fig. 3.19b. The unknown resultants at the cut section are P_α, M_α, and V_α. Using the appropriate equilibrium equation, it will be found that

$$M_\alpha = V_1 (r - r \cos \alpha) + H_1 r \sin \alpha - Q(r \sin \alpha - r \sin 30°)$$

$$V_\alpha = Q \cos \alpha - V_1 \sin \alpha - H_1 \cos \alpha$$

and

$$P_\alpha = Q \sin \alpha + V_1 \cos \alpha - H_1 \sin \alpha$$

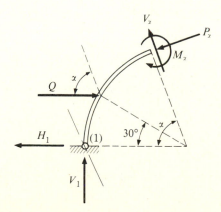

Figure 3.19b

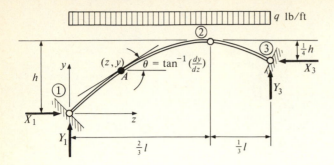

Figure 3.20a

Example 3.6 Consider next the three-hinged arch shown in fig. 3.20a. The arch axis is defined as a parabola whose geometry varies as

$$y = 3h\left[\frac{z}{l} - \frac{3}{4}\left(\frac{z}{l}\right)^2\right]$$

y and z are measured from point 1. Letting

$$\eta = \frac{y}{h} \quad \text{and} \quad \xi = \frac{z}{l}$$

the shape of the arch can be described in the following dimensionless form

$$\eta = 3(\xi - \tfrac{3}{4}\xi^2)$$

The slope at any location would be

$$\theta = \frac{d\eta}{d\xi} = 3\left(1 - \frac{3}{2}\xi\right)$$

Following the procedures used in the preceding example, the reaction components are

$$X_1 = X_3 = \frac{2ql^2}{9h} \quad \text{and} \quad Y_1 = \frac{2}{3}ql$$

$$Y_3 = \frac{1}{3}ql$$

(The vertical component of force at the internal hinge 2 is found to be zero.) The moment and horizontal and vertical component of force at a cut section in member 1-2 would be found from a consideration of fig. 3.20b.

$$M_A = Y_1 z - X_1 y - qz\frac{z}{2}$$

$$= Y_1 z - q\frac{z^2}{2} - X_1(3h)\left[\frac{z}{l} - \frac{3}{4}\left(\frac{z}{l}\right)^2\right]$$

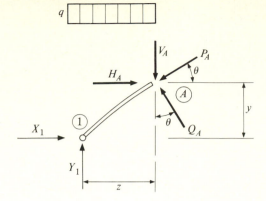

Figure 3.20b

$$V_A = Y_1 - qz = \frac{2}{3}q - qz$$

$$H_A = X_1 = \frac{2ql^2}{9h}$$

To facilitate analysis and design, these horizontal and vertical force components should be expressed as shear and axial thrust components. The appropriate transformation equations are

$$Q_A = V_A \cos \theta + H_A \sin \theta$$

$$P_A = -V_A \sin \theta + H_A \cos \theta$$

If $h = 20$ ft, $l = 40$ ft, $q = 500$ lb/ft, and $z = 10$ ft, the following are obtained:

$$V_A = 8{,}333.33 \text{ lb} \downarrow \qquad H_A = 8{,}888.89 \text{ lb} \leftarrow \qquad M_A = 66.5 \text{ ft} \cdot \text{lb} \circlearrowright$$

At $z = 10$ ft, $\theta = 43°9'$, and

$$Q_A = 5.18 \text{ lb} \searrow \qquad P_A = 785.85 \text{ lb} \swarrow \text{ (compression)}$$

3.5 TRUSSES

A truss is a configuration in which the individual members composing the structure are straight. The members are connected at their points of intersection by means of frictionless pins or hinges. Moreover, it is assumed that loading is applied only at these points of connection. (It is, of course, also expected that the bars are joined in such a way that the system is stable.)

Based on the above definitional restrictions, it can be demonstrated that each bar in a truss is a two-force member—one that can transmit only axial forces.

One further assumption is normally made—that the deflections are small and equilibrium can be defined in the undeformed position.

Trusses are widely used in bridges, buildings, towers, ships, aircraft, and many other structural forms. They are efficient structures for certain applications and can

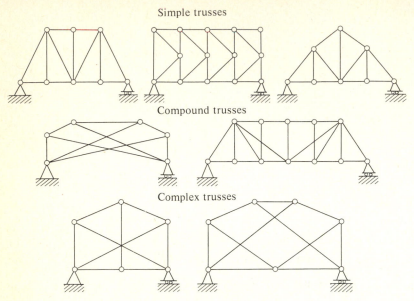

Simple trusses

Compound trusses

Complex trusses

Figure 3.21 Typical statically determinate trusses.

sustain relatively heavy loads and can be used for relatively long spans with comparatively small dead weights. Like frames, trusses can be determinate or indeterminate. Moreover, again like frames, indeterminate trusses can be either internally or externally indeterminate, or both. In this chapter attention will be restricted to determinate trusses.

Statically determinate trusses are classified as one of three types: simple trusses, compound trusses, and complex trusses. A simple two-dimensional truss is constructed by first joining together three bars in a triangular pattern, and then adding two more bars (an additional triangle) for each additional joint. A compound truss is a stable system that is formed by interconnecting two or more simple trusses. The rigidity of such a system is assured by providing sufficient independent constraints at the points where the trusses attach. A complex truss also can be statically determinate. It does not satisfy, however, the definitions of either the simple or the compound cases given above. Complex trusses are therefore those that are neither simple nor compound. Several examples of these types of trusses are shown in fig. 3.21.

Determinacy and stability of trusses The simplest stable configuration that a planar truss can take is a triangle. For a space truss it would be a tetrahedron. These are illustrated in fig. 3.22. The possible combinations of simple triangles (or tetrahedrons) that can be used to form a truss are almost endless. The choice of the particular form used to accomplish a desired end is for most cases at the discretion of the designer.

A given truss configuration is determinate or indeterminate depending upon how many bars there are and how they are joined together. (It is to be understood that depending on how the bars are joined, and in some cases how the loads are applied,

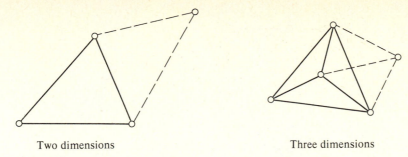

Two dimensions   Three dimensions

Figure 3.22 Typical stable configurations of two-dimensional trusses.

trusses also can be stable or unstable.) Defining

$$N_r = \text{number of external reactions}$$

$$N_b = \text{number of bars (bar forces)}$$

$$N_j = \text{number of joints}$$

and noting that at each joint in a planar truss there exist two independent equations of static equilibrium

$$2N_j = \text{total number of static equilibrium equations available}$$

(For a three-dimensional truss, there would be $3N_j$ available equations.) From a purely mathematical point of view, the following can be stated:

If $N_r + N_b < 2N_j$, the truss is unstable!

If $N_r + N_b > 2N_j$, the truss is stable and indeterminate!

If $N_r + N_b = 2N_j$, the truss is stable and determinate!

(For a space truss—simply substitute $3N_j$ for $2N_j$ in these relationships.) Unfortunately, these rules do not cover some special cases; for example, the following:

1. If a portion of a structure is indeterminate and part of that structure is unstable, such as the one illustrated in fig. 3.23a, the structure is unstable! The formula, however, indicates that it is determinate.

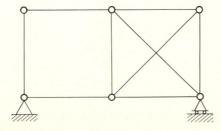

Figure 3.23a Unstable truss.

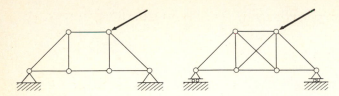

Figure 3.23b Unstable truss.

2. A structure that is externally indeterminate and internally unstable, or vice versa, as is shown in fig. 3.23b, still satisfies the rules listed above for a determinate structure. It is obviously not determinate.

In most cases the question of external indeterminacy can be determined almost by observation. Internally, however, it is often not so obvious. In complicated situations, the following rules may be used:

For plane trusses: If $3 + N_b = 2N_j$, the truss is determinate!

For space trusses: If $6 + N_b = 3N_j$, the truss is determinate!

Several additional examples are shown in fig. 3.24. The reader should determine if these trusses are stable or unstable, and whether or not they are determinate.

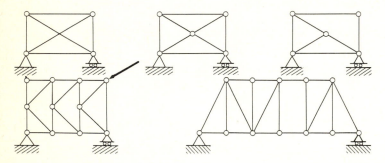

Figure 3.24

A truss is defined to be a collection of two-force members joined together by frictionless pins. At each of these joints there exists a concurrent force system that must be in equilibrium. The analysis of such a system can be simplified by noting the following:

1. If two forces are concurrent, the forces must be either collinear or zero (see fig. 3.25). From $\sum F_{horiz} = \sum F_{vert} = 0$

If $\theta \neq 0$, then $F_1 \equiv F_2 \equiv 0$

If $\theta = 0$, then $F_1 = F_2$

Figure 3.25

2. If three forces are concurrent and only two are collinear, the third force must be equal to zero (see fig. 3.26).

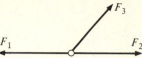

From $\sum F_{\text{vert}} = 0$

$$F_3 = 0$$

Figure 3.26

3. If in a three-dimensional system all concurrent forces lie in a plane except one, then that one force must be zero. As indicated, it is assumed that F_1, F_2, and F_3 lie in the xy plane. F_4 has a component in the z direction. From $\sum F_z = 0$

$$F_4 = 0$$

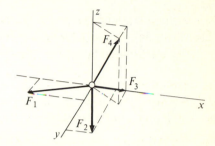

Figure 3.27

Two types of free bodies normally are used in truss analysis. These are (*a*) joints, per se, and (*b*) selection of a part of the truss—with the remainder removed. These procedures are most often referred to as the *method of joints* and the *method of sections*, respectively. Often the two methods are combined when solving a given problem.

When analyzing a joint of a plane truss as a free body, two independent relationships can be defined. These will be based on the relationships

$$\sum F_x = 0 \quad \text{and} \quad \sum F_y = 0$$

For example, consider the joint illustrated in fig. 3.28.

$$\sum F_x = 0 \qquad F_2 + F_1 \cos \theta = Q$$
$$\sum F_y = 0 \qquad F_1 \sin \theta = Q$$

Two equations: two unknowns. Therefore, F_1 and F_2 can be solved in terms of Q and θ.

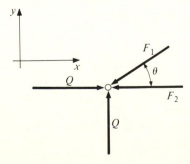

Figure 3.28

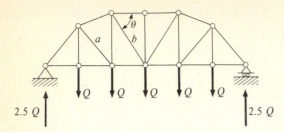

Figure 3.29a Determinate planar truss.

When using the method of sections, the additional, third equilibrium condition, that is, $\sum M = 0$, can be added. Consider the plane truss shown in fig. 3.29a. The force in the diagonal (or web) member labeled b is desired. It is to be recognized that a cut through the third panels from the left will involve the member in question. This is shown in fig. 3.29b. Since member b is the only member at the cut section that has a vertical component, this fact can be used to determine the force in the member.

$$\sum F_y = 0 \qquad F_b \sin \theta = 2.5Q - Q - Q$$

or

$$F_b = \frac{0.5Q}{\sin \theta}$$

To ascertain the bar force in member a in fig. 3.29a, a section would be selected through the second panel. The top chord member at the cut section has a component of force in the vertical direction. Therefore, the simple, one-equation procedure used above will not be sufficient. However, if it is recognized that the unknown forces associated with the top and bottom chord members intersect at a point, and if summation of moments about that point is taken, then there will be contained only one unknown in the resulting equation (see fig. 3.29c). Thus, from the single expression $\sum M_0 = 0$

$$F_a(d) = 2.5Qb - Q(a + b)$$

or

$$F_a = 2.5Q\frac{b}{a} - Q\frac{a + b}{d}$$

The distances a, b, and d are determined from the geometry of the truss.

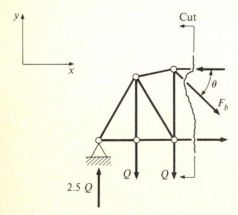

Figure 3.29b Using method of sections to determine F_b.

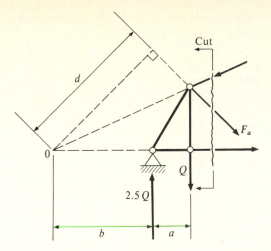

Figure 3.29c Using method of sections to determine F_a.

It should be evident to the reader that the sequence in which various free bodies—either joints, sections, or combinations—are considered is important. A more advantageous sequence can reduce considerably the time required to analyze a structure.

The first step in any truss analysis is to determine whether or not the structure in question is stable, and further whether it is determinate. It also is useful to attempt to identify any bars that will have zero force (for the particular loading specified). Some members will have bar forces that are not zero, but are obvious, and these also should be identified. The joint method is most often used for these purposes, although this is by no means a rigid rule.

Example 3.7 Consider the roof truss shown in fig. 3.30a. It is required that the forces in the various members be determined for the idealized loading given. There are three reaction components, twenty-one bars, and twelve joints.

$$N_r + N_b = 24 \qquad \text{and} \qquad 2N_j = 24$$

Therefore, the structure should be both stable and determinate.

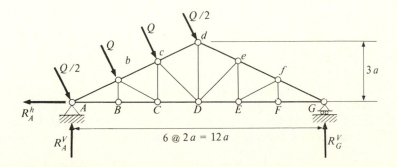

Figure 3.30a

If joint F is taken as a free body, there are only three possible forces that can develop. Of these, F_{EF} and F_{FG} are collinear. Therefore $F_{fF} = 0$, and $F_{EF} = F_{FG}$. Considering next joint f as a free body, bar fE also must have a zero force. Using this procedure at various joints throughout the structure results in a determination that the following members will develop zero forces.

$$F_{Bb} = F_{De} = F_{Ee} = F_{Ef} = F_{Ff} = 0$$

The members associated with these zero forces have been shown dashed in fig. 3.30b. They will not contribute to the resistance of the structure to the loading shown. (It is to be recognized that the absence of a force in a particular member of a structure subjected to a given loading is insufficient reason, in and of itself, to remove that member. Most structures are subjected to a variety of loadings. The design of an individual member depends on the greatest induced force under any one or more of these possibilities.)

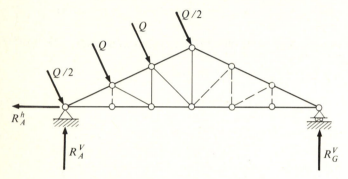

Figure 3.30b Truss with zero-force members removed.

For the truss as a whole, three equilibrium equations are available to establish the three reaction components.

$$\sum F_x = 0 \qquad R_A^h = \frac{3Q}{\sqrt{5}} = \frac{3\sqrt{5}Q}{5} \quad \leftarrow$$

$$\sum M_A = 0 \qquad R_G^v(12a) = Q\sqrt{5}\,a + 2Q\sqrt{5}\,a + \tfrac{3}{2}Q\sqrt{5}\,a$$

$$= \frac{3\sqrt{5}}{8}Q \quad \uparrow$$

$$\sum F_y = 0 \qquad R_A^v = \frac{6}{\sqrt{5}}Q - \frac{3\sqrt{5}}{4}Q = \frac{33\sqrt{5}}{40}Q \quad \uparrow$$

The individual joints of the truss, each in turn, are used as free bodies to determine the remaining bar forces. Since only two equilibrium equations are available at each joint, it is necessary to select as a starting point a joint that has no more than two unknowns. Joint G will be selected. This is shown in fig. 3.30c.

$$\sum F_y = 0 \qquad F_{fG}\left[\frac{1}{\sqrt{5}}\right] = R_G^v = \frac{3\sqrt{5}}{8}Q$$

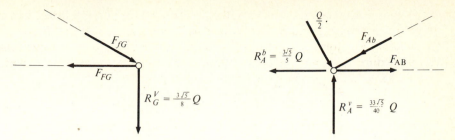

Figure 3.30c   **Figure 3.30d**

or

$$F_{fG} = \tfrac{15}{8}Q \qquad \text{(compression)}$$

$$\sum F_x = 0 \qquad F_{FG} = F_{fG}\left[\frac{2}{\sqrt{5}}\right] = \frac{3\sqrt{5}}{4}Q \qquad \text{(tension)}$$

The forces in the members were assumed in the correct direction as indicated by the plus values of the solution.

For joints f, e, F, and E to be in equilibrium,

$$F_{DE} = F_{EF} = F_{FG} = \frac{3\sqrt{5}}{4}Q \qquad \text{(tension)}$$

and

$$F_{de} = F_{ef} = F_{fG} = \frac{3\sqrt{5}}{8}Q \qquad \text{(compression)}$$

Proceeding in the same fashion, starting at joint A, the forces in the members Ab and AB are obtained (fig. 3.30d).

$$\sum F_y = 0 \qquad F_{Ab}^v = \frac{5}{8}\sqrt{5}Q$$

or

$$F_{Ab} = \sqrt{5}\,F_{Ab}^v = \tfrac{25}{8}Q \qquad \text{(compression)}$$

$$\sum F_x = 0 \qquad F_{AB} + \frac{1}{\sqrt{5}}\frac{Q}{2} - F_{Ab}^h - \frac{3\sqrt{5}}{5}Q = 0$$

or

$$F_{AB} = \tfrac{7}{4}\sqrt{5}Q \qquad \text{(tension)}$$

The section shown in fig. 3.30e is next used to compute F_{bc} and F_{CD}.

$$\sum M_A = 0 \qquad F_{Cc}(4a) - Q\sqrt{5}\,a = 0$$

or

$$F_{Cc} = \frac{\sqrt{5}}{4}Q \qquad \text{(tension)}$$

$$\sum F_y = 0 \qquad F_{bc}^v + \frac{2}{\sqrt{5}}Q + \frac{2}{\sqrt{5}}\frac{Q}{2} - F_{Cc} - \frac{33}{40}\sqrt{5}Q = 0$$

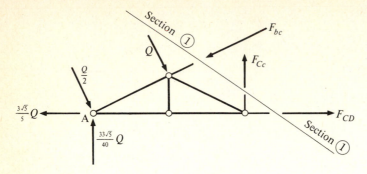

Figure 3.30e

or $\qquad F_{bc} = \frac{19}{8}Q \qquad$ (compression)

$$\sum F_x = 0 \qquad F_{CD} + \frac{1}{\sqrt{5}}\frac{Q}{2} + \frac{1}{\sqrt{5}}Q - F_{bc}^h - \frac{3}{5}\sqrt{5}Q = 0$$

or $\qquad F_{CD} = \frac{5}{4}\sqrt{5}Q \qquad$ (tension)

Continuing with joint C, then with section 2-2, then with joint d, and finally with joint D, as shown in fig. 3.30f, 3.30g, 3.30h, and 3.30i, the following values are obtained for the various members of the truss:

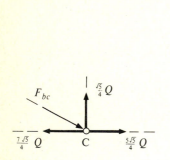

Figure 3.30f

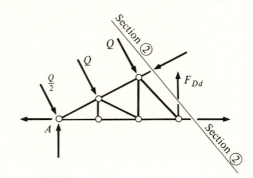

Figure 3.30g

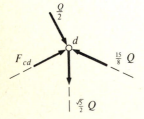

Figure 3.30h

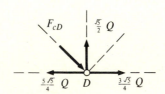

Figure 3.30i

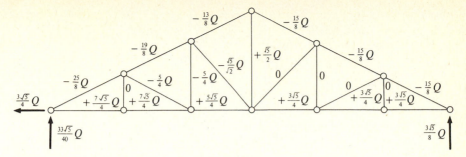

Figure 3.30j Axial forces in the various truss members.

$F_{AB} = \frac{7}{4}\sqrt{5}\,Q$ (tension) $\quad F_{ab} = \frac{25}{8}Q$ (compression) $\quad F_{bB} = 0$

$F_{BC} = \frac{7}{4}\sqrt{5}\,Q$ (tension) $\quad F_{bc} = \frac{19}{8}Q$ (compression) $\quad F_{cC} = \dfrac{\sqrt{5}}{4}\,Q$ (tension)

$F_{CD} = \frac{5}{4}\sqrt{5}\,Q$ (tension) $\quad F_{cd} = \frac{13}{8}Q$ (compression) $\quad F_{dD} = \dfrac{\sqrt{5}}{2}\,Q$ (tension)

$F_{DE} = \frac{3}{4}\sqrt{5}\,Q$ (tension) $\quad F_{de} = \frac{15}{8}Q$ (compression) $\quad F_{eE} = 0$

$F_{EF} = \frac{3}{4}\sqrt{5}\,Q$ (tension) $\quad F_{ef} = \frac{15}{8}Q$ (compression) $\quad F_{fF} = 0$

$F_{FG} = \frac{3}{4}\sqrt{5}\,Q$ (tension) $\quad F_{fg} = \frac{15}{8}Q$ (compression)

$$F_{bC} = \tfrac{5}{4}Q \text{ (compression)} \qquad F_{De} = 0$$

$$F_{cD} = \sqrt{\tfrac{5}{2}}\,Q \text{ (compression)} \qquad F_{Ef} = 0$$

These results are shown on a sketch of the truss in fig. 3.30j, where compression is denoted by a minus sign, and tension by a plus.

Let it be assumed that $Q = 10,000$ lb and that a (in fig. 3.30a) is equal to 2 ft. Determine the least weight single-angle steel section of A36 steel that can be used for the lower chord member DE ($\frac{3}{4}$-in connecting bolts are to be used to attach the members one to the other).

The tensile force to be sustained by the member in question is

$$F_{DE} = \frac{3}{4}\sqrt{5}(10) = \frac{(3)(2.236)}{4}(10) = 16.77 \text{ kips (tension)}$$

The normal stress associated with the load is assumed to be uniformly distributed over the cross section, and equals

$$\sigma = \frac{F_{DE}}{A}$$

Therefore, the required cross-sectional area is

$$A_{\text{rqd}} = \frac{F_{DE}}{\sigma_{\text{all}}}$$

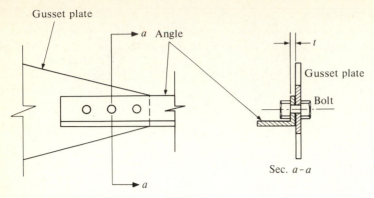

Figure 3.30k Single line-bolted connection

In this equation σ_{all} is the allowable tensile stress of the material. For A36 steel,† $\sigma_{all} = 22$ ksi, except at pinholes, and $\sigma_{all} = 16$ ksi on net sections at pinholes. The required *net* area will be

$$A_{rqd} = \frac{16.77}{22} = 0.762 \text{ in}^2$$

But the net area in relation to the gross area depends on the arrangement of the fasteners. Figure 3.30k, for example, presumed that the bolts are placed in a single line. Figure 3.30l, on the other hand, assumes a staggered arrangement. For the first case $A_{net} = A_{gross} - (\frac{3}{4} + \frac{1}{8})t$, where t is the thickness of the angle leg. (It is to be noted that $\frac{1}{8}$-in has been added to the bolt diameter of $\frac{3}{4}$-in to facilitate erection.) When staggered hole arrangements are used, the net section corresponds to the smallest area over which failure can occur. For the case in

† See AISC *Manual for Steel Construction.*

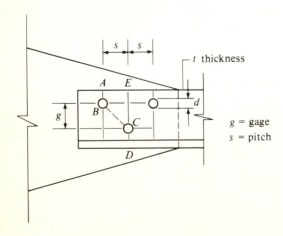

Figure 3.30l Staggered bolted connection.

question, the possibility of several lines of fracture exists. For example, in fig. 3.30*l* the net section could occur along either path *E-C-D* or *A-B-C-D*, as shown. For either case, the net section is the gross area, A_{gross} (that is, the actual cross-sectional area of the member) less the area of any holes that might participate in the fracture, plus the effect of the stagger. (It has been suggested that a factor of $s^2/4g$ be added for each slanting segment.) The net area for the two indicated possible paths is therefore:

For path *E-C-D*: $\qquad A_{net} = A_{gross} - dt$

For path *A-B-C-D*: $\qquad A_{net} = A_{gross} - 2dt + \dfrac{s^2}{4g}t$

where t is the thickness of the angle, and d is the diameter of the connector plus the desired amount of overage to facilitate construction.

It will be assumed that a connection of the type shown in fig. 3.30*k* will be used. As a starting point assume $t = \frac{1}{4}$-in. The gross area of the member must then be

$$A_{gross} = A_{rqd} + A_{hole} = 0.762 + (\tfrac{7}{8})(\tfrac{1}{4}) = 0.982 \text{ in}^2$$

A $2\frac{1}{2} \times 2 \times \frac{1}{4}$ in angle has an area of 1.06 in² and a weight of 3.62 lb/ft. It is the lightest $\frac{1}{4}$-inch thick angle section that supplies a sufficient area. If t is assumed as $\frac{5}{16}$-in, then $A_{gross} = 0.983$ in² and the resulting section would be $2\frac{1}{2} \times 1\frac{1}{2} \times \frac{5}{16}$ in, which has an area of 1.15 in² and a weight per foot of 3.92 lb. The $\frac{1}{4}$-in thick design results in the lesser weight and therefore normally would be selected.

(This example is intended merely to illustrate a procedure that could be used to define the size of a tensile member needed for a given bar force. In actuality, the bar force of 16.77 kips for which the member was designed represents only part of the load that the member would actually be required to sustain. To this force—which resulted from an idealized wind loading—there would have to be added the effects of the other loads that act on the truss: dead weight of the truss itself, the weight of the roofing material and purlins, any gravity loads such as snow and ice, loads that might be hung on the lower chord of the truss, etc.)

Example 3.8 As a second truss example, consider the two-dimensional bridge shown in fig. 3.31*a*. Only the bar forces in members *BC*, *bC*, *cC*, and *cD* are desired.

Considering the entire structure as a free body, it is determined that the reactions are

$$R_A = Q$$

$$R_G = 2Q$$

Since most of the forces in the members connecting an interior joint *C* are desired, it is probably most efficient to use the method of sections. For example,

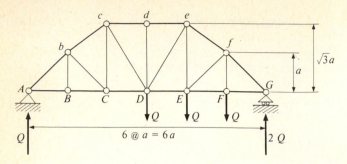

Figure 3.31a

consider a section through the second panel as indicated in fig. 3.31b.

$$\sum M_b = 0 \qquad - F_{BC}(a) + Qa = 0$$

or
$$F_{BC} = Q \quad \text{(tension)}$$

To compute the bar force in member bC, the free body shown in fig. 3.31c will be used. Moments will be summed about point c. The force in member bC will be assumed in its horizontal and vertical component form.

$$\sum M_c = 0 \qquad Q(2a) - F_{BC}(\sqrt{3}\,a) - F_{bc}^h(\sqrt{3}\,a) = 0$$

or
$$F_{bc}^h = Q\frac{2 - \sqrt{3}}{\sqrt{3}} = 0.219Q$$

This corresponds to a force in the member of

$$F_{bc} = \sqrt{2}\,F_{bc}^h = 0.310Q \quad \text{(tension)}$$

Since we now know two of the bar forces at joint C, the remaining two can be established by a consideration of the equilibrium of the joint (see fig. 3.31d).

$$\sum F_h = 0 \qquad F_{BC} - F_{bc}^h - F_{CD} = 0$$

or
$$F_{CD} = Q - 0.219Q = 0.781Q \quad \text{(compression)}$$

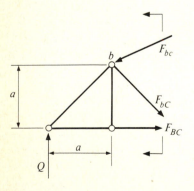

Figure 3.31b

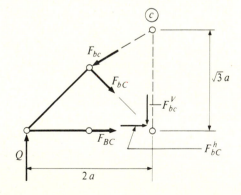

Figure 3.31c

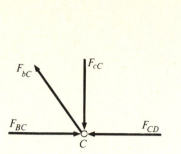

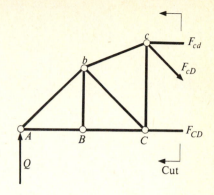

Figure 3.31d   **Figure 3.31e**

$$\sum F_v = 0 \qquad F_{bC}^v - F_{cC} = 0$$

or
$$F_{cC} = 0.219Q \quad \text{(compression)}$$

The force in bar cD can be determined using the method of sections as indicated in fig. 3.31e.

$$\sum F_v = 0 \qquad \frac{\sqrt{3}}{2} F_{cD} - Q = 0$$

or
$$F_{cD} = \frac{2}{\sqrt{3}} Q \quad \text{(tension)}$$

Compound trusses For relatively long-span truss bridges, the three-hinged form shown in fig. 3.32a is frequently used. The analysis of such a structure is essentially the same as for the simpler cases just considered.

The truss in question is a compound truss. Six components of reaction are indicated, however, two of those can be presumed to be zero—since bar eE can transmit no horizontal component of force, and no applied horizontal forces act anywhere on the truss. Moreover, again from the center truss, the forces F_{eE} and V_I

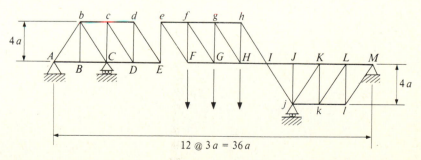

Figure 3.32a Compound truss.

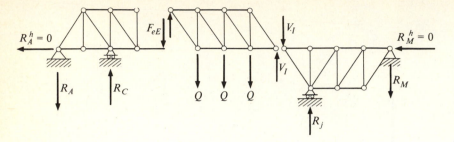

Figure 3.32b Separation of compound truss into three simple trusses.

can be established from equilibrium considerations: $F_{eE} = \frac{3}{2}Q$ (compression), and $V_I = \frac{3}{2}Q$ (in the direction shown). The reactions of the two simple end trusses can therefore be readily established (see fig. 3.32b).

$$R_A = \tfrac{3}{2}Q \quad \text{(downward)}$$

$$R_C = 3Q \quad \text{(upward)}$$

$$R_M = \tfrac{1}{2}Q \quad \text{(downward)}$$

$$R_J = 2Q \quad \text{(upward)}$$

With these values, the analysis of the structure for the individual bar forces would proceed as previously described.

A different type of compound truss system occurs when two or more simple trusses are connected by one, two, or three bars and hinges as illustrated in fig. 3.33a. Here, it is not readily obvious that a determinate (stable) situation exists. Assuming stability, the reactions are readily obtained:

$$R_A^v = \tfrac{3}{8}Q \quad \text{(upward)} \quad R_A^h = 0$$

$$R_F = \tfrac{5}{8}Q \quad \text{(upward)}$$

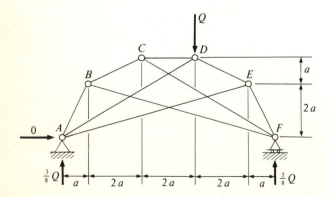

Figure 3.33a Compound truss.

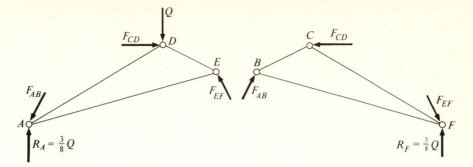

Figure 3.33b Separation of compound truss into simple trusses.

Next, the truss is broken up into simple trusses, each acted upon by the applied loading, external reactions, and the unknown bar forces in the presumed cut members (fig. 3.33*b*). Each of these simple trusses taken as a free body provides sufficient equations of statics to allow determination of the unknown connecting bar forces. Once these are calculated, the truss can be analyzed in a simple manner. Consider the two simple trusses *ADE* and *BCF* shown in fig. 3.33*b* connected by the bars *AB*, *CD*, and *EF*. Each free body contains only three unknowns. Using the equations $\sum F_x = \sum F_y = \sum M_i = 0$, the required bar forces F_{AB}, F_{CD}, and F_{EF} can be established.

The roof truss shown in fig. 3.34*a* is another type of compound truss. It consists of two simple trusses connected by bar *CD* and a hinge at 0. A free body of one of the parts is shown in fig. 3.34*b*. The unknown internal forces are F_{CD}, H_0, and V_0.

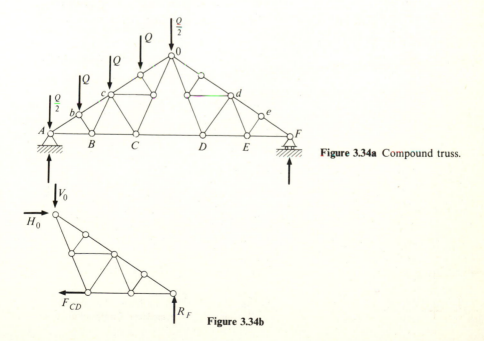

Figure 3.34a Compound truss.

Figure 3.34b

These readily can be established from a consideration of static equilibrium of either one of the simple trusses.

Complex trusses Complex trusses, because of their geometry, are not analyzed using the same procedures that were used for simple and compound trusses. For these types of structures, special methods are employed.† Since a complex truss can be statically determinate, it is possible to write two equations of equilibrium at each of the joints giving a set of linear algebraic equations that can be solved simultaneously for the forces in all of the members. (These methods will be considered later in the sections dealing with matrix and computer solutions.)

Three-dimensional trusses While nearly all engineering structures are three-dimensional, most can be idealized into equivalent two-dimensional systems. There are, however, some configurations such as a tower, or a dome, etc., where the members are arranged in such a way that it is impossible to divide them into a series of planar structures. These three-dimensional structures therefore must be analyzed as complete units.

The methods of analysis for three-dimensional trusses are essentially the same as those for planar trusses. First, the truss is examined to determine its stability and determinacy. Next, reactions are calculated. Finally, using either the method of joints, or the method of sections, or a combination of the two, forces in the various members would be determined.

In a space truss, if it is to be externally determinate, there can be no more than six components of reaction. The types of reaction normally presumed are:

1. Hinge (a ball in a socket)—three reaction components
2. Slotted roller (movement free in one direction)—two reaction components
3. Ball support (free to move in two directions)—one reaction component

A fixed-support would contain all six possible components at that one location.

As was true in the two-dimensional case, it is possible (almost by inspection) to establish the forces in certain members of three-dimensional trusses. This is especially so for zero force members. Consider the three situations illustrated in fig. 3.35. In case c, where it has been assumed that all of the bars have components in all of the coordinate directions, it is, of course, possible for two of the bars to be coplanar. Using the information given above in case a, the force in the third bar must necessarily equal zero—leaving two bars for further examination. The problem then becomes one for which $\sum F_\xi = \sum F_\eta = 0$, where ξ and η are appropriate axes in the plane of the two members in question. From two-dimensional considerations these can have values other than zero only if the forces are collinear—for that special case, they must be equal and opposite.

The analysis of three-dimensional trusses can become rather complicated. This is not because the concepts are unusual or that individual equation formulations at the

---

† S. P. Timoshenko and D. H. Young, *Theory of Structures*, 2d ed., 1965, McGraw-Hill Book Company. These structures are quite well handled by the methods described in chap. 9 of this book.

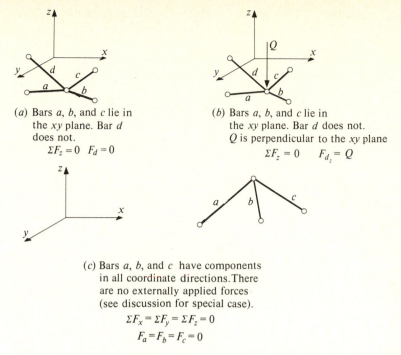

(a) Bars a, b, and c lie in
the xy plane. Bar d
does not.

$$\Sigma F_z = 0 \quad F_d = 0$$

(b) Bars a, b, and c lie in
the xy plane. Bar d does not.
Q is perpendicular to the xy plane

$$\Sigma F_z = 0 \quad F_{d_z} = Q$$

(c) Bars a, b, and c have components
in all coordinate directions. There
are no externally applied forces
(see discussion for special case).

$$\Sigma F_x = \Sigma F_y = \Sigma F_z = 0$$
$$F_a = F_b = F_c = 0$$

Figure 3.35 Rules for establishing zero-force members in three-dimensional trusses.

various joints are difficult. Rather, it is owing to the fact that the resulting large number of equations must be solved simultaneously. With the availability of digital computers, however, these solutions become routine. For this reason space trusses are considered in more detail in the chapter dealing with matrix methods.

3.6 RIGID SPACE FRAMES

Two different types of space frames are commonly encountered. In the one type the structure itself is two-dimensional, but the applied loads have components out of the plane of the structure. The second type involves structures that are truly three-dimensional. In both cases, the number of internal stress resultants at a cut section will equal six.

Figure 3.36a shows a horizontally curved beam. It is fully fixed at end A and free at B. A uniformly distributed load q is presumed applied along the member and acts in the y direction. In addition, a concentrated load Q is applied at the free end at an angle α to the horizontal (xz) plane. The centerline axis of the curved member is specified as a circle having a radius r and a total central angle β. (As before, the dimensions of the cross section are assumed to be small when compared to the radius r.) The six reaction components at fixed end A are determined from the equations for static equilibrium of the structure as a whole.

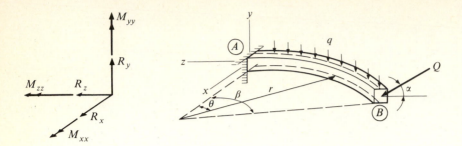

Possible reaction
components at A

Figure 3.36a Horizontally curved beam subjected to uniformly distributed vertical loading.

Considering equilibrium of forces in the x direction,

$$\sum F_x = 0$$

or

$$R_x = Q \cos \alpha \cos \beta$$

For equilibrium in the y direction,

$$R_y = Q \sin \alpha + \int_A^B q \, ds = Q \sin \alpha + \int_A^B qr \, d\theta$$

For $\sum F_z = 0$,

$$R_z = Q \cos \alpha \sin \beta$$

Next, consider the bending moments about the x axis at point A. From $\sum M_{xx} = 0$,

$$M_{xx} = bQ \sin \alpha + \int_A^B aq \, ds$$

where, from sketch a of fig. 3.36b, it is noted that

$$a = r \sin \theta$$

$$b = r \sin \beta$$

$$ds = r \, d\theta$$

For moment about the y axis, $\sum M_{yy} = 0$ or

$$M_{yy} = cQ \cos \alpha = Qr \cos \alpha \sin \beta$$

since $c = r \sin \beta$.

For moment about the z axis, $\sum M_{zz} = 0$ or

$$M_{zz} = eQ \sin \alpha + \int_A^B dq \, ds$$

or

$$M_{zz} = Qr(1 - \cos \beta) \sin \alpha + \int_A^B qr^2(1 - \cos \theta) \, d\theta$$

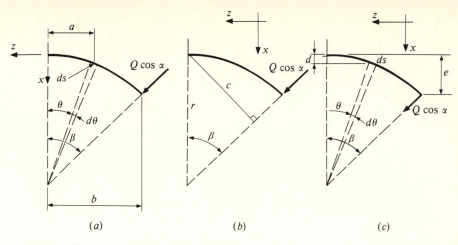

(a)   (b)   (c)

Figure 3.36b Component dimensions of beam.

since

$$d = r(1 - \cos \theta)$$
$$e = r(1 - \cos \beta)$$
$$ds = r \, d\theta$$

The twisting moment at any cross section can be defined using the dimensions identified in fig. 3.37.

$$T_\phi = (Q \sin \alpha) \overline{CD} + \int_0^\phi (q \, ds) \overline{CE}$$

where

$$\overline{CD} = r(1 - \cos \phi)$$
$$\overline{CE} = r[1 - \cos (\phi - \theta)]$$
$$ds = r \, d\theta$$

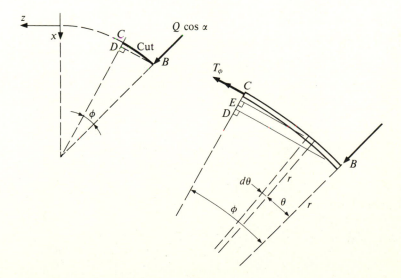

Figure 3.37 Consideration of element lengths.

Therefore

$$T_\phi = Qr(1 - \cos \phi) \sin \alpha + qr^2 \int_0^\phi [1 - \cos (\phi - \theta)] \, d\theta$$

or $\qquad T_\phi = Qr(1 - \cos \phi) \sin \alpha + qr^2(\phi - \sin \phi) \qquad 0 \le \phi \le \beta$

To facilitate graphing and to make the particular solution more general, the non-dimensional form in fig. 3.38 will be used.

$$\frac{T_\phi}{Qr \sin \alpha} = 1 - \cos \phi + \frac{qr}{Q \sin \alpha}(\phi - \sin \phi)$$

This figure is a plot of the equation for two values of $qr/Q \sin \alpha$. (It should be noted that a value of β was not needed to obtain these solutions.) Additional corresponding sets of curves for bending moments and shear could be developed.

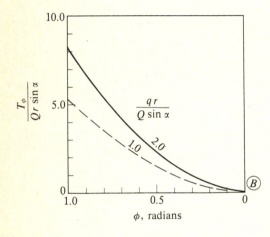

Figure 3.38 Variation in developed torsional moment as a function of ϕ.

Consider next the space structure and loading shown in fig. 3.39. The configuration consists of two straight members that are rigidly connected at corner B. The angle of connection is 90° and both members lie in the xz plane. (It is to be recalled that a rigid joint is one that transmits all of the stress resultants. Moreover, the angle of connection between the two members remains constant throughout the loading.)

A quick examination of the structure as a whole reveals that there will be no reactive forces in either the z or x direction at point A. (There are no applied forces in

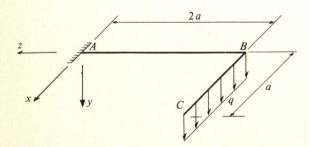

Figure 3.39 Three-dimensional space structure.

these directions.) Furthermore, there will be no bending about the y axis. Using the remaining equilibrium equations:

$$\sum F_y = 0 \qquad R_{y_A} = qa$$

$$\sum M_{xx_A} = 0 \qquad M_{xx} = (qa)(2a) = 2qa^2$$

$$\sum M_{zz_A} = 0 \qquad M_{zz} = (qa)\left(\frac{a}{2}\right) = \frac{1}{2}qa^2$$

Member AB resists bending, twisting, and shear. Member BC is subjected only to bending and shear.

3.7 CABLES

Cables are flexible structural members that can resist only axial tension. They are used primarily in long-span suspension bridges, guyed masts, cable-supported roofs, etc. Because of their flexible nature, cables develop horizontal reactions whenever they are subjected to vertical loads. (This, of course, would not be true when the cable is used as a hanger and is suspended vertically.) The horizontal component of the cable tension is constant at every point along the length of the member in question.

A typical cable is shown in fig. 3.40. It is connected at points 1 and 2 and is presumed subjected to a uniformly distributed vertical force of q pounds per horizontal foot of projection. The reactions are $V_1, H_1, V_2,$ and H_2. The available equations of static equilibrium are

$$\sum F_z = 0$$

$$\sum F_y = 0$$

$$\sum M_1 = 0 \qquad (\text{or} \sum M_2 = 0)$$

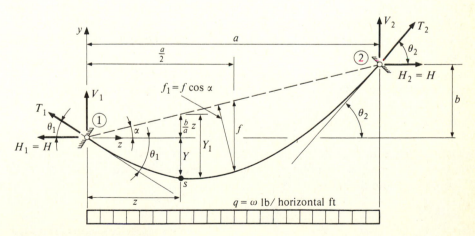

Figure 3.40 Geometry of cables.

It is further known that at any point on the cable $\sum M = 0$. However, this condition is not helpful unless the deflected shape of the cable can be established.

The deformed cable position for the given load is shown as a solid line in fig. 3.40. The sum of the moments of all of the forces to the left of the indicated point $S(z, y)$, taken about S, must be zero—since, by definition, the cable has no bending resistance. That is,

$$V_1 z - H\left(y_1 - \frac{b}{a} z\right) - \frac{w}{2} z^2 = 0 \tag{3.10}$$

where y is the vertical distance at the point S between the cable and the straight line (chord) joining the two ends. A second relationship involving V_1 and H can be written by using the required relationship $\sum M_2 = 0$.

$$Hb + V_1 a - \frac{w}{2} a^2 = 0 \tag{3.11}$$

Eliminating V_1 from these two equations, yields

$$\left(\frac{w}{2} a - \frac{Hb}{a}\right) z - H\left(y_1 - \frac{a}{b} z\right) - \frac{w}{2} z^2 = 0$$

or

$$Hy_1 = \frac{w}{2} az - \frac{w}{2} z^2$$

At $z = a/2$ the cable sag at the midpoint is $y_1 = f$. Therefore

$$H = \frac{wa^2}{8f} \qquad \text{or} \qquad f = \frac{wa^2}{8H}$$

Substituting this into eq. (3.11) gives

$$\frac{wa^2}{8f} y_1 = \frac{wa}{2} z - \frac{wz^2}{2}$$

or

$$y_1 = \frac{4fz}{a^2} (a - z) \tag{3.12}$$

Using the relationship $y = (b/a)z - y_1$, the general expression for the geometry of the cable is

$$y = \frac{b}{a} z - \frac{4fz}{a^2} (a - z) \tag{3.13}$$

(It should be recognized that this is an equation of a parabola.)

To determine the reaction forces T_1 and T_2—acting in the directions of the tangents at the ends of the cables—the end slopes, θ_1 and θ_2, must first be determined. The slope at any point along the cable can be obtained by differentiating with respect to z the deflection y. That is,

$$\frac{dy}{dz} = \frac{b}{a} - \frac{4f}{a^2} (a - 2z) \tag{3.14}$$

At end 1, $z = 0$

$$\left.\frac{dy}{dz}\right|_{z=0} = \tan \theta_1 = \frac{b}{a} - \frac{4f}{a} = \frac{1}{a}(b - 4f) \tag{3.15}$$

At end 2, $z = a$

$$\left.\frac{dy}{dz}\right|_{z=a} = \tan \theta_2 = \frac{b}{a} + \frac{4f}{a} = \frac{1}{a}(b + 4f)$$

T_1 and T_2 are therefore

$$T_1 = H \sec \theta_1$$

$$T_2 = H \sec \theta_2 \tag{3.16}$$

where

$$H = \frac{w}{8f}a^2$$

Since the horizontal force H has a constant value along the entire length of the cable, the cable tension T at point S is

$$T = H \sec \theta \tag{3.17}$$

where θ is the slope at the point in question (see fig. 3.40). From eq. (3.17) it is evident that the maximum tension in the cable occurs at the point of maximum slope—in this case, at one of the ends of the cable.

The lowest point on the cable is determined by setting $dy/dz = 0$. That is,

$$\frac{dy}{dz} = \frac{b}{a} - \frac{4f}{a^2}(a - 2z) = 0$$

or

$$z_0 = \frac{a}{2}\left(1 - \frac{b}{4f}\right)$$

(It is interesting to note that the slope at the midpoint $z = a/2$ is always parallel to the line joining the end points 1 and 2.)

The total length of the cable S can be determined using the geometry defined. For a small length of arc, ds,

$$(ds)^2 = (dz)^2 + (dy)^2$$

or

$$ds = \sqrt{(dz)^2 + (dy)^2} = dz\sqrt{1 + \left(\frac{dy}{dz}\right)^2} \tag{3.18}$$

The total length is therefore

$$S = \int_0^a ds = \int_0^a \sqrt{1 + \left(\frac{dy}{dz}\right)^2}\, dz \tag{3.19}$$

But the square-root term can be expanded using the binomial equation.

$$(1 + x)^n = 1 + nx + \frac{n(n-1)x^2}{2!} + \frac{n(n-1)(n-2)x^3}{3!} + \cdots$$

That is,

$$\sqrt{1 + \left(\frac{dy}{dz}\right)^2} = (1 + y'^2)^{1/2}$$

$$= 1 + \tfrac{1}{2}y'^2 + (\tfrac{1}{2})(\tfrac{1}{2})(-\tfrac{1}{2})(y'^2)^2 + \cdots \qquad (3.20)$$

Introducing this into eq. (3.19), and noting that

$$y' = \frac{dy}{dz} = \frac{b}{a} - \frac{4f}{a}(a - 2z)$$

yields

$$S = \int_0^a \left\{ 1 + \frac{1}{2}\left[\frac{b}{a} - \frac{4f}{a^2}(a - 2z)\right]^2 - \frac{1}{8}\left[\frac{b}{a} - \frac{4f}{a^2}(a - 2z)\right]^2 + \cdots \right\} dz \qquad (3.21)$$

Taking only the first two terms in this series, the length of the cable is given by the expression

$$S = a\left(\sec \alpha + \frac{8}{3}\frac{f^2}{a^2}\frac{1}{\sec^3 \alpha}\right) \qquad (3.22)$$

(This equation is considered to be sufficiently accurate for cables whose sag is not excessive. For greater accuracy, more terms in the series must be included.)

Since the tensile force varies over the length of the cable, the unit strain also varies. Therefore, to compute the elongation of the cable, this must be taken into account. Defining the total change in length as ΔS, each differential length ds extends $d(\Delta S) = (T/AE)\,ds$, where T is the tension at any point S, and

$$T = H \sec \theta = H\frac{ds}{dz}$$

Therefore the total change in length of the cable is

$$\Delta S = \int d(\Delta S) = \int \frac{T}{AE}\,ds = \int \frac{H\,ds/dz}{AE}\,ds = \frac{H}{AE}\int_0^a \frac{(ds)^2}{dz} \qquad (3.23)$$

But since

$$ds = \left[1 + \left(\frac{dy}{dz}\right)^2\right]^{1/2} dz$$

$$\Delta S = \frac{H}{AE}\int_0^a \left[1 + \left(\frac{dy}{dz}\right)^2\right] dz$$

or

$$\Delta S = \frac{H}{AE}\int_0^a \left\{1 + \left[\frac{b}{a} - \frac{4f}{a^2}(a - 2z)\right]^2\right\} dz$$

or

$$\Delta S = \frac{Ha}{AE}\left[1 + \left(\frac{b}{a}\right)^2 + \frac{16}{3}\left(\frac{f}{a}\right)^3\right] \qquad (3.24)$$

This can be written in terms of the average tensile force in the cable, $\bar{T}$:

$$\Delta S = \frac{\bar{T}S}{AE}$$

where
$$\bar{T} = \frac{Ha}{S}\left[1 + \left(\frac{b}{a}\right)^2 + \frac{16}{3}\left(\frac{f}{a}\right)^3\right] \tag{3.25}$$

A cable is designed in the same way as any other structural component. Based on the initial geometry and the loads to be carried, the maximum tensile force is computed. Then using the allowable unit stress stipulated[†] for the type and kind of cable desired the required size is determined.

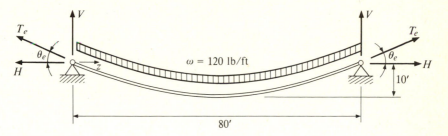

Figure 3.41

Example 3.9 The cable shown in fig. 3.41 spans an 80-ft horizontal span. It is subjected to a uniformly distributed vertical load of 120 lb/ft. (Because the cable weight is small in comparison to this value, it is assumed to be included in the w value.) The end supports are at the same elevation and the sag at the midpoint of the span is 10 ft. A 6×7 zinc-coated prestretched bridge rope having an allowable stress of 100,000 lb/in² and a modulus of elasticity of $E = 20 \times 10^6$ lb/in² is to be selected.[‡]

It is to be recognized that the system is symmetrical, and $b = 0$. H is determined from the relationship

$$H = \frac{wa^2}{8f} = \frac{(120)(80)^2}{(8)(10)} = 9600 \text{ lb}$$

---

† The idea of a unit stress in a cable is not absolutely correct. The cable is made up of many separate wires spun together to form a wire rope. The allowable stress, then, is an average stress for the whole rope, voids included. The same is true for the modulus of elasticity E.

‡ The notation 6×7 indicates the placement of wires within the rope—six strands, seven wires in each strand. (For details see any wire rope manual.)

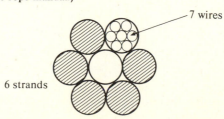

7 wires

6 strands

The slope of the cable at either support is θ_e.

$$\theta_e = \tan^{-1} \frac{1}{a}(b - 4f) = \tan^{-1} (0.500) = 26°34'$$

The maximum tensile force is therefore

$$T_{\max} = T_e = H \sec \theta_e = (9600)(1.1181) = 10{,}734 \text{ lb}$$

For an average allowable stress of 100 ksi, the required cable area is $A_{\text{rqd}} = 0.1073 \text{ in}^2$. A $\frac{1}{2}$-in round 6×7 bridge rope, which has an area of 0.119 in^2 and weighs 0.42 lb/ft, will be more than adequate.

The length of cable required is determined from eq. (3.22).

$$S = a\left(\sec \alpha + \frac{8}{3}\frac{f^2}{a^2 \sec^3 \alpha}\right)$$

$$= 80\left[1.000 + \frac{8}{3}\left(\frac{10}{80}\right)^2\right] = 83.336 \text{ ft}$$

(Note that since $b = 0$, $\alpha = 0$.) The total weight of the cable is $(0.42) \times (83.336) = 35.0 \text{ lb}$. The average cable tension $\bar{T}$ is given by

$$\bar{T} = \frac{Ha}{S}\left[1 + \frac{16}{3}\left(\frac{f}{a}\right)^3\right] = \frac{(9600)(80)}{(83.336)}\left[1.000 + \frac{16}{3}\left(\frac{10}{80}\right)^3\right] = 9312 \text{ lb}$$

The total elongation is therefore

$$\Delta S = \frac{\bar{T}S}{AE} = \frac{(9312)(83.336)(12)}{(0.119)(20{,}000)} = 3.913 \text{ in} = 0.326 \text{ ft}$$

and the cable would be cut to a length

$$L = S - \Delta S = 83.336 - 0.326 = 83.010 \text{ ft}$$

3.8 INFLUENCE LINES

Thus far in this chapter, attention has been directed toward the determination of stress resultants, when the applied loading is specified in magnitude and is position fixed. But, as was pointed out in chap. 1, certain live loads are not fixed in location: automobiles travel across bridges, winds traverse buildings, materials or goods stored in a warehouse are placed in different locations at different times, etc. These moving or movable live loads create for the designer a problem—because the movements produce changes in the magnitude (and sometimes even in the sense) of the maximum stresses in the various members. Since all components of the structure must be selected to sustain the greatest induced force, the designer must carefully examine the effects of such loads. The concept of *influence lines* has proven to be quite useful in studying the effects of these types of loads.

An influence line is simply a graphic or pictorial representation of the magnitude of a particular stress resultant at one point in the structure, when a unit load is placed

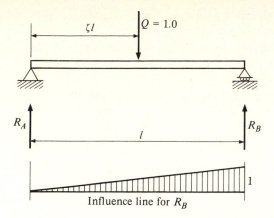

Figure 3.42 Influence line for simple-beam end reaction.

at various other locations on that structure. Consider the simply supported beam shown in fig. 3.42. It is desired to determine the influence line for the reaction at B. Since, by definition, this influence line represents the magnitude of R_B for any arbitrary placement of the unit load, it will be presumed that the load is placed ζl from the left-hand end of the span, as shown. The reaction R_B then can be determined from the equilibrium equation $\sum M_A = 0$.

$$Q(\zeta l) - R_B(l) = 0 \qquad \text{or} \qquad \frac{R_B}{Q} = \zeta$$

By selecting various locations for the load; that is, by choosing various values of ζ from 0 to 1.0, the influence line for R_B is obtained. (It should be emphasized that the ordinate of the influence line at ζ represents the value of the reaction at B for an applied unit load at ζ.) The units of the influence line ordinate for reactions, shears, and axial forces are, for example, lb/lb, while the ordinate of the influence line for moment is ft·lb/lb.

Since influence lines describe the effect on a particular resultant of a unit load in any location, and since in elastic, first-order systems superposition (that is, algebraic summing) holds, the effect of a series of simultaneously applied loads can be readily established by multiplying each load by its corresponding influence line ordinate and summing the results. It follows that the value of the stress resultant that results from a uniformly distributed load is equal to the load intensity multiplied by the area under the influence line for that particular stress resultant. Assume, for example, that there are three concentrated loads, each of magnitude Q, applied at the quarter points of the simple beam shown in fig. 3.42. The vertical reaction at end B would be

$$R_B = Q(\tfrac{1}{4}) + Q(\tfrac{1}{2}) + Q(\tfrac{3}{4}) = \tfrac{3}{2}Q$$

(The values in the parentheses are from the influence line diagram in question.) If the beam were subjected to a uniformly distributed vertical load of q lb/ft, the reaction at B would be

$$R_B = \int_0^l (\text{ordinate of IFL})q \; d(\zeta l) = \tfrac{1}{2}(1)(l)q = \tfrac{1}{2}ql$$

For a partially distributed load (or patch loading), only the area under the influence line between the limits of the partial loading would be considered.

In the several examples that follow, the construction of influence lines corresponding to different types and locations of stress resultants is illustrated. These examples should be studied carefully, with special attention being paid to the difference between an influence line and a shear and bending-moment diagram.

In statically determinate structures, the influence lines for stress resultants will consist of straight-line segments. This is due to the linear relationship between the loads and stress resultants. Consequently, when constructing an influence line, values at several controlling locations will be computed, and these then will be connected by straight lines.

Example 3.10 For the simply supported beam with overhang shown in fig. 3.43, influence lines for R_A, R_B, M_B, M_D, and V_D are required. D is located midway between the two supports. To determine the influence line for the vertical reaction at A, three placements of the unit load will be considered:

$$\text{Unit load at } A \qquad R_A = 1 \quad \uparrow$$

$$\text{Unit load at } B \qquad R_A = 0$$

$$\text{Unit load at } C \qquad R_A = \tfrac{1}{2} \quad \downarrow$$

The corresponding influence line is shown as sketch a of fig. 3.43.

For the reaction at B:

$$\text{Unit load at } A \qquad R_B = 0$$

$$\text{Unit load at } B \qquad R_B = 1 \quad \uparrow$$

$$\text{Unit load at } C \qquad R_B = \tfrac{3}{2} \quad \uparrow$$

The influence line for R_B is shown in sketch b. (Note that these computed singular values have been joined by a straight line.)

To construct the influence line for bending moment at location B, the free body diagram shown in fig. 3.44 will be used. A section just to the right of the support is chosen for consideration. When the unit load is applied anywhere between A and B, the moment at B must equal zero. When it is between B and C, however, values are obtained. For a unit load at C, the moment becomes

$$M_B = (1)(a) = a$$

and acts in the direction shown on the free body diagram. The influence line for the bending moment at B is given in fig. 3.43c.

For the moment at the interior point D, the free body diagram shown in fig. 3.45 will be used. More than that—for a major portion of the diagram, the influence line for the reaction at A can be used. When the unit load is between D and C, the only force governing the value of M_D will be the reaction R_A. Therefore, in that range

$$M_D = R_A(a) \qquad \text{(unit load between } D \text{ and } C\text{)}$$

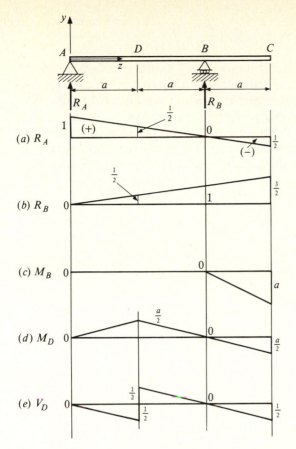

Figure 3.43 Influence lines for a simple beam with an overhanging portion.

When the unit load is in the region AD, a free body diagram of the right-hand portion of the beam could be considered and this would give

$$M_D = R_B(a) \qquad \text{(unit load between } A \text{ and } D\text{)}$$

The resulting influence line is shown in sketch d of fig. 3.43.

The free body diagram of fig. 3.45 also could be used to establish the influence line values for shear at section D. With the unit load between A and D,

$$V_D = R_B$$

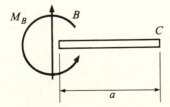

Figure 3.44

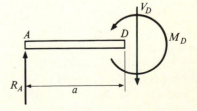

Figure 3.45

When it is between D and C,

$$V_D = R_A$$

Sketch e of fig. 3.43 is the desired influence line. The signs correspond to the assumed plus convention shown in the free body diagrams.

Let it now be assumed that the beam in question is subjected to a movable, distributed, live load of q, a moving, concentrated, live load of Q, and its own dead weight w. The presumed controlling condition for design will be maximum bending. Since the live loads can occur anywhere on the span, they should be placed in such a way that maximum moment values are realized. Recalling that the magnitude of a particular stress resultant is given by the magnitude of the applied load multiplied by the influence line ordinate at the point of application of the load, it is evident that a maximum value for M_B is produced when the distributed load q is placed over the entire range from B to C, and the concentrated load Q is located at end C. Maximum positive values of M_D occur when only span AB is loaded with q, and the concentrated load Q is placed at D.

At B (for the maximum negative moment):

$$M_B = -\left[(q+w)\frac{a^2}{2} + Qa\right]$$

At D (for the maximum positive moment):

$$M_D = +q\frac{1}{2}\frac{a}{2}2a + w\left[\frac{1}{2}\frac{a}{2}2a + \frac{1}{2}a\left(-\frac{a}{2}\right)\right] + Q\frac{a}{2}$$

$$= q\frac{a^2}{2} + w\left(\frac{a^2}{2} - \frac{a^2}{4}\right) + Q\frac{a}{2}$$

At D (for the maximum negative moment):

$$M_D = q\frac{1}{2}a\left(-\frac{a}{2}\right) + w\left[\frac{1}{2}\frac{a}{2}2a + \frac{1}{2}a\left(-\frac{a}{2}\right)\right] + Q\left(-\frac{a}{2}\right)$$

$$= -q\frac{a^2}{4} + w\left(\frac{a^2}{2} - \frac{a^2}{4}\right) - Q\frac{a}{2}$$

Maximum positive moment at B occurs when both q and Q are placed outside of segment BC.

When a single moving load is applied to a beam, it is obvious from the influence line diagram where that particular load should be placed to produce maximum (or minimum) effects. However, when several interconnected loads are applied, proper placement may not be readily apparent.

In fig. 3.46 a series of concentrated loads $Q_1, Q_2, \ldots, Q_8$ move from right to left across the beam. They are spaced by the distance, $a, b, \ldots, g$ as shown. It is desired to place the loads on the span in such a fashion that maximum shear and maximum bending moment at interior location C will be produced. Qualitatively, the appropriate influence line diagrams are shown in the figure.

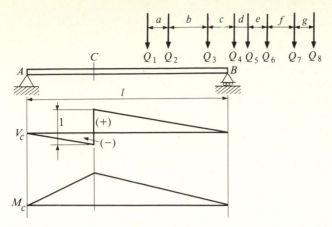

Figure 3.46 Simple beam subjected to a series of space concentrated loads.

Consider, first, the question of shear at C. As the loads move from right to left, before the first load Q_1 reaches C, the shear at C will equal the left-hand reaction R_A. When Q_1 is just at C, V_C is a maximum (but not necessarily the absolute maximum for all of the possible placements of the loads). As Q_1 passes C, and moves further to the left, V_C decreases because $V_C = R_A - Q_1$. (This can also be deduced from the influence line diagram.) As the second load Q_2 reaches point C, a second maximum value of V_C is realized. As soon as Q_2 passes, V_C decreases, etc. Therefore, it can be concluded that the absolute maximum of V_C occurs when one of the loads in the series is located at point C. From the shape of the influence line, it is comparatively easy to tell approximately the proper location of the loads to produce the absolute maximum value. Examination of two or three cases is usually sufficient to verify the choice. Moreover, with the ordinates of the influence line known, relatively little time is consumed in these examinations.

To ascertain the largest possible moment at point C, for the train of loads indicated, the same procedure is followed. In turn, each load is placed at point C and the moment value computed for the loads on the span. In general, the maximum moment will occur when as many of the larger loads as are possible are located near the peak of the diagram. (Precise methods have been developed which will give the exact placement of series of loads of the type shown. The purpose of this section is to develop concepts, and those methods and criteria have therefore not been included.)

If the loading is a moving, uniformly distributed live load, rather than a series of concentrated forces, then from the influence line for V_C it can be concluded that maximum shear will be realized when that uniform load is placed over either the entire positive or the entire negative portion of the diagram—whichever has the larger area. Similarly, from the influence line for M_C, maximum moment at C is realized when the entire simple span is subjected to the distributed load.

Example 3.11 The second influence line example will deal with a compound beam, specifically, one that contains an internal hinge. The beam in question is shown in fig. 3.47. Influence lines for R_A, R_B, R_D, M_E, V_E, and M_B are desired. E

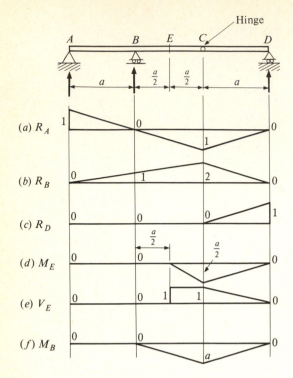

Figure 3.47 Influence lines.

is at the midpoint of span *BC*. Diagrams *a* and *b* are determined in the same manner as those previously developed. The process will not here be repeated. It is to be recognized, however, that for the reaction at *D*, with the unit load applied anywhere between *A* and *C*, $R_D = 0$. With the unit load at *D*, $R_D = 1.0$. Thus, the influence line shown as *c* of fig. 3.47.

The influence line for moment at *E* would be developed in several stages. For the unit load between *A* and *E*, and using *EC* as a free body, $R_D = V_C = 0$. Therefore $M_E = 0$. For the unit load between *E* and *C*, and using *CE* as a free body, $M_E = (1)$ (moment arm). For the unit load between *C* and *D*, and using *CD* and *CE* as free bodies, $M_E = V_C(a/2)$. The resulting influence line is given as (*d*).

The influence line for shear also requires special consideration. For the unit load between *A* and *E*, and using *CD* and *EC* as free bodies,

$$R_D = V_C = V_E = 0$$

For the unit load between *E* and *C*, and using *CD* and *EC* as free bodies, $V_C = 0$ and $V_E = $ unity. For the unit load between *C* and *D*, and using *CD* and *EC* as free bodies, $V_C = 1.0 - R_D$ and $V_E = V_C$. This is shown in sketch *e*.

The influence line for moment at *B* would be developed in a fashion similar to that used for case *c* of fig. 3.43.

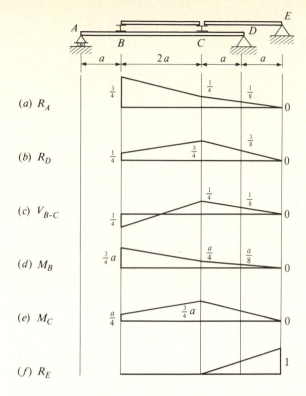

Figure 3.48 Influence lines.

Example 3.12 (see fig. 3.48) It is specified that loads can be applied only along the upper simple beams BC and CE. Influence lines for certain stress resultants in the lower beam AD are desired: R_A, R_D, V_{BC}, M_B and M_C. For completeness, the influence line for the end reaction R_E (in the upper beam) also will be determined.

First, since the loads can be applied between B and E only, the influence lines will extend over only that range. Secondly, forces are transmitted to the lower beam AD only at points B and C. These equal the end reactions of beams BC and CE which were specified to be simply supported at their end points.

For the unit load at B, the vertical reaction at A is $\frac{3}{4}$, while that at D is $\frac{1}{4}$. For the unit load at C, $R_A = \frac{1}{4}$ and $R_D = \frac{3}{4}$. For the unit load at E, $R_A = R_D = 0$ and $R_E = 1.0$. Thus, influence lines a, b, and f are obtained.

The moment at B equals the reaction R_A times the distance a. Therefore, sketch d is readily apparent. In similar fashion, the moment at C equals the reaction R_D times the distance a. The influence line is as shown in sketch e.

The shear in panel BC depends both on the value of the reaction R_A of beam AD, and the vertical force transmitted from the top to the lower beam at point B. If the unit load were placed at B, the shear would equal the value of the influence line ordinate for R_A minus the unit force from BC, that is, $\frac{3}{4} - 1 = -\frac{1}{4}$. For the unit load at C, no forces are transmitted from BC at B. Therefore $V_{BC} = R_A = +\frac{1}{4}$. The influence line is shown in c.

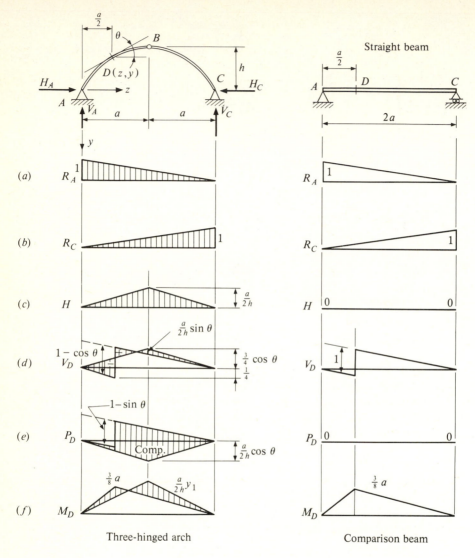

Figure 3.49 Influence lines for a three-hinged arch compared with those of a simple beam.

Example 3.13: Three-hinged arch The structure in question has a span of $2a$. It is symmetrical, and the height at the center point B, which also is the interior hinge point, equals h. (For comparison purposes there also will be shown in the right-hand portion of fig. 3.49 influence lines for a simple beam of span $2a$.)

The vertical components of reaction at ends A and C of the arch are identical to those for the simple beam—when each is presumed subjected to a unit vertical force.

The horizontal components of reactions in the arch at A and C are identical:
$$H_A = H_C = H.$$

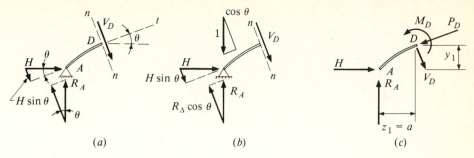

(a)   (b)   (c)

Figure 3.50

If the unit load were to the left of the interior hinge B, and BC were taken as a free body, summation of moments about hinge B would yield

$$H = \frac{R_C a}{h}$$

If the load were to the right

$$H = \frac{R_A a}{h}$$

The influence line for H is shown in sketch c. (The simple beam subjected to vertical loads has no horizontal reaction component.)

For shear at section D, consider the unit load between D and C, and select as the free body AD. From fig. 3.50a, it is evident by summing forces in the n direction that

$$V_D = R_A \cos\theta - H \sin\theta$$

Using the influence lines for R_A and H, multiplying them by $\cos\theta$ and $\sin\theta$, respectively, yields the appropriate portion of the influence line for shear shown in sketch d of fig. 3.49. When the unit load is between A and D (see fig. 3.50b), the component of the unit load must be subtracted, that is,

$$V_D = R_A \cos\theta - H \sin\theta - 1 \cos\theta$$

The influence line for the axial thrust at section D is obtained in the same manner. For example, when the unit load is between D and C and AD is selected as a free body

$$P_D = H \cos\theta + R_A \sin\theta$$

When the load is between A and D

$$P_D = H \cos\theta + R_A \sin\theta - 1 \sin\theta$$

For moment at section D, the dimensions indicated in fig. 3.50c are significant. For the unit load between D and C,

$$M_D = R_A(a) - Hy_1$$

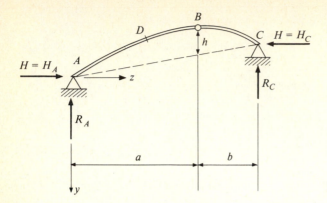

Figure 3.51a

For the unit load between A and D,

$$M_D = R_B(3a) - Hy_1$$

The influence line in question is shown in sketch f of fig. 3.49.

Example 3.14 For an unsymmetrical three-hinged arch, the reactions H_A and H_C do not lie on the same horizontal line (see fig. 3.51a). This creates computational problems. The arch can be more conveniently studied if the reactions at A and C are resolved into components acting in the direction of the line joining the points A and C and vertically as shown in fig. 3.51b. (Note that the new system of axes are not orthogonal.) The influence line for H can be obtained from the influence line for H'.

To determine the influence line for H', consider the unit vertical load between B and C, and use AB as a free body. Summing moments about B

$$R'_A(a) - H'(h \cos \phi) = 0$$

or

$$H' = \frac{R'_A a}{h \cos \phi}$$

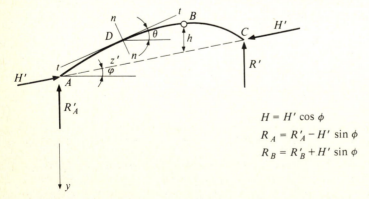

$$H = H' \cos \phi$$
$$R_A = R'_A - H' \sin \phi$$
$$R_B = R'_B + H' \sin \phi$$

Figure 3.51b Expression of the reactions in nonrectangular coordinates.

But $H' = H/\cos \phi$, therefore

$$H = R'_A \frac{a}{h}$$

When the unit load is between A and B, and BC is used as a free body

$$H = R'_B \frac{b}{h}$$

H_{max} occurs at B and equals ab/h. Straight lines would be drawn between this value at B and zero at A and C.

For shear at section D, assuming the unit vertical load is applied to the right of D, and using AD as the free body,

$$V_D = R'_A \cos \theta - H' \sin (\theta - \phi) = R'_A \cos \theta - H \frac{\sin (\theta - \phi)}{\cos \phi}$$

Similarly, the axial force would be

$$P_D = R'_A \sin \theta + H' \cos (\theta - \phi) = R'_A \sin \theta + H \frac{\cos (\theta - \phi)}{\cos \phi}$$

(It is to be recognized that R'_A and R'_B are described by straight lines.)

Example 3.15 For the truss shown in fig. 3.52, it is specified that vertical loads can be applied only along the bottom chord members. Moreover, these loads will be transmitted to the truss through floor beams only at the panel points.

It should be recognized that the influence lines for the reactions are the same as those that would be obtained for a simple beam. These are shown in sketches a and b. For the force in bar Bb, a free body diagram of joint B will be selected for examination, and the unit load will be placed, in turn, at $A, B, C, \ldots, G$. A value is obtained only when the load is at B, and for that case $F_{Bb} = 1.0$ (tension). The influence line is, therefore, as shown in sketch c.

For the force in member bC and the unit load between C and G, the free body shown in fig. 3.53 is selected. Summing moments about 0,

$$F_{bC} = \sqrt{\tfrac{2}{3}} R_A \quad \text{(tension)}$$

When the unit load is between A and B, a free body of the remainder of the structure is used. Summing moments about 0,

$$F_{bC} = \frac{14}{3\sqrt{2}} R_G \quad \text{(compression)}$$

The influence line for F_{bC} is therefore as shown in sketch d of fig. 3.52. (By selecting for detailed examination several points within panel BC, the reader should verify the shape and magnitude of the influence line shown in that region.)

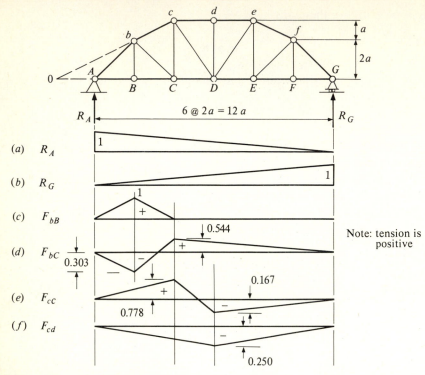

6 @ 2a = 12a

(a) R_A

(b) R_G

(c) F_{bB}

(d) F_{bC}

(e) F_{cC}

(f) F_{cd}

Note: tension is positive

Figure 3.52 Influence lines for a truss.

For the force in member cC, a section will be selected that cuts bc, cC, and CD and moments will be taken about point 0. If the unit load is between D and G, the controlling equation is

$$F_{cC} = \tfrac{1}{3}R_A \quad \text{(compression)}$$

If it is between A and C

$$F_{cC} = \tfrac{7}{3}R_G \quad \text{(tension)}$$

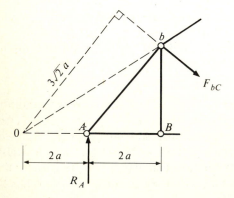

Figure 3.53

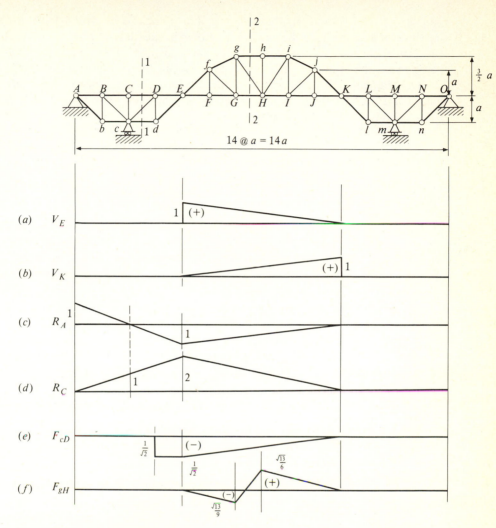

Figure 3.54 Influence lines for a truss.

A section obtained by cutting members cd, cD, and CD would be used to determine the force F_{cd}. Moments would be taken about point D.

It should be evident that the influence line for F_{dD} will be zero throughout the entire length of the span.

Example 3.16 The structure shown in fig. 3.54 is a compound truss having four external reactions. Since internal frictionless hinges exist at E and K, the system is determine. Considering the simple segment AE, it is evident that a unit load placed anywhere between A and E will not contribute to the internal vertical reaction (shear) at point E. (That is, the influence line for V_E will be zero for that entire range.) Similarly, when the unit load is between K and 0, $V_E = 0$. When the load is between E and K, the influence line varies as it would for the end reactions of a simple span beam. This is shown in sketch a. The influence line for the vertical shear at K is obtained in the same fashion. It is shown in sketch b. Using these, the influence lines for R_A and R_C can be readily obtained.

To determine the influence line for the force in diagonal member cD, a section 1-1 is cut, the chosen free body is from that cut to point E, and $\sum F_{\text{vert}} = 0$ is used. For the bar force F_{gh}, a cut is made at section 2-2, as shown in the figure. Calculations would proceed as for member cd of fig. 3.52. For F_{gH} the same free body would be selected, but the control condition would be $\sum F_{\text{vert}} = 0$.

3.9 DESIGN ENVELOPES

It should be evident from the preceding sections of this chapter that a knowledge of the size, or shape, or variation of the cross section is not required for analysis of stress resultants. This is true for all determinate structures, but, as will be demonstrated later, this is not the case for indeterminate ones. It also should be apparent that when dealing with moving or movable loads, or with several different static loads, the question exists as to which of the various possibilities are the most significant ones for design. Influence lines provide insight into this later question. But even there, it must be remembered that these diagrams are constructed for particular locations in the structure. It is also true that normal stress may control the design in one region of the structure, but shear could well be critical in another region. The applied loads associated with these two cases may be quite different.

To facilitate design, use frequently is made of what is termed a *design envelope*. This is nothing more than a pictorial or graphical representation of the controlling design conditions at each point in the structure. Depending upon the stress resultant being considered, an envelope could be developed which would define, for example, the maximum moment at each point in each member of the structure in question. Similarly, another envelope could indicate the maximum shear values—under all possible conditions of loading. An envelope describing thrust or one showing the maximum possible values of torsional moment also could be constructed. These diagrams would provide useful information for selecting the actual sizes and variations along the members.

It must, of course, be understood that not infrequently the costs of fabrication, or construction, or the ready availability of certain discrete and separate prefabricated member sizes dictate the use of particular prismatic members. This is especially true for standard metal shapes and timbers. Even for concrete members, forming costs may somewhat restrict variations in the dimensions of the cross section, although the amounts of steel reinforcement used provide a certain degree of flexibility.

To illustrate the concept of design envelopes, consider the 30-ft-long bridge girder shown in fig. 3.55a. This girder is to be used as an overhead crane, capable of lifting 5 tons. It is a simple beam, and variations in depth along the member are allowed.

A possible, critical design condition would be the determination of the depth and thickness of the web of the girder at its ends. Shearing forces (and therefore shearing

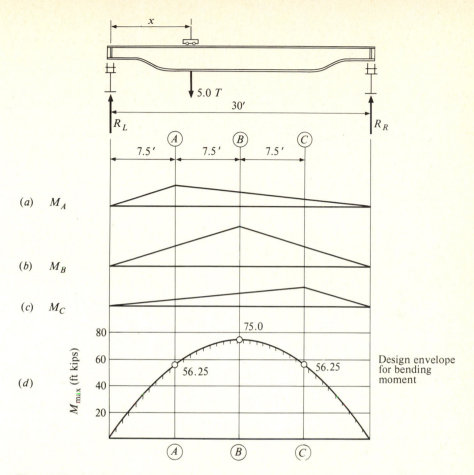

Figure 3.55a–d Influence lines and design envelope for a crane girder.

stresses) are high in these regions. A second, probably more critical design consideration, however, is the variation in the maximum live-load bending moment within the
span. A design envelope will first be constructed for this condition.

Three points along the member will be selected for examination: the one-quarter
point, the one-half point, and the three-quarter point. These have been designated in
fig. 3.55 as sections A, B, and C, respectively. The influence lines for bending
moments at each of these points have the same general form: each will have a
maximum value at the point in question, and will decrease linearly to zero at the end
supports. Therefore, maximum conditions will be realized by placing the 5-ton load
at the points in question. The maximum bending moments at points A and C are

$$M_{A,\,max} = M_{C,\,max} = \underbrace{(10 \text{ kips})}_{\substack{\text{applied} \\ \text{load}}} \underbrace{(\tfrac{3}{4})(7.5)}_{\substack{\text{influence} \\ \text{line} \\ \text{ordinate}}} = 56.25 \text{ ft} \cdot \text{kips}$$

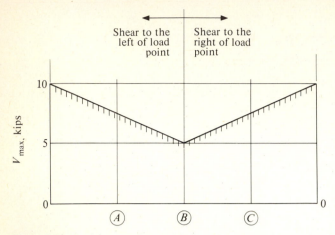

Figure 3.55e Design envelope considering shearing forces.

At the midpoint

$$M_{B, \max} = (10 \text{ kips})(\tfrac{1}{2})(15) = 75.0 \text{ ft} \cdot \text{kips}$$

Using these values, the live-load moment design envelope shown in sketch d of fig. 3.55 is obtained. The girder cross sectional dimensions would be selected to sustain this live-load moment plus the dead-load moment due to the weight of the girder, plus any required impact factors.

As indicated in fig. 3.46, the influence line for vertical shear in a single beam has two possible maximum values: one just to the left of the unit load, and the other just to the right. Moreover, the sum of these two values will equal one. For the girder in question, the design envelope for shearing forces would be that shown in fig. 3.55e.

3.10 PROBLEMS

3.1 Draw the shear and bending moment diagrams for the beams shown:

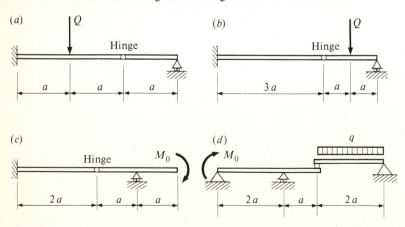

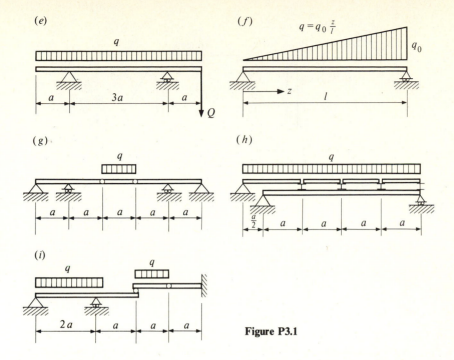

Figure P3.1

3.2 For the rigid frames shown, draw the shear, bending moment, and axial force diagrams.

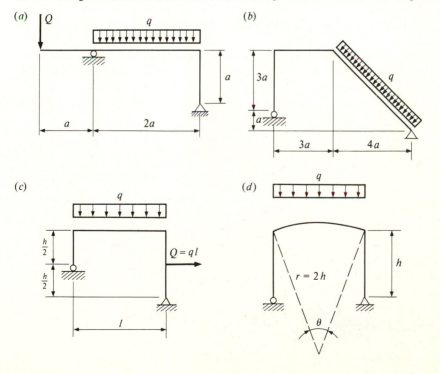

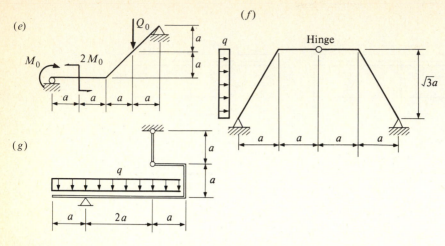

Figure P3.2

3.3 For the three hinged arches shown, draw the shear, bending moment, and axial force diagrams.

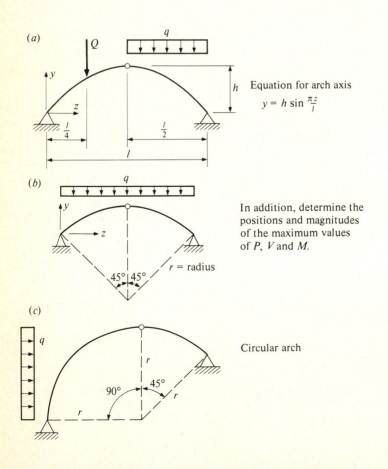

(a)

h Equation for arch axis

$$y = h \sin \frac{\pi z}{l}$$

(b)

In addition, determine the
positions and magnitudes
of the maximum values
of P, V and M.

r = radius

(c)

Circular arch

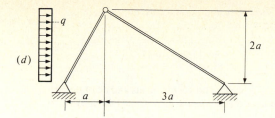

Figure P3.3

3.4 By the method of joints, determine the forces in all of the bars of the trusses shown. (Advantage should be taken of symmetry, zero force members, etc.)

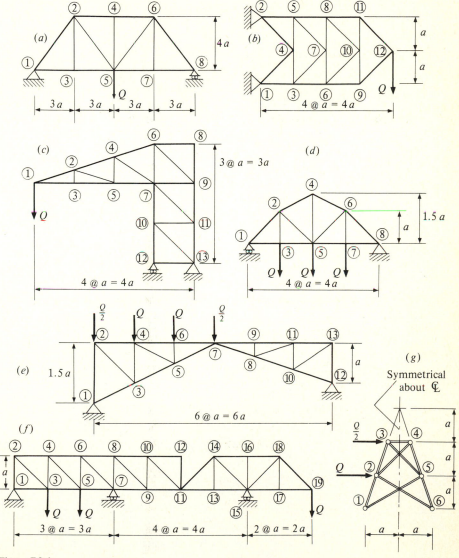

Figure P3.4

3.5 Determine the following bar forces in the trusses shown in prob. 3.4 by section method.

(a) 2-5	(b) 3-4	(c) 4-6	(d) 2-3	(e) 2-3	(f) 4-5
3-5	3-7	6-7	2-4	2-4	4-6
	6-7	7-10	2-5	3-4	10-11
		7-11		8-10	14-16

3.6 For the three-dimensional truss shown, determine the reaction components and all of the forces in all of the bars. All members are of length a, have the same cross-sectional area A, and have the same modulus of elasticity E.

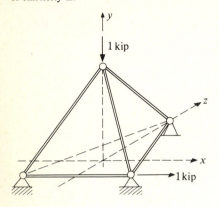

Figure P3.6

3.7 Draw the shear, thrust, bending moment, and torsional moment diagrams for the curved bar shown.

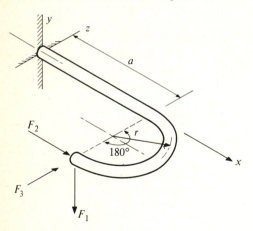

Figure P3.7

3.8 Determine the value of the maximum tension in the cable under its own weight. How long should it be? Where is the low point on the cable, and what is the deflection?

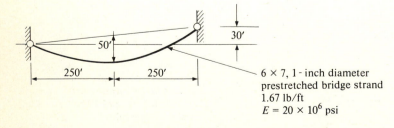

6 × 7, 1-inch diameter
prestretched bridge strand
1.67 lb/ft
$E = 20 \times 10^6$ psi

Figure P3.8

3.9 For the beams shown, draw the influence lines for the quantities indicated.

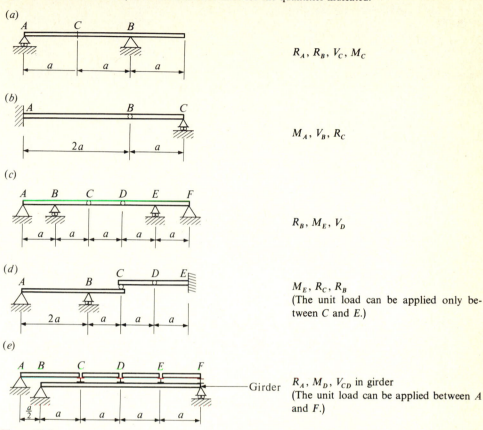

(a)

R_A, R_B, V_C, M_C

(b)

M_A, V_B, R_C

(c)

R_B, M_E, V_D

(d)

M_E, R_C, R_B
(The unit load can be applied only between C and E.)

(e)

Girder R_A, M_D, V_{CD} in girder
(The unit load can be applied between A
and F.)

Figure P3.9

3.10 Draw the indicated influence lines for the frames and arches shown. All unit loads are to be assumed vertical unless otherwise specified.

(a)

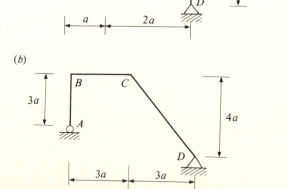

1. Axial force in column CD
2. Moment at B
3. Shear just to the right of B
(The load can be applied only from A to
C.)

(b)

1. Moment at C
2. Shear just to the right of C
3. Axial force in member CD
(The load can be applied between A and
B, and is horizontal.)

(c)

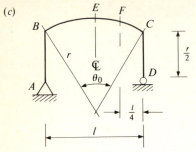

1. Moment at E and F
2. Shear at E and F
3. Axial force at E and F
(The load can be applied only along BC.)

(d)

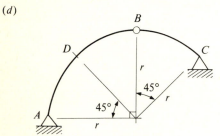

1. Horizontal reaction at A
2. Shear force at hinge B
3. Moment, shear, and axial force at D
(The load can be applied from A to C.)

Figure P3.10

3.11 Draw the influence lines for the plane trusses shown.

(a) Pratt truss

Bar forces in members a, b, c, and d.

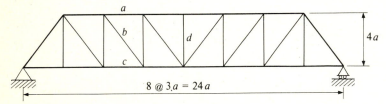

(b) Parker truss

Bar forces in members a, b, c, and d.

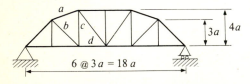

(c) Baltimore truss

Bar forces in members a, b, c, d, and e.

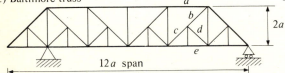

(d)

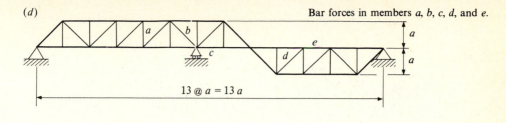

Bar forces in members a, b, c, d, and e.

$13 @ a = 13 a$

(e)

Bar forces in members a, b, c, d, and e.

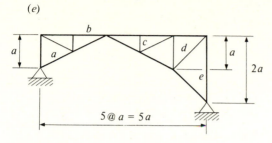

$5 @ a = 5a$

Figure P3.11

DEFORMATIONS OF DETERMINATE STRUCTURES—SOLUTIONS TO DIFFERENTIAL EQUATIONS

Load-deformation relationships are needed for a variety of purposes. Deflections, in and of themselves, may be limiting design factors. But as important, methods for solving indeterminate structures presuppose a knowledge of, or an ability to determine, deformations.

Chapter 2 dealt with methods for determining stresses in structural members. In chap. 3, the relationships that must exist between stress resultants and applied loads in determinate structures were examined. In both of those chapters, it was repeatedly emphasized that a basic assumption was being made—that equilibrium was formulated in the undeformed state. This is not to say that it was assumed that the structure would not deflect. Rather, the first-order formulation presupposed that the resulting deformations would be of sufficiently small magnitude that their effect on the equilibrium equations could be neglected.

In many instances first-order procedures are adequate and can be justified. There are other cases, however, where such simplification and idealization are unwarranted and unrealistic. It is one of the purposes of this (and the next) chapter to give a clearer insight into just when or under what conditions this might be the case.

The primary purpose of this chapter, however, is to develop methods for predicting the deformational response of individual slender members or determinate structures composed of such members. These methods will be based upon the solution of the appropriate differential equations. In chap. 5 deformational solutions will be obtained using energy methods.

4.1 LINEAR AND NONLINEAR RESPONSE OF STRUCTURES

A variety of classifications may be used to describe the deformational response of structures: small as opposed to large, cyclic as opposed to steady, deterministic as opposed to random, elastic as opposed to inelastic, etc. One of the more important groups describes the general interrelationship that exists between the applied loads and the overall deformational response: *linear* as opposed to *nonlinear*. Many factors influence whether the former or the latter will be the case—and correspondingly, whether more complicated analyses will be required. For these and other reasons, it is desirable to review the fundamental definitions and assumptions that are involved.

Nonlinear material behavior Many, if not most, methods of structural analysis prescribe at the outset that a linear relationship exists between stresses and strains. While most materials exhibit such behavior, at least in their initial ranges of loading, it is to be recognized that linearity of stress-strain response is not synonymous with or a necessary condition for *elasticity*. Elasticity requires that deformations due to applied forces be fully recovered when the loads are removed. Inelastic behavior, on the other hand, has associated with it the realization of a permanent set upon removal of the loads.

On a unit basis, inelastic behavior has associated with it an unloading path that differs from that of loading. The loop so formed is referred to as a hysteresis loop, and is a measure of the *energy loss* in the cycle of loading and unloading.

Inelastic stress-strain relationships are observed in a variety of forms. In some, for example, when there is a "sudden" repositioning of atoms or molecules in the material in question, a piecewise linear type of stress-strain response may be observed. In others, a more continuous nonlinear variation takes place. A number of examples of typical (idealized) uniaxial stress-strain relationships are shown in fig. 4.1. (Note that in all cases, both the loading and the unloading paths have been indicated.)

Most of this chapter will be restricted to a consideration of materials that respond in a linear, elastic fashion, that is, stresses will be presumed to be proportional to strains. This is a necessary requirement for the linear theory of structures—but it is not sufficient. Examples will be given where structures composed of materials which exhibit linear, elastic stress-strain behavior, respond to certain particular types of loading in a nonlinear fashion.

Large displacements In developing the equations leading to eq. (2.18), it was assumed—almost by definition—that the curvatures at a particular location in a beam could be equated to the second derivative of the deflection with respect to the independent variable z. As now will be demonstrated, this is not an absolutely correct statement. Consider the deflection curve shown in fig. 4.2. Points P and P' along the beam are of particular interest. At point P the slope (with respect to the z axis) is θ. At point P', which is Δs further along the deflection curve, the slope has a value $\theta + \Delta\theta$. The change in slope in the distance Δs is therefore $\Delta\theta$. Or, in the limit, $d\theta$ will

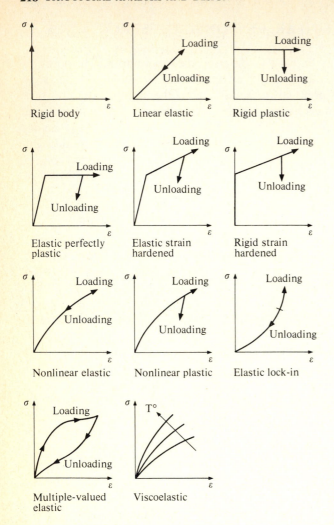

Fig. 4.1 Typical models of uniaxial stress-strain laws.

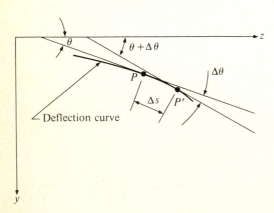

Figure 4.2 Beam-deflection curve.

be the change over the differential arc element ds. Since curvature is the first derivative of the slope with respect to distance along the member, that is

$$\phi = \frac{d\theta}{ds} \tag{4.1}$$

and since

$$\tan \theta = \frac{dy}{dz} \quad \text{or} \quad \theta = \tan^{-1} \frac{dy}{dz} \tag{4.2}$$

$$\phi = \frac{d\theta}{ds} = \frac{d}{ds}\left(\tan^{-1} \frac{dy}{dz}\right) = \frac{d}{dz}\left(\tan^{-1} \frac{dy}{dz}\right)\frac{dz}{ds}$$

$$= \frac{1}{1 + (dy/dz)^2} \frac{d}{dz} \frac{dy}{dz} \frac{dz}{ds}$$

$$\phi = \frac{1}{1 + (dy/dz)^2} \frac{d^2y}{dz^2} \frac{dz}{ds} \tag{4.3}$$

The element length ds can be related to the coordinate dimensions dz and dy by the relationship

$$ds = \sqrt{(dz)^2 + (dy)^2} = dz\sqrt{1 + (dy/dz)^2}$$

Therefore

$$\frac{dz}{ds} = \frac{1}{\sqrt{1 + (dy/dz)^2}} \tag{4.4}$$

Substituting eq. (4.4) into eq. (4.3) yields

$$\phi = \frac{d\theta}{ds} = \frac{d^2y/dz^2}{[1 + (dy/dz)^2]^{3/2}} = -\frac{M}{EI} \tag{4.5}$$

(It is to be noted that curvature has been equated to $-M/EI$, the same as was done in chap. 2.)

For the case where the slope of the beam is small,

$$\tan \theta \doteq \theta \doteq \frac{dy}{dz}$$

$$\left(\frac{dy}{dz}\right)^2 \ll 1.0$$

Equation (4.5) then reduces to

$$\frac{d^2y}{dz^2} = -\frac{M}{EI} \tag{4.6}$$

the value determined earlier.

To illustrate the relative importance of large deflections in beam bending, consider the end deflection of the cantilever beam shown in fig. 4.3. The member is

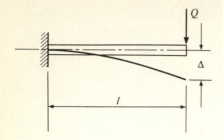

Figure 4.3 Cantilever beam subjected to concentrated end load.

presumed to be of uniform cross section throughout its length, and the material is linearly elastic and has a modulus of elasticity of E. Solving eq. (4.6), and taking into account the appropriate boundary conditions,

$$\Delta = \frac{Ql^3}{3EI} \tag{4.7}$$

Solving eq. (4.5)[†], the corresponding relationship would be

$$\Delta = \frac{2Ql^3}{6EI + 3Ql^2} \tag{4.8}$$

This can be written in the same form as eq. (4.7):

$$\Delta = \frac{Ql^3}{3EI} \frac{1}{1 + \frac{1}{2}(Ql^2/EI)} \tag{4.9}$$

Recalling from chap. 2 that if the section in question is symmetrical, the extreme fiber stress (normal stress) is given by the equation

$$\sigma = \frac{M}{I} \frac{d}{2}$$

where d is the depth of the section. At the fixed-support point where the moment is maximum, the stress also would be maximum. Therefore

$$\sigma_{max} = \frac{Ql}{I} \frac{d}{2} \qquad \text{or} \qquad Ql = \frac{\sigma_{max} I}{d/2}$$

This can be rewritten as

$$\frac{1}{2} \frac{Ql^2}{EI} = \frac{\sigma_{max}}{E} \frac{l}{d} \tag{4.10}$$

Substituting this into eq. (4.9) yields

$$\Delta = \frac{Ql^3}{3EI} \left[\frac{1}{1 + \sigma_{max} l / Ed} \right] \tag{4.11}$$

---

† The particular procedure that was used to determine this equation is described in R. Frisch-Fay, *Flexible Bars*, Butterworth & Co. (Publishers), Ltd., London, 1962.

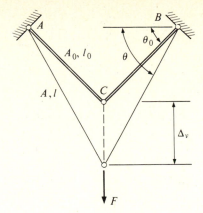

Figure 4.4a Large deformations of a two-member truss.

Since for most structural materials the ratio of σ_{max} to E is seldom greater than 10^{-3}, and since l/d is normally in the range of from 5 to 40, the influence of large displacements on this type of beam-bending problem can reasonably be ignored.

As a second example of large-deformation theory, consider the two-bar truss system shown in fig. 4.4a. Each of the bars has an undeformed cross-sectional area of A_0, and a length of l_0. The angle θ_0 describes the inclination with respect to the horizontal of each of the members. It is desired to determine the vertical deflection at point C, that is, Δ_v. (The assumption will be made that the stress-strain behavior of the material in question is linearly elastic.) As loads are applied, the geometry of the structure changes; the cross-sectional areas become smaller, and the angle θ_0 changes to θ. Taking into account these effects, the solution given in fig. 4.4b is obtained.† It is

† For a detailed discussion of the theory of finite deformations, see Y. C. Fung, *Foundations of Solid Mechanics*, Prentice-Hall, Inc., Englewood Cliffs, N.J., 1965.

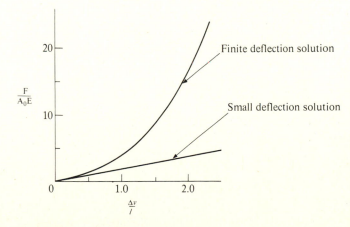

Figure 4.4b Load-deflection relationships.

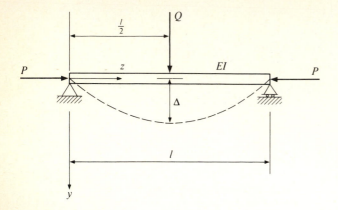

Fig. 4.5a Laterally loaded beam subjected to axial thrust.

to be noted that so long as the ratio Δ_v/l_0 is not large, say, 0.1 or 0.2, the influence of formulating the problem in the undeformed, as opposed to the deformed, state is not significant.

Geometrical nonlinearities From the above two examples it is evident that the use of large-deflection theory results in nonlinear load-deformational response. Under certain types of loading, however, even when small deformations are presumed, nonlinear behavior can be predicted.

Consider for illustration of this basic problem type the uniform cross-section member load as shown in fig. 4.5a. It is presumed that the member deflects in the plane of the applied transverse load Q. To establish the differential equation, the free body shown in fig. 4.5b will be selected. The known end reactions are $Q/2$ and P. The moment at the cut section is

$$M = \frac{Q}{2}z + Py$$

Therefore, since $d^2y/dz^2 = -M/EI$

$$EI\,\frac{d^2y}{dz^2} = -\left(\frac{Q}{2}z + Py\right)$$

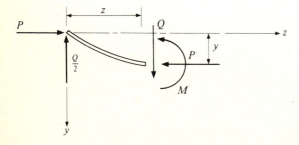

Figure 4.5b

or

$$\frac{d^2y}{dz^2} + \frac{P}{EI}y + \frac{Q}{2EI}z = 0 \tag{4.12}$$

Taking into account the appropriate boundary conditions—as will be demonstrated later in this chapter—the load versus centerline lateral-deflection relationship is

$$\Delta = \frac{Q}{2P\sqrt{P/EI}}\left(\tan\frac{l}{2}\sqrt{\frac{P}{EI}} - \frac{l}{2}\sqrt{\frac{P}{EI}}\right)$$

or

$$\Delta = \frac{Ql^3}{48EI}\left[\frac{3[\tan(l/2)\sqrt{P/EI} - (l/2)\sqrt{P/EI}]}{(l^3/8)(P/EI)^{3/2}}\right] \tag{4.13}$$

If the applied axial thrust P is held constant, there exists a linear relationship between Q and Δ. For the special case where $P = 0$, the relationship becomes

$$\Delta = \frac{Ql^3}{48EI} \tag{4.14}$$

Where P is not constant—independent of whether or not Q is held constant—the applied-load versus lateral-deflection relationship is nonlinear. This is shown in fig. 4.5c.

This class of problems, where overall deformations are observed to be small, but where equilibrium must be formulated using an assumed deformed position of the member, is often referred to as *problems involving geometrical nonlinearities*. As contrasted with *first-order* formulations, which were described in chap. 2 and also earlier in this chapter, these problems are said to result from *second-order* considerations.

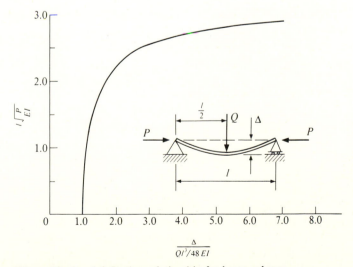

Figure 4.5c Load-deflection relationship for beam column.

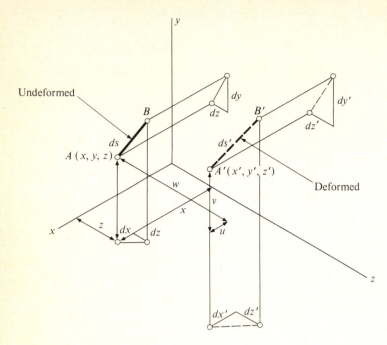

Figure 4.6 Unit deformational response.

Higher-order formulations—unit behavior It is to be recognized that the concept of second-order formulations is not one that applies only to overall behavior of members. *Unit response* also can be examined in this fashion. Consider, for example, the development of the basic relationships that must hold between strains and displacements in differential elements. Presume that a *ds* length element (which initially was located between points *A* and *B* as shown in fig. 4.6) assumes a totally new orientation due to the application of loading. Point $A(x, y, z)$ moves to a new location $A'(x', y', z')$, where $x' = x + u$, $y' = y + v$, and $z' = z + w$.

The length of the element in the undeformed state is given by

$$ds = [(dx)^2 + (dy)^2 + (dz)^2]^{1/2} \tag{4.15}$$

The deformed length $A'B'$ would be

$$ds' = [(dx')^2 + (dy')^2 + (dz')^2]^{1/2} \tag{4.16}$$

If the undeformed length *ds* and the deformed length *ds'* are the same, then the element is said to have undergone a *rigid body translation*. If, however, they differ, strain has occurred. A measure of this strain is the change in the length in *AB*, that is, $ds' - ds$. These are defined by eqs. (4.15) and (4.16) in the square root form. A measure of the strain could just as well be obtained, and with greater ease, by squaring both sides of both equations and taking the difference $(ds')^2 - (ds)^2$. Since by definition

$$x' = x + u \qquad y' = y + v \qquad \text{and} \qquad z' = z + w$$

$$(ds')^2 = (du)^2 + 2du\ dx + (dx)^2 + (dv)^2 + 2dv\ dy + (dy)^2$$
$$+ (dw)^2 + 2dw\ dz + (dz)^2$$

and

$$(ds')^2 - (ds)^2 = 2(du\ dx + dv\ dy + dw\ dz)$$
$$+ (du)^2 + (dv)^2 + (dw)^2 \qquad (4.17)$$

the total differentials du, dv, and dw can be expressed as follows:

$$du = \frac{\partial u}{\partial x}\ dx + \frac{\partial u}{\partial y}\ dy + \frac{\partial u}{\partial z}\ dz$$

$$dv = \frac{\partial v}{\partial x}\ dx + \frac{\partial v}{\partial y}\ dy + \frac{\partial v}{\partial z}\ dz$$

$$dw = \frac{\partial w}{\partial x}\ dx + \frac{\partial w}{\partial y}\ dy + \frac{\partial w}{\partial z}\ dz$$

Substituting these into eq. (4.17) yields

$$
\begin{aligned}
(ds')^2 - (ds)^2 = {} & 2\left\{\frac{\partial u}{\partial x} + \frac{1}{2}\left[\left(\frac{\partial u}{\partial x}\right)^2 + \left(\frac{\partial v}{\partial x}\right)^2 + \left(\frac{\partial w}{\partial x}\right)^2\right]\right\}(dx)^2 \\
& + 2\left\{\frac{\partial v}{\partial y} + \frac{1}{2}\left[\left(\frac{\partial u}{\partial y}\right)^2 + \left(\frac{\partial v}{\partial y}\right)^2 + \left(\frac{\partial w}{\partial y}\right)^2\right]\right\}(dy)^2 \\
& + 2\left\{\frac{\partial w}{\partial z} + \frac{1}{2}\left[\left(\frac{\partial u}{\partial z}\right)^2 + \left(\frac{\partial v}{\partial z}\right)^2 + \left(\frac{\partial w}{\partial z}\right)^2\right]\right\}(dz)^2 \\
& + 2\left\{\frac{\partial v}{\partial x} + \frac{\partial u}{\partial y} + \frac{\partial u}{\partial x}\frac{\partial u}{\partial y} + \frac{\partial v}{\partial x}\frac{\partial v}{\partial y} + \frac{\partial w}{\partial x}\frac{\partial w}{\partial y}\right\}dx\ dy \\
& + 2\left\{\frac{\partial w}{\partial x} + \frac{\partial u}{\partial z} + \frac{\partial u}{\partial x}\frac{\partial u}{\partial z} + \frac{\partial v}{\partial x}\frac{\partial v}{\partial z} + \frac{\partial w}{\partial x}\frac{\partial w}{\partial z}\right\}dx\ dz \\
& + 2\left\{\frac{\partial w}{\partial y} + \frac{\partial v}{\partial z} + \frac{\partial u}{\partial y}\frac{\partial u}{\partial z} + \frac{\partial v}{\partial y}\frac{\partial v}{\partial z} + \frac{\partial w}{\partial y}\frac{\partial w}{\partial z}\right\}dy\ dz \qquad (4.18)
\end{aligned}
$$

or

$$\tfrac{1}{2}[(ds')^2 - (ds)^2] = \varepsilon_x(dx)^2 + \varepsilon_y(dy)^2 + \varepsilon_z(dz)^2 +$$
$$+ \gamma_{xy}\ dx\ dy + \gamma_{xz}\ dx\ dz + \gamma_{yz}\ dy\ dz \qquad (4.19)$$

where ε_x, ε_y, and ε_z are normal (extensional) strains. γ_{xy}, γ_{xz}, and γ_{yz} are shearing strains.

The strain-deformational relationships are the following:

$$\varepsilon_x = \frac{\partial u}{\partial x} + \frac{1}{2}\left[\left(\frac{\partial u}{\partial x}\right)^2 + \left(\frac{\partial v}{\partial x}\right)^2 + \left(\frac{\partial w}{\partial x}\right)^2\right]$$

$$\varepsilon_y = \frac{\partial v}{\partial y} + \frac{1}{2}\left[\left(\frac{\partial u}{\partial y}\right)^2 + \left(\frac{\partial v}{\partial y}\right)^2 + \left(\frac{\partial w}{\partial y}\right)^2\right]$$

$$\varepsilon_z = \frac{\partial w}{\partial z} + \frac{1}{2}\left[\left(\frac{\partial u}{\partial z}\right)^2 + \left(\frac{\partial v}{\partial z}\right)^2 + \left(\frac{\partial w}{\partial z}\right)^2\right] \tag{4.20}$$

$$\gamma_{xy} = \frac{\partial v}{\partial x} + \frac{\partial u}{\partial y} + \frac{\partial u}{\partial x}\frac{\partial u}{\partial y} + \frac{\partial v}{\partial x}\frac{\partial v}{\partial y} + \frac{\partial w}{\partial x}\frac{\partial w}{\partial y}$$

$$\gamma_{xz} = \frac{\partial w}{\partial x} + \frac{\partial u}{\partial z} + \frac{\partial u}{\partial x}\frac{\partial u}{\partial z} + \frac{\partial v}{\partial x}\frac{\partial v}{\partial z} + \frac{\partial w}{\partial x}\frac{\partial w}{\partial z}$$

$$\gamma_{yz} = \frac{\partial w}{\partial y} + \frac{\partial v}{\partial z} + \frac{\partial u}{\partial y}\frac{\partial u}{\partial z} + \frac{\partial v}{\partial y}\frac{\partial v}{\partial z} + \frac{\partial w}{\partial y}\frac{\partial w}{\partial z}$$

For a two-dimensional case, it can be assumed that the *derivatives of the* deformation in the third direction need not be considered. For example, in the *yz* plane, the appropriate equations would be

$$\varepsilon_y = \frac{\partial v}{\partial y} + \frac{1}{2}\left[\left(\frac{\partial v}{\partial y}\right)^2 + \left(\frac{\partial w}{\partial y}\right)^2\right]$$

$$\varepsilon_z = \frac{\partial w}{\partial z} + \frac{1}{2}\left[\left(\frac{\partial v}{\partial z}\right)^2 + \left(\frac{\partial w}{\partial z}\right)^2\right] \tag{4.21}$$

$$\gamma_{yz} = \frac{\partial w}{\partial y} + \frac{\partial v}{\partial z} + \frac{\partial v}{\partial y}\frac{\partial v}{\partial z} + \frac{\partial w}{\partial y}\frac{\partial w}{\partial z}$$

For a one-dimensional case,

$$\varepsilon_z = \frac{\partial w}{\partial z} + \frac{1}{2}\left(\frac{\partial w}{\partial z}\right)^2 \tag{4.22}$$

If it now is assumed that the displacements *u*, *v*, and *w* are small with respect to the size of the body in question, and further that all products of partial derivatives are negligibly small, the six-equation set becomes

$$\varepsilon_x = \frac{\partial u}{\partial x} \qquad \varepsilon_y = \frac{\partial v}{\partial y} \qquad \varepsilon_z = \frac{\partial w}{\partial z}$$

$$\gamma_{xy} = \frac{\partial v}{\partial x} + \frac{\partial u}{\partial y} \qquad \gamma_{xz} = \frac{\partial w}{\partial x} + \frac{\partial u}{\partial z} \qquad \gamma_{yz} = \frac{\partial w}{\partial y} + \frac{\partial v}{\partial z} \tag{4.23}$$

These equations describe the strain-displacement relationships for *small-displacement theory*. It is to be noted that under this theory normal strain ε_i depends only upon the displacement in the *i*th direction. Shearing strains γ_{ij} involve only the *i*th and *j*th planes. When considering large displacements the cross-term effects become important and must be included.

4.2 DEFORMATIONAL RESPONSE OF UNIT LENGTH ELEMENTS TO INDIVIDUAL STRESS RESULTANTS

Problems concerning the prediction of deflections generally occur in one of two forms: a loading is specified, and deflections throughout the structure are desired, or deflections at a given location are sought for a variety of different loading possibilities. This second category is particularly important when solving problems using *flexibility methods*. Generally speaking, problems concerned with the definition of particular deformations are more readily solved using energy methods (to be described in chap. 5). When the entire deflected shape is desired, however, solution of the appropriate differential equation(s) would have its advantages. (It must be understood, of course, that a differential equation solution can be used to determine deformations at a particular location.)

Individual structural members may be subjected to single stress resultants or combinations of them. If a first-order formulation is used, superposition of the individual effects can be assumed. When the problem is nonlinear, this cannot be done, and all stress resultant components must be handled simultaneously.

Four basically different types of stress resultants must be considered: axial force, bending moment, shearing force, and twisting moment. These are illustrated in figs. 4.7 to 4.10. In each of the cases shown, it has been assumed that deformation of the member can occur only within the indicated dz length. The remaining portions are rigid. The total deformation of a real member, then, could be obtained by integrating the derived unit response over the entire length of the member.

For two-dimensional structures composed of slender structural members of the type normally encountered in civil engineering construction, it has been observed that deformations due to the following individual and combination of stress resultants are the more important ones to examine:

1. Deformations due to bending (such as in beams)
2. Deformations due to combined bending and axial thrust (such as in beam columns, columns in frames, etc.)
3. Deformations in trusses due to the combination of axial forces

Shearing deformations for slender members are usually insignificant, as will be subsequently demonstrated.

The assumptions that will be made in the remaining parts of this section are (a) first-order formulation, (b) the unit material response to imposed loading is linearly elastic, (c) the material is both isotropic and homogeneous, (d) the forces and deformations increase gradually, and (e) the deflections and deformations are small. Emphasis will be directed toward the solution of two-dimensional problems.

Deformations due to axial forces If it is assumed that the applied force acts along a z axis that is coincident with the centroid of the cross section, and further that a linear distribution of strains across a section occurs, then the case shown in fig. 4.7 is realized. The differential length dz increases an amount $d\delta$, where

$$d\delta = \varepsilon \, dz$$

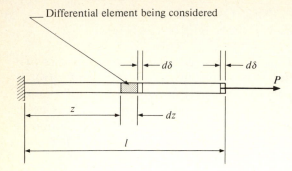

Figure 4.7 Structural member subjected to axial tension.

ε is the strain which is presumed to be uniform across the cross section and can be related to the applied force

$$\varepsilon = \frac{\sigma}{E} = \frac{P}{AE}$$

Therefore, the axial elongation of the member due to the unit element response is

$$d\delta = \frac{P}{AE}\, dz \tag{4.24}$$

where A is the cross-sectional area, and E is Young's modulus of elasticity. Assuming an elastically deformable material over the entire length

$$\delta = \int_0^l d\delta = \int_0^l \frac{P}{AE}\, dz \tag{4.25}$$

For a prismatic member subjected to a constant axial force

$$\delta = \frac{Pl}{AE} \tag{4.26}$$

It should be recognized that the geometrically defined differential relationship, $d\delta = \varepsilon\, ds$, could just as well have been obtained using the third of the normal strain equations developed earlier and given as eq. (4.23):

$$\varepsilon_z = \frac{\partial w}{\partial z}$$

Deformations due to bending moments The governing differential equation is given as eq. (4.6). It can be derived, however, from first principles: Presume that the beam in question is composed of longitudinal fibers each of which responds individually to its applied stress. The shortening over the element length dz of the particular fiber that is located y below the neutral axis is

$$d\delta = \varepsilon\, dz$$

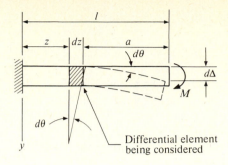

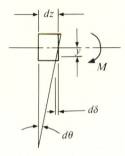

Figure 4.8 Structural member subjected to bending moment.

as was shown in the preceding example. From chap. 2, however, the stress at that location is

$$\sigma = \frac{My}{I}$$

Since $\sigma = E\varepsilon$ (or $\varepsilon = \sigma/E$),

$$d\delta = \frac{My}{EI}\,dz$$

Using the sign convention of chap. 2 (tension positive), it can be seen from the geometry of fig. 4.8 that

$$\frac{-d\delta}{y} = d\theta$$

Therefore

$$d\theta = \frac{-M}{EI}\,dz \tag{4.27}$$

The deflection of the span at the free end, due to the deformational response of the element, would be

$$d\Delta = a\,d\theta$$

where $a = l - z$. Therefore

$$\Delta = \int_0^l (l - z)\, d\theta = \int_0^l -\frac{M}{EI}(l - z)\, dz \tag{4.28}$$

Depending on the particular loading, M may or may not be a function of z. If the beam is prismatic, I will be a constant. If not, its variation with z will have to be described. If a uniform material is used throughout the length, E is constant. (The term EI is frequently referred to as the *bending rigidity* or bending stiffness of the section, and is an important quantity when examining stability and natural frequencies of vibration.)

Deformations due to shearing forces The shearing stress-strain relationship is given by the equation

$$\tau = \gamma G$$

where G is the *shearing modulus of elasticity*. If the unit deformational response is defined by (see fig. 4.9)

$$d\Delta = \gamma_{max}\, dz$$

then

$$d\Delta = \frac{\tau_{max}}{G}\, dz$$

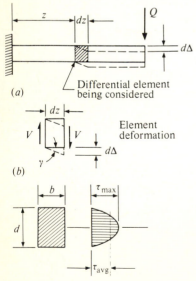

(a)

Differential element being considered

(b)

Element deformation

(c)

Cross section   Shearing stress

Figure 4.9 Structural member subjected to transverse shear.

But from eq. (2.23)

$$\tau_{max} = \frac{VQ_{max}}{Ib} = \frac{3}{2}\frac{V}{A} = \frac{3}{2}\tau_{avg}$$

or

$$d\Delta = \frac{3}{2}\frac{V}{AG}dz \tag{4.29}$$

If the entire span is elastic

$$\Delta = \frac{3}{2}\int_{0}^{l}\frac{V}{AG}dz \tag{4.30}$$

AG is defined as the *shear rigidity* of the member.

(It is to be recognized that the maximum shearing stress, which occurs at the neutral axis of the cross section, is related to the average differently depending upon the specific shape of the cross section. For the rectangular shape $\tau_{max} = 1.5\tau_{avg}$. For the circular shape, $\tau_{max} \doteq 1.33_{avg}$. For the wide-flange or I shape, $\tau_{max} \doteq 1.0\tau_{avg}$. Frequently, these are written as $\tau_{max} = \lambda_v \tau_{avg}$.)

Deformations (of circular sections) due to twisting moments In the preceding discussions concerning bending and shear, it was assumed that the cross section had at least one axis of symmetry. Moreover, it was further presumed that the applied loads and the resulting deformations acted within that plane. In both cases the cross-sectional form strongly influenced the behavior of the member.

As shown in chap. 2, members subjected to twisting moments resist deformation in two distinct and separate ways: by uniform torsion (St. Venant type torsion), and by warping of the cross section. Certain cross-sectional forms resist the applied twisting moment primarily by warping, while others realize little stiffness from this source. The circular cross section resists twisting moment entirely by uniform

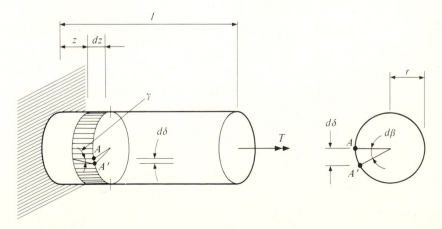

Figure 4.10 Circular cross-section member subjected to twisting moment.

torsion. Since a later section of this chapter will be devoted entirely to a discussion of torsional rigidities of sections which warp, discussion (and derivations) at this stage will be restricted to a consideration of members of circular cross section.

Again, a differential length element will be presumed elastic, with the remainder of the member rigid (see fig. 4.10). Due to the applied torque, point A moves to A' and

$$d\delta = \gamma \, dz$$

where γ is the shearing strain on the outside surface of the cylindrical member. Shearing stress on this surface (from chap. 2) equals

$$\tau_{max} = \frac{Tr}{\kappa_T}$$

Therefore, using the stress-strain relationship $\gamma = \tau/G$,

$$d\delta = \frac{\tau}{G} \, dz = \frac{Tr}{G\kappa_T} \, dz$$

The change in angle at the end of the member due to this rotation of the unit element is

$$d\beta = \frac{d\delta}{r} = \frac{T}{G\kappa_T} \, dz \tag{4.31}$$

The total angle of twist then becomes

$$\beta = \int_0^l \frac{T}{G\kappa_T} \, dz \tag{4.32}$$

Deformations due to distributed stress resultants If the applied loads are distributed rather than concentrated as presumed above, the following integral and differential relationships are to be used.

For distributed axial thrust (fig. 4.11a):

$$P = \int_0^z p \, dz \qquad \text{or} \qquad \frac{dP}{dz} = p$$

For distributed transverse loads—in shear (fig. 4.11b):

$$V = \int_0^z q \, dz \qquad \text{or} \qquad \frac{dV}{dz} = q$$

For distributed transverse loads—in bending (fig. 4.11c):

$$M = \int_0^z V \, dz = \iint q \, dz$$

or

$$\frac{d^2M}{dz^2} = \frac{dV}{dz} = q$$

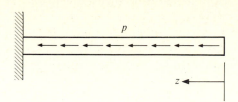

Figure 4.11a Structural members subjected to distributed loads.

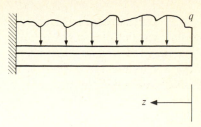

Figure 4.11b Structural members subjected to distributed loads.

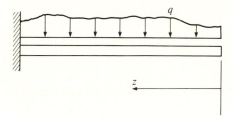

Figure 4.11c Structural members subjected to distributed loads.

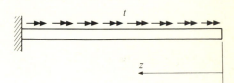

Figure 4.11d Structural members subjected to distributed loads.

For distributed twisting moments t (fig. 4.11d):

$$T = \int_0^z t \, dz \qquad \text{or} \qquad \frac{dT}{dz} = t$$

It is to be understood that p, q, and t can be constants or functions of z.

4.3 BENDING DEFORMATIONS

Within the assumptions of elementary beam theory, the deformed shape of a member subjected to lateral loads and bending moments can be described in terms of the location of its centroidal axis. This is usually referred to as the elastic curve of the deformed beam (fig. 4.12).

It is assumed that an elastic curve, $y = f(z)$, is continuous in the interval $a < z < b$, the length of the member, and that its derivatives exist. From eqs. (2.16), (2.17), (4.6), and (4.17) and the use of elementary geometrical principles, the governing differential equations for beam bending are those listed in table 4.1. The sign convention assumes plus loads and deformations in the direction of plus coordinate axes. Slopes which increase y with z are plus, as are curvatures which increase θ with z.

If a particular beam and lateral load are specified, it should be evident from table 4.1 that integration of $q(z)$ four times, considering the appropriate boundary conditions, results in an expression for the deflected shape of the member. This sequence of

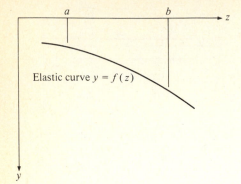

Figure 4.12

steps for a member subjected to a uniformly distributed lateral loading is illustrated in fig. 4.13.

When concentrated lateral loads are applied to the beam, shears are readily defined. For these cases, three integrations are sufficient to determine y. Similarly, for applied moments only two integrations need be carried out.

In fig. 4.13 it was assumed that the cross section remained constant over the length of the beam. Had EI varied, the relative values of ϕ, θ, and y would have been somewhat different from those shown. The values for q, V, and M, however, would not have been affected. (This later statement is true for statically determinate structures, but will not be the case for indeterminate ones.)

Integration results in the introduction of arbitrary constants, one for each integration. These are evaluated from a consideration of the boundary conditions at the

Table 4.1 Governing differential equations for beam bending

Distributed load	$q(z) = -\dfrac{dV}{dz} = EI\,\dfrac{d^4y}{dz^4}$
Shear	$V(z) = \dfrac{dM}{dz} = -EI\,\dfrac{d^3y}{dz^3}$
Moment	$M(z) = -EI\,\dfrac{d^2y}{dz^2}$
Curvature†	$\dfrac{d^2y}{dz^2} = \dfrac{d\theta}{dz} = \phi(z)$
Slope	$\dfrac{dy}{dz} = \theta(z)$
Displacement	$y = \Delta(z)$

† Note: $\phi = -\dfrac{M}{EI}$

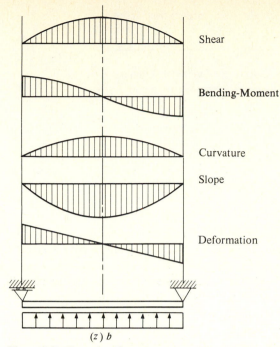

Shear

Bending-Moment

Curvature

Slope

Deformation

$(z) b$

Figure 4.13 Shear, bending-moment, curvature, slope, and deformation diagrams for a beam subjected to a uniformly distributed lateral load.

ends of the length intervals in question. Since most beam-bending problems are formulated either at the fourth derivative (lateral load) or second derivative (moment) stage, boundary conditions associated with the various supports must be specified for these two cases. These are listed in tables 4.2 and 4.3. To illustrate how these elementary boundary situations are used in "real" structural problems consider the following several examples.

Example 4.1 The beam ABC is composed of two different size members rigidly connected at their point of attachment. At A there is a simple support; at B, a simple support; and at C, a free end. The entire length is subjected to a uniformly distributed lateral load q. At free end C, there also is applied a concentrated force Q. Presuming solution from a fourth-order formulation, indicate all boundary conditions required to define the deformed shape of the beam. The structure and loading are shown in fig. 4.14.

Each of the spans AB and BC require the specification of four boundary conditions. A total of eight boundary and continuity conditions must therefore be defined. These are given in fig. 4.14.

Table 4.2 Boundary and continuity conditions: Integration from the fourth derivative

Pictorial representation	Mathematical expressions†	Verbal descriptions
Simple support	$y = 0$ $EIy'' = 0$ (or just $y'' = 0$)	No deflection No moment
Fixed support	$y = 0$ $y' = 0$	No deflection No slope
Free end	$EIy'' = 0$ (or $y'' = 0$) $EIy''' = 0$ (or $y''' = 0$)	No moment No shear
Interior hinge Hinge L ← → R	$y_L = y_R$ $y_L'' = 0$ $y_R'' = 0$ $(EI)_L y_L''' = (EI)_R y_R'''$	Deflections equal No moment No moment Shear forces equal
Interior roller support L ← → R	$y_L = 0$ $y_R = 0$ $y_L' = y_R'$ $(EI)_L y_L'' = (EI)_R y_R''$	No deflection No deflection Slopes equal Moments equal
Interior load point Q L ← → R	$y_L = y_R$ $y_L' = y_R'$ $(EI)_L y_L'' = (EI)_R y_R''$ $(EI)_L y_L''' - (EI)_R y_R''' = Q$	Deflections equal Slopes equal Moments equal Total shear $= Q$

† $y' = \dfrac{dy}{dz}$, $y'' = \dfrac{d^2y}{dz^2}$, $y''' = \dfrac{d^3y}{dz^3}$

Table 4.3 Boundary and continuity conditions: Integration from the second derivative

Pictorial representation	Mathematical expressions	Verbal descriptions
Simple support	$y = 0$	No deflection
Fixed support	$y = 0$ $y' = 0$	No deflection No slope
Free end	No meaningful boundary conditions can be specified	
Interior hinge	$y_L = y_R$	Deflections equal
Interior roller support	$y_L = y_R = 0$ $y'_L = y'_R$	No deflections Slopes equal
Interior load point	$y_L = y_R$ $y'_L = y'_R$	Deflections equal Slopes equal

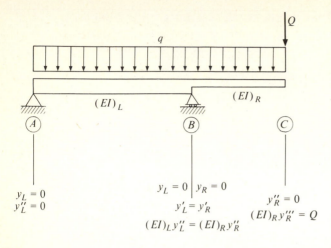

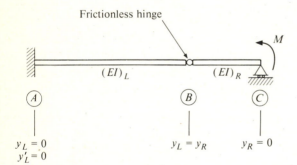

$y_L = 0$
$y_L'' = 0$

$y_L = 0 \mid y_R = 0$
$y_L' = y_R'$
$(EI)_L y_L'' = (EI)_R y_R''$

$y_R'' = 0$
$(EI)_R y_R''' = Q$

Figure 4.14 Boundary and continuity conditions.

Example 4.2 The span ABC shown in fig. 4.15 contains an interior frictionless hinge at point B. Presuming a second-order differential equation formulation, define the appropriate boundary and continuity conditions.

Again, as in example 4.1, it is assumed that two separate spans are involved. Since a second-order equation is integrated for each span, four distinct and separate boundary conditions must be established. These are shown in fig. 4.15.

$y_L = 0$
$y_L' = 0$

$y_L = y_R$

$y_R = 0$

Figure 4.15 Boundary and continuity conditions.

Example 4.3 Presuming a fourth-order formulation for the structure and loading shown in fig. 4.16, the sixteen required boundary and continuity conditions are those listed.

From these examples it should be clear that boundary or continuity conditions are introduced at

1. All interior support points
2. The ends of beams
3. All interior-hinge locations
4. All points where there is an abrupt change in cross-sectional shape
5. All points of discontinuity in the loading function

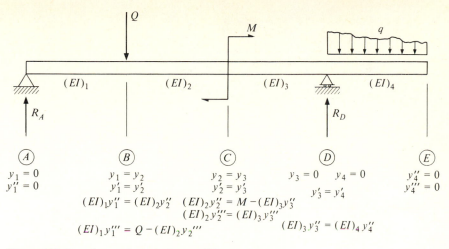

Figure 4.16 Boundary and continuity conditions.

When relatively complicated loadings are imposed, as was the case in fig. 4.16, the number of integration constants to be evaluated becomes large. The amount of work required to solve such a problem increases appreciably. Fortunately, most of the computational difficulties are those associated with solving the simultaneous equations that result—a task ideally suited to a digital computer. In many complicated loading situations successive integration of singularity functions† also can be used to advantage.

The order of the appropriate boundary or continuity conditions must always be less than the order of the differential equation governing the solution. When using a second-order formulation, boundary and continuity conditions can be defined only in terms of the function (that is, y), and its first derivative. Similarly, should integration proceed from a third-order equation, the appropriate conditions for the evaluation of the constants of integration can include only second derivatives, first derivatives, and the function itself. If a fourth-order equation is used, the allowable equations can include y, y', y'', and y'''. In those special cases where it is possible to prescribe more than the required number of boundary conditions, the equations containing the lowest-order derivatives take precedence.

4.3.1 Method of Direct Integration

Example 4.4 As an illustration of the direct integration method, consider the prismatic (uniform cross section) beams shown in fig. 4.17. This simply supported member is subjected to a single concentrated lateral load of Q. It is

---

† See, for example, S. H. Crandall, N. C. Dahl, and T. J. Lardner, *An Introduction to the Mechanics of Solids*, 2d ed., p. 164, McGraw-Hill Book Company, New York, 1972.

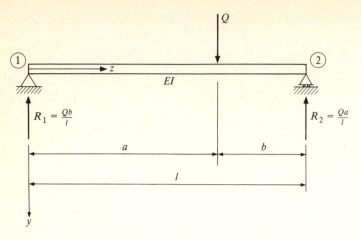

Figure 4.17 Simple beam subjected to a concentrated load.

desired to determine the equation of the elastic curve, and from it the maximum deflection of the beam.

Boundary conditions

At $z = 0$ $y = 0$

At $z = l$ $y = 0$

For the range $0 \le z \le a$, the bending moment can be determined from the free body diagram shown in fig. 4.18. Summing moments about the cut section results in

$$+M - \frac{Qb}{l}z = 0$$

or

$$M = \frac{Qb}{l}z$$

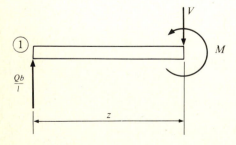

Figure 4.18

Integrating from the second-order differential equation $d^2y/dz^2 = -M/EI$ gives the following:

$$EI\,\frac{d^2y}{dz^2} = -\frac{Qb}{l}\,z \qquad \text{for } 0 \le z \le a$$

$$EI\,\frac{dy}{dz} = -\frac{Qbz^2}{2l} + C_1 \tag{4.33}$$

$$EI(y) = -\frac{Qbz^3}{6l} + C_1 z + C_2$$

From the boundary conditions $y = 0$ at $z = 0$, $C_2 = 0$.

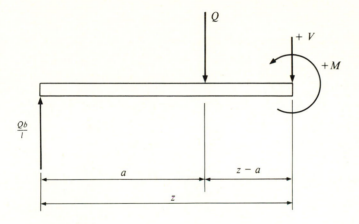

Figure 4.19

For the range $a \le z \le l$, the free body diagram would be that shown in fig. 4.19. For equilibrium of moments

$$+M - \frac{Qb}{l}\,z + Q(z - a) = 0$$

Therefore

$$EI\,\frac{d^2y}{dz^2} = Q(z - a) - \frac{Qb}{l}\,z \qquad \text{for } a \le z \le l$$

$$EI\,\frac{dy}{dz} = Q\,\frac{z^2}{2} - Qaz - \frac{Qbz^2}{2l} + C_3 \tag{4.34}$$

$$EI(y) = \frac{Qz^3}{6} - \frac{Qaz^2}{2} - \frac{Qbz^3}{6l} + C_3 z + C_4$$

The boundary condition at $z = l$ yields the relationship

$$C_4 = -\tfrac{1}{3}Qal^2 - C_3 l$$

In addition, the slopes and the deflections at the load point must be equal—whether integration is from the left or to the right of Q.

	$0 \leq z \leq a$	$0 \leq z \leq l$
$EI\dfrac{dy}{dz}$	$-\dfrac{Qbz^2}{2l} + C_1$	$\dfrac{Qz^2}{2} - Qaz - \dfrac{Qbz^2}{2l} + C_3$
$EI(y)$	$-\dfrac{Qbz^3}{6l} + C_1 z$	$\dfrac{Qz^3}{6} - \dfrac{Qaz^2}{2} - \dfrac{Qbz^3}{6l} + C_3 z + C_4$

Equating the slopes at $z = a$ yields

$$C_1 = C_3 - \frac{Qa^2}{2}$$

For the deflections

$$C_1 = -\frac{Qa^2}{3} + C_3 - \frac{1}{a}C_4$$

Eliminating C_1 between the two of these equations gives

$$C_4 = -\tfrac{1}{6}Qa^3$$

In turn, back substitution yields

$$C_3 = \frac{1}{3}Qal + \frac{1}{6}Q\frac{a^3}{l}$$

and

$$C_1 = \frac{1}{3}Qal + \frac{1}{6}\frac{Qa^3}{l} - \frac{1}{2}Qa^2$$

The equations of the elastic curve are therefore

$$EI(y) = -\frac{Qbz^3}{6l} + \left(\frac{1}{3}Qal + \frac{1}{6}Q\frac{a^3}{l} - \frac{1}{2}Qa^2\right)z \qquad 0 \leq z \leq a$$

$$EI(y) = \frac{Qz^3}{6}\frac{a}{l} - \frac{Qaz^2}{2} + \left(\frac{Qal}{3} - \frac{Qa^3}{6l}\right)z + \frac{Qa^3}{6} \qquad a \leq z \leq l$$

(4.35)

If it is assumed that $a \geq \frac{1}{2}l$, the maximum deflection will occur to the left of the applied load. The exact location of the maximum deflection can be determined by maximizing the deflection relationship.

$$\frac{d}{dz}[EI(y)] = 0$$

or

$$-\frac{Qbz^2}{2l} + \frac{Qal}{3} + \frac{Qa^3}{6l} - \frac{Qa^2}{2} = 0 \qquad (4.36)$$

The distance to the point of maximum deflection is therefore

$$z_{cr} = \sqrt{\frac{1}{3}\frac{a}{b}(2l^2 - 3al + a^2)} \qquad (4.37)$$

But since $b = l - a$, z_{cr} can be written in terms of a and l as

$$z_{cr} = \sqrt{\tfrac{1}{3}a(2l - a)}$$

This value substituted into the equation of the elastic curve [the first of eqs. (4.35)] gives the maximum deflection. For example, if the lateral load were applied at the midpoint of the span (that is, $a = b = l/2$),

$$z_{cr} = \sqrt{\frac{1}{3}\left(\frac{l}{2}\right)\left(2l - \frac{l}{2}\right)} = \frac{l}{2}$$

and

$$y_{max} = +\frac{Ql^3}{48EI} \qquad (4.38)$$

Example 4.5 As another example, consider the determination of the rotation of the rigid joint 2 of the structure shown in fig. 4.20. The maximum horizontal deflection of the column also is desired.

Considering the entire structure, the components of reaction can be determined from equilibrium considerations. These are

$$R_1^H = qa \quad \leftarrow$$
$$R_1^V = 0$$
$$R_3^V = qa \quad \uparrow$$

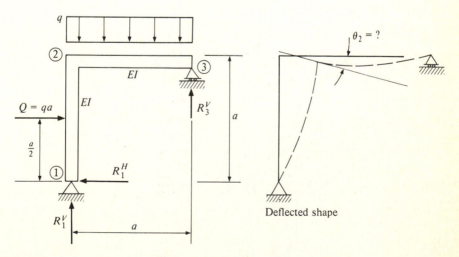

Figure 4.20

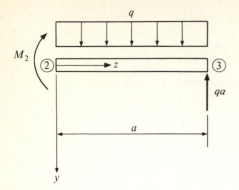

Figure 4.21

The moment at the joint in question (that is, M_2) is defined from a consideration of the equilibrium of the top beam. Considering the free body shown in fig. 4.21,

$$M_2 = \frac{qa^2}{2}$$

(It is to be noted that the origin of the coordinate axes for this member 2–3 is assumed located at point 2. In all likelihood, point 2 will move to the right—because of the horizontal load on the column. The location of the reference at point 2 provides a relative reference that is not affected by the horizontal translation. This added z value is the same horizontal displacement of the top of the column, and would be determined after all the boundary conditions had been satisfied.)

Proceeding as before, the free body diagram shown in fig. 4.22 is selected.

$$M - \frac{qa^2}{2} + q\frac{z^2}{2} = 0$$

or

$$M = q\frac{a^2}{2} - q\frac{z^2}{2} \qquad (4.39)$$

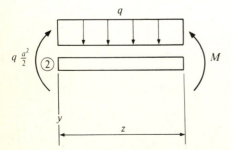

Figure 4.22

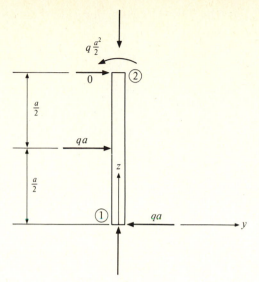

Figure 4.23

Therefore

$$EI\left(\frac{d^2y}{dz^2}\right)_B = -\frac{qa^2}{2} + \frac{qz^2}{2}$$

$$EI\left(\frac{dy}{dz}\right)_B = -\frac{qa^2z}{2} + \frac{qz^3}{6} + C_1$$

$$EI(y)_B = -\frac{qa^2z^2}{4} + \frac{qz^4}{24} + C_1z + C_2 \tag{4.40}$$

Assuming that the column does not shorten, at $z = 0$, $y = 0$, or $C_2 = 0$. At $z = a$, the boundary condition requires that $y = 0$. This gives

$$C_1a = \tfrac{5}{24}qa^4 \qquad \text{or} \qquad C_1 = \tfrac{5}{24}qa^3 \tag{4.41}$$

(Note that this also is the value of the slope at point 2, referred to the local coordinate system for member 2-3.)

For column 1-2, between the point 1 and the concentrated lateral load, and presuming a new local coordinate system, as shown in fig. 4.23, equilibrium is described by the following:

$$EI\left(\frac{d^2y}{dz^2}\right)_C = -qaz \qquad 0 \le z \le \frac{a}{2}$$

$$EI\left(\frac{dy}{dz}\right)_C = -\frac{qaz^2}{2} + C_3$$

$$EI(y)_C = -\frac{qaz^3}{6} + C_3z + C_4 \tag{4.42}$$

At $z = 0$, $y = 0$, and therefore $C_4 = 0$. The slope and deflection at the load point are

$$EI\left(\frac{dy}{dz}\right)_{C,\,z=a/2} = -\frac{qa^3}{8} + C_3$$

$$EI(y)_{C,\,z=a/2} = -\frac{qa^4}{48} + \frac{C_3 a}{2} \tag{4.43}$$

For the upper half of the column, the equations would be

$$EI\left(\frac{d^2y}{dz^2}\right)_C = -\frac{qa^2}{2} \qquad \frac{a}{2} \le z \le a$$

$$EI\left(\frac{dy}{dz}\right)_C = -\frac{qa^2}{2}z + C_5$$

$$EI(y)_C = -\frac{qa^2z^2}{4} + C_5 z + C_6 \tag{4.44}$$

At the lateral load point $(z_{\text{local}} = \tfrac{1}{2}a)$, the slope and deflection are

$$EI\left(\frac{dy}{dz}\right)_{C,\,z=a/2} = -\frac{qa^3}{4} + C_5$$

$$EI(y)_{C,\,z=a/2} = -\frac{qa^4}{16} + C_5\frac{a}{2} + C_6 \tag{4.45}$$

Equating these to the values obtained from the integration for the lower half of the column yields

$$-\frac{qa^3}{8} + C_3 = -\frac{qa^3}{4} + C_5 \tag{4.46}$$

and

$$-\frac{qa^4}{48} + C_3\frac{a}{2} = -\frac{qa^4}{16} + C_5\frac{a}{2} + C_6$$

If these are solved simultaneously

$$C_6 = -\tfrac{1}{48}qa^4$$

At the top of the column, the slope and deflection are given by the equations

$$EI\left(\frac{dy}{dz}\right)_{C,\,z=a} = -\frac{qa^3}{2} + C_5$$

$$EI(y)_{C,\,z=a} = -\tfrac{1}{4}qa^4 + C_5 a + C_6 \tag{4.47}$$

The slope of the top beam at the same point (that is, at point 2) was given by the equation

$$EI\left(\frac{dy}{dz}\right)_{B,\,z=0} = C_1 = \frac{5}{24}qa^3$$

Therefore

$$-\frac{qa^3}{2} + C_5 = \frac{5}{24}qa^3$$

or

$$C_5 = \tfrac{17}{24}qa^3 \tag{4.48}$$

For completeness, it also should be noted that substitution of this value into eq. (4.46) yields

$$C_3 = \tfrac{14}{24}qa^3 \tag{4.49}$$

With these constants of integration, it now is possible to determine the point of maximum horizontal deflection of the column.

$$\frac{d}{dz}[EI(y)_c] = 0 = -\frac{qa^2z}{2} + \frac{17qa^3}{24}$$

or

$$z_{cr} = \tfrac{17}{12}a \tag{4.50}$$

This point is not within the length of the actual column. The maximum deflection must therefore occur at the end of the member, at $z = a$. The maximum value of y is therefore

$$EI(y)_{C,\,z=a} = -\frac{qa^4}{4} + \frac{17qa^4}{24} - \frac{qa^4}{48} = \frac{21}{48}qa^4$$

or

$$y_{C,\,z=a} = \frac{21}{48}\frac{qa^4}{EI} \tag{4.51}$$

Example 4.6 The elastic curve of a nonprismatic beam is determined in exactly the same way. The tapered beam shown in fig. 4.24 is an example of such a member. As shown, the cross section is rectangular having a constant width of b,

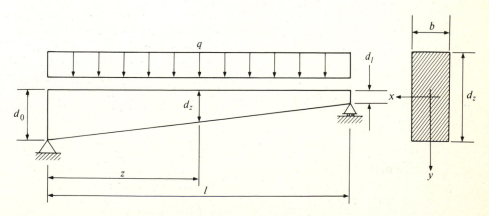

Figure 4.24 Linearly tapered rectangular beam subjected to a uniformly distributed lateral load.

and a linearly varying depth d_z. A uniformly distributed load of q is applied over the entire length of the member.

If the angle of taper is small, ordinary beam theory has been shown to yield sufficiently accurate results.† Thus the governing differential equation is

$$\frac{d^2y}{dz^2} = -\frac{M_z}{EI_z}$$

where

$$M_z = \frac{qlz}{2} - \frac{qz^2}{2}$$

$$I_z = \tfrac{1}{12}bd_z^3$$

$$d_z = d_0 - \frac{d_0 - d_l}{l}z = d_0 - \alpha z$$

$$\alpha = \frac{d_0 - d_l}{l}$$

The elastic curve, therefore, is defined by the equation

$$\frac{d^2y}{dz^2} = \frac{qz^2/2 - qlz/2}{(Eb/12)(d_0 - \alpha z)^3} \qquad \begin{array}{ll} \textit{Boundary conditions} \\ \text{At } z = 0 & y = 0 \\ \text{At } z = l & y = 0 \end{array} \qquad (4.52)$$

4.3.2 Moment-Area Method

The moment-area method is a relatively easy to use, semigeometrical method for determining the deformational response of beams. To facilitate the development of the basic concepts of the method, consider the behavior of a member of length l, subjected to a loading which has associated with it an M/EI (that is, a curvature) diagram of the type shown in fig. 4.25a. The resulting elastic curve has been shown in figure 4.25b. Consider, specifically, locations i and j on that elastic curve. These are separated by a distance dz. The change in the slopes of tangents drawn through these two points equals $d\theta$. However

$$\frac{d\theta}{dz} = \phi_z$$

or

$$d\theta = \phi_z \, dz = -\left(\frac{M}{EI}\right)_z dz \qquad (4.53)$$

The following first moment-area theorem, therefore, can be stated:

† See G. C. Lee, M. Morrell, and R. L. Ketter, Design of Tapered Members, *Weld. Res. Counc. Bull.*, no. 173, June, 1972.

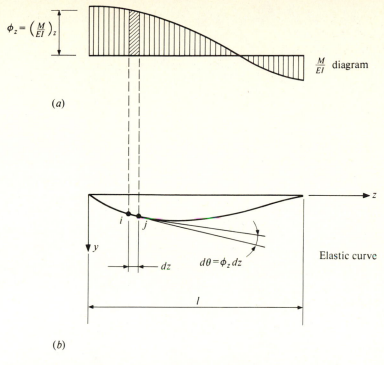

Figure 4.25 Bending moment and deflections diagrams.

For beams in bending, the change in slope between the tangents at two points on the elastic curve is equal to the area under the curvature (that is, $-M/EI$) diagram between the points in question.

In equation form:

$$\theta_B - \theta_A = \int_A^B \left(-\frac{M}{EI}\right)_z dz \tag{4.54}$$

Remembering that, by definition, deflections and slopes are small such that $\tan \theta = \sin \theta = \theta$, another useful relationship also can be defined. Consider fig. 4.26, where i and j are the same points defined in fig. 4.25. Extending the tangent lines to some point, say B, the *tangential deviation* (that is, the y distance between the two tangent lines) at B due to this single dz length element will equal the change in slope $d\theta$ times the distance from that section to B.

$$d\Delta = m \, d\theta = m(\phi_z) \, dz = m\left(-\frac{M}{EI}\right)_z dz$$

If the influence of all other intermediary dz elements between i and B are considered, noting that both M/EI and m change for each element, then the second moment-area theorem can be stated:

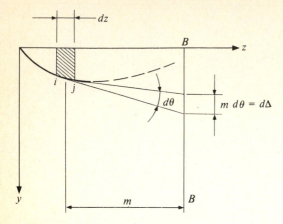

Figure 4.26 Tangential deviation.

The deviation from a point on the elastic curve, say, point B, to a tangent line drawn through the elastic curve at point A equals the first moment with respect to B of the $-M/EI$ diagram between points A and B.

In equation form:

$$\Delta_{BA} = \int_A^B \left(-\frac{M}{EI} \right)_z z_B \, dz \qquad (4.55)$$

where z_B is the distance measured from location B. This is shown in fig. 4.27.

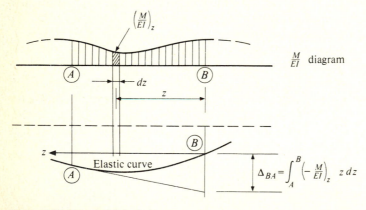

Figure 4.27 Tangential deviation.

Example 4.7 Consider the determination of the end slope and deflection of a cantilever beam subjected to a concentrated lateral load, as shown in fig. 4.28. It should be recognized that the slope and the deflection of the cantilever beam at its fixed end (that is, at B) are zero. Therefore, using the first of the moment-area

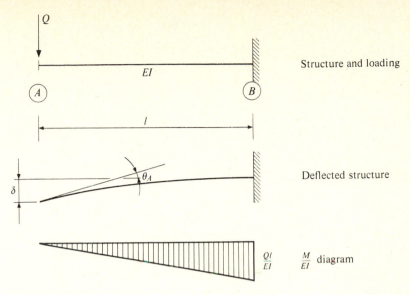

Figure 4.28

theorems, the area under the M/EI diagram between end B and the point in question will yield the slope of the beam at that point. For the slope at end A

$$\theta_A = -\frac{1}{2}\left(-\frac{Ql}{EI}\right)l = \frac{1}{2}\frac{Ql^2}{EI} \tag{4.56}$$

(It should be recognized that the M/EI diagram has been plotted on the compression side of the beam—as was the case for all of the examples heretofore considered. By the earlier stated convention this results in a negative value for the moment along the entire length of the beam. Frequently, it is just as easy to mentally ascertain the general directions of deflection and to work from that rather than to adhere vigorously to a predescribed sign convention at each stage of integration. For this reason, it is usual practice to include a sketch of the anticipated deflected shape of the structure as part of the problem formulation.) The deviation of the elastic curve at end A from a tangent drawn through end B yields the desired deflection δ. Using the second moment-area theorem and taking the first moment of the entire M/EI diagram about A

$$\delta = -\frac{1}{2}\left(-\frac{Ql}{EI}\right)(l)\left(\frac{2}{3}l\right) = \frac{1}{3}\frac{Ql^3}{EI} \tag{4.57}$$

Example 4.8 The member shown in fig. 4.29 is presumed to be of uniform cross section. It is desired to determine (1) the deflection at the end of the overhang δ_C, (2) the maximum upward deflection in span AB, and (3) the slope at the overhang end of the beam, θ_C.

While there may be easier formulations for the determination of any one particular quantity in question, because of the desire to obtain several different answers a more general approach will be used. First there will be established a

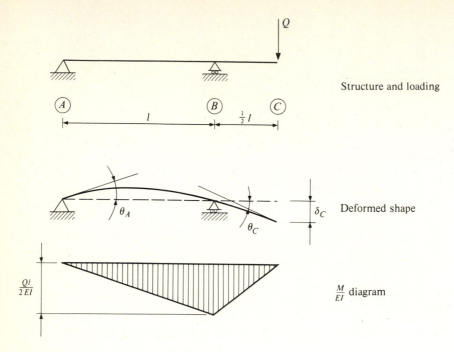

Structure and loading

Deformed shape

$\frac{M}{EI}$ diagram

Figure 4.29

horizontal datum through the point of greatest upward deflection in span AB (see fig. 4.30).

θ_A (the slope at the left-hand end of the member) can be determined by considering the entire M/EI diagram between points A and B.

$$\theta_A = \frac{\Delta_{BA}}{l} = \frac{1}{l}\left[\frac{1}{2}\left(\frac{Ql}{2EI}\right)(l)\left(\frac{l}{3}\right)\right] = \frac{1}{12}\frac{Ql^2}{EI} \tag{4.58}$$

Between endpoint A and a yet to be defined interior location D, the slope of the member continuously decreases in magnitude from θ_A to zero. From the first of the moment-area theorems, this change in slope equals the area under the M/EI diagram. Therefore

$$\frac{Ql^2}{12EI} - \left[\frac{1}{2}\left(\frac{1}{2}\frac{Qm}{EI}\right)(m)\right] = 0$$

or

$$m = \frac{l}{\sqrt{3}} = 0.576l \tag{4.59}$$

The (upward) deflection of that point is equal to the tangential deviation Δ_{BD} (or Δ_{AD}).

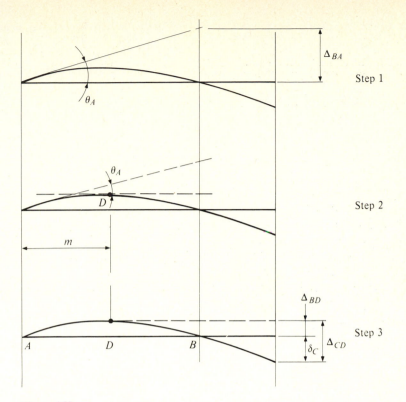

Figure 4.30

$$\Delta_{BD} = \left(\frac{0.576Ql}{2EI}\right)(0.424l)\left(\frac{0.424l}{2}\right) + \frac{1}{2}\left(\frac{0.424Ql}{2EI}\right)(0.424l)\left(\frac{0.424l}{3}\right)$$

$$= 0.026\frac{Ql^3}{EI} + 0.0063\frac{Ql^3}{EI} = 0.0323\frac{Ql^3}{EI} \tag{4.60}$$

Correspondingly, Δ_{CD} would be obtained by taking the first moment of the M/EI diagram between D and C about point C.

$$\Delta_{CD} = \left(\frac{0.576Ql}{2EI}\right)(0.424l)(0.712l) + \frac{1}{2}\left(\frac{0.424Ql}{2EI}\right)(0.424l)(0.641l)$$

$$+ \frac{1}{2}\left(\frac{Ql}{2EI}\right)\left(\frac{l}{2}\right)\left(\frac{2l}{6}\right)$$

$$= 0.0869\frac{Ql^3}{EI} + 0.0288\frac{Ql^3}{EI} + 0.0416\frac{Ql^3}{EI} = 0.1573\frac{Ql^3}{EI} \tag{4.61}$$

The deflection at C is therefore the difference between the values given in eqs. (4.61) and (4.60).

$$\delta_C = \Delta_{CD} - \Delta_{BD} = 0.1573\frac{Ql^3}{EI} - 0.0323\frac{Ql^3}{EI} = 0.1250\frac{Ql^3}{EI} \tag{4.62}$$

The slope at the end of the member is the area under the M/EI diagram between points D and C

$$\theta_C = \left(\frac{0.576Ql}{2EI}\right)(0.424l) + \frac{1}{2}\left(\frac{0.424Ql}{2EI}\right)(0.424Ql) + \frac{1}{2}\left(\frac{Ql}{2EI}\right)(0.500l)$$

$$= 0.1221\frac{Ql^2}{EI} + 0.0449\frac{Ql^2}{EI} + 0.1250\frac{Ql^2}{EI} = 0.2920\frac{Ql^2}{EI} \tag{4.63}$$

4.3.3 Superposition

The deformation of structures subjected to several simultaneously applied distinct and separate loading systems may be obtained by algebraically summing the deformations due to the individual cases. This, of course, presumes that the single-loading deformations are known, or can be readily obtained, and that the system being considered is a linear one. For example, the centerline deflection of a beam subjected to both a uniformly distributed and a concentrated lateral load applied at the midpoint of the beam would be obtained from the information given in fig. 4.31.

$$\delta^C_{\mathbb{C}} = \delta_{\mathbb{C}_1} + \delta_{\mathbb{C}_2} = \frac{Ql^3}{48EI} + \frac{5}{384}\frac{ql^4}{EI}$$

It must be emphasized that this type of superposition holds only for linear system.

Several typical beam-deformation relationships are given in table 4.4 for use in subsequent beam-deflection computations.

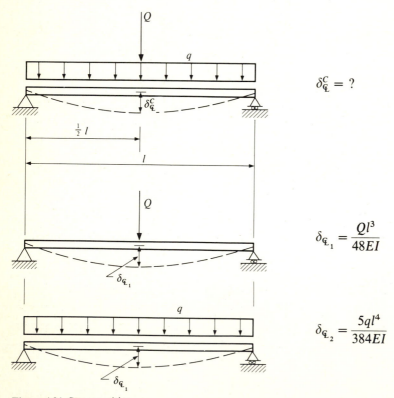

$$\delta^C_{\mathbb{C}} = ?$$

$$\delta_{\mathbb{C}_1} = \frac{Ql^3}{48EI}$$

$$\delta_{\mathbb{C}_2} = \frac{5ql^4}{384EI}$$

Figure 4.31 Superposition.

Table 4.4 Beam-deformation relationships

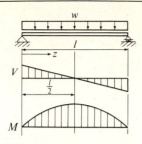

$$V_z = w\left(\frac{l}{2} - z\right)$$

$$M_z = \frac{wl}{2}z - \frac{w}{2}z^2$$

$$\delta_z = \frac{wz}{24EI}(l^3 - 2lz^2 + z^3)$$

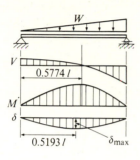

$$V_z = \frac{W}{3} - \frac{Wz^2}{l^2}$$

$$M_z = \frac{Wz}{3l^2}(l^2 - z^2)$$

$$\delta_z = \frac{Wz}{180EIl^2}(3z^4 - 10l^2z^2 + 7l^4)$$

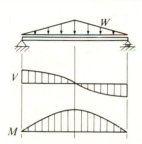

$$V_z = \frac{W}{2l^2}(l^2 - 4z^2) \qquad z \le \frac{l}{2}$$

$$M_z = Wz\left(\frac{1}{2} - \frac{2z^2}{3l^2}\right) \qquad z \le \frac{l}{2}$$

$$\delta_z = \frac{Wz}{480EIl^2}(5l^2 - 4z^2)^2$$

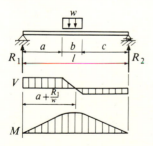

$$V_z = R_1 - w(z - a) \qquad a \le z \le a + b$$

$$M_z = R_1 z \qquad\qquad z \ge a$$

$$= R_1 z - \frac{w}{2}(z - a)^2 \qquad a \le z \le a + b$$

$$= R_2(l - z) \qquad\qquad z \ge a + b$$

Table 4.4 (continued)

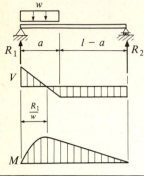

$$V_z = R_1 - wz \qquad z \leq a$$

$$M_z = R_1 z - \frac{wz^2}{2} \qquad z \leq a$$

$$= R_2(l - z) \qquad z \geq a$$

$$\delta_z = \frac{wz}{24EIl}[a^2(2l - a)^2 - 2az^2(2l - a) + lz^3]$$
$$\qquad z \leq a$$

$$= \frac{wa^2(l - z)}{24EIl}(4zl - 2z^2 - a^2)$$
$$\qquad z \geq a$$

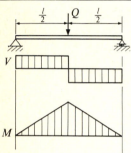

$$V_z = \frac{Q}{2}$$

$$M_z = \frac{Q}{2}z \qquad z \leq \frac{l}{2}$$

$$\delta_z = \frac{Qz}{48EI}(3l^2 - 4z^2) \qquad z \leq \frac{l}{2}$$

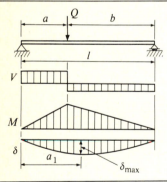

$$M_z = \frac{Qbz}{l} \qquad z \leq a$$

$$\delta_z = \frac{Qbz}{6EIl}(l^2 - b^2 - z^2) \qquad z \leq a$$

$$\delta_{max} = \frac{Qab(a + 2b)\sqrt{3a(a + 2b)}}{27EIl}$$

$$a_1 = \sqrt{\frac{a(a + 2b)}{3}} \qquad \text{when} \qquad a \geq b$$

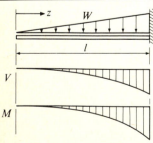

$$V_z = -W\frac{z^2}{l^2}$$

$$M_z = -\frac{W}{3}\frac{z^3}{l^2}$$

$$\delta_z = \frac{W}{60EIl^2}(z^5 - 5l^4z + 4l^5)$$

Table 4.4 (continued)

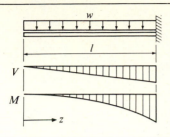

$$V_z = -wz$$

$$M_z = -\frac{w}{2}z^2$$

$$\delta_z = \frac{w}{24EI}(z^4 - 4l^3z + 3l^4)$$

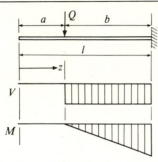

$$V_z = -Q \qquad\qquad z \geq a$$

$$M_z = -Q(z - a) \qquad z \geq a$$

$$\delta_z = \frac{Qb^2}{6EI}(3l - 3z - b) \qquad z \leq a$$

$$\quad = \frac{Q(l - z)^2}{6EI}(3b - l + z) \qquad z \geq a$$

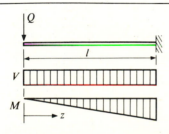

$$V_z = -Q$$

$$M_z = -Qz$$

$$\delta_z = \frac{Q}{6EI}(2l^3 - 3l^2z + z^3)$$

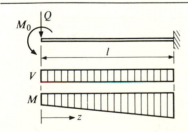

$$V_z = -Q$$

$$M_z = -(Qz + M_0)$$

$$\delta_z = \frac{Q}{6EI}(2l^3 - 3l^2z + z^3) + \frac{M_0z^2}{2EI}$$

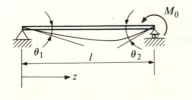

$$\delta_z = \frac{M_0 lz}{6EI}\left(1 - \frac{z^2}{l^2}\right)$$

$$\delta_{max} = \frac{M_0 l^2}{9\sqrt{3}\,EI} \qquad \text{at } z = \frac{l}{\sqrt{3}}$$

$$\theta_1 = \frac{M_0 l}{6EI} \qquad\qquad \theta_2 = \frac{M_0 l}{3EI}$$

Table 4.4 (continued)

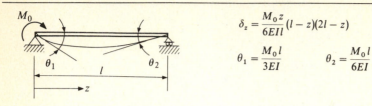

$$\delta_z = \frac{M_0 z}{6EIl}(l - z)(2l - z)$$

$$\theta_1 = \frac{M_0 l}{3EI} \qquad \theta_2 = \frac{M_0 l}{6EI}$$

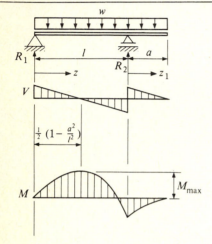

$V_z = R_1 - wz$ between supports

$= w(a - z_1)$ overhang

$M_z = \frac{wz}{2l}(l^2 - a^2 - zl)$ between supports

$= \frac{w}{2}(a - z_1)^2$ overhang

$\delta_z = \frac{wz}{24EIl}(l^4 - 2l^2 z^2 + lz^3 - 2a^2 l^2 + 2a^2 z^2)$

between supports

$= \frac{wz_1}{24EI}(4a^2 l - l^3 + 6a^2 z_1 - 4az_1^2 + z_1^3)$

overhang

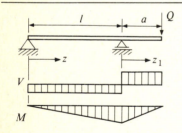

$\delta_z = \frac{Qaz}{6EIl}(l^2 - z^2)$ between supports

$= \frac{Qz_1}{6EI}(2al + 3az_1 - z_1^2)$

overhang

4.3.4 Deflections and Design

In chap. 1 several criteria, each of which constituted a potentially limiting design condition, were discussed. One of those was deflection. Most building codes contain some provision specifying the maximum deformation that can be tolerated. As an illustration of how these might apply consider the following example.

Example 4.9 The beam in question is 24 ft long and is subjected to a uniform live load of 500 lb/ft. The maximum allowable bending stress is 22 ksi, and the maximum deflection (in this case, at the center of the span) is limited to 1/360 of the span length. E for the material is 30×10^3 ksi.

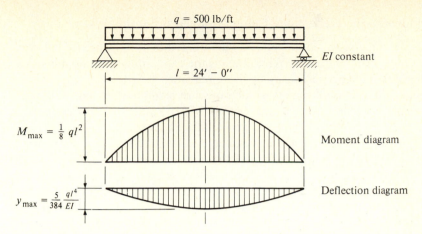

Figure 4.32

The design limitations are

$$\sigma_{max} = \frac{M_{max}}{S_{rqd}} = \frac{\frac{1}{8}ql^2}{S_{rqd}} = 22 \text{ ksi}$$

and

$$\Delta_{max} = \frac{5}{384}\frac{ql^4}{EI_{rqd}} = \left(\frac{1}{360}\right)(24)(12) = 0.8 \text{ in}$$

From the first of these,

$$S_{rqd} = \frac{M}{22} = \frac{(0.5)(24)^2(12)}{(8)(22)} = 19.636 \text{ in}^3$$

From the second,

$$I_{rqd} = \frac{5}{384}\frac{ql^4}{E\Delta_{max}} = \frac{(5)(0.5)(24 \times 12)^4}{(12)(384)(0.8)(30 \times 10^3)} = 220.32 \text{ in}^4$$

A 12-in-deep, 19 lb/ft wide-flange shape ($S = 21.1 \text{ in}^3$) is sufficient for the allowable stress requirement. For deflections to be within the prescribed limit, however, a 14-in-deep, 26 lb/ft wide-flange ($S = 35.1 \text{ in}^3$, $I = 244 \text{ in}^4$) is required. It is to be recognized that the weight of the member, itself, has not, thus far, been taken into account. Since it also is uniformly distributed, the actual distributed load on the beam will be 526 lb/ft. The derived equations are linear with respect to q; therefore

$$I_{rqd} = 220.32(\tfrac{526}{500}) = 231.77 \text{ in}^4$$

Since this is less than that supplied, the member selected is more than adequate.

4.3.5 Determination of Deflections Using Numerical Methods

In some cases it is not convenient to perform the formal integrations needed to define the deflected shape of the member. When such is the case, integration by numerical methods using a digital computer may be more realistic.

Several basically different numerical methods are available for the solution of problems of this type. One known as the method of *finite differences* (or *collocation method*) presumes the application of the given differential equation (in difference form) at preselected locations along the member. Another method assumes forward numerical integration from one boundary of the member to the other. In this later category, a method that is popular with numerical analysts is the fourth-order Runge-Kutta type process known as the Kutta-Simpson one-third method.† Because of the generality of its application, this will be the method first described. [It is to be recognized that the differential equations of beam bending are not first-order equations, as is required normally for application of the Runge-Kutta methods. Therefore, to put them in this form, equivalent sets of simultaneous first-order differential equations must be developed. As an elementary example of how this might be done, consider the second-order differential equation $y'' = f(z, y)$. If a new variable t is introduced such that $y' = t$, then the two first-order equations to be solved would be $t' = f(z, y)$ and $y' = t$.]

Kutta-Simpson One-third Method For a set of n first-order differential equations, the Kutta-Simpson method presumes the form

$$y_i' = f_i(z, y_j) \qquad \begin{cases} i = 1, 2, \ldots, n \\ j = 1, 2, \ldots, n \end{cases} \qquad (4.64)$$

with the prescribed initial conditions

$$y_{i,0} = \alpha_i \qquad i = 1, 2, \ldots, n$$

Assuming an integration interval of h

$$y_{i,z+h} = y_{i,z} + \Delta y_i \qquad (4.65)$$

where

$$\Delta y_i = \tfrac{1}{6}(k1_i + 2k2_i + 2k3_i + k4_i) \qquad (4.66)$$

$$k1_i = hf_i(z, y_1, y_2, \ldots, y_n)$$

$$k2_i = hf_i\left(z + \frac{h}{2}, y_1 + \frac{k1_1}{2}, y_2 + \frac{k1_2}{2}, \ldots, y_n + \frac{k1_n}{2}\right)$$

$$k3_i = hf_i\left(z + \frac{h}{2}, y_1 + \frac{k2_1}{2}, y_2 + \frac{k2_2}{2}, \ldots, y_n + \frac{k2_n}{2}\right) \qquad (4.67)$$

$$k4_i = hf_i(z + h, y_1 + k3_1, y_2 + k3_2, \ldots, y_n + k3_n)$$

For all of these equations $i = 1, 2, \ldots, n$.

---

† For details of the method and its derivation see R. L. Ketter and S. Prawel, *Modern Methods of Engineering Computation*, p. 274, McGraw-Hill Book Company, New York, 1969.

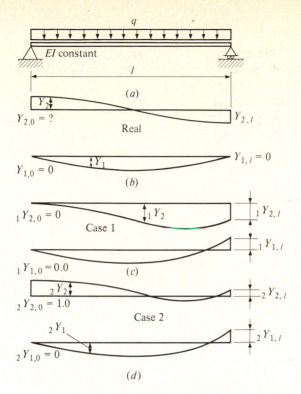

Figure 4.33 General deflection relationships for numerical solutions.

It is presumed that a value for each of the $y_{i,z}$'s exists at the point z. Then, using these, values for each $y_{i,z+h}$ can be computed. Here, as mentioned above, h is the integration interval, or step size, or increment in the independent variable z. If $k1_i$, $k2_i$, $k3_i$, and $k4_i$ are evaluated—in that order—Δy_i can be computed. In turn, f_i is the function defined by the ith first-order differential equation.

Consider, for example, the uniformly loaded, simply supported beam-bending problem defined in fig. 4.33. The governing differential equation is

$$\frac{d^2y}{dz^2} = -\frac{M}{EI} = -\frac{q}{2EI}(lz - z^2) \tag{4.68}$$

and the boundary conditions are at $z = 0$, $y = 0$; at $z = l$, $y = 0$. Since the basic procedure requires a set of first-order equations,

$$\frac{dy}{dz} = T \quad \text{and} \quad \frac{dT}{dz} = -\frac{q}{2EI}(lz - z^2) \tag{4.69}$$

In the notation of eqs. (4.64) to (4.67)

$$y_1 = y \qquad f_i = T = y_2$$

$$y_2 = T \qquad f_2 = -\frac{q}{2EI}(lz - z^2)$$

and

$$y_{1, z+h} = y_{1, z} + \Delta y_1$$

$$\Delta y_1 = \tfrac{1}{6}(k1_1 + 2k2_1 + 2k3_1 + k4_1)$$

$$k1_1 = hf_1(z, y_{1, z}, y_{2, z})$$

$$k2_1 = hf_1\left(z + \frac{h}{2}, y_{1, z} + \frac{k1_1}{2}, y_{2, z} + \frac{k1_2}{2}\right)$$

$$k3_1 = hf_1\left(z + \frac{h}{2}, y_{1, z} + \frac{k2_1}{2}, y_{2, z} + \frac{k2_2}{2}\right)$$

$$k4_1 = hf_1(z + h, y_{1, z} + k3_1, y_{2, z} + k3_2)$$

$$y_{2, z+h} = y_{2, z} + \Delta y_2$$

$$\Delta y_2 = \tfrac{1}{6}(k1_2 + 2k2_2 + 2k3_2 + k4_2)$$

$$k1_2 = hf_2(z, y_{1, z}, y_{2, z})$$

$$k2_2 = hf_2\left(z + \frac{h}{2}, y_{1, z} + \frac{k1_1}{2}, y_{2, z} + \frac{k1_2}{2}\right)$$

$$k3_2 = hf_2\left(z + \frac{h}{2}, y_{1, z} + \frac{k2_1}{2}, y_{2, z} + \frac{k2_2}{2}\right)$$

$$k4_2 = hf_2(z + h, y_{1, z} + k3_1, y_{2, z} + k3_2)$$

$$(4.70)$$

Since $f_1 = y_2$ and $f_2 = -q/2EI(lz - z^2)$,

$$k1_1 = h(y_{2, z})$$

$$k1_2 = -\frac{qh}{2EI}(lz - z^2)$$

$$k2_1 = h\left(y_{2, z} + \frac{k1_2}{2}\right)$$

$$k2_2 = -\frac{qh}{2EI}\left[l\left(z + \frac{h}{2}\right) - \left(z + \frac{h}{2}\right)^2\right]$$

$$k3_1 = h\left(y_{2, z} + \frac{k2_2}{2}\right)$$

$$k3_2 = k2_2$$

$$k4_1 = h(y_{2, z} + k3_2)$$

$$k4_2 = -\frac{qh}{2EI}[l(z + h) - (z + h)^2]$$

$$(4.71)$$

If values were available for both $y_{1, 0}$ and $y_{2, 0}$ (from known initial or boundary conditions), then the process would be started by selecting a value for the interval h, say, $l/20$. Values for $y_{1, 1}$ and $y_{2, 1}$ would then be computed. Using these as the initial conditions for the second interval, $y_{1, 2}$ and $y_{2, 2}$ would be determined. The process continues in this fashion, step by step, until values for $y_{1, l}$ and $y_{2, l}$ have been obtained.

Unfortunately, for the problem defined, only $y_{1, 0}$ is known at the outset. Modification of the process therefore is required. Assume, for example, that values for both $y_{1, 0}$ and $y_{2, 0}$ are given, say, 0.0 and 0.0, respectively. Using these values and the above defined integrations, solutions for y_1 and y_2 (shown as case 1 in fig. 4.33c) can be obtained. The terminal values at $z = l$ for this case are $_1y_{1, l}$ and $_1y_{2, l}$.

Next, the entire process is repeated; this time, however, using as values for $y_{1, 0}$ and $y_{2, 0}$ the assumed quantities 0.0 and 1.0. The solution shown as case 2 in fig. 4.33d is obtained. Here terminal values at $z = l$ are $_2y_{1, l}$ and $_2y_{2, l}$ as indicated. (Note that the homogeneous form must be used for this case.)

Since the problem is a linear one, and since both of the solutions (labeled case 1 and case 2) are independent and satisfy the governing differential equation, the sum of the two situations also will be a solution. The following can therefore be written:

$$y_{1, z} = (_1y_{1, z}) + B(_2 y_{1, z})$$

$$y_{2, z} = (_1y_{2, z}) + B(_2 y_{2, z})$$

$$(4.72)$$

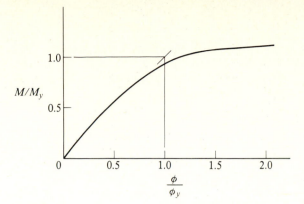

Figure 4.34 Nondimensional moment-curvature relationship.

where the constant B is determined from the equation

$$y_{1,l} = (_1 y_{1,l}) + B(_2 y_{1,l}) \tag{4.73}$$

The same integrating device also can be used to solve nonlinear bending problems. In such applications, however, because of the nonlinearity, the superposition process just described cannot be used. Instead, a one-dimensional search for the correct value of the unknown initial slope $y_{2,0}$ would be undertaken.

As an example of a nonlinear bending problem, assume that the beam shown in fig. 4.33 is made from a material which possesses a nonlinear-elastic (see fig. 4.1) stress-strain relationship of the form

$$\frac{\varepsilon}{\varepsilon_y} = A \frac{\sigma}{\sigma_y} + B \left(\frac{\sigma}{\sigma_y} \right)^n \tag{4.74}$$

ε_y and σ_y represent yield strain and stress values.

It should be recognized that the relationship between bending moment and curvature is no longer linear. Instead the two will be related in a manner similar to that shown in fig. 4.34. M_y and ϕ_y are the values of moment and curvature which correspond to initial yielding in the cross section.

If the constants A, B, and n in the stress-strain relationship are taken as 1.0, 0.429, and 100.0, respectively, and if it is assumed that the cross section in question is an I shape, the curve shown in fig. 4.34 can be described by the equation[†]

$$\frac{\phi}{\phi_y} = - \left[\frac{M}{M_y} + 0.255 \left(\frac{M}{M_y} \right)^{18.5} \right] \tag{4.75}$$

Because of this nonlinearity, it is more convenient to describe the equilibrium of the beam by the two second-order relationships

$$\frac{d^2 M}{dz^2} = - \frac{q}{EI}$$

[†] For details, see S. Prawel and R. L. Ketter, Deformation of Pile Supported Structures, *ASCE Civil Engineering in the Oceans 2*, 1969, p. 147.

and

$$\frac{d^2y}{dz^2} = -(M + 0.255M^{18.5}) \tag{4.76}$$

rather than use the more conventional fourth-order formulation. The known boundary conditions are

$$\text{At } z = 0 \quad \begin{cases} y = 0 \\ y'' = 0 \end{cases} \qquad \text{At } z = l \quad \begin{cases} y = 0 \\ y'' = 0 \end{cases}$$

As in the previous example, these two second-order equations now must be reduced to a set of four first-order ones. Letting $V = dM/dz$, and $\theta = dy/dz$, the desired equations would be

$$\frac{dV}{dz} = -\frac{q}{EI}$$

$$\frac{dM}{dz} = V$$

$$\frac{d\theta}{dz} = -(M + 0.255M^{18.5}) \tag{4.77}$$

$$\frac{dy}{dz} = \theta$$

with the known boundary conditions

$$\text{At } z = 0 \quad \begin{cases} y_0 = 0 \\ M_0 = 0 \end{cases} \qquad \text{At } z = l \quad \begin{cases} y_l = 0 \\ M_l = 0 \end{cases}$$

In terms of the Runge-Kutta equations (4.64),

$$V = y_1 \qquad M = y_2 \qquad \theta = y_3 \qquad y = y_4$$

and

$$f_1 = -\frac{q}{EI} \qquad f_2 = V = y_1 \qquad f_3 = -(y_2 + 0.255y_2^{18.5}) \qquad \text{and} \qquad f_y = y_3$$

It is necessary that there be available an initial value for each of the variables y_1 through y_4. In this instance, however, only initial values of y_1 and y_4 are known. Therefore, the remaining two unknown initial conditions must be assumed, and a procedure must be developed to ensure that the correct values have been selected. (As stated previously, the nonlinear nature of this problem precludes the use of superposition as employed in the previous example.)

Multivariant searching methods have been successfully used to resolve problems of this type. The one most often encountered is the *Method of steepest descent.*† This

---

† See R. L. Ketter and S. Prawel, *Modern Methods of Engineering Computations*, p. 467, McGraw-Hill Book Company, New York, 1969.

method involves the minimization of a function of the error in the known *final conditions*, that is, the boundary conditions at $z = l$. This error is related to the unknown initial conditions

$$E = f(y_{1,0}, y_{3,0}) \tag{4.78}$$

For this problem, a suitable error function would be

$$E = |y_{2,l}| + |y_{4,l}| \tag{4.79}$$

(It is to be understood that while the unknown initial conditions do not, as such, appear in the proposed error expression, the value of that expression is dependent upon the values chosen for these unknown initial conditions. If the exact values were available for $y_{1,0}$ and $y_{3,0}$, the error term would vanish, and the computed values for $y_{2,l}$ and $y_{4,l}$ would be zero.)

Application of the method of steepest decents involves the use of the iterative equations

$$_{k+1}y_{1,0} = {_k}y_{1,0} - (\Delta T) {_k}\left[\frac{\partial f}{\partial y_{1,0}}\right]$$

$$_{k+1}y_{3,0} = {_k}y_{3,0} - (\Delta T) {_k}\left[\frac{\partial f}{\partial y_{3,0}}\right] \tag{4.80}$$

Here ΔT is a step size that can be taken either as a constant or as a variable. The pre-subscripts $k + 1$ and k refer to the to-be-computed values and the on-hand values, respectively, of $y_{1,0}$ and $y_{3,0}$.

The computational procedure requires that an initial guess ($k = 0$) be made for both of the variables $y_{1,0}$ and $y_{3,0}$. Then, denoting these as $_0 y_{1,0}$ and $_0 y_{3,0}$, respectively, values of $y_{2,l}$ and $y_{4,l}$ are computed using the four-variable Runge-Kutta algorithm. A value for E can then be determined. Next, a small variation is made in $y_{1,0}$ and the equations reintegrated. Denoting this change in $y_{1,0}$ as $\Delta y_{1,0}$ and the resulting change in E as $\Delta E y_1$, the partial derivative $_0[\partial f/\partial y_{1,0}]$ can be approximated by the expression

$$_0\left[\frac{\partial f}{\partial y_{1,0}}\right] \simeq {_0}\left[\frac{\Delta E y_1}{\Delta y_{1,0}}\right] \tag{4.81}$$

Repeating this process for a small change in $y_{3,0}$ to produce $\Delta E_{y,3}$, then $_1 y_{1,0}$ and $_1 y_{3,0}$ can be obtained from

$$_1 y_{1,0} = {_0}y_{1,0} - \Delta T {_0}\left[\frac{\partial f}{\partial y_{1,0}}\right]$$

$$_1 y_{3,0} = {_0}y_{3,0} - \Delta T {_0}\left[\frac{\partial f}{\partial y_{3,0}}\right]$$

or

$$_1 y_{1,0} = {_0}y_{1,0} - \Delta T {_0}\left[\frac{\Delta E_{y1}}{\Delta y_{1,0}}\right]$$

$$_1 y_{3,0} = {_0}y_{3,0} - \Delta T {_0}\left[\frac{\Delta E_{y3}}{\Delta y_{3,0}}\right]$$

$$\tag{4.82}$$

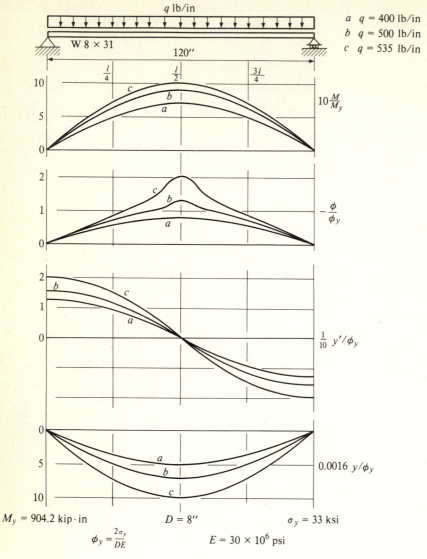

Figure 4.35 Moment, curvature, slope and deflection diagrams for several values of lateral loads (above the elastic limit).

The whole process is repeated using the just computed ($k = 1$) values to compute the next ($k = 2$) values of $y_{1,0}$ and $y_{3,0}$ and so on, until the difference between the kth and kth plus one values of each is acceptably small.

The final results using this method for a W 8 × 31 (wide-flange shape, 8 in deep by 31 lb/ft) section 120 in long for various values of q are shown in fig. 4.35. Note the pronounced difference in shape between the moment and curvature diagrams.

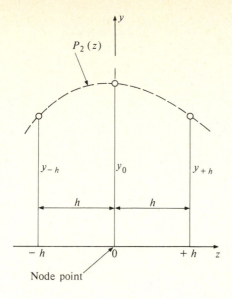

Figure 4.36 Assumed deflection notation for finite differences.

Finite-difference method Another numerical procedure that is well adapted to the computation of beam deflections is the finite difference method.† This method presumes that within a given interval the true function can be represented by a polynomial of order n. That is,

$$y = f(z) \cong P_n(z) = a_n z^n + a_{n-1} z^{n-1} + a_{n-2} z^{n-2} + \cdots a_2 z^2 + a_1 z + a_0 \quad (4.83)$$

A second-order approximation, for example, would be

$$y \cong P_2(z) = a_2 z^2 + a_1 z + a_0$$

Using the local coordinate system shown in fig. 4.36, and assuming expansion about the reference point 0, the coefficient of the second-order polynomial would be

$$a_0 = y_0$$

$$a_1 = \frac{1}{2h}(-y_{-h} + y_{+h})$$

$$a_2 = \frac{1}{2h^2}(y_{-h} - 2y_0 + y_{+h})$$

(The subscript 0 in these expressions denotes the reference, or node, point. The subscripts $-h$ and $+h$ refer to points one increment of h to the left and to the right of the node, respectively, as shown in the figure.)

---

† See R. L. Ketter and S. Prawel, *Modern Methods of Engineering Computations*, chaps. 9 and 10, McGraw-Hill Book Company, New York, 1969.

An approximation to the first and second derivatives of y with respect to z is obtained by differentiating the polynomial. That is,

$$\frac{dy}{dz} = y' \cong \frac{d}{dz} P_2(z) = 2a_2 z + a_1$$

and

$$\frac{d^2 y}{dz^2} = y'' \cong \frac{d^2}{dz^2} P_2(z) = 2a_2$$

(4.84)

At the node point

$$y_0' \cong a_1 = \frac{1}{2h}(-y_{-h} + y_{+h})$$

$$y_0'' \cong 2a_2 = \frac{1}{h^2}(y_{-h} - 2y_0 + y_{+h})$$

(4.85)

These are referred to as the *first difference* and the *second difference*, respectively.

Assume, for illustration of the method, that a solution is desired to the differential equation

$$y'' + Ay' + B = 0$$

(4.86)

in the range $0 \le z \le k$. The boundary conditions are specified as: at $z = 0$, $y = a$, and at $z = k$, $y = b$.

First, the range k is divided into m equal increments, each of size h; that is, $h = k/m$. A node point is placed at the end of each of these segments. Figure 4.37 illustrates an example where four increments (or five nodes, including the ends) are used.

The unknown values of y at each of the interior node points are y_1, y_2, and y_3, as indicated. The known zero values at the ends of the range in question are also shown.

The *difference equation* corresponding to the given differential equation would be obtained by substituting the values for the various derivatives,

$$y'' + Ay' + B \cong \frac{1}{h^2}(y_{-h} - 2y_0 + y_{+h}) + \frac{A}{2h}(-y_{-h} + y_{+h}) + B = 0 \quad (4.87)$$

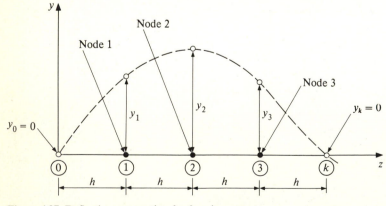

Figure 4.37 Deflection assumption for four increments.

(Note that the difference equation is written only in terms of the y's at the nodes.) This gives at points 1, 2, and 3 the following linear simultaneous equations in the unknowns y_1, y_2, and y_3.

At 1:
$$\frac{1}{h^2}(0 - 2y_1 + y_2) + \frac{A}{2h}(0 + y_2) + B = 0$$

At 2:
$$\frac{1}{h^2}(y_1 - 2y_2 + y_3) + \frac{A}{2h}(-y_1 + y_3) + B = 0 \qquad (4.88)$$

At 3:
$$\frac{1}{h^2}(y_2 - 2y_3 + 0) + \frac{A}{2h}(-y_2 + 0) + B = 0$$

In the more normal form for solution, these would be

$$(-2)y_1 + \left(1 + \frac{Ah}{2}\right)y_2 + (0)y_3 = -Bh^2$$

$$\left(1 - \frac{Ah}{2}\right)y_1 + (-2)y_2 + \left(1 + \frac{Ah}{2}\right)y_3 = -Bh^2$$

$$(0)y_1 + \left(1 - \frac{Ah}{2}\right)y_2 + (-2)y_3 = -Bh^2$$

To illustrate further the use of the finite-difference method, consider the solution of the differential equation (4.52), which defined the elastic curve for a linearly tapered beam subjected to a uniformly distributed lateral load.

$$y'' = \frac{qz^2/2 - qlz/2}{(Eb/12)(d_0 - \alpha z)^3} = \frac{6qz}{Eb}\frac{z - l}{(d_0 - \alpha z)^3}$$

where

$$\alpha = \frac{d_0 - d_l}{l}$$

The known boundary conditions are $y_0 = 0$, and $y_l = 0$.

Presume that the following are given:

$$b = 12 \text{ in}$$
$$d_0 = 18 \text{ in}$$
$$d_l = 12 \text{ in}$$
$$l = 18 \text{ ft}$$
$$E = 3 \times 10^6 \text{ lb/in}^2$$
$$q = 150 \text{ lb/ft}$$

The value of α is then

$$\alpha = \frac{d_0 - d_l}{l} = \frac{18 - 12}{(12)(18)} = 0.0278$$

Using this data, the finite-difference equation becomes

$$y_{-h} - 2y_0 + y_{+h} = \frac{kh^2z(z - 18)}{f_1(z)}$$

where

$$k = \frac{6q}{Eb} = 2.0833 \times 10^{-6}$$

and

$$f_1(z) = d_0^3 - 3d_0^2\alpha z + 3d_0\alpha^2z^2 - \alpha^3z^3$$

(z is measured in feet.) To facilitate examination, several values of $f_1(z)$ are here listed:

z, ft	$f_1(z)$	$\dfrac{z(z - 18)}{f_1(z)}$
4.5	2.59904	−23.3740
6.0	2.36966	−30.3841
9.0	1.95214	−41.4929
12.0	1.58698	−45.3692
13.5	1.42285	−42.6960

If only one interior node point is presumed (as shown in fig. 4.38a),

$$y\!\!\Big/_0^{\,0} - 2y_1 + y\!\!\Big/_1^{\,0} = \frac{kh^2z(z - 18)}{f_1(z)}$$

Since z (at point 1) = 9 ft,

$$-2y_1 = kh^2(-41.4929)$$

or

$$y_1 = \frac{kh^2}{2}(41.4929) = 0.0035 \text{ ft} = 0.0420 \text{ in}$$

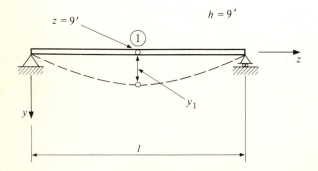

Figure 4.38a

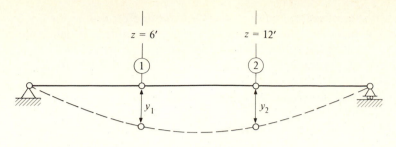

Figure 4.38b

The same problem, but with two assumed equally spaced interior node points, is illustrated in fig. 4.38b. The unknown displacements are y_1 and y_2, at $z = 6$ ft and 12 ft, respectively. Writing the difference equation at each of these points yields

At 1: $$0 - 2y_1 + y_2 = kh^2(-30.3841)$$

At 2: $$y_1 - 2y_2 + 0 = kh^2(-45.3692)$$

Solving these simultaneously gives as values of y_1 and y_2

$$y_1 = 0.00265 \text{ ft} = 0.0318 \text{ in}$$

$$y_2 = 0.00303 \text{ ft} = 0.0364 \text{ in}$$

Using three interior node points (that is, $h = 4.5$ ft), the three governing equations would be (see fig. 4.38c):

At 1: $$0 - 2y_1 + y_2 = kh^2(-23.3740)$$

At 2: $$y_1 - 2y_2 + y_3 = kh^2(-41.4929)$$

At 3: $$y_2 - 2y_3 + 0 = kh^2(-42.6960)$$

The resulting values of the deflection are

$$y_1 = 0.002065 \text{ ft} = 0.0248 \text{ in}$$

$$y_2 = 0.003144 \text{ ft} = 0.0377 \text{ in}$$

$$y_3 = 0.002473 \text{ ft} = 0.0297 \text{ in}$$

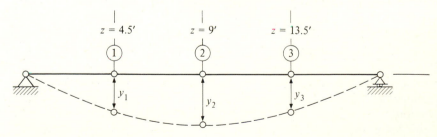

Figure 4.38c

Greater accuracy can, of course, be achieved by assuming more interior node points. It is interesting, however, to compare the values computed for the deflection at $z = 9$ ft for the one node and the three node cases. For the one node case, $y_{z=9'} = 0.0420$ in. For the three node case, $y_{z=9'} = 0.0377$ in. The difference is 0.0043 in or only an 11 percent improvement. This is typical of finite-difference solutions. For most linear problems, large numbers of nodes are usually not necessary.

It should be evident that when the differential equation is fourth-order rather than second, the difference expressions thus far derived are insufficient. For fourth-order cases, it is necessary that there be presumed at the outset a fourth-order polynomial:

$$y \cong P_4(z) = a_4 z^4 + a_3 z^3 + a_2 z^2 + a_1 z + a_0$$

Moreover, five reference points are required for evaluation. Assuming these are symmetrically placed about the central node point, the resulting difference expressions for the various derivatives are

$$y = y_0$$

$$y' = \frac{dy}{dz} \cong \frac{1}{12h}(y_{-2h} - 8y_{-h} + 8y_{+h} - y_{+2h})$$

$$y'' = \frac{d^2y}{dz^2} \cong \frac{1}{12h^2}(-y_{-2h} + 16y_{-h} - 30y_0 + 16y_{+h} - y_{+2h}) \qquad (4.89)$$

$$y''' = \frac{d^3y}{dz^3} \cong \frac{1}{2h^3}(-y_{-2h} + 2y_{-h} - 2y_{+h} + y_{+2h})$$

$$y'''' = \frac{d^4y}{dz^4} \cong \frac{1}{h^4}(y_{-2h} - 4y_{-h} + 6y_0 - 4y_{+h} + y_{+2h})$$

(It is to be understood that, depending on the problem, boundary conditions also may need to be expressed in this expanded fourth-order form.)

4.4 DEFORMATIONS DUE TO COMBINED BENDING AND SHEAR

In the preceding section, deformations were computed on the basis of bending moments alone. The effects of other stress resultants, for example shear, were not considered. Strictly speaking, then, the solutions obtained represent only the bending portion of the total deflection, and correspond to the actual deflections only for the special case of a beam subjected to constant moment. For the more general case where moments vary, shearing-stress resultants are present and will have an effect on the deformational response. In many cases the effect of shear is small. In certain special situations, however, omitting these types of deformation can lead to significant errors. For linear-elastic cases, the total deformation of the beam can be obtained by superimposing the separately computed effects of shear and bending. In this section the differential equations governing the shearing deformation of beams

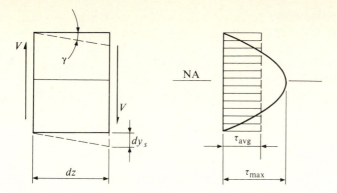

Figure 4.39 Unit length of member subjected to transverse shear.

will be developed, and a typical problem will be solved to illustrate the relative importance of this type of behavior.

Consider the dz length element shown in fig. 4.39. The loading is a vertical shear V acting on each of the cut faces. These forces produce the parallelogram type of displacement pattern shown by the dashed lines. If it is assumed that the cross section is rectangular, the distribution of shearing stress over the section will be parabolic and will have its maximum value at the neutral axis of the cross section. The average shear stress is given by the expression

$$\tau_{avg} = \frac{V}{A} \tag{4.90}$$

which is related to the maximum value by the relationship

$$\tau_{max} = \tfrac{3}{2}\tau_{avg} \tag{4.91}$$

(If the section were other than rectangular, the equation would be

$$\tau_{max} = \lambda_V \, \tau_{avg}$$

where λ_V is the *shear shape factor*.)

The vertical displacement of the right-hand face of the element with respect to the left is given by

$$dy_s = \gamma \, dz$$

where γ is the shearing strain, and y_s is the shear deformation. If the shearing modulus of elasticity is G, then

$$\gamma = \frac{\tau}{G}$$

or

$$\frac{dy_s}{dz} = \frac{\tau_{max}}{G} = \frac{\lambda_V V}{GA} \tag{4.92}$$

Equation (4.92) described the shear deformation y_s as a function of the total shear at the point in question, z. If the function $V(z)$ is known, the equation can be

Table 4.5 Governing differential equations for beam deformations due to shear

Distributed load	$-q(z) = \dfrac{dV}{dz} = \dfrac{GA}{\lambda_V} \dfrac{d^2 y_s}{dz^2}$
Shear	$V(z) = \dfrac{GA}{\lambda_V} \dfrac{dy_s}{dz}$

integrated directly to yield the desired deformation. This is generally the case for concentrated loads since shear is constant over large segments of the beam. For a beam subjected to a distributed load,

$$-q(z) = \frac{dV}{dz} \tag{4.93}$$

Therefore, the more general second-order relationship can be written

$$\frac{GA}{\lambda_V} \frac{d^2 y_s}{dz^2} = -q(z) \tag{4.94}$$

The governing differential equations for shear deformations are summarized in table 4.5.

As an example of the use of these relationships, consider the cantilever beam shown in fig. 4.40. The deflection at the free end is desired. The member is rectangular in cross section, and supports a uniformly distributed load q.

Considering first the deformations due to bending, the governing differential equation is

$$EI \frac{d^2 y_b}{dz^2} = -M_z = -\frac{1}{2} q z^2$$

The boundary conditions are: at $z = l$, $y_b = 0$ and $y_b' = 0$. The solution is given by eq. (4.95).

$$y_b = \frac{q}{EI} \left(\frac{z^4}{24} - \frac{l^3}{6} z + \frac{l^4}{8} \right) \tag{4.95}$$

To determine the deformations due to shear, the second of the differential equations listed in table 4.5 would be used.

$$\frac{GA}{\lambda_V} \frac{dy_s}{dz} = V(z) = -qz$$

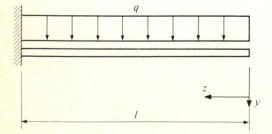

Figure 4.40

The boundary condition at $z = l$ is $y_s = 0$. Integrating the above expression and evaluating the constant of integration from the known boundary yields

$$y_s = \frac{\lambda_V}{2} \frac{q}{GA} (l^2 - z^2) \tag{4.96}$$

The total deformation of the beam at location z is the sum of eqs. (4.95) and (4.96). That is,

$$y = \frac{q}{EI} \left(\frac{z^4}{24} - \frac{l^3}{6} z + \frac{l^4}{8} \right) + \frac{\lambda_V}{2} \frac{q}{GA} (l^2 - z^2) \tag{4.97}$$

If $\bar{y} = y/l$ and $\xi = z/l$, this can be written as

$$\bar{y} = \frac{ql^3}{EI} \left(\frac{\xi^4}{24} - \frac{1}{6} \xi + \frac{1}{8} \right) + \frac{\lambda_V ql}{2GA} (1 - \xi^2)$$

At the free end (where $\xi = 0$)

$$\bar{y}_{\xi=0} = \frac{ql^3}{8EI} \left(1 + \frac{4\lambda_V}{l^2} \frac{EI}{GA} \right) \tag{4.98}$$

The term $ql^3/8EI$ is the deflection due to bending alone. The term $4\lambda_V EI/l^2 GA$ represents the added deflection due to the influence of shear. Assuming a rectangular shape and $E/G = 3.0$, this reduces to

$$y_{\xi=0} = \frac{ql^4}{8EI} \left[1 + \frac{3}{2} \left(\frac{d}{l} \right)^2 \right] \tag{4.99}$$

where d is the depth of the cross section in question. The solution to this equation is shown in fig. 4.41. For normal bending members, the d/l ratio is usually much

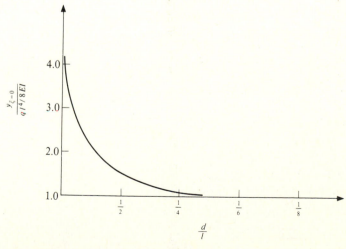

Figure 4.41 Influence of shear on lateral deflection of a uniformly loaded cantilever beam.

smaller than one-tenth. Since the contribution of shear to deformations at that particular d/l value is only 1.5 percent of the total, shear deformations in beam bending are usually ignored.

4.5 LATERAL DEFORMATIONS DUE TO THE COMBINED ACTION OF BENDING AND AXIAL FORCE

It is to be understood at the outset that the deformations being examined in this section are lateral deformations, that is, those perpendicular to the longitudinal axis of the member.

In the preceding section, which dealt with shearing deformations, the total deformation of a beam could be defined by superimposing the independent shear and bending effects. This was because the problem was a linear one. Further, it was concluded that, except for very short beams, shearing deformations are usually quite small. In this section it will be demonstrated that the influence of axial force on bending deformations can be most important, and in many cases must be seriously considered. It will also be observed that the heretofore described general superposition procedure is inappropriate.

The interaction between bending moments and axial forces is probably best illustrated from a consideration of fig. 4.42. Here, a simply supported beam is sub-

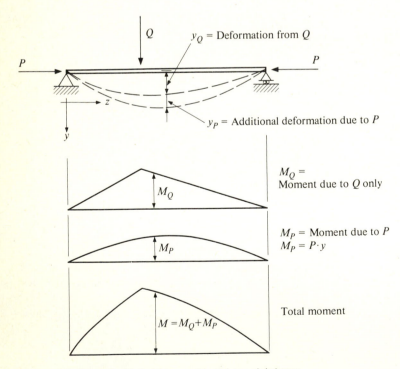

Figure 4.42 Laterally loaded beam subjected to axial thrust.

jected to a single lateral load Q and axial compressive forces P. (It is assumed that the axial forces are less than those which would produce buckling.) The moment diagram due to the lateral load Q is linearly varying as indicated in the first of the moment sketches. This moment produces deflections, which, in turn, cause additional moments along the member due to P times y. Still further deflections result. Finally, a stable situation is reached where the deflections correspond to the bending moments due to both Q and $P \times y$ (see fig. 4.42).

It should be evident that the iterative process just described actually need not be carried out to obtain a solution. The influence of the axial force on the bending moment can be incorporated directly into the differential equation.

$$EI \frac{d^2y}{dz^2} = -M_z = -(M_0 + Py)$$

where M_0 is the moment due to the lateral forces, end moments, etc., which can and probably will vary with z, and Py takes into account the added influence of the axial force and deflection. This can be written in the more standard form

$$\frac{d^2y}{dz^2} + \frac{P}{EI} y = -\frac{M_0}{EI} \tag{4.100}$$

Shear force equilibrium of beam-column elements can be obtained by differentiating once with respect to z the moment-equilibrium equation. A second differentiation yields the equilibrium equation for lateral loads. The governing equations are therefore those listed in table 4.6.

Table 4.6 Governing differential equations for combined bending and thrust: Axial force in compression

$\dfrac{d^4y}{dz^4} + k^2 \dfrac{d^2y}{dz^2} = \dfrac{1}{EI} q(z)$	Transverse load
$\dfrac{d^3y}{dz^3} + k^2 \dfrac{dy}{dz} = \dfrac{1}{EI} V(z)$	Shear
$\dfrac{d^2y}{dz^2} + k^2 y = -\dfrac{1}{EI} M(z)$	Moment
where $k^2 = \dfrac{P}{EI}$	

Because of the interaction between the axial force and the moments, the deformations of beam columns must be determined before any of the stress resultants can be computed. These deformations are obtained from the solution to the appropriate differential equation(s).

As an example, consider the uniformly loaded beam column shown in fig. 4.43. The fourth-order form of the governing equation will be used.

$$y'''' + k^2 y'' = \frac{q}{EI} \tag{4.101}$$

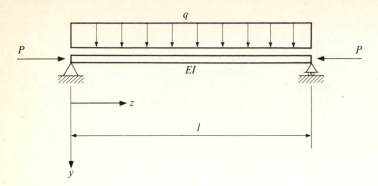

Figure 4.43

The boundary conditions for the beam column shown are

$$\text{At } z = 0 \quad \begin{cases} y = 0 \\ y'' = 0 \end{cases} \quad \text{and} \quad \text{At } x = l \quad \begin{cases} y = 0 \\ y'' = 0 \end{cases}$$

The general solution is

$$y_n = A \cos kz + B \sin kz + Cz + D \tag{4.102}$$

The particular solution is

$$y_p = \frac{q}{2k^2 EI} z^2$$

Therefore

$$y = A \cos kz + B \sin kz + Cz + D + \frac{q}{2k^2 EI} z^2 \tag{4.103}$$

From the boundary conditions, the constants are determined to be

$$A = -D = \frac{q}{k^4 EI}$$

$$B = \frac{q}{k^4 EI} \frac{1 - \cos kl}{\sin kl}$$

$$C = -\frac{ql}{2k^2 EI}$$

The deflection is therefore

$$y(z) = \frac{ql^4}{16EIu^4} \left[\frac{\cos (u - 2uz/l)}{\cos u} - 1 \right] - \frac{ql^2}{8EIu^2} [z(l - z)] \tag{4.104}$$

where $u = kl/2$.

The maximum deflection occurs at midspan, and has a value of

$$y_{max} = \frac{5}{384} \frac{ql^4}{EI} \eta(u) \tag{4.105}$$

where the function $\eta(u)$ is given by the expression

$$\eta(u) = \frac{12(2 \sec u - u^2 - 2)}{5u^4} \tag{4.106}$$

and

$$u = \frac{kl}{2} = \frac{l}{2}\sqrt{\frac{P}{EI}} = \frac{1}{2}\sqrt{\frac{Pl^2}{EI}} = \frac{\pi}{2}\sqrt{\frac{P}{\pi^2 EI/l^2}} = \frac{\pi}{2}\sqrt{\frac{P}{P_e}} \tag{4.107}$$

It is to be recognized that the midspan deflection is the same as that for a uniformly loaded beam—except that it now is multiplied by a factor $\eta(u)$ which is a function of P, EI, and l.

In general, the slope of a beam column can be obtained by differentiating the deflection expression. For the particular member under consideration, the maximum slope occurs at the ends and has a value equal to

$$\theta_{max} = \theta_0 = \frac{dy(z)}{dz}\bigg|_{z=0} = \frac{ql^3}{24EI}\left[\frac{3(\tan u - u)}{u^3}\right]$$

or

$$\theta_0 = \frac{ql^3}{24EI}\chi(u) \tag{4.108}$$

where

$$\chi(u) = \frac{3(\tan u - u)}{u^3}$$

The maximum moment is given by the expression

$$M_{max} = -EI\left[\frac{d^2 y(z)}{dz^2}\right]_{z=l/2} = \frac{ql^2}{8}\frac{2(1 - \cos u)}{u^2 \cos u}$$

or

$$M_{max} = \frac{ql^2}{8}\lambda(u) \tag{4.109}$$

where

$$\lambda(u) = \frac{2(1 - \cos u)}{u^2 \cos u}$$

Example 4.10 For the beam column shown in fig. 4.43 assume that the member is a W 8 × 31 shape ($A = 9.12$ in^2, $I = 110.0$ in^4, $S = 27.4$ in^3), and is 15 ft long. The lateral load is 1 kip/ft and the axial thrust is $P = 130$ kips. It is desired to determine the maximum displacement, slope, and normal stress.

Using the terms defined in eq. (4.107)

$$P_e = \frac{\pi^2 EI}{l^2} = 320 \text{ kips}$$

Therefore

$$\frac{P}{P_e} = \frac{130}{320} = 0.406$$

and

$$u = \frac{\pi}{2}\sqrt{\frac{P}{P_e}} = 1.00$$

The corresponding values of the magnification factors are

$$\eta(u) = 1.690 \qquad \chi(u) = 1.672 \qquad \text{and} \qquad \lambda(u) = 1.704$$

Therefore

$$y_{max} = \frac{5}{384} \frac{ql^4}{EI} \eta(u) = 0.583 \text{ in}$$

$$\theta_0 = \frac{ql^3}{24EI} \chi(u) = 0.0103 \text{ rad}$$

and

$$M_{max} = \frac{ql^2}{8} \lambda(u) = 47.925 \text{ ft} \cdot \text{kips}$$

The maximum normal stress is therefore

$$\sigma_{max} = \frac{M_{max}}{S} = \frac{(47.925)(12)}{(27.4)} = 20.99 \text{ ksi}$$

As a second illustration, consider the member shown in fig. 4.44. The lateral loading is a generally placed concentrated load of Q and the axial thrust is P.

Because of the discontinuity at Q, the problem is better dealt with in two parts: one considering the beam to the left of Q; the other to the right. As before, moment equilibrium is defined by taking moments about an arbitrary section a distance z from the left-hand end of the member. (It is to be understood that the member is assumed in the deformed position.) Thus, for the left-hand segment of the beam

$$EI \frac{d^2y}{dz^2} + Py = -\frac{Qc}{l} z \qquad\qquad \text{left}$$

For the right-hand portion

$$EI \frac{d^2y}{dz^2} + Py = -\frac{Q(l-c)(l-z)}{l} \qquad \text{right}$$

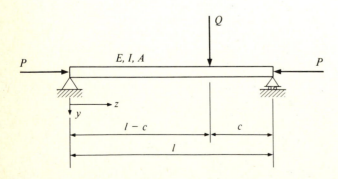

Figure 4.44

Letting $k^2 = P/EI$, these become

$$y'' + k^2 y = -\frac{Qc}{EIl}z \qquad \text{left}$$

$$y'' + k^2 y = -\frac{Q(l-c)}{EIl}(l-z) \qquad \text{right}$$

The solutions are

$$y = A\cos kz + B\sin kz - \frac{Qc}{Pl}z \qquad \text{left}$$

$$y = C\cos kz + D\sin kz - \frac{Q(l-c)}{Pl}(l-z) \qquad \text{right}$$

To evaluate these constants the boundary conditions must be taken into account:

$$\text{At } z = 0,\ y = 0$$
$$\text{At } z = l,\ y = 0$$
$$\text{At } z = (l-c),\ y_{\text{left}} = y_{\text{right}}$$
$$\text{At } z = (l-c),\ y'_{\text{left}} = y'_{\text{right}}$$

The constants of integration are therefore

$$A = 0$$

$$B = \frac{Q\sin kc}{Pk\sin kl}$$

$$C = \frac{Q\sin [k(l-c)]}{Pk}$$

$$D = -\frac{Q\sin [k(l-c)]}{Pk\tan kl}$$

and the equations for the elastic curve are

$$y = \frac{Q\sin kc}{Pk\sin kl}\sin kz - \frac{Qc}{Pl}z \qquad\qquad 0 \le z \le (l-c)$$

$$y = \frac{Q\sin [k(l-c)]}{Pk\sin kl}\sin [k(l-z)] - \frac{Q(l-c)(l-z)}{Pl} \qquad (l-c) \le z \le l \tag{4.110}$$

It follows that

$$y' = \frac{Q\sin kc}{P\sin kl}\cos kz - \frac{Qc}{Pl} \qquad\qquad \text{left}$$

$$y' = -\frac{Q\sin [k(l-c)]}{P\sin kl}\cos [k(l-c)] + \frac{Q(l-c)}{Pl} \qquad \text{right} \tag{4.111}$$

and

$$y'' = -\frac{Qk \sin kc}{P \sin kl} \sin kz \qquad\qquad \text{left}$$

$$y'' = -\frac{Qk \sin [k(l-c)]}{P \sin kl} \sin [k(l-z)] \qquad \text{right}$$

(4.112)

If the load Q is applied at midspan, the centerline deflection also will be the maximum deflection.

$$y_{max} = y_{z=l/2} = \frac{Q}{2Pk}\left(\tan \frac{kl}{2} - \frac{kl}{2}\right)$$

Again using the u notation

$$y_{max} = \frac{Ql^3}{48EI}\frac{3(\tan u - u)}{u^3} = \frac{Ql^3}{48EI}\chi(u)$$

(4.113)

where, again,

$$\chi(u) = \frac{3(\tan u - u)}{u^3}$$

For this same case, where Q is located at midspan, the maximum slope (at the ends of the member) is

$$\theta_{max} = \frac{Ql^2}{16EI}\frac{2(1-\cos u)}{u^2 \cos u} = \frac{Ql^2}{16EI}\lambda(u)$$

(4.114)

where

$$\lambda(u) = \frac{2(1-\cos u)}{u^2 \cos u}$$

The maximum moment at midspan is given by the expression

$$M_{max} = EI \left.\frac{d^2y}{dz^2}\right|_{z=l/2} = \frac{Ql}{4}\frac{\tan u}{u}$$

(4.115)

Next, consider the bending of a beam column due to the application of an end couple (M_B, in fig. 4.45). The governing differential equation is

$$y'' + k^2 y = -\frac{M_B z}{EIl}$$

(4.116)

where $k^2 = P/EI$. The boundary conditions are $y_0 = 0$ and $y_l = 0$. The general solution to the equation is

$$y = A \cos kz + B \sin kz + \frac{M_B}{Pl}z$$

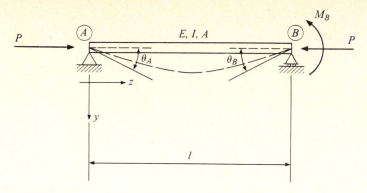

Figure 4.45

Substituting this into the boundary conditions yields the values of the constants. And, using these,

$$y = \frac{M_B}{P}\left(\frac{\sin kz}{\sin kl} - \frac{z}{l}\right) \tag{4.117}$$

Of special interest are the end slopes θ_A and θ_B.

$$\theta_A = \frac{dy}{dz}\bigg|_{z=0} = \frac{M_B l}{6EI}\phi(u)$$

and

$$\theta_B = \frac{dy}{dz}\bigg|_{z=l} = \frac{M_B l}{3EI}\psi(u) \tag{4.118}$$

where

$$\phi(u) = \frac{3}{u}\left(\frac{1}{\sin 2u} - \frac{1}{2u}\right)$$

$$\psi(u) = \frac{3}{2u}\left(\frac{1}{2u} - \frac{1}{\tan 2u}\right) \tag{4.119}$$

It should be noted that for $P = 0$, $\phi(u) = \psi(u) = 1.0$, and the solution reduces to the case of simple beam bending.

When moments are applied at both ends of the member and it is assumed that P is held constant, superposition of the relationships derived above can be presumed. For this reason, that case will not be derived here. Table 4.7 summarizes for future reference a number of deformational relationships for several typical beam columns.

For convenience, the functions $\eta(u)$, $\lambda(u)$, $\chi(u)$, $\phi(u)$, and $\psi(u)$ have been computed and are available in tabular form. Table 4.8 lists values at 0.2 increments in $2u$ from 0 to π. Values, as functions of P/P_e, also are listed in the Appendix.

Figure 4.46 contains plots of the latter three variables $\chi(u)$, $\phi(u)$, and $\psi(u)$ as functions of $2u$. The range $0 \le 2u \le 2\pi$ is shown.

To further illustrate the effect of axial thrust on bending, consider the case of a prismatic member subjected to equal end moments that produce a single curvature

Table 4.7 Beam-column deformation relationships

$$y = \frac{ql^4}{16EIu^4}\left[\frac{\cos(u - 2uz/l)}{\cos u} - 1\right] - \frac{ql^2}{8EIu^2}[z(l - z)]$$

$$y_{max} = y_{l/2} = \frac{5}{384}\frac{ql^4}{EI}\eta(u)$$

$$\theta_{max} = \theta_0 = \theta_l = \frac{ql^3}{24EI}\chi(u)$$

$$M_{max} = M_{l/2} = \frac{ql^2}{8}\lambda(u)$$

EI constant

$$y = \frac{Ql}{2P}\left[\frac{\sin(2uz/l)}{2u\cos u} - \frac{z}{l}\right]$$

$$y_{max} = y_{l/2} = \frac{Ql^3}{48EI}\chi(u)$$

$$\theta_{max} = \theta_0 = \theta_l = \frac{Ql^2}{16EI}\lambda(u)$$

$$M_{max} = M_{l/2} = \frac{Ql}{4}\frac{\tan u}{u}$$

EI constant

$$y = \frac{M_B}{P}\left[\frac{\sin(2uz/l)}{\sin 2u} - \frac{z}{l}\right]$$

$$\theta_A = \frac{M_B l}{6EI}\phi(u)$$

$$\theta_B = \frac{M_B l}{3EI}\psi(u)$$

EI constant

$$y = \frac{M_B}{P}\left[\frac{\sin(2uz/l)}{\sin 2u} - \frac{z}{l}\right] + \frac{M_A}{P}\left[\frac{\sin 2u(1 - z/l)}{\sin 2u} - \left(1 - \frac{z}{l}\right)\right]$$

$$\theta_A = \frac{M_A l}{3EI}\psi(u) + \frac{M_B l}{6EI}\phi(u)$$

$$\theta_B = \frac{M_A l}{6EI}\phi(u) + \frac{M_B l}{3EI}\psi(u)$$

EI constant

where $2u = kl = \sqrt{\dfrac{P}{EI}}\, l = \pi\sqrt{\dfrac{P}{P_e}}$

$$P_e = \frac{\pi^2 EI}{l^2}$$

$$\eta(u) = \frac{12(2\sec u - u^2 - 2)}{5u^4}$$

$$\lambda(u) = \frac{2(1 - \cos u)}{u^2\cos u}$$

$$\chi(u) = \frac{3(\tan u - u)}{u^3}$$

$$\phi(u) = \frac{3}{u}\left(\frac{1}{\sin 2u} - \frac{1}{2u}\right)$$

$$\psi(u) = \frac{3}{2u}\left(\frac{1}{2u} - \frac{1}{\tan 2u}\right)$$

Table 4.8 Table of functions $\eta(u)$, $\lambda(u)$, $\phi(u)$, $\psi(u)$, **and** $\chi(u)$†

$2u$	$\eta(u)$	$\lambda(u)$	$\phi(u)$	$\psi(u)$	$\chi(u)$
0	1.000	1.000	1.000	1.000	1.000
0.2	1.004	1.004	1.005	1.003	1.004
0.4	1.016	1.016	1.019	1.011	1.016
0.6	1.037	1.038	1.044	1.025	1.037
0.8	1.070	1.073	1.080	1.045	1.068
1.0	1.114	1.117	1.130	1.074	1.111
1.2	1.173	1.176	1.198	1.111	1.169
1.4	1.250	1.255	1.288	1.161	1.245
1.6	1.354	1.361	1.408	1.227	1.346
1.8	1.494	1.504	1.571	1.315	1.482
2.0	1.690	1.704	1.799	1.437	1.672
2.2	1.962	1.989	2.134	1.612	1.949
2.4	2.400	2.441	2.660	1.885	2.382
2.6	3.181	3.240	3.589	2.362	3.144
2.8	4.822	4.938	5.632	3.396	4.808
2.9	6.790	6.940	7.934	4.555	6.680
3.0	11.490	11.670	13.506	7.349	11.201
π	∞	∞	∞	∞	∞

† The values shown in table 4.8 are taken from S. P. Timoshenko and J. Gere, *Theory of Elastic Stability*, 2d ed., pp. 521–529, McGraw-Hill Book Company, New York, 1961. They are reproduced by permission of the publisher. Values of these quantities, as a function of P/P_e, are listed in the Appendix of this volume.

type of deformation; that is, $M_A = M_B = M_0$. The elastic curve for such a loading is given by the equation

$$y = \frac{M_0}{P \cos (kl/2)} \left[\cos \left(\frac{kl}{2} - kz \right) - \cos \frac{kl}{2} \right]$$

or

$$y = \frac{M_0 l^2}{8EI} \frac{2}{u^2 \cos u} \left[\cos \left(u - \frac{2uz}{l} \right) - \cos u \right]$$

The maximum deflection occurs at $z = l/2$ and equals

$$y_{max} = \frac{M_0 l^2}{8EI} \lambda(u) \tag{4.120}$$

The end slope is

$$\theta_0 = \frac{M_0 l}{2EI} \frac{\tan u}{u} \tag{4.121}$$

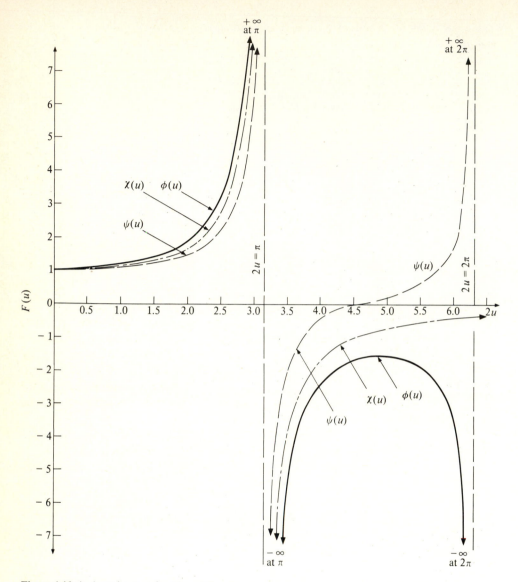

Figure 4.46 ψ, ϕ, and χ as a function of $2u$.

The maximum bending moment is

$$M_{\text{max}} = M_0(\sec u) \tag{4.122}$$

This later equation can be put in a more convenient form for examination by nondimensionalizing.

$$\frac{M_{\text{max}}}{M_0} = \sec u = \sec \left[\frac{\pi}{2} \sqrt{\frac{P}{P_e}} \right] \tag{4.123}$$

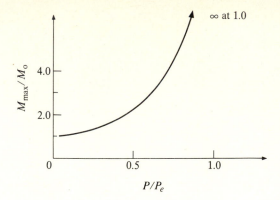

Figure 4.47 Influence of axial thrust on the maximum bending moment.

where

$$P_e = \frac{\pi^2 EI}{l^2}$$

P_e is known as the Euler buckling load. A plot of eq. (4.123) is shown in fig. 4.47. For $P/P_e = 0$, the beam column problem reduces to one of pure bending with $M_{max} = M_0$. At $P/P_e = 0.25$, $M_{max} = 1.4M_0$. This corresponds to an increase in moment over that at zero axial thrust of 40 percent. At $P/P_e = 0.5$, the increase is 220 percent. Finally at $P/P_e = 1.0$, the axial-load multiplying factor becomes infinitely large. From this example, it is clear that axial force can have a profound effect upon maximum moments in beam columns. It can be readily shown that it has similar effects upon all of the beam column deformations.

As a somewhat different looking determinate structural problem consider the behavior of the column and beam system shown in fig. 4.48. The upper end of the column is free, and the lower end is rigidly attached to a horizontal beam.

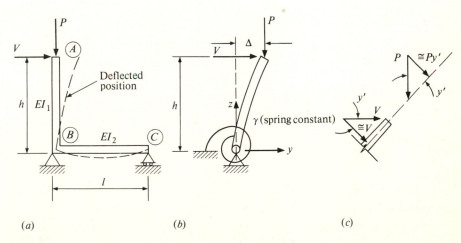

Figure 4.48a–c Structure subjected to axial thrust and bending.

The applied loading is an axial thrust P and a horizontal shear V as shown. It is desired to determine the moment at rigid joint B. (It is to be noted that the two member sizes are different, although constant sections are presumed for each.)

If member AB is presumed to be an elastically restrained beam column, then the structure to be analyzed is that shown in fig. 4.48b. The governing differential equation is

$$y'''' + k^2 y'' = 0 \tag{4.124}$$

which has the following solution:

$$y = A \sin kz + B \cos kz + Cz + D \tag{4.125}$$

The obvious boundary conditions are

$$\text{At } z = 0 \qquad y = 0$$
$$\text{At } z = 0 \qquad \gamma y' = (EI_1)y''$$

and

$$\text{At } z = h \qquad y'' = 0$$

At the upper end of the member, however, there is a second, somewhat special condition that must be defined: the relationship between the applied loads and the end shear in the member. If the deformations are presumed to be small, the condition described in fig. 4.48c will hold. Since plus shears are assumed to be directed in the y direction on the most positive face,

$$\text{At } z = h \qquad V + Py' + (EI_1)y''' = 0$$

Substitution of the general solution into each of these yields the following simultaneous set of equations:

$$(0)A + \qquad (1.0)B + \qquad (0)C + \qquad (1.0)D = 0$$
$$(k\xi)A + \qquad (k^2)B + \qquad (\xi)C + \qquad (0)D = 0$$
$$(\sin kh)A + \qquad (\cos kh)B + \qquad (0)C + \qquad (0)D = 0 \qquad (4.126)$$
$$(0)A + \qquad (0)B + \qquad (1.0)C + \qquad (0)D = -\frac{V}{P}$$

where $\xi = \gamma/EI_1$. Since only the moment at the rigid joint is desired, and since the moment is a function of the constant B, that is,

$$M_B = M_{z=0} = (EI_1)y''_{z=0} = -EI_1 k^2(B)$$

only this one factor need be determined. Unfortunately, this cannot be directly accomplished, and A and C must also be established. Carrying out the necessary operations it is determined that

$$B = -\frac{V}{Pk} \frac{\xi \sin kh}{-k \sin kh + \xi \cos kh}$$

The moment at the base of the column is therefore

$$M_B = Vh \frac{(\sin 2u)/2u}{\cos 2u - (2u/\xi h) \sin 2u} \tag{4.127}$$

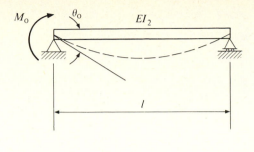

Figure 4.48d

The spring constant γ can be directly related to the end bending stiffness of the lower beam (see fig. 4.48d).

$$\gamma = \frac{M_0}{\theta_0} = 3 \frac{EI_2}{l}$$

Since

$$\xi = \frac{\gamma}{EI_1} = \frac{3EI_2}{lEI_1}$$

$$\xi h = \frac{3h}{l} \frac{EI_2}{EI_1} = 3\zeta$$

Substituting this into eq. (4.127) gives the desired equation.

$$M_B = Vh \frac{(\sin 2u)/2u}{(\cos 2u) - (2u/3\zeta) \sin 2u} \tag{4.128}$$

where $\zeta = I_2 h/I_1 l$ and $2u = \pi\sqrt{P/P_e}$. A plot of this relationship is shown in fig. 4.49 for two values of $I_2 h/I_1 l$.

In all of the beam column illustrations thus far considered, it has been assumed that the axial force is in compression. Consider now the influence of a tensile axial force. For such a case, the governing differential equations are given in table 4.9.

Table 4.9 Governing differential equations for combined bending and thrust: Axial force in tension

$\dfrac{d^4y}{dz^4} - k^2 \dfrac{d^2y}{dz^2} = \dfrac{1}{EI} q(z)$	Transverse load
$\dfrac{d^3y}{dz^3} - k^2 \dfrac{dy}{dz} = \dfrac{1}{EI} V(z)$	Shear
$\dfrac{d^2y}{dz^2} - k^2y = -\dfrac{1}{EI} M(z)$	Moment
where $k^2 = \dfrac{P}{EI}$	

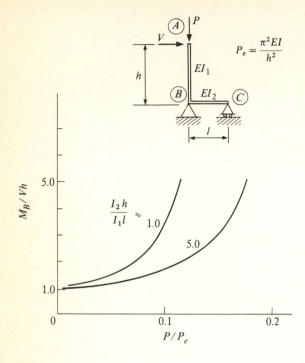

Figure 4.49 Influence of axial thrust on bending moment at B.

Assuming that q is a uniformly distributed lateral load as shown in fig. 4.43, the general solution to the first of the listed differential equations is

$$y = A \sinh kz + B \cosh kz + Cz + D + \frac{qz^2}{2k^2EI} \tag{4.129}$$

Using the boundary conditions: at $z = 0$ and l, $y = y'' = 0$, the equation of the elastic curve is

$$y = \frac{ql^4}{16EIu^4}\left[\frac{\cosh\,(u - 2uz/l)}{\cosh u} - 1\right] + \frac{ql^2}{8EIu^2}\,z(l - z) \tag{4.130}$$

At the midspan

$$y_{max} = \frac{5}{384}\frac{ql^4}{EI}\left[\frac{12[(2/\cosh u) + u^2 - 2]}{5u^4}\right] = \frac{5}{384}\frac{ql^4}{EI}\,[\eta^+(u)] \tag{4.131}$$

Since $\eta^+(u) < 1.0$, for all values of u, axial tension always reduces the deflection.

The axial-load amplification factors that were developed for compressive forces can be readily modified for axial tension. The transformation conditions would be $ki = k\sqrt{-1}$ for k, $ui = u\sqrt{-1}$ for u, and $-P$ for P. Since $\sin iu = i \sinh u$, $\cos iu = \cosh u$ and $\tan iu = i \tanh u$, the functions† become

---

† These functions are tabulated in the Appendix of this volume.

$$\phi^+ = \frac{3}{u}\left(\frac{1}{2u} - \frac{1}{\sinh 2u}\right)$$

$$\psi^+ = \frac{3}{2u}\left(\frac{1}{\tanh 2u} - \frac{1}{2u}\right)$$

$$\eta^+ = \frac{12(2 \operatorname{sech} u + u^2 - 2)}{5u^4} \tag{4.132}$$

$$\lambda^+ = \frac{2(\cosh u - 1)}{u^2 \cosh u}$$

$$\chi^+ = \frac{3(u - \tanh u)}{u^3}$$

Throughout this discussion, it has been presumed that the axial thrust in question was applied through the centroid of the cross section at each of the ends of the member. Bending moments, lateral loads, etc., also were assumed, but these were not necessarily tied to the magnitude of the thrust. There are cases where eccentric thrusts are applied to the ends of the member (see fig. 4.50). In those cases $M_A = Pe_A$; $M_B = Pe_B$, and the elastic curve is defined by the equation

$$y = e_B\left(\frac{\sin kz}{\sin kl} - \frac{z}{l}\right) + e_A\left[\frac{\sin [k(l-z)]}{\sin kl} - \frac{l-z}{l}\right] \tag{4.133}$$

the end rotations would be those listed as the last case in table 4.7. There $M_A = Pe_A$, and $M_B = Pe_B$.

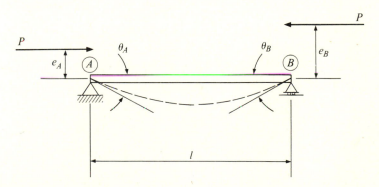

Figure 4.50

4.5.1 Interaction Curves

It was noted in chap. 1 that a frequently specified structural design criterion is the attainment of an allowable normal stress in the member. In chap. 2 this definition of failure was repeatedly used in the design examples. But in all those cases it was stated (or assumed) that the deflection had no influence on the induced bending moment.

From figs. 4.47 and 4.49, as well as from an examination of the maximum moment equations listed in table 4.7, it should be evident that axial thrust can significantly increase *primary* bending moments (that is, those which result from lateral loads, applied end moments, etc.). An assumption of no (or limited) influence of deflections may be most unreasonable for such situations.

Assume that a limiting allowable normal stress will be the design criterion. Further, assume that an analytical expression can be written which defines the bending moment at any point along the member. The limiting stress, then, will be related to the axial thrust and bending moment by the equation (see sec. 2.7)

$$\sigma_{max} = \sigma_{all} = \frac{P}{A} + \frac{M_{max}}{S} \tag{4.134}$$

where A and S are properties of the cross section in question. Since P is constant along the member, the solution depends on the location and magnitude of the maximum bending moment. (It is to be recognized that at a cut section along the member, the axial thrust acts through the centroid, and the bending moment is about an axis which passes through that point.)

Consider the beam column shown in fig. 4.51. It is of length L and has a constant cross section. The loading consists of two equal and opposite axial thrusts, and two end moments; M_0 at the left, and βM_0 at the right. The deflection curve for this case is

$$y = \frac{M_0}{P} \left[\beta \left(\frac{\sin kz}{\sin kL} - \frac{z}{L} \right) + \frac{\sin k(L-z)}{\sin kL} - \frac{L-z}{L} \right] \tag{4.135}$$

The slope and the moment are

$$\theta = \frac{M_0}{P} \left[\beta \left(\frac{k \cos kz}{\sin kL} - \frac{1}{L} \right) - \frac{k \cos k(L-z)}{\sin kL} + \frac{1}{L} \right] \tag{4.136}$$

and

$$M = -\frac{EIM_0}{P} \frac{k^2}{\sin kL} [\beta \sin kz + \sin k(L-z)] \tag{4.137}$$

The maximum moment occurs where $\partial M / \partial z = 0$, or

$$\beta \cos kz - \cos k(L-z) = 0$$

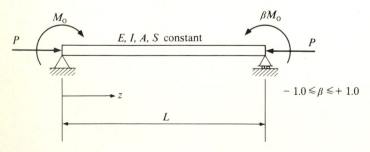

$-1.0 \leqslant \beta \leqslant +1.0$

Figure 4.51

This can be written as

$$\beta = \frac{\cos 2u(1 - z/L)}{\cos 2u(z/L)} \tag{4.138a}$$

or

$$\cot 2u\left(\frac{z}{l}\right) = \frac{\sin 2u}{\beta - \cos 2u} = \frac{\sin \pi\sqrt{P/P_e}}{\beta - \cos \pi\sqrt{P/P_e}} \tag{4.138b}$$

(It must, of course, be understood that for a particular value of β, eqs. (4.138a) or (4.138b) may predict a location of the maximum moment that is off the end of the member. Where such is the case, the maximum value occurs at the end.)

Consider the special case where $\beta = +1.0$; that is, the member is subjected to a constant (single-curvature type) primary bending moment. From eq. (4.138), it is determined that the maximum moment occurs at the midspan (that is, $z = L/2$) and has a value equal to that given in eq. (4.139).

$$M_{max} = M_0 \sec\left(\frac{\pi}{2}\sqrt{\frac{P}{P_e}}\right) \tag{4.139}$$

where

$$P_e = \frac{\pi^2 EI}{L^2}$$

Substituting eq. (4.139) into (4.134), and assuming that the allowable normal stress is the yield stress of the material σ_y,

$$\sigma_y = \frac{P}{A} + \frac{M_0(\zeta)}{S}$$

where

$$\zeta = \sec\left(\frac{\pi}{2}\sqrt{\frac{P}{P_e}}\right)$$

Dividing both sides of the equation by σ_y yields

$$\frac{P}{P_y} + \frac{M_0}{M_y}(\zeta) = 1.0 \tag{4.140}$$

where $P_y = A\sigma_y$, the maximum yield value of the axial thrust when bending moment is zero; and $M_y = S\sigma_y$, the initial yield value of the bending moment when axial thrust is zero.

If the end moment is due to the eccentric application of the axial thrust, eq. (4.140) can be written as

$$1.0 = \frac{P}{P_y} + \frac{M_0}{M_y}(\zeta) = \frac{P}{P_y} + \frac{Pe}{S\sigma_y}(\zeta) = \frac{P}{P_y} + \frac{Pe}{(Ar^2/c)\sigma_y}(\zeta)$$

or

$$\frac{P}{P_y}\left[1 + \frac{ec}{r^2}(\zeta)\right] = 1.0 \tag{4.141}$$

where r is the radius of gyration of the cross section about the centroidal axis, and c is the distance from centroid to the extreme fiber of the cross section.

As indicated above, it is possible using this procedure to predict a location of maximum moment off the end of the span. Consider, for illustration, the case where $\beta = 0$. From eq. (4.138a) the section of maximum moment will be determined from the relationship

$$0 = \frac{\cos 2u(1 - z/L)}{\cos 2u(z/L)}$$

or

$$\cos 2u\left(1 - \frac{z}{L}\right) = 0$$

Since the cosine is first equal to zero at $\pi/2$,

$$2u\left(1 - \frac{z}{L}\right) = \frac{\pi}{2}$$

But $2u = \pi\sqrt{P/P_e}$. Therefore

$$\frac{z}{L} = 1 - \frac{\frac{1}{2}}{\sqrt{P/P_e}} \tag{4.142}$$

For P/P_e less than $\frac{1}{4}$, z/L will have a negative value—which is impossible. In that range the maximum occurs at the end of the member, at $z = 0$.

Table 4.10 summarizes the multiplication (or amplification) factor ζ for a variety of different loading conditions. Since some specifications in this country, as well as a number abroad, have introduced the conservative approximation

$$\zeta_{\text{approx}} = \frac{1}{1 - P/P_e}$$

values for that factor also have been tabulated.

Fundamentally, bending moments produce deflections. These, in turn, cause the amplification factors that have been developed. The five cases of primary bending moments illustrated in table 4.10 are quite varied. They probably are sufficient to allow approximation of a number of other loading conditions.

Many structural members are primarily subjected to axial thrusts and end-bending moments. To facilitate solution of these, fig. 4.52 presents in graphic form ζ values for β ranging from $+1.0$ to -1.0, and P/P_e from 0 to 1.0.

It is to be recognized that the above developed interaction equation (4.140) presumes initial yielding of the extreme fiber. Rather than this limiting condition, one somewhat less might be more appropriately chosen. Defining the desired allowable values as

$$P_{\text{all}} = C_P P_y = C_P A\sigma_y$$

and

$$M_{\text{all}} = C_B M_y = C_B S\sigma_y$$

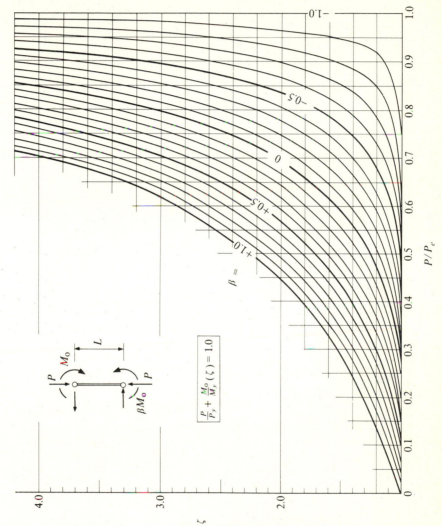

Figure 4.52 Moment-magnification factors for beam columns subjected to end moments.

Table 4.10 Moment magnification factors for various loadings:

$$\frac{P}{P_y} + \frac{M_0}{M_y}(\zeta) = 1.0$$

Loading				
	q P	Q P	P M_0	
Primary moment diagram				
M_0 $\beta = +1.0$	M_0	M_0	M_0 $\beta = 0$	M_0 / M_0 $\beta = -1.0$
Equation for ζ				
$\dfrac{1}{\cos u}$	$\dfrac{2(1 - \cos u)}{u^2 \cos u}$	$\dfrac{\tan u}{u}$	1.0 for $\dfrac{P}{P_e} \le \dfrac{1}{4}$ $\dfrac{1}{\sin 2u}$ for $\dfrac{P}{P_e} \ge \dfrac{1}{4}$	1.0 $\dfrac{1}{1 - P/P_e}$ for $\dfrac{P}{P_e} \le 1.0$

296

Values for ζ

$\dfrac{P}{P_e} \downarrow$						
0	1.000	1.000	1.000	1.000	1.000	1.000
0.1	1.137	1.098	1.114	1.000	1.000	1.111
0.2	1.310	1.205	1.256	1.000	1.000	1.250
0.3	1.534	1.352	1.442	1.011	1.000	1.423
0.4	1.831	1.545	1.685	1.093	1.000	1.667
0.5	2.253	1.817	2.031	1.257	1.000	2.000
0.6	2.886	2.233	2.547	1.538	1.000	2.500
0.7	3.937	2.899	3.402	2.034	1.000	3.333
0.8	6.061	4.258	5.127	3.068	1.000	5.000
0.9	12.392	8.289	10.260	6.200	1.000	10.000
1.0	∞	∞	∞	∞	1.000, ∞	∞

$$P_e = \frac{\pi^2 EI}{L^2}, \quad 2u = \pi\sqrt{\frac{P}{P_e}}$$

where C_P and C_B are factors chosen (or prescribed) to provide a sufficient margin for safety against yielding, the modified interaction equation would be

$$\frac{P}{P_{all}} + \frac{M_0}{M_{all}} (\zeta) = 1.0 \tag{4.143}$$

Many specifications write eq. (4.143) in the following allowable stress form

$$\frac{f_a}{F_a} + \frac{f_b}{F_b} (\zeta) = 1.0 \tag{4.144}$$

where

f_a = actual axial stress = P/A

F_a = allowable axial stress if there is no bending = $C_P \sigma_y$

f_b = actual bending stress = M/S

F_b = allowable bending stress if there is no axial load = $C_B \sigma_y$

It is important to note, however, that while eqs. (4.140), (4.143), and (4.144) may look somewhat different, in essence they are all based on the same general assumptions: (a) "failure" corresponds to the first realization of yield stress at the most highly strained cross section; (b) the member is constrained to deform in the plane of the applied moments, lateral loads, etc.; and (c) no element—flange, web, etc.—of the member buckles locally.

In all of the beam-column examples that have been considered, a particular quantity continually appears and reappears: $2u$. It is listed in this form as well as others.

$$2u = kl = l\sqrt{\frac{P}{EI}} = \pi\sqrt{\frac{P}{P_e}}$$

where $P_e = \pi^2 EI/l^2$. The important point to note is that it not only contains a property of the cross-section EI, and the length of the member l, but it also contains the applied thrust P. But the first term of the interaction equation also contains the term P. A direct plot of P/P_y versus M/M_y versus P/P_e, therefore, is not the most helpful graph. However, by noting that

$$\frac{P}{P_e} = \frac{P}{P_e} \times \frac{P_y}{P_y} = \frac{P}{P_y}\frac{P_y}{P_e} = \frac{P}{P_y}\frac{A\sigma_y}{\pi^2 EI/l^2}$$

or

$$\frac{P}{P_e} = \frac{P}{P_y}\frac{(l/r)^2}{\pi^2 E/\sigma_y} \tag{4.145}$$

a meaningful nondimensional term can be established. Here l is the length of the member, r is the radius of gyration about the centroidal axis perpendicular to the plane of bending, E is the modulus of elasticity, and σ_y is the yield stress of the material. The ratio l/r is frequently referred to as the *slenderness ratio*.

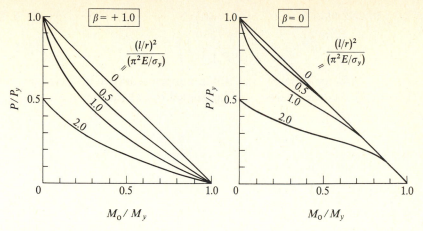

Figure 4.53 Elastic-limit interaction curves.

Figure 4.53 is a plot of P/P_y versus M_0/M_y versus $(l/r)^2/(\pi^2 E/\sigma_y)$. Two ratios of end moment are shown; $\beta = +1.0$, and $\beta = 0$. It is to be noted that the maximum values of P/P_y for both cases are the same, regardless of the slenderness ratio. For smaller values of l/r the axial yield value governs. For larger values, buckling occurs. This will be discussed in detail in a later chapter.

Example 4.11 A structural aluminum round tube of 5-in external diameter and $\frac{1}{4}$-in thickness is 8 ft 0 in long. It is subjected to an eccentrically applied compressive axial thrust of 10 kips. At one end of the member the eccentricity is e. At the other end, it is zero. (This corresponds to $\beta = 0$.) If the yield strength of the material is 35 ksi and it is desired to have a load factor of safety of 1.9 against attainment of this yield-strength value, determine the limiting value of e that can be accepted.

For the shape in question

$$A = 3.731 \text{ in}^2$$

$$S = 4.220 \text{ in}^2$$

$$I = 10.55 \text{ in}^4$$

$$r = 1.682 \text{ in}$$

$$c = 2.5 \text{ in}$$

weight per foot = 4.388 lb

Since it is the maximum eccentricity that is desired, the interaction equation form (4.141) will be used. That is,

$$\frac{P}{P_y}\left(1 + \frac{ec}{r^2}\zeta\right) = 1.0$$

In terms of e, this would be

$$e = \frac{r^2}{c\zeta}\left(\frac{1}{(P/P_y)} - 1\right) \tag{4.146}$$

To ensure the load factor of safety, the 10-kip applied load will be increased to 19 kips (LF $= 1.9$), and the analysis will be carried out at that prorated value.

$$\frac{P}{P_y} = \frac{19}{(3.731)(35)} = 0.145$$

$$\frac{l}{r} = \frac{(8)(12)}{1.682} = 57.07$$

$$\frac{\pi^2 E}{\sigma_y} = \frac{(3.142)^2(10,000)}{35} = 2820.6$$

$$\frac{P}{P_e} = \frac{P}{P_y}\frac{(l/r)^2}{\pi^2 E/\sigma_y} = (0.145)\left[\frac{(57.07)^2}{(2820.6)}\right] = 0.167$$

It is to be noted that since $P/P_e < 0.25$, the maximum moment occurs at the end of the member, and $\zeta = 1.0$. Therefore

$$e = \frac{r^2}{c\zeta}\left[\frac{1}{(P/P_y)} - 1\right] = \left[\frac{(1.682)^2}{(2.5)(1.0)}\right]\left[\frac{1}{(0.145)} - 1\right]$$

or

$$e = 6.68 \text{ in}$$

Example 4.12 A rectangular timber beam column of 15-ft length is subject to 10-kip tensile force at each end and a concentrated lateral load of $Q = 5$ kips at its midpoint. If $\sigma_{all} = 1700$ psi, and $E = 1200$ ksi, determine a permissible design envelope assuming that $P \cdot y$ does not contribute to a decrease in the deflection of the member. (It is to be recognized that this is a conservative assumption.)

From the above

$$P_{app} = 10 \text{ kips}$$

$$M_{app}\Big|_{z=l/2} = \frac{Ql}{4} - P\frac{Ql^3}{48EI} = \frac{Ql}{4}\left(1 - \frac{Pl^2}{12EI}\right)$$

$$M_{all} = S\sigma_{all} = \tfrac{1}{6}bd^2(1.70) \qquad \text{in} \cdot \text{kips}$$

$$P_{all} = A\sigma_{all} = bd(1.70) \qquad \text{kips}$$

Since

$$\frac{P}{P_{all}} + \frac{M}{M_{all}} = 1.0$$

$$\frac{(10)}{bd(1.70)} + \frac{\dfrac{(5)(180)}{4}\left[1 - \dfrac{(10)(180)^2}{(12)(1200)(bd^3/12)}\right]}{\tfrac{1}{6}bd^2(1.70)} = 1.0$$

Table 4.11 Comparative design solutions

	b, in	
d, in	Disregarding second term in bracket	Including all terms in equation
5.00	32.94	30.85
10.00	8.53	8.28
15.00	3.92	3.85
20.00	2.28	2.25

or

$$\frac{5.882}{bd} + \frac{794.118}{bd^2}\left(1 - \frac{270.00}{bd^3}\right) = 1.0$$

Possible design values are given in table 4.11.

Example 4.13 An 8-in deep steel wide-flange shape, 24 ft 0 in in length, is to be selected to support an axial force of $P = 120$ kips and a single-end bending moment $(\beta = 0)$ of $M_0 = 800$ in·kips. This moment is applied about the strong axis of the section. The material has a Young's modulus of elasticity of 30,000 ksi and a yield strength of $\sigma_y = 50$ ksi. It is specified that the allowable axial stress (in the absence of bending) should be $(0.5\sigma_y)$, and the allowable in bending (in the absence of axial force) should be $0.6\sigma_y$. That is,

$$P_{\text{all}} = C_P A \sigma_y = (0.5)(50)A$$
$$M_{\text{all}} = C_B S \sigma_y = (0.6)(50)S$$

The interaction equation to be used is eq. (4.143)

$$\frac{P}{P_{\text{all}}} + \frac{M_0}{M_{\text{all}}}(\zeta) = 1.0$$

ζ is determined from fig. 4.52 and depends on the value of $P/P_{e,\,\text{all}}$

$$\frac{P}{P_{e,\,\text{all}}} = \frac{P}{(0.5)[(\pi^2)(30,000)I/l^2]} = \frac{67.232}{I}$$

In tabular form, the various terms in the equation are listed in table 4.12. The values listed in column 10 of table 4.12 are the sums of the two terms of the left-hand side of the interaction equation. The section whose value is closest to 1.0 (but slightly under) will be the most desirable section. For the case illustrated that would be the W 8 × 48 section.

Table 4.12 Interaction values for trial sections

(1) Section	(2) A	(3) I_x	(4) S_x	(5) $\dfrac{P}{P_{e,\,all}}$	(6) ζ	(7) $\dfrac{M_0}{M_{all}}$	(8) $(6) \times (7)$	(9) $\dfrac{P}{P_{all}}$	(10) $(8)+(9)$
W 8 × 35	10.3	126	31.1	0.53	1.32	0.858	1.13	0.47	1.60
W 8 × 40	11.8	146	35.5	0.46	1.18	0.751	0.89	0.41	1.30
W 8 × 48	14.1	184	43.2	0.37	1.06	0.618	0.66	0.34	1.00
W 8 × 58	17.1	227	52.0	0.30	1.01	0.513	0.52	0.28	0.80
W 8 × 67	19.7	272	60.4	0.25	1.00	0.442	0.44	0.24	0.68

4.5.2 Reinforced Concrete Beam Columns

The analysis and design of short† reinforced concrete beam columns is not greatly different from that of other beam columns. The only real difference lies in the inability of the concrete to withstand anything but very low tensile stresses.

Consider for illustration of the above concept the rectangular cross section and loading shown in fig. 4.54. Steel reinforcing bars are presumed in both the compression and tension sides of the member and have values of A_s' and A_s, respectively. For various values of e, ranging from 0 to ∞, typical strain distributions are shown in fig. 4.55. In all cases it has been presumed that P has been increased to the value where $\varepsilon_{c,\,max}$ is realized. In case a, where $e = 0$, the response of the member is clearly dependent upon the strengths of the concrete and steel in compression. On the other hand, since most members are underreinforced, case d is governed by the tensile strength of the steel reinforcing bars labeled A_s. Between these two extremes,

† Reinforced concrete columns in buildings are generally short columns—ones that are not subject to buckling.

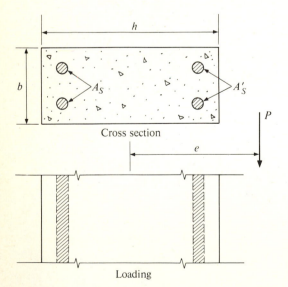

Cross section

Loading

Figure 4.54 Doubly reinforced concrete column subjected to an eccentrically applied axial thrust.

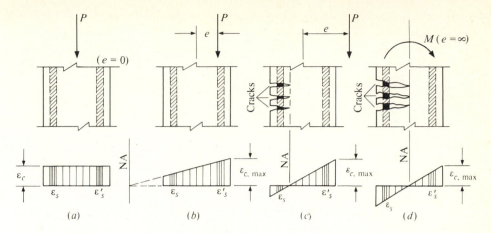

Figure 4.55 Influence of eccentricity on behavior.

one or the other of these types of failure occurs. There is, however, a particular value of P and e for which a *balanced* situation exists, and both conditions are realized simultaneously. This is usually denoted as P_b and e_b, and corresponds to $\varepsilon_{c,\,max} = \varepsilon_{c,\,ult}$ and $\varepsilon_s = \varepsilon_y$. If $P < P_b$ (or $e > e_b$), a tension failure occurs. If, on the other hand, $P > P_b$ or $e < e_b$, the failure is compressive. These three possibilities are shown in fig. 4.56, where the corresponding assumed stress distributions also are given.

The axial thrust and the eccentricity corresponding to the balanced case can be determined from force and moment equilibrium considerations (see case b of fig. 4.56). The maximum concrete strain will be taken as 0.003, as in chap. 2. The stress in the tensile steel A_s will be the yield value. The value of the stress in the steel A_s', however, is unknown—at this stage. Since it has been indicated in the sketch that the strain is greater than ε_y, it will be assumed that the stress is equal to the yield value. (This assumption must be subsequently verified as part of the analysis.)

From the indicated linear strain distribution, and a presumed value for E_s of 29×10^6 lb/in², the neutral axis will be defined by the term c_b, where

$$c_b = \frac{87,000}{87,000 + \sigma_y}\,d \qquad (4.147)$$

[Note that this equation is identical to (2.75).] Similarly $a_b = k_1 c_b$, where k_1 is as defined in chap. 2.

The balanced load P_b is obtained by summing forces in the axial direction.

$$P_b = 0.85\sigma_c' b a_b + A_s'\sigma_y - A_s'(0.85\sigma_c') - A_s\sigma_y \qquad (4.148a)$$

[Frequently the term $A_s'(0.85\sigma_c')$, which represents the volume of the concrete stress block occupied by the steel bars A_s', is omitted.] The balanced moment $M_b = P_b e_b$ is defined by summing moments of the forces about the steel bars A_s.

$$M_b = P_b e_b = (0.85\sigma_c')b a_b\left(d - \frac{a_b}{2}\right) + A_s'\sigma_y(d - d') - A_s'(0.85\sigma_c')(d - d') \qquad (4.148b)$$

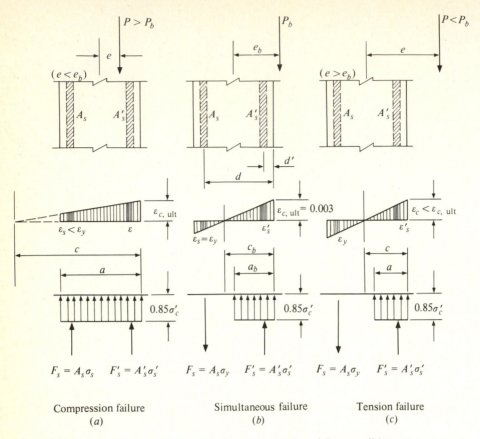

Figure 4.56 Strain and (assumed) stress distributions for various failure conditions.

As noted earlier, both of these equations presume $\varepsilon'_s \geq \varepsilon_y$. If this is not the case, then the actual value of σ'_s would have to be ascertained and used in the second terms of both equations.

Interaction diagrams are developed for concrete beam columns by repeated solution of the axial-force and moment-equilibrium equations. While in most cases the results are presented in terms of dimensionless parameters, it must be emphasized that these are valid only for the geometry specified. The process is illustrated diagrammatically in fig. 4.57. The section in question is b wide and h deep, and contains the reinforcing steel A_s and A'_s. Assuming a constant value of e for increasing values of P, the corresponding values of M are computed and plotted as shown. Finally a value of P_u is realized for which the member fails, either in tension or compression. This provides one point on the ultimate-strength interaction diagram. Next, a second value of e is selected, and the process is repeated. A third value is chosen, etc., until the entire range has been established. A typical resulting curve is shown in fig. 4.58.

To facilitate usage, nondimensional diagrams of the type just described have been developed and are available. (See, for example, the charts contained in the American Concrete Institute publication SP-7, published in 1964; or those given by

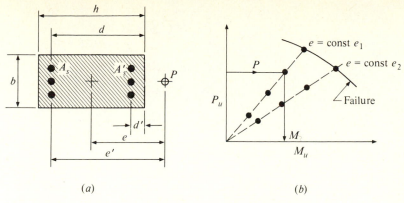

(a)   (b)

Figure 4.57 Interaction relationships for various values of thrust.

C. Winter et al., *Design of Concrete Structures*, 8th ed., McGraw-Hill Book Company, New York, 1972. They normally are of the general form shown in fig. 4.59.)

Interaction formulas for reinforced concrete beam columns can be obtained by numerically approximating the interaction diagrams. For example, the range governed by compression failure could be approximated by a straight line of the form

$$\frac{M_b}{M_u} = \frac{P_0 - P_b}{P_0 - P_u}$$

or

$$P_u = \frac{P_0}{1 + (P_0/P_b - 1)(e/e_b)} \tag{4.149}$$

For the range where yielding in tension defines failure

$$P_u = (0.85\sigma'_c)bh\left\{-p + 1 - \frac{e'}{d} + \left(1 - \frac{e'}{d}\right)^2 + 2p\left[\mu'\left(1 - \frac{d'}{d}\right) + \frac{e'}{d}\right]\right\} \tag{4.150}$$

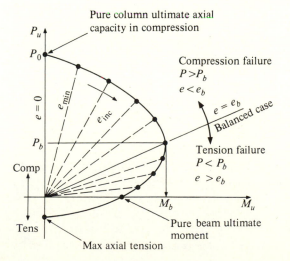

Figure 4.58 Axial-thrust bending-moment interaction relationship.

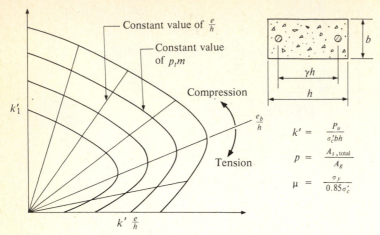

Figure 4.59 Nondimensional interaction diagram for a symmetrical, rectangular-cross-section reinforced concrete beam column. (All curves shown correspond to one given value of γ.)

can be used. In this equation,

$$\mu' = \mu - 1$$

$$\mu = \frac{\sigma_y}{(0.85\sigma_c')}$$

$$p = \frac{A_{s,\,total}}{bh}$$

e' = eccentricity of the applied load, measured from the tension steel A_s (see fig. 4.57)

d' = the distance from the compression face of the cross section to the compression steel A_s' (see fig. 4.57)

These approximations are illustrated in fig. 4.60.

In fig. 4.61a to d and 4.62a to d there are given beam-column interaction diagrams† for reinforced concrete columns with *short*, rectangular and circular cross

† From G. Winter et al., *Design of Concrete Structures*, 8th ed., McGraw-Hill Book Company, New York, 1972. They are reproduced with the permission of the publisher.

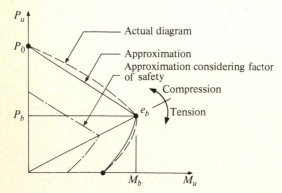

Figure 4.60

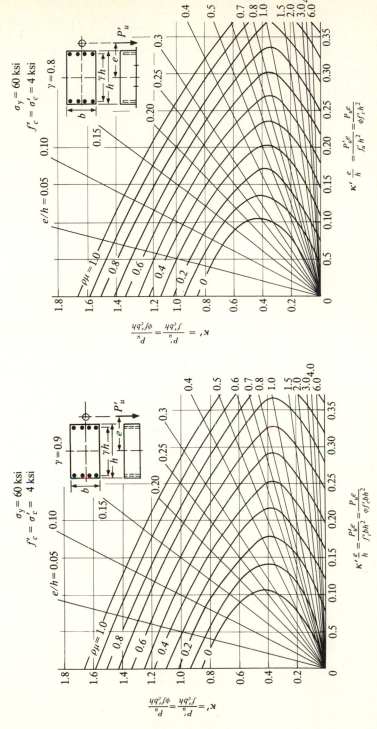

Figure 4.61a Beam-column interaction diagram—rectangular section, $\gamma = 0.9$, $\alpha_c' = 4$ k ksi, $\alpha_y = 60$ ksi. (*From G. Winter et al., Design of Concrete Structures, 8th ed., New York, McGraw-Hill, Copyright © 1972; reproduced with the permission of the publisher.*)

Figure 4.61b Beam-column interaction diagram—rectangular section $\gamma = 0.8$, $\alpha_c' = 4$ ksi, $\alpha_y = 60$ ksi. (*From G. Winter et al., Design of Concrete Structures, 8th ed., New York, McGraw-Hill, Copyright © 1972; reproduced with the permission of the publisher.*)

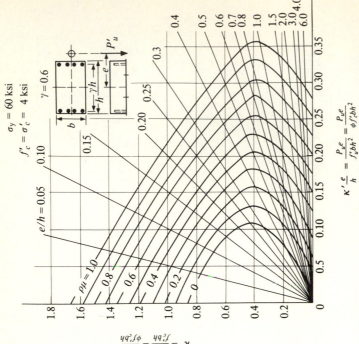

Figure 4.61c Beam-column interaction diagram—rectangular section $\gamma = 0.7$, $\alpha'_c = 4$ ksi, $\alpha_y = 60$ ksi. *(From G. Winter et al., Design of Concrete Structures, 8th ed., New York, McGraw-Hill, Copyright © 1972: reproduced with the permission of the publisher.)*

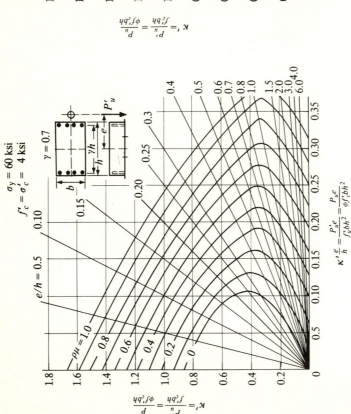

Figure 4.61d Beam-column interaction disgram—rectangular section $\gamma = 0.6$, $\alpha'_c = 4$ ksi, $\alpha_y = 60$ ksi. *(From G. Winter et al., Design of Concrete Structures, 8th ed., New York, McGraw-Hill, Copyright © 1972: reproduced with the permission of the publisher.)*

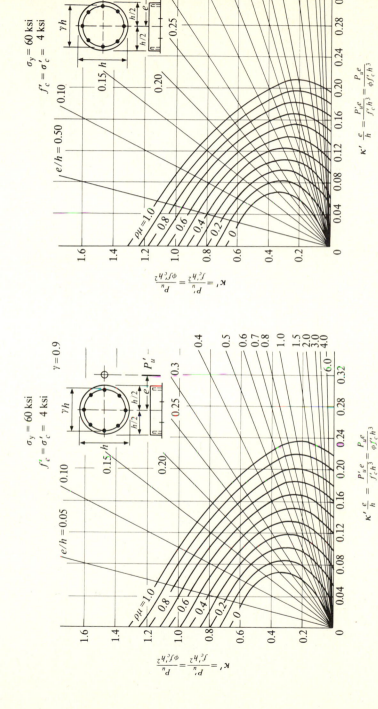

Figure 4.62a Beam-column interaction diagram—circular section $\gamma = 0.9$, $\alpha'_c = 4$ ksi, $\alpha_y = 60$ ksi. (*From G. Winter et al., Design of Concrete Structures, 8th ed., New York, McGraw-Hill, Copyright © 1972; reproduced with the permission of the publisher.*)

Figure 4.62b Beam-column interaction diagram—circular section $\gamma = 0.8$, $\alpha'_c = 4$ ksi, $\alpha_y = 60$ ksi. (*From G. Winter et al., Design of Concrete Structures, 8th ed., New York, McGraw-Hill, Copyright © 1972; reproduced with the permission of the publisher.*)

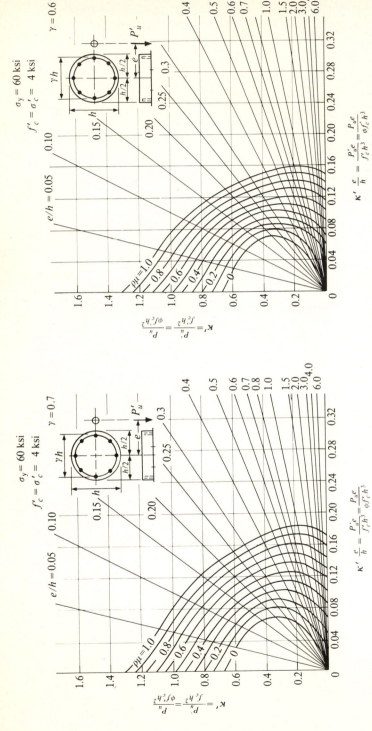

Figure 4.62c Beam-column interaction diagram—circular section $\gamma = 0.7$, $\alpha'_c = 4$ ksi, $\alpha_y = 60$ ksi. (*From G. Winter et al., Design of Concrete Structures, 8th ed., New York, McGraw-Hill, Copyright © 1972; reproduced with the permission of the publisher.*)

Figure 4.62d Beam-column interaction diagram—circular section $\gamma = 0.6$, $\alpha'_c = 4$ ksi, $\alpha_y = 60$ ksi. (*From G. Winter et al., Design of Concrete Structures, 8th ed., New York, McGraw-Hill, Copyright © 1972; reproduced with the permission of the publisher.*)

sections. Values of γ of 0.9, 0.8, 0.7, and 0.6 have been presumed. In all the diagrams, σ_c' (f_c' in the figures) is 4 ksi, and σ_y (f_y in the figures) is 60 ksi. Since k_1 is 0.85 for values of σ_c' up to and including 4 ksi, these curves would hold for lesser values of σ_c'. For values of σ_y other than 60 ksi, or for $\sigma_c' > 4$ ksi, other diagrams would have to be used.

Example 4.14 As an example of an analysis type of problem, determine the maximum axial load carrying capacity of the circular column shown in fig. 4.63.

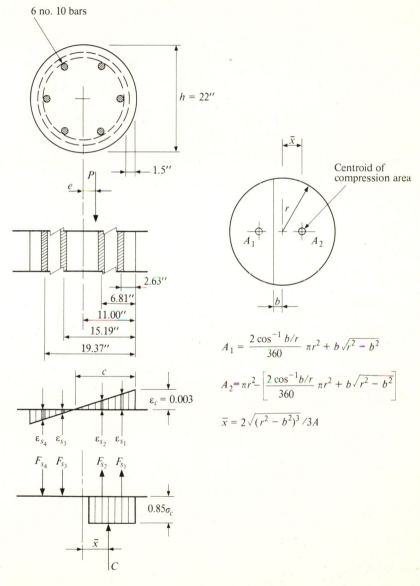

$$A_1 = \frac{2\cos^{-1} b/r}{360}\, \pi r^2 + b\sqrt{r^2 - b^2}$$

$$A_2 = \pi r^2 - \left[\frac{2\cos^{-1} b/r}{360}\, \pi r^2 + b\sqrt{r^2 - b^2}\right]$$

$$\bar{x} = 2\sqrt{(r^2 - b^2)^3}\,/3A$$

Figure 4.63 Circular reinforced concrete column subjected to an eccentrically applied thrust.

The overall diameter of the member is $h = 22$ in, and six symmetrically placed no. 10 bars provide the reinforcing. The spiral is $\frac{1}{2}$ in in diameter, and the concrete cover over the spiral is $1\frac{1}{2}$ in. The ultimate strength of the concrete in compression is $\sigma'_c = 3500$ lb/in², and the yield strength of the steel is $\sigma_y = 60$ ksi. As is required in many specifications, it is to be assumed that the load is placed at a minimum eccentricity of 1.0 in or $0.05h$, whichever is the larger value.

For the prescribed eccentricity,

$$e = 1.0 \text{ in} \quad \text{or} \quad 0.05h = (0.05)(22) = 1.1 \text{ in}$$

Therefore $e = 1.1$ in. For a balanced design situation

$$\varepsilon_{s4} = \varepsilon_y = \frac{\sigma_y}{E} = \frac{60}{29{,}000} = 0.0021$$

and

$$c_b = \left(\frac{87{,}000}{87{,}000 + 60{,}000}\right)(19.37) = 11.46 \text{ in}$$

Thus,

$$a_b = 0.85c_b = 9.74 \text{ in}$$

The area of the compressive stress block and the location of its centroid are defined by the equations shown in the right-hand sketch of fig. 4.63. For the balanced design case, $b = 1.26$ in, $A_2 = 162.4$ in², and $\bar{x} = 5.35$ in. Thus,

$$C_b = (0.85\sigma'_c)(162.4) = 483.1 \text{ kips}$$

and

$$\varepsilon_{s1} = 0.002312 \qquad \sigma_{s1} = 60 \text{ ksi (compression)}$$
$$\varepsilon_{s2} = 0.001217 \qquad \sigma_{s2} = 35.3 \text{ ksi (compression)}$$
$$\varepsilon_{s3} = 0.000976 \qquad \sigma_{s3} = 28.3 \text{ ksi (tension)}$$
$$\varepsilon_{s4} = 0.00207 \qquad \sigma_{s4} = 60.0 \text{ ksi (tension)}$$

Whenever a steel bar lies within the area of the compressive stress block, the steel stress should be corrected by a factor of $-0.85\sigma'_c$. The forces in the various bars are therefore

$$F_{s1} = 72.42 \text{ kips (compression)}$$
$$F_{s2} = 82.10 \text{ kips (compression)}$$
$$F_{s3} = 73.66 \text{ kips (tension)}$$
$$F_{s4} = 76.2 \text{ kips (tension)}$$

and the balanced axial force is

$$P_b = -(483.1 + 82.10 + 72.42) + (73.66 + 76.2) = -487.78 \text{ kips}$$

The balanced moment is

$$M_b = (76.2)(8.37) + (73.66)(4.19) + (82.10)(4.19) + (72.42)(8.37)$$
$$+ (483.1)(5.35) = 4,481.22 \text{ in·kips}$$

Therefore

$$e_b = \frac{M_b}{P_b} = \frac{4481.22}{487.78} = 9.19 \text{ in}$$

Since $e < e_b$, the member will fail in compression. The actual location of the neutral axis will be found by repeated application of the above procedure for various trial values of c. This is shown in table 4.13. Interpolating between the values computed, yields for $e = 1.1$ in

$$c = 22.85 \text{ in}$$
$$P_u = 1358 \text{ kips}$$
$$M_u = 1518 \text{ in·kips}$$

Example 4.15 To find P_u for an eccentricity of $e = 20.0$ in, the same process would be repeated using trial values of $c = 8, 9, 10,$ and 11 in. This is shown in table 4.14.

For $e = 20$ in, interpolating between the computed values gives

$$c = 8.60 \text{ in}$$
$$P_u = 216 \text{ kips}$$
$$M_u = 4262 \text{ in·kips}$$

It is to be noted that if the design charts had been used and appropriate adjustments were made for the incorporated ϕ modification factor, the values would have been:

Example 4.14: $e = 1.1$ in, $P_u = 1369.01$ kips, $M_u = 1520.53$ in·kips
Example 4.15: $e = 20$ in, $P_u = 217.09$ kips, $M_u = 4269.13$ in·kips

Example 4.16 The design process is the reverse of that just described. Suppose that a symmetrical rectangular column is to be selected to support a service load of

$$\text{DL} = 80 \text{ kips}$$
$$\text{LL} = 60 \text{ kips}$$

at an eccentricity of 4.0 in. σ'_c for the concrete is to be 3500 lb/in^2 and $\sigma_y = 60$ ksi. In addition it is specified that the steel percentage be kept as close as possible to 0.015, with a minimum value allowed of 0.01.

The ultimate loads are defined by using a load factor of the type described in chap. 2.

$$P_u = 1.5(\text{DL}) + 1.8(\text{LL}) = (1.5)(80) + (1.8)(60) = 228 \text{ kips}$$

and

$$M_u = P_u e = (228)(4.0) = 912 \text{ in·kips}$$

Table 4.13 Solutions for various trial values of c, example 4.14

c, in	a, in	A_{comp}, in²	x, in	σ_{s_1}	σ_{s_2}	σ_{s_3}	σ_{s_4}	F_{s_1}	F_{s_2}	F_{s_3}	F_{s_4}	C, kips	P_u, kips	M_u, in·kips	e, in
11.46	9.74	162.40	5.35	−60.0	−35.3	+28.3	+60.0	−72.42	−82.10	+71.88	+76.20	−483.14	−487.77	4481.22	9.19
18.0	15.3	282.19	2.45	−60.0	−54.09	−13.57	+6.61	−72.42	−129.83	−26.91	+8.39	−839.52	−1060.29	3164.44	2.98
20.0	17.0	315.20	1.66	−60.0	−57.38	−20.92	−2.74	−72.42	−138.19	−45.58	−3.48	−937.72	−1197.39	2521.69	2.11
21.0	17.85	330.37	1.29	−60.0	−58.79	−24.07	−6.75	−72.42	−141.77	−53.58	−8.57	−982.85	−1259.19	2171.83	1.72
23.0	19.55	356.99	0.619	−60.0	−60.0	−29.54	−13.73	−72.42	−144.84	−67.47	−13.66	−1062.02	−1360.43	1473.41	1.08
22.75	19.34	353.99	0.695	−60.0	−60.0	−28.91	−12.92	−72.42	−144.84	−65.87	−16.41	−1053.11	−1352.65	1531.60	1.13

(+) indicates tension, (−) indicates compression

Table 4.14 Solutions for various trial values of c, example 4.15

c, in	a, in	A_{comp}, in²	x, in	F_{s_1}	F_{s_2}	F_{s_3}	F_{s_4}	C, kips	P_u, kips	M_u, in·kips	e, in
8.0	6.80	99.86	7.02	−70.39	−32.87	+152.40	+76.20	−297.08	−446.50	4088.77	23.81
9.0	7.65	117.53	6.52	−72.42	−46.21	+151.98	+76.20	−349.65	−347.75	4354.10	18.13
10.0	8.50	135.56	6.05	−72.42	−62.93	+114.46	+76.20	−403.29	−240.10	4428.09	12.73
11.0	9.35	153.90	5.57	−72.42	−76.61	+84.18	+76.20	−457.85	−171.74	4467.90	10.01

The design charts given as fig. 4.51a to d will be used.

Assuming that $d' = 2.0$ in, the parameter γ is given by the equation

$$\gamma = \frac{h - 4.0}{h}$$

For entering the charts,

$$k' = \frac{P}{f'_c bh} = \frac{228}{3.5bh} = \frac{65.14}{bh}$$

and

$$\frac{k'e}{h} = \frac{260.56}{bh^2}$$

Further

$$\mu = \frac{f_y}{0.85f'_c} = \frac{\sigma_y}{0.85\sigma'_c} = \frac{60}{(0.85)(3.5)} = 20.17$$

Since it is required that the percentage of reinforcing steel be greater than 0.01, and optimally at 0.015, $p > 0.01$, and desirably $p = 0.015$. This gives for $p\mu$, $0.20 \le p\mu$, and optimally $p\mu = 0.30$. Two trial designs are shown in the following:

		Assume					From charts		Inter-polation
							$\gamma = 0.6$	$\gamma = 0.7$	$\gamma = 0.67$
	b	h	$\dfrac{e}{h}$	k'	$k'\dfrac{e}{h}$	γ	$p\mu$	$p\mu$	$p\mu$
Trial 1	12″	12″	0.33	0.452	0.151	0.67	0.18	0.18	0.18
Trial 2	10″	12″	0.33	0.543	0.181	0.67	0.35	0.35	0.35

Since $p\mu$ computed in trial 1 was less than that allowed, a smaller section was chosen for examination in trial 2. This is within the permissible range and will be further considered.

Selecting the 10×12-in section,

$$A_g = (10.0)(12.0) = 120 \text{ in}^2$$

Since $p = A_s/A_g$, and $p\mu = 0.35$

$$A_s = pA_g = \left(\frac{0.35}{20.17}\right)(120) = (0.017)(120) = 2.04 \text{ in}^2$$

This steel area is to be divided equally between the two faces of the beam column. For a tied column of this type, the minimum number of bars that can be used is four. The following are possibilities

Four no. 6 bars	$A_s = 1.76 \text{ in}^2$
Four no. 7 bars	$A_s = 2.40 \text{ in}^2$
Six no. 6 bars	$A_s = 1.86 \text{ in}^2$

The appropriate choice, therefore, would be four no. 7 bars, two in each face. The selection is shown in fig. 4.64.

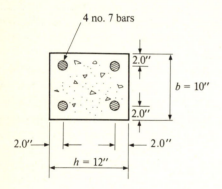

Figure 4.64 Rectangular reinforced concrete column.

To complete the design, an investigation would be carried out to check the capacity of the selected section. Referring to fig. 4.65, the values shown in table 4.15 are obtained. A graph of c versus e is shown in fig. 4.66.

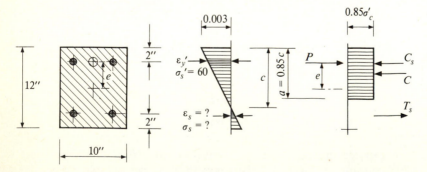

Figure 4.65 Strain and stress distributions for an eccentrically applied axial thrust.

Table 4.15 Solutions for various trial values of c, example 4.16

c	a	C	T_s	C_s	P_u	M_u	$e = \dfrac{M_u}{P_u}$
6.0	5.10	-151.73	69.60	-66.03	-148.16	1065.99	7.19
7.0	5.95	-177.01	48.37	-68.43	-197.07	1002.66	5.09
8.0	6.80	-202.30	26.10	-68.43	-244.63	904.10	3.70
9.0	7.65	-227.59	11.48	-68.43	-284.54	814.65	2.86
10.0	8.5	-252.88	0	-68.43	-321.31	716.26	2.23
7.8	6.63	-197.24	29.58	-68.43	-236.09	921.63	3.91

It is to be observed that for the given cross section the value of c corresponding to $e = 4.0$ in is $c = 0.78$. (Computed values of P_u, M_u, and e corresponding to $c = 0.78$ are given in the last line of the table.) Since the computed value of e is 3.91 in rather than 4.0 in, a slightly smaller value of c is indicated. From the tabulated results, a smaller c yields a higher value of M_u. Since the value of M_u is already sufficiently large, there is no problem here. For P_u, however, the opposite is true as is indicated by the dashed line in fig. 4.66. (P_u for $e = 4.0$ in is 230 kips.) Since the indicated capacity is greater than the applied load, the section is satisfactory.

The design is completed by adding lateral reinforcement in the form of closed hoops or ties. [As noted in chap. 2, the tie spacing must be such that the vertical bars are effectively restrained from buckling outward under load. Since this is primarily a function of the unsupported length of the reinforcing bars, codes contain rules governing tie (and spiral) spacing. A typical stipulation is at least a no. 3 tie for no. 10 or smaller reinforcing bars; spacing at the least distance given by—16 bar diameters, 48 tie diameters, or the least dimension of

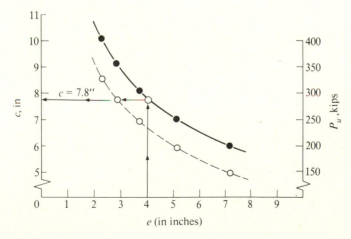

Figure 4.66 Distance to the neutral axis as a function of eccentricity.

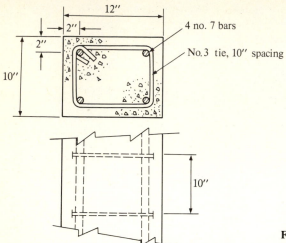

12″

2″

2″

10″

4 no. 7 bars

No. 3 tie, 10″ spacing

10″

Figure 4.67 Tied reinforced concrete column.

the column.] The following will be used: a no. 3 tie ($d = 0.375$ inches) spaced at the smallest of 10 in, $(16)(0.875) = 14$ in or $(48)(0.375) = 18$ in. The final column design is shown in fig. 4.67. (It is to be remembered that this is a design for a short column, one that does not buckle.)

4.6 DEFORMATIONS DUE TO TORSION

It was shown in chap. 2 that the ability of a slender structural member to resist torsional moments depends on two phenomena: St. Venant type torsional resistance, and warping resistance. The developed differential equation of equilibrium was

$$\frac{d^4\theta}{dz^4} - \frac{G\kappa_T}{EI_\omega}\frac{d^2\theta}{dz^2} = -\frac{m_z}{EI_\omega} \tag{4.151}$$

where θ is the total angle of twist at the point z, and m_z is the magnitude of the distributed torque at that point. The solution of this equation, subject to the boundary conditions, defines the torsional deformation of the member in question.

Writing eq. (4.151) in the normal form,

$$\theta'''' - \lambda^2\theta'' = -\frac{m_z}{EI_\omega} \tag{4.152}$$

where

$$\lambda^2 = \frac{G\kappa_T}{EI_\omega}$$

The solution to the homogeneous equation (that is, $m_z = 0$) is

$$\theta = A\sinh \lambda z + B\cosh \lambda z + Cz + D$$

$$\theta' = A\lambda \cosh \lambda z + B\lambda \sinh \lambda z + C$$

$$\theta'' = A\lambda^2 \sinh \lambda z + B\lambda^2 \cosh \lambda z \tag{4.153}$$

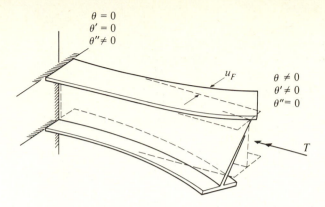

$\theta = 0$
$\theta' = 0$
$\theta'' \neq 0$

u_F

$\theta \neq 0$
$\theta' \neq 0$
$\theta'' = 0$

T

Figure 4.68 Wide-flange-cross-section cantilever beam subjected to a twisting moment.

The constants of integration A, B, C, and D are defined by considering the appropriate boundary conditions.

While it is necessary that the boundary conditions employed be in terms of the angle of twist θ, in many cases the conditions are more easily described by considering, instead, the lateral flange displacement u_F. (Obviously this holds only for flanged types of cross sections.) In chap. 2 it was shown that for small values of θ the lateral flange displacement u_F could be described by

$$u_F = \frac{d}{2}\theta$$

Thus, for first and second derivatives of θ, the corresponding values of u_F can be used (see fig. 4.68). The member in question is fixed at one end and free at the other, as shown. A torque T is applied at the free end. At the fixed end, since the member is not free to rotate, θ must be zero. At the free end, θ is clearly not zero. At the fixed end, the slope of the flange in the z direction (u_F') is seen to be zero. Since θ is proportional to u_F, it also must be equal to zero. At the free end, however, u_F' (and thus θ') is not zero. The second derivative of the angle of twist is zero at the free end (because there is no reaction to moment in the flanges, and u_F'' is zero). At the fixed end, $\theta'' \neq 0$.

Real world boundary conditions are illustrated in fig. 4.69. The assumptions normally made for these cases are indicated.

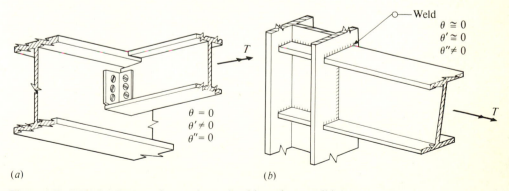

Weld
$\theta \cong 0$
$\theta' \cong 0$
$\theta'' \neq 0$

T

T

$\theta = 0$
$\theta' \neq 0$
$\theta'' = 0$

(a) (b)

Figure 4.69 Typical end connections and associated boundary conditions.

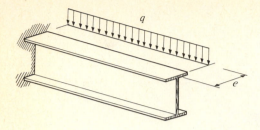

Figure 4.70 Eccentrically applied lateral load.

As is true of all differential equations, the complete solution is made up of the homogeneous and the particular cases. The particular solution depends upon how the torsional loading is applied. If the torque is concentrated, there is no particular solution. If, however, it is distributed, a particular solution exists. Consider, for example, the member subjected to the eccentric loading shown in fig. 4.70. For this case, the distributed torque $m_z = qe$, where q is the magnitude of the distributed load, and e is its eccentricity, the particular solution would be

$$\theta_P = F \frac{qe}{EI_\omega} z^2 \tag{4.154}$$

Substituting this into the governing equation gives for the constant

$$F = \frac{1}{2\lambda^2}$$

The complete solution for this problem is therefore

$$\theta = A \sinh \lambda z + B \cosh \lambda z + Cz + D + \frac{qe}{2G\kappa_T} z$$

$$\theta' = A\lambda \cosh \lambda z + B\lambda \sinh \lambda z + C + \frac{qe}{G\kappa_T} z \tag{4.155}$$

$$\theta'' = A\lambda^2 \sinh \lambda z + B\lambda^2 \cosh \lambda z + \frac{qe}{G\kappa_T}$$

To illustrate the use of these equations, consider the behavior of the simply supported beam shown in fig. 4.71. It is presumed that the member has an I-type cross section, and that it is supported laterally all along its centroidal axis. As shown, the load q is vertical and is uniformly distributed. It is applied at an eccentricity e from the vertical axis of the member.

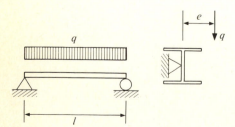

Figure 4.71

The torsional boundary conditions for the case shown are

$$\text{At } z = 0 \qquad \theta = 0 \text{ and } \theta'' = 0$$

$$\text{At } z = l \qquad \theta = 0 \text{ and } \theta'' = 0$$

Substituting eq. (4.155) into these equations yields the following values for the integration constants

$$A = \frac{qe}{\lambda^2 G\kappa_T} \frac{\cosh \lambda l - 1}{\sinh \lambda l}$$

$$B = -\frac{qe}{\lambda^2 G\kappa_T}$$

$$C = -\frac{qel}{2G\kappa_T}$$

$$D = \frac{qe}{\lambda^2 G\kappa_T}$$

The solution is therefore

$$\theta = \frac{qe}{G\kappa_T} \left(\frac{\cosh \lambda l - 1}{\lambda^2 \sinh \lambda l} \sinh \lambda z - \frac{1}{\lambda^2} \cosh \lambda z - \frac{l}{2} z + \frac{1}{\lambda^2} + \frac{z^2}{2} \right)$$

$$\theta' = \frac{qe}{G\kappa_T} \left(\frac{\cosh \lambda l - 1}{\lambda \sinh \lambda l} \cosh \lambda z - \frac{1}{\lambda} \sinh \lambda z - \frac{l}{2} + z \right)$$

$$\theta'' = \frac{qe}{G\kappa_T} \left(\frac{\cosh \lambda l - 1}{\sinh \lambda l} \sinh \lambda z - \cosh \lambda z + 1 \right)$$

$$\theta''' = \frac{qe}{G\kappa_T} \left(\frac{\lambda \cosh \lambda l - 1}{\sinh \lambda l} \cosh \lambda z - \lambda \sinh \lambda z \right)$$

In chap. 2 it was shown that the total torque at any point along a member is the sum of the St. Venant and warping torsions. That is,

$$T_z = T_{SV} + T_\omega$$

where

$$T_{SV} = G\kappa_T \theta'$$

and

$$T_\omega = -EI_\omega \theta'''$$

Thus the total torque at point z is given by

$$T_z = \frac{qe}{\lambda} \left(\frac{\cosh \lambda l - 1}{\sinh \lambda l} \cosh \lambda z - \sinh \lambda z + \lambda z - \frac{\lambda l}{2} \right)$$

$$- \frac{EI_\omega qe}{G\kappa_T} \left[\frac{\lambda(\cosh \lambda l - 1)}{\sinh \lambda l} \cosh \lambda z - \lambda \sinh \lambda z \right]$$

or

$$T_z = qe\left(z - \frac{l}{2} \right)$$

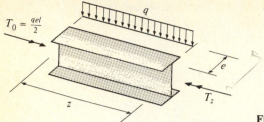

$T_0 = \dfrac{qel}{2}$

Figure 4.72

This is the same result that would have been obtained by considering the equilibrium of the free body shown in fig. 4.72.

The maximum warping normal stress in the flanges could be determined from eq. (2.45),

$$\sigma_{\omega,\,max} = -\frac{Edb}{4}\frac{d^2\theta}{dz^2}$$

That is

$$\sigma_{\omega,\,z,\,max} = -\frac{Edb}{4}\frac{qe}{G\kappa_T}\left(\frac{\cosh\lambda l - 1}{\sinh\lambda l}\sinh\lambda z - \cosh\lambda z + 1\right)$$

The largest normal stress will occur at the point $z = l/2$.

$$\sigma_{\omega,\,max} = -\frac{Edb}{4G\kappa_T}qe\left[\frac{\cosh(\lambda l/2) - (\sinh\lambda l/2)}{\sinh\lambda l} + 1\right]$$

To this must be added the maximum normal stress due to bending. For this type of loading, at the point $z = l/2$,

$$\sigma_{b,\,max} = \frac{M(d/2)}{I} = \frac{M}{S} = \frac{ql^2}{8S}$$

The distribution of the resulting normal stress at the center of the member is shown in fig. 4.73.

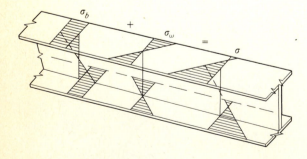

Figure 4.73 Normal stresses due to combined bending and twisting.

Example 4.17 Determine the maximum uniformly distributed load that can be carried by a A-36 steel W 12 × 85 beam, 20 ft long and braced laterally as indicated in fig. 4.71, if it is simply supported† and the vertical applied load is eccentric by 6 in.

The section constants and material properties for this member (as listed by the American Institute of Steel Construction) are as follows:

$$E = 30 \times 10^6 \text{ lb/in}^2 \qquad G = 11 \times 10^6 \text{ lb/in}^2$$

$$b = 12.105 \text{ in} \qquad d = 12.5 \text{ in}$$

$$S_x = 116 \text{ in}^3 \qquad \lambda = 0.0152/\text{in} \qquad \lambda^2 = \frac{G K_T}{E I_w}$$

$$\kappa_T = 4.80 \text{ in}^4 \qquad l = 20 \text{ ft} = 240 \text{ in} \qquad I_w = \frac{d^2}{4}(I_y)$$

$$\sigma_{\text{all}} = (36.0)(0.66) = 24 \text{ ksi}$$

$$e = 6 \text{ in (defined by problem)}$$

For bending:

$$\sigma_{b,\,\text{max}} = \frac{M}{S} = \frac{ql^2}{8S} = \frac{(240)^2}{(8)(116)} q = 5.172q \qquad \text{leaves } 8 \text{ in } \frac{lb}{ft}$$

$$\frac{2 = 62.068}{12} = 5.17$$

For torsion:

$$\sigma_{\omega,\,\text{max}} = \frac{Edbqe}{4G\kappa_T}\left(\frac{\cosh(\lambda l/2) - \sinh(\lambda l/2)}{\sinh \lambda l} + 1\right) \qquad 8 \text{ in } \frac{k}{in}$$

Since $\lambda l = 3.648$,

NOTE: SINCE THIS
IS IN K/ft

$$\sigma_{\omega,\,\text{max}} = \frac{Ebdqe}{4G\kappa_T}(1.0084) = 10.837q \qquad \text{THE EQUN ÷ BY}$$

G UNITS =
12 in/ft WHICH IS

$$10.837\frac{lb}{in^2} \qquad \text{NOT SHOWN TO}$$

The maximum normal stress is then

get 10.837 q

$$\sigma_{\text{max}} = \sigma_{b,\,\text{max}} + \sigma_{\omega,\,\text{max}} = 5.172q + 10.837q = 16.009q$$

where q is expressed in lb/ft.

The maximum q is therefore given by

$$q_{\text{all}} = q_{\text{max}} = \frac{\sigma_{\text{all}}}{16.009} = \frac{24.0}{16.009} = 1.499 \text{ kips/ft}$$

Had only primary bending been considered, the maximum value of q would have been

$$q_b = \frac{24.0}{5.172} = 4.640 \text{ kips/ft}$$

(The actual normal stress under such a loading would have been

$$\sigma_{\text{max}} = 16.009q = (16.009)(4.640) = 74.282 \text{ ksi},$$

clearly a disastrous overloading for the material with $\sigma_y = 36$ ksi.)

† Simply supported here means both in bending and in torsion.

4.7 TIME-DEPENDENT DEFORMATIONS

In all the cases that have been considered thus far, the structures were subjected to *static forces*. This was true even for the case of a *rolling load*. Actually, with the exception of the dead weight of a structure, all loads are more or less dynamic in nature because their application involves some variation with time. Common examples of structures subjected to dynamic loading are: (*a*) buildings subjected to gusting winds or blasts, (*b*) bridges or crane girders supporting moving loads, (*c*) structures subjected to sudden uneven changes in temperature, etc.

The question of whether for design or analysis purposes a particular loading should be considered as dynamic as opposed to static is not a simple one. It is a relative matter and not simply based upon how fast or how slowly the load is applied to the structure. As will be discussed later in this section, the most important parameter in this respect is the natural period of free vibration of the structure. The load may be regarded as static when its variation with time is slow with respect to the natural frequency of the structure, and dynamic when it is not.

Free vibration of the structure is defined as the general oscillative motion that occurs after the force causing the deformation has been removed. Free vibrations in structures are therefore independent of loading characteristics and depend only upon properties of the structure itself. *Forced vibrations*, on the other hand, occur in conjunction with loads that vary with time. In addition to being dependent upon the structure, they also are dependent upon the loading.

The *natural period* of free vibration is the time required for the entire structure to go through one complete cycle of free vibration. Since all parts of the structure are involved, implicit in this definition is that each part of the structure is moving at the same frequency. The natural frequency is a property of the structure. More specifically, the natural frequency depends upon the distribution of mass and stiffness of the structure in question.

All the principles developed for static analysis are applicable when the response is dynamic. In dynamic analysis, however, it is first necessary to determine the relationship that exists between the deflection of the structure and time—and this requires a consideration of the effects of inertial forces. Knowing the variation in deformation with time, stresses can be computed using the exact same methods that were used for the static cases. In many respects the calculations of stresses in a dynamically loaded structure follow the same processes used to analyze beam columns or for nonuniform torsion. In both of those cases it was first necessary to determine the deflection of the structure. Then, using those computed deformations (and their derivatives), stresses were calculated.

The dynamic deformation (that is, the dynamic response) of a structural system is dependent upon the response of each of the component parts that make up the structure. These can be any one or a combination of three fundamental response modes, as illustrated in fig. 4.74. In that figure it has been presumed that the structural element in question is a cantilever beam subjected to a loading at its free end, which load has associated with it a mass very much larger than the mass of the beam itself.

If the free end of the beam is displaced in the *y* direction and then suddenly released, a time-dependent up-and-down motion, shown in fig. 4.74*b*, is observed. This oscillation is resisted by the bending strength of the beam, and the mode is

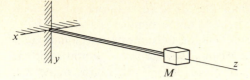

(*a*) Mass-beam system

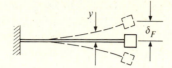

(*b*) Flexural mode

(*c*) Extensional mode

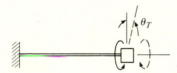

(*d*) Torsional mode

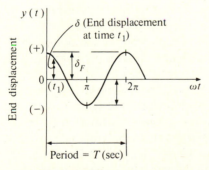

(*e*) Note: (1) One cycle occurs in one period of time

(2) The frequency is the number of cycles that occur in one unit of time

Figure 4.74 Fundamental dynamic response modes.

referred to as the *flexural mode* of vibration. The *axial* (or extensional) *mode* is illustrated in fig. 4.74c and results when the free end of the member is displaced in the axial direction, and suddenly released. If the free end is twisted about its longitudinal axis (z axis) and released, the motion of the beam is described as vibration in the *torsional mode*. States of motion involving combinations of these basic modes are referred to as *coupled modes* of vibration.

Assuming an elastic behavior, upon removal of the forced displacement the beam begins to vibrate periodically between specific limits as indicated in fig. 4.74e. The maximum displacement of the beam in either direction is defined as the *amplitude* of the vibration. Since there are no forces applied after the initial displacement, the case shown represents free vibration. At any time after the initial disturbance, the displacement of the end of the beam is described by an equation having the form

$$\delta = \delta_F \cos \omega t \qquad (4.156)$$

where ω is the *circular frequency* of vibration (expressed in radians per second) and δ_F is the amplitude.

As indicated in the note on fig. 4.74, the period T is the time required for one cycle of periodic motion to occur. The frequency f defines the number of cycles that occur in one second. The frequency is therefore the inverse of the period. That is

$$f = \frac{1}{T}$$

Since for this case $T = 2\pi/\omega$, the circular frequency ω is related to the frequency by the relationship

$$\omega = 2\pi f$$

Thus far in this discussion, it has been more or less inferred that the amplitude δ_F does not change with time. This will be true *only* if there are *no* forces involved other than those due to the mass and the spring (the elastic restraint of the beam). In reality, this situation can never arise because there will always be forces of one type or another that tend to resist the motion and reduce the amplitude as time progresses. Collectively, these resisting forces that are present in vibrating systems are called *damping forces*.

Damping forces can be either external to the vibrating system or they can be internal. An example of an external damping force would be the drag resistance that is offered by the air as the vibrating structure moves through it. Internal damping results from various frictional forces that develop within the structure itself—such as that caused by slippage at joints. When these types of damping forces exist, the amplitude of free vibration is gradually reduced, and there exists what is referred to as *damped free vibration*. In such cases, the vibrating body will eventually come to rest in its static equilibrium position. The free and "damped" cases are illustrated diagrammatically in fig. 4.75.

As was pointed out earlier, forced vibrations are those that occur when the disturbing force is not removed from the vibrating structure. If frictional (or damping) forces are included in the analysis of such cases, the problem is said to be one of *damped forced vibration*. Correspondingly, if damping is not included—the fictitious case of *undamped forced vibration* is obtained.

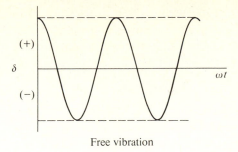

Free vibration

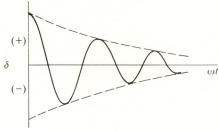

Damped free vibration

Figure 4.75 Deformational response as a function of time.

The loading applied to a structure can be due to a number of causes and can occur in an almost endless number of varieties. Generally, however, they can be classified in a dynamic sense. For example, a load could be suddenly applied and then removed producing an *impact* type of dynamic loading. Or, it could be a load that has a motion entirely of its own and is resting upon the structure producing an *impulsive* type of loading. Still another case would correspond to that situation where the load is *periodic* in nature and forces the structure to vibrate—at least at the point of load application—at the frequency and amplitude of the load. A similar situation could occur in conjunction with a *nonperiodic* load. An unsteady wind load is an example of a nonperiodic dynamic load causing a damped forced vibration. When the wind stops, the motion is of the damped free type, and the structure gradually returns to its at rest position. An earthquake also produces a forced vibration. When the ground motion ceases, the structure comes to rest after going through a number of cycles of damped free vibrations.

To describe the motion of a member such as a simply supported beam, it must be known how the displacement of each point along the member responds with time. In such a case, the displacement y is the dependent variable. The independent variables are the distance along the beam z and the time t. To model such a beam, it is assumed that each differential element of the member is connected to its adjacent element by a spring K, and a *dashpot c*, as indicated in fig. 4.76. The spring constant K represents the bending stiffness of the beam, and the value of c represents the internal damping. The mass of the element is dM.

If the number of degrees of freedom possessed by the system is defined as the number of displacements required to completely define the response of the beam, this model clearly has an infinite number of degrees of freedom. This is because a value of y is required for each of the elements, and there are an infinite number of them.

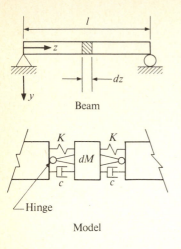

Beam

Model

Figure 4.76 Assumed dynamic model of a beam.

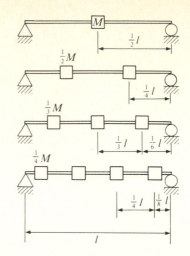

Figure 4.77 Multiple-degree-of-freedom models.

In many cases it is sufficient to describe the response at only a limited number of certain selected points along the beam. In such a case the model would have a finite number of degrees of freedom, the actual number corresponding to the number of points selected. This type of model is usually referred to as a *lumped* or *concentrated-mass* system. While it is assumed that the mass is concentrated at certain specific locations and the remainder of the member is considered to be weightless, the beam is presumed to retain its full structural strength and stiffness.

The member shown in fig. 4.76 could be modeled in a number of ways, depending upon the degree of sophistication desired. Some of these are shown in fig. 4.77. The first one shown involves the motion of only one point. The entire mass of the beam is presumed to be concentrated at the center, and the displacement at this point is the only dependent variable. This is, therefore, an example of a single-degree-of-freedom system. The remaining cases represent two-, three-, and four-degree-of-freedom models. In each case the beam has been divided into the desired number of parts and the mass of that part has been concentrated at the center of that segment. Clearly, as the number of degrees of freedom employed increases, the error involved in the approximation decreases, and the solutions approach the infinite-degree-of-freedom case.

For many structural systems the error associated with assuming a limited number of degrees of freedom is relatively small. For example, in fig. 4.74 it was specified that the mass at the end of the cantilever was large in comparison to the mass of the beam. In this instance, the error resulting from a single-degree-of-freedom assumption would in all likelihood not be excessive. For the three-story rigid frame shown in fig. 4.78, it would not be unreasonable to assume that by far the greater part of the total mass is located in the floor systems and in the roof. The three-degree-of-freedom model shown would thus seem to be appropriate. In any event, the particular model chosen to represent a given structure will depend upon the accuracy desired in the solution.

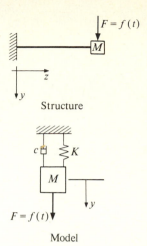

Structure

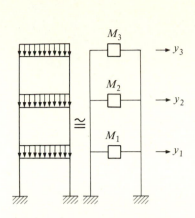

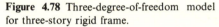

Figure 4.78 Three-degree-of-freedom model for three-story rigid frame.

Model

Figure 4.79 Single-degree-of-freedom model.

4.7.1 Single-Degree-of-Freedom Systems

It is common in a dynamic analysis to represent the structural system being studied by a series of interconnected lumped masses, springs, and dashpots. These are joined in such a way that the equation of motion of the mass accurately reflects the motion of the system being modeled. The cantilever beam in fig. 4.74, for example, would be idealized by the spring-mass model shown in fig. 4.79.

The force F represents an external force acting at the free end of the member that varies with time according to the function $f(t)$. It is assumed that there is an internal damping force resulting from slippage at the connection and from the internal resistance of the atoms or molecules of the material. Internal damping of this type is generally presumed to be viscous, and the resulting damping force thus acts in a direction opposite to the direction in which the mass is moving and is proportional to the velocity of the mass. In equation form, this would be

$$\text{Damping force} = F_D = -c\,\frac{dy}{dt} \qquad (4.157)$$

where c is the constant of proportionality (known as the damping constant), dy/dt is the velocity of the mass, and the negative sign reflects the deceleration of the damping force. The spring constant K represents the bending stiffness of the beam, and is presumed to be linear. (A detailed discussion of the evaluation of this constant is given in a later chapter of this volume.)

The equilibrium of the spring-mass model of fig. 4.79 can be defined directly from the free body shown in fig. 4.80, using Newton's second law of motion, or D'Alembert's principle of dynamic equilibrium. The applied force is $F = f(t)$. The resisting forces are the force in the spring, Ky, the damping force $c\dot{y}$, and inertia force

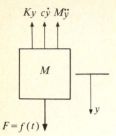

Ky $c\dot{y}$ $M\ddot{y}$

M

y

$F = f(t)$

Figure 4.80

$M\ddot{y}$. (The dots over the y's in fig. 4.80 denote derivatives with respect to t.) The resulting equilibrium equation is

$$M\ddot{y} + c\dot{y} + Ky = f(t) \tag{4.158}$$

Undamped free vibration When the force F is zero and there is no damping ($c = 0$), motion can occur only if the system is in some way disturbed. Otherwise it simply remains at rest in its static equilibrium state—that is, the displaced state resulting from the weight of the beam and mass. An initial displacement y_0, or an initial velocity $\dot{y}_0$, or a combination of the two would provide such a disturbance. Since in this case the motion is unaffected by any external force, it is a free-vibration problem and the equation of motion is simply

$$M\ddot{y} + Ky = 0$$

If ω^2 is defined as

$$\omega^2 = \frac{K}{M}$$

this could be written in the standard harmonic equation form

$$\ddot{y} + \omega^2 y = 0 \tag{4.159}$$

which has a solution

$$y = C_1 \sin \omega t + C_2 \cos \omega t \tag{4.160}$$

(It is to be recognized that as previously defined ω is the circular frequency.) C_1 and C_2 are the constants of integration and depend on the particular set of initial conditions specified for the problem in question. For a given initial displacement y_0, and an initial velocity $\dot{y}_0$ (both at $t = 0$), the governing equations would be

$$y_0 = C_1 \sin \omega(0) + C_2 \cos \omega(0)$$

$$\dot{y}_0 = C_1 \omega \cos \omega(0) + C_2 \omega \sin \omega(0)$$

From these

$$C_1 = \frac{\dot{y}_0}{\omega} \quad \text{and} \quad C_2 = y_0$$

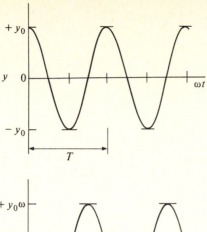

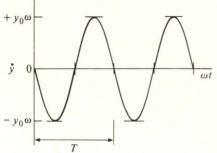

Figure 4.81 Deformational response as a function of time.

The equation of motion for the system is therefore

$$y = \frac{\dot{y}_0}{\omega} \sin \omega t + y_0 \cos \omega t \qquad (4.161)$$

If it is presumed that the motion is initiated from rest with a displacement of y_0 (with $\dot{y}_0 = 0$), then the displacement of the mass at any time t is given by the equation

$$y = y_0 \cos \omega t \qquad (4.162a)$$

and the velocity would be

$$\dot{y} = -y_0 \omega \sin \omega t \qquad (4.162b)$$

These are shown graphically in fig. 4.81. It is to be recognized that the deformation is cyclic—it repeats itself over and over. This will always be the case for an undamped system.

It also should be noted that while the undamped system is not a real one, it does provide the information necessary to determine the maximum stresses in the member. (This is because of the fact that the effect of damping is to reduce the amplitude of vibration with time. The maximum amplitude is thus the same for both the undamped and the damped systems, and corresponds to the initial displacement y_0. The maximum stress—and thus the design—therefore would be governed by this quantity.) For the cantilever beam in question, the displacement at the free end is given by the equation

$$y_{\text{end}} = y_0 = \frac{Pl^3}{3EI} \qquad (4.163)$$

The load necessary to produce this deformation is therefore

$$P = \frac{3y_0 EI}{l^3}$$

The corresponding maximum moment (at the fixed end) is

$$M_{max} = Pl = \frac{3y_0 EI}{l^2}$$

and the maximum normal stress would be

$$\sigma_{max} = \frac{M_{max}}{S} = \frac{3y_0 EI}{Sl^2} \tag{4.164}$$

where S is the section modulus.

The period T, that is the time required to complete one cycle of vibration, is given by

$$T = \frac{2\pi}{\omega} = 2\pi \sqrt{\frac{M}{K}}$$

The natural frequency is

$$f = \frac{1}{T} = \frac{1}{2\pi} \sqrt{\frac{K}{M}} \tag{4.165}$$

Undamped forced vibration Forced vibration, as previously defined, is motion occurring in conjunction with an applied force $F = f(t)$. Assume that $f(t)$ is constant at a value of F_1, is applied suddenly, and remains for an indefinite period of time. For such a case the differential equation is

$$\ddot{y} + \frac{K}{M} y = \frac{F_1}{M} \tag{4.166}$$

and the solution is

$$y = C_1 \sin \omega t + C_2 \cos \omega t + \frac{F_1}{K} \tag{4.167}$$

If it is assumed that $y_0 = \dot{y}_0 = 0$ at $t = 0$

$$C_1 = 0 \quad \text{and} \quad C_2 = -\frac{F_1}{K}$$

The final solution is therefore

$$y = \frac{F_1}{K}(1 - \cos \omega t) \tag{4.168}$$

which is shown graphically in fig. 4.82.

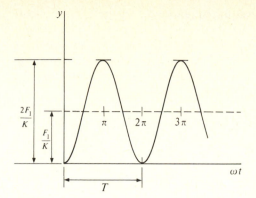

Figure 4.82

From this figure it is to be observed that this forced case is similar to that for free vibration. The only difference is that the horizontal (time) axis has been shifted vertically by an amount F_1/K.

The maximum normal stress develops in such a system each time the value of y reaches its maximum value of $2(F_1/K)$. This is exactly twice the displacement and therefore twice the stress that would have developed had the load been applied "statically." The maximum stress for this case is therefore obtained from

$$\frac{M_{max}l^2}{3EI} = 2\frac{F_1}{K}$$

or

$$\sigma_{max} = \frac{6F_1 EI}{Sl^2 K} \tag{4.169}$$

One convenient way in which this dynamic load increase can be described is through the use of the *dynamic load factor* (DLF). The dynamic load factor is defined as the ratio of the dynamic deformation (caused by the load F) to the static deformation (produced by the same load). For the example just considered, the static deformation is F_1/K. The dynamic load factor is therefore

$$\text{DLF} = \frac{y_{dyn}}{y_{stat}} = \frac{(F_1/K)(1 - \cos \omega t)}{F_1/K} = 1 - \cos \omega t \tag{4.170}$$

This is shown in fig. 4.83.

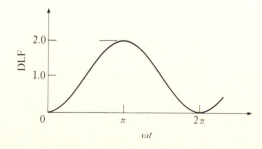

Figure 4.83 Dynamic load factor as a function of time.

In a large number of instances only the maximum value of the dynamic load factor is of practical interest since this defines the worst possible state of deflection and, correspondingly, stress. In the preceding example, the applied load was specified as constant with time. This represents the most elementary case of dynamic loading. In more complex cases in which the applied load is not constant, or there are several loads, the dynamic load factor is based upon some arbitrarily selected load value.

Next, consider the case where the load F_1 is applied for a limited time only, say, from $t = 0$ to $t = t_d$. During this time period, the motion of the system is given by the equation

$$y = \frac{F_1}{K}(1 - \cos \omega t) \qquad 0 \le t \le t_d \tag{4.171}$$

At $t = t_d$

$$y_{t_d} = \frac{F_1}{K}(1 - \cos \omega t_d)$$

and

$$\dot{y}_{t_d} = \frac{F_1}{K}\omega(\sin \omega t_d)$$

Since the load is removed at $t = t_d$, motion from this time on is free and thus described by the relationship

$$y = \frac{\dot{y}_0}{\omega}(\sin \omega t) + y_0(\cos \omega t) \qquad t_d \le t \le \infty \tag{4.172}$$

Substituting into this expression the initial conditions

$$y_0 = y_{t_d} = \frac{F_1}{K}(1 - \cos \omega t_d)$$

$$\dot{y}_0 = \dot{y}_{t_d} = \frac{F_1}{K}\omega(\sin \omega t_d)$$

Replacing t by t_d, there is obtained

$$y = \frac{F_1}{K}\{(\sin \omega t_d)[\sin \omega(t - t_d)] + (1 - \cos \omega t_d)[\cos \omega(t - t_d)]\}$$

or

$$y = \frac{F_1}{K}[\cos \omega(t - t_d) - \cos \omega t] \tag{4.173}$$

Since F_1/K is the static deformation under the load F_1, the dynamic load factor is given by

$$\text{DLF} = 1 - \cos \omega t = 1 - \cos 2\pi \frac{t}{T} \qquad t \le t_d$$

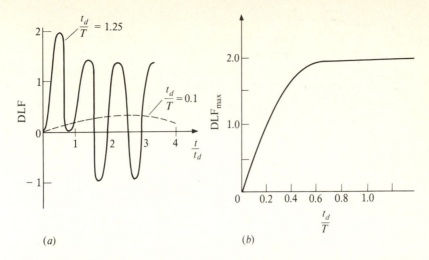

Figure 4.84 Dynamic load factor as a function of the ratio of the time of load application to the natural period of the system.

and

$$DLF = \cos \omega(t - t_d) - \cos \omega t$$

$$= \cos 2\pi \frac{t - t_d}{T} - \cos 2\pi \frac{t}{T} \qquad t \geq t_d \qquad (4.174)$$

From this it should be apparent that the important parameter in defining the dynamic load factor is really the ratio of the actual time t to the natural period T of the system—and not either one separately. Since the breakpoint in the solution is the time t_d, it is convenient in this case to use the ratio of t_d/T to describe the DLF. Figure 4.84a shows the dynamic load factor versus t for the particular ratios $t_d/T = 1.25$ and 0.10. Part b of the figure illustrates the relationship between the maximum dynamic load factor and the ratio t_d/T. It is to be noted that for this case of rectangular loading, the maximum value of the dynamic load factor remains constant at a value of 2.0 for loads that are applied for more than one-half of the natural period, and decrease rapidly for loads applied for a shorter period of time.

Example 4.18 To illustrate these concepts, consider the simply supported steel beam shown in fig. 4.85a. The beam is 24 ft long, and supports at its midspan a mass which weight is 10 kips and a single concentrated force at the same point of 10 kips. The concentrated load is suddenly applied and then removed after an elapsed time of t_d as shown in fig. 4.85c. The weight of the beam is to be neglected. The spring constant for the single-degree-of-freedom model K would be determined as follows: The centerline deflection for a simple beam subjected to a concentrated load Q at the centerline would be

$$\Delta_{\mathbb{C}} = \frac{Ql^3}{48EI}$$

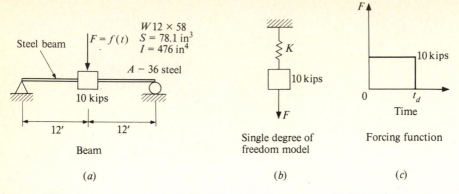

Figure 4.85

Therefore, the stiffness is given by:

$$K = \frac{Q}{\Delta_{\mathbb{C}}} = \frac{48EI}{l^3} = 28.69 \text{ kips/in}$$

The natural period of the beam-mass system, in seconds, is

$$T = 2\pi \sqrt{\frac{M}{K}} = 2\pi \sqrt{\frac{10}{(386)(28.69)}} = 0.189 \text{ s}$$

For the static load at midspan; that is, for the 10 kips + 10 kips = 20 kips load; the maximum normal stress would be

$$\sigma_{\max} = \frac{Ql}{4S} = \frac{(20)(24)(12)}{(4)(78.1)} = 18.44 \text{ ksi}$$

If t_d is taken equal to 0.05 s, $t_d/T = 0.264$ and the dynamic load factor is 1.43. The corresponding normal stress is therefore

$$\sigma_{\max} = 9.22 + (1.43)(9.22) = 22.40 \text{ ksi}$$

If, however, $t_d = 0.1$ s, $t_d/T = 0.529$, DLF $= 2.0$, and

$$\sigma_{\max} = 9.22 + (2.0)(9.22) = 27.66 \text{ ksi}$$

For any value of t_d greater than 0.1 s, the same value of maximum normal stress would be realized. (Thus, for load durations of over 0.1 s, the A-36 steel beam would be overstressed.)

Considering initial yielding as the failure criterion, the proper section to be used for this loading would be determined by trial and error. From a section modulus table† it is determined that the next lightest section that will fail by

---

† See American Institute of Steel Construction, *Manual of Steel Construction.*

yielding (as opposed to buckling) is a W 14 × 68 ($S = 103$ in^3, $I = 724$ in^4). For this section,

$$K = \frac{48EI}{l^3} = \frac{(48)(30,000)(724)}{(24 \times 12)^3} = 43.64 \text{ kips/in}$$

$$T = 2\pi \sqrt{\frac{10}{(386)(43.64)}} = 0.153 \text{ s}$$

For $t_d = 0.1$ s

$$\frac{t_d}{T} = \frac{0.1}{0.153} = 0.653$$

which gives DLF = 2.0. The static stress is

$$\sigma_{max} = \frac{QL}{4S} = \frac{(10 + 10)(24)(12)}{4(103)} = 13.98 \text{ kips}$$

The dynamic stress is

$$\sigma_{max} = \frac{13.98}{2} + (2.00)\left(\frac{13.98}{2}\right) = 20.97 \text{ ksi}$$

which is satisfactory.

Damped free vibration As was noted earlier, if concern is only for the establishment of the maximum deformation, examination of the undamped system is sufficient. If, however, it is desired to describe the oscillations after the initial disturbances, the effect of damping must be included.

For the one-degree-of-freedom system shown in fig. 4.79 with viscous damping, equilibrium is defined by the relationship

$$M\ddot{y} + c\dot{y} + Ky = f(t) \tag{4.175}$$

For free vibration where $f(t) = 0$, the response would be

$$y = e^{-\beta t}(C_1 \sin \omega_d t + C_2 \cos \omega_d t)$$

where

$$\beta = \frac{c}{2M}$$

and

$$\omega_d = \sqrt{\omega^2 - \beta^2} \tag{4.176}$$

ω_d is the natural frequency of the damped system, and is a measure of the amount of damping present. Depending upon the relative magnitudes of the circular frequency ω and the quantity β, this can be real, zero, or even imaginary. If $\omega < \beta$, the system is said to be overdamped, and ω_d is not real. Such a case of limited practical importance. The limit of real response is the case of $\omega = \beta$. This particular case is referred to as one of critical damping, and while it is of relatively little importance in itself, it does serve as a convenient reference point for describing the amount of damping that is actually present. For example, it has been observed by studying real structures that damping typically exists between 5 and 10 percent of critical damping c_{cr}. For critical damping

$$\omega = \beta = \frac{c_{cr}}{2M}$$

or

$$c_{cr} = 2M\omega = 2\sqrt{KM} \tag{4.177}$$

To determine the integration constants C_1 and C_2, use must be made of the initial conditions (at $t = 0$).

$$y = y_0 \quad \text{and} \quad \dot{y} = \dot{y}_0$$

Substituting the general solution into these expressions yields

$$C_1 = \frac{\dot{y}_0 + y_0\beta}{\omega_d}$$

and

$$C_2 = y_0$$

The solution is therefore

$$y = e^{-\beta t}\left(\frac{\dot{y}_0 + \beta y_0}{\omega_d} \sin \omega_d t + y_0 \cos \omega_d t\right) \tag{4.178}$$

This is shown graphically for the case where $\dot{y}_0 = 0$ in fig. 4.86. Note that the exponential term $e^{-\beta t}$ defines an envelope of decreasing amplitudes within which the oscillations must remain.

Introducing critical damping into the solution (that is, $\omega = \beta$)

$$y = e^{-\omega t}[\dot{y}_0 t + (1 + \omega t)y_0] \tag{4.179}$$

which is clearly nonperiodic. For this case, and assuming an initial displacement y_0, the system merely returns to its at-rest position with no oscillation at all.

As noted earlier, all real systems are damped to some degree. Since the natural frequency of the damped system ω_d is different from that of the undamped system, it

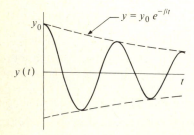

Figure 4.86 Envelope of amplitudes.

would be helpful to know how much error is introduced by neglecting the damping in computing the natural frequency. To bound the error, it will be assumed that the system in question has 15 percent of critical damping. (Note that this is not a typical structure, since it does not fall within the 5 to 10 percent range.) For $\beta = 0.15\omega$

$$\omega_d = \sqrt{\omega^2 - \beta^2} = \sqrt{0.9775\omega^2} = 0.989\omega$$

The difference is clearly not large, even for this extreme case. For this reason, it is common practice to ignore the effects of damping when computing natural frequencies.

Periodic loading A periodic loading is one that forces the structure to vibrate, at least at the point of application, at the same frequency and at the same amplitude as that of the load. A common source of such a disturbance would be a machine having a heavy flywheel rotating out of balance. A time-dependent loading of this type is usually described by an equation having the form

$$F = f(t) = F_M \sin \alpha t \tag{4.180}$$

where F_M is the amplitude of the load F, and α is the circular frequency of the load.

If such a load is applied to an undamped, single-degree-of-freedom system for an indefinite period of time, the equation of motion is

$$M\ddot{y} + Ky = F_M \sin \alpha t \tag{4.181}$$

and its solution is

$$y = C_1 \sin \omega t + C_2 \cos \omega t + \frac{F_M}{M}\left(\frac{\sin \omega t}{\omega^2 - \alpha^2}\right) \tag{4.182}$$

where ω, as previously defined, is the circular frequency. Presuming that the system starts from rest (that is, $y_0 = 0$ and $\dot{y}_0 = 0$), the constants of integration are

$$C_1 = -\frac{F_M}{M}\frac{\alpha}{\omega}\frac{1}{\omega^2 - \alpha^2}$$

and

$$C_2 = 0$$

This gives for the dynamic displacement

$$y = \frac{F_M}{M(\omega^2 - \alpha^2)}\left(\sin \alpha t - \frac{\alpha}{\omega}\sin \omega t\right) \tag{4.183}$$

The dynamic load factor would be

$$\text{DLF} = \frac{y_{\text{dyn}}}{y_{\text{stat}}} = \frac{1}{1 - (\alpha/\omega)^2}\left(\sin \alpha t - \frac{\alpha}{\omega}\sin \omega t\right) \tag{4.184}$$

It is to be recognized that the overall response is dependent upon (a) the natural frequency of the system, and (b) the frequency of the load.

The maximum value of the DLF, for a given value of α/ω, can be obtained by graphing the two separate parts of eq. (4.184), and then adding them together. The

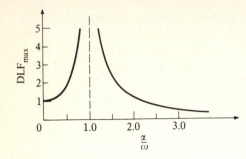

Figure 4.87 The maximum dynamic load factor as a function of the ratio of the circular frequency of an applied periodic loading to the natural frequency of the system.

maximum values so obtained are shown in fig. 4.87. If the frequency of the load α is very small when compared to the natural frequency ω, the dynamic load factor approaches 1.0, and there is little dynamic effect. If α is very large compared to ω, the dynamic load factor approaches zero. Here, for very high frequencies of the loading, the beam is unable to respond to the rapid oscillations and so remains stationary.

From the figure it is evident that something significant happens when $\alpha = \omega$, that is, when the load and the structures are in resonance one with the other. Substituting these values into eq. (1.184) the indeterminate case 0/0 is obtained, and the solution must be found using l'Hospital's rule

$$\lim \text{DLF}_{\alpha \to \omega} = \frac{t \cos \alpha t - (1/\omega) \sin \omega t}{-2(\alpha/\omega^2)}$$

or

$$\text{DLF}_{\alpha = \omega} = \tfrac{1}{2}(\sin \omega t - \omega t \cos \omega t) \tag{4.185}$$

This indicates that at resonance ($\alpha = \omega$) the dynamic load factor increases steadily with time and becomes infinite at $t = \infty$. The significance, from a theoretical point of view, should be obvious: When the pulsating load is at or near the natural frequency of the system, displacements and stresses grow larger and larger, and the structure destroys itself.

From a practical standpoint, the effect of damping must be considered, for it clearly plays a role in the response of the system at resonance. The equation of motion for damped single-degree-of-freedom systems subjected to pulsating loads is given by eq. (4.175), or

$$M\ddot{y} + c\dot{y} + Ky = F_M \sin \alpha t \tag{4.186}$$

Its solution is

$$y = e^{-\beta t}(C_1 \sin \omega_d t + C_2 \cos \omega_d t) + \frac{F_M}{K} \left[\frac{(1 - \alpha^2/\omega^2) \sin \alpha t - 2(\alpha\beta/\omega^2) \cos \alpha t}{(1 - \alpha^2/\omega^2) + 4(\alpha\beta/\omega^2)^2} \right] \tag{4.187}$$

where

$$\beta = \frac{c}{2M} \quad \text{and} \quad \omega_d = \sqrt{\omega^2 - \beta^2}$$

Since primary concern is with the damped system, the displacement associated with the free part of the vibration will disappear after a few cycles, and attention need be focused only on the forced part of the vibration. This forced displacement can be expressed in the form

$$y_f = \frac{(F_M/K)[(1 - \alpha^2/\omega^2)^2 + 4(\alpha\beta/\omega^2)^2]^{1/2} \sin(\alpha t + \phi)}{(1 + \alpha^2/\omega^2)^2 + 4(\alpha\beta/\omega^2)^2} \qquad (4.188)$$

where ϕ is the phase angle. The maximum displacement occurs when $\sin(\alpha t + \phi) = 1.0$. For that case the dynamic load factor is given by

$$\text{DLF}_{max} = \left[\left(1 - \frac{\alpha^2}{\omega^2}\right)^2 + 4\left(\frac{\alpha\beta}{\omega^2}\right)^2\right]^{-1/2} \qquad (4.189)$$

A graph of this relationship is shown in fig. 4.88. As indicated, even a small amount of damping prevents the occurrence of infinite amplitudes at resonance. However, the amplitudes may be large, remembering that normal structural damping is 5 to 10 percent of critical damping. It is clear that if at all possible, resonant excitation should be avoided.

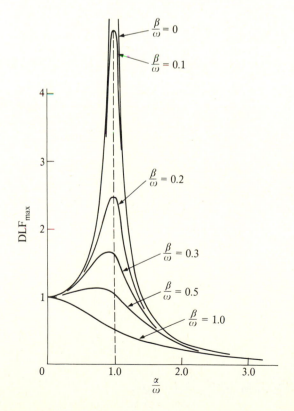

Figure 4.88 Maximum dynamic load factors for periodic loadings.

4.7.2 Multidegree-of-Freedom Systems

The procedures used to analyze multidegree-of-freedom systems are direct extensions of those developed for the single-degree-of-freedom case. For illustration, consider the two-degree-of-freedom structure shown in fig. 4.89. It is assumed that the cantilever in question carries a dynamic load F_T and a mass M at its end, and a mass of $2M$ at its center. Sketches of fig. 4.89b describe the two-degree model, as well as the corresponding free body diagrams of each mass.

For mass 1

$$2M\ddot{y}_1 + Ky_1 - K(y_2 - y_1) = 0$$

For mass 2 $\qquad\qquad\qquad\qquad\qquad\qquad\qquad\qquad\qquad\qquad$ (4.190)

$$M\ddot{y}_2 - Ky_1 + \tfrac{5}{4}K(y_2) = F_T$$

The natural frequency would be obtained from a consideration of the homogeneous equations

$$2M\ddot{y}_1 + Ky_1 - K(y_2 - y_1) = 0$$

$$M\ddot{y}_2 - Ky_1 + \tfrac{5}{4}Ky_2 = 0$$

(4.190a)

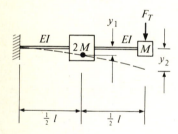

(a) Structure and loading

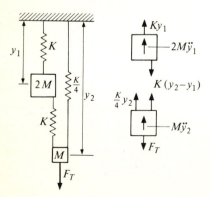

(b) Assumed model $\qquad\qquad\qquad\qquad\qquad$ **Figure 4.89** Multidegree-of-freedom beams.

Moreover, since (by definition of natural frequency) both masses will oscillate with the same frequency

$$y_1 = c_1 \sin \omega t$$

$$y_2 = c_2 \sin \omega t \qquad (4.191)$$

where c_1 and c_2 are constants to be determined and ω is the natural circular frequency. Substituting eq. (4.191) into (4.190a) yields

$$(2K - 2M\omega^2)c_1 - Kc_2 = 0$$

$$-Kc_1 + (\tfrac{5}{4} - M\omega^2)c_2 = 0 \qquad (4.192)$$

Since values other than zero for c_1 and c_2 can exist only if the value of the determinant of the coefficients is zero,

$$\begin{vmatrix} 2K - 2M\omega^2 & -K \\ -K & \tfrac{5}{4}K - M\omega^2 \end{vmatrix} = 0$$

or

$$(\omega^2)^2 - \frac{9}{4}\frac{K}{M}\omega^2 + \frac{3}{4}\frac{K^2}{M^2} = 0 \qquad (4.193)$$

Solving this for ω^2 gives

$$\omega^2 = (1.125 \pm 0.718)\frac{K}{M}$$

The natural circular frequencies are therefore

$$\omega_1 = 1.358\sqrt{\frac{K}{M}} \qquad \text{and} \qquad \omega_2 = 0.638\sqrt{\frac{K}{M}} \qquad (4.194)$$

The fundamental mode shapes (not magnitudes) are obtained by substituting these values back into eq. (4.192).

For the higher mode, $\omega_1 = 1.358\sqrt{K/M}$

$$-1.686Kc_1 - Kc_2 = 0$$

$$-Kc_1 - 0.593Kc_2 = 0$$

or

$$c_1 = -0.593c_2$$

Setting $c_2 = 1.0$,

$$y_1 = -0.593 \sin \omega t$$

$$y_2 = 1.0 \sin \omega t$$

which is shown graphically in fig. 4.90.

For the lowest mode, where $\omega_2 = 0.638\sqrt{K/M}$

$$1.186Kc_1 - Kc_2 = 0$$

$$-Kc_1 + 0.843c_2 = 0$$

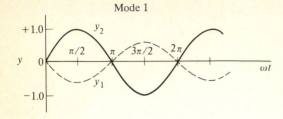

Figure 4.90

which gives

$$c_1 = 0.843c_2$$

If c_2 is selected as unity

$$y_1 = 0.843 \sin \omega t$$

$$y_2 = 1.0 \sin \omega t$$

and the graph is shown in fig. 4.91.

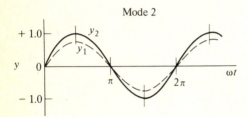

Figure 4.91

To define the actual response of such a system, the governing differential equations (4.190) must be solved. For illustration, assume that F_T is defined as shown in fig. 4.92 (which is typical of an impact type of loading). Further, assume that in fig. 4.89 $M = 5$ kips (including the weight of the beam), and the member is a W 14 × 87 shape ($S = 138$ in^3, $I = 967$ in^4) and is made of an ASTM type A-440 steel ($\sigma_y = 50$ ksi). If the length is $l = 20$ ft, the spring constant is

$$K = 4\frac{3EI}{l^3} = 25.182 \text{ kips/in}$$

The differential equations then become

$$\ddot{y}_1 + 1946.06y_1 - 973.03y_2 = 0$$

and   (4.195)

$$\ddot{y}_2 - 1946.06y_1 + 2432.58y_2 = 77.28F_T$$

If, for example, the structure were at rest when the load is applied, the initial condition to be taken into account would be at $t = 0$

$$y_1 = \dot{y}_1 = 0 \quad \text{and} \quad y_2 = \dot{y}_2 = 0 \quad\quad (4.196)$$

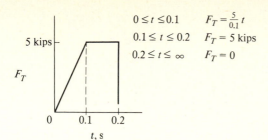

$$0 \leq t \leq 0.1 \qquad F_T = \frac{5}{0.1}t$$
$$0.1 \leq t \leq 0.2 \qquad F_T = 5 \text{ kips}$$
$$0.2 \leq t \leq \infty \qquad F_T = 0$$

Figure 4.92 Imposed loading as a function of time.

While a variety of procedures could be used to solve this problem, the numerical procedure defined previously as eq. (4.67) will be used. This fourth-order Runge-Kutta method requires that each equation to be solved be of the first order. Since the set in question is second order, the following substitutes will be introduced:

$$y_3 = \dot{y}_1 \qquad \text{and} \qquad y_4 = \dot{y}_2$$

The four first-order equations are then

$$\dot{y}_1 = y_3 \qquad\qquad\qquad = f_1$$
$$\dot{y}_2 = y_4 \qquad\qquad\qquad = f_2 \qquad\qquad (4.197)$$
$$\dot{y}_3 = 973.03 y_2 - 1946.06 y_1 \qquad = f_3$$
$$\dot{y}_4 = 77.28 F_T - 2432.58 y_2 + 1946.06 y_1 = f_4$$

Presuming a time increment $h = 0.0005$ s, and the initial conditions indicated in eq. (4.196), the Runge-Kutta formulation is

$$\Delta y_i = \tfrac{1}{6}(K_{1i} + 2K_{2i} + 2K_{3i} + K_{4i})$$

$$K_{1i} = 0.0005 f_i(t_0, y_{10}, y_{20}, y_{30}, y_{40})$$

$$K_{2i} = 0.0005 f_i\left(t_0 + \frac{0.0005}{2}, y_{10} + \frac{K_{11}}{2}, y_{20} + \frac{K_{12}}{2}, y_{30} + \frac{K_{13}}{2}, K_{40} + \frac{K_{14}}{2}\right)$$

$$K_{3i} = 0.0005 f_i\left(t_0 + \frac{0.0005}{2}, y_{10} + \frac{K_{21}}{2}, y_{20} + \frac{K_{22}}{2}, y_{30} + \frac{K_{23}}{2}, K_{40} + \frac{K_{24}}{2}\right)$$

$$K_{4i} = 0.0005 f_i(t_0 + 0.0005, y_{10} + K_{31}, y_{20} + K_{32}, y_{30} + K_{33}, y_{40} + K_{34})$$
$$(4.198)$$

where y_{i0} is the initial value of y_i, and for each step in time takes on the value of y_i computed in the previous step.

The process is begun by setting $t = 0$ and computing y_1, y_2, y_3, and y_4 at the end of the first increment of time (0.0005 s). Since f_i, $i = 1, 2, 3$, and 4, must be computed at t_0, $t_0 + \tfrac{1}{2}h$ and at $t_0 + h$, the values of F_T at those particular times must be known. In this range

$$F_T = \frac{5.0}{0.1}t$$

or

$$F_0 = 0, \quad F_{0.00025} = 0.125 \text{ kips} \qquad \text{and} \qquad F_{0.0005} = 0.25 \text{ kips}$$

Thus,

$$K_{11} = hy_{30} \qquad K_{12} = hy_{40} \qquad K_{13} = h(973.03y_{20} - 1946.06y_{10})$$

$$K_{14} = h(0 - 2432.58y_{20} + 1946.06y_{10})$$

$$K_{21} = h\left(y_{30} + \frac{K_{13}}{2}\right) \qquad K_{22} = h\left(y_{40} + \frac{K_{14}}{2}\right)$$

$$K_{23} = h\left[973.03\left(y_{20} + \frac{K_{12}}{2}\right) - 1946.06\left(y_{10} + \frac{K_{11}}{2}\right)\right]$$

$$K_{24} = h\left[(0.125)(77.28) - 2432.58\left(y_{20} + \frac{K_{12}}{2}\right) + 1946.06\left(y_{10} + \frac{K_{11}}{2}\right)\right]$$

$$K_{31} = h\left(y_{30} + \frac{K_{23}}{2}\right) \qquad K_{32} = h\left(y_{40} + \frac{K_{24}}{2}\right)$$

$$K_{33} = h\left[973.03\left(y_{20} + \frac{K_{22}}{2}\right) - 1946.06\left(y_{10} + \frac{K_{21}}{2}\right)\right]$$

$$K_{34} = h\left[(0.125)(77.28) - 2432.58\left(y_{20} + \frac{K_{22}}{2}\right) + 1946.06\left(y_{10} + \frac{K_{21}}{2}\right)\right]$$

$$K_{41} = h(y_{30} + K_{33}) \qquad K_{42} = h(y_{40} + K_{34})$$

$$K_{43} = h[973.03(y_{20} + K_{32}) - 1946.06(y_{10} + K_{31})]$$

$$K_{44} = h[(0.250)(77.28) - 2432.58(y_{20} + K_{32}) + 1946.06(y_{10} + K_{31})]$$

Solving these and introducing the resulting values into the first of the equations listed as (4.198) yields

$$y_{1_{0.0005}} = 0 \text{ in}$$

$$y_{2_{0.0005}} = 8.0500 \times 10^{-8} \text{ in}$$

$$y_{3_{0.0005}} = 9.7911 \times 10^{-9} \text{ in/s}$$

$$y_{4_{0.0005}} = 4.8298 \times 10^{-4} \text{ in/s}$$

The values are then used as the initial conditions for the next step.

The above described procedure is repeated, over and over, until the desired time limit is reached. Figure 4.93 is a plot of y_1 and y_2 obtained using time increments of $h = 0.0005$ s up to an elapsed time of 2.0 s.

In many, if not most cases, it is the stress caused by the dynamics loading that is of primary interest. But since normal stresses depend on bending moments and these, in turn, vary as the curvature, the values heretofore calculated are directly applicable. The only problem is the reconversion of the lumped-mass system to a continuous one. This can be accomplished in an approximate way by presuming that the displacement y at any time t can be represented by an nth order polynomial. The order should be such that each of the previously selected degrees of freedom can be accommodated as well as all of the boundary conditions.

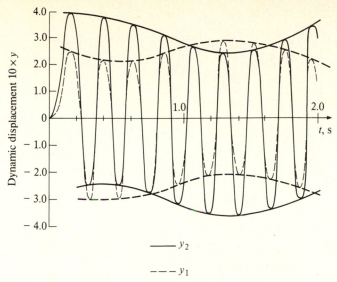

Figure 4.93 Displacement as a function of time.

For the case illustrated where there were selected two degrees of freedom and there were specified two initial conditions, a fourth-order expression is required (see fig. 4.94a).

$$y = a_0 + a_1 z + a_2 z^2 + a_3 z^3 + a_4 z^4$$

Since $y_0 = 0$, and $y_0' = 0$, $a_0 = a_1 = 0$. The remaining constants are

$$a_2 = \frac{1}{l^2}\left(12y_1 - \frac{9}{4}y_2\right)$$

$$a_3 = \frac{1}{l^3}\left(-20y_1 + \frac{23}{4}y_2\right)$$

$$a_4 = \frac{1}{l^4}\left(8y_1 - \frac{5}{2}y_2\right)$$

The moment is therefore

$$M = -EIy'' = -EI\left[\frac{2}{l^2}\left(12y_1 - \frac{9}{4}y_2\right) + \frac{6}{l^3}\left(-20y_1 + \frac{23}{4}y_2\right)z\right.$$
$$\left. + \frac{12}{l^4}\left(8y_1 - \frac{5}{2}y_2\right)z^2\right] \tag{4.199}$$

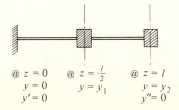

@ z = 0   @ z = $\frac{l}{2}$ @ z = l
 y = 0   y = y_1   y = y_2
 y' = 0   y'' = 0   **Figure 4.94a** Initial conditions.

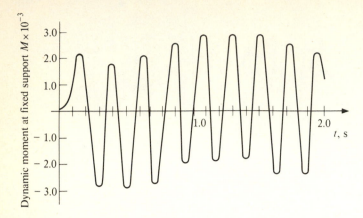

Figure 4.94b Moment at the support as a function of time.

The moment at the fixed end of the beam, as a function of time, is thus given by the expression

$$M_0 = -EI\left[\frac{2}{l^2}\left(2y_1 - \frac{9}{4}y_2\right)\right]$$

This is shown graphically in fig. 4.94b.

4.7.3 Distributed Mass and Load

A continuous formulation for the time-dependent elastic curve of a beam results when the inertia term is added to the formulation of chap. 2. That is,

$$EIy'''' = q(z) = w(t, z) - m\ddot{y}$$

where $w(t, z)$ is a distributed lateral load that is both a function of time and distance, and $m\ddot{y}$ is the inertia term with m the mass per unit of length. In its more normal form

$$EI\,\frac{\partial^4 y}{\partial z^4} + m\,\frac{\partial^2 y}{\partial t^2} = w(t, z) \tag{4.200}$$

This is a partial differential equation of the hyperbolic type and can be solved for the displacements y as a function of distance and time in a number of ways.[†]

One method of solution that is particularly well-adapted to a digital computer and problems involving beam bending is based upon the finite-difference method described earlier in this chapter. The procedure is to replace the term $\partial^4 y/\partial z^4$ in the differential equation by the fourth-order difference

$$\frac{\partial^4 y}{\partial z^4} \cong \frac{1}{h^4}\left(y_{-2h} - 4y_{-h} + 6y_0 - 4y_{+h} + y_{+2h}\right)$$

† For closed form solutions see C. R. Wylie, Jr., *Advanced Engineering Mathematics*, McGraw-Hill Book Company, New York, 1966. For numerical procedures, see R. L. Ketter and S. Prawel, *Modern Methods of Engineering Computations*, pp. 401–421, McGraw-Hill Book Company, New York, 1969.

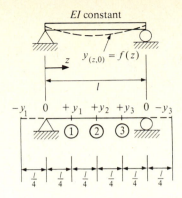

Figure 4.95 Four-segment finite-difference assumption.

The partial differential equation then becomes

$$\frac{EI}{h^4}(y_{-2h} - 4y_{-h} + 6y_0 - 4y_{+h} + y_{+2h}) + m\,\frac{d^2 y_0}{dt^2} = w(t_1 z_0) \qquad (4.201)$$

a second-order differential equation of the initial value type.

To illustrate the process, consider the simply supported beam shown in fig. 4.95, subjected to an initial displacement $y_0 = f(z)$. The boundary conditions are

$$\text{At } z = 0, \; y_{(0,\,t)} = 0, \qquad \text{At } z = l, \; y_{(l,\,t)} = 0$$

$$y''_{(0,\,t)} = 0 \qquad\qquad\qquad y''_{(l,\,t)} = 0$$

Therefore, at each of the node points 1, 2, and 3, eq. (4.201) can be written

$$\frac{d^2 y_1}{dt^2} + \frac{EI}{mh^4}(-y_1 + 0 + 6y_1 - 4y_2 + y_3) = 0$$

$$\frac{d^2 y_2}{dt^2} + \frac{EI}{mh^4}(0 - 4y_1 + 6y_2 - 4y_3 + 0) = 0$$

$$\frac{d^2 y_3}{dt^2} + \frac{EI}{mh^4}(y_1 - 4y_2 + 6y_3 + 0 - y_3) = 0$$

This set can be solved for displacements at 1, 2, and 3 as a function of time using the Runge-Kutta method illustrated for the two-degree-of-freedom system.

4.8 PROBLEMS

4.1 Starting from a statement of the governing differential equation and a listing of the appropriate boundary conditions, determine expressions for the elastic curve for each of the following:

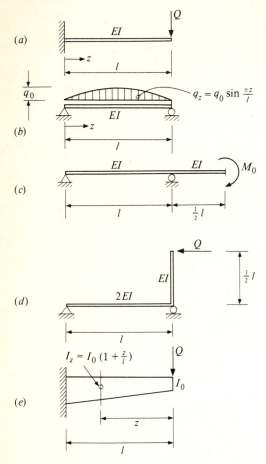

Figure P4.1

4.2 For each of the cases shown, find the indicated deformations:

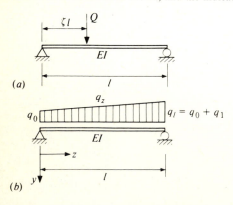

1. The elastic curve
2. The maximum deflection
3. The maximum deflection if $\zeta = \frac{1}{2}$

1. The elastic curve
2. The elastic curve, if $q_0 = 0$
3. The elastic curve, if $q_1 = 0$

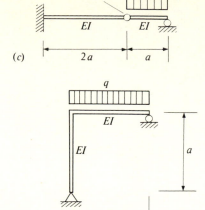

1. The elastic curve
2. The deflection at the interior hinge

(c)

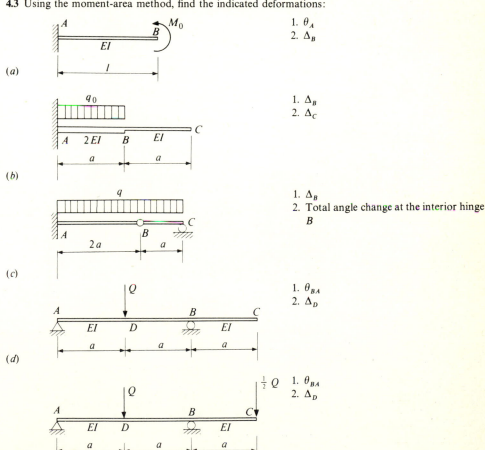

1. The elastic curve of the beam
2. The maximum horizontal deflection of the column
3. The rotation at the base of the column

(d)

Figure P4.2

4.3 Using the moment-area method, find the indicated deformations:

1. θ_A
2. Δ_B

(a)

1. Δ_B
2. Δ_C

(b)

1. Δ_B
2. Total angle change at the interior hinge B

(c)

1. θ_{BA}
2. Δ_D

(d)

1. θ_{BA}
2. Δ_D

(e)

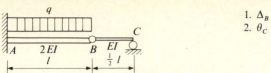

1. Δ_B
2. θ_C

Figure P4.3

(*f*)

4.4 Using the principle of superposition, find the indicated deformations.

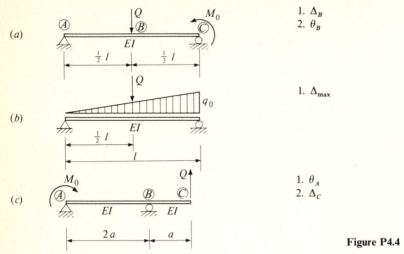

(*a*)

1. Δ_B
2. θ_B

(*b*)

1. Δ_{max}

(*c*)

1. θ_A
2. Δ_C

Figure P4.4

4.5 Using the Runge-Kutta method, establish the elastic curve for the following cases:

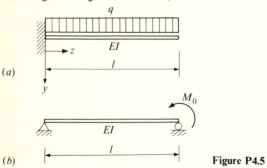

(*a*)

(*b*)

Figure P4.5

4.6 Using the finite difference method, determine points on the elastic curve for the following cases:

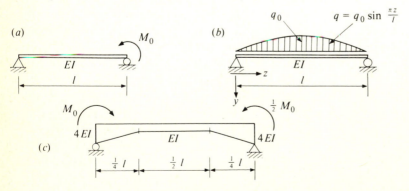

(*a*)

(*b*) $q = q_0 \sin \dfrac{\pi z}{l}$

(*c*)

Figure P4.6

4.7 Using any method, find the indicated deformations:

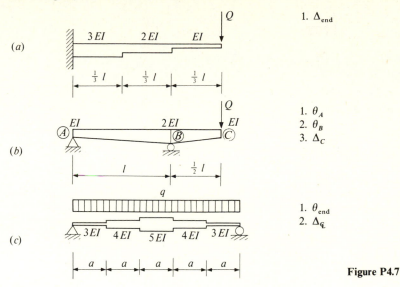

(a)

1. Δ_{end}

(b)

1. θ_A
2. θ_B
3. Δ_C

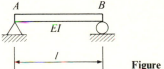

(c)

1. θ_{end}
2. $\Delta_{\mathbb{C}}$

Figure P4.7

4.8 For the beam shown, draw the influence line for θ_A.

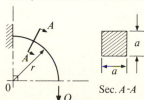

Figure P4.8

4.9 Considering only bending deformations, determine θ_B and Δ_C.

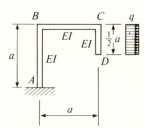

Figure P4.9

4.10 Repeat prob. 4.9, but include bending, shear, and axial deformations.

4.11 Determine the location of the centerline at the end of the curved beam shown, after loading. Bending, shear, and axial deformations are to be included.

Sec. A-A

Figure P4.11

4.12 For the beam shown, find the ratio of midspan bending deformation to midspan shear deformation.

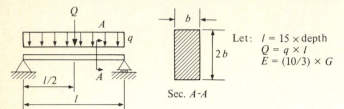

Let: $l = 15 \times \text{depth}$
$Q = q \times l$
$E = (10/3) \times G$

Sec. A-A

Figure P 4.12

4.13 For the beam column shown, derive the following:

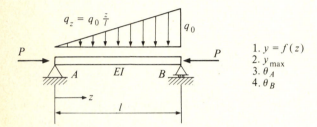

$q_z = q_0 \frac{z}{l}$

q_0

1. $y = f(z)$
2. y_{max}
3. θ_A
4. θ_B

Figure P4.13

4.14 A cantilever beam of length l, bending stiffness EI, and cross-sectional area A is subjected to an axial tensile force P, and a lateral load at its end of Q. Determine the equation of the elastic curve. If Δ is the deflection at the free end of the member, plot the variation in Δ as a function of several values of P/Q.

4.15 If the axial force P in prob. 4.14 is compressive, find the equation of the elastic curve. What are the upper limiting values for the force P?

4.16 For the member shown, determine θ_A and Δ_C.

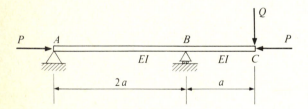

Figure P4.16

4.17 A rectangular beam (1 by 4 in in cross section) is subjected to an eccentric tensile force of 20 kips. Determine the normal stresses at the top and bottom of the beam.

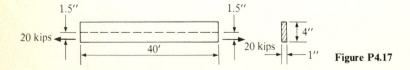

Figure P4.17

4.18 Including the interaction effects, determine the midspan deflection of the beam shown in prob. 4.17, assuming that $E = 30,000$ ksi, that 20-kip axial force is in compression, and that the beam column does not buckle.

4.19 Combined axial and bending stresses are defined by the equation

$$\sigma_{max} = \sigma_A + \sigma_B$$

where $\sigma_A = P/A$, and $\sigma_B = Mc/I$. If the maximum stress is specified as the yield stress, the interaction equation becomes

$$\frac{\sigma_A}{\sigma_y} + \frac{\sigma_B}{\sigma_y} = 1.0$$

and the interaction curve is that shown. For the member defined in prob. 4.17 determine the points on the interaction curve which correspond to (a) $e = 1.5$ in, (b) $e = 0$ in, and (c) $e = 15$ in.

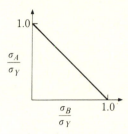

Figure P4.19

4.20 What is the maximum value of the compressive axial force P that can be carried by the column shown in prob. 2.34, considering minimum eccentricity as defined in ACI 318-71?

4.21 Select a rectangular reinforced concrete section to carry a compressive force of 100 kips at an eccentricity of 10 in. Let $\sigma_c' = 3500$ lb/in², $\sigma_y = 60$ ksi, and p minimum.

4.22 Determine P and M if $e = 25$ in, for the section shown in fig. 4.63.

4.23 Select a round reinforced concrete column to carry a compressive axial force of 100 kips at a minimum e, $\sigma_c' = 4000$ lb/in², and $\sigma_y = 60$ ksi.

4.24 A 12 by 18 in reinforced concrete column is reinforced by eight no. 7 bars. The bars are located as shown. The load P is compressive and is eccentric to the x-x axis 3 in and to the y-y axis 12 in. If $\sigma_c' = 4000$ lb/in² and $\sigma_y = 60$ ksi, what is the maximum value of P that the column can sustain?

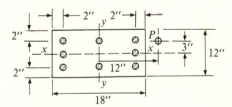

Figure P4.24

4.25 In steel frame buildings the exterior wall (brick or block) normally is carried by spandrel beams with the loads acting eccentric to the web of the beam. The spandrel beam therefore is subjected to a case of combined bending and torsion, as shown. If the beam in question is made of A-242 steel, and the ends are simply supported in both torsion and bending, find the maximum normal and shear stress in the beam.

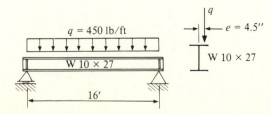

Figure P4.25

4.26 The beam section shown carries a crane rail with maximum crane wheel loads as indicated. Determine the maximum normal and shear stress assuming that the ends of the crane girder are simply supported and are free to warp. $L = 30$ ft.

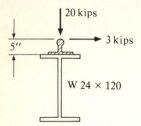

Figure P4.26

4.27 For the beam in prob. 4.26 let the vertical load of 20 kips be located at midspan. By lumping $\frac{1}{2}$ of the mass of the beam at midspan and $\frac{1}{4}$ at each of the simple supports, find the displacement and velocity of the midspan at $t = 10$ s if the 20-kip load is suddenly removed at $t = 2$ s. Assume that the beam is at rest just prior to the load removal.

4.28 It is assumed that the columns of the rigid frame shown have negligible weight when compared with the weight of the girder and the vertical loading. Moreover, it is assumed that the girder is rigid. Under these conditions, it is reasonable to presume that the system can be approximated by a single-degree-of-freedom model. Sketch that spring-mass model, and determine the natural frequency of vibration of the frame.

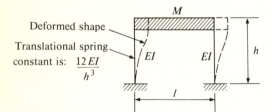

Figure P4.28

4.29 In a conventional reinforced concrete beam, steel reinforcing bars are placed in the concrete to resist the tension. The cracks that result in the concrete on the tension side of the member can be eliminated if these steel bars or tendons are installed under tension; that is, if the member is *prestressed*. This prestressing force in the steel causes compression in the concrete with the result that the cross section is in compression before any bending (due to any externally applied force) occurs. Thus,

$$\sigma = \sigma_{\text{prestress}} \pm \frac{My}{I}$$

Show that when $e = d/6$ the bending stress that can be applied without causing tension in the concrete is twice as large as for the case where $e = 0$, for the same maximum compressive stress in the beam.

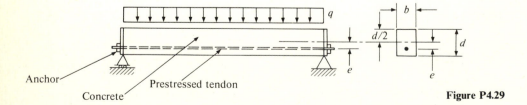

Figure P4.29

FIVE

DEFORMATIONS—ENERGY METHODS

In general, materials (and therefore structures) deform under the action of forces. As these deformations occur, work is done. Internally, energy is stored in the members. Externally, the applied forces move in the directions of the force vectors.

For most structural applications it is reasonable to presume that loads are applied gradually, and that an adiabatic condition of no significant heat exchange between the structure in question and its surroundings exists. For such cases there will be an equalization of work done by the applied forces and the strain energy stored in the body. (This, of course, presumes that appropriate account has been taken of the kinetic energy of any resulting rigid body motion.)

In the preceding chapter, deformations were determined by solving appropriate (equilibrium defined) differential equations subject to particular boundary conditions. For many real cases, it was demonstrated that such solutions are, at best, long and tedious. For these types of problems, as well as for other less difficult ones, deformational solutions obtained by minimizing energy expressions are much more readily obtained. It is the primary purpose of this chapter to develop such procedures.

5.1 STRAIN ENERGY VERSUS COMPLEMENTARY ENERGY

In structural mechanics, it is the energy stored in the individual members and the work done by the applied forces (or stresses) as they move through displacements (or strains) that are of primary concern. Typical examples of these are shown in fig. 5.1.

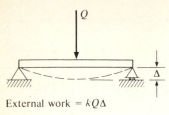

External work = $kQ\Delta$

(a)

Internal work = $\int_v c(\sigma)(\varepsilon)\, dv$

(b)   **Figure 5.1** Examples of external and internal work.

In equation form, they would be written as follows:

$$W_{int} = \int_V c(\sigma)(\varepsilon)\, dV$$

$$W_{ext} = \sum_{i=1}^{n} kQ_i\Delta_i$$

(5.1)

where c and k are constants of proportionality; Q_i is the applied external load; Δ_i is the displacement of the structure at the load point, in the direction of Q_i; n is the number of applied forces; σ is the developed stress in the body (both normal and shearing stress), which can and probably will vary throughout the structure; and ε is the corresponding strain.

Unlike the previously discussed concept of equilibrium—which could be applied to particular subportions of a structure, even a unit element—energy methods are based on the concept of equality of internal energy stored and external work done for the entire system in question. It should be recognized that these two concepts are different, though in many situations one can serve as a direct substitute for the other.

Since it is equality of energy and work that is necessary for solution, it is important that there be a clear understanding of the types of systems that are *conservative* and which ones are not. For the purposes of this text, a system is conservative when all of the internal and external energies are retained and can be recovered upon removal of the applied loads. (This in no way requires linearity of load-deformational response. It is only required that the system respond elastically to the loads imposed.) A nonconservative system, on the other hand, is one in which energy is "lost" (that is, changed in form) during deformation. Such losses could be the result of friction, inelastic behavior, etc. Conservative systems will be presumed throughout this chapter.

Energy that is stored in a structure as a result of the action of applied loads, accelerations, temperature changes, etc., is referred to as *strain energy*. Consider the

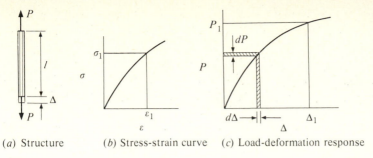

(*a*) Structure   (*b*) Stress-strain curve   (*c*) Load-deformation response

Figure 5.2 Load-deformation response of tension member.

single tension bar structure shown in fig. 5.2. The member is presumed to be of uniform cross section, and have a length *l*. Further, it is subjected to an axial tensile force which acts through the centroid of the cross section.

As P increases from zero, the member elongates. The actual magnitude of that elongation depends on the stress-strain properties of the material, the cross-sectional area A, the length l, and the applied load P. Assuming that the load is applied gradually, a curve similar to that shown in fig. 5.2*c* is obtained. Since the strain energy stored in the bar must be the same as the work done by the applied load

$$W_{\text{int}} = W_{\text{ext}}$$

or

$$W_{\text{int}} = \int dW_{\text{ext}} = \int_0^{\Delta_1} P \, d\Delta = \text{strain energy} \tag{5.2}$$

The total strain energy, then, is equal to the area *under* the load-deformation curve. If linearity of stress-strain response (and therefore load-deformation response) is presumed,

$$W_i = \tfrac{1}{2}P_1\Delta_1$$

Complementary energy, by definition, is the area *above* the load-deformation curve. Denoting this quantity by the symbol W_i^*,

$$W_i^* = \int_0^{P_1} \Delta \, dP = \text{complementary energy} \tag{5.3}$$

Again, presuming linearity of response,

$$W_i^* = \tfrac{1}{2}P_1\Delta_1$$

Assuming that both strain energy and complementary energy can be analytically defined, it should be evident from eqs. (5.2) and (5.3) that

$$\frac{dW_i}{d\Delta} = P$$

and

$$\frac{dW_i^*}{dP} = \Delta$$

$$(5.4)$$

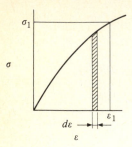

Figure 5.3 Stress-strain relationships.

This would suggest—and it is true—that strain energy formulations will be more useful in defining stress resultants; whereas, complementary energy is better suited to the calculation of deformations.

Strain energy also can be expressed in terms of stresses and strains. For example, if the stress-strain relationship is as shown in fig. 5.3, the response of a differential element would be

$$dW_i = \int_0^{\varepsilon_1} \sigma \, d\varepsilon \, dV$$

For the linear case, assuming only one component acting,

$$dW_i = \tfrac{1}{2}\sigma_1 \varepsilon_1 \, dV \tag{5.5}$$

It should be noted that eq. (5.5) expresses the strain energy for a differential element of volume at a point in the structure. To obtain the total strain energy, this would have to be integrated over the entire volume, taking into account the variation in σ and ε.

There are six stress components possible at any point in a three-dimensional system. The more general strain energy expression, therefore, would be

$$W_i = \sum_{j=1}^{6} \int_V \int_0^{\varepsilon} \sigma_j \, d\varepsilon_j \, dV$$

For the linear case,

$$W_i = \frac{1}{2} \sum_{j=1}^{6} \int_V \sigma_j \varepsilon_j \, dV \tag{5.6}$$

Similar complementary energy expressions could be defined.

Strain-Energy Components For linear systems, strain-energy components can be superimposed. Since most structural problems are of this type, consideration of the individual effects is both appropriate and desirable.

(a) Consider, for example, the axially loaded tension member shown in fig. 5.4a. The corresponding differential element is given in fig. 5.4b. The change in length of the element, due to the applied axial thrust, is given by eq. (4.24); that is,

$$d(\Delta L) = \varepsilon \, dz = \frac{P}{AE} \, dz$$

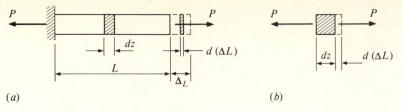

(*a*)   (*b*)

Figure 5.4 Deformation due to tensile loading.

or

$$P = AE \, \frac{d(\Delta L)}{dz} \tag{5.7}$$

The strain energy stored in the element is therefore

$$dW_i = \frac{1}{2} P \, d(\Delta L) = \frac{1}{2} \frac{P^2}{AE} \, dz = \frac{1}{2} AE \left[\frac{d(\Delta L)}{dz} \right]^2 dz$$

For the full member

$$W_i = \int_0^L \frac{P^2}{2AE} \, dz = \int_0^L \frac{AE}{2} \left[\frac{d(\Delta L)}{dz} \right]^2 dz \tag{5.8}$$

(*b*) The strain energy associated with beam bending is determined in a similar fashion, considering the element shown in fig. 5.5.† From eq. (4.27),

$$d\theta = \frac{M}{EI} \, dz$$

For the differential length element, then, the strain energy stored would be

$$dW_i = \frac{1}{2} M \, d\theta = \frac{1}{2} \frac{M^2}{EI} \, dz$$

For the entire member

$$W_i = \int_0^L \frac{M^2}{2EI} \, dz = \int_0^L \frac{EI}{2} \left(\frac{d^2 y}{dz^2} \right)^2 dz \tag{5.9}$$

† It should be noted that M has been assumed in the same direction as the rotation $d\theta$. This is not the same convention defined in the differential equation solutions of chap. 4.

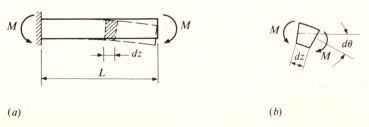

(*a*)   (*b*)

Figure 5.5 Deformation due to bending.

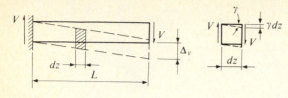

(a)

(b)

Figure 5.6 Deformation due to transverse shear.

(c) The influence of transverse shear is shown in fig. 5.6. Since

$$\gamma = \frac{\tau}{G} = \frac{\lambda V}{AG}$$

and

$$dW_i = \frac{1}{2} V\gamma \, dz = \frac{1}{2} \lambda \frac{V^2}{AG} \, dz$$

the strain energy stored in the member is

$$W_i = \int_0^L \frac{\lambda}{2} \frac{V^2}{AG} \, dz = \int_0^L \frac{AG}{2\lambda} \left(\frac{d\Delta_v}{dz}\right)^2 dz \tag{5.10}$$

(d) When computing the strain energy associated with members subjected to torsional moments, it is desirable to separate the effects of St. Venant and warping. That is,

$$W_i = W_{i,\,sv} + W_{i,\,\omega}$$

From eq. (4.31)

$$\frac{d\theta}{dz} = \frac{T_{sv}}{G\kappa_T}$$

The St. Venant strain energy associated with a unit length element is

$$dW_{i,\,sv} = \frac{1}{2} T_{sv} \, d\theta = \frac{1}{2} \frac{(T_{sv})^2}{G\kappa_T} \, dz$$

or

$$W_{i,\,sv} = \int_0^L \frac{1}{2} \frac{(T_{sv})^2}{G\kappa_T} \, dz = \int_0^L \frac{G\kappa_T}{2} \left(\frac{d\theta}{dz}\right)^2 dz \tag{5.11a}$$

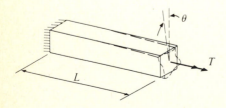

Figure 5.7 Deformation due to twisting moment.

For warping torsion

$$dW_{i,\,\omega} = 2(\tfrac{1}{2}M_F\, d\theta_F)$$

But from eq. (2.37),

$$M_F = EI_F\, \frac{d^2 u_F}{dz^2} = \frac{EI_y}{2}\, \frac{d\theta_F}{dz}$$

The warping strain energy associated with a unit length element is, then,

$$dW_{i,\,\omega} = M_F\, \frac{2M_F}{EI_y}\, dz = \frac{2}{EI_y}\, M_F^2\, dz$$

or

$$W_{i,\,\omega} = \int_0^L \frac{2}{EI_y}\, M_F^2\, dz = \int_0^L \frac{EI_y}{2}\, \frac{d^2 u_F}{dz^2}\, dz \qquad (5.11b)$$

This gives for the total strain energy

$$W_i = \int_0^L \left(\frac{1}{2}\, \frac{T_{sv}^2}{G\kappa_T} + \frac{2}{EI_y}\, M_F^2 \right) dz = \int_0^L \left[\frac{G\kappa_T}{2} \left(\frac{d\theta}{dz} \right)^2 + \frac{EI_\omega}{2} \left(\frac{d^2\theta}{dz^2} \right)^2 \right] dz \qquad (5.12)$$

where $I_\omega = \tfrac{1}{2}\, d^2 I_y$ for I and W sections.

For all the cases considered, it is to be understood that strain energy is a positive, linearly additive, scalar quantity. It is independent of the chosen sign system.

5.2 REAL WORK AND VIRTUAL WORK

When studying the energy-related behavior of structures, it is often convenient to presume deformations and forces that are not real. These fictitious quantities are referred to as *virtual displacements* and *virtual forces*. For illustration of the concept, consider the load-deformation curves shown in fig. 5.8. (The behavior illustrated in the left-hand sketch corresponds to a nonlinear, elastic situation. The right-hand graph is for the linear case.) As indicated in the figures, the areas *under* the curves to the left of the vertical line $\Delta = \Delta_1$ represent the strain energy that will be stored in the

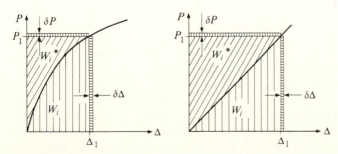

Figure 5.8 Load-deformation curves.

structure. Since this work is associated with real forces and real displacements, it is referred to as *real work*. Correspondingly, the *real complementary energy* W_i^* is represented by the area *above* the curve and below the horizontal line $P = P_1$. Since both W_i and W_i^* depend upon the actual $P - \Delta$ relationship realized, they are path-dependent, and the history of loading can be significant.

If, however, it is presumed that the structure is already deformed to the points P_1 and Δ_1, and is in equilibrium, and if there is then introduced a very small, additional (*virtual*) displacement or force, there will be additional work done. This additional virtual work, however, will not depend upon the previous history of loading.

Virtual strain energy, by definition, is

$$W_{iv} = P_1 \, \delta\Delta \tag{5.13a}$$

Correspondingly, *virtual complementary energy* is

$$W_{iv}^* = \Delta_1 \, \delta P \tag{5.13b}$$

To generate virtual strain energy there would be introduced into the system a set of virtual displacements that are compatible with the geometric boundary conditions. The virtual work would be the product of the real forces times these virtual displacements. If, on the other hand, virtual forces are introduced—which forces are in equilibrium—and if the real deformations are multiplied by these virtual forces, then there is generated virtual complementary energy.

For the linear case shown in the second graph of fig. 5.8, the virtual work done by both the virtual force δP and the virtual displacement $\delta\Delta$ is the same. (Also to be noted is the absence of the $\frac{1}{2}$ factor that is normally present in real work expressions.) Virtual strain energy and virtual complementary energy can therefore be used interchangeably in linearly elastic systems.

It is to be recognized that virtual work does not depend on the load-deformation relationship as does real work. The concept of virtual work, therefore, is much more general and can be used for any type of elastic or inelastic material. Real work is limited in application to elastic (conservative) systems.

The principle of virtual displacements The principle of virtual displacements follows directly from the preceding discussion, and can be stated as follows:

> *If a system is in equilibrium under a set of real external loads, then for any given virtual displacement, the work done by the external loads is the same as the internal energy stored.*

Implied in this statement are the following:

1. The principle, in effect, replaces the requirement of equilibrium, since it presumes that the system in question is in equilibrium when the virtual displacement is imposed.
2. The principle is completely general. It is as valid for nonconservative systems as it is for conservative ones. (This is because the virtual displacement is applied to a system that is, by definition, in a deformed state that is in equilibrium. The particular history of loading need not be known.)

3. The virtual displacement can be any geometrically acceptable displacement. (That is, it must satisfy the given geometrical boundary conditions, but need not necessarily satisfy the governing equations of equilibrium.)
4. The virtual displacement causes no changes in the stresses that are contained within the structure.

In equation form, the principle of virtual displacements is

Virtual work = virtual strain energy

$$\sum_{i=1}^{n} F_i D_i = \int_V \sigma \varepsilon \, dV \tag{5.14}$$

or

$$\sum_{i=1}^{n} (\text{real forces at } i)(\text{virtual displacements at } i)$$

$$= \int_V (\text{real internal stress at a point})(\text{virtual strain at the same point}) \, dV$$

The principle of virtual forces The complement to the principle of virtual displacements is the principle of virtual forces. This states

If the deformed shape of a given body is compatible with the specified constraints, then for any given virtual force the virtual complementary work will equal the virtual complementary energy.

This implies:

1. The system of virtual forces must, in themselves, satisfy equilibrium.
2. The virtual forces have associated with them virtual stresses. (These, however, cause no changes in the deformations within the structure.)

In equation form

$$\sum_{i=1}^{n} F_i D_i = \int_V \sigma \varepsilon \, dV \tag{5.15}$$

or

$$\sum_{i=1}^{n} (\text{virtual force at } i)(\text{real displacement at } i)$$

$$= \int_V (\text{virtual stress at a point})(\text{real strain at the same point}) \, dV$$

5.3 THE PRINCIPLES OF MINIMUM POTENTIAL ENERGY AND MINIMUM COMPLEMENTARY ENERGY

In differential calculus, when $y = f(z)$, where z is the independent variable, the first derivative of y with respect to z represents the rate of change of y as z increases. A vanishing first derivative indicates a maximum or a minimum value of the dependent variable. If, for example, y is the shape of the elastic curve of a structural member subjected to external loading, the following can be said:

> *Of all possible values of z, there is at least one value z_1 such that when $z = z_1$ the function $y = f(z)$, that is, the deflection of the beam, has its maximum (or minimum) value.*

In variational calculus, the concept is more generalized but the idea of minimization remains the same. If $x = f(y)$ and $y = f_1(z)$, the maximum or minimum value of x depends upon how the entire function $y = f_1(z)$ changes—rather than upon the value of a single variable. Consider, for example, the case of a member that is deformed due to the application of certain applied loads, and presume that the deflection curve associated with those loads is given by the function $y = f_1(z)$. The strain energy stored in the member W_i is a function of the deflection y, as is the external work done by the applied loads. If the total potential energy of the system (represented by the symbol Π) is defined as the sum of the strain energy W_i and the work done by the externally applied loads W_e, then

$$\Pi = W_i - W_e = f(y) \tag{5.16}$$

where y is the deflection. (It is to be recognized that as the applied force(s) move through the displacement(s), there is a decrease in total potential realized. Thus the negative sign preceding W_e.) Equation (5.16) is an expression of the total potential energy of the system in question and is a function of y. y, in turn, is a function of the distance z along the member. Clearly, any change in Π is dependent upon how the entire function y changes, rather than upon how y varies at a given point.

The rate of change of one function with respect to another is defined as the *variation*. *Variational calculus* involves the study of such changes. For the case referred to above, the potential energy function of a member subjected to external loading is given by the relationship

$$\Pi = W_i - W_e \tag{5.17}$$

For a stationary (maximum or minimum) value of Π, the first variation of the function, defined as

$$\delta\Pi = \delta(W_i - W_e)$$

will be equal to zero. That is,

$$\delta\Pi = \delta(W_i - W_e) = \delta W_i - \delta W_e = 0 \tag{5.18}$$

Using both the principle of virtual displacements and the concept of stationary potential energy, the very important principle of *minimum potential energy* can be stated:

> *Of all of the possible deformations that fulfill the prescribed geometrical constraints, the actual one will make the total potential energy Π a minimum.*

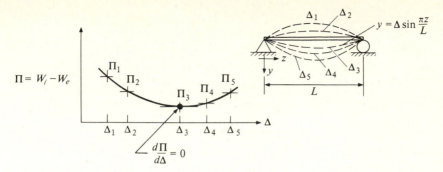

Figure 5.9 Illustration of minimum potential energy.

Since the variation referred to is with respect to deformations, the principle of minimum potential energy yields results in the form of loads or stress resultants. The principle is directly (and readily) applicable, therefore, to the analysis of redundant systems, since equilibrium of stress resultants need not be considered.

Consider for illustration of the general procedure the simple determinate beam shown in the inset sketch of fig. 5.9. It is assumed that $y = \Delta_j \sin (\pi z/L)$, where Δ_j can be any function or constant. If Π is to have a stationary value (that is, a maximum or a minimum value), then

$$\frac{d\Pi}{d\Delta_j} = 0$$

since the magnitude of y depends upon Δ_j.

The principle of *minimum complementary energy* is similar to that of minimum potential energy. The basic difference is that it is derived from the principle of virtual forces. It can be stated as follows:

> *Of all of the possible states of stress in equilibrium with the applied loads and consequent boundary forces, the actual one will make the total complementary energy Π^* a minimum.*

That is,

$$\Pi^* = W_i^* - W_e^* \tag{5.19}$$

and

$$\delta\Pi^* = \delta(W_i^* - W_e^*) = 0 \tag{5.20}$$

5.3.1 The Equality of Minimum Energy and Differential Equation Solutions

It was pointed out earlier in this chapter that minimum energy procedures are frequently used to obtain approximations to actual solutions. It is possible, however, by keeping all terms in their general form, to demonstrate that the two methods yield the same results. Consider for illustration the simple beam shown in fig. 5.10. It is

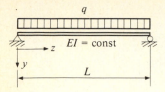

Figure 5.10

subjected to a uniformly distributed lateral load q and has a constant cross section. The total potential of the system is $\Pi = W_i - W_e$, where

$$W_i = \int_0^L \frac{EI}{2} (y'')^2 \, dz \tag{5.21a}$$

and

$$W_e = \int_0^L qy \, dz \tag{5.21b}$$

Therefore,

$$\Pi = \int_0^L \left[\frac{EI}{2} (y'')^2 - qy \right] dz \tag{5.22}$$

In *functional form*, this would be

$$\Pi = \int_a^b F(z, y, y'') \, dz$$

where

$$F(z, y, y'') = \frac{EI}{2} (y'')^2 - qy \tag{5.23}$$

and

$$y'' = \frac{d^2 y}{dz^2}$$

From the calculus of variation,† it is known that the condition for minimizing the function

$$\Pi = \int_a^b F(z, y, y', y'') \, dz \tag{5.24}$$

is

$$F_y - \frac{d}{dz} F_{y'} + \frac{d^2}{dz^2} F_{y''} = 0, \tag{5.25}$$

---

† See R. L. Ketter and S. Prawel, *Modern Methods of Engineering Computations*, p. 440, eq. (12.62), McGraw-Hill Book Company, New York, 1969.

where $y = f_1(z)$ and y, y', and y'' are continuous functions in the interval $a \leq z \leq b$, with y_a and y_b known. Therefore, from eq. (5.23)

$$F_y = \frac{\partial F}{\partial y} = -q$$

$$F_{y'} = \frac{\partial F}{\partial y'} = 0$$

$$F_{y''} = \frac{\partial F}{\partial y''} = EI y''$$

Substituting these values into eq. (5.25) yields

$$-q - 0 + \frac{d^2}{dz^2}(EI y'') = 0$$

or

$$EI \frac{d^4 y}{dz^4} = q$$

This is the same as the first of the differential equations listed in table 4.1.

From the above it is evident that the principle of minimum potential energy can be viewed as a substitute for the equilibrium requirement. It is an alternate form of the principle of virtual displacements and as such takes on the advantages and disadvantages of that procedure.

5.3.2 Approximate Deformations Using the Principle of Minimum Potential Energy

Energy methods are used most frequently to obtain approximate solutions. The procedure would be as follows:

1. Assume a displacement function; for example,

$$y = a_1 \varphi_1 + a_2 \varphi_2 + a_3 \varphi_3 + \cdots + a_n \varphi_n$$

where the φ's are functions of z which satisfy the geometrical constraints. (It should be understood that the selected function need not necessarily be of the form of the actual solution. It must, however, be written in terms of one or more undetermined parameters.)
2. The total potential energy Π of the system under consideration is next defined in terms of the presumed displacement function. Π, therefore, becomes a function of the undefined parameters $a_1, a_2, \ldots, a_n$.
3. The first variation of Π with respect to the displacement yields a set of linear, simultaneous equations that can be solved for the displacement function coefficients. Back substitution of the resulting values yields the desired solution.

The above described procedure is known as the *Ritz* method, and is very useful in a large number of instances. The key feature of the method is the choosing of an appropriate displacement function. Not only must it satisfy the specified geometric constraints, it also must be capable of yielding useful results. Since the actual solution is not known at the outset, one can only guess at its form or perhaps employ a convergent series type of approximation. Clearly, the error involved depends on the quality of the chosen function.

It is to be emphasized that strain energy is *always* a positive quantity—independent of the chosen deformation relationship or sign convention. This is important to remember when equating internal and external energies.

Example 5.1 To illustrate the procedure, consider the beam shown in fig. 5.11. It will be assumed that the deflections can be approximated by the displacement function

$$y = \Delta \sin \frac{\pi z}{L}$$

The second derivative of this expression would be

$$y'' = -\Delta \left(\frac{\pi}{L}\right)^2 \sin \frac{\pi z}{L}$$

Therefore

$$W_i = \frac{1}{2} \int_0^L EI(y'')^2 \, dz = + \frac{EI}{2} \int_0^L \Delta^2 \left(\frac{\pi}{L}\right)^4 \sin^2 \left(\frac{\pi z}{L}\right) dz$$

or

$$W_i = + \frac{\Delta^2 \pi^4 EI}{4L^3} \tag{5.26}$$

The first variation of W_i (that is, the influence of a small disturbance in y) would be

$$\delta W_i = \frac{\partial W_i}{\partial \Delta} d\Delta = + \frac{\Delta \pi^4 EI}{2L^3} d\Delta \tag{5.27}$$

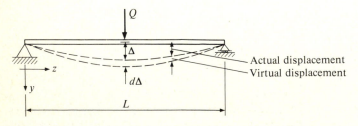

Figure 5.11 Virtual displacement.

Correspondingly, the change in the external potential energy of the system would be

$$\delta W_e = Q \ d\Delta \tag{5.28}$$

For equilibrium to be satisfied

$$\delta W_i = \delta W_e$$

or

$$\frac{\Delta \pi^4 EI}{2L^3} \ d\Delta = Q \ d\Delta$$

which gives the following value for Δ

$$\Delta = \frac{2L^3 Q}{\pi^4 EI}$$

Substituting this value into the original expression for the deflection yields

$$y = \frac{2QL^3}{\pi^4 EI} \sin \frac{\pi z}{L}$$

At $z = \frac{1}{2}L$

$$y = \frac{QL^3}{48.7EI}$$

which compares quite well with the exact value of $QL^3/48EI$.

To improve the accuracy of the approximation, more terms could be added to the displacement function. For example, y could be assumed as the series

$$y = a_1 \sin \frac{\pi z}{L} + a_2 \sin \frac{2\pi z}{L} + \cdots = \sum_n^\infty a_n \sin \frac{n\pi z}{L} \tag{5.29}$$

The second derivative of y with respect to z would then be

$$\frac{d^2 y}{dz^2} = -a_1 \left(\frac{\pi}{L}\right)^2 \sin \frac{\pi z}{L} - a_2 \left(\frac{2\pi}{L}\right)^2 \sin \frac{2\pi z}{L} - a_3 \left(\frac{3\pi}{L}\right)^2 \sin \frac{3\pi z}{L} - \cdots$$

$$= -\sum_n^\infty a_n \left(\frac{n\pi}{L}\right)^2 \sin \frac{n\pi z}{L} \tag{5.30}$$

and the strain energy would be

$$W_i = \frac{1}{2} EI \int_0^L \left[-a_1 \left(\frac{\pi}{L}\right)^2 \sin \frac{\pi z}{L} - a_2 \left(\frac{2\pi}{L}\right)^2 \sin \frac{2\pi z}{L} - \cdots \right]^2 dz$$

Expanding the integrand, using the formula

$$(-\zeta_1 - \zeta_2 - \zeta_3 - \cdots)^2 = (\zeta_1^2 + \zeta_2^2 + \cdots + 2\zeta_1\zeta_2 + 2\zeta_2\zeta_3 + \cdots)$$

it is to be noted that two kinds of terms will be involved: (a) squares of the same quantity, and (b) products of two different quantities. The strain energy therefore can be written as in eq. (5.31).

$$W_i = \frac{1}{2}\left[EI \sum_{n=1}^{\infty} a_n^2 \left(\frac{n\pi}{L}\right)^4 \int_0^L \sin^2 \frac{n\pi z}{L}\, dz \right.$$

$$\left. + \sum_{\substack{n=1 \\ m=1,\, n\neq m}}^{\infty} a_n a_m \left(\frac{n^2 m^2 \pi^4}{L^4}\right) \int_0^L \sin \frac{n\pi z}{L} \sin \frac{m\pi z}{L}\, dz \right] \quad (5.31)$$

It can be shown that for $n = m$ the definite integral is

$$\int_0^L \sin^2 \frac{n\pi z}{L}\, dz = \frac{L}{2}$$

For $n \neq m$,

$$\int_0^L \sin \frac{n\pi z}{L} \sin \frac{m\pi z}{L}\, dz = 0$$

Thus,

$$W_i = \frac{\pi^4 EI}{4L^3}(a_1^2 + 2^4 a_2^2 + 3^4 a_3^2 + \cdots)$$

or

$$W_i = \frac{\pi^4 EI}{4L^3} \sum_n n^4 a_n^2 \quad (5.32)$$

If the displacement function y is given a small variation δy, that is,

$$y + \delta y = \sum_n^{\infty} (a_n + da_n) \sin \frac{n\pi z}{L}$$

then

$$\delta y = \sum_n^{\infty} da_n \sin \frac{n\pi z}{L}$$

and the variation in strain energy becomes

$$\delta W_i = \frac{\partial W_i}{\partial a_n} da_n = \frac{\pi^4 EI}{2L^3} \sum n^4 a_n\, da_n \quad (5.33)$$

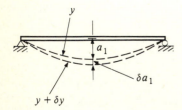

Figure 5.12

The change in external work done due to the assumed variation in y equals Q times the distance traveled. If the load is located at $z = c$

$$\delta W_e = \left[\sum da_n \sin \frac{n\pi c}{L} \right] Q \tag{5.34}$$

For the entire system,

$$\delta \Pi = \delta W_i - \delta W_e = 0$$

or

$$\delta W_i = \delta W_e$$

Therefore,

$$\frac{\pi^4 EI}{2L^3} \sum n^4 a_n \, da_n = Q \sum da_n \sin \frac{n\pi c}{L}$$

This gives

$$a_n = \sum \frac{2QL^3}{\pi^4 EI n^4} \sin \frac{n\pi c}{L} \qquad n = 1, 2, 3, \ldots$$

which, when substituted into eq. (5.29), yields the deflection expression

$$y = \frac{2QL^3}{\pi^4 EI} \sum_{n=1}^{\infty} \frac{1}{n^4} \sin \frac{n\pi c}{L} \sin \frac{n\pi z}{L}$$

If the load is applied at the midspan,

$$y = \frac{2QL^3}{\pi^4 EI} \sum_{n=1}^{\infty} \frac{1}{n^4} \sin \frac{n\pi}{2} \sin \frac{n\pi z}{L}$$

Since $\sin n\pi/2 = 0$ for $n = 2, 4, 6 \ldots$, the deflection at the midspan becomes

$$y_{\mathbb{C}} = \frac{2QL^3}{\pi^4 EI} \sum_{n=1}^{\infty} \frac{1}{n^4} \sin^2 \frac{n\pi}{2} = \frac{2QL^3}{\pi^4 EI} \sum \frac{1}{n^4} \qquad n = 1, 3, 5, \ldots \text{ odd}$$

or

$$y_{\mathbb{C}} = \frac{2QL^3}{\pi^4 EI} \left(1 + \frac{1}{3^4} + \frac{1}{5^4} + \frac{1}{7^4} + \cdots \right)$$

$$= \frac{QL^3}{48.7 EI} \left(1 + \frac{1}{3^4} + \frac{1}{5^4} + \cdots \right)^{\dagger} \tag{5.35}$$

Example 5.2 This same example (fig. 5.11) could be solved assuming a polynomial displacement function. For example, assume that

$$y = a_0 + a_1 z + a_2 z^2 + a_3 z^3 \tag{5.36}$$

---

† Note that in this particular example the assumption of more terms in the sine series does not improve the accuracy of the solution. This is one of the interesting and difficult aspects of energy methods.

The geometrical constraints to be satisfied by this function are

$$\text{At } z = 0 \qquad y = 0$$
$$\text{At } z = \tfrac{1}{2}L \qquad y' = 0$$

(5.37)

From the first of these, $a_0 = 0$. From the second,

$$a_1 = -a_2 L - \tfrac{3}{4}a_3 L^2$$

Therefore the initially chosen displacement function can be written as

$$y = -(a_2 L + \tfrac{3}{4}a_3 L^2)z + a_2 z^2 + a_3 z^3 \tag{5.38}$$

The associated internal strain energy will be

$$W_i = 2(\tfrac{1}{2}) \int_0^{1/2L} EI(y'')^2 \, dz = EI \int_0^{1/2L} (4a_2^2 + 24a_2 a_3 z + 36a_3^2 z^2) \, dz$$

or

$$W_i = EI(2a_2^2 L + 3a_2 a_3 L^2 + \tfrac{3}{2}a_3^2 L^3) \tag{5.39}$$

The external work done will be

$$W_e = Q(y_{\mathbb{C}}) = Q\left[-\frac{L^2}{4}(a_2 + a_3 L) \right]$$

Therefore, from the principle of minimum potential energy,

$$\delta W_i = \delta W_e$$

or

$$\delta[2a_2^2 L + 3a_2 a_3 L^2 + \tfrac{3}{2}a_3^2 L^3] = \frac{Q}{EI} \delta\left[-\frac{L}{4}(a_2 + a_3 L) \right]$$

From this

$$\frac{\partial W_i}{\partial a_2} = 4La_2 + 3L^2 a_3 \qquad \text{and} \qquad \frac{\partial W_e}{\partial a_2} = \frac{Q}{EI}\left(-\frac{L^2}{4} \right)$$

$$\frac{\partial W_i}{\partial a_3} = 3L^2 a_2 + 3L^3 a_3 \qquad \text{and} \qquad \frac{\partial W_e}{\partial a_3} = \frac{Q}{EI}\left(-\frac{L^3}{4} \right)$$

or

$$4La_2 + 3L^2 a_3 = -\frac{QL^2}{4EI}$$

$$3L^2 a_2 + 3L^3 a_3 = -\frac{QL^3}{4EI}$$

Solving these two equations simultaneously yields

$$a_2 = 0$$

$$a_3 = -\frac{Q}{12EI}$$

Substitution into the expression for a_1 gives

$$a_1 = \frac{3QL^2}{48EI}$$

The general deflected shape is therefore

$$y = \frac{QL^3}{48EI}\left[3\frac{z}{L} - 4\left(\frac{z}{L}\right)^3\right] \tag{5.40}$$

and the centerline deflection is

$$y_{\mathbb{C}} = \frac{QL^3}{48EI}$$

It should be noted that this is the exact solution to the problem defined. [When it is considered that the true deflected shape will be a cubic equation, and when such a form is contained within the general family of assumed cubic polynomial displacement functions given as eq. (5.36), this is not surprising.]

In general, if the true deflected shape is contained within the assumed displacement function, the minimum potential energy process will lead to the exact solution of the problem. If, however, the exact solution is not within the family selected (as was the case for the first solution where a sine variation was presumed), then the particular member of that family that satisfies equilibrium will be the result. The quality of the answer depends almost entirely upon the quality of the presumed displacement function.

Example 5.3 As another example, consider the determination of the deflection of the beam column shown in fig. 5.13 using the Ritz method. It will be assumed that the deflection can be represented by the series

$$y = \sum a_n \sin \frac{n\pi z}{L} \qquad n = 1, 2, 3, \dots$$

The strain energy therefore will be

$$W_i = \int_0^L \frac{EI}{2}\left(\frac{d^2 y}{dz^2}\right)^2 dz = \frac{\pi^4 EI}{4L^3}\sum n^4 a_n^2$$

For small changes in the deflected shape, the variation in the strain energy is

$$\delta W_i = \frac{\pi^4 EI}{2L^3}\sum n^4 a_n \, da_n \tag{5.41}$$

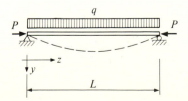

Figure 5.13

as previously derived. (This is because the strain energy is a function of the assumed deformation, and not of the applied loading. Since the displacement functions are the same in both examples 5.1 and 5.2, the internal strain energies and their associated variations also will be the same in both cases.)

The external work done by the applied loads in this example is different from that of example 5.1. First, the lateral load is uniformly distributed. Secondly, axial force is present and contributes to the external work done as the member bends.

The external work done by the uniformly distributed lateral load due to a small change in y is given by

$$\delta W_e^q = \int_0^L (q\ dz)\ \delta y = \int_0^L q \sum \left(\sin \frac{n\pi z}{L}\ dz \right) da_n$$

or

$$\delta W_e^q = q \sum \left(\frac{L}{n\pi} \right)(1 - \cos n\pi)\ da_n \qquad n = 1, 2, 3, \ldots$$

This reduces to

$$\delta W_e^q = q \sum \left(\frac{2L}{n\pi} \right) da_n \qquad n = 1, 3, 5, \ldots \tag{5.42}$$

The external work done by the axial force is

$$\delta W_e^P = P\ \delta(\Delta L)$$

where (ΔL) is the total distance through which P moves as the member bends. As shown in fig. 5.14,

$$\Delta L = \int_0^L (ds - dz)\ dz$$

where s is measured along the elastic curve. Since

$$ds - dz = \sqrt{(dz)^2 + (dy)^2} - dz = dz \left[1 + \left(\frac{dy}{dz} \right)^2 \right]^{1/2} - dz$$

$$= dz \left[1 + \frac{1}{2}\left(\frac{dy}{dz} \right)^2 - \frac{1}{8}\left(\frac{dy}{dz} \right)^4 + \cdots - 1 \right] \cong \frac{1}{2}\left(\frac{dy}{dz} \right)^2$$

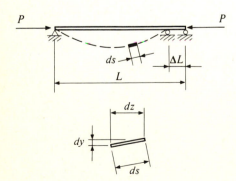

Figure 5.14

the axial shortening will be

$$\Delta L = \int_0^L \frac{1}{2}\left(\frac{dy}{dz}\right)^2 dz \tag{5.43}$$

For

$$y = \sum a_n \sin \frac{n\pi z}{L}$$

$$y' = \sum a_n \frac{n\pi}{L} \cos \frac{n\pi z}{L}$$

Therefore

$$\Delta L = \int_0^L \frac{1}{2}\left(a_1 \frac{\pi}{L} \cos \frac{\pi z}{L} + a_2 \frac{2\pi}{L} \cos \frac{2\pi z}{L} + \cdots\right)^2 dz$$

Since the presumed displacement function is an orthogonal function, it can be shown that

$$\int_0^L \cos^2\left(\frac{n\pi z}{L}\right) dz = \frac{L}{2} \qquad \text{for } m = n$$

and

$$\int_0^L \cos\left(\frac{n\pi z}{L}\right) \cos\left(\frac{m\pi z}{L}\right) dz = 0 \qquad \text{for } m \neq n$$

Therefore

$$\Delta L = \int \frac{1}{2}(y')^2 \, dz = \frac{\pi^2}{4L} \sum n^2 a_n^2 \tag{5.44}$$

The change in ΔL due to a small variation in y is

$$\delta(\Delta L) = \frac{\partial(\Delta L)}{\partial a_n} da_n = \sum \frac{\pi^2 n^2}{2L} a_n \, da_n$$

The work done by P during this variation δy is therefore

$$\delta W_e^P = P \,\delta(\Delta L) = P \sum \frac{\pi^2 n^2}{2L} a_n \, da_n \tag{5.45}$$

The total external work done by the applied forces during the variation δy is

$$\delta W_e = \delta W_e^q + \delta W_e^P$$

or

$$\delta W_e = q \sum \left(\frac{L}{n\pi}\right)(1 - \cos n\pi) \, da_n + P \sum \frac{n^2 \pi^2}{2L} a_n \, da_n$$

Using the principle of minimum potential energy, $\delta W_i = \delta W_e$, or

$$\frac{\pi^4 EI}{2L^3} \sum n^4 a_n \, da_n = qL \sum \left(\frac{1}{n\pi}\right)(1 - \cos n\pi) \, da_n + P \sum \frac{n^2 \pi^2}{2L} a_n \, da_n$$

Solving for a_n,

$$a_n = \frac{qL \sum (1/n\pi)(1 - \cos n\pi)}{\sum (\pi^4 EI/2L^3)n^4 - P \sum (n^2\pi^2/2L)} \qquad \text{where } n = 1, 2, 3, \ldots$$

Noting that when $n = 2, 4, 6 \ldots$, $(1 - \cos n\pi) = 0$ and therefore $a_n = 0$, this can be rewritten as

$$a_n = \frac{qL \sum (1/n\pi)}{\sum (\pi^4 EI/2L^3)n^4 - P \sum (n^2\pi^2/2L)} \qquad \text{where } n = 1, 3, 5, \ldots \quad (5.46)$$

If $P_e = \pi^2 EI/L^2 =$ the Euler buckling load, and if $\alpha = P/P_e$, eq. (5.46) can be written in the form

$$a_n = \frac{4qL^4}{\pi^5 EI} \sum \frac{1}{n^3(n^2 - \alpha)} \qquad n = 1, 3, 5, \ldots$$

The deflection curve is therefore

$$y = \frac{4qL^4}{\pi^5 EI} \sum \frac{1}{n^3(n^2 - \alpha)} \sin \frac{n\pi z}{L} \qquad n = 1, 3, 5, \ldots \quad (5.47)$$

Since

$$\frac{4}{\pi^5} \cong \frac{5.019}{384}$$

$$y \cong \frac{5.019}{384} \frac{qL^4}{EI} \sum \frac{1}{n^3(n^2 - \alpha)} \sin \frac{n\pi z}{L} \qquad n = 1, 3, 5, \ldots$$

Considering only the first term in this series, and setting the axial load P equal to zero, the solution is well within 1 percent of the exact solution. As P approaches the Euler load (that is, as $\alpha \to 1.0$) the magnification term approaches infinity. The Euler buckling load therefore represents the upper limit of the axial-force capacity for the beam column in question. (This is the same conclusion that was reached in chap. 4 using the exact differential equation solution.)

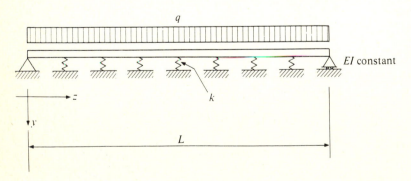

Figure 5.15

Example 5.4 As a fourth example, consider the determination of the deflected shape as a uniformly loaded beam resting on an elastic foundation. It is assumed that the spring constant of the foundation is k and is constant for the entire span. (k has the dimensions of pounds per inch per inch.) As in the previous example, it will be assumed that the deflected shape can be represented by the function

$$y = \sum a_n \sin \frac{n\pi z}{L} \qquad n = 1, 2, 3, \ldots$$

For this problem, the strain energy can be considered in two parts: one, the bending energy stored in the beam, W_i^B; the other, the energy stored in the foundation, W_i^F. Since the assumed deflected shape has the same general form as examples 5.1 and 5.3, the strain energy due to bending of the beam will be the same as in those previous cases.

$$W_i^B = \frac{\pi^4 EI}{4L^3} \sum_{n=1}^{\infty} a_n^2 n^4 \tag{5.48}$$

The strain energy stored in the foundation is given by the expression

$$W_i^F = \frac{1}{2} \int_0^L (ky\,dz)y = \frac{1}{2}k \int_0^L y^2\,dz$$

(The $\frac{1}{2}$ is required because q acts on the beam from zero to its full value.) Substituting into this equation the assumed value of y,

$$W_i^F = \frac{1}{2}k \int_0^L \left[a_1 \sin \frac{\pi z}{L} + a_2 \sin \frac{2\pi z}{L} + \cdots \right]^2 dz$$

Again, noting that

$$\int_0^L \sin^2 \frac{n\pi z}{L}\,dz = \frac{L}{2} \qquad \text{for } m = n$$

and

$$\int_0^L \sin \frac{m\pi z}{L} \sin \frac{n\pi z}{L}\,dz = 0 \qquad \text{for } m \neq n$$

the strain energy in the foundation is

$$W_i^F = \frac{1}{2}k \int_0^L \sum a_n^2 \sin^2 \frac{n\pi z}{L}\,dz = \frac{1}{4}kL \sum a_n^2$$

The total strain energy stored in the system is therefore

$$W_i = W_i^B + W_i^F$$

or

$$W_i = \frac{\pi^4 EI}{4L^3} \sum a_n^2 n^4 + \frac{kL}{4} \sum a_n^2 \tag{5.49}$$

The first variation of W_i is given by

$$\delta W_i = \left(\frac{\pi^4 EI}{4L^3} \sum n^4 + \frac{kL}{2}\right) a_n \, da_n \tag{5.50}$$

From eq. (5.42), the corresponding first variation of the external work done by the uniformly distributed lateral load q is

$$\delta W_e = \int q \, dz \, \delta y = \frac{2qL}{\pi} \sum \frac{1}{n} \, da_n \qquad n = 1, 3, 5, \ldots \tag{5.51}$$

For $\delta \Pi = 0$,

$$\delta W_i = \delta W_e$$

which gives

$$a_n = \frac{4qL^4}{\pi^5 EI} \sum \frac{1}{n(n^4 + kL^4/\pi^4 EI)} \qquad n = 1, 3, 5, \ldots \tag{5.52}$$

The equation of the elastic curve is therefore

$$y = \frac{4qL^4}{\pi^5 EI} \sum \frac{1}{n(n^4 + kL^4/\pi^4 EI)} \sin \frac{n\pi z}{L} \qquad n = 1, 3, 5, \ldots \tag{5.53}$$

Considering only the first term of this series,

$$y = \frac{4qL^4}{\pi^5 EI} \left[\frac{1}{1 + kL^4/\pi^4 EI}\right] \sin \frac{\pi z}{L}$$

this gives as a midspan deflection

$$y_{\textcentoldstyle} = \frac{4qL^4}{\pi^5 EI} \left(\frac{1}{1 + kL^4/\pi^4 EI}\right)$$

Since for any value of k greater than zero the parentheticised term will always have a value less than one, the deflections will be less than those of a non-supported beam. For $k = 0$, the solution is identical to the simple beam solution obtained in example 5.3.

5.4 MINIMUM COMPLEMENTARY ENERGY

In the preceding section, equilibrium solutions were obtained by minimizing potential energy. In those cases variation in Π was with respect to displacement functions. As indicated by eq. (5.20), similar procedures can be developed which minimize complementary energies. For those, variation will be with respect to prescribed equilibrium force functions. The basic equation is

$$\delta \Pi^* = \delta W_i^* - \delta W_e^* = 0$$

or

$$\delta W_i^* = \delta W_e^* \tag{5.54}$$

In many instances it is convenient to stipulate a force function in which only one load or stress resultant has the ability to change. For such cases

$$dW_i^* = dW_e^* = \Delta \, dP$$

or

$$\Delta = \frac{dW_i^*}{dP} \tag{5.55}$$

For linear cases, complementary energy is equal to strain energy, that is, $W_i^* = W_i$, and

$$\Delta = \frac{dW_i}{dP} \tag{5.56}$$

This expression and the resulting concept is most often referred to as *Castigliano's theorem*. In words:

> *The rate of change of the strain energy associated with a given structural system with respect to a given load is equal to the deflection of that system at the load point in question in the direction of the load.*

In still other words: To determine a component of deformation at a given point in a structure, all that is needed is the rate of change in W_i with respect to a load located at the point in question, acting in the direction of the desired deformation.

In the previous section various expressions for strain-energy components were developed. For example, for beam bending

$$W_i = \int_0^L \frac{M^2}{2EI} \, dz$$

If more than one bending member were present in the system in question

$$W_i = \sum_j \int_0^{L_j} \frac{M_j^2}{2EI_j} \, dz_j \tag{5.57}$$

and Castigliano's theorem leads to

$$\Delta = \frac{dW_i}{dP} = \sum_j \int \frac{M_j}{EI_j} \frac{\partial M_j}{\partial P} \, dz_j \tag{5.58}$$

In these, the subscript j denotes a particular member in the structure.

Axial thrust, shear, and torsion also have associated with them strain energies. Since all these terms are positive and are scalar quantities, they can be added to obtain the total strain energy of the complete system. It also is to be noted that deformational response can be one of two types: displacements or rotations. For the first of these, the loading with respect to which variation occurs is an applied force. For the second, it would be an applied moment. In summary then, the following would be the governing equations for the determination of deformations of slender bars:

Displacements:

$$\Delta_0 = \sum \left(\int \frac{M}{EI} \frac{\partial M}{\partial Q_0} dz + \int \frac{P}{AE} \frac{\partial P}{\partial Q_0} dz + \int \frac{\lambda V}{AG} \frac{\partial V}{\partial Q_0} dz \right.$$
$$\left. + \int \frac{T_{SV}}{G\kappa_T} \frac{\partial T_{SV}}{\partial Q_0} dz + \int \frac{4M_F}{EI_y} \frac{\partial M_F}{\partial Q_0} dz \right) \tag{5.59}$$

Bending rotations:

$$\theta_0 = \sum \left(\int \frac{M}{EI} \frac{\partial M}{\partial M_0} dz + \int \frac{P}{AE} \frac{\partial P}{\partial M_0} dz + \int \frac{\lambda V}{AG} \frac{\partial V}{\partial M_0} dz \right.$$
$$\left. + \int \frac{T_{SV}}{G\kappa_T} \frac{\partial T_{SV}}{\partial M_0} dz + \int \frac{4M_F}{EI_y} \frac{\partial M_F}{\partial M_0} dz \right) \tag{5.60}$$

Torsional rotations:

$$\beta_0 = \sum \left(\int \frac{M}{EI} \frac{\partial M}{\partial T_0} dz + \int \frac{P}{AE} \frac{\partial P}{\partial T_0} dz + \int \frac{\lambda V}{AG} \frac{\partial V}{\partial T_0} dz \right.$$
$$\left. + \int \frac{T_{SV}}{G\kappa_T} \frac{\partial T_{SV}}{\partial T_0} dz + \int \frac{4M_F}{EI_y} \frac{\partial M_F}{\partial T_0} dz \right) \tag{5.61}$$

M, P, V, and T represent, respectively, the bending moment, axial thrust, shear, and twisting moment as functions of the applied loads. Δ_0 is the displacement under the load Q_0, in the direction of Q_0. θ_0 is the rotation of the member at the point of the applied moment M_0, in the sense of the moment. β_0 is the rotation about a longitudinal axis at the point of an applied torque T_0, in the sense of that torque.

It is not unusual to require deflections or rotations at locations other than those at which loads or moments are applied. When such is the case, fictitious loads are presumed at those points and the necessary integrations are carried out. After these operations have been completed, the fictitious quantities are equated to zero.

For emphasis, it should here again be stated that while the method now being described is complementary to that of minimum potential energy—$\delta \Pi = \delta W_i - \delta W_e = 0$, and $\delta \Pi^* = \delta W_i^* - \delta W_e^* = 0$—due to the consequence of the assumption that $W_i^* = W_i$, Castigliano's theorem is restricted to the solution of linear elastic structures.

Example 5.5 For the horizontally curved beam shown in fig. 5.16, the vertical deflection at the load point in the direction of Q is desired.

Neglecting the transverse shear term, the total complementary energy of the system shown is

$$W_i^* = W_i = \int_0^{\frac{1}{2}\pi} \frac{M_\theta^2}{2EI} r \, d\theta + \int_0^{\frac{1}{2}\pi} \frac{T_\theta^2}{2G\kappa_T} r \, d\theta$$

where

$$M_\theta = Qr \sin \theta$$

and

$$T_\theta = Qr(1 - \cos \theta)$$

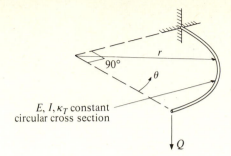

Figure 5.16

From Castigliano's theorem

$$\Delta = \frac{\partial W_i^*}{\partial Q} = \frac{r}{EI} \int_0^{\frac{1}{2}\pi} M_\theta \frac{\partial M_\theta}{\partial Q} d\theta + \frac{r}{G\kappa_T} \int_0^{\frac{1}{2}\pi} T_\theta \frac{\partial T_\theta}{\partial Q} d\theta$$

Since

$$\frac{\partial M_\theta}{\partial Q} = r \sin \theta \quad \text{and} \quad \frac{\partial T_\theta}{\partial Q} = r(1 - \cos \theta)$$

$$\Delta = \frac{\pi Q r^3}{4EI} + \frac{Q r^3}{4G\kappa_T}(3\pi - 8)$$

or

$$\Delta = \frac{\pi Q r^3}{4EI}\left(1 + 0.454 \frac{EI}{G\kappa_T}\right) \tag{5.62}$$

Example 5.6 As a second example using Castigliano's theorem, consider the determination of the centerline deflection of a uniformly loaded, simply supported, uniform cross-section beam. Since no concentrated load Q is specified at the point where deflection is desired, it will be necessary to introduce a fictitious load Q_0 at that location. The beam and loading are symmetrical about the centerline, therefore the total complementary energy—considering both the real, distributed load q and the fictitious concentrated load Q_0—is given by

$$W_i^* = 2 \int_0^{\frac{1}{2}L} \frac{M^2}{2EI} dz$$

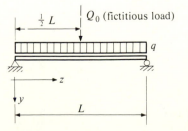

Figure 5.17

The midpoint deflection is

$$\Delta_{\mathbb{Q}} = \frac{dW_i^*}{dQ_0} = 2 \int_0^{\frac{1}{2}L} \frac{M}{EI} \frac{\partial M}{\partial Q_0} dz$$

where

$$M = \left(\frac{qL}{2} - \frac{qz}{2} + \frac{Q_0}{2}\right) z \quad \text{and} \quad \frac{\partial M}{\partial Q_0} = \frac{1}{2}z$$

Upon substitution and integration

$$\Delta_{\mathbb{Q}} = \frac{5}{384} \frac{qL^4}{EI} + \frac{Q_0}{48} \frac{L^3}{EI} \tag{5.63a}$$

But, $Q_0 = 0$. Therefore,

$$\Delta_{\mathbb{Q}} = \frac{5}{384} \frac{qL^4}{EI} \tag{5.63b}$$

This is the exact solution.

Example 5.7 In many cases it is the deformation of a structure under its own weight that is desired. Consider, for example, the vertical elongation of a bar suspended as shown in fig. 5.18. The complementary energy is given by

$$W_i^* = \int_0^L \frac{P^2}{2AE} dz = \frac{1}{2AE} \int_0^L (Q_0 + \gamma Az)^2 dz$$

Δ is then

$$\Delta = \frac{\partial W_i^*}{\partial Q_0} = \frac{\gamma L^2}{2AE}$$

where γ is the unit weight of the material.

Example 5.8 The uniform cross-section rigid frame structure $ABCD$ in sketch a of fig. 5.19, is rigidly attached at point A. It is free at point D. The loading to which it is subjected is a uniformly distributed vertical force acting along

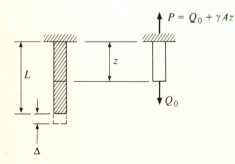

Figure 5.18

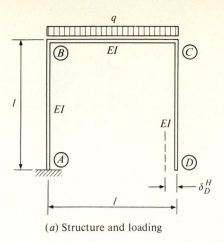

(*a*) Structure and loading

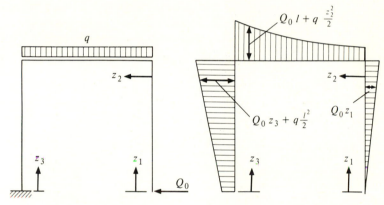

(*b*) Loading and coordinate system
(*c*) Bending moment diagram

Figure 5.19

member BC. The horizontal displacement of point D due to the loading q is desired. (This is shown as δ_D^H.) It will be assumed that the problem can be considered as one of bending alone, that is, the influence of axial thrust and shear will be neglected. The complementary energy can therefore be expressed as

$$
\begin{aligned}
W_i^* &= \sum \int \frac{M^2}{2EI}\, dz \\
&= \int_0^l \frac{(Q_0 z_1)^2}{2EI}\, dz_1 + \int_0^l \frac{(Q_0 l + \frac{1}{2}q z_2^2)^2}{2EI}\, dz_2 + \int_0^l \frac{(Q_0 z_3 + \frac{1}{2}q l^2)^2}{2EI}\, dz_3
\end{aligned}
$$

The desired deformation is obtained using the relationship

$$
\delta_D^H = \frac{\partial W_i^*}{\partial Q_0} = \frac{Q_0 l^3}{EI}\left(\frac{5}{3}\right) + \frac{q l^4}{EI}\left(\frac{5}{12}\right)
$$

But $Q_0 = 0$. Therefore

$$\delta_D^H = \frac{5}{12} \frac{ql^4}{EI}$$

The vertical deformation and the rotation of point D, had these been desired, would be found in a similar fashion. For the vertical deformation, Q_0 would be assumed in the vertical direction. To determine θ_D, a fictitious moment M_0 would be applied at point D.

Castigliano's theorem can be used to calculate the deformations of trusses. Since, by definition, trusses are composed of pin-connected, axial force members,

$$\Delta = \sum_{i=1}^{n} \left(\frac{P_i L_i}{A_i E_i}\right)\left(\frac{\partial P_i}{\partial Q_0}\right) \tag{5.64}$$

where n is the number of members in the structure. If an applied load acted at the point and in the direction of the desired deformation, then a fictitious load Q_0 need not be introduced. Differentiation would be with respect to that given load.

Relative movement between two points also can be ascertained using Castigliano's theorem. For example, to determine the change in distance between joints a and b of the truss shown in fig. 5.20a, due to any given external (or for that matter, internal) loading, the location of the fictitious Q_0 loads would be as shown. Similarly, to determine the rotation of a bar, say, member cd in fig. 5.20b,

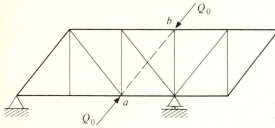

Figure 5.20a Loading to determine relative movement between a and b.

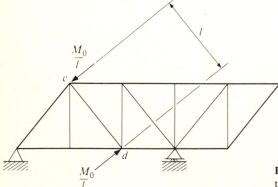

Figure 5.20b Loading to determine rotation of member cd.

the pair of imaginary forces indicated would be applied. The corresponding equation would be

$$\theta_{cd} = \Sigma \frac{P_i L_i}{A_i E_i} \frac{\partial P_i}{\partial M_0}$$

In both of these cases, after the necessary operations have been carried out, the fictitious forces and moments would be set equal to zero.

Example 5.9 Determine the vertical deflection of the point of load application of the crane shown in fig. 5.21. Cross-sectional areas are indicated on each of the members. The effective moduli of the booms and the stranded cables are $E_{booms} = 30 \times 10^6$ lb/in², and $E_{cables} = 15 \times 10^6$ lb/in².
    The solution is summarized in the following table.

(1)	(2)	(3)	(4)	(5)	(6)	(7)
Member	Length, ft	$AE \times 10^{-6}$, lb	P†, lb	$\dfrac{PL}{AE} \times 10^6$	$\dfrac{\partial P}{\partial Q}$	$\dfrac{PL}{AE}\dfrac{\partial P}{\partial Q} \times 10^6$, ft
AC	50.00	15.0	$+2.50Q$	$+8.333Q$	$+2.50$	$20.833Q$
BC	40.00	150.0	$-1.50Q$	$-0.400Q$	-1.50	$0.600Q$
CD	63.24	15.0	$+1.581Q$	$+6.665Q$	$+1.581$	$10.537Q$
BD	84.84	150.0	$-2.121Q$	$-1.200Q$	-2.121	$2.545Q$
						$\Sigma = 34.515Q$

† Tension positive.

Therefore

$$\delta_D = 0.0000034515Q$$

where δ_D is in feet and Q is in pounds.

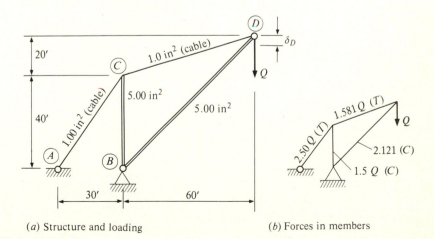

(a) Structure and loading   (b) Forces in members

Figure 5.21

5.5 THE UNIT LOAD METHOD

A direct, strict application of the theorem of minimum complementary energy [eq. (5.56)] leads to a number of cumbersome "bookkeeping" techniques ill-suited to the solution of large complex structures. A more flexible approach that is readily adapted and which can be shown to be a natural interpretation of that theorem is known as the *unit load*, or *dummy unit load*, method. (It also is known variously as the Maxwell-Mohr method, the method of Virtual Velocities, the Dummy Load method, etc.)

It should be noted at the outset that the unit load method can be derived directly from the principle of virtual work, and therefore there are no restrictions on the type of material behavior presumed when using this method of solution. The most common applications, however, are for linearly elastic materials, and for these the process can be derived directly from Castigliano's theorem.

In the preceding section it was shown that the total complementary energy of, for example, a single beam bending system can be expressed as

$$W_i^* = \int \frac{M^2}{2EI} \, dz$$

From Castigliano's formulation of the problem, it was shown that

$$\Delta_0 = \int \frac{M}{EI} \frac{\partial M}{\partial Q_0} \, dz$$

where 0 is the location of desired deflection Δ_0, and Q_0 is a concentrated load applied at that point, in the direction of Δ_0. It is now assumed that the bending moment in question is due to a general set of applied loads Q plus the particular load Q_0, that is,

$$M = f_1(Q) + f_2(Q_0)$$

where f_1 and f_2 are functions of z. For such a case,

$$\frac{\partial M}{\partial Q_0} = \frac{\partial f_1(Q)}{\partial Q_0} + \frac{\partial f_2(Q_0)}{\partial Q_0}$$

Since, by definition, $f_1(Q)$ is not a function of Q_0,

$$\frac{\partial f_1(Q)}{\partial Q_0} = 0$$

Therefore

$$\Delta_0 = \int \frac{M}{EI} \frac{\partial f_2(Q_0)}{\partial Q_0} \, dz \qquad (5.65)$$

But $\partial f_2(Q_0)/\partial Q_0$ is nothing more than the moment diagram in the member in question due to the application of a unit (1 lb) load at location 0 in the direction of desired deflection Δ_0. Denoting this moment as m, eq. (5.65) becomes

$$\Delta_0 = \int \frac{Mm}{EI} \, dz \qquad (5.66)$$

In a similar fashion it can be shown that the various partial derivatives contained in eqs. (5.59), (5.60), and (5.61) have the values

$$\frac{\partial M}{\partial Q_0} = m = \text{bending moment due to } Q_0 = 1$$

$$\frac{\partial P}{\partial Q_0} = p = \text{axial thrust due to } Q_0 = 1$$

$$\frac{\partial V}{\partial Q_0} = v = \text{shear due to } Q_0 = 1$$

$$\frac{\partial T_{\text{sv}}}{\partial Q_0} = t_{\text{sv}} = \text{torsional moment due to } Q_0 = 1$$

$$\frac{\partial M_F}{\partial Q_0} = m_F = \text{flange bending moment due to } Q_0 = 1$$

$$\frac{\partial M}{\partial M_0} = m = \text{bending moment due to } M_0 = 1$$

$$\frac{\partial P}{\partial M_0} = p = \text{axial thrust due to } M_0 = 1$$

$$\frac{\partial V}{\partial M_0} = v = \text{shear due to } M_0 = 1$$

$$\frac{\partial T_{\text{sv}}}{\partial M_0} = t_{\text{sv}} = \text{torsional moment due to } M_0 = 1$$

$$\frac{\partial M_F}{\partial M_0} = m_F = \text{flange bending moment due to } M_0 = 1$$

$$\frac{\partial M}{\partial T_0} = m = \text{bending moment due to } T_0 = 1$$

$$\frac{\partial P}{\partial T_0} = p = \text{axial thrust due to } T_0 = 1$$

$$\frac{\partial V}{\partial T_0} = v = \text{shear due to } T_0 = 1$$

$$\frac{\partial T_{\text{sv}}}{\partial T_0} = t_{\text{sv}} = \text{torsional moment due to } T_0 = 1$$

$$\frac{\partial M_F}{\partial T_0} = m_F = \text{flange bending moment due to } T_0 = 1$$

The general deformation equation therefore becomes

$$
\left.\begin{array}{c}\Delta_0 \\ \theta_0 \\ \text{or} \\ \beta_0\end{array}\right\} = \Sigma \left(\int \frac{Mm}{EI}\,dz + \int \frac{Pp}{AE}\,dz + \int \frac{\lambda Vv}{AG}\,dz \right.
$$
$$
\left. + \int \frac{T_{SV}}{G\kappa_T}\,t_{SV}\,dz + \int \frac{4M_F}{EI_y}\,m_F\,dz \right) \tag{5.67}
$$

In eq. (5.67) it is assumed that each of the terms—M, m, P, p, V, v, T_{SV}, t_{SV}, M_F, and m_F; as well as E, G, A, I, I_y, and κ_T—can vary with z. For most real structures, however, it is not unusual to find members of constant cross section. For those cases the material and cross-sectional properties can be taken outside the integral. Moreover, for axial force (truss) members, the axial thrust terms are constants over the lengths of the various members in question and integration is readily accomplished.

$$
\Sigma \int_0^L \frac{Pp}{AE}\,dz = \Sigma \frac{PL}{AE}\,p
$$

Considering just the bending situation, assuming a member of constant cross section,

$$
\int_0^L \frac{Mm}{EI}\,dz = \frac{1}{EI} \int_0^L Mm\,dz
$$

Assuming given distributions of M and m (as functions of z) it is possible to obtain algebraic expressions for the integral in question. For example, if M and m vary linearly as indicated in fig. 5.22, the algebraic expression for the integral would be

$$
\frac{1}{EI} \int_0^L Mm\,dz = \frac{L}{3EI}\left[M_L m_L + M_R m_R + \tfrac{1}{2}(M_L m_R + M_R m_L)\right] \tag{5.68}
$$

For other assumed variations, other expressions would be obtained.

It should be recognized that due to the nature of the unit load, m will always vary linearly. M, however, may vary in a variety of fashions. Table 5.1 summarizes a number of the cases frequently encountered. In the left-hand column are shown various forms for M. Across the top are various m diagrams. Each table entry is the integral of the corresponding product moment.

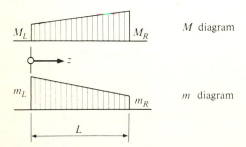

M diagram

m diagram

Figure 5.22 Applied and unit moment diagrams.

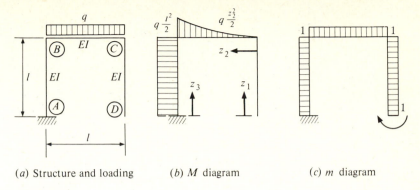

(a) Structure and loading   (b) M diagram   (c) m diagram

Figure 5.23

Example 5.10 For the rigid frame shown in fig. 5.23a, determine the rotation at point D. The M moment diagram is shown in sketch b. The m diagram is in sketch c. It is to be noted that both of these have been indicated on the *tension side* of the member. The signs are therefore consistent between the two sketches. The unit rotation at D has been chosen in the clockwise direction. A computed plus value for θ_D will therefore correspond to a clockwise rotation at that point.

Assuming only bending deformations, the rotation at point D can be determined from the following equations

$$\theta_D = \int_0^l \frac{(0)(1)}{EI} dz_1 + \int_0^l \frac{q(z_2)^2(1)}{EI} dz_2 + \int_0^l \frac{ql^2(1)}{2EI} dz_3$$

$$\text{Member } CD \qquad\qquad \text{Member } BC \qquad\qquad \text{Member } AB$$

or using the values listed in table 5.1

$$\theta_D = 0 + \frac{l}{6EI}\left[(1)\left(\frac{ql^2}{2} + \frac{4ql^2}{8}\right)\right] + \frac{l}{EI}\left[\frac{ql^2}{2}(1)\right] = \frac{2}{3}\frac{ql^3}{EI}$$

Example 5.11 Member CF, in the truss structure shown in fig. 5.24, has been fabricated 2 in too short. What will be the horizontal displacement of point D? In eqs. (5.59) to (5.61) and (5.67) the terms

$$\int \frac{P}{AE} dz \qquad \text{and} \qquad \frac{PL}{AE}$$

represent the elongation or shortening of an axial force member due to the "primary condition of load"—be that an applied external force, a change in length due to temperature change, a fabrication error in length, etc. Therefore, for the problem indicated, $(PL/AE)_{CF} = -2$ in, where it is assumed that axial elongation (due to a tensile force) is positive.

Sketch b of fig. 5.24 gives the forces in the various truss members due to the application of a unit force at point D (the point of desired deflection). It is to be noted that the force in member CF is 0.500 and is in compression.

Table 5.1 Evaluation of the integral $\displaystyle\int_0^L Mm\,dz$

m diagram $\rightarrow$ M diagram $\downarrow$	m_L $\boxed{+}$ $m_R = m_L$	m_L ◿ m_R	m_L ◺ 0	m_L ◹ $m_R = -m_L$	0 ◺ m_R
Linear variation in M					
M_L $\boxed{+}$ $M_R = M_L$	$L(M_L m_L)$	$\dfrac{L}{2}[M_L(m_L + m_R)]$	$\dfrac{L}{2}(M_L m_L)$	0	$\dfrac{L}{2}(M_L m_R)$
M_L ◿ M_R	$\dfrac{L}{2}[(M_L + M_R)m_L]$	$\dfrac{L}{3}\big[M_L(m_L + \tfrac{1}{2}m_R) + M_R(m_R + \tfrac{1}{2}m_L)\big]$	$\dfrac{L}{3}[(M_L + \tfrac{1}{2}M_R)m_L]$	$\dfrac{L}{6}(M_L m_L - M_R m_L)$	$\dfrac{L}{3}[(M_R + \tfrac{1}{2}M_L)m_R]$
M_L ◺ 0	$\dfrac{L}{2}(M_L m_L)$	$\dfrac{L}{3}[M_L(m_L + \tfrac{1}{2}m_R)]$	$\dfrac{L}{3}(M_L m_L)$	$\dfrac{L}{6}(M_L m_L)$	$\dfrac{L}{6}(M_L m_R)$
M_L ◹ $M_R = -M_L$	0	$\dfrac{L}{6}[M_L(m_L - m_R)]$	$\dfrac{L}{6}(M_L m_L)$	$\dfrac{L}{3}(M_L m_L)$	$-\dfrac{L}{6}(M_L m_R)$
0 ◺ M_R	$\dfrac{L}{2}(M_R m_L)$	$\dfrac{L}{3}[M_R(m_R + \tfrac{1}{2}m_L)]$	$\dfrac{L}{6}(M_R m_L)$	$-\dfrac{L}{6}(M_R m_L)$	$\dfrac{L}{3}(M_R m_R)$

Parabolic variation in M

M_L ▦ M_R $(M_C$ = central ordinate$)$	$\frac{L}{6}[(M_L + 4M_C + M_R)m_L]$	$\frac{L}{6}[(M_L + 2M_C)m_L + (2M_C + M_R)m_R]$	$\frac{L}{6}[(M_L + 2M_C)m_L]$	$\frac{L}{6}[(M_L + M_R)m_L]$	$\frac{L}{6}[(2M_C + M_R)m_R]$
0 ⌢ M_C ⌢ 0	$\frac{2L}{3}[M_C m_L]$	$\frac{L}{3}[M_C(m_L + m_R)]$	$\frac{L}{3}(M_C m_L)$	0	$\frac{L}{3}(M_C m_L)$
M_L ⌢ M_C ⌢ 0	$\frac{L}{6}[(M_L + 4M_C)m_L]$	$\frac{L}{6}[(M_L + 2M_C)m_L + 2M_C m_R]$	$\frac{L}{6}[(M_L + 2M_C)m_L]$	$\frac{L}{6}(M_L m_L)$	$\frac{L}{3}(M_C m_R)$
0 ⌢ M_C ⌢ M_R	$\frac{L}{6}[(4M_C + M_R)m_L]$	$\frac{L}{6}[(2M_C)m_L + (2M_C + M_R)m_R]$	$\frac{L}{3}(M_C m_L)$	$\frac{L}{6}(M_R m_L)$	$\frac{L}{6}[(2M_C + M_R)m_R]$

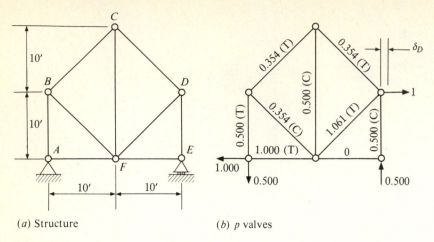

(a) Structure

(b) p values

Figure 5.24

Since CF is the only member to have a change in length,

$$\delta_D = \Sigma \left(\frac{PL}{AE}\right) p = (\Delta L)_{CF} p_{CF} = (-2)(-0.500) = 1 \text{ in}$$

The answer is positive. Therefore, the deflection is in the direction assumed for the unit load.

Example 5.12 The three-dimensional truss shown in fig. 5.25 is subjected to a 10-kip vertical force at point D. The possible components of reaction have been indicated. The modulus of elasticity is 30×10^6 lb/in^2. The cross-sectional areas of the various members are those listed in the following table. Determine the deflection of D in the z direction.

Member	Length, ft	Area, in^2	P, kips	p	$\frac{PL}{AE}p$ ($\times 10^3$), ft
AB	14.14	1.00	+0.884	+0.354	+0.147
BC	14.14	1.00	+0.884	+0.354	+0.147
AC	20.00	1.00	+0.625	−0.750	−0.313
AD	22.91	2.00	−2.863	+1.145	−0.436
BD	20.62	2.00	−5.154	−2.062	+3.652
CD	22.91	2.00	−2.863	+1.145	−1.252
				$\Sigma =$	+1.945

Therefore

$$\delta_D = +0.001945 \text{ ft}$$

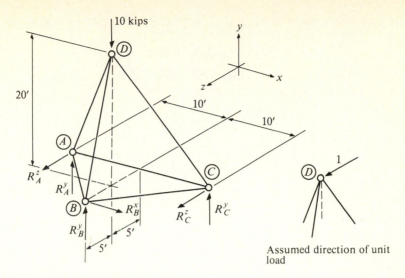

Figure 5.25

5.6 MAXWELL-BETTI'S RECIPROCAL THEOREMS

In this section, to facilitate discussion, a double subscript notation will be introduced. The first of the subscripts will refer to the point at which deformation occurs. The second indicates the location of the applied external forces. For example, Δ_{ij} would stand for the deflection at location i due to the application of a force at j. In the terms used in the preceding section

$$\Delta_{ij} = \sum \left(\int \frac{M_j m_i}{EI}\, ds + \int \frac{P_j p_i}{AE}\, ds + \cdots \right)$$

If the script delta is used to denote the unit response of the member, that is, δ_{ij} would stand for the deflection of point i due to the application of a unit force at j, then

$$\Delta_{ij} = Q_j \delta_{ij} \qquad (5.69a)$$

where Q_j is the magnitude of the force applied at j. Correspondingly,

$$\Delta_{ji} = Q_i \delta_{ji} \qquad (5.69b)$$

Assume, now, that a given linear elastic system is subjected to two externally applied force systems Q_i and Q_j, and further let it be assumed that the load Q_i is first applied. If that load is increased continuously from zero to its maximum value, the external work done during that loading would be equal to $\frac{1}{2}Q_i \Delta_{ii}$. If at that stage a second loading Q_j were introduced (keeping Q_i constant at its full value), the external work done would be

$$(W_e)_{ij} = \tfrac{1}{2}Q_i \Delta_{ii} + \tfrac{1}{2}Q_j \Delta_{jj} + \underbrace{Q_i \Delta_{ij}}_{\substack{\text{additional work done} \\ \text{when } Q_j \text{ is added}}}$$

If the loading sequence were reversed and Q_j were first applied, the external work done would be

$$(W_e)_{ji} = \tfrac{1}{2}Q_j\Delta_{jj} + \tfrac{1}{2}Q_i\Delta_{ii} + Q_j\Delta_{ji}$$

For linear elastic systems, however, the work done—independent of the path chosen—must be the same. Therefore

$$(W_e)_{ij} = (W_e)_{ji}$$

or

$$Q_i\Delta_{ij} = Q_j\Delta_{ji} \tag{5.70}$$

This equation has become known as *Betti's law*. Verbally, it can be stated as follows:

> *The work done by a first load Q_i acting through the displacement due to a second load Q_j is equal to the work done by the second load acting through the displacement of the first load.*

Substituting (5.69a) and (5.69b) into eq. (5.70) yields

$$Q_i(Q_j\delta_{ij}) = Q_j(Q_i\delta_{ji})$$

or

$$\delta_{ij} = \delta_{ji} \tag{5.71}$$

which can be stated as follows:

> *The deformation at i due to a unit load at j is equal to the deformation at j due to a unit load at i.*

This is known as *Maxwell's reciprocal theorem*, and has wide application in the solution of many determinate and indeterminate structural problems.

The reciprocal theorem derived above has direct application in the construction of influence lines for deformations. Recalling that an influence line describes a particular behavior (deflection, for example) at a given point due to the application of unit forces at various locations in the structure, and noting the reciprocal relationship given above as (5.71), the deflected curve of a beam caused by a unit load at the point in question is in fact the influence line for the deflection at that same point in the beam. For rotations, the imposed loading would be a unit moment, and the deflected shape would be the influence line.

5.7 PROBLEMS

5.1 For the space frame shown in prob. 3.7, assume that $a = r$, $F_2 = 0$, and $F_1 = Q$. If the cross section is circular and solid, determine the deflection of the structure at the load point Q.

5.2 The cantilever beam described in prob. 4.1e is subjected to a concentrated vertical force at its free end. Using the principle of minimum complementary energy determine the vertical deflection of the end of the beam.

5.3 A load Q is supported by the two-bar truss system shown. If the stress-strain relationship is elastic but nonlinear as described, determine the total strain energy and the total complementary energy of the system.

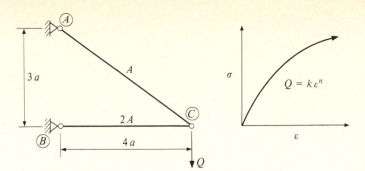

$$Q = k\varepsilon^n$$

Figure P5.3

5.4 Using Castigliano's theorem, obtain the indicated deformations:

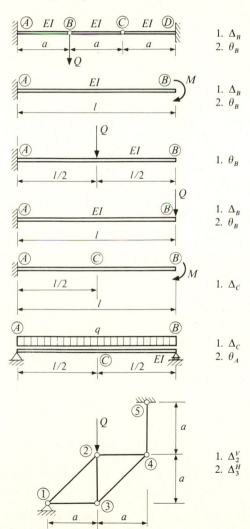

1. Δ_B
2. θ_B

1. Δ_B
2. θ_B

1. θ_B

1. Δ_B
2. θ_B

1. Δ_C

1. Δ_C
2. θ_A

1. Δ_2^V
2. Δ_3^H

All bars have the same A's and E's

Figure P5.4

5.5 If in the truss shown in prob. 5.4 there is a uniform temperature rise of 100°F, and if there are no external loads acting, what will be the location of joint 4. (Assume that the coefficient of linear expansion is $\alpha = 0.001$ per unit length per degree.)

5.6 By the unit load method, determine the indicated deformations:

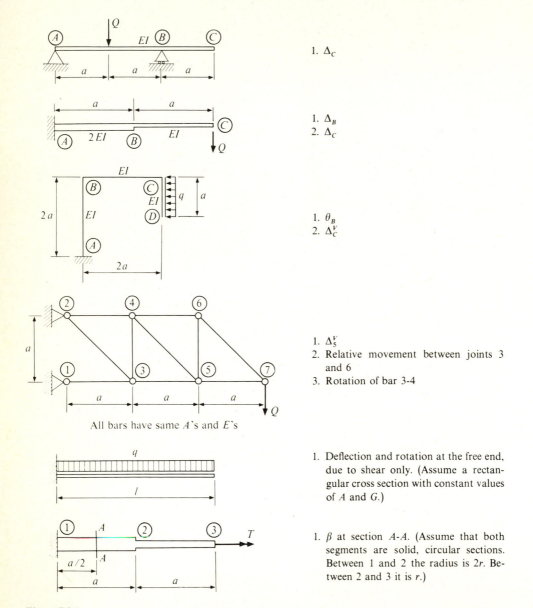

1. Δ_C

1. Δ_B
2. Δ_C

1. θ_B
2. Δ_C^V

1. Δ_5^V
2. Relative movement between joints 3 and 6
3. Rotation of bar 3-4

All bars have same A's and E's

1. Deflection and rotation at the free end, due to shear only. (Assume a rectangular cross section with constant values of A and G.)

1. β at section A-A. (Assume that both segments are solid, circular sections. Between 1 and 2 the radius is $2r$. Between 2 and 3 it is r.)

Figure P5.6

5.7 Using the indicated trigonometric series, obtain approximations for the deflected shapes based on the principle of minimum potential energy. Discuss the accuracy of the solution versus the number of terms in the series.

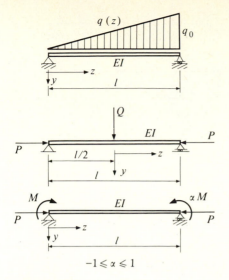

$$y = \sum a_n \sin \frac{n\pi z}{l} \qquad n = 1, 3, 5, \ldots$$

(Origin at the midspan.)

$$y = \sum b_n \cos \frac{n\pi z}{l} \qquad n = 0, 1, 2, \ldots$$

$$y = \sum a_n \sin \frac{n\pi z}{l} \qquad n = 1, 2, 3, \ldots$$

For $\alpha = +1, 0$ and -1, compare the solutions with eq. (4.135).

Figure P5.7

STABILITY OF STRUCTURAL MEMBERS

6.1 INTRODUCTION

Assume that an initially straight, uniform cross section, "flagpole"-type column is subjected to a concentrated vertical force at its uppermost end. Further, assume that that load acts along the centroidal axis of the column. Still further, assume that the applied force is continuously increased from zero. Using the methods heretofore described, it is possible to define the axial shortening of this column as a function of the magnitude of the applied load, the cross-sectional dimensions, the length of the member, and the modulus of elasticity of the material.

As the load P is increased there eventually will be reached a particular critical value of the load for which the straight member finds it equally easy to sustain the load in a laterally bent configuration. This is illustrated in fig. 6.1. (Δ is the total axial shortening of the column. δ is the lateral displacement of the free end.) At this value of P, which is noted on the diagram as "buckling," the member finds it possible to either remain in the straight position or to deflect laterally. Below the critical value, the member will be straight. Above, it will be bent. At the critical value, then, a condition of *bifurcation* exists—with two adjacent, equilibrium positions being possible for the same external force. More than that, the deformational modes of resistance for these two cases are totally different. Buckling—a condition of bifurcation—constitutes one of the ways in which a structural member can become unstable.

From fig. 6.1 it should be evident that buckling does not necessarily correspond to the ultimate load-carrying capacity of the member. For the case illustrated, there is indicated considerable strength above the buckling load—though deformations associated with those increased loads may be appreciable. As P is increased, however, a

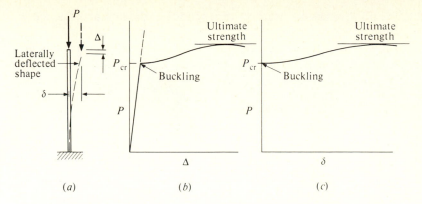

Figure 6.1 Load-deformation behavior of members subjected to axial compression.

particular value of thrust will be reached beyond which further increases are impossible, the member continues to deflect, and at least two adjacent equilibrium positions can be found for the same value of P. It should be recognized that both of these positions have the same laterally bent configuration. Such a situation also describes a condition of general structural instability—not one of buckling, but rather one of *progressive deflection*.

In summary: Instability occurs when two or more adjacent equilibrium positions exist for the same value of P. Buckling differs from progressive deflection in that the two adjacent equilibrium positions correspond to two fundamentally different deformational modes. (It is, of course, to be understood that under very special circumstances the two cases could occur simultaneously.)

Consider now as a second example the deformational response of the laterally loaded beam column shown in fig. 6.2. It will be assumed that the axial thrust and the lateral load increase proportionally and that the member deflects laterally in the plane of the applied forces, which is further presumed to be the most flexible of the cross section's bending directions. Again, two load-deformation curves will be

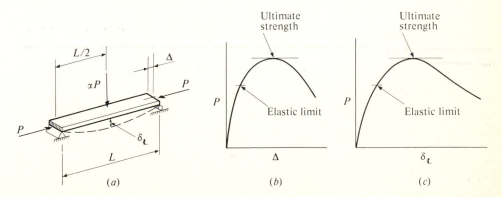

Figure 6.2 Load-deformation behavior of beam columns—bending about the minor principal axis.

shown: P versus Δ, and P versus $\delta_{\mathbb{C}}$. Here $\delta_{\mathbb{C}}$ corresponds to the lateral deflection at the lateral load point.

Unlike the flagpole column of fig. 6.1, both axial and lateral deflections are observed from the outset of load application. Assuming a linear elastic behavior,

$$\Delta = \int_0^L \frac{P}{AE}\, dz + \int_0^L \tfrac{1}{2}(y')^2\, dz \qquad (6.1)$$

The first of the two terms indicated in eq. (6.1) defines the axial shortening of the member due to the thrust P. The second is due to the lateral bending of the member.† (It should be noted that for most cases, the second term is appreciably greater than the first.) From eq. (4.113) the lateral deflection $\delta_{\mathbb{C}}$ varies with P according to the relationship

$$\delta_{\mathbb{C}} = \frac{\alpha PL^3}{48EI}\frac{3(\tan u - u)}{u^3}$$

where $u = (\pi/2)\sqrt{P/P_e}$, and $P_e = \pi^2 EI/L^2$. From this equation it is evident that as P approaches P_e, $\delta_{\mathbb{C}}$ goes to ∞. But as the centerline deflection increases, the moment at that section also increases. Therefore, for all *real materials*, a particular value of P will be reached less than P_e which corresponds to initial yielding at the centerline section. This has been shown in the sketches as the "elastic limit." Above the elastic limit value the member responds in a progressively more flexible fashion to increased values of P. Finally there is reached a constant value of P (below P_e) for which the member continues to deflect, and adjacent equilibrium positions under the same load can be established. The member is indifferent with regard to which of these equally possible positions it assumes, and a condition of instability exists.

It should be understood that for this second case no buckling occurred. Instability (at ultimate strength) corresponded to progressive deflection.

Consider, now, the behavior of a laterally loaded beam column, where the axis of initial bending is the strong axis of the cross section. This is illustrated in fig. 6.3. (It will be assumed that the member is constrained to remain vertical at its points of end support; however, between those points the member is unrestrained.)

As the load P is increased from zero there is realized from the outset both vertical and axial deformations. (The vertical deflection is initially in the direction of the applied load αP.) There is reached a value of thrust, however, for which the member finds it equally possible to sustain the loading in a laterally bent and twisted configuration. That is, between the two end supports, the member bends and twists out of its initial plane of loading. One possible geometric indicator of that type of movement is the indicated twisting angle β. As noted in the figure, this situation also corresponds to buckling: There are two adjacent equilibrium positions for the same load, and these two positions correspond to two distinctly different deformational modes.

---

† From fig. 5.13, the axial deformation due to lateral bending is

$$\Delta L = ds - dz = [\sqrt{1 + (y')^2} - 1]\, dz$$
$$= [1 + \tfrac{1}{2}(y')^2 - \tfrac{1}{8}(y')^4 + \tfrac{1}{48}(y')^6 - \cdots - 1]\, dz \cong \tfrac{1}{2}(y')^2\, dz$$

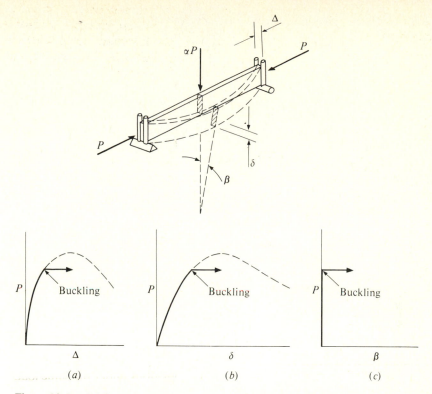

Figure 6.3 Load-deformation behavior of beam columns—bending about the major principal axis.

Lateral-torsional buckling in fig. 6.3 does not necessarily correspond to the ultimate carrying capacity of the member in question. As was true for the first case described, there could be increased values of P beyond this critical value, depending upon the cross-sectional properties, the length of the member, the end constraints, the modulus of elasticity, the shearing modulus, etc. Eventually, however, a maximum value of P will be reached for which deformations (in the bent and twisted configuration) progressively increase, instability occurs, and the maximum load-carrying capacity is realized.

From the above it is evident that deformational modes are, at least, of the following three types: (*a*) pure axial shortening, (*b*) lateral bending (about any given cross-sectional axis), and (*c*) combined bending and twist. It should be noted, however, that other modes also are possible: (*d*) pure twisting (about a longitudinal axis) and (*e*) local buckling (for example, wrinkling of a flange in a region of high stress). For particular types of structures still other modes could occur: "oil-canning" of longitudinally loaded, thin, circular cross-section shells, "snap-through" of shallow, laterally loaded, cylindrical shells, etc.

Stability, and its counterpart instability, are often explained by analogy to the behavior of a rigid ball on the three surfaces shown in fig. 6.4. In each of the cases shown, the ball is assumed to be in equilibrium in the position indicated; however, if the balls are slightly disturbed from these positions, the responses are quite different.

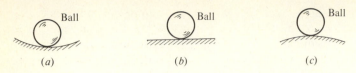

Figure 6.4

For case *a*, positive work will be required, and the ball will return to its original position upon removal of the disturbance. (This case corresponds to points on the ascendency sides of the load-deformation curves given in figs. 6.1, 6.2, and 6.3.) This equilibrium position is stable. Case *c*, on the other hand, represents a state of instability (or unstable equilibrium) where the disturbance results in the giving up of energy, and movement becomes progressive. (Such would be the case for all descendency portions of the illustrated load-deformation curves.) The neutral equilibrium position modeled in case *b* neither returns to its original position nor continues to move when the disturbance ceases. Such would be the case when structural members buckle, or when they reach their maximum load-carrying capacity. This condition of neutral equilibrium is frequently stated as the necessary condition for structural instability.

Buckling can occur in both the elastic and the inelastic ranges of material behavior, and procedures for determining these will be developed in this chapter. For real materials, ultimate carrying capacity is realized in the inelastic range.

Four distinctly different classical methods are available for the solution of buckling problems. The first of these, the so-called *statical* method, requires that a second, infinitesimally near-equilibrium position be found which will sustain the same load. *Energy* methods (virtual work, minimum potential energy, minimum complementary energy, etc.) as defined in chap. 5 also can be used to establish neutrality of a given equilibrium state. A third method, known as the *work* method, requires that zero force (load or moment) causes the member to remain in the deformed position. In the *kinetic* or fourth method, the equations of motion are formulated and the load is established which results in deformation with zero frequency of vibration. The first two methods will be the primary ones considered in this chapter.

Ultimate strength solutions are available for only a few problems, and for these the work involved is extensive. For most cases, it is necessary to know the history of loading of the particular member and to establish the deformational response of the member as a function of that history. Exceptions to this are the several approximate methods that have been developed to answer the very specific question of load carrying capacity, and for which it is possible to demonstrate the relative insensitivity of that particular load to the deflected shape of the member.

6.2 BUCKLING OF CENTRALLY LOADED COLUMNS

It will be assumed, as given, that the member in question is initially straight, slender, of solid cross section, and subjected to an axial compressive force applied along the centroidal axis of the member. Moreover, it will be presumed in this part of this chapter that the member is constructed of a homogeneous, perfectly elastic material.

In chap. 4 equations were derived which define the response of such structural members to various types and kinds of beam-column loading. For example, if it is assumed that a pin-connected member is subjected to an axial thrust plus two equal and opposite end moments that produce single curvature deformations, the equations defining the centerline deflection, the end slopes, and the maximum bending moment are eqs. (4.120), (4.121), and (4.123). For the case of a uniformly distributed lateral load plus an applied axial thrust, the equations are (4.105), (4.108) and (4.109). For a concentrated lateral load at midspan, the equations are (4.113), (4.114) and (4.115). It should be noted that corresponding sets of equations could be obtained using the energy methods described in chap. 5. For example, for a member subjected to a uniformly distributed lateral load plus axial thrust applied at its ends, the centerline deflection is that defined by eq. (5.47). Values for the slope and bending moment could be obtained by differentiation.

The equations derived in chaps. 4 and 5 can be used to establish the critical loads associated with column buckling. For example, for the beam-column subjected to axial thrust plus equal end moments, M_0, the maximum bending moment occurs at the centerline, and is given by eq. (4.123).

$$\frac{M_{max}}{M_0} = \sec\left(\frac{\pi}{2}\sqrt{\frac{P}{P_e}}\right)$$

(In this equation, and as will be true throughout this chapter, P_e is the Euler load and is defined by the relationship

$$P_e = \frac{\pi^2 EI}{L^2}$$

EI is the bending stiffness about a centroidal axis of the cross section which is perpendicular to the plane of lateral deformation. It is assumed that the member is of constant cross section throughout its length.) As P approaches P_e, $\sec\left[(\pi/2)\sqrt{P/P_e}\right]$ approaches infinity and, therefore, M_{max} increases without limit. P_e, then, clearly defines the value of P at which lateral deflections are possible with M_0 approaching zero. For the assumed pin-ended conditions, this critical value of the axial thrust is known as the Euler buckling load, and is denoted as P_e. (It is to be recognized that similar conclusions concerning the buckling load could have been reached considering any one of the other loading conditions referred to above.)

6.2.1 Elastic Buckling as an Eigenvalue Problem

Buckling loads can be and most often are derived directly from a consideration of the differential equations for beam-column equilibrium that were developed in chap. 4. Assume the second-order formulation of the problem:

$$\frac{d^2 y}{dz^2} + k^2 y = -\frac{1}{EI}M(z) \tag{6.2}$$

where

$$k^2 = \frac{P}{EI}$$

and $M(z)$ is the bending moment due to the primary loading. If $M(z) = 0$, the homogeneous differential equation to be solved is

$$\frac{d^2 y}{dz^2} + k^2 y = 0 \tag{6.3}$$

which has as a general solution

$$y = A \sin kz + B \cos kz \tag{6.4}$$

To evaluate the arbitrary constants A and B, the boundary conditions associated with the end supports must be used. Consider, for illustration, a column pinned at both ends. The boundary conditions would be

$$\text{At } z = 0 \qquad y = 0$$

$$\text{At } z = L \qquad y = 0$$

Substituting the general solution given as eq. (6.4) into these boundary conditions yields the following two, homogeneous, algebraic equations, which must be simultaneously satisfied.

$$A \sin 0 + B \cos 0 = 0$$

$$A \sin kL + B \cos kL = 0$$

From the first of these,

$$B = 0$$

From the second equation,

$$A \sin kL = 0$$

This condition can be satisfied in either one of two ways: A can be zero, or $\sin kL$ can be zero. If A is selected, y will be identically zero for all values of P. In other words, for that assumption, the member cannot buckle. If, on the other hand, $\sin kL$ is set equal to zero, deflections can occur. For $\sin kL = 0$, it is necessary that

$$kL = n\pi \qquad n = 1, 2, 3, \ldots \tag{6.5}$$

But since

$$k^2 = \frac{P}{EI}$$

this can be written as

$$P_{cr} = \frac{n^2 \pi^2 EI}{L^2} \tag{6.6}$$

The smallest P_{cr} value corresponds to the case where $n = 1$, for which

$$P_{cr} = P_e = \frac{\pi^2 EI}{L^2} \tag{6.7}$$

the Euler buckling load.

The Euler buckling load also can be derived from a consideration of the fourth-order formulation of the general bending problem. From table 4.6, setting $q(z) = 0$, the governing homogeneous differential equation is

$$\frac{d^4y}{dz^4} + k^2\,\frac{d^2y}{dz^2} = 0 \tag{6.8}$$

where

$$k^2 = \frac{P}{EI}$$

The general solution to eq. (6.8) is

$$y = A \sin kz + B \cos kz + C\frac{z}{L} + D \tag{6.9}$$

Again, for the simply supported end conditions, the boundary values to be met are

$$
\begin{aligned}
&\text{At } z = 0 && y = 0 \\
&\text{At } z = 0 && y'' = 0 \\
&\text{At } z = L && y = 0 \\
&\text{At } z = L && y'' = 0
\end{aligned}
$$

Considering only the first and second of these requirements,

$$B + D = 0$$

and

$$B = 0$$

Therefore

$$B = D = 0 \tag{6.10}$$

From the boundary conditions at $z = L$, two more independent equations can be written:

$$A \sin kL + C(1.0) = 0$$

$$A(-k^2 \sin kL) + C(0) = 0$$

Since $k^2 = P/EI = 0$ is a trivial solution, as is the case where $A \equiv C \equiv 0$, the only nontrivial case exists when $C = 0$ and

$$\sin kL = 0$$

This yields the identical Euler buckling load expression that was obtained as eq. (6.7); that is,

$$P_{cr} = P_e = \frac{\pi^2 EI}{L^2}$$

This procedure for determining the buckling load of an axially loaded column can be generalized to handle any known boundary conditions. For example, if both ends of the member are "fixed" against both bending rotations and lateral translation, the boundary conditions to be met are

$$\text{At } z = 0 \qquad y = 0$$
$$\text{At } z = 0 \qquad y' = 0$$
$$\text{At } z = L \qquad y = 0$$
$$\text{At } z = L \qquad y' = 0$$

and the simultaneous, linear, homogeneous, algebraic equations to be solved are

$$(0)A + \qquad (1.0)B + \quad (0)C + (1.0)D = 0$$

$$(k)A + \qquad (0)B + \left(\frac{1}{L}\right)C + \quad (0)D = 0$$

$$(\sin kL)A + \quad (\cos kL)B + (1.0)C + (1.0)D = 0$$

$$(k \cos kL)A + (-k \sin kL)B + \left(\frac{1}{L}\right)C + \quad (0)D = 0$$

For a given structural member, the unknowns are A, B, C, D, and k. If $A \equiv B \equiv C \equiv D \equiv 0$, the member will remain straight for all values of P. If, however, a solution is forced by equating the determinant of the coefficients to zero, a value of k_{cr} (and thereby P_{cr}) can be established at which lateral deflections are possible.† For the particular set of conditions defined above the buckling solution would be

$$P_{cr} = \frac{\pi^2 EI}{(0.25)L^2} \tag{6.11}$$

Had the member in question been fixed at its base and pin-connected at its uppermost end, the boundary conditions would be

$$\text{At } z = 0 \qquad y = 0$$
$$\text{At } z = 0 \qquad y' = 0$$
$$\text{At } z = L \qquad y = 0$$
$$\text{At } z = L \qquad y'' = 0$$

and the critical load listed as eq. (6.12) would result.

$$P_{cr} = \frac{\pi^2 EI}{(0.49)L^2} \tag{6.12}$$

† For a detailed discussion of a number of different methods for solving eigenvalue problems, see chap. 7 of R. L. Ketter and S. Prawel, *Modern Methods of Engineering Computation*, McGraw-Hill Book Company, New York, 1969.

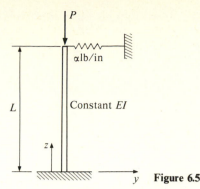

P

$\alpha\,\mathrm{lb/in}$

L

Constant EI

z

y **Figure 6.5**

Example 6.1 To illustrate the generality of the procedure just described, consider the solution of the buckling problem defined in fig. 6.5. The member is of constant cross section, of length L, and is subjected to an axial thrust at its uppermost end. The base of the column is fixed against both rotation and translation. At the upper end, the member is free to rotate, but it is constrained against translation by a spring of stiffness α lb/in. The differential equation which must be used to solve this problem—since at least one of the boundary conditions involves shear, which is a third-derivative consideration—is the fourth-order equation:

$$\frac{d^4y}{dz^4} + k^2\,\frac{d^2y}{dz^2} = 0$$

where

$$k^2 = \frac{P}{EI}$$

The boundary conditions are

$$\text{At } z = 0 \qquad y = 0$$
$$\text{At } z = 0 \qquad y' = 0$$
$$\text{At } z = L \qquad y'' = 0$$
$$\text{At } z = L \qquad y''' + k^2 y' - \gamma y = 0$$

where $\gamma = \alpha/EI$. (The fourth equation in this set states that the end shear developed in the member at $z = L$ is resisted by the force in the spring due to the lateral deflection.)

Substituting the general solution to the differential equation, that is,

$$y = A \sin kz + B \cos kz + C\frac{z}{L} + D$$

into each of these boundary conditions yields the four, simultaneous, homogeneous, linear, algebraic equations listed as eq. (6.13). The unknowns are A, B, C

and D, as well as the critical axial thrust value P_{cr}—which, as was indicated in the previous example, is contained in the term k.

$$(0)A + (1.0)B + (0)C + (1.0)D = 0$$

$$(k)A + (0)B + \left(\frac{1}{L}\right)C + (0)D = 0$$

$$(-k^2 \sin kL)A + (-k^2 \cos kL)B + (0)C + (0)D = 0$$

$$(-\gamma \sin kL)A + (-\gamma \cos kL)B + \left(\frac{k^2}{L} - \gamma\right)C + (-\gamma)D = 0$$

(6.13)

For these to have a solution other than the trivial one $A = B = C = D = 0$, the determinant of the coefficients must equal zero. That is,

$$\begin{vmatrix} (0) & (1.0) & (0) & (1.0) \\ (kL) & (0) & (1.0) & (0) \\ (-\sin kL) & (-\cos kL) & (0) & (0) \\ (-\gamma L \sin kL) & (-\gamma L \cos kL) & (k^2 - \gamma L) & (-\gamma L) \end{vmatrix} = 0 \qquad (6.14)$$

This yields the buckling solution (that is, the characteristic equation)

$$\gamma L^3 = \frac{\alpha L^3}{EI} = \frac{(kL)^3}{kL - \tan kL} \qquad (6.15)$$

which is shown graphically in fig. 6.6.

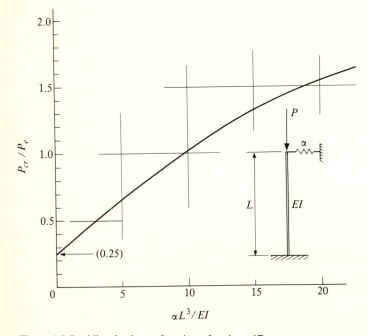

Figure 6.6 Buckling load as a function of spring stiffness.

Example 6.2 The uniform cross section, two-span, continuous column shown in fig. 6.7 is subjected to an axial force of P in segment A-B. The axial force in segment BC is presumed to be equal to zero. Determine the flexural buckling load for the structure.

In essence the column under consideration is member AB shown in sketch b. The boundary conditions at end A of that member are $y = 0$ and $y'' = 0$. At end B, they are $y = 0$ and $EIy'' = \gamma y'$. γ is the spring constant associated with span BC. That is,

$$\gamma = \frac{M_0}{\theta_0}$$

It is to be noted that member BC is pin-connected at B and fixed at C, and is subjected to an applied end moment M_0 at B, as indicated. For this particular member, the boundary conditions to be met are

$$\text{At } z = 0 \qquad y = 0$$

$$\text{At } z = 0 \qquad y'' = -\frac{M_0}{EI}$$

$$\text{At } z = l \qquad y = 0$$

$$\text{At } z = l \qquad y' = 0$$

Since there are no lateral loads acting on this member, the differential equation of equilibrium is

$$\frac{d^4 y}{dz^4} = 0$$

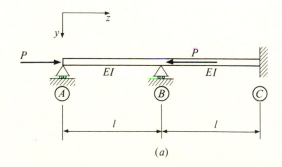

(a)

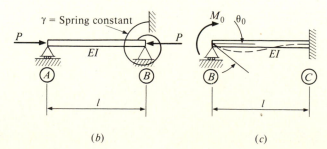

(b)   (c)   **Figure 6.7**

which has as a general solution

$$y = az^3 + bz^2 + cz + d$$

Substituting this equation into the boundary conditions, and solving for the ratio M_0/θ_0, the spring constant becomes

$$\gamma = \frac{M_0}{\theta_0} = \frac{4EI}{l}$$

Consider now the column segment AB. The governing differential equation is

$$\frac{d^4 y}{dz^4} + k^2 \frac{d^2 y}{dz^2} = 0$$

which has as a general solution

$$y = A \sin kz + B \cos kz + C \frac{z}{L} + D$$

The boundary conditions are

$$\text{At } z = 0 \qquad y = 0$$
$$\text{At } z = 0 \qquad y'' = 0$$
$$\text{At } z = l \qquad y = 0$$
$$\text{At } z = l \qquad EIy'' = \gamma y' = \frac{4EI}{l} y'$$

Substituting into these the general solution, and equating the determinant of the resulting coefficient matrix to zero, yields the following characteristic equation for the buckling load:

$$\tan kl = \frac{4kl}{(kl)^4 + 4} \tag{6.16}$$

where $kl = \pi\sqrt{P_{cr}/P_e}$. A graphical solution of this equation is shown in fig. 6.8. Since $(kL)_{cr} = 1.22\pi$,

$$(kL)_{cr} = \sqrt{\frac{P_{cr}}{EI}} \, l = 1.22\pi$$

or

$$\frac{P_{cr}}{P_e} = 1.49 \tag{6.17}$$

Example 6.3 The rigid beam and column structure shown in fig. 6.9 is the *inverted version* of the structure shown as fig. 4.48 in chap. 4. The loading, however, is different from that previously considered: only P is assumed to act.

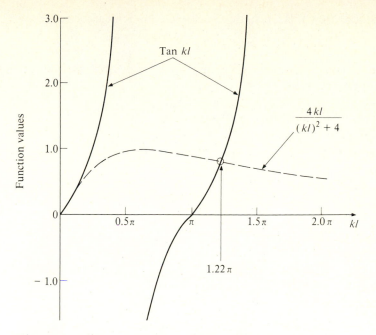

Figure 6.8 Graphical determination of buckling load.

The algebraic set of homogeneous equations to be solved is given as eq. (4.126)—with $V = 0$. For these to have other than the trivial solution $A = B = C = D = 0$, the determinant of the coefficient matrix must equal zero:

$$
\begin{vmatrix}
0 & 1.0 & 0 & 1.0 \\
k\xi & k^2 & \xi & 0 \\
\sin kh & \cos kh & 0 & 0 \\
0 & 0 & 1.0 & 0
\end{vmatrix} = 0
$$

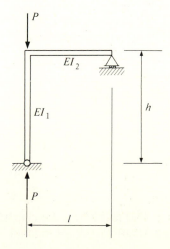

Figure 6.9

Here, $\xi = (3/l)(EI_2/EI_1)$. The resulting buckling equation can be reduced to the expression

$$kh \tan kh = 3\frac{I_2}{I_1}\frac{h}{l} \tag{6.18}$$

which is shown graphically in fig. 6.10.

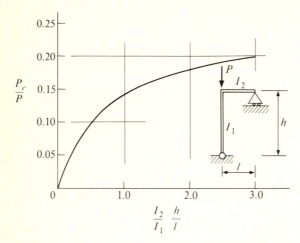

Figure 6.10 Buckling load as a function of relative bending stiffnesses.

In each of the above examples, ratios of the critical load to the Euler buckling load for the structure in question were determined; where, by definition, the Euler load is the buckling load of a pin-ended member of length L. The solutions obtained could equally well have been written in the form

$$P_{cr} = \frac{\pi^2 EI}{(KL)^2} \tag{6.19}$$

where K is a so-called *effective length factor*. Using this form of representation, the critical buckling load would be described in terms of the Euler load of a hypothetical pin-ended member of length KL. For the various solutions obtained earlier in this section, the corresponding effective length factors could be obtained from the expression

$$K = \sqrt{\frac{P_e}{P_{cr}}} \tag{6.20}$$

K values for a number of simple cases are given in Table 6.1. Elaborate charts or nomographs have been developed which give effective length factors as functions of

Table 6.1 Effective length factor

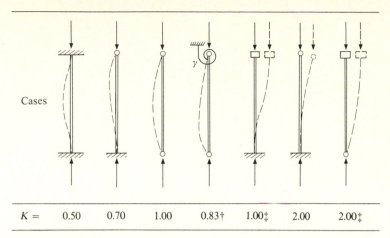

Cases							
$K =$	0.50	0.70	1.00	0.83†	1.00‡	2.00	2.00‡

† For this case the assumed spring constant is $\gamma = 4EI/l$ (see example 6.2).

‡ The slope is maintained zero at the upper end of the column.

both translational and rotational end restraints. They are not included here, but are available in the literature.†

When a column is an integral part of a structure, its ends are connected to other beams or columns, and these provide rotational as well as translational restraint. To determine the true buckling load of a particular column in a given structure, a stability analysis of the entire structure must be carried out. This is a tedious and often completely unmanageable process for multistory, multibay frames. Based on past experience or knowledge of limiting values associated with idealized boundary conditions, engineers generally use their judgment and estimate the effective length factor K for a given design situation. Frequently, it is sufficient to estimate the end restraints from a consideration of the members immediately connected to the column in question. The alignment charts developed by Jackson and Moreland, are particularly useful in this regard. They are reproduced here as fig. 6.11.

† Bruce G. Johnston, The Column Research Council *Guide to Design Criteria for Metal Compression Members*, 3d ed., John Wiley & Sons, Inc., New York, 1976.

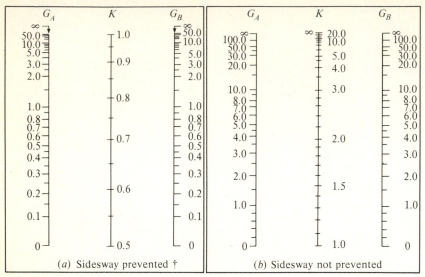

| G_A | K | G_B | | G_A | K | G_B |

(a) Sidesway prevented †

(b) Sidesway not prevented

† No relative displacement of the column ends in the plane of buckling.

The subscripts A and B refer to the joints at the two ends of the column section being considered. G is defined as

$$G = \frac{\sum I_c/L_c}{\sum I_g/L_g} \qquad (6.21)$$

in which $\sum$ indicates a summation of all members rigidly connected to that joint and lying in the plane in which buckling of the column is being considered. I_c is the moment-of-inertia and L_c is the corresponding unbraced length of the column section, and I_g is the moment-of-inertia and L_g the corresponding unbraced length of the girder or other restraining member. I_c and I_g are taken about axes perpendicular to the plane of buckling.

For a column base connected to a footing by a frictionless hinge, G is theoretically infinite but should be taken as 10 in design practice. If the column base is rigidly attached to a properly designed footing, G approaches a theoretical value of zero but should be taken as 1.0. Other values may be used if justified by analysis.

The girder stiffness I_g/L_g should be multiplied by a factor when certain conditions at the far end are known to exist. For the case with sidesway prevented, the appropriate multiplying factors are as follows:

    1.5 for far end of girder hinged, and
    2.0 for far end of girder fixed against rotation

For the case with sidesway not prevented, the multiplying factors are:

    0.5 for far end of girder hinged, and
    0.67 for far end of girder fixed

Having determined G_A and G_B for a column section, K is obtained by constructing a straight line between the appropriate points on the scales for G_A and G_B.

Figure 6.11 Alignment charts for effective length of columns in continuous frames. (*Courtesy of Jackson and Moreland, Division of United Engineers and Contractors, Inc.*)

Noting that

$$\sigma_{cr} = \frac{P_{cr}}{A} \qquad \text{and} \qquad I = Ar^2$$

where A is the cross-sectional area and r is the radius of gyration about the cross-sectional axis which governs buckling, the critical buckling stress can be determined from the relationship

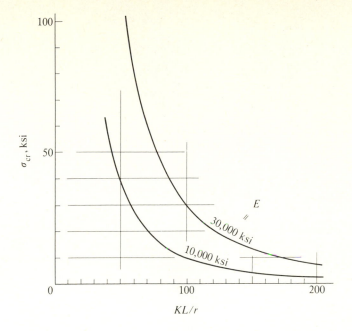

Figure 6.12a Buckling stress as a function of effective slenderness ratio.

$$\sigma_{cr} = \frac{\pi^2 E}{(KL/r)^2} \tag{6.22}$$

KL/r is referred to as the *effective slenderness ratio* of the column. This relationship is shown in fig. 6.12a for two different values of E.

It should be recalled that one of the fundamental assumptions made throughout these entire derivations was that of a linear, elastic material. The solutions shown in fig. 6.12a, then, correspond only to those cases where buckling occurs in this range of behavior. Defining the limiting value as the yield point stress σ_y, and nondimensionalizing with respect to that value, the solutions given in fig. 6.12b are obtained.

If instead of a pronounced yield value the material displayed a proportional limit followed by a gradually increasing, more flexible stress-strain behavior, then the curves shown in fig. 6.12b would be applicable only in the range $\sigma_{cr} \leq \sigma_{proportional\ limit}$ with the substitution of the proportional limit stress for the yield stress. Above the proportional limit, inelastic procedures, as will hereafter be described, would have to be used to define the buckling load.

In summary: The solution of buckling problems as classical eigenvalue problems requires that there first be obtained a general solution to the governing differential equation of equilibrium. Secondly, that solution would then be substituted into each of the known boundary conditions—the total number of equations being equal to the order of the initially defined differential equation. This set of homogeneous, linear, algebraic equations must be solved simultaneously. Unfortunately, the set in question contains as unknowns not only the constants contained in the general solution, but also the unknown axial force term P. In strict mathematical terms, then, the set is indeterminate—there are more unknowns than there are independent equations. On the other hand, all equations are equal to zero, and this condition allows a nontrivial solution to be forced. If the determinant of the coefficients of the algebraic set is equated to zero, an expression will be obtained which contains not only P_{cr}, but

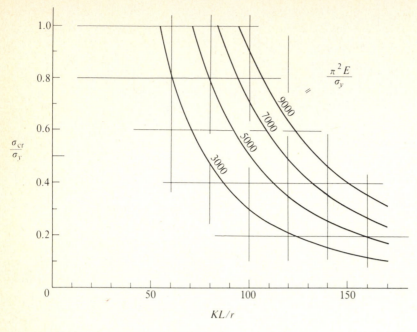

Figure 6.12b Column-buckling curves.

the length of the member, the bending stiffness, various spring stiffness terms (when these are present), etc. When the conditions specified by this equation are met, deflections can occur. Solving the equation for P_{cr} gives the critical values of thrust for which buckling occurs. Since the problem is mathematically indeterminate, it is impossible to ascertain the actual magnitudes of the deflections at buckling. The general deflected shape, however, can be established by assuming a value of 1.0 for any one of the constants and solving for the others in terms of it.

Finally, the results of these types of analysis can be described in the form $P_{cr} = f(\pi, E, I, L)$, as the ratio P_e/P_{cr}, or in terms of an effective length factor K, where

$$P_{cr} = \frac{\pi^2 EI}{(KL)^2}$$

6.2.2 Energy Methods for Determining the Elastic, Flexural Buckling Load

Eigenvalue solutions as defined in the preceding section are exact solutions for buckling loads. For complicated cases—for example, where unusual boundary conditions exist, where the cross-sectional properties vary along the length of the member, etc.—exact solutions may be difficult if not impossible to obtain. For such cases approximate solutions are necessary. In general, these are of the following types: energy methods, finite differences, and numerical integration. A fourth extremely useful approximate method is known as the *finite-element method*.

In chap. 5 it was demonstrated that an elastic body is in equilibrium if the first variation of the total potential of the system in question is equal to zero. If it is assumed that under a particular set of applied forces a given body is in equilibrium in a first position, say, a straight position, and if that position is used as a reference for the establishment of a second, adjacent equilibrium position, say, a laterally bent position, then that new position will be governed by the relationship

$$\delta\Pi^+ = \delta(W_i^+ - W_e^+) = \delta W_i^+ - \delta W_e^+ = 0 \tag{6.23}$$

(The + superscript notation is used to denote only the changes associated with going from the first to the second equilibrium positions.) Since bending is the phenomenon under consideration,

$$\delta W_i^+ = \int_0^l \frac{M^2}{2EI}\, dz = \int_0^l \frac{EI}{2}\, (y'')^2\, dz \tag{6.24}$$

and [from eq. (5.43)]

$$\delta W_e^+ = P_{cr}\, \Delta L = \frac{P_{cr}}{2} \int_0^l (y')^2\, dz \tag{6.25}$$

In these equations y corresponds to the lateral deflection that occurs during buckling. (It is to be noted that the full value of the axial load, P_{cr}, is assumed to act during the entire deflection process.)

From eqs. (6.23), (6.24) and (6.25)

$$P_{cr} = \frac{EI \int_0^l (y'')^2\, dz}{\int_0^l (y')^2\, dz} \tag{6.26}$$

This is known as *Rayleigh's quotient*. The accuracy of the solution, using this method, largely depends upon the accuracy of the assumed buckling mode. It is particularly important to note that the selected configuration should be consistent with the given geometrical boundary conditions of the column.

Consider the buckling of a pin-ended column of length L for which the boundary conditions are: at $z = 0$, $y = 0$, and at $z = l$, $y = 0$. Assuming the polynomial

$$y = a_0 + a_1 z + a_2 z^2 \tag{6.27}$$

then from the boundary conditions

$$a_0 = 0 \quad \text{and} \quad a_1 + a_2 l = 0$$

Equation (6.27) then becomes

$$y = a_3 z(z - l) \tag{6.28}$$

(What initially appeared to be a three-arbitrary-constant selection actually turned out to be a single-variable equation.) Substituting eq. (6.28) into eq. (6.26) yields the critical load

$$P_{cr} = \frac{12EI}{l^2} \tag{6.29}$$

(This is to be compared to the exact solution $P_{cr} = P_e = \pi^2 EI/l^2$.) If, on the other hand, the single sine wave were selected as the assumed deflection mode; that is,

$$y = \Delta \sin \frac{\pi z}{l} \tag{6.30}$$

then substitution into eq. (6.26) would predict the exact buckling solution

$$P_{cr} = P_e = \frac{\pi^2 EI}{l^2}$$

This is due to the fact that the single sine wave represents the exact buckling mode as was previously demonstrated.

One question immediately arises: Would the result be improved if more terms of the power series had been assumed in eq. (6.27)? In general, this is true. However, it should be recognized that for such cases the deflected shape would contain more than one arbitrary parameter and these would appear in the expression for P_{cr}. Solution for P_{cr} would then require minimization with respect to each of these parameters. Such minimization, however, can become cumbersome and can lead to rather complex expressions. A more direct application of the principle of minimum potential energy can facilitate solutions.

As pointed out in chap. 5, the minimum potential energy theory is based on the principle of virtual displacements and has all of the advantages and limitations of that principle. It replaces the equations of equilibrium, but it does not guarantee geometrical compatability of the assumed deflected shape with the prescribed geometrical conditions of the structure. This must be handled separately.

The total potential associated with a bending change can be defined by eq. (6.31)

$$\Pi^+ = W_i^+ - W_e^+ = \tfrac{1}{2}E \int_0^l I(y'')^2 \, dz - \tfrac{1}{2}P \int_0^l (y')^2 \, dz \tag{6.31}$$

(Here it has been assumed that the moment of inertia of the cross section varies along the length of the member. Had this not been the case, I could have been taken outside the integral sign, as has been previously shown.) If it is assumed that the deflected configuration can be represented by the series

$$y = a_1 \varphi_1 + a_2 \varphi_2 + a_3 \varphi_3 + \cdots + a_n \varphi_n \tag{6.32}$$

where $\varphi_1, \varphi_2, \varphi_3, \ldots, \varphi_n$ are functions of z, then substitution of (6.32) into (6.31) yields

$$\Pi^+ = F_1(a_1, a_2, a_3, \ldots, a_n) + PF_2(a_1, a_2, a_3, \ldots, a_n) \tag{6.33}$$

In this expression F_1 and F_2 will be quadratic functions of the arbitrary parameters $a_1, a_2, a_3, \ldots, a_n$. The problem of obtaining expressions for the first variation of Π^+ therefore becomes nothing more than an ordinary maximum-minimum problem in which the variables are $a_1, a_2, a_3, \ldots, a_n$. That is

$$\frac{\partial \Pi^+}{\partial a_i} = \frac{\partial}{\partial a_i}(W_i^+ - W_e^+) = 0 \qquad i = 1, 2, \ldots, n \tag{6.34}$$

Since first derivatives of quadratic functions are linear functions, eq. (6.34) will be linear, homogeneous equations in the unknown $a_1, a_2, a_3, \ldots, a_n$. They can be displayed in the following form:

$$\begin{aligned}
\beta_{11}a_1 + \beta_{12}a_2 + \beta_{13}a_3 + \cdots \beta_{1n}a_n &= 0 \\
\beta_{21}a_1 + \beta_{22}a_2 + \beta_{23}a_3 + \cdots \beta_{2n}a_n &= 0 \\
\cdots \cdots \cdots \cdots \cdots \cdots \cdots \cdots \cdots \\
\beta_{n1}a_1 + \beta_{n2}a_2 + \beta_{n3}a_3 + \cdots \beta_{nn}a_n &= 0
\end{aligned} \tag{6.35}$$

It is to be recognized that the β's contain P, E, I, and L terms, as well as numbers.

For a homogeneous set of linear, algebraic equations to have other than a trivial solution, the determinant of the coefficient matrix must equal zero. That is,

$$\begin{vmatrix}
\beta_{11} & \beta_{12} & \beta_{13} & \cdots & \beta_{1n} \\
\beta_{21} & \beta_{22} & \beta_{23} & \cdots & \beta_{2n} \\
\cdots & \cdots & \cdots & \cdots & \cdots \\
\beta_{n1} & \beta_{n2} & \beta_{n3} & \cdots & \beta_{nn}
\end{vmatrix} = 0 \tag{6.36}$$

This results in an nth order equation that must be solved for the smallest value of P_{cr}. This method of solution is generally known as the Rayleigh-Ritz method, or just Ritz method. It is frequently used to obtain approximate solutions to buckling problems.

Consider for illustration of the Ritz method of solution the flagpole-type column shown in fig. 6.13. The member is originally straight, of uniform cross section, and has an axial thrust of P applied at the uppermost free end.

It will be assumed that the deflected configuration can be represented by the polynomial

$$y = az + bz^2 + cz^3 \tag{6.37}$$

Since the geometrical boundary conditions to be satisfied are

$$\text{At } z = 0 \qquad y = 0$$

$$\text{At } z = 0 \qquad y' = 0$$

a must be equal to zero. Therefore

$$y = bz^2 + cz^3 \tag{6.37a}$$

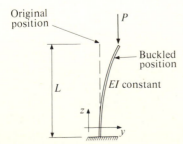

Original position

P

Buckled position

EI constant

L

z

y

Figure 6.13

The internal strain energy and the external work done as the member deflects (during buckling) will be

$$W_i^+ = \frac{EI}{2} \int_0^L (y'')^2 \, dz = \frac{EI}{2} \int_0^L (2b + 6cz)^2 \, dz = EI(2b^2L + 6bcL^2 + 6c^2L^3)$$

and

$$W_e^+ = \frac{P}{2} \int_0^L (y')^2 \, dz = \frac{P}{2} \int_0^L (2bz + 3cz^2)^2 \, dz = \frac{P}{2}\left(\frac{4}{3}b^2L^3 + 3bcL^4 + \frac{9}{5}c^2L^5\right)$$

Therefore

$$\Pi^+ = W_i^+ - W_e^+ = EI(2b^2L + 6bcL^2 + 6c^2L^3) - \frac{P}{2}\left(\frac{4}{3}b^2L^3 + 3bcL^4 + \frac{9}{5}c^2L^5\right) \tag{6.38}$$

Differentiating partially with respect to each of the unknowns b and c yields the equations

$$\frac{\partial \Pi^+}{\partial b} = EI(4bL + 6cL^2) - \frac{P}{2}\left(\frac{8}{3}bL^3 + 3cL^4\right) = 0$$

$$\frac{\partial \Pi^+}{\partial c} = EI(6bL^2 + 12cL^3) - \frac{P}{2}\left(3bL^4 + \frac{18}{5}cL^5\right) = 0 \tag{6.39}$$

Collecting terms

$$b\left(\frac{4}{3}\frac{PL^2}{EI} - 4\right) + c\left(\frac{3}{2}\frac{PL^3}{EI} - 6L\right) = 0$$

$$b\left(\frac{3}{2}\frac{PL^2}{EI} - 6\right) + c\left(\frac{9}{5}\frac{PL^3}{EI} - 12L\right) = 0 \tag{6.40}$$

For a nontrivial solution the determinant of the coefficients must equal zero, or

$$\begin{vmatrix} \left(\frac{4}{3}\frac{PL^2}{EI} - 4\right) & \left(\frac{3}{2}\frac{PL^3}{EI} - 6L\right) \\ \left(\frac{3}{2}\frac{PL^2}{EI} - 6\right) & \left(\frac{9}{5}\frac{PL^3}{EI} - 12L\right) \end{vmatrix} = 0 \tag{6.41}$$

This gives the quadratic equation

$$3\left(\frac{PL^2}{EI}\right)^2 - 104\left(\frac{PL^2}{EI}\right) + 240 = 0 \tag{6.42}$$

which has as a smallest root

$$\frac{P_{cr}L^2}{EI} = 2.487$$

The critical load is therefore

$$P_{cr} = 2.487 \frac{EI}{L^2}$$

This value should be compared with the exact solution $P_{cr} = 2.467EI/L^2$ (It should be noted that the predicted solution using this approximate method is larger than the exact solution. In general, the Ritz method will always result in a critical load larger than the true buckling load.)

The assumed deflected configuration [eq. (6.37)] need not be of polynomial form. For example, for the same structure and loading shown in fig. 6.14, and assuming a moving coordinate system that passes through the uppermost end of the deflected column but has zero for z at the base, a possible cosine series approximation would be

$$y = a_1 \cos \frac{\pi z}{2L} + a_3 \cos \frac{3\pi z}{2L} + a_5 \cos \frac{5\pi z}{2L} + \cdots$$

or

$$y = \sum_{n = 1, 2, 3, \ldots} a_{2n-1} \cos \frac{(2n - 1)\pi z}{2L} \tag{6.43}$$

The first and second derivatives with respect to z of this equation are

$$y' = - \sum_{n = 1, 2, 3, \ldots} a_{2n-1} \left(\frac{2n - 1}{2L} \pi \right) \sin \frac{(2n - 1)\pi}{2L}$$

and

$$y'' = - \sum_{n = 1, 2, 3, \ldots} a_{2n-1} \left(\frac{2n - 1}{2L} \pi \right)^2 \cos \frac{(2n - 1)\pi}{2L}$$

Therefore

$$W_i^+ = \frac{EI}{2} \int_0^L (y'')^2 \, dz = \frac{EI}{2} \int_0^L \left[-\sum a_{2n-1} \left(\frac{2n - 1}{2L} \right)^2 \pi^2 \cos \left(\frac{2n - 1}{2L} \right) \pi z \right]^2 dz$$

$$= \frac{EI}{2} \frac{\pi^4}{32L^3} \sum_{n = 1, 2, 3, \ldots} (2n - 1)^4 a_{2n-1}^2 \tag{6.44}$$

and

$$W_e^+ = \frac{P}{2} \int_0^L (y')^2 \, dz = \frac{P}{2} \int_0^L \left[-a_{2n-1} \left(\frac{2n - 1}{2L} \right) \pi \sin \left(\frac{2n - 1}{2L} \right) \pi z \right]^2 dz$$

$$= \frac{P\pi^2}{16L} \left[\sum_{n = 1, 2, 3, \ldots} (2n - 1)^2 a^2 (2n - 1) \right] \tag{6.45}$$

Assuming only three terms in the series, the total potential is

$$\Pi^+ = W_i^+ - W_e^+ = -\frac{\pi^4 EI}{64L^3} (a_1^2 + 3^4 a_3^2 + 5^4 a_5^2) + \frac{P\pi^2}{16L} (a_1^2 + 3^2 a_3^2 + 5^2 a_5^2)$$

or

$$\Pi^+ = a_1^2 \left(\frac{P\pi^2}{16L} - \frac{\pi^4 EI}{64L^3} \right) + a_3^2 \left(\frac{3^2 P\pi^2}{16L} - \frac{3^4 \pi^4 EI}{64L^3} \right) + a_5^2 \left(\frac{5^2 P\pi^2}{16L} - \frac{5^4 \pi^4 EI}{64L^3} \right) \tag{6.46}$$

Differentiating the total potential with respect to each of the unknowns a_1, a_3, and a_5 and equating the resulting expressions to zero yields the following three, independent equations:

$$\frac{\partial \Pi^+}{\partial a_1} = a_1 \left(\frac{P\pi^2}{8L} - \frac{\pi^4 EI}{32L^3} \right) = 0$$

$$\frac{\partial \Pi^+}{\partial a_2} = a_2 \left(\frac{3^2 P\pi^2}{8L} - \frac{3^4 \pi^4 EI}{32L^3} \right) = 0 \qquad (6.47)$$

$$\frac{\partial \Pi^+}{\partial a_3} = a_3 \left(\frac{5^2 P\pi^2}{8L} - \frac{5^4 \pi^4 EI}{32L^3} \right) = 0$$

For the first of these to be satisfied

$$\frac{P\pi^2}{8L} = \frac{\pi^4 EI}{32L^3}$$

or

$$P_{cr} = \frac{\pi^2 EI}{4L^2} \qquad (6.48a)$$

For the second,

$$P_{cr} = \frac{3^2 \pi^2 EI}{4L^2} \qquad (6.48b)$$

For the third,

$$P_{cr} = \frac{5^2 \pi^2 EI}{4L^2} \qquad (6.48c)$$

The buckling value predicted by eq. (6.48a) is the smallest, and therefore would be assumed to be the solution to the problem in question. (Equation 6.48a is, in fact, the exact solution to the problem. The first term of the assumed cosine series corresponds identically to the true deflected shape of the member at buckling.)

It should be observed that unlike the previous case, partial differentiation with respect to the parameters a_1, a_2, and a_3 did not result in the definition of a set of simultaneous algebraic equations. Rather, each differentiation yielded a separate equation containing the single unknown a_1, a_2, or a_3, and a parenthesized term which included the unknown critical axial load value P_{cr}. This was because the chosen set of functions are "orthogonal" over the interval of integration. Fourier series are frequently used for Ritz solutions because of this property.

When solving buckling problems by the Rayleigh-Ritz methods, it is useful to consider the assumed displacement function of the following general form:

$$y(z) = \sum_{n=0}^{\infty} a_n f(z) z^n \qquad (6.49)$$

where $f(z)$ is a specifically chosen function which satisfies the geometrically prescribed boundary conditions and $\sum a_n z^n$ is a power series.

For fixed-fixed ends: $\qquad f(z) = z^2(z-l)^2$

For fixed-hinged ends: $\qquad f(z) = z^2(z-l)$

For hinged-hinged ends: $\qquad f(z) = z(z-l)$ $\qquad\qquad$ (6.50)

For fixed-free ends: $\qquad f(z) = z^2$

Note that the polynomial functions of eqs. (6.28) and (6.37a) are the simplest possible forms which satisfy both eqs. (6.49) and (6.50).

Example 6.4 The linearly tapered member shown in fig. 6.14 is subjected to an axial thrust along its centroidal axis. It is specified that the member is constrained in a fashion that physically prevents it from buckling in the weak direction. In the strong direction, the member is pinned at each of its ends, but is unrestrained between those points. The critical load associated with buckling about the strong axis of the member is desired.

The variation in cross-sectional depth is defined by the relationship

$$d_z = d_0\left(1 + \frac{z}{L}\gamma\right) \qquad\qquad (6.51)$$

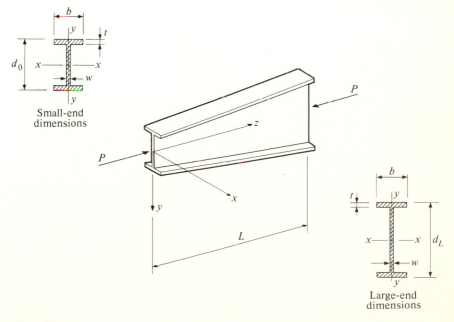

Figure 6.14

where d_0 is the smaller end depth, and γ is a measure of the magnitude of taper of the member. γ is defined by the relationship

$$\gamma = \frac{d_L}{d_0} - 1$$

Thus, for a prismatic member, $\gamma = 0$. For a member whose depth at the larger end is three times that at the smaller end, $\gamma = 2.0$. For all variations in γ, it is specified that the flange widths and thicknesses are constant, as is the thickness of the web of the member. The equation for $\delta\Pi = 0$ is

$$\delta\Pi = \delta\left[\int_0^L EI_x(y'')^2\,dz - P\int_0^L (y')^2\,dz\right] = 0$$

or

$$\int_0^L EI_x y''\,\delta y''\,dz - P\int_0^L y'\,\delta y'\,dz = 0 \tag{6.52}$$

where I_x is the moment of inertia about the strong axis of the cross section which varies with z.

For simply supported ends, the displacement function can be assumed to be of the following general form, according to eqs. (6.49) and (6.50):

$$y = \sum_{m=0}^{\infty} a_m z(z - L)z^m \tag{6.53}$$

Substituting this into eq. (6.52) yields

$$\sum_{m=0}^{\infty} a_m\left(EI_x y''_m y''_n\,dz - P\int_0^L y'_m y'_n\,dz\right) = 0 \tag{6.54}$$

It is to be noted that the summation index for the variational form has been changed from m to n to distinguish it from the displacement function.

To facilitate generalization, a nondimensional variable ζ will be introduced, where

$$\zeta = \frac{z}{L} \qquad 0 \le \zeta \le 1.0 \tag{6.55}$$

The dimensional variables along the length of the member, then, become

$$d_z = d_0(1 + \zeta\gamma)$$
$$I_x = I_{x0}(1 + \gamma\zeta)(1 + \mu\gamma\zeta)$$

where

$$\mu = \frac{d_0^3 w}{12I_{x0}}$$

and

$$I_{x0} = \text{moment of inertia of the smallest end of the}$$
$$\text{member } (\zeta = 0), \text{ about the } xx \text{ axis}$$

In this form, eq. (6.54) can be written as

$$\sum_{m=0}^{\infty} a_m \left(\int_0^{1.0} I_x y_m'' y_n'' \, d\zeta - \lambda \pi^2 \int_0^{1.0} y_m' y_n' \, d\zeta \right) = 0 \qquad (6.56)$$

where

$$\lambda = \frac{P_{cr}}{P_{ex0}}$$

and

$$P_{ex0} = \frac{\pi^2 E I_{x0}}{L^2}$$

Solution of eq. (6.56), presuming deflection function (6.53), yields a set of homogeneous, algebraic equations—one equation for each of the m terms. To have other than a trivial solution, the determinant of the coefficient matrix of that set must equal zero. It should be obvious to the reader that for any realistic number of terms, solutions can best be achieved using the computer. This has been done for a variety of cross sections and member lengths.[†] Only three of those sections will be listed here, and only one length of member will be presumed: $L = 144$ in. The cross-sectional dimensions for the three sections considered are given in table 6.2. For all of the members $E = 30{,}000$ ksi. Assuming 15 terms in the assumed deflection function, the results are given in table 6.3.

[†] For a detailed discussion of the solution of linearly tapered member problems, see G. C. Lee, M. L. Morrell and R. L. Ketter, Design of Tapered Members, *Weld. Res. Counc. Bull. no. 173*, June 1972.

Table 6.2 Geometry of cross sections at small end of tapered member

	Member A, in	Member B, in	Member C, in
d_0	6.00	6.00	6.00
b	4.00	4.00	12.00
t	0.25	0.75	0.75
w	0.10	0.25	0.25

Table 6.3 Influence of the degree of taper on the magnitude of the buckling load

	$\dfrac{P_{cr}}{P_{exo}}$		
γ	Member A, in	Member B, in	Member C, in
0	1.0000	1.0000	1.0000
0.5	1.5571	1.5505	1.5352
1.0	2.1832	2.1671	2.1276
1.5	2.8783	2.8462	2.7759
2.0	3.6419	3.5945	3.4642
2.5	4.4668	4.3998	4.2354
3.0	5.3498	5.2747	5.0458
3.5	6.3291	6.2025	5.8859
4.0	7.3201	7.1677	6.7976
4.5	8.4257	8.2707	7.7635
5.0	9.6224	9.3750	8.8067
6.0	12.1054	11.8031	10.9478

6.2.3 Torsional Buckling of Axially Loaded Columns

Most real structural members have different properties about each of their principal cross-sectional axes. Also, such members are joined at their ends to other members that have different effective bending stiffness in the two major directions. The possibility of buckling about either of the two principal axes of the cross section therefore is real and must be examined. There is, however, another possible independent buckling mode that in a few cases also must be examined: torsional buckling about the longitudinal axis.

The torsional behavior of uniform cross-section members subjected to axial forces is governed by the differential equation

$$\frac{d^4\beta}{dz^4} + \frac{P\bar{r}_0^2 - G\kappa_T}{EI_\omega}\frac{d^2\beta}{dz^2} = 0 \tag{6.57}$$

where P = the applied axial thrust

β = the angle of twist of the cross section

I_x = the moment of inertia of the cross section about the major principal axis xx

I_y = the moment of inertia of the cross section about the minor principal axis yy

A = the area of the cross section

$\bar{r}_0$ = the polar radius of gyration of the section about the shear center

$\bar{r}_0^2 = \dfrac{I_x + I_y}{A} + x_0^2 + y_0^2$

x_0 = the x distance from the shear center to the centroid

y_0 = the y distance from the shear center to the centroid

κ_T = the St. Venant torsion constant of the cross section

I_ω = the warping constant of the cross section
E = Young's modulus of elasticity of the material
G = shearing modulus of elasticity of the material

Equation (6.57) is of the standard form

$$\frac{d^4\beta}{dz^4} + \lambda^2 \frac{d^2\beta}{dz^2} = 0 \tag{6.58}$$

It has as a general solution

$$\beta = A \sin \lambda z + B \cos \lambda z + C\left(\frac{z}{L}\right) + D \tag{5.59}$$

Eigenvalue solutions for this case can be obtained in the same fashion that was used earlier in this chapter for determining flexural buckling.

If, for example, the four boundary conditions to be satisfied are those associated with simple end supports, that is,

$$\text{At } z = 0 \qquad \beta = 0$$
$$\text{At } z = 0 \qquad \beta'' = 0$$
$$\text{At } z = L \qquad \beta = 0$$
$$\text{At } z = L \qquad \beta'' = 0$$

and if the cross section is assumed to be doubly symmetric, then the value of the torsional buckling load will be found to be

$$P_{z,\,cr} = \left(\frac{\pi^2 E I_\omega}{L^2} + G\kappa_T\right)\frac{A}{I_x + I_y} \tag{6.60}$$

For other end conditions, suitably modified equations can be derived.

When both flexural and torsional movements occur at the instant of buckling, three coupled differential equations (with all of their associated boundary conditions) must be satisfied. The three equations are[†]

$$EI_x \frac{d^4v}{dz^4} + P\left(\frac{d^2v}{dz^2} + y_0 \frac{d^2\beta}{dz^2}\right) = 0$$

$$EI_y \frac{d^4u}{dz^4} + P\left(\frac{d^2u}{dz^2} + x_0 \frac{d^2\beta}{dz^2}\right) = 0 \tag{6.61}$$

$$EI_\omega \frac{d^4\beta}{dz^4} + (P\bar{r}_0^2 - G\kappa_T)\frac{d^2\beta}{dz^2} + Py_0 \frac{d^2u}{dz^2} - Px_0 \frac{d^2v}{dz^2} = 0$$

where u is the deformation of the member in the x direction, v is the deformation in the y direction, and β is the rotation about the longitudinal axis z.

[†] For a derivation of these equations see T. V. Galambos, *Structural Members and Frames*, chap. 4, Prentice-Hall, Inc., Englewood Cliffs, N.J., 1968.

For simply supported end conditions for each of the deformation variables, it can be demonstrated that eq. (6.61) will be satisfied by the assumed deformation functions

$$v = C_1 \sin \frac{\pi z}{L} \qquad u = C_2 \sin \frac{\pi z}{L} \qquad \text{and} \qquad \beta = C_3 \sin \frac{\pi z}{L}$$

Substituting these into eq. (6.61) yields the following set of linear, simultaneous, homogeneous algebraic equations:

$$C_1 \left(\frac{\pi^2 EI_x}{L^2} - P \right) + C_2(0) \qquad + C_3(Px_0) \qquad = 0$$

$$C_1(0) \qquad + C_2 \left(\frac{\pi^2 EI_y}{L^2} - P \right) + C_3(-Py_0) \qquad = 0$$

$$C_1(Px_0) \qquad + C_2(-Py_0) \qquad + C_3 \left[\frac{\pi^2 EI_\omega}{L^2} + G\kappa_T - P\bar{r}_0^2 \right] = 0$$

For a nontrivial solution, the determinant of the coefficients of C_1, C_2, and C_3 must equal zero. This gives the eigenvalue equation

$$(P - P_x)(P - P_y)(P - P_z) - P^2(P - P_y)\frac{x_0^2}{\bar{r}_0^2} - P^2(P - P_x)\frac{y_0^2}{\bar{r}_0^2} = 0 \qquad (6.62)$$

where, in the terms of the notation previously used,

$$P_x = P_{ex} \quad = \text{the Euler buckling load about the } xx \text{ axis}$$
$$P_y = P_{ey} \quad = \text{the Euler buckling load about the } yy \text{ axis}$$
$$P_z = P_{z, cr} = \text{the torsional buckling load}$$

The smallest of the roots of eq. (6.62) will be the critical buckling load for the column in question.

Had the boundary conditions been other than the simple ones assumed, different (and in all likelihood, more complicated) deformation functions would have had to be selected. In turn, the resulting eigenvalue equation would have been more complicated.

Example 6.5 A 280-in long, 14 × 43 lb/ft wide-flange steel shape is to be used as a column. It is specified that all the end conditions of the member are of the simple support type. What will be the buckling load of the column?
The properties of the member in question are

$$L = 280 \text{ in}$$
$$A = 12.6 \text{ in}^2$$
$$I_x = 429 \text{ in}^4$$
$$I_y = 45.1 \text{ in}^4$$
$$E = 30,000 \text{ ksi}$$
$$G\kappa_T = 0.349E$$
$$EI_\omega = 1950E$$

Using the equations previously derived, the three independent, elastic, column buckling loads are

$$P_{e_x} = \frac{\pi^2 E I_x}{L^2} = 1620 \text{ kips}$$

$$P_{e_y} = \frac{\pi^2 E I_y}{L^2} = 170 \text{ kips}$$

$$P_{z, \text{cr}} = \left(\frac{\pi^2 E I_\omega}{L^2} + G\kappa_T\right)\frac{A}{I_x + I_y} = 474 \text{ kips}$$

The critical condition, therefore, is flexural buckling about the minor axis of the member at a value of axial thrust of 170 kips. (The stress associated with this load is $170/12.6 = 13.49$ ksi, which is well within the elastic limit of all structural steels.)

If the column in question had been supported in the x direction such that bending about the yy axis was prevented, but twisting about the longitudinal axis was allowed, then the column would buckle torsionally at the next higher critical value, $P_{\text{cr}} = 474$ kips. For this case, $\sigma_{\text{cr}} = 37.6$ ksi. Depending upon whether or not this value is above or below the yield level of the particular steel specified, the solution computed will not or will be valid. (The stress associated with buckling about the xx axis equals 128.6 ksi, a value higher than the yield value of almost all structural steels. Therefore, independent of whether or not the cross section is further constrained and thereby prevented from twisting, the strong-axis elastic buckling solution is only of academic interest.) Methods for solving inelastic buckling problems will be discussed in the next section of this chapter.

6.2.4 Inelastic Column Buckling—Influence of Material Properties

Real columns in real structures are frequently constrained against deformation in their weak direction. Moreover, it is not unusual to have them so braced at their ends or along their lengths that twisting is successfully prevented. The result is a prediction of elastic buckling about the strong axis of the section. For many of these cases there results an elastic flexural buckling prediction that is larger than that associated with the yield stress or proportional limit stress of the material. This is especially true for columns in building-type construction, where ceiling heights, applied loads or other architectural or fabrication requirements result in a prescription of column proportions that make them relatively short or "stocky." For these types of members, prorated design stresses, that is, design stresses increased to account for the desired margin of safety, frequently fall in the range of inelastic behavior. In fact, for steel building structures, many if not most real designs occur in that range. Almost without exception, concrete building columns are short.

As indicated in chap. 1, certain materials exhibit stress-strain properties of the type illustrated in fig. 6.15a. In the initial loading range, the material responds linearly to imposed stresses. There is reached, however, a proportional limit above which strains increase at an ever increasing rate. This continues until there is reached a value of stress, denoted in the figure as σ_y, for which strains increase as stress

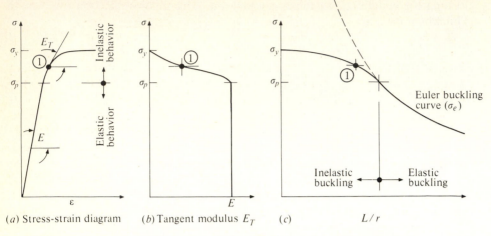

(a) Stress-strain diagram   (b) Tangent modulus E_T (c) L/r

Figure 6.15 Tangent modulus buckling of an idealized axially loaded compression member.

remains constant. (It is, of course, to be recognized that the stress-strain diagram given is not necessarily a complete curve to fracture. Depending on the material, there subsequently could be realized increases in stress associated with strain-hardening.)

In the linear elastic range of material behavior, the slope of the stress-strain curve equals E, Young's modulus of elasticity. Above the proportional limit, the slope E_T varies from E to zero with increasing stress. Figure 6.15b is a plot of the variation in E_T as a function of σ for the case shown.

Within the elastic range of material behavior, assuming pin-ended conditions for the column in question, the flexural buckling load is defined by the equation

$$P_{cr} = P_e = \frac{\pi^2 EI}{L^2} \tag{6.63}$$

In terms of stress, this would be

$$\sigma_{cr} = \sigma_e = \frac{\pi^2 E}{(L/r)^2} \tag{6.64}$$

which is shown in fig. 6.15c.

It is to be recalled that, independent of whether or not stresses are elastic or inelastic, just prior to buckling a column is subjected to a uniform distribution of stress over the entire cross section; that is, all longitudinal fibers of the member are at the same point on the stress-strain curve. It would seem reasonable, therefore, to assume that in the inelastic range of applied axial stresses the member, at buckling, will respond as if it were composed of a material whose modulus of elasticity is the tangent modulus at the stress value in question. That is,

$$(\sigma_{cr})_T = \frac{\pi^2 E_T}{(L/r)^2} \tag{6.65}$$

Unfortunately, as will be hereafter demonstrated, such as assumption is not strictly valid.

As lateral deformation occurs, bending moments develop and compressive stresses increase on the concave side of the member. On the convex side, they decrease. For increasing values of stress, the changes due to bending are related to the bending strains by the tangent modulus E_T. However, for those segments of the cross section where unloading occurs, as illustrated in fig. 1.19, the full elastic modulus E governs the behavior of each of the fibers. The effective unit bending stiffness of the member—the quantity that governs lateral bending deformations, and therefore buckling—is as a result not just a function of E_T, but a function of both E_T and E.

Consider the symmetrical cross section shown in fig. 6.16a. It is assumed that there is applied to the member in question, through the centroid, an axial thrust that causes a uniformly distributed normal stress of σ_1 in the longitudinal direction. The corresponding normal strains are ε_1. (These values have been indicated in sketches a and b of fig. 6.15.) At this particular value of load, the member is assumed to buckle, and therefore deflect laterally an infinitesimal amount. Assuming that at the most highly strained cross section along the member, plane sections before bending remain plane, a linear variation in bending strains can be presumed across that section, as indicated in fig. 6.16b. These, in turn, result in the stress distribution shown in fig. 6.16c.

For these assumptions, recalling that—by definition—P remains constant at

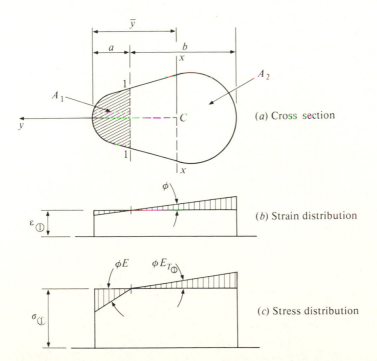

(a) Cross section

(b) Strain distribution

(c) Stress distribution

Figure 6.16 Strain and stress distributions at onset of bending—reduced modulus concept.

buckling, there will be within the cross section in question an axis, parallel to the principal axis of bending, along which no change in strain occurs. This is shown as axis 1-1 in fig. 6.16a. (It is assumed that this axis is b distance from the concave side of the member, and a from the convex side.) The entire area A_1, shown cross-hatched, experiences unloading. For P to remain constant

$$\int_{A_2} \Delta\sigma \, dA - \int_{A_1} \Delta\sigma \, dA = 0 \tag{6.66}$$

where $\Delta\sigma$ is the change in stress due to bending. If the cross section is rectangular, this would give

$$\frac{E_T}{E} = \left(\frac{a}{b}\right)^2 \tag{6.67}$$

For the more general case,

$$\frac{E_T}{E} = \frac{\int_{A_2} [(\bar{y} - a) - y] \, dA}{\int_{A_1} [y - (\bar{y} - a)] \, dA} \tag{6.68}$$

The effective bending stiffness of the member can be determined by the relationship

$$M = E_{\text{eff}} I \varphi$$

or

$$\frac{M}{\varphi} = E_R I \tag{6.69}$$

where E_R is the reduced modulus for the cross section and I is the moment of inertia about the indicated xx axis. From

$$M = \int_{A_2} [(\bar{y} - a) - y]^2 E \varphi \, dA + \int_{A_1} [y - (\bar{y} - a)]^2 E_T \varphi \, dA$$

E_R can be determined in terms of E_T and E:

$$E_R = E\left(\frac{I_1}{I} + \frac{E_T}{E}\frac{I_2}{I}\right) \tag{6.70}$$

where

I_1 = moment of inertia of area A_1 about 1-1 axis

I_2 = moment of inertia of area A_2 about 1-1 axis

I = total moment of inertia of cross section about xx axis

For a rectangular cross section, substituting in the values for a and b from eq. (6.67), eq. (6.70) becomes

$$E_R = E_T \frac{4}{1 + 2\sqrt{E_T/E} + E_T/E} \tag{6.71}$$

The corresponding, inelastic buckling loads and stresses are

$$(P_{cr})_R = \frac{\pi^2 E_R I}{L^2} \tag{6.72a}$$

and

$$(\sigma_{cr})_R = \frac{\pi^2 E_R}{(L/r)^2} \tag{6.72b}$$

[Since the denominator of the bracketed terms of eq. (6.71) is always less than 4.0 for $E_T/E < 1.0$, E_R will always be greater than E_T, and the *reduced modulus buckling load* will always be greater than the *tangent modulus buckling load*. The actual magnitude of the increase depends on the stress-strain relationship, the magnitude of the average stress prior to buckling, and the cross section of the member.]

The above developed analyses for the tangent modulus and the reduced modulus buckling loads were proposed originally by Engesser (1889 and 1895). Using the definitions commonly accepted, there seems to be little question but that the buckling load in the inelastic range should be the reduced modulus load. Unfortunately, experimentally determined values from carefully conducted tests provided a much greater degree of correspondence with the tangent modulus load than they did with the reduced modulus load. For over 50 years this presented an unexplained paradox, and various reasons for the discrepancy were advanced—for example, eccentricity of load application, out-of-straightness of the columns, etc. It was not until 1948 that Shanley and in 1950 Duberg and Wilder provided the explanation.

Consider the behavior of the model column examined by Duberg and Wilder,† which is here shown as fig. 6.17a. The member in question is of length l and is connected at its base to two symmetrically placed springs of stiffness C_L and C_R. It is presumed that the member itself is rigid, and that deformations occur only in the springs. (This is the same as presuming a simply supported column of twice this length, where a deformable "cell" exists at the center section of the member, and where the member is rigid everywhere else.) It is presumed that the load-deformation curves of the springs are governed by the diagram shown in fig. 6.17b.

When the applied load is below P_1, the system is in the elastic range. When $P > P_1$ inelastic behavior is observed and the *loading* spring follows the curve whose slope is α_2. *Unloading* follows the path shown, with a slope of α_1.

Elastic buckling Elastic buckling occurs in the range $P \leq P_1$, with $C_L = C_R = C_1$. Considering small deflections (see fig. 6.18),

$$\delta = \frac{e_R - e_L}{b} l \tag{6.73}$$

For equilibrium,

$$\sum M_0 = 0$$

---

† J. E. Duberg and T. W. Wilder, Column Behavior in the Plastic Range, *J. Aeronaut. Sci.*, vol. 17, no. 6, June 1950.

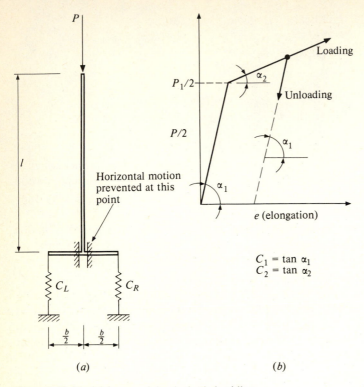

$$C_1 = \tan \alpha_1$$
$$C_2 = \tan \alpha_2$$

(a) (b)

Figure 6.17 Model for examining inelastic buckling.

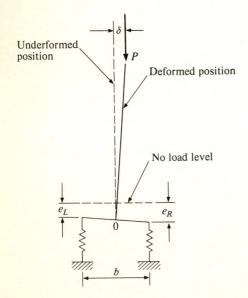

Figure 6.18 Deformation of model column.

This gives

$$P\delta = \tfrac{1}{2}b(C_1 e_R) - \tfrac{1}{2}b(C_1 e_L)$$

or $\qquad\qquad\qquad\qquad\qquad\qquad\qquad\qquad\qquad\qquad\qquad$ (6.74)

$$P\delta = \tfrac{1}{2}bC_1(e_R - e_L)$$

From eqs. (6.73) and (6.74)

$$\left(P - \frac{C_1 b^2}{2l}\right)(e_R - e_L) = 0$$

If $e_R = e_L$, a trivial (nonbuckling) solution is obtained. If, however, the other parenthesized term is equated to zero,

$$P_{cr} = P_e = \frac{C_1 b^2}{2l} \qquad\qquad\qquad (6.75)$$

This is the elastic critical buckling load for the model column in question.

Inelastic "buckling"—no strain reversal It is assumed that at the time of buckling
the compressive strain in the left-hand spring remains unchanged, while that in the
right-hand spring increases an amount Δe_R. Therefore, $\delta = (l/b)\,\Delta e_R$. The spring
constant for increased deformations in the inelastic range is C_2. Summing forces in
the vertical direction and moments about point 0 yield

$$\sum F_v = 0 \qquad\qquad\qquad \Delta P = \Delta e_R\, C_2$$

$$\sum M_0 = 0 \qquad (P + \Delta P)\frac{l}{b}\,\Delta e_R = \frac{1}{2}b\,\Delta e_R\, C_2$$

Assuming that at the time of buckling the increment of axial force is small; that is
$\Delta P \to 0$; the buckling equation becomes

$$P_{cr} = P_T = \frac{C_2 b^2}{2l} \qquad\qquad\qquad (6.76)$$

where P_T is the tangent modulus buckling load.

For this case, consider now the rate of change of ΔP with respect to the lateral
deflection δ:

$$\Delta P = \Delta e_R\, C_2 = \frac{b}{l}\,\delta C_2 = 2P_T\frac{\delta}{b}$$

or

$$\frac{\partial(\Delta P)}{\partial\delta} = \frac{2P_T}{b} > 0$$

The rate is positive; that is, increased load is required for deformation to occur. P_T,
therefore, is a possible point of bifurcation on the load-deflection curve, but it is not a
point of neutral equilibrium. It can therefore be concluded that the tangent modulus
load is smaller than the true buckling load of the member.

Inelastic "buckling"—no change in load If it is assumed that $\Delta P \equiv 0$, then from equilibrium considerations

$$\sum F_v = 0 \qquad \Delta P = \Delta e_L \, C_1 + \Delta e_R \, C_2 = 0 \qquad (6.77)$$

and

$$\sum M_0 = 0 \qquad \frac{Pl}{b}(\Delta e_R - \Delta e_L) = \frac{1}{2} b(\Delta e_R \, C_2 - \Delta e_L \, C_1) \qquad (6.78)$$

Substituting (6.78) into (6.77) yields

$$P \, \Delta e_R \left(1 + \frac{C_2}{C_1}\right) = \frac{b^2}{l}\frac{C_1 C_2}{C_1 + C_2} = \frac{b^2 C_1}{l}\frac{b^2 C_2}{l}\frac{1}{C_1 + C_2}\frac{l}{b^2}$$

$$= 4P_T P_e \frac{1}{C_1 + C_2}\frac{l}{b^2}$$

or

$$P_{cr} = P_R = \frac{4P_T P_e}{(1/b^2)(2lP_e + 2lP_T)}\frac{l}{b^2} = \frac{2P_T P_e}{P_T + P_e}$$

or

$$P_R = \frac{2}{1 + P_T/P_e} P_T \qquad (6.79)$$

Since $P_T/P_e \leq 1.0$,

$$1 + \frac{P_T}{P_e} \leq 2$$

or

$$P_R \geq P_T$$

The rate of change of ΔP with respect to δ is zero, since by definition $\Delta P \equiv 0$. Hence P_R is the buckling load corresponding to neutral equilibrium.

Complete strain reversal—unloading only For this case, the left-hand spring is assumed to unload while the right-hand spring remains constant. That is,

$$\Delta e_R = 0 \qquad \text{and} \qquad \Delta e_L = f(C_1)$$

Summing forces in the vertical direction, and moments about 0

$$\sum F = 0 \qquad \Delta P = -\Delta e_L \, C_1$$

$$\sum M_0 = 0 \qquad (P + \Delta P)\frac{l}{b} \Delta e_L = \frac{b}{2} C_1 \, \Delta e_L$$

For $\Delta P \to 0$,

$$P_{cr} = P_e = \frac{b^2 C_1}{2l}$$

which is the same general elastic expression obtained previously. The rate of change of ΔP with respect to δ is therefore

$$\Delta P = -\Delta e_L\, C_1 = -\frac{\delta}{l}\, bC_1$$

or

$$\frac{\partial(\Delta P)}{\partial \delta} = -2\frac{P_e}{b} < 0$$

The predicted critical load, then, is in unstable equilibrium. The load can never be reached.

Figures 6.19a and b summarize the general buckling behavior to be expected for both *short columns*, that is, for those where buckling takes place in the inelastic range, and *long columns*, where elastic buckling occurs. In the inelastic range, the load corresponding to true buckling is somewhere between the tangent modulus load P_T and the reduced modulus load P_R. In the elastic range, the Euler solution governs.

It is important to note the basic differences between elastic and inelastic buckling theories. For elastic buckling the bending stiffness of the member is constant. There is one unique buckling load associated with a given bending stiffness and a prescribed set of member geometrical parameters. For inelastic buckling, the bending stiffness is reduced from that in the elastic range, but the extent of that reduction depends upon the magnitude of the applied load at the time of buckling. The later is related not only to geometry but to the reduced stiffness at the time in question. To determine the inelastic buckling load, therefore, at least one assumption has to be made in order

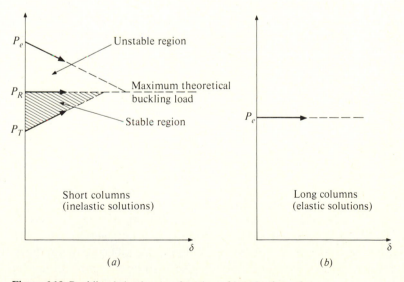

Figure 6.19 Buckling behavior as a function of length of member.

to provide sufficient conditions for solution. Theoretically, an infinite number of inelastic buckling loads, all bounded by P_T and P_R, can be determined, depending upon, for example, the assumed strain distribution.

The difference between the tangent modulus and the reduced modulus buckling loads is relatively small for most real construction members. Moreover, members are seldom if ever straight, and applied thrusts are not precisely located along the centroidal axis of the member—both of which effects tend to reduce the critical load of the column. Test results also tend to confirm the reasonableness of a prediction of buckling strengths based on the tangent modulus theory. For these and other reasons, the Column Research Council of Engineering Foundation issued the following official position statement in May 1952:

> It is the considered opinion of the Column Research Council that the tangent modulus formula affords a proper basis for the establishment of working-load formulas.

The accepted governing equations for column buckling are therefore

Elastic buckling: $$P_{cr} = \frac{\pi^2 E I}{(KL)^2}$$

Inelastic buckling: $$P_{cr} = \frac{\pi^2 E_T I}{(KL)^2} = \frac{\pi^2 E I m}{(KL)^2} \tag{6.80}$$

where m is a modifying function or factor which relates the tangent-modulus E_T to the elastic modulus E.

6.2.5 Residual Stresses and the Inelastic Buckling Load of Steel Columns

Certain structural members, due to the manner of their forming or fabrication, contain initial, internal, residual stresses, which stresses are in equilibrium within themselves at each and every section along the member. They correspond to zero applied external loading. It is to be understood, however, that these locked-in stresses are additive to the stresses which result from applied forces, and their combination can cause yielding of portions of the cross section below those values that would be predicted neglecting these stresses.

In general, residual stresses exist in structural members as a result of plastic deformations. These deformations may be due to differential cooling after hot-rolling or welding, or they may be the result of fabrication operations such as cold-bending or cambering. In hot-rolled steel members, deformations occur during the process of cooling from a very high temperature to room temperature. As the member passes through the last rolling mill, temperature is essentially constant across the cross section. As it rests on the cooling bed, however, certain parts cool faster than others, and those parts, initially, shrink at a faster rate. It also is to be recognized that the modulus of elasticity, as well as other material properties, vary with temperature—

the higher the temperature, the smaller the value of the various material strength and stiffness parameters.

Consider, for example, the development of cooling residual stresses in a typical, hot-rolled, wide-flange, steel cross section. As the member sits on the cooling bed, the tips of the flanges and the center of the web cool at a faster rate than the areas near the flange-web junction. Differential shrinkage in the section therefore is resisted by the hotter portions of the member. Internal stresses develop. The hotter portions, however, because of more flexible material properties at the higher temperatures, flow to accommodate to these stresses. A point is reached, however, when the areas near the flange-web junctions are cooling at the faster rate, thereby subjecting the flange tips to compressive forces and the flange-web junctions to tension. Depending upon the rate of heat exchange, and the material properties at elevated temperatures, there results an eventual, stable, room-temperature state which has a distribution of internal stresses of the general form shown in fig. 6.20a. The maximum compressive residual stress for that particular condition occurs at the flange tips and these have, on the average, an intensity of $0.3\sigma_y$. Across the flanges the stress distribution will vary from this maximum compressive value to a tensile stress at the flange-web junction. Depending upon the relative dimensions of the cross section, the web may be entirely subjected to tensile stresses, or, as indicated in the figure, a part may be in compression. It is assumed that residual stresses are constant through the thicknesses of the various elements.

For welded shapes, the approximate maximum compressive residual stress is $0.5\sigma_y$. The maximum tensile residual stress, which occurs at the weld, however, is at the yield value. (It is to be understood that with this set of internal stresses, yielding will occur the moment the cross section is subjected to applied tensile forces.)

Since residual stresses exist in the cross section without associated, externally

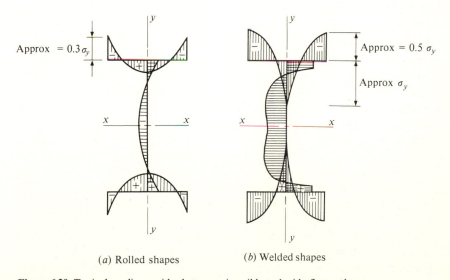

(a) Rolled shapes   (b) Welded shapes

Figure 6.20 Typical cooling residual stresses in mild steel wide-flange shapes.

applied forces, they must—within themselves—satisfy the statical equilibrium conditions at each and every section along the member. This requires that

$$\int_A \sigma_r \, dA = 0$$

$$\int_A \sigma_r x \, dA = 0 \qquad\qquad (6.81)$$

$$\int_A \sigma_r y \, dA = 0$$

where σ_r denotes the residual stress, and A is the area of the cross section.

It is to be recognized that if a compressive force corresponding to a uniformly distributed compressive stress over the entire cross section is applied to the member, the resulting stress distribution will be governed by the equation

$$\sigma = \sigma_r + \sigma_a$$

where $\sigma_a = P/A$—providing σ is less than σ_y at the most highly stressed location. As σ_a is increased, however, a value will be reached for which this combined stress equals the yield value, and the particular fiber in question can sustain no further increase in stress (though it can continue to deform in a constrained fashion). The total cross section, however, can support a greater load—but with less of the cross section resisting the increase. This is illustrated in fig. 6.21.

The assumed cross section is an idealized wide-flange shape, with the web area equal to zero. The residual stresses in the flanges have a maximum compressive value of $0.3\sigma_y$ at the tips and a value of $0.3\sigma_y$ in tension at the flange-web junction. Variation between these points is assumed to be linear.

The externally applied force is along the centroidal axis of the member. Therefore, the stress distribution associated with the load is symmetrical. If the combined stress is everywhere below the yield value, σ_a is uniformly distributed over the cross section.

In fig. 6.21c six different stress distributions across the flanges are shown. These correspond to different values of σ_a/σ_y, as indicated. At stage 4 the combined stress at the flange tips just equals the yield value. As the external load is increased beyond that value, yielding progresses across the section. Finally, the full value $\sigma_a/\sigma_y = 1.0$ is reached. (It is to be recognized that since the residual stresses are in equilibrium within themselves, the external load associated with stage 6 is the same as that which would have been realized had there been no residual stresses present in the member.)

Summarizing: In the initial stages of response to imposed external loading, the member behaves as if the cross section contained no residual stresses whatsoever. There is reached, however, for the assumed residual stress pattern, at $P_p = A(\sigma_a) = A(0.7\sigma_y)$, an axial thrust for which initial yielding occurs at the outermost points of the flanges. For $P > P_p$, yielding progresses across the section with strains increasing at an ever-increasing rate with intensified average stress.

If a column contained such residual stresses, and if that member were to be analyzed for buckling, the residual stresses would have no effect so long as $\sigma_a \leq 0.7\sigma_y$. If, however, the member buckles after a part of the cross section has

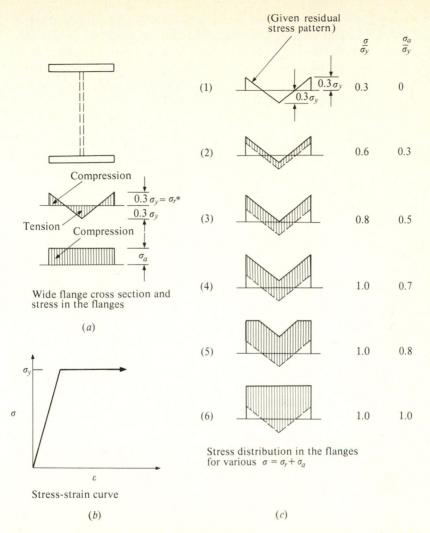

Figure 6.21 Influence of initial residual stresses on the applied stress distribution in wide-flange members.

yielded, the effective bending rigidity will be less than the fully elastic value EI. This is due to the fact that in the yielded regions, the modulus of elasticity is zero (see fig. 6.21b), and the effective bending stiffness is the bending stiffness of that part of the cross section which remains elastic. Thus, the inelastic buckling load and buckling stress of hot-rolled steel members containing residual stresses can be written as

$$P_{cr} = \frac{\pi^2 EI}{(KL)^2} \frac{I_e}{I}$$

and   (6.82)

$$\sigma_{cr} = \frac{\pi^2 E}{(KL/r)^2} \frac{I_e}{I}$$

where I is the moment of inertia of the entire cross section about the assumed axis of bending, and I_e is the moment of inertia of that part which remains elastic. It is to be recognized that the modification factor I_e/I depends not only upon the magnitude and distribution of the residual stress but also upon the geometrical properties of the particular cross-sectional shape in question. The axis of bending also is significant.

For the idealized two-flange cross section of fig. 6.21a,

$$\frac{I_e}{I} = \left(\frac{A_e}{A}\right)^3 \qquad \text{for weak-axis bending}$$

and

$$\frac{I_e}{I} = \frac{A_e}{A} \qquad \text{for strong-axis bending} \tag{6.83}$$

where A = area of the entire cross section = $2bt$

A_e = effective cross-sectional area = $2b_e t$

b = flange width

b_e = width of the elastic portion of the cross section

t = flange thickness

Therefore eq. (6.82) can be written as

$$\sigma_{\text{cr}} = \frac{\pi^2 E}{(KL/r)^2_{yy}} \tau^3 \qquad \text{for weak-axis buckling} \tag{6.84}$$

and

$$\sigma_{\text{cr}} = \frac{\pi^2 E}{(KL/r)^2_{xx}} \tau \qquad \text{for strong-axis buckling} \tag{6.85}$$

where

$$\tau = \frac{A_e}{A}$$

Since τ is less than unity, the reduction is more pronounced for weak-axis buckling.

To construct column-buckling curves, it is first necessary that there be determined the variation of τ with respect to the average stress acting on the cross section. This can be obtained by considering case 5 of fig. 6.21c. The average stress is defined by the relationship

$$\sigma_{\text{avg}} = \frac{P}{A}$$

where P is the total compressive force acting on the partially yielded section, and A is the original cross-sectional area. Again neglecting the web,

$$P = \sigma_y(A - A_e) + \left\{\frac{1}{2}\left[\left(\sigma_y - 2\sigma_r^* \frac{b_e}{b}\right) + \sigma_y\right]\right\} A_e$$

The average stress therefore will be

$$\sigma_{\text{avg}} = \frac{P}{A} = \sigma_y - \sigma_r^* \left(\frac{A_e}{A}\right)^2$$

or

$$\frac{\sigma_{\text{avg}}}{\sigma_y} = 1 - \left(\frac{\sigma_r^*}{\sigma_y}\right)\tau^2 \tag{6.86}$$

This is shown in fig. 6.22a, assuming a value of $\sigma_r^* = 0.3\sigma_y$. The variation in τ as a function of the average stress is shown in sketch b of that figure.

The effective (average) stress-strain curve for the entire cross section also can be determined experimentally. If a short segment of the column (a *stub column*) is subjected to an applied load over the entire cross section, if an average stress on that member is obtained by dividing the total applied force by the original cross-sectional area, and further, if strains are determined from a consideration of the axial shortening of the entire stub column, then—for buckling purposes—the average stress-strain curve so obtained will correspond to the effective stress-strain relationship.

$$E_T = \frac{d\sigma_{\text{avg}}}{d\varepsilon} = \frac{d(P/A)}{d(P/A_e E)} = E\frac{A_e}{A} = E\tau \tag{6.87}$$

The variation in τ associated with this measured value of average stress should be the same as that computed using eq. (6.86).

Having defined τ, eqs. (6.84) and (6.85) can be plotted. This is shown schematically in fig. 6.23. λ is the nondimensional parameter

$$\lambda = \sqrt{\frac{\sigma_y}{\sigma_e}} = \frac{KL}{r}\frac{1}{\pi}\sqrt{\frac{\sigma_y}{E}}$$

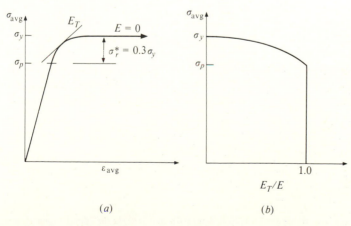

(a)   (b)

Figure 6.22 Buckling stress as a function of the tangent modulus.

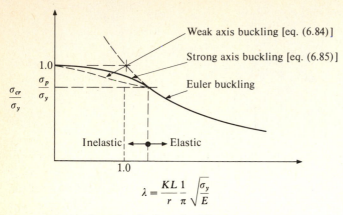

Figure 6.23 Nondimensional column-buckling curves.

It is most important to recognize that both of the inelastic buckling curves shown in fig. 6.23 are based on the tangent modulus concept of first possible lateral movement. As such, they represent a lower bound to the true solution—a safe approximation for design purposes.

Because the variations in observed buckling values for a variety of shapes is a matter of only a few percent, and (on the nondimensional bases) this is also true for strong-axis as opposed to weak-axis buckling, the Column Research Council has recommended a single inelastic column-buckling curve for all cases.† The curve is a parabola of the form

$$\sigma_{cr,inelas} = \sigma_y - \frac{\sigma_r^*(\sigma_y - \sigma_r^*)}{\pi^2 E}\left(\frac{KL}{r}\right)^2$$

or

$$\sigma_{cr,inelas} = \sigma_y - \frac{\sigma_r^*(\sigma_y - \sigma_r^*)}{\sigma_{cr,elas}} \tag{6.88}$$

where σ_r^* = maximum compressive residual stress. Furthermore, because of the various possibilities of distribution and magnitude of the residual stresses (for example, hot-rolled sections versus welded shapes), the Column Research Council has recommended for design that $\sigma_r^* = 0.5\sigma_y$. For this particular value,

$$\sigma_{cr} = \sigma_y\left[1 - \frac{\sigma_y}{4\pi^2 E}\left(\frac{KL}{r}\right)^2\right] \tag{6.89}$$

† Bruce G. Johnston, The Column Research Council *Guide to Design Criteria for Metal Compression Members*, 3d ed., chap. 2, John Wiley & Sons, Inc., New York, 1976.

6.3 LATERAL BUCKLING OF UNSUPPORTED, DOUBLY SYMMETRIC CROSS-SECTION BEAMS

Many beam-type members are braced by other elements of the structure in such a manner that they are constrained to deflect only in the direction of the applied lateral loads. For example, concrete-slab floor systems are extremely rigid in their own plane, and normally these slabs are continuously attached to the supporting beams. The beams, therefore, can deflect only in a direction perpendicular to the slab. Horizontal and rotational motions of the beams are prevented by the floor construction. There are instances, however, where beams and girders have no lateral support or bracing over parts or all their spans. A typical case is a crane-runway girder. Such members are usually devoid of any lateral bracing over the entire length between supporting columns. Such laterally unbraced beams can buckle in a manner somewhat similar to columns and must be safeguarded by appropriate adjustment of design stresses.

Figure 6.24 shows a beam in pure bending. It is simply supported at its ends, and also is held against tipping at these locations. If the beam is of the I or wide-flange type, its top flange is under uniform compression and would tend to buckle in the weak direction—namely, downward—if the web were not there to prevent it. If the force in the top flange were large enough, however, the flange would buckle in the direction in which it is free to move, that is, horizontally. On the other hand, the bottom flange is in tension and tends to remain straight. Since the two flanges and the web are tied together and represent one rigid unit, buckling—if it is to occur—can take place only in the manner shown in fig. 6.24c: The top flange bends farther than the bottom flange, and in consequence the entire cross section rotates. The same holds true for a beam of rectangular cross section as shown in sketch d.

The nature of this *lateral-torsional* buckling phenomenon is the same as for

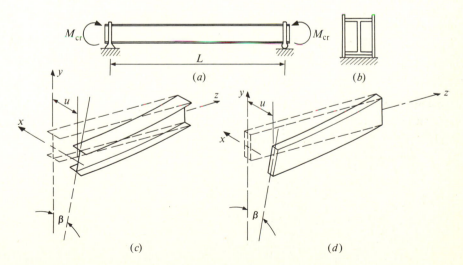

Figure 6.24 Lateral-torsional buckling of a beam subjected to end moments.

columns. The beam is stable with no tendency toward lateral-torsional motion until the bending moment reaches a certain critical magnitude M_{cr}. At that value, however, the member becomes unstable and can undergo rotations and lateral deflections which may lead to collapse. Bifurcation occurs!

Description of the motion referred to above requires the use of two geometrical parameters: one for the horizontal translation, and the other for the twisting of the cross section. The deformation in the plane of the applied moment is assumed to be unrelated to the lateral and torsional displacements, due to the assumption of small deformation theory.

6.3.1 Elastic Buckling—Differential Equation Solution

Because more than one dependent variable is involved, it is desirable to introduce a set of variable axes for the buckled state of the beam. The ones that will be used here are those suggested by Timoshenko and Gere.† They are summarized in table 6.4. The variable coordinates are related to the fixed coordinates (described in the undeformed position) by the direction cosines. u and v are the displacements in the x direction and y direction, respectively. β is the angle of twist with respect to the longitudinal axis z.‡ These relationships can best be visualized from a consideration of fig. 6.25. Sketch a shows a cross section in the deformed, buckled position $(-u, -v, \beta)$. The side view of the beam is shown in sketch b. A top view is in sketch c.

For small values of u, v, and β, it can be assumed that

$$I_x = I_\xi$$

$$I_y = I_\eta$$

† S. P. Timoshenko and J. M. Gere, *Theory of Elastic Stability*, 2d ed., McGraw-Hill Book Company, New York, 1961.

‡ Strictly speaking, this should be the shear center axis in the longitudinal direction. For doubly symmetrical cross sections, the centroidal longitudinal axis and the shear center longitudinal axis coincide, as indicated in fig. 6.25a.

Table 6.4 Cosines between axes

Variable coordinates	Fixed coordinates		
	x	y	z
ξ	1	β	$-\dfrac{du}{dz}$
η	$-\beta$	1	$-\dfrac{dv}{dz}$
ζ	$\dfrac{du}{dz}$	$\dfrac{dv}{dz}$	1

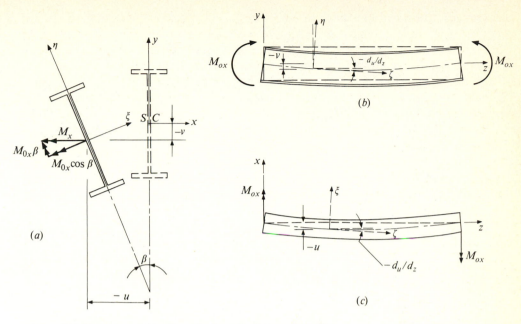

Figure 6.25 Geometry and loading of a laterally deformed member.

For the wide-flange cross section, and for many rectangular shapes, it can be further assumed that the bending rigidity about the x axis is quite large when compared to that about the y axis. The deflection in the plane of the applied moment v, therefore, is small when compared to u and β.[†]

With respect to the variable axes, the components of applied moment at any point on the beam are

$$M_\xi = M_{0x}$$

$$M_\eta = M_{0x}\beta \tag{6.90}$$

$$M_\zeta = -\frac{du}{dz}M_{0x}$$

The resistance of the beam to these components can be written as

$$M_\xi = EI_\xi \frac{d^2v}{dz^2} = EI_x \frac{d^2v}{dz^2}$$

$$M_\eta = EI_\eta \frac{d^2u}{dz^2} = EI_y \frac{d^2u}{dz^2} \tag{6.91}$$

$$M_\zeta = G\kappa_T \frac{d\beta}{dz} - EI_\omega \frac{d^3\beta}{dz^3}$$

[†] N. S. Trahair and S. T. Woolcock, Effect of Major Axis Curvature on I-Beam Stability, *C. E. Res. Rep. no. R 190*, University of Sydney, Sydney, Australia, March 1972.

From eqs. (6.90) and (6.91),

$$EI_x \frac{d^2v}{dz^2} = M_{0x}$$

$$EI_y \frac{d^2u}{dz^2} - M_{0x}\beta = 0 \qquad (6.92)$$

$$EI_\omega \frac{d^3\beta}{dz^3} - G\kappa_T \frac{d\beta}{dz} - M_{0x}\frac{du}{dz} = 0$$

The first of these equations describes bending about the strong axis of the cross section, as was done in chap. 4. It should be noted that it is independent of the lateral and torsional displacements. The last two equations, however, are coupled and must be solved simultaneously. More than that, they are homogeneous equations. The problem is therefore the classical eigenvalue problem—where the characteristic equation contains the critical value of the applied bending moment.

Solution to the last two equations of (6.92) can be facilitated by first differentiating the torsional equilibrium equation once with respect to z and then eliminating the d^2u/dz^2 term between the two. This gives the single fourth-order differential equation

$$EI_\omega \frac{d^4\beta}{dz^4} - G\kappa_T \frac{d^2\beta}{dz^2} - \frac{M_{0x}^2}{EI_y}\beta = 0 \qquad (6.93)$$

which is of the form

$$\beta'''' - \lambda_1 \beta'' - \lambda_2 \beta = 0$$

where

$$\lambda_1 = \frac{G\kappa_T}{EI_\omega} \qquad \text{and} \qquad \lambda_2 = \frac{M_{0x}^2}{E^2 I_y I_\omega}$$

The general solution to eq. (6.93) is

$$\beta = C_1 \cosh \alpha_1 z + C_2 \sinh \alpha_1 z + C_3 \sin \alpha_2 z + C_4 \cos \alpha_2 z \qquad (6.94)$$

C_1, C_2, C_3, and C_4 are constants of integration and

$$\alpha_1 = \sqrt{\frac{\lambda_1 + \sqrt{\lambda_1^2 + 4\lambda_2}}{2}} \qquad \alpha_2 = \sqrt{\frac{-\lambda_1 + \sqrt{\lambda_1^2 + 4\lambda_2}}{2}} \qquad (6.95)$$

Substituting the general solution into the boundary conditions

$$\beta(0) = 0 \qquad \beta(L) = 0 \qquad \beta''(0) = 0 \qquad \text{and} \qquad \beta''(L) = 0$$

yields the following four simultaneous homogeneous algebraic equations:

$$C_1(1) \qquad\quad + C_2(0) \qquad\qquad + C_3(0) \qquad\qquad + C_4(1) \qquad\qquad\qquad = 0$$

$$C_1(\alpha_1^2) \qquad\quad + C_2(0) \qquad\qquad + C_3(0) \qquad\qquad + C_4(-\alpha_2^2) \qquad\qquad\quad = 0$$

$$C_1(\cosh \alpha_1 L) \quad + C_2(\sinh \alpha_1 L) \quad + C_3(\sin \alpha_2 L) \quad + C_4(\cos \alpha_2 L) \qquad = 0$$

$$C_1(\alpha_1^2 \cosh \alpha_1 L) + C_2(\alpha_1^2 \sinh \alpha_1 L) + C_3(-\alpha_2^2 \sin \alpha_2 L) + C_4(-\alpha_2^2 \cos \alpha_2 L) = 0$$

For a nontrivial solution the determinant of the coefficient matrix must equal zero. This gives as the eigenvalue (or characteristic value) equation

$$(\alpha_1^2 + \alpha_2^2)^2 \sinh \alpha_1 L \sin \alpha_2 L = 0 \tag{6.96}$$

Since the parenthesized term is the sum of two positive numbers, it cannot equal zero. Sinh $\alpha_1 L$ can equal zero only if $(\alpha_1 L) = 0$. But this is a trivial solution. Therefore, eq. (6.96) can have a solution only if

$$\sin \alpha_2 L = 0 \tag{6.97}$$

This requires that

$$\alpha_2 L = n\pi \tag{6.98}$$

Substituting the value for α_2,

$$(M_{0x})_{\text{cr}} = \frac{n\pi}{L} \sqrt{EI_y G\kappa_T \left(1 + \frac{n^2\pi^2 EI_\omega}{G\kappa_T L^2}\right)} \tag{6.99}$$

The smallest critical moment occurs for $n = 1$ and has a value

$$(M_{0x})_{\text{cr}} = \frac{\pi}{L} \sqrt{EI_y G\kappa_T + EI_y \frac{\pi^2}{L^2} EI_\omega} \tag{6.100}$$

The first term under the square root, $EI_y G\kappa_T$, represents the combined resistance of the beam to lateral bending and St. Venant's torsion. The second term is the contribution of the combined lateral bending and warping torsional rigidities. The relative importance of these two groupings of cross-sectional and material properties depends upon the unsupported length of the beam L.

The warping constant I_ω is negligibly small for rectangular cross sections. For such members, the second set of terms can be reasonably neglected. That is,

$$(M_{0x})_{\text{cr}} = \frac{\pi}{L} \sqrt{EI_y G\kappa_T} \tag{6.100a}$$

In fact, even for thin-walled, open cross-section I beams, if the length of the member is relatively long, eq. (6.100a) will be adequate for engineering design purposes.

It is to be emphasized that the critical moment obtained using either eq. (6.100) or (6.100a) corresponds to buckling of a member whose end supports are, by definition, simple for both lateral bending and twisting. That is, at both ends of the member $u = u'' = \beta = \beta'' = 0$. In real structures the ends of the member are usually restrained, and the buckling load is greater. For such cases, the critical buckling moment can be expressed in terms of effective length factors—in the same fashion as was done for columns.

$$(M_{0x})_{\text{cr}} = \frac{\pi}{K_y L} \sqrt{EI_y G\kappa_T + \frac{\pi^2}{(K_z L)^2} EI_y EI_\omega} \tag{6.101}$$

where K_y and K_z are the effective length factors for lateral bending and twisting,

Table 6.5 Effective length factors for various end conditions†,‡

Boundary conditions		K_y	K_z
At $z = 0$	At $z = L$		
$u = u'' = \beta = \beta'' = 0$	$u = u'' = \beta = \beta'' = 0$	1.000	1.000
$u = u'' = \beta = \beta'' = 0$	$u = u'' = \beta = \beta' = 0$	0.904	0.693
$u = u'' = \beta = \beta'' = 0$	$u = u' = \beta = \beta'' = 0$	0.626	1.000
$u = u'' = \beta = \beta'' = 0$	$u = u' = \beta = \beta' = 0$	0.693	0.693
$u = u'' = \beta = \beta' = 0$	$u = u'' = \beta = \beta' = 0$	0.883	0.492
$u = u' = \beta = \beta'' = 0$	$u = u' = \beta = \beta' = 0$	0.431	0.693
$u = u' = \beta = \beta' = 0$	$u = u' = \beta = \beta' = 0$	0.492	0.492
$u = u' = \beta = \beta'' = 0$	$u = u' = \beta = \beta'' = 0$	0.434	1.000
$u = u' = \beta = \beta'' = 0$	$u = u'' = \beta = \beta' = 0$	0.606	0.492

† T. V. Galambos, *Structural Members and Frames*, table 3.2, p. 108, Prentice-Hall, Inc., Englewood Cliffs, N.J., 1968.

‡ V. Z. Vlasov, *Thin Walled Elastic Beams*, Y. Schechtman, trans. chap. 5, Moscow, 1959; Israel Program for Scientific Translation, Jerusalem, 1961.

respectively. Galambos, based on work by Vlasov, has given a number of these values for a variety of end conditions. They are here repeated in table 6.5.

The lateral buckling moment of an unsupported beam rigidly attached to two adjacent spans will be bounded by the two idealized values predicted by eq. (6.101), with $K = 1.00$ and $K = 0.5$. To actually determine the lateral buckling load for continuous beams, continuity conditions must be established and introduced into the boundary conditions for the member in question. This problem is better handled using the finite-element approach.

6.3.2 Elastic Buckling—Energy Method Solution

The critical lateral-torsional buckling moment also can be obtained using the Rayleigh-Ritz approach described earlier in this chapter. Consider the member in the buckled configuration shown in fig. 6.26. For ease of visualization, only the web of the wide-flange member is pictured. Moreover, the indicated lateral and axial deformations are assumed to be due only to lateral-torsional buckling; that is, deformations in the plane of the applied moments, due to primary bending in that plane, have not been shown. The displacements of the top and bottom flanges of the buckled position can be expressed in terms of u and β of the centroid of the cross section by the relationships

Top flange:
$$u_T = u + \beta \frac{d}{2}$$

$$(6.102)$$

Bottom flange:
$$u_B = u - \beta \frac{d}{2}$$

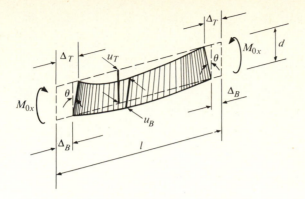

Figure 6.26 Bending member in laterally buckled configuration.

where d is the depth of the member. At buckling, the shortening and lengthening of the longitudinal fibers of the flanges due to their lateral displacement

$$2\Delta_T = \frac{1}{2} \int_0^l \left(\frac{du_T}{dz} \right)^2 dz$$

and (6.103)

$$2\Delta_B = \frac{1}{2} \int_0^l \left(\frac{du_B}{dz} \right)^2 dz$$

From fig. 6.26 it is seen that the angle of rotation through which M_{0x} travels is

$$\theta = \frac{\Delta_T - \Delta_B}{d}$$

Using eqs. (6.102) and (6.103)

$$\theta = \frac{1}{2} \int_0^l \frac{du}{dz} \frac{d\beta}{dz} dz \qquad (6.104)$$

The potential energy due to the applied end moments is therefore

$$W_e = 2M_{0x}\theta$$

or

$$W_e = M_{0x} \int_0^l \frac{du}{dz} \frac{d\beta}{dz} dz \qquad (6.105)$$

The strain energy of the beam due to the combined lateral and torsional displacements is given by

$$W_i = \frac{1}{2} EI_y \int_0^l \left(\frac{d^2u}{dz^2} \right)^2 dz + \frac{1}{2} G\kappa_T \int_0^l \left(\frac{d\beta}{dz} \right)^2 dz + \frac{1}{2} EI_\omega \int_0^l \left(\frac{d^2\beta}{dz^2} \right)^2 dz \quad (6.106)$$

Thus the total potential associated with the buckled shape is

$$\Pi = W_i - W_e$$

$$= \frac{1}{2} EI_y \int_0^l (u'')^2 dz + \frac{1}{2} G\kappa_T \int_0^l (\beta')^2 dz + \frac{1}{2} EI_\omega \int_0^l (\beta'')^2 dz - M_{0x} \int_0^l (u')(\beta') dz$$

$$(6.107)$$

If it is assumed that

$$u = u_0 \sin \frac{\pi z}{l}$$

$$\beta = \beta_0 \sin \frac{\pi z}{l} \tag{6.108}$$

which satisfies simply supported end conditions, then the Rayleigh-Ritz procedure requires that

$$\delta \Pi = \frac{\partial \Pi}{\partial u_0} \delta u_0 + \frac{\partial \Pi}{\partial \beta_0} \delta \beta_0 = 0$$

or

$$\frac{\partial \Pi}{\partial u_0} = 0$$

$$\frac{\partial \Pi}{\partial \beta_0} = 0 \tag{6.109}$$

From eqs. (6.107), (6.108), and (6.109), and noting that

$$\int_0^l \sin^2 \left(\frac{\pi z}{l} \right) dz = \int_0^l \cos^2 \left(\frac{\pi z}{l} \right) dz = \frac{l}{2}$$

there results the following characteristic equation:

$$EI_\omega \frac{\pi^4}{l^4} + G\kappa_T \frac{\pi^2}{l^2} - \frac{(M_{0x})_{cr}^2}{EI_y} = 0$$

This gives

$$(M_{0x})_{cr} = \frac{\pi}{l} \sqrt{EI_y G\kappa_T + EI_y \frac{\pi^2}{l^2} EI_\omega} \tag{6.110}$$

which is identical to eq. (6.100). [This is to be expected since the assumed functions (6.108) are the exact functions.]

If a polynomial function were presumed for the displacement, an approximate answer would be obtained. A possible four-term function for the simply supported case might be

$$u = z(z - l)(b_0 + b_1 z + b_2 z^2 + b_3 z^3)$$

$$\beta = z(z - l)(c_0 + c_1 z + c_2 z^2 + c_3 z^3) \tag{6.111}$$

From eqs. (6.107), (6.109), and (6.111), and presuming a beam of constant cross section whose dimensions are identical with the smaller end dimensions of the tapered member defined as Member A in table 6.2, the critical lateral buckling moment presuming this polynomial representation is approximately 5 percent higher than the value predicted by the exact solution [eqs. (6.100) and (6.110)].

It should here be noted that solutions have been obtained for a large number of lateral buckling problems—using both the differential equation approach and energy methods. Of major importance are those which presumed transverse loading other than the uniform moment case previously described. Typical loadings on beams are distributed loads, concentrated loads, unequal end moments, etc.

Table 6.6 Lateral buckling coefficients

Beam and loading	Boundary conditions	$K = K_y = K_z$	C_1
q lb/ft $\;\;$ $ql^2/8$	$u = u'' = \beta = \beta'' = 0$ $u = u' = \beta = \beta' = 0$	1.00 0.50	1.13 0.97
q $\;\;$ $ql^2/12$ $\;\;$ $ql^2/24$	$u = u'' = \beta = \beta'' = 0$ $u = u' = \beta = \beta' = 0$	1.00 0.50	1.30 0.86
Q $\;\;$ $Ql/4$	$u = u'' = \beta = \beta'' = 0$ $u = u' = \beta = \beta' = 0$	1.00 0.50	1.35 1.07
Q $\;\;$ $Ql/8$ $\;\;$ $Ql/8$	$u = u'' = \beta = \beta'' = 0$ $u = u' = \beta = \beta' = 0$	1.00 0.50	1.70 1.04
Q $\;\;$ Ql	$u = u' = \beta = \beta' = 0$	1.00	1.30
q $\;\;$ $ql^2/2$	$u = u' = \beta = \beta' = 0$	1.00	2.05
M $\qquad$ αM $-1.0 \le \alpha \le +1.0$	$u = u'' = \beta = \beta'' = 0$	1.00	[†](1) $C_1 \le 2.3$ [†](2) $C_1 = 1.75 - 1.05\alpha$ $+ 0.3\alpha^2$

† M. G. Salvadori, Lateral Buckling of Eccentrically Loaded I-Columns, *Trans. Am. Soc. Civ. Eng.*, vol. 121, p. 1163, 1956.

As discussed in chap. 6 of the Column Research Council Guide,† it is possible to define the elastic lateral buckling strength of beams subjected to varying moments using eq. (6.112).

$$(M_{x,\,max})_{cr} = C_1 \frac{\pi}{K_y L} \sqrt{EI_y GK_T + \left(\frac{\pi}{K_z L}\right)^2 EI_y EI_\omega} \qquad (6.112)$$

where $(M_{x,\,max})_{cr}$ = maximum primary bending moment along the beam
C_1 = a modification factor (or function) which accounts for different conditions of loading
K_y = effective length factor for lateral bending
K_z = effective length factor for twisting about the longitudinal axis

Values for C_1 for a variety of loading conditions are given in table 6.6.

The critical stress corresponding to the lateral buckling moment can be obtained by dividing the critical maximum moment value given in eq. (6.112) by the section modulus of the cross section about the strong xx axis, S_x.

$$\sigma_{cr} = C_1 \frac{\pi}{S_x(K_y L)} \sqrt{EI_y GK_T + \left(\frac{\pi}{K_z L}\right)^2 EI_y EI_\omega} \qquad (6.113)$$

It is to be understood that this equation is applicable for elastic lateral buckling only. Although it was derived based on an assumed I-beam type cross section, it will provide approximate answers to sections with only one axis of symmetry, providing that axis is in the plane of the applied loads. An I shape of unequal flange widths is one such example, where the values of κ_T, I_y, and I_ω may be reasonably substituted into the equation. For rectangular shapes, I_ω is negligible and the second grouping of terms in eq. (6.113) can be neglected.

6.3.3 Inelastic Lateral Buckling

The theories of inelastic buckling discussed in sec. 6.2.4 apply to lateral buckling in exactly the same manner as they do for axial column buckling. However, in this case the reduction in rigidities must be considered for EI_y, GK_T, and EI_ω. The exact reduction factors depend upon the nonlinear nature of the stress-strain relationship of the material. In the case of the monotonically increasing type of curve, such as for aluminum alloys, the tangent modulus concept can be readily adopted to modify eq. (6.113). That is,

$$\sigma_{cr} = C_1 \frac{\pi}{S_x(K_y L)} \sqrt{(EI_y)_{eff}(GK_T)_{eff} + \left(\frac{\pi}{K_z L}\right)^2 (EI_y)_{eff}(EI_\omega)_{eff}} \qquad (6.114)$$

---

† Bruce C. Johnston, The Column Research Council *Guide to Design Criteria for Metal Compression Members*, 3d ed., John Wiley & Sons, Inc., New York, 1976.

To obtain $(EI_y)_{eff}$ and $(EI_\omega)_{eff}$ the same principles that were discussed in the column case may be used. For $(G\kappa_T)_{eff}$, however, it has been shown that for rectangular sections the full elastic value $G\kappa_T$ may be used to determine the critical moment.† For I beams the term $(G\kappa_T)_{eff}$ is relatively unimportant; a very small difference in the critical buckling moment exists using either the full elastic value or a reduced value.

For steel I beams the influence of residual stresses must be considered. This can be done in the same fashion as for the column case (see fig. 6.27). In sketch *a* the residual stress and an applied stress σ_a due to bending about the strong axis of the section are shown. It is to be noted that the applied stress in the upper flange is in compression, whereas that in the lower flange is in tension. As the applied moment is increased, a particular value of M will be reached beyond which yielding occurs. In the upper flange this yielding starts from the flange tips and progresses inward toward the web. In the bottom flange yielding starts at the web-flange junction since there the combined tensile stress is greatest. A generalized yielded case is shown in sketch *b*. It is evident, then, that the effective rigidities are functions of the applied loading—as was true for the inelastic column buckling case, including residual stresses. A semi-inverse process of assuming a strain distribution and computing the associated applied bending moment and the bending rigidities is therefore appropriate. In general, the difference between the critical load predicted by the reduced modulus theory and that based on the tangent modulus theory is quite small. Since the tangent modulus approach is much simpler, that procedure is most often used. In fact, analytical results have demonstrated that the parabolic approximation that was

---

† B. G. Neal, The Lateral Instability of Yielded Mild Steel Beams of Rectangular Cross-Section, *Philos. Trans. R. Soc., London*, vol. 242, 1950.

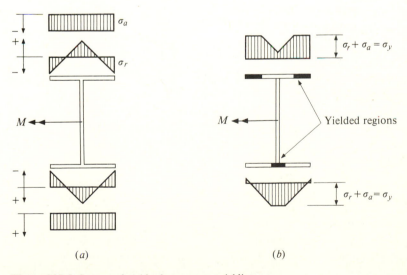

Figure 6.27 Influence of residual stresses on yielding.

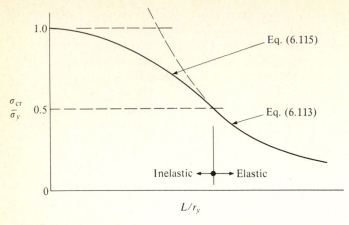

Figure 6.28 Elastic and inelastic buckling.

used for the inelastic column buckling case [eq. (6.88)] is an appropriate design representation for inelastic lateral buckling.

$$(\sigma_{cr})_{inelas} = \sigma_y - \frac{\sigma_r^*(\sigma_y - \sigma_r^*)}{(\sigma_{cr})_{elas}} \tag{6.115}$$

where $(\sigma_{cr})_{elas}$ is defined by eq. (6.113).

For the entire length spectrum, assuming an idealized stress-strain curve and a maximum residual stress of $\sigma_r^* = 0.5\sigma_y$, the design curves would be those shown in fig. 6.28. (Note that a factor of safety of 1.0 has been presumed.)

6.3.4 General Observations Concerning Lateral Buckling of Steel I Beams

The lateral buckling equation (6.100) can be written in the following form:

$$(M_{0x})_{cr} = \frac{\pi}{L}\sqrt{EI_y G\kappa_T \left(1 + \frac{\pi^2}{L^2}\frac{EI_\omega}{G\kappa_T}\right)} \tag{6.116}$$

The relative importance of warping to St. Venant type torsion therefore can be determined by comparing to 1.0 the term

$$R = \left(\frac{\pi}{L}\right)^2 \frac{EI_\omega}{G\kappa_T}$$

For steel, $G \doteq \frac{3}{8}E$. Therefore

$$R \doteq 3.7\frac{I_\omega}{L^2\kappa_T} \tag{6.117}$$

In general, most rolled wide-flange shapes have a ratio I_ω/κ_T of between 10^3 and 10^4. Moreover, these types of members usually have spans of between 8 and 40 ft. R may therefore range between 0.02 and 5. For long, unsupported spans R may be as low as the 0.02 value given and for such cases the St. Venant term will be adequate.

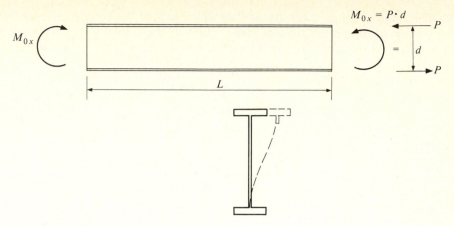

Figure 6.29 Lateral buckling of the top flange.

On the other hand, for I shapes with very thin plate elements and very short spans, R may be sufficiently large that only warping need be considered.

Welded or riveted plate girders of the type and size used in civil engineering construction usually have very thin webs and narrow flanges. For these types of members the webs may be as much as 8 ft deep for spans of only 60 ft. For such members, the theories heretofore developed may not lead to appropriate predictions of strength. The individual elements of the cross section may be so flexible that the members cannot retain the cross-sectional form up to the calculated lateral buckling loads. For such cases, lateral buckling strengths can be more reasonably predicted by assuming that the compression flanges act as columns. Depending upon the proportions of the various plate elements, the effective cross section that may be presumed in such an analysis will contain a portion of the web; that is, the column section will have a T form. Since buckling downward is prevented by the web, buckling must be lateral and the critical moment value will be

$$(M_{0x})_{cr} = P_{cr}d = \frac{\pi^2 E(I_y/2)}{L^2}d \qquad (6.118)$$

Conceptually, this is illustrated in fig. 6.29.

6.4 BEAM COLUMNS, COMBINED BENDING AND AXIAL THRUST

Seldom are structural members subjected to only axial thrust or to only end-bending moments. More often than not the critical design condition is a combination of these two extremes.

In chap. 4 methods were developed which allow the prediction of the elastic, in-plane, load-deformational response curves for members subjected to combined bending and thrust. Extension of those methods to include nonlinear stress-strain properties also was introduced and illustrated.

Actually, beam columns may respond in a variety of fashions to imposed loading—depending upon the type of cross section; upon whether or not the member remains in the plane of the applied moments, that is, whether lateral-torsional buckling occurs; upon the stress-strain properties of the material, etc. For example, consider the wide-flange shape shown in fig. 6.30. The member in question is of length L and is subjected to an axial thrust P plus two equal but opposite end moments that cause a single curvature type of deformation. The applied end moments are about the x axis of the cross section, which is further presumed to be the major bending axis of the section.

Assume that a constant axial thrust less than the buckling value is first applied to the member and that the end-bending moments are then increased from zero. The straight-line, elastic end moment versus end-rotation response curve shown in sketch a is obtained. The equation governing the solution is listed in table 4.7. (It should be noted that had P increased with M_{0x}, the relationship would not have been linear.) As end moments increase eventually a particular value will be reached for which initial yielding occurs in the member. This has been shown as point A on the graph. For increases in M_{0x} beyond this value, the member responds in a more flexible fashion as indicated by the dashed curve a.

Assuming that the member continues to deflect in the plane of the applied moment, it is possible to define the entire load-deformational response curve in the inelastic range. However, for most real members, at some point along the dashed curve a, say, at point B, the member will laterally deflect. The load-deformational response curve beyond that point will then be of the type shown as curve b. (It must be understood that while it has been assumed in this illustrative case that lateral-torsional buckling occurred in the inelastic range, this need not have been the case.

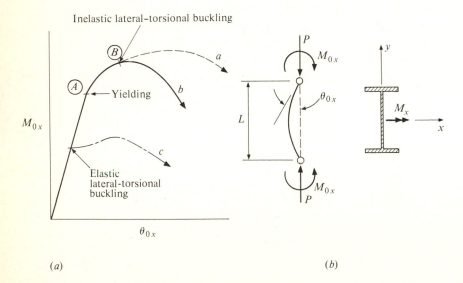

(a)   (b)

Figure 6.30 Load-deformational response of a beam column.

For certain members, lateral-torsional buckling occurs early in the elastic response range. Such a case is illustrated by curve c.) In general, all members realize their ultimate carrying capacity in the inelastic range.

6.4.1 Elastic Lateral-Torsional Buckling of Beam Columns

For doubly symmetric cross-section members subjected to axial thrust plus equal end-bending moments applied about the strong axis of the member, the differential equations of equilibrium may be written as:

$$EI_x \frac{d^2 v}{dz^2} + Pv = M_{0x} \tag{6.119}$$

$$EI_y \frac{d^2 u}{dz^2} + Pu - M_{0x}\beta = 0 \tag{6.120}$$

$$EI_\omega \frac{d^3 \beta}{dz^3} - (G\kappa_T - Pr_0^2) \frac{d\beta}{dz} - M_{0x} \frac{du}{dz} = 0 \tag{6.121}$$

where all of the definitions listed in sec. 6.3.1 are valid, and in addition

$$P = \text{axial force}$$

$$r_0 = \text{polar radius of gyration}$$

$$r_0^2 = \frac{I_x + I_y}{A}$$

Equation (6.119) defines the "in-plane," beam-column bending behavior of the member in question. Equations (6.120) and (6.121) govern the lateral-torsional deformational behavior. [When $P = 0$, these equations reduce, identically, to (6.92).]

Recalling that the single sine curve displacement function is the exact solution for both the axially loaded column and the uniform-moment beam-buckling cases, it is reasonable to assume such a buckled configuration for a hinged-end beam column. That is,

$$u = u_0 \sin \frac{\pi z}{L}$$

$$\beta = \beta_0 \sin \frac{\pi z}{L}$$

Substituting these into (6.120) and (6.121) yields the two equations

$$\left[\left(\frac{\pi^2 EI_y}{L^2} - P \right) u_0 - (M_{0x})\beta_0 \right] \frac{\pi^2}{L^2} \sin \frac{\pi z}{L} = 0$$

$$\left[(-M_{0x}) u_0 + \left(\frac{\pi^2 EI_\omega}{L^2} + G\kappa_T - Pr_0^2 \right) \beta_0 \right] \frac{\pi^2}{L^2} \sin \frac{\pi z}{L} = 0$$

For a nontrivial solution,

$$\begin{vmatrix} \left(\dfrac{\pi^2 EI_y}{L^2} - P\right) & (-M_{0x}) \\[3mm] (-M_{0x}) & \left(\dfrac{\pi^2 EI_\omega}{L^2} + G\kappa_T - Pr_0^2\right) \end{vmatrix} = 0$$

from which the characteristic equation (6.122) is obtained:

$$(P_{e_y} - P)(r_0^2 P_{e_z} - Pr_0^2) = (M_{0x})^2 \tag{6.122}$$

Here

$$P_{e_y} = \frac{\pi^2 EI_y}{L^2} = \text{Euler buckling load in the weak direction}$$

and

$$P_{e_z} = \frac{1}{r_0^2}\left(\frac{\pi^2 EI_\omega}{L^2} + G\kappa_T\right) = \begin{array}{l}\text{elastic torsional buckling load} \\ \text{under pure axial compression}\end{array}$$

Equation (6.122) can be rewritten in the more usable form

$$M_{0x} = r_0\sqrt{(P_{e_y} - P)(P_{e_z} - P)} \tag{6.123}$$

Several observations can be made concerning this equation: If M_{0x} is equal to zero, that is, the member is axially loaded, then

$$P_{cr}\Bigg\langle \begin{array}{l} P_{e_y} = \dfrac{\pi^2 EI_y}{L^2} \\[4mm] \text{or} \\[4mm] P_{e_z} = \dfrac{1}{r_0^2}\left(\dfrac{\pi^2 EI_\omega}{L^2} + G\kappa_T\right) \end{array}$$

If, on the other hand, $P = 0$,

$$(M_{0x})_{cr} = r_0\sqrt{P_{e_y} P_{e_z}}$$

$$= r_0\sqrt{\frac{\pi^2 EI_y}{L^2}\frac{1}{r_0^2}\left(\frac{\pi^2 EI_\omega}{L^2} + G\kappa_T\right)}$$

$$= \frac{\pi}{L}\sqrt{EI_y G\kappa_T + \frac{\pi^2}{L^2} EI_y EI_\omega}$$

This is the same expression as was obtained for the pure bending case.

Another special characteristic can be observed from eq. (6.123). The quantity under the square root sign must be positive for M_{0x} to have any realistic value. This means that P must be either greater than both P_{e_y} and P_{e_z}, or less than both of them. It therefore can be concluded that the lateral-torsional buckling load must be smaller than either of the individual buckling loads P_{e_y} or P_{e_z} of the column.

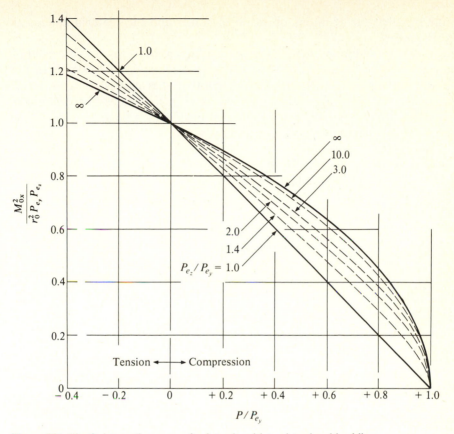

Figure 6.31 Elastic interaction curves for lateral and lateral-torsional buckling.

Equation (6.123) can be nondimensionalized, as given in eq. (6.124):

$$\frac{M_{0x}^2}{r_0^2 P_{e_y} P_{e_z}} = \left(1 - \frac{P}{P_{e_y}}\right)\left(1 - \frac{P}{P_{e_y}}\frac{P_{e_y}}{P_{e_z}}\right) \qquad (6.124)$$

Interaction curves based on this equation are given in fig. 6.31.

6.4.2 Inelastic Lateral-Torsional Buckling of Beam Columns

Inelastic lateral-torsional buckling solutions can be obtained using any of the general inelastic theories already described. The tangent modulus approach offers the easiest, conservative solution, and the results are sufficiently accurate for design purposes. For hot-rolled or welded structural steel members, members that contain significant residual stresses, the relationships between the various stiffness parameters and the average stress acting on the partially yielded cross section must first be determined. This could be accomplished in the same manner as previously described. However, for partially yielded beam-column cross sections, the situation is more complicated due to the fact that the positions of the shear center and the centroid change as yielding progresses.

The increase in ultimate load-carrying capacity above the inelastic lateral-torsional buckling load depends on the cross-sectional dimensions, length of the member, stress-strain properties, etc. Experimental studies have shown, however, that for columns of relatively low slenderness ratio in the plane of bending (for example, $L/r_x < 40$), the inelastic lateral-torsional buckling load corresponds, essentially, to the ultimate carrying capacity of the member.

For design purposes, the inelastic lateral-torsional buckling load can be approximated using the same procedures that were used for determining the inelastic column buckling and beam buckling loads. That is,

$$\left(\sigma_{cr}\right)_{inelas} = \sigma_y - \frac{\sigma_r^*\left(\sigma_y - \sigma_r^*\right)}{\left(\sigma_{cr}\right)_{elas}} \tag{6.125}$$

where $\left(\sigma_{cr}\right)_{elas}$ is the elastic lateral-torsional buckling stress.

6.4.3 Ultimate Carrying Capacity of Beam Columns

As previously has been stated, beam columns that are constrained to deflect in the plane of the applied moment—due either to bending about the weak axis of the cross section, or because bracing is provided in the perpendicular direction—reach a stage in their loading history where instability occurs with deflections continuing to increase even though the applied load is held constant. Without exception, this phenomena occurs in the inelastic range of material behavior. It is the purpose of this section to describe a method for determining this ultimate load-carrying capacity.

The following are the steps that will be followed to obtain the solution in question:

1. A particular type of cross section and stress-strain curve will be assumed.
2. For the beam column in question, a loading will be prescribed which allows a single variable deformation function to be reasonably presumed.
3. A general linear strain distribution across the most highly strained section along the member will be presumed, and, using the assumed stress-strain relationship, this strain distribution will be described in terms of the corresponding stresses.
4. The stress distributions, in turn, will be translated into expressions for thrust, bending moment, and curvature at the section in question.
5. Using a single-point collocation approach and taking into account the assumed deflection function, the external loading and deflection at the assumed control section will be related to the internal resultants defined in stage 4.
6. Finally, this load-deflection relationship will be maximized to determine the ultimate load-carrying capacity of the member.

This procedure was first applied to the rectangular cross section by Jezek.†

---

† K. Jezek, Die Tragfähigkeit axial gebrückter und auf Biegung beanspruchter Stahlstäbe, *Stahlbau*, vol. 9, 1936.

The cross section that will be assumed will be the rectangle, and bending will be prescribed about the weak axis of the section. It will be presumed that the material has an elastic–perfectly plastic stress-strain curve of the type shown in fig. 4.1.

The beam column in question is shown in fig. 6.32. Equal but opposite bending moments m are applied to each end of the member, and there also is imposed an axial thrust P. The member is of length L. As shown in the figure, the assumed coordinate system will have its center located at the center of the member, and it further will be presumed that the deflection is governed by the equation

$$y = \delta \cos \frac{\pi z}{L} \tag{6.126}$$

where δ is the lateral deflection at the centerline (that is, at $z = 0$). (Once δ is known, the entire deflection curve is prescribed.)

It should be evident from the loading and presumed deflected configuration that the most highly strained cross section along the member will be the centerline section. Correspondingly, the stress distribution at that section will be the greatest. Assuming a free body diagram of the upper (or lower) half of the member, equilibrium requires that

$$P_{\mathtext{\$}} = P$$

$$M_{\mathrm{\textcent}} = m + P\delta \tag{6.127}$$

$M_{\mathrm{\textcent}}$ and $P_{\mathrm{\textcent}}$ at the cut section depend upon the stress distribution, which, in turn, depends upon the strain distribution, which, in turn, is governed by the presumed amount of yielding and the curvature at the section in question.

Consider fig. 6.33. If it is assumed that "plane sections remain plane," and that loading has progressed beyond the point of initial compressive yielding in the member, one of two stress distributions across the section is possible at the centerline section. As shown for case A: compressive yielding has penetrated into the member on the concave side an amount a, but the convex, tension side remains

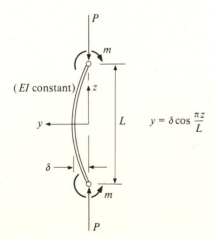

Figure 6.32 Assumed beam column.

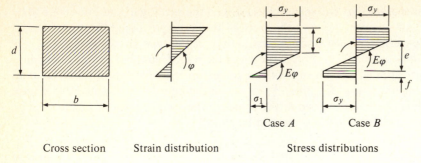

Cross section	Strain distribution	Stress distributions

Figure 6.33 Generalized strain and stress distributions at the length of the member.

elastic. For case *B*, however, it has been assumed that both tensile and compressive yielding have occurred. It is to be noted that in both cases, stresses in the elastic region vary linearly with distance from the neutral axis.

Case A For stress distribution *A* in fig. 6.33, the axial thrust acting on the member will be

$$P = \int_A \sigma \, dA = P_y - \tfrac{1}{2}(d - a)(\sigma_y + \sigma_1)b$$

or

$$\frac{P}{P_y} = 1 - \frac{1}{2}\left(1 - \frac{a}{d}\right)\left(1 + \frac{\sigma_1}{\sigma_y}\right) \tag{6.128}$$

where $P_y = bd\,\sigma_y$ = the axial force corresponding to yielding in compression over the entire cross section

The bending moment corresponding to the case *A* stress distribution is

$$M = \int_A \sigma y \, dA = \frac{1}{2}(d - a)(\sigma_y + \sigma_1)b\left(\frac{d}{2} - \frac{d - a}{3}\right)$$

or

$$\frac{M}{M_y} = \left(1 - \frac{a}{d}\right)\left(1 + \frac{\sigma_1}{\sigma_y}\right)\left(\frac{1}{2} + \frac{a}{d}\right) \tag{6.129}$$

where

$$M_y = \frac{bd^2}{6}\,\sigma_y = \text{the bending moment corresponding to initial yielding}$$
(in the absence of any axial force)

It also is possible to describe the curvature φ in terms of the unknowns σ_1 and a.

$$\tan E\varphi \doteq E\varphi = \frac{\sigma_y + \sigma_1}{d - a}$$

Nondimensionalizing this equation gives

$$\frac{\varphi}{\varphi_y} = \frac{1}{2}\left(\frac{1 + \sigma_1/\sigma_y}{1 - a/d}\right)$$ (6.130)

where

$$\varphi_y = \frac{2\sigma_y}{Ed} = \text{the curvature corresponding to } M_y.$$

$$1 + \frac{\sigma_1}{\sigma_y} = \frac{2(1 - P/P_y)}{1 - a/d}$$ (6.131)

If this is substituted into eq. (6.129)

$$\frac{a}{d} = \frac{9}{8}\left(\frac{M/M_y}{1 - P/P_y}\right) - \frac{1}{2}$$ (6.132)

$$\frac{\varphi}{\varphi_y} = \frac{4}{9}\frac{(1 - P/P_y)^3}{[(1 - P/P_y) - \frac{3}{4}(M/M_y)]^2}$$ (6.133)

Case B For stress distribution *B*:

$$\frac{P}{P_y} = 1 - \frac{e}{d} - 2\frac{f}{d}$$ (6.134)

$$\frac{M}{M_y} = 6\left[\frac{f}{d}\left(1 - \frac{f}{d}\right) + \frac{e}{d}\left(\frac{1}{2} - \frac{f}{d} - \frac{e}{3d}\right)\right]$$ (6.135)

and

$$\frac{\varphi}{\varphi_y} = \frac{1}{e/d}$$ (6.136)

Eliminating e/d and f/d between these equations yields

$$\frac{\varphi}{\varphi_y} = \frac{1}{\sqrt{3[1 - (P/P_y)^2 - \frac{3}{2}(M/M_y)]}}$$ (6.137)

where the various nondimensionalizing parameters are as previously defined.

Consider now what happens externally. The assumed deflected configuration is given by eq. (6.126), which can be differentiated to relate the curvature at any section along the member to the centerline deflection

$$y'' = \varphi = \delta\frac{\pi^2}{L^2}\cos\frac{\pi z}{L}$$

At the centerline,

$$\varphi_\mathbb{C} = \delta\frac{\pi^2}{L^2}$$ (6.138)

This also can be nondimensionalized by the factor φ_y, where $\varphi_y = M_y/EI$. That is,

$$\frac{\varphi_{\mathcal{C}}}{\varphi_y} = \frac{\delta(\pi^2/L^2)}{M_y/EI} = \delta \frac{\pi^2 EI}{L^2} \frac{c}{A\sigma_y r^2}$$

or

$$\frac{\varphi_{\mathcal{C}}}{\varphi_y} = \delta \frac{P_e}{P_y} \frac{c}{r^2} \tag{6.139}$$

c is the distance from the centroid to the extreme fiber of the cross section, and r is the radius of gyration about the axis of bending.

Equating (6.133) and (6.139) and keeping in mind that the moment at the center-line section is equal to the primary applied bending moment m plus the axial thrust times the midpoint deflection δ, we arrive at the expression

$$\delta\left(9\frac{P_e}{P_y}\frac{c}{r^2}\right)\left(1 - \frac{P}{P_y} - \frac{3}{4}\frac{m+P\delta}{M_y}\right)^2 = 4\left(1 - \frac{P}{P_y}\right)^3 \tag{6.140}$$

For a constant value of m and a defined member size (that is, P_e and P_y are known), equation (6.140) is an expression of the axial thrust–centerline lateral-deflection relationship for the member. The critical deflection associated with the ultimate load-carrying capacity, therefore, can be obtained from the condition

$$\frac{dP}{d\delta} = 0 \tag{6.141}$$

This gives

$$\delta_{cr} = \frac{1 - P/P_y - \frac{3}{4}(m/M_y)}{P/M_y} \tag{6.142}$$

Substituting (6.142) into (6.140) and simplifying yields

$$\frac{m}{M_y} = 3\left(1 - \frac{P}{P_y}\right)\left(1 - \sqrt[3]{\frac{P}{P_e}}\right) \qquad \text{case } A \tag{6.143}$$

Proceeding in exactly the same fashion, case B would have the following interaction relationship:

$$\frac{m}{M_y} = \frac{3}{2}\left[1 - \left(\frac{P}{P_y}\right)^2 - \sqrt[3]{\left(\frac{P}{P_e}\right)^2}\right] \qquad \text{case } B \tag{6.144}$$

To determine the boundaries of applicability of eqs. (6.143) and (6.144), equate the two. This gives

$$\sqrt[3]{\frac{P}{P_e}} = 1 - \frac{P}{P_y} \tag{6.145}$$

which when substituted into eq. (6.143) yields the boundary equation

$$\frac{m}{M_y} = 3\left(1 - \frac{P}{P_y}\right)\frac{P}{P_y} \tag{6.146}$$

It should be evident from eq. (6.144) that if the quantity P/P_e is held constant, the interaction relationship between P/P_y and m/M_y is linear in the range where case A holds. Where B governs, the interaction relationship is parabolic. A set of curves can therefore be readily constructed which have as coordinate values P/P_y and m/M_y and as cross curves various values of P/P_e. As was indicated in sec. 4.5.1 [eq. (4.145)] such a representation is not the most usable one for design purposes, since the third variable contains a mixture of the applied axial thrust, the geometrical properties of the member, and the material properties. Rather, a more usable representation can be obtained by introducing a modification to the P/P_e term of the following type:

$$\frac{P}{P_e} = \frac{P}{P_e}\frac{P_y}{P_y} = \frac{P}{P_y}\frac{P_y}{P_e} = \frac{P}{P_y}\gamma \tag{6.147a}$$

where

$$\gamma = \frac{\sigma_y}{\pi^2 E}\left(\frac{L}{r}\right)^2 \tag{6.147b}$$

From this it is possible to construct curves on the coordinate axes P/P_y and m/M_y for various values of γ—where γ is a function only of the slenderness ratio in the direction of bending and the material properties.

In fig. 6.34 ultimate-strength interaction curves have been shown for various values of γ: 0, 0.4, 1.0, and 2.0. These are the solid lines. (They have as extremes $P/P_y = 1.0$ for $m/M_y = 0$, and $m/M_y = 1.5$ for $P/P_y = 0$.) Also shown in the figure by the dash-dot curves are the corresponding initial yield solutions. They were obtained using the equations and tabulated values listed in table 4.10. It should be understood that these solutions hold only for a rectangular cross-section member that is sub-

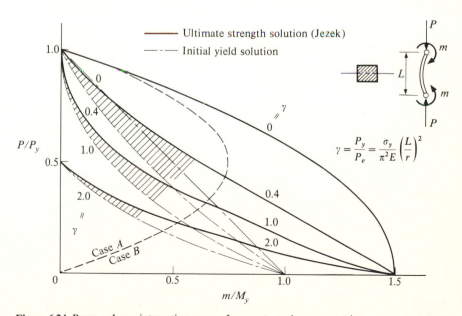

Figure 6.34 Beam-column interaction curves for a rectangular cross section.

jected to equal and opposite end moments and axial thrusts that produce single curvature deformations. Moreover, it is to be understood that the ultimate-strength solutions are approximations based on the assumption of a single-variable displacement function. (Such an assumption, however, is a reasonable one for the type of loading assumed. Experimental evidence supports the results.)

When these methods are extended to the case of a wide-flange member, several troublesome features arise. Whereas for the rectangle only two general inelastic cases had to be considered, for the wide-flange there will be four: (*a*) yielding into the compression flange, tension flange elastic; (*b*) yielding into the web on the compression side, tension flange elastic; (*c*) yielding into the web on the compression side, yielding in the flange on the tension side; and (*d*) yielding into the web on the compression side, yielding into the web on the tension side. Solutions for each of these cases would have to be obtained, and boundaries established between each. Moreover, as has been repeatedly pointed out in this chapter, most real wide-flange members contain residual stresses, and these must be taken into account.

To facilitate the ultimate strength solution of as-rolled, wide-flange steel beam columns, procedures were developed (which include the influence of residual stress) which allow the definition of M-P-φ relationships.[†] [These would be comparable to eqs. (6.133) and (6.137) for the rectangle.] These relationships have been numerically integrated[‡] to obtain ultimate-strength solutions comparable to the approximate ones described above. They are shown in fig. 6.35 for an assumed W 8 × 31 where

† R. L. Ketter, E. L. Kaminsky, and L. S. Beedle, Plastic Deformations of Wide-Flange Beam-Columns, *Trans. Am. Soc. Civ. Eng.*, vol. 120, p. 1070, 1955.

‡ T. V. Galambos and R. L. Ketter, Column under Combined Bending and Thrust, *Proc. Am. Soc. Civ. Eng.*, vol. 85, no. EM2, April, 1959.

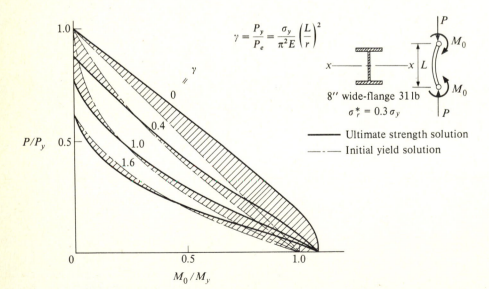

$$\gamma = \frac{P_y}{P_e} = \frac{\sigma_y}{\pi^2 E}\left(\frac{L}{r}\right)^2$$

8″ wide-flange 31 lb
$\sigma_r^* = 0.3\sigma_y$

——— Ultimate strength solution
—·—· Initial yield solution

Figure 6.35 Beam-column interaction curves for a W 8 × 31 member containing residual stresses.

$\sigma_r^* = 0.3\sigma_y$. Again, for comparison purposes, elastic limit solutions are indicated. (It should be noted that the selected wide-flange shape has one of the smaller shape-factors: M_{ult}/M_y, for $P/P_y = 0$. Therefore, these results should be adequate for design purposes.)

Several interesting observations can be made by comparing figs. 6.34 and 6.35. For the rectangular shape (fig. 6.34), where no residual stresses were presumed, the ultimate-strength solutions are everywhere greater than the initial-yield solutions—as would be expected. When residual stresses are included for the wide-flange shape (fig. 6.35), such is not always the case. For the lower values of P/P_y, it is true. For the higher values, however, an elastic strength prediction neglecting residual stress will be an overestimate of the ultimate carrying capacity of the member.

The two sets of curves also provide a visual comparison of the relative reserve in strength that exists above the initial yield solutions. For the rectangle, for $P = 0$, there is a 50 percent reserve in strength. For the W 8×31 ($P = 0$), it is only 10 percent. For various values of P/P_y the reserve is indicated by the shading.

Beam columns are subjected to loading conditions other than those assumed. In fact, for building-type construction, a much more frequent design case consists of an applied moment at one end of the member with zero moment at the other end. W 8×31 solutions (including the influence of residual stresses) also have been obtained for this case (see fig. 6.36) as well as for many other cases.†

---

† R. L. Ketter, Further Studies of the Strength of Beam-Columns, *Proc. Am. Soc. Civ. Eng.*, vol. 87, no. ST 6, p. 135, August, 1961.

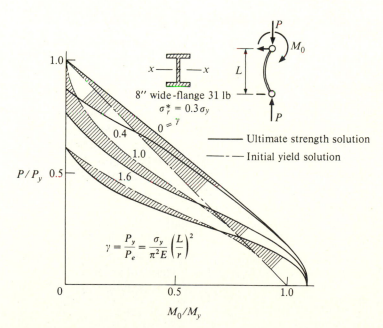

Figure 6.36 Beam-column interaction curves for a W 8×31 member subjected to moments at one end only.

As indicated earlier, the interaction curves shown in figs. 6.35 and 6.36 for ultimate-load carrying capacity were obtained by numerically integrating known, nonlinear M-P-φ relationships. The process used was long and tedious and each point on each of the curves required that there be constructed the actual load-deformation curve for an assumed, given-length member, starting from the elastic limit case and progressing until a condition of "continued deflection" under constant external loading was realized.

It is to be recognized that the M-P-φ relationships referred to above can be approximated (to the degree desired) by nonlinear algebraic expressions using standard curve-fitting techniques. These equations, in turn, can then be incorporated into standard computer solutions of nonlinear differential equations, and the desired interaction curves obtained. For a W 8 × 31 shape ($\sigma_r^* = 0.3\sigma_y$), a suitable algebraic equation would be†

$$\frac{\varphi}{\varphi_y} = \frac{M}{M_y} + a(10)^b \left(\frac{M}{M_y}\right)^c \qquad (6.148)$$

where

$$a = 0.255$$

$$b = 4.46 \frac{P}{P_y}$$

$$c = \left\{ \left[\left(142.19 \frac{P}{P_y} - 304.58 \right) \frac{P}{P_y} + 231.19 \right] \frac{P}{P_y} - 8.19 \right\} \frac{P}{P_y} + 18.50$$

(Note that this equation is the same as (4.76), when the axial thrust is set equal to zero.)

6.4.4 The Interaction Equation for Beam Columns

In this section all of the various possible failure modes for beam columns have been examined, and methods have been developed for determining the associated critical loads. Considering the fact that most real members are subjected to a variety of types and kinds of loading, each of which may cause a different type of failure, and recognizing that the effort required to obtain a solution for just one given set of conditions is a formidable task, it is obvious that approximate design equations are required.

In general, it can be said that structural members of the type being considered are subjected to two basically different kinds of loading: axial-thrust and bending-moment. An expression relating these two, then, is desired. The expression could be linear or powered in either of the selected variables; however, the desired form should be the least complicated one that provides the degree of accuracy desired—since the equation will be used repetitively. The interaction equation that has been

---

† S. P. Prawel, Jr., and R. L. Ketter, Deformations of Pile Supported Structures, *Proc. (Civil Engineering in the Ocean II) Am. Soc. Civ. Eng.*, p. 417, 1970.

shown to give a reasonable prediction of structural strength, and yet is relatively simple, is the one given as eq. (6.149):

$$\frac{P}{P_u} + \frac{M}{M_u(1/C_m)}\frac{1}{1 - P/P_e} \le 1.0 \tag{6.149}$$

Here P = applied axial load

M = maximum bending moment due to applied transverse loading or applied end moments. (M does not include the secondary moment due to axial thrust times deflection. Furthermore, M is considered to act in the plane of symmetry of the cross section only.)

P_u = the axial force carrying capacity of the member when axial force alone exists. (P_u may be the elastic or the inelastic buckling load, depending upon the slenderness of the member. Furthermore, it may be either the flexural buckling load about the xx axis or about the yy axis, depending upon the end conditions or intermediate lateral supports along the member. For very short length members, P_u approaches the yield load $A\sigma_y$.)

M_u = the bending-moment carrying capacity of the member when moment alone exists. (M_u may be either the elastic or inelastic lateral buckling moment, when the member is laterally unsupported. When the member is laterally supported, M_u will be the maximum moment that the cross section can sustain, that is, the fully plastic moment.)

C_m = a modification factor to M_u, which accounts for the corrections for moment gradient, different loading conditions, support conditions, etc. (In general, C_m is smaller than unity but greater than 0.4. Suggested values are given in most specifications.)

$1 - \dfrac{P}{P_e}$ = an amplification factor that yields a conservative approximation of the effect of axial thrust times deflection on the maximum bending moment. (P_e is the Euler buckling load of the member in the plane of the applied bending moments. $P_e = P_u$, if the factor of safety is 1.0, if the member buckles in the elastic range of stress and if deformations are constrained to occur in the plane of the applied moments. $P_u \ne P_e$ if the member fails in the inelastic range, or if axial buckling occurs in a plane other than the applied moment plane. Regardless of P_u, P_e is the elastic axial buckling load—the Euler load—in the plane of the applied moments, independent of whether or not the member is laterally constrained.)

Equation (6.149) can be expanded to handle those situations where the member is subjected to a condition of biaxial bending.

$$\frac{P}{P_u} + \frac{M_x}{(M_{u_x}/C_{m_x})(1 - P/P_{e_x})} + \frac{M_y}{(M_{u_y}/C_{m_y})(1 - P/P_{e_y})} \le 1.0 \tag{6.150}$$

For this case, P_u is computed based upon the larger effective slenderness ratio of the member. The subscripts x and y refer to the two principal directions of bending.

For thin-walled, open cross-section members, eq. (6.150) neglects the warping normal stress that is induced into the member due to the combination of axial force and biaxial bending. The modification factors C_{m_x} and C_{m_y} therefore should probably be adjusted to take care of the torsional moment effects. For members with strong torsional resistance (for example, box-type columns), eq. (6.150) predicts adequately the biaxial strength of the member.

6.5 LOCAL PLATE BUCKLING OF STRUCTURAL MEMBERS

As noted in several of the earlier chapters of this text, one of the major concerns of design is cost. In many, if not most cases, the cost of a structural member is dependent upon its weight. This, in turn, can be related to the cross-sectional area.

When stability is the design consideration, the parameter of greatest significance is the effective slenderness ratio. The smaller the value of that quantity, the greater will be the load that the member can sustain. The cross-sectional form selected, then, should be the one that gives, per unit of cross-sectional area, the greatest radius of gyration about an axis perpendicular to the direction of anticipated buckling. Presuming a known buckling direction, a cross section composed of two flange-type, exceedingly thin elements, located as far apart as possible, would be the desired form. Unfortunately, the two elements would have to be held together—at least at a certain number of points along the length—to make them work together, and this would substract from the optimum nature of this suggested solution. Moreover, as was demonstrated in several of the considered examples, the direction of buckling is not always obvious. If the end constraints are equal in all directions, the "best" solution would be a thin-walled, cylindrical shape. For different end conditions in different directions, the optimum form will take a different shape.

In each of the cases alluded to above, the desirability of a thin element was noted. More than that, from an overall-stability point of view, it can be reasoned that the thinner the element, the more efficient the solution. However, depending upon the loading, the material properties, the type of cross section, etc., a ratio of the element width to thickness will be reached above which the cross section refuses to maintain its original form, and local wrinkling (that is, local buckling) of that individual element occurs before the anticipated solution is realized. When this happens, the equations previously developed are no longer valid. It is the primary purpose of this section to develop criteria for the prevention of local buckling of cross-sectional elements subjected to typical loadings.

Figure 6.37 shows a number of cross-sectional shapes that are used for metal compression and bending members. Except for the hollow circular section, all are composed, essentially, of flat-plate elements. These elements are subjected to direct compression, bending, shear, or any combination of these—in the plane of the plate element itself.

Thin-walled sections of the types illustrated in fig. 6.37a, b, c, and d are particularly efficient as columns, since they have approximately equal strengths and rigidities in all directions. This is exactly so for the hollow cylinder shown in (a). Depending upon the material and size desired, type (b) cross sections can be manufactured by extrusion or by fabrication, where welding or other means of rigid attachment of the edges of the four plate elements is used. Type (c) cross sections are

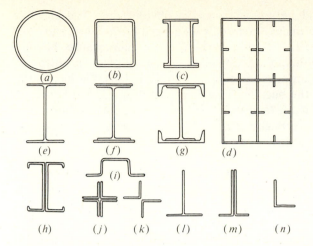

(a) (b) (c)

(e) (f) (g) (d)

(i)

(h) (j) (k) (l) (m) (n)

Figure 6.37 Typical cross-sectional shapes.

composed of two channel shapes attached to two flat-plate elements. The individual parts may be welded or "glued," or they may be intermittently connected by rivets, bolts, etc., along the length of the member. Type (d) is a most elaborate cross-sectional form, and normally would be built up by welding. (When fabricating such a member, the sequence of welding, as well as the amount and location of preheat required to ensure that the member is of the desired, straight form after cooling, can be a major undertaking. Unless the members are subsequently annealed, the resulting residual stress pattern may also constitute a troublesome design problem.)

Wide-flange or similar types of open cross sections (fig. 6.27e, f, g, and h) are commonly used in building and bridge construction. Where several different elements are joined, the methods of attachment described above would be used.

The two elements shown as (h) are manufactured by cold-forming steel sheets. Normally, they would be attached one to the other by spot welding. Cross section (i) could be either a single, cold-formed member, or two Z-type elements could be welded to provide the hat shape.

Cross sections of the type illustrated in (i) to (n) are not the most suitable ones for resisting significant amounts of compression. They are generally used as secondary members.

The response of the individual plate elements of a cross section is not unlike the response of the overall member to the imposed external loading: equilibrium must be maintained, deformations will occur, and there is the possibility that for a given set of conditions bifurcation also can happen.

While there are particular situations where transverse loads are applied directly to given cross-sectional elements, at most of these locations elaborate care is taken (by the introduction of stiffness, cover plates, etc.) to ensure that the cross section maintains its form. Away from these points, the individual plate elements of the cross section are subjected to in-plane forces, and it is this loading condition that is of major concern in the selection of the proportions of the various elements which constitute the cross section.

Plate buckling differs from column buckling in several ways, and the more important ones of these should be noted. First, the obvious one: The buckling strength of an individual plate element is markedly influenced by the edge conditions along the length of the member. Secondly, and of greater importance, but not obvious from a visual examination of the problem: A considerably larger postbuckling strength exists for plates than does for columns. For columns, the increase in load-carrying capacity beyond the buckling load is negligibly small and hence the tangent modulus load is a reasonable (and slightly conservative) criterion for design purposes. For many plates, the postbuckling strength must be taken into consideration, or extremely conservative designs will result.

If increasing compressive forces are applied to opposite edges of a flat plate, a critical stress will be reached at which stress the plate buckles out of its plane. Such a critical stress may be in the elastic or the inelastic range of material behavior. For the plate, as for the column, small geometrical imperfections and residual stresses may cause initial bending at a load that is less than the buckling load.

Plate elements have length and width dimensions as well as thickness. If it is specified that the loading to which the element is subjected is in the direction of the longer of the two plan dimensions, that is, the length, it can be shown that if that longer dimension is at least several times the width, the buckling load is essentially independent of the actual length dimension, and the buckling deformations will be of a wave form. The elastic buckling strength of such a long plate element, then, is primarily determined by the plate width-to-thickness ratio b/t and by the restraint conditions that exist along the longitudinal boundaries of the element. It can be shown that the following equation may be used to approximate the critical buckling stress of flat plate elements in columns that are subjected to uniform compressive forces:

$$\sigma_{\mathrm{cr}} = k \; \frac{\pi^2 E \sqrt{\tau}}{12(1 - \mu^2)(b/t)^2} \tag{6.151}$$

where $\tau = E_T/E$, $\mu = $ Poisson's ratio, b and t are the width and thickness of the plate, respectively; and k is a factor depending upon the longitudinal boundary conditions as defined in table 6.7.†

When a column cross section is composed of various connected plate elements, it is possible (actually, probable) that one or more of the elements will be more susceptible to buckling than will be the others. For such a case, the less critical parts provide edge constraint to the more flexible elements. A lower bound to the actual solution therefore can be obtained by presuming that all the elements are simultaneously at their buckling point and provide no bending resistance to their adjacent counterpart elements. Simply supported edge conditions (or free edges where such is the case) would then be presumed.

As indicated earlier in this section, critical stresses associated with local buckling are significant when analyzing a given structure for its strength. But probably more

---

† The values are taken from Bruce G. Johnston, The Column Research Council *Guide to Design Criteria for Metal Compression Members*, 3d ed., chap. 3, John Wiley & Sons, Inc., New York, 1976.

Table 6.7 Factors for plate buckling

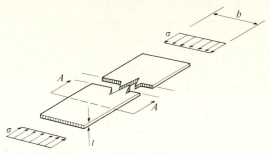

Case	Description of edge support	k value	Sketch of section A-A
1	One edge simply supported, the other edge free	0.42	
2	One edge fixed, the other edge free	1.28	
3	Both edges simply supported	4.00	
4	One edge simply supported, the other edge fixed	5.42	
5	Both edges fixed	6.97	

important for design is the definition of the limiting value of b/t for which it can be presumed that local buckling will not be a problem. In many bridge and building codes, the basic requirement with respect to local buckling is that the proportions of the elements of the cross section be such that the element is capable of reaching the yield stress of the material prior to the attainment of the buckling load. (It is to be recognized that such a requirement does not provide any additional safety factor; that is, FS = 1.00.) For such cases

$$\frac{b}{t} = \sqrt{\frac{\pi^2 k E}{12(1 - \mu^2)\sigma_y}} \tag{6.152}$$

or

$$\left(\frac{b}{t}\right)_{\text{max}} \leq C \sqrt{\frac{k}{\sigma_y}} \tag{6.153}$$

where C is a numerical constant depending upon the properties of the material. For structural steels where $E = 30 \times 10^6$ lb/in^2, and $\mu = 0.3$, $C = 5210$.

Both compression and bending exist in certain cross-sectional elements of structural members; for example, in the web of an I-shape beam column. Coefficients to take into account this variation in stress across the plate element also have been developed—but only for elastic buckling. The governing equation is

$$\sigma_{cr} = k_C \frac{\pi^2 E}{12(1 - \mu^2)(b/t)^2} \tag{6.154}$$

with the values of k_C given in table 6.8.

Table 6.8 Values of k_C†

$\dfrac{\sigma_2}{\sigma_1}$	Both edges simply supported	Both edges fixed	Top edge free		Bottom edge free	
			Bottom edge simple support	Bottom edge fixed	Top edge simple support	Top edge fixed
$+1.00$ (Pure compression)	4.00	6.97	0.42	1.33	0.42	1.33
$+0.33$	5.8					
0	7.8	13.6	0.57	1.61	1.70	5.93
-0.33	11.0					
-0.67	15.7					
-1.00 (Pure bending)	23.9	39.6	0.85	2.15		

† From Bruce G. Johnston, The Column Research Council *Guide to Design Criteria for Metal Compression Members*, 3d ed., John Wiley & Sons, Inc., New York, 1976.

When a plate element is subjected to a uniformly applied shearing stress along the four edges, the critical buckling stress can be computed from the relationship

$$\sigma_{cr} = k_S \frac{\pi^2 E}{12(1 - \mu^2)(b/t)^2} \tag{6.155}$$

where k_S is a so-called shear-buckling coefficient.

For all edges simply supported:

$$k_S = 5.34 + \frac{4.0}{(a/b)^2} \tag{6.155a}$$

For all edges fixed:

$$k_S = 8.98 + \frac{5.6}{(a/b)^2} \tag{6.155b}$$

a is the length of the plate element, and b is the width. a/b is the aspect ratio, and must be greater than or equal to 1.0.

Inelastic solutions corresponding to eqs. (6.154) and (6.155) are currently lacking. One of the reasons for this is that in the inelastic range it is necessary to postulate a yield criterion to take into account the contributions due to normal stress and shearing stress. Even for the pure shear case, the condition is not clear, since the yield stress is not a well-established value. If the von Mises yield criterion is used, it can be shown that the modifying factor to be applied to the elastic case is $\sqrt{\tau}$, the term shown in eq. (6.151). Use of it will give a conservative answer in the inelastic range.

A number of comprehensive summaries of known plate-buckling solutions are available in the literature. The reader is referred to the list of references given at the end of this chapter for further material.

6.6 SUMMARY

In this chapter, there have been developed methods for determining the stability (or lack of stability) of structural members. This is an important, and frequently governing, design consideration. Therefore, for ease of reference, the most important equations that have been developed will be here summarized. The equation numbers are those originally defined.

Axial compression
Elastic flexural buckling:

$$P_{cr} = \frac{\pi^2 EI}{(KL)^2} \tag{6.19}$$

or

$$\sigma_{cr} = \frac{\pi^2 E}{(KL/r)^2} \tag{6.22}$$

K is the effective length factor of the column. For idealized end conditions, K values are given in table 6.1. When the column is part of a frame, the K values can be estimated from the alignment charts given as fig. 6.11.

Elastic torsional buckling:

$$(P_z)_{\text{cr}} = \left(\frac{\pi^2 E I_\omega}{L^2} + G \kappa_T\right) \frac{A}{I_x + I_y} \tag{6.60}$$

Inelastic buckling:

$$(\sigma_{\text{cr}})_{\text{inelas}} = \sigma_y - \frac{\sigma_r^*(\sigma_y - \sigma_r^*)}{(\sigma_{\text{cr}})_{\text{elas}}} \tag{6.88}$$

Bending
Elastic lateral buckling of I beams:

$$\sigma_{\text{cr}} = \frac{(M_x)_{\text{cr}}}{S_x} = C_1 \frac{\pi}{S_x(K_y L)} \sqrt{E I_y G \kappa_T + \left(\frac{\pi}{K_z L}\right)^2 E I_y E I_\omega} \tag{6.113}$$

C_1 is a modification factor which accounts for different conditions of loading. K_y and K_z are effective length factors. Numerical values for these are listed in tables 6.5 and 6.6.

Inelastic lateral buckling of I beams:

$$(\sigma_{\text{cr}})_{\text{inelas}} = \sigma_y - \frac{\sigma_r^*(\sigma_y - \sigma_r^*)}{(\sigma_{\text{cr}})_{\text{elas}}} \tag{6.115}$$

Combined bending and axial compression
Elastic lateral-torsional buckling:

$$(P_{e_y} - P)(r_0^2 P_{e_z} - P r_0^2) = (M_{0x})^2 \tag{6.122}$$

(Nondimensional curves of this equation are given in fig. 6.31.)

Inelastic lateral-torsional buckling:

$$(\sigma_{\text{cr}})_{\text{inelas}} = \sigma_y - \frac{\sigma_r^*(\sigma_y - \sigma_r^*)}{(\sigma_{\text{cr}})_{\text{elas}}} \tag{6.125}$$

Elastic, in-plane bending solutions: [See chaps. 4 and 5 (particularly table 4.10 and fig. 4.52).]

Ultimate strength, in-plane bending solutions: (See figs. 6.34, 6.35 and 6.36, as well as the references listed.)

The interaction equation:

$$\frac{P}{P_u} + \frac{M}{M_u(1/C_m)} \frac{1}{1 - P/P_e} \leq 1.0 \tag{6.149}$$

For biaxial bending:

$$\frac{P}{P_u} + \frac{M_x}{(M_{u_x}/C_{m_x})(1 - P/P_{e_x})} + \frac{M_y}{(M_{u_y}/C_{m_y})(1 - P/P_{e_y})} \le 1.0 \quad (6.150)$$

Local (plate element) buckling

$$\sigma_{\text{cr}} = k \frac{\pi^2 E \sqrt{\tau}}{12(1 - \mu^2)(b/t)^2}$$

$$(6.151)$$
$$(6.154)$$
$$(6.155)$$

Values of k are given in tables 6.7 and 6.8, and in eqs. (6.155a and b) for different loadings and boundary conditions.

Limiting (b/t) values for plate elements stressed to the elastic limit:

$$\frac{b}{t} \le C \sqrt{\frac{k}{\sigma_y}} \quad (6.153)$$

where for structural steel $C = 5210$.

6.7 PROBLEMS

6.1 A uniform cross-section column of length l is subjected to an applied axial thrust at its ends of P. The member is pin-connected at its uppermost end (that is, $y = y'' = 0$), and restrained at its base against bending rotations by a spring of stiffness γ in·lb/rad (that is, $y = 0$, and $EI_y'' = \gamma y'$). Determine the flexural buckling load of the column as a function of l, EI, and γ. Graphically, describe the effective length factor K of the column as a function of the ratio γ/EI.

6.2 A uniform cross-section column of length l is subjected to an applied axial thrust at its ends of P. At its base, the member is fixed (that is, $y = y' = 0$), and at its uppermost end the member is prevented from translation, and is restrained against bending rotation by a spring of stiffness γ (that is, $y = 0$, and $EIy'' = \gamma y'$). Determine the flexural buckling load of the column as a function of l, EI, and γ. Graphically, describe the effective length factor K of the column as a function of the ratio γ/EI.

6.3 A fully continuous column of length $a + b$ is subjected to an applied axial thrust of P as shown in the sketch. If the member is fixed at A (that is, $y = y' = 0$) and constrained against lateral deflection at B (that is, $y = 0$):

(a) Determine the effective length factor of segment AB as a function of the ratio a/b.

(b) Determine the effective length factor of segment BC as a function of the ratio a/b.

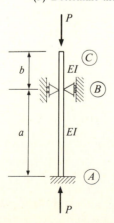

Figure P6.3

6.4 If the column illustrated in prob. 6.3 were pin-connected at its base A (that is, $y = y'' = 0$), what would be the corresponding effective length factors, as functions of a/b?

6.5 Assume that the rigid frame structure shown in fig. 6.9 is fixed at its base, rather than being pin-ended as shown. Determine the critical buckling load as a function of the ratio $I_2 h/I_1 l$, and graphically describe the effective length factor of the column as a function of that ratio.

6.6 The step column shown in the sketch is fixed at its base and free at its uppermost end. Assuming that buckling will be of the flexural type, define the appropriate differential equations and boundary conditions required to solve the eigenvalue problem. What will be the order of the resulting eigenvalue equation? Define the determinant to be solved.

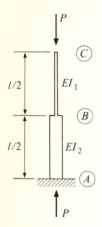

Figure P6.6

6.7 Solve prob. 6.6 using an energy method, assuming as a deflection function

$$y = az + bz^2 + cz^3$$

6.8 A W 24 × 61 member of 24 ft 0 in length is used to support an axial thrust applied at its ends. The member is nowhere restrained along its length. Assuming an elastic–fully plastic type of material with no residual stresses, determine the buckling load of the member, given the following end conditions:

(a) Simple supports at both ends:

$$u(0) = u(l) = u''(0) = u''(l) = 0$$
$$v(0) = v(l) = v''(0) = v''(l) = 0$$
$$\beta(0) = \beta(l) = \beta''(0) = \beta''(l) = 0$$

(b) Fixed supports at both ends:

$$u(0) = u(l) = u'(0) = u'(l) = 0$$
$$v(0) = v(l) = v'(0) = v'(l) = 0$$
$$\beta(0) = \beta(l) = \beta'(0) = \beta'(l) = 0$$

(c) Simple supports in the strong direction and fixed both in the weak direction and for torsional behavior:

$$u(0) = u(l) = u''(0) = u''(l) = 0$$
$$v(0) = v(l) = v'(0) = v'(l) = 0$$
$$\beta(0) = \beta(l) = \beta'(0) = \beta'(l) = 0$$

6.9 Using the alignment charts of fig. 6.11, determine the flexural buckling load of column AB of the frame shown.

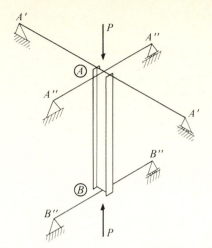

Figure P6.9

The column is a W 8×31 length 14 ft 0 in. The yield stress of the material is $\sigma_y = 50$ ksi, and $E = 30,000$ ksi.

At the uppermost end of the column, the beams are rigidly attached. The beams sizes and lengths are as follows:

$$AA' \quad W \; 10 \times 29 \quad L = 36 \text{ ft 0 in}$$
$$AA'' \quad W \; 8 \times 17 \quad L = 12 \text{ ft 0 in}$$

The column is presumed pin-connected in the strong direction at its base. In the weak direction:

$$BB'' \quad W \; 8 \times 17 \quad L = 12 \text{ ft 0 in}$$

6.10 Determine the flexural buckling load of an axially loaded column 15 ft long. The member is a W 14×74, and has a yield stress level of $\sigma_y = 42$ ksi. The following end conditions are to be assumed:

(a) Both ends hinged in both the strong and weak directions.

(b) Both ends hinged in the strong direction, and fixed in the weak direction.

(c) The top of the column is fixed in both directions. The bottom is fixed in the weak direction, and hinged in the strong direction.

(d) Both ends are hinged in both the strong and the weak directions, but a lateral support is provided at the midheight of the column in the weak direction.

6.11 Obtain the Euler buckling load for a hinged-end column using energy methods, assuming as a deflection function

$$y = A \sin \frac{\pi z}{l}$$

6.12 Using the Rayleigh-Ritz procedure, determine the buckling load for an axially loaded column, assuming

$$y = Az^2(z - l)$$

6.13 What would be the critical KL/r value that divides the regions of elastic and inelastic buckling for a column with $\sigma_r^* = 0.5\sigma_y$. (The answer should be given in terms of E and σ_y.)

6.14 Based on eqs. (6.84) and (6.85) and fig. 6.23, determine the average stress-strain curve for a W 14×136 column with $\sigma_r^* = 0.5\sigma_y$. Then plot the strong and weak axis inelastic column buckling curves that are tangent to the elastic buckling curve.

6.15 Assuming a material with stress-strain properties of the type

$$\frac{\varepsilon}{\varepsilon_y} = \frac{\sigma}{\sigma_y} + \frac{3}{7}\left(\frac{\sigma}{\sigma_y}\right)^n$$

where σ_y = yield point stress, and ε_y = yield point strain; determine the tangent modulus buckling curve for a rectangular, constant cross-section column.

6.16 Due to conditions existing during cooling, a hot-rolled, rectangular cross-section steel member is obtained which is not straight and has a more or less uniform curvature. A decision is made to straighten the member by subjecting it to applied moments at each of its ends. These moments are increased to such an extent that inelastic behavior is observed and the resulting permanent-set, when the moments are removed, causes the member to be straight. It is to be understood that due to the loading prescribed, each and every cross section along the length of the member now contains internal residual stresses. It is specified that they are of the type and magnitude shown in the sketch.

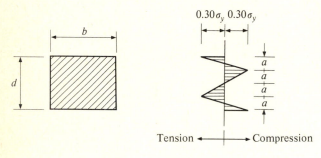

$$0.30\sigma_y \quad 0.30\sigma_y$$

Tension ←——|——→ Compression

Cross section   Residual stress pattern   **Figure P6.16**

    If such a member is to be used as a column, determine the column (flexural) buckling curve, assuming an idealized elastic–perfectly plastic stress-strain curve and the tangent modulus concept of inelastic buckling. What will be the critical slenderness ratio which divides the elastic and the inelastic ranges of column behavior?

6.17 Using the Rayleigh-Ritz method, and assuming that

$$u = u_0\left(1 - \cos\frac{2\pi z}{l}\right)$$

$$\beta = \beta_0\left(1 - \cos\frac{2\pi z}{l}\right)$$

show that the characteristic equation for lateral buckling of a fixed-end beam ($u = u' = \beta = \beta' = 0$) subjected to a uniform applied moment M_{0x} is

$$4EI_y \frac{\pi^2}{l^2}\left(G\kappa_T + 4EI_\omega \frac{\pi^2}{l^2}\right) - M_{0x}^2 = 0$$

6.18 For a simply supported beam of rectangular cross section, derive a relationship between the lateral buckling stress and the aspect ratio d/b of the cross section. (Make your own assumptions when necessary.)

6.19 In prob. 6.18, what would be the aspect ratio for which the critical buckling stress is equal to an allowed maximum buckling stress of $\frac{2}{3}\sigma_y$. (Use: $E = 30{,}000$ ksi, $G = \frac{3}{8}E$, and $\sigma_y = 36$ ksi and $l/d = 20$.)

6.20 Plot the elastic and inelastic lateral buckling curves (σ_{cr}/σ_y versus l/r_y) for a W 27×94 beam, with $\sigma_r^* = 0.4\sigma_y$. (Use: $E = 30{,}000$ ksi, $G = \frac{3}{8}E$, and $\sigma_y = 36$ ksi. It is to be noted that for an I-type cross section

$$I_\omega = \tfrac{1}{4}d^2 I_y$$

where d is the depth of the cross section.)

6.21 If the geometrical properties of an I-type cross section are approximated by the following:

$$I_x = 2bt\left(\frac{d}{2}\right)^2 + \tfrac{1}{12}wd^3$$

$$I_y = \tfrac{1}{6}tb^3$$

$$I_\omega = \tfrac{1}{4}d^2I_y$$

$$\kappa_T = \tfrac{2}{3}bt^3$$

$$G = \tfrac{3}{8}E$$

where

$$t = \text{flange thickness}$$

$$w = \text{web thickness}$$

$$b = \text{flange width}$$

$$d = \text{section depth}$$

show that eq. (6.113) can be written as

$$\sigma_{cr} = C_1\sqrt{\left(\frac{K_1}{(ld/bt)}\right)^2 + \left(\frac{K_2}{(l/r_T)^2}\right)^2}$$

where K_1 and K_2 are constants involving E and G, and

r_T = the radius of gyration about the y axis of the compression flange of the beam plus one-sixth of the web

$$r_T^2 = \frac{b^2}{12(1 + A_w/A_F)^2}$$

$A_w = \tfrac{1}{6}(\text{area of the web})$

$A_F = \text{area of the flanges}$

6.22 Using the relationship given in prob. 6.21, determine for two specific cross sections the relative contribution of the two terms under the square root sign to the total lateral buckling stress, as a function of the unsupported length l. The two cross sections are
  (a) W 16 × 50   (a typical beam section)
  (b) W 8 × 31   (a typical column section)
It should be assumed that $K_2 = 15K_1$ and $K_z = K_y$.

6.23 Describe the curves to the left of the vertical axis $P/P_{e_y} = 0$ in fig. 6.31. Do these curves represent lateral-torsional buckling solutions? Why?

6.24 A laterally supported, W 8 × 31 member, 201 in in length, is subjected to an axial thrust of $0.4P_y$ and equal end moments M_0. Using eq. (6.149), what will be the maximum value of M_0 that the member can sustain, assuming

$$C_m = 0.85 \qquad \text{and} \qquad \sigma_y = 54 \text{ ksi}$$

Check this against the ultimate-strength solutions given in fig. 6.35. If the member were laterally unsupported, what would be the value of M_0? [Check this against eq. (6.122).]

6.25 C_m in eq. (6.149) is a correction factor for M_u. For a W 8 × 31 beam column of length $l = 15$ ft, plot interaction curves of P/P_y versus M/M_u, assuming (a) a uniform primary moment diagram ($\alpha = +1.0$), and (b) a triangular primary moment diagram ($\alpha = 0$). For both cases, it should be assumed that $C_m = 0.40$, 0.85, and 1.00.

6.8 SELECTED GENERAL REFERENCES

Bleich, H.: *Buckling Strength of Metal Structures*, McGraw-Hill Book Company, New York, 1952.

Galambos, T. V.: *Structural Members and Frames*, Prentice-Hall, Inc., Englewood Cliffs, N.J., 1968.

Johnston, B. G.: The Column Research Council *Guide to Design Criteria for Metal Compression Members*, 3d ed., John Wiley & Sons, Inc., New York, 1976.

Timoshenko, S. P. and J. M. Gere: *Theory of Elastic Stability*, 2d ed., McGraw-Hill Book Company, New York, 1961.

Valasov, V. Z.: *Thin Walled Elastic Beams*, Y. Schechtman, trans. (Moscow, 1959, Israel Program for Scientific Translation, Jerusalem, 1961.)

Yu, W. W.: *Cold-Formed Steel Structures*, McGraw-Hill Book Company, New York, 1973.

6.9 APPENDIX

Considering only the possibility of flexural buckling, it should be evident from the earlier sections of this chapter that the relative proportions of cross-sectional elements can be significant factors in determining the axis of buckling. Of course, end restraints are of major importance in determining effective length factors; however, even if these are known or reasonably can be approximated, the problem of selecting the overall proportions of a desired cross section so that the resulting buckling loads about the two principal axes are of about the same magnitude is still present.

For the rectangular cross section, the problem is relatively simple. Given a rectangular cross section of depth d and width b, the ratio of the radii of gyration about the two principal axes is

$$\frac{r_x}{r_y} = \sqrt{\frac{I_x}{I_y}} = \frac{d}{b}$$

That is

$$\frac{r_x}{r_y} = \zeta \frac{d}{b}$$

where

$$\zeta = 1.0$$

For a wide-flange type shape of over all depth d, flange width b, flange thickness t, and web thickness w, the ratio becomes

$$\frac{r_x}{r_y} = \zeta \frac{d}{b}$$

where

$$\zeta = \sqrt{\frac{1 - (1 - \beta)(1 - \alpha)^3}{\alpha + (1 - \alpha)(\beta)^3}}$$

$$\alpha = \frac{2t}{d}$$

$$\beta = \frac{w}{b}$$

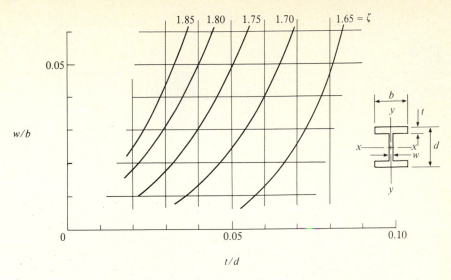

Figure 6.38 ζ as a function of cross-sectional properties.

The interrelationship among and between these variables is shown graphically in fig. 6.38. The range of variables indicated encompasses most rolled, wide-flange, and I-type shapes.

SEVEN

DESIGN OF STRUCTURAL MEMBERS

7.1 INTRODUCTION

It was noted in chap. 1 that, in the broadest sense, structural engineering encompasses the following: (*a*) planning the layout for the system in question; (*b*) estimating the nature, magnitude, and distribution of forces that reasonably can be expected to exist during the structure's anticipated life; (*c*) resolving the manner in which these applied forces are transmitted among and between the various components of the system; (*d*) establishing the desired margin of safety for a given situation; (*f*) proportioning the individual members and their connections so that there will be sustained the anticipated loading(s); and (*g*) preparing the necessary detail, fabrication, and construction drawings, contracts, and specifications, as well as determining construction or erection procedures. A structural designer not only must be fully aware of all of these aspects of design but also must be capable of arriving at a best solution based on insufficient information—taking into account many (often conflicting) objectives, such as safety, economy, practicality, etc.

   In a narrow sense, structural design is concerned with the determination of the minimum allowable geometries of structural elements and their connections based upon the results of structural analyses, using acceptable performance criteria such as allowable stress, ultimate strength, maximum deflection, stiffness, etc. Performance limitations for real design situations are given in recognized codes and specifications. Unfortunately, these manuals of acceptable design practice vary from region to region and from country to country. Moreover, they change with time. Therefore, for most of the examples that will be considered, this chapter will deal with the basic concepts and procedures of structural member design—not necessarily those design specifications of present-day engineering practice. It must be understood that not all aspects of design details will be covered here. The latter can best be learned by carrying out complete design projects as homework, based on current design codes—or in subsequent courses, for those students who wish to pursue design as a career upon graduation.

It will be assumed in this chapter that a given structure has been analyzed, and the forces or moments acting on the various component members are known. While in the preceding chapters only determinate structures were examined, the methods and procedures for design that will be developed here apply equally well in indeterminate cases—once the resultant forces and moments acting on the various elements have been established.

There are three basic criteria that govern the design of most structural members: (a) stress or strength, (b) deformation, and (c) stiffness or stability. Chapters 2 and 3 dealt with stress and stress resultants. Chapters 4 and 5 were concerned with methods for determining deformations. In Chapter 6, stability was considered. Natural frequencies of vibration were discussed in Chapter 4. All this earlier information provides the basis for the design procedures that will be presented in this chapter. Additional principles and design details will be introduced as required.

For the first two of the design criteria, stress or strength, and deformation, it is possible to develop a straightforward design procedure which will ensure the realization of a particular desired state for a given loading. Such is very seldom the case when stability is the controlling condition. There, the objective is to have more than a given reserve in capacity to ensure that instability at the given loading is not realized.

7.1.1 Connections

Component members which make up completed structures are fastened at their ends and at various intermediate points to other members. Depending upon the chosen material and design detail selected, these connections can be made to reproduce to a significant degree the presumptions of the analysis—pin connections, fully rigid connections, connections for partial restraint, etc.

Connections in reinforced concrete construction are more or less rigid unless particular care is exercised to produce other conditions. For example, if a beam is doubly reinforced, and if those reinforcing rods are bent and crossed at the neutral axis, the member will behave as if it contained a hinge at that location. On the other hand, continuing reinforcing rods from a column into an adjacent beam—assuming that the concrete is poured continuously—produces a rigid connection.

Metal structures are permanently fastened together by means of rivets, bolts, or welds. Rivets may be " placed " in the shop or in the field, and these types of connections are used in a large number of different situations. Unfinished, or turned, common bolts, on the other hand, are used only when the members are not subjected to shock or vibration. High-strength bolts are relatively recent introductions to civil engineering structural construction, and are used in a variety of applications. Welding, the third general method of fastening members listed above, is a process of joining metal parts by means of heat and pressure, which causes fusion. Resistance welding is said to occur when the heat of fusion is due to the passage of an electric current through the pieces being joined. Fusion temperatures also can be reached using an electric arc or an oxyacetylene flame. At this time, most structural welding is accomplished using the electric arc as the temperature source.

Welds are classified according to their relative position as flat, horizontal, vertical, and overhead. They also are categorized as groove, fillet, plug, and slot. By type of joint, they would be known as butt, lap, T, corner, or edge.

Glues and epoxy and other resins, as well as other means, are used to connect polymers and silicones, and in certain cases, metals. Joints of this kind are usually of the lap type, although other configurations are sometimes observed.

7.1.2 Main Member Design

When approaching the main member design problem, it is common practice to presume at the outset that the selected cross section will behave as if it is a "fixed-shape geometric unit" that maintains its proportions throughout the loading history. For example, if the cross section in question is of the wide-flange or T type, it would be presumed that the web and the flanges remain at the same angles, one to the other, as loading occurs. Moreover, local wrinkling (that is, plate buckling) of the flanges or the web will not occur. Still another assumption probably would be made: The connections will be proportioned so that the loads are transmitted to the member in question over the entire depth of the cross section. If bending is about the principal axis of the cross section, it also might be initially assumed that adequate lateral support will be provided to ensure that lateral-torsional buckling does not control the design.

Normally, for reinforced concrete members, it would be presumed at the outset that sufficient bond can develop between the reinforcing rods and the concrete in order that the two act together. It also usually is assumed that adequate provision will be made (later) in the design process—through the introduction of vertical bars, inclined bars or stirrups—to ensure that premature failure due to diagonal tension will not take place.

After the general shape and size of the cross section in question have been established, attention must be given to the provision of details which will ensure the desired behaviors. It should be recognized that the particular details vary, depending upon the material being used. For these and other reasons, the basic organization and sequence of the presentation in this chapter will be according to the material being used:

Design of steel members and structural components
Design of reinforced concrete members
Design of composite members

It is to be understood that although a particular material is being considered in each of these sections, the general methods discussed are applicable to other materials. For example, the methods developed in the section on steel design apply equally well to design in aluminum—although the relative importance of the various types of behavior may not be the same for the two cases.

At a number of previous places in this volume, design procedures or information were presented. It is advantageous to recall some of these:

Chapter 1. Discussion of structural design in general; establishment of design loadings; failure criteria; factors of safety, and pertinent design codes and specifications.

Chapter 2. Simple beams in bending; cross-sectional properties; permissible design ranges— for bending members (example 2.12), for bending and shears (examples 2.4 and 2.5), for short axially loaded concrete columns (example 2.14), and for concrete beams (example 2.16).

Chapter 3. Design of simple beams in a bridge floor system (example 3.4); design of a steel tension member, considering both the net and gross areas of the cross section (example 3.7); design of cables (example 3.9); and moment and shear design envelopes for bridge girders (sec. 3.9).

Chapter 4. Design of bending members using deformational failure criterion (example 4.9); interaction equations or diagrams for combined bending and thrust; design for thrust and bending (examples 4.12, 4.13, and 4.16).

7.2 DESIGN OF STEEL MEMBERS

Stress or *strength*, and *stiffness* are the primary controlling criteria for most structural metal designs. *Deformations* are important when certain types of entire structures are being examined, but seldom does this mode of failure control the design of individual structural members in civil engineering practice.

For the most part in steel design, the control criteria is dependent upon the type of loading to which the member in question is subjected. For example, if the member is subjected only to tensile forces, the design condition of failure will be the strength of the member—or, as this is sometimes expressed, the attainment of a particular average stress across the cross section. For shear, allowable stress or strength governs. When the loading is compression, stability (and, therefore, stiffness) more often than not is the controlling condition. The design of bending members is governed by strength (or stress) or stability, depending upon the type of cross section in question, the plane of loading, the length of the member, the kind and amount of end and lateral support, etc. When combined conditions of loading are considered, strength or stiffness must be examined.

Structural steel has both a high yield strength and a high modulus of elasticity (see table 1.4). Therefore, the resulting designed members are relatively slender. Stability is more often than not the design criterion for all but tension members. Therefore, increasing the load-carrying capacity of most steel members requires that additional lateral support be provided. At the one extreme, if the member in question is continuously supported along its entire length, buckling cannot occur (except for local "wrinkling" of the cross-sectional elements). In this case the maximum stress in the member governs the design and equals that allowed by the material. (Of course, an appropriate margin for safety would have to be taken into account.) At the other extreme, where bending is about the strong axis of the cross section and no lateral supports exist along the member, instability determines the load-carrying capacity. Stresses may be exceedingly small, but the ability of the member to sustain additional loading is nonexistent.

In the design of structural steel members, the laterally unsupported length along the member is significant. If this length is large, most likely the design criterion will be lateral buckling. If, on the other hand, the member is adequately braced against lateral or torsional motion(s), that is, the unsupported length is relatively small, maximum stress will probably control the design.

A steel member subjected to axial compression will fail due to column buckling. The member could be part of a rigid frame, or it could be a truss member. An axial force member, however, may be subjected to tension. When such is the case, buckling will not occur, and more often than not design will be based upon maximum stress.

The concept of an *effective length factor* is of importance when considering the possibility of buckling. Individual members in structures are connected to foundations and to other members, and these provide deformational and rotational restraint to the ends of the member in question. The effective length factor takes into account the influence of these adjacent members and foundations [see, for example, table 6.1, or fig. 6.11, or eq. (6.21)]. Because the various members are not ideally connected, designers are often faced with the necessity of estimating appropriate values for K, the effective length factor.

In general, when buckling is the control factor, the designer is presented with two options. Additional supporting or bracing members can be added, which will increase the buckling load of the main member, or a sufficiently large main member can be selected, thereby eliminating buckling as a real problem. The total cost of the bracing and main member should be taken into account when comparing these two possible design solutions.

7.2.1 Tension Members

In theory, the determination of the required area of a tension member is relatively simple. In practice, the proportioning of the cross section so that it is both economical and readily connects to adjacent members is somewhat more involved. The form of the cross section is significantly dependent upon the type of structure of which it is a part, and upon the connection that will be used.

In the selection of tension members, two conditions govern. The first of these has to do with the net cross-sectional area of the member. Longitudinal normal stresses at that section should not exceed a prespecified allowable stress. (Current practice places that limiting normal stress at $0.6\sigma_y$.) The second condition is an upper limit on the effective slenderness ratio that the member should possess. This is not done to prevent buckling—for by definition a member subjected to tensile loading cannot buckle—but rather to control the lateral bending stiffness of the member and, thereby, its natural frequency of vibration. (Current practice requires that $(KL/r)_{min} \leq 240$ for main members of trusses and 300 for secondary members such as diagonal bracing.)

Example 7.1 To illustrate the determination of the required net cross-sectional area of steel tension members, consider the $\frac{1}{2}$-in-thick 11 in-wide plate strip shown in fig. 7.1. Three holes for $\frac{3}{4}$-in-diam rivets—at locations B, C, and,

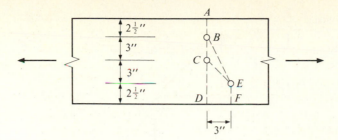

Figure 7.1

E—exist in the plate. (For $\frac{3}{4}$-in-diam rivets, holes are generally made $\frac{1}{8}$ in larger in diameter—to facilitate driving. Thus the diameter of the holes is $\frac{7}{8}$ in.)

There are three possible *net sections* that must be considered: *ABCD*, *ABCEF*, and *ABEF*. The net width for *ABCD* is 9.25 in. For *ABCEF*, the net width is 9.125 in. For *ABEF*, it is 9.625 in. The net cross-sectional area of the steel plate, therefore, is $9.125 \times 0.50 = 4.56$ in². (To ensure an adequate margin of safety, it is common practice for civil engineering type construction to further require that no more than 85 percent of the gross area be counted in tension member design. For the case under consideration, this maximum permitted net area is $0.85 \times 11 \times 0.50 = 4.68$ in², which is greater than that calculated above as the net area.)

Example 7.2 As a second illustration of tension member design, assume that a tensile force of 39,000 lb is resisted by a single steel angle, 9 ft long. The allowable stress is 20,000 lb/in². The member in question is connected at its ends to other members by $\frac{3}{4}$-in bolts, only one of which is to be placed at any one section along the length of the member.

For such problems it is convenient to tabulate permissible trials, somewhat as shown in table 7.1:

Table 7.1 Possible design solutions, example 7.2

Angle thickness, in	Area of one $\frac{7}{8}$-in-diam hole, in²	Gross area required, in²	Lightest angle and cross-sectional area, in²
$\frac{1}{4}$	0.219	2.169	$6 \times 3\frac{1}{2} \times \frac{1}{4}$ (2.31)
$\frac{5}{16}$	0.273	2.223	$4 \times 4 \times \frac{5}{16}$ (2.40)
$\frac{3}{8}$	0.328	2.278	$3\frac{1}{2} \times 3 \times \frac{3}{8}$ (2.30)
$\frac{7}{16}$	0.383	2.333	$3 \times 3 \times \frac{7}{16}$ (2.43)
$\frac{1}{2}$	0.438	2.388	$3 \times 2\frac{1}{2} \times \frac{1}{2}$ (2.50)

All the sections listed in the table are satisfactory. The lightest is the $3\frac{1}{2} \times 3 \times \frac{3}{8}$ angle, which has a minimum slenderness ratio of $(9 \times 12)/0.62 = 174$ (less than 240). It has a gross area of 2.30 in²; 85 percent of 2.30 in² is 1.96 in², which is greater than the required net cross-sectional area of $39,000/20,000 = 1.950$ in².

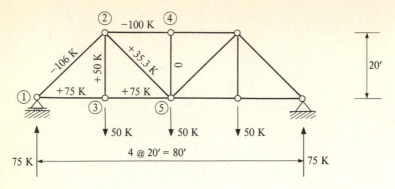

Figure 7.2

Example 7.3 The truss shown in fig. 7.2 is four panels in length and is subjected to 50-kip vertical loads at each of the three interior panel points. The forces acting are shown on each of the members, and these were determined using the methods in chap. 3.

It is specified that the steel from which the truss is to be fabricated has a yield stress of $\sigma_y = 42$ ksi, and the connectors are to be $\frac{3}{4}$-in-diam rivets. The 7th edition of the AISC Specification is to be the basis for the design.

Design of member 2-3 Member 2-3 is 20 ft long, and is subjected to 50 kips in tension. Using the specification in question, the allowable stress is $0.6\sigma_y = 25.2$ ksi. (This implies a load factor of safety against yielding of 1.67.) The required net area, therefore, is $50/25.2 = 1.98$ in^2.

The cross section in question may be a single rolled shape, or a built up section based upon a rolled shape, or it may be a more generalized form using various arrangements of plate elements. Because of the relatively small area required, it will be assumed that the cross section is composed of two angles, arranged as shown in fig. 7.3. The type of cross section chosen is not doubly symmetric. The effect of load eccentricity at the ends of the member, therefore, must be taken into account. (The AISC Specification requires that the slender-

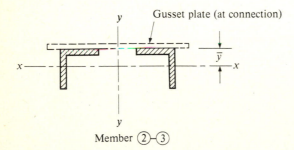

Member ②-③

Figure 7.3

ness ratio of tension members be limited to 240. This also may constitute a design limitation.) Try two angles $4 \times 4 \times \frac{5}{16}$. For each of these

$$A_x = 2.40 \text{ in}^2$$

$$I_x = 3.71 \text{ in}^4$$

$$r_x = 1.24 \text{ in}$$

$$\bar{y} = 1.12 \text{ in}$$

Assuming for the entire cross section two rows of $\frac{3}{4}$-in-diam rivets for connection purposes, two $\frac{7}{8}$-in-diam holes must be subtracted from the gross area to obtain the net area at the ends.

$$A' = 2[2.40 - (\tfrac{5}{16})(\tfrac{7}{8})] = 4.25 \text{ in}^2$$

This is greater than the minimum required value indicated above of 1.98 in². Also, based on the net section, the two angles have

$$I'_x = 6.72 \text{ in}^4 \qquad \text{and} \qquad \bar{y} = 1.24 \text{ in}$$

To check the loading eccentricity induced stresses, the simplest form of the interaction equation will be used:†

$$\frac{f_a}{F_a} + \frac{f_b}{F_b} \leq 1.0 \tag{7.1}$$

where f_a = actual axial tensile stress
f_b = actual bending stress due to loading eccentricity
$F_a = F_b = 0.6\sigma_y$—because the member in question is a tension member, there is no possibility of failure due to buckling or instability.

To determine f_b, it is necessary to know the eccentricity of load application at the end of the member (that is, the distance from the indicated x-x axis in fig. 7.3 to the center of the gusset plate). If a $\frac{1}{2}$-in-thick gusset plate is used, the eccentricity is $0.25 + 1.24 = 1.49$ in. Therefore the bending stress in tension is given by

$$f_b = \frac{Mc}{I_x} = \frac{(50)(1.49)(1.24)}{6.72} = 13.7 \text{ ksi}$$

From the interaction equation, the adequacy of the assumed cross section can be determined. Since $f_a = 50/4.25 = 11.76$,

$$\frac{f_a + f_b}{0.6\sigma_y} = \frac{11.8 + 13.7}{25.2} = 1.01 > 1.0$$

The selected cross section almost meets the specification interaction equation requirement.

---

† This corresponds to eq. (4.144) with $\zeta = 1.0$.

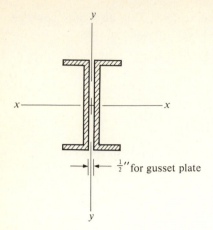

$\frac{1}{2}''$ for gusset plate

Figure 7.4

Design of member 3-5 Member 3-5 also is a tension member. Its design load is 75 kips. The required area, therefore, is $75/25.2 = 2.98$ in^2. For purposes of illustration, assume that the cross section is made up of two channels, back-to-back, as shown in fig. 7.4.

Because the chosen configuration is doubly symmetric, there will be no load eccentricity—and, therefore, no bending stresses. The design will be based upon the net cross-sectional area and the slenderness requirements.

Assume two C 8 × 11.5 channels make up the cross section. Their properties would then be

$$A = 6.76 \text{ in}^2 \qquad A' = 6.76 - 4[(\tfrac{7}{8})(0.220)] = 5.99 \text{ in}^2$$

$$I_x = 65.2 \text{ in}^4$$

$$I_y = 7.20 \text{ in}^4$$

$$r_x = 3.11 \text{ in}$$

$$r_y = 1.032 \text{ in}$$

The largest slenderness ratio is about the weak axis:

$$\frac{L}{r_y} = \frac{(20)(12)}{1.032} = 233 < 240$$

Since the slenderness ratio is less than the limiting value specified, and the net cross-sectional area is greater than 2.98 in^2, the section is more than adequate.

Connection of members to gusset plate For illustration, consider joint no. 3. In the vertical direction, the force in member 2-3 is 50 kips. In the horizontal direction, the force to be transmitted is 75 kips.

Using $\frac{3}{4}$-in-diam rivets (assuming ASTM A502 material for the rivets), the cross-sectional area of each rivet is 0.44 in^2, and the allowable shearing stress

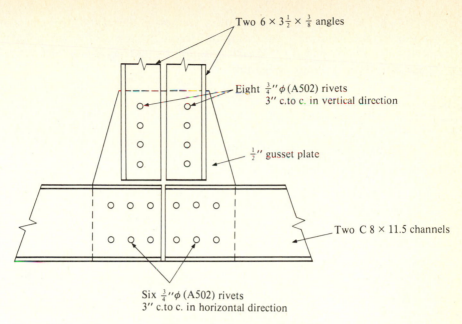

Two $6 \times 3\frac{1}{2} \times \frac{3}{8}$ angles

Eight $\frac{3}{4}'' \phi$ (A502) rivets
3″ c.to c. in vertical direction

$\frac{1}{2}''$ gusset plate

Two C 8 × 11.5 channels

Six $\frac{3}{4}'' \phi$ (A502) rivets
3″ c.to c. in horizontal direction

Figure 7.5 Typical connection of bottom chord to vertical member.

is 15 ksi.† Thus, for the 50-kip load in member 2-3, the required number of rivets in *single shear* is

$$\frac{50}{(15)(0.44)} = 7.6 \qquad \text{use 8 rivets}$$

For the 75-kip horizontal load, 12 rivets in single shear, or 6 rivets in double shear, are required. Because of the symmetrical arrangement of channels, the double-shear case holds.

A typical connection arrangement for joint 2 is illustrated in fig. 7.5.

7.2.2 Centrally Loaded Columns

Example 7.4 The axial buckling load of a given wide-flange cross-section column depends upon the end-boundary conditions. Consider the elastic buckling loads of two different length columns; one with a slenderness ratio of $L/r_x = 20$, and the other with $L/r_x = 120$.

The end conditions for the columns in question are‡

1. At the top: $v = v'' = u = u' = \beta = \beta'' = 0$
2. At the bottom: $v = v'' = u = u' = \beta = \beta' = 0$

---

† From the AISC Specification.
‡ Equations (6.61) are the governing differential equations.

The column, then, at both ends is simply supported about the strong axis and fixed about the weak axis, with warping free at the top and restrained at the bottom. For these defined end conditions, approximate, conservative, effective-length factors suitable for design are

$$K_x = 1.00$$

$$K_y = 0.50$$

$$K_z = 0.70$$

From tables 6.1 and 6.5, more exact values for these conditions would be

$$K_x = 1.000$$

$$K_y = 0.431$$

$$K_z = 0.693$$

For a particular chosen 18-in wide-flange structural steel column, the following geometrical properties are given:

$$E = 30,000 \text{ ksi} \qquad I_\omega = 3860 \text{ in}^6$$

$$G = 11,000 \text{ ksi} \qquad \kappa_T = 2.17 \text{ in}^4$$

$$A = 17.7 \text{ in}^2 \qquad r_x = 7.47 \text{ in}$$

$$I_x = 986 \text{ in}^4 \qquad r_y = 1.68 \text{ in}$$

$$I_y = 50.1 \text{ in}^4$$

The elastic axial buckling loads, therefore, are

1. For $L/r_x = 20$ (assuming $K_x = 1.00$, $K_y = 0.50$, $K_z = 0.70$)

$$(P_{cr})_x = \frac{\pi^2 E I_x}{(K_x L)^2} = 13,080 \text{ kips} \tag{7.2}$$

$$(P_{cr})_y = \frac{\pi^2 E I_y}{(K_y L)^2} = 2658 \text{ kips} \tag{7.3}$$

$$(P_{cr})_z = \frac{1}{r_0^2}\left(G\kappa_T + \frac{\pi^2 E I_\omega}{(K_z L)^2}\right) = 407.8 + 1785.2 = 2193 \text{ kips} \tag{7.4}$$

The critical load for this length member is 2193 kips, and the mode of buckling is twisting about the longitudinal axis.

2. For $L/r_x = 120$ (assuming $K_x = 1.00$, $K_y = 0.50$, $K_z = 0.70$)

$$(P_{cr})_x = 363 \text{ kips}$$

$$(P_{cr})_y = 73.8 \text{ kips}$$

$$(P_{cr})_z = 407.8 + 49.6 = 457.4 \text{ kips}$$

For this length member, weak-axis lateral buckling controls. (Note that when $L/r_x = 20$, $(P_{cr})_x > (P_{cr})_z$, but when $L/r_x = 120$, $(P_{cr})_z > (P_{cr})_x$. This is understandable. As indicated, length does not influence the St. Venant torsional rigidity contribution to the torsional buckling load.)

It is to be understood that example 7.4 was for the purpose of demonstrating the relative influences of end conditions on the buckling mode. Only elastic buckling behavior was examined. For a real situation, it would have to be determined whether elastic or inelastic behavior controlled.

Example 7.5 When the column in question is an integral part of a structure, the allowable column load not only depends upon the properties of the particular member in question but also upon the restraint provided by the adjacent, connecting members. To illustrate this, consider the in-plane buckling behavior of columns 1-2 of the rigid frame shown in fig. 7.6. It will be presumed that the members are initially straight, the connections to other members are rigid, and the vertical loads are applied along the centroidal axes of the vertical members. Both fixed and pinned bases for the lower columns are to be considered.

Figure 6.11 is a set of nomographs (or alignment charts) which define effective-length factors. Two distinct and separate cases are given, depending upon whether or not it is possible for the ends of the member in question to translate. Rotational restraints at the ends are taken into account by the G factors—which are the ratios of the bending stiffness of the member in question to the sums of the stiffness of the adjoining members [eq. (6.21)]. Theoretically, G for a hinged connection is ∞. In the design specification, however, it is suggested for this case that a value of 10 be used. If the column is rigidly attached at its base to a properly designed footing, G would have a theoretical value of 0. Again, a suggested limiting value of $G = 1.0$ is specified.

Table 7.2 summarizes various design possibilities for the structure and loading specified.

It is evident from table 7.2 that the buckling loads differ considerably, depending upon the translational and rotational restraint at the ends of the column.

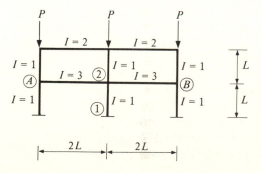

Figure 7.6

Table 7.2 Possible buckling loads for column 1-2

	G_1	G_2	K	$P_{cr} = \dfrac{\pi^2 EI}{(KL)^2}$
Column bases hinged				
a. Sidesway prevented				
1. Full continuity at A and B	Use 10	0.667	0.83	$14.3\dfrac{EI}{L^2}$
2. Hinge connection at A and B	Use 10	0.445	0.80	$15.4\dfrac{EI}{L^2}$
b. Sidesway permitted				
1. Full continuity at A and B	Use 10	0.667	2.25	$1.95\dfrac{EI}{L^2}$
2. Hinge connection at A and B	Use 10	1.33	1.90	$2.73\dfrac{EI}{L^2}$
Columns bases fixed				
a. Sidesway prevented				
1. Full continuity at A and B	1.0	0.667	0.75	$17.5\dfrac{EI}{L^2}$
b. Sidesway permitted				
1. Full continuity at A and B	1.0	0.667	1.25	$6.32\dfrac{EI}{L^2}$

Example 7.6 Assuming a desired load factor of safety of 1.8, determine the allowable axial loads that two different length steel W 12×65 columns can sustain if their effective slenderness ratios are $KL/r_y = 50$ and $KL/r_y = 150$. It is defined that weak-axis flexural buckling governs.

The following conditions are specified:

$$\sigma_y = 36 \text{ ksi}$$

$$\sigma_r^* = 0.3\sigma_y$$

$$E = 30,000 \text{ ksi}$$

$$FS = 1.8$$

1. For $KL/r_y = 50$, the elastic buckling stress is

$$(\sigma_{cr})_{elas} = \frac{\pi^2 E}{(KL/r_y)^2} = 118 \text{ ksi} \tag{7.5}$$

which is considerably more than $\sigma_y - \sigma_r^*$ or 25.2 ksi. Therefore, buckling will occur in the inelastic range. From eq. (6.88)

$$(\sigma_{cr})_{inelas} = \sigma_y - \frac{(\sigma_y - \sigma_r^*)\sigma_r^*}{(\sigma_{cr})_{elas}} = 33.69 \text{ ksi} \tag{7.6}$$

The allowable design stress, therefore, would be

$$\sigma_{all} = \frac{\sigma_{cr}}{FS} = \frac{33.69}{1.8} = 18.7 \text{ ksi}$$

and the allowable axial load would be

$$P_{all} = \sigma_{all} \times A = (18.7)(19.1) = 357 \text{ kips}$$

2. For $KL/r_y = 150$,

$$(\sigma_{cr})_{elas} = \frac{\pi^2 E}{(KL/r_y)^2} = 13.2 \text{ ksi} < 25.2 \text{ ksi}$$

Therefore, the load-carrying capacity is based upon elastic buckling, and the allowable load is

$$P_{all} = \frac{13.2}{1.8} \times 19.1 = 140 \text{ kips}$$

Example 7.7 A 12-ft-long column, pin-connected at both of its ends, is subjected to a compressive axial thrust of 200 kips. It is part of a school building. Assuming $\sigma_y = 50$ ksi (A-441 steel) and $E = 29,000$ ksi, and taking into account the seriousness of potential failure, select a wide-flange column section to sustain the load ($\sigma_r^* = 0.3\sigma_y$).

Referring to sec. 1.6 (specifically table 1.6) safety factors may be determined considering the two factors—probability of failure and seriousness of failure. For

$$A = \text{very good (vg)}$$
$$B = \text{good (g)}$$
$$C = \text{very good (vg)}$$
$$D = \text{very serious (vs)}$$
$$E = \text{serious (s)}$$

the load factor of safety for this situation should be

$$FS = (1.3)(1.5) = 1.95$$

Obviously, a number of different design solutions exist and could be examined. Here, only the W 8×35 will be considered. For that shape,

$$A = 10.3 \text{ in}^2 \qquad I_y = 42.5 \text{ in}^4 \qquad r_y = 2.03 \text{ in}$$

Therefore,

$$\frac{KL}{r_y} = 71$$

From eq. (6.88), the buckling stress is

$$\sigma_{cr} = \sigma_y - \frac{\sigma_r^*(\sigma_y - \sigma_r^*)}{\pi^2 E}\left(\frac{KL}{r}\right)^2 = 40.75 \text{ ksi}$$

The allowable axial thrust is then

$$P_{all} = \frac{(40.75)(10.3)}{1.95} = 215.2 \text{ kips} > 200 \text{ kips}$$

The W 8×35 cross section is adequate.

Example 7.8 Select a pin-connected, wide-flange column of 15-ft length to resist an axial compressive force of 2500 kips. It is to be assumed that $E = 30,000$ ksi, $\sigma_y = 42$ ksi, $\sigma_r^* = \frac{1}{2}\sigma_y$, and FS $= 1.95$. An additional, special architectural requirement for this design is that the cross-sectional dimensions must be equal to or less than 20×18 in.

It is to be recognized that the applied axial force is quite large. (Assuming that the allowable stress is one-half the yield stress, that is, 21 ksi, the required area is $2500/21 = 119$ in^2.) One of the largest rolled sections is the W 14×426, which has a cross-sectional area of $A = 125$ in^2, and a radius of gyration in the weak direction of $r_y = 4.34$ in. A 15-ft-length column of this cross section would have a slenderness ratio about the weak axis of

$$\frac{KL}{r_y} = \frac{(1.0)(15)(12)}{4.34} = 41.5$$

The elastic axial buckling stress, therefore, will be

$$(\sigma_{cr})_{elas} = \frac{\pi^2 E}{(KL/r_y)^2} = 172 \text{ ksi}$$

But this value is considerably higher than the effective proportional limit stress of the material:

$$\sigma_p = \sigma_y - \sigma_r^* = 42 - (\tfrac{1}{2})(42) = 21 \text{ ksi}$$

Therefore, the member will buckle in the inelastic range.† Using eq. (6.89), the allowable stress in the inelastic range for the W 14×426 member is

$$\sigma_{all} = \frac{\sigma_{cr}}{FS} = \frac{\sigma_y}{FS}\left[1 - \frac{\sigma_y}{4\pi^2 E}\left(\frac{KL}{r_y}\right)^2\right] = 20.22 \text{ ksi}$$

---

† This fact could have been determined in a somewhat different fashion. Solving the elastic buckling equation for the slenderness ratio, that particular value that results in a stress equal to the proportional limit is

$$\left(\frac{KL}{r}\right)_p = \sqrt{\frac{\pi^2 E}{\sigma_p}}$$

If the slenderness ratio for a given column is above this transition value, buckling will be in the elastic range. For smaller length members, inelastic buckling occurs. For the case where $\sigma_p = \frac{1}{2}\sigma_y$,

$$\left(\frac{KL}{r}\right)_p = \sqrt{\frac{2\pi^2 E}{\sigma_y}} = 119$$

For most practical situations, the transition slenderness ratio occurs between 100 and 130.

This is less than the initially assumed stress of 21 kips. Hence the W 14 × 426 is adequate. Actually, the member will resist—with a factor of safety of 1.95—an applied axial thrust of 20.22 × 125 = 2528 kips.

Example 7.9 When axial forces are very large, it is desirable and at times necessary to use built-up sections. Even at lower load levels, when the member in question is a part of, for example, a truss, built-up sections may be required.

The various elements of built-up cross sections must be connected so that they act together, rather than as individual parts. They can be directly welded, bolted, or riveted when they adjoin one another. When they are separated, however, *stay plates*, *batten plates*, or *lacing* are normally added to maintain the proper spacing and interaction. Lacing and stay plates are important details of design and must be proportioned to resist the most severe conditions of loading—normally the shearing component of stress. Their spacing along the length of the main elements is determined by the bending or buckling characteristics of the elements.

Perhaps the simplest built-up column type member is a rolled wide-flange shape to which *cover plates* have been attached to the flanges. Such a design will here be discussed.

It is specified that the axial load to be sustained is 2500 kips, the same as example 7.8. Again, it will be assumed that the member is 15-ft long and the ends are pin-connected. This time, however, an A-36 type steel ($\sigma_y = 36$ ksi) is prescribed. Assuming that the allowable stress is 19 ksi, the required cross-sectional area becomes $2500/19 = 132$ in². A W 14 × 320 with two 20 in × $1\frac{1}{4}$ cover plates will provide the following:

$$A = 94 + (2)(25) = 144 \text{ in}^2$$

$$I_y = 1635 + 2[(\tfrac{1}{12})(1.25)(20)^3] = 3302 \text{ in}^4$$

$$r_y = \sqrt{\frac{3302}{144}} = 4.79 \text{ in}$$

The effective slenderness ratio is

$$\frac{KL}{r_y} = 37.6$$

and the allowable buckling stress (using FS = 1.95) is

$$\sigma_{all} = \frac{36}{1.95}\left[1 - \frac{36}{4\pi^2 E}(37.6)^2\right] = 17.67 \text{ ksi}$$

The cross-sectional design, therefore, is adequate. Actually the built-up member in question could support—with a factor of safety of 1.95—an axial thrust of

$$P_{all} = (17.67)(144) = 2544 \text{ kips}$$

Example 7.10 As noted in example 7.9, built-up columns are frequently used in truss bridge construction. Here, connection details and the desirability of having loads transferred along centroidal axes without the introduction of large eccentricities more or less prescribe "spread" types of open cross sections. The particular choice of elements for a given set of loads is based upon a number of practical considerations: architectural requirements, availability of materials, appearance, economy, etc.

Compression chord members of trusses are not necessarily doubly symmetric. Often the cross sections are symmetrical only about the vertical axis, which more often than not is the plane of loading. A typical compression chord member is shown in fig. 7.7. The problem is to determine the axial thrust that the member can sustain if its length is 19 ft, $K = 1.00$, $\sigma_y = 36$ ksi, and FS = 1.80.

Built-up members of the type in question can be fabricated either by riveting or welding. It will be assumed for this particular case that welding is used. Therefore, maximum compression residual stresses equal to $\sigma_r^* = 0.5\sigma_y$ must be presumed in the analysis.

The cross-sectional area of the section in question is

$$A = (2)(12.48) + (20)(0.50) = 34.96 \text{ in}^2$$

Measuring down from the top of the built-up cross section, as indicated in fig. 7.7, the horizontal centroidal axis is located at

$$\bar{y} = \frac{(2 \times 12.6)(9.5) + (10)(0.25)}{34.96} = 6.92 \text{ in}$$

The moments of inertia are

$$I_x = 1704.4 \text{ in}^4$$

$$I_y = 1553 \text{ in}^4$$

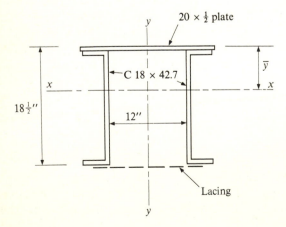

Figure 7.7

The least radius of gyration is

$$r_y = \sqrt{\frac{I_y}{A}} = \sqrt{\frac{1553}{34.96}} = 6.66 \text{ in}$$

The governing slenderness ratio therefore is

$$\frac{KL}{r_y} = \frac{(1.00)(19)(12)}{6.66} = 34.23,$$

and the allowable stress (in the inelastic range) is

$$\sigma_{\text{all}} = \frac{36}{1.8} \left[1 - \frac{36}{4\pi^2 E} (34.23)^2 \right] = 19.29 \text{ ksi}$$

The allowable load is $19.29 \times 34.96 = 674$ kips.

The lacing indicated in fig. 7.7 also must be designed. It can be either single or double, that is, overlapping, in pattern, and can be made up of rectangular bars or angles or channels depending upon the load requirements of the design. Most specifications provide specific requirements for the proportioning of these details—requirements that are based upon the probability of imperfections in the fabrication, load application, etc. These are most often expressed in terms of a design lateral shear that must be resisted, which is a percentage of the axial thrust. For example, the American Institute of Steel Construction specification requires that the member be able to sustain a lateral shear equal to 2 percent of the maximum compressive load of the column. Moreover, that specification requires that the individual lacing members, under the assumed loading, have a buckling stress greater than that of the column as a whole. (It is not uncommon to have required a slenderness ratio for the lacing that will result in buckling of those elements at the yield stress—assuming no residual stress.) Of course, the lacing bars must be able to sustain any forces that may be imposed for any loading.

While it may not have been obvious from the carefully selected sets of example problems discussed in this section, the design of structural members subjected to axial compressive loads is necessarily a trial-and-error process. This is true whether the member buckles in the elastic or the inelastic range of behavior. The allowable stress depends upon the magnitude and distribution of the cross-sectional area. The required area depends upon the allowable stress. One cannot be obtained without having first selected a trial value for the other. For these types of problems, experience has no substitute.

7.2.3 Beams

The relative ease with which steel beams are designed depends upon a number of factors: the shape of the cross section, its orientation with respect to the plane of loading, the end constraints, the lateral support conditions, the width-thickness ratios of the various elements of the cross section, etc. In most real situations, beams are subjected to loadings which produce bending moments in one plane significantly greater than those in all other planes. It is common practice to select cross sections which take this fact into account—that is, ones which have markedly different bending properties in the two principal directions. For these types of shapes, there is always the possibility that the member will buckle laterally and twist as the loading is increased. The member's strength, then, is determined in large part by the lateral support provided to the beam.

For beams constrained to deform in the plane of primary bending—due either to lateral loading about the weak axis of the member, or due to the provision of adequate lateral support—the maximum stress (either due to bending or shear) or at times the maximum deflection governs the design. Design procedures for these cases are straightforward, and tables can be developed which require little more than looking up safe-load values for the various standard shapes. If, however, the beam does not have sufficient lateral support, the design procedure is somewhat more complicated, and a trial-and-error approach not unlike that described for compressive column loads is required.

In chap. 2, examples were given of the design of simple beams based upon maximum allowable bending stress and maximum allowable shearing stress. In chaps. 4 and 5, deformation was considered as the design limitation. It also is to be recalled that in chap. 4 nonuniform torsional stresses were determined for thin-walled beams. In the following series of examples, emphasis will be placed upon those aspects of design not covered in the earlier chapters.

Example 7.11 A 100-ft-long W 12 × 190 beam is laterally unsupported over its entire length. Assuming that the plane of the web of the member is in the plane of the applied bending moments and neglecting the weight of the beam itself, determine for each of the loading and end-support conditions shown in fig. 7.8 the maximum normal stress in the beam corresponding to lateral-torsional buckling. Assume $\sigma_y = 65$ ksi, $E = 30,000$ ksi, and $G = 11,000$ ksi. The cross-sectional properties for the W 12 × 190 are

$$I_y = 590 \text{ in}^4 \qquad I_\omega = 23,600 \text{ in}^6$$

$$\kappa_T = 48.9 \text{ in}^4 \qquad S_x = 263 \text{ in}^3$$

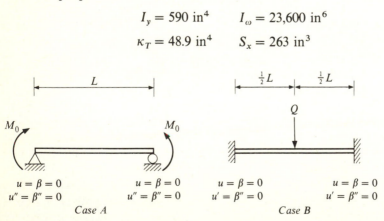

$$u = \beta = 0 \qquad\qquad u = \beta = 0$$
$$u'' = \beta'' = 0 \qquad\qquad u'' = \beta'' = 0$$
Case A

$$u = \beta = 0 \qquad\qquad u = \beta = 0$$
$$u' = \beta'' = 0 \qquad\qquad u' = \beta'' = 0$$
Case B

Figure 7.8

Case A: For the uniform moment case, the lateral buckling moment is given by eq. (6.100), which is here reproduced as (7.7).

$$(M_0)_{cr} = \frac{\pi}{L} \sqrt{EI_y GK_T + \frac{\pi^2}{L^2} EI_y EI_\omega} = 8114 \text{ in. kips} \qquad (7.7)$$

The maximum normal stress, which occurs at the extreme fibers of the beam, is

$$(\sigma_x)_{cr} = \frac{(M_0)_{cr}}{S_x} = \frac{8114}{263} = 30.9 \text{ ksi}$$

This is less than the yield stress value minus any compressive residual stress. Therefore, the member buckles in the elastic range.

$$(\sigma_x)_{cr} = 30.9 \text{ ksi}$$

Case B: For the concentrated load case, the buckling moment can be determined using eq. (6.112) and the values given in table 6.6. For this case (where $K_y = K_z = 1.00$), for the given boundary conditions in the lateral and torsional directions,

$$C_1 = 1.70$$

Therefore,

$$(M_x)_{cr} = (1.70)(8114) = 13,793.8 \text{ in} \cdot \text{kips}$$

and (assuming elastic behavior) the maximum normal stress is

$$(\sigma_x)_{cr} = (1.70)(30.9) = 52.5 \text{ ksi}$$

If the cross section is fabricated by welding and the maximum compressive residual stress is $\sigma_r^* = \frac{1}{2}\sigma_y$, the proportional limit stress is

$$\sigma_p = \sigma_y - \frac{1}{2}\sigma_y = \frac{1}{2}\sigma_y = 32.5 \text{ ksi}$$

The cross section, then, will buckle in the inelastic range. The corresponding buckling stress, from eq. (6.115), is

$$(\sigma_{cr})_{inelas} = \sigma_y - \frac{\sigma_r^*(\sigma_y - \sigma_r^*)}{(\sigma_{cr})_{elas}} = 65 - 20.1 = 44.9 \text{ ksi}$$

Example 7.12 The cross section shown in fig. 7.9 is subjected to bending moments about the strong axis. The member is braced laterally at discrete points along its length. If it is assumed that the strong-axis bending moment between the points of lateral bracing is constant and the lateral and torsional effective length factors for these segments are 1.00, and further if it is presumed that there should be a load factor of safety of 1.65, determine the allowable buckling stress for the member. The lateral unbraced lengths of the members in question are 12 ft and 60 ft, respectively.

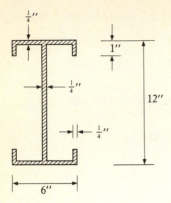

Figure 7.9

The material and geometric properties for the section are as follows:

$E = 30,000$ ksi $A = 6.63$ in^2

$I_y = 15.22$ in^4

$v = 0.3$ $I_x = 156.8$ in^4

$S_x = 26.1$ in^3

$G = \dfrac{E}{2(1+v)} = 11,540$ ksi $\kappa_T = 0.138$ in^4

$I_\omega = 525$ in^6

However,

$$\sigma_{cr} = \frac{(M_x)_{cr}}{S_x} \tag{7.8}$$

where

$$(M_x)_{cr} = \frac{\pi}{L}\sqrt{EI_y G\kappa_T + \left(\frac{\pi}{L}\right)^2 EI_y EI_\omega}$$

The critical normal stress, then, can be written as

$$\sigma_{cr} = \frac{\pi\sqrt{EI_y G\kappa_T}}{S_x L}\left[\sqrt{1 + \frac{\pi^2}{L^2}\frac{EI_\omega}{G\kappa_T}}\right] \tag{7.9}$$

For the shape specified,

$$\sigma_{cr} = \frac{1970}{L}\sqrt{1 + \frac{97,600}{L^2}}$$

For $L = 12$ ft $= 144$ in

$$\sigma_{cr} = 13.68\sqrt{1 + 4.71} = 32.70 \text{ ksi}$$

For $L = 60$ ft $= 720$ in

$$\sigma_{cr} = 2.74\sqrt{1 + 0.188} = 2.99 \text{ ksi}$$

For structural members which have relatively short unsupported length; that is, for those where

$$\frac{\pi^2}{L^2} E I_\omega \gg G \kappa_T$$

the critical bending moment can be approximated by

$$M_{cr} \approx \frac{\pi^2 E}{L^2} \sqrt{I_y I_\omega} \tag{7.10}$$

For members with relatively long unsupported lengths,

$$\frac{\pi^2}{L^2} E I_\omega \ll G \kappa_T$$

and

$$M_{cr} \approx \frac{\pi}{L} \sqrt{E I_y G \kappa_T} \tag{7.11}$$

Since for the short length case, the equation is of the same form as the Euler's column buckling formula (indeed, warping torsion is directly associated with normal stress), longitudinal residual stresses such as those due to cooling or welding will have a direct influence on the critical moment. Therefore, for only the warping torsional term there would have to be considered the possibility of two separate cases: elastic behavior and inelastic (transition) behavior. Equation (6.115) would govern in the inelastic range.

It is interesting to observe that by ignoring either one of the terms in eq. (6.100), the maximum error introduced will occur when

$$G \kappa_T = \frac{\pi^2}{L^2} E I_\omega \tag{7.12}$$

and approximately will be equal to $\sqrt{2} - 1$, or 41.4 percent.

Example 7.13 A two-span, constant cross-section beam of 40-ft length (16 ft for one span and 24 ft for the other) is used to support a uniformly distributed vertical load of 3 kips/ft and a concentrated end moment of 75 ft · kips, as shown in fig. 7.10. Assuming $\sigma_y = 65$ ksi, $\sigma_r^* = 0.3\sigma_y$, and FS = 1.85, select an acceptable wide-flange rolled shape. In fig. 7.10 also is shown the bending-moment diagram. Methods for determining such moment diagrams are discussed in chap. 8.

If the beam is laterally supported continuously along its length, the 145.5 ft · kips maximum bending moment governs the design. Since the allowable stress is

$$\sigma_{all} = \frac{65}{1.85} = 35.1 \text{ ksi}$$

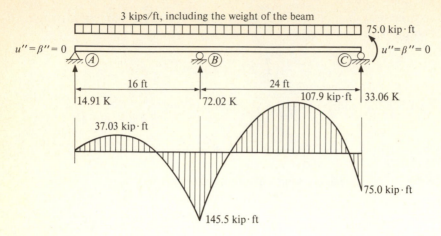

Figure 7.10

the required section modulus is

$$S_{rqd} = \frac{145.5}{35.1} \times 12 = 49.74 \text{ in}^3$$

A W 18×35 provides $S_x = 57.9 \text{ in}^3$.

Consider, now, the case where the beam is laterally supported *only* at the three vertical support locations A, B, and C. For this case, the 24-ft span governs the design, and the lateral-buckling stress equation is given by eq. (6.113):

$$\sigma_{cr} = C_1 \frac{\pi}{S_x(K_y L)} \sqrt{EI_y G\kappa_T + \left(\frac{\pi}{K_z L}\right)^2 EI_y EI_\omega} \qquad (7.13)$$

where C_1 is the modification factor to take into account other than uniform-moment situations (see table 6.6). L is the unsupported length in the weak direction and equals 24 ft. The moment diagram for the right-hand span approximates that for a beam fixed at both ends and subjected to a uniformly distributed lateral load. Therefore, it is appropriate to select $C_1 = 1.30$. Try a W 16×58:

$$S_x = 94.4 \text{ in}^3$$

$$I_y = 65.3 \text{ in}^4$$

$$\kappa_T = 1.98 \text{ in}^4$$

$$I_\omega = 3780 \text{ in}^6$$

Substituting these into the above critical stress equation, (7.13) equals

$$\sigma_{cr} = 40.1 \text{ ksi}$$

(It is to be noted that this value is less than $0.7\sigma_y$. Therefore, the member buckles in the elastic range.) Taking into account the desired design factor of safety, the allowable stress for the W 16×58 member is

$$\sigma_{all} = \frac{40.1}{1.85} = 21.68 \text{ ksi}$$

The actual maximum normal stress in the span in question occurs at the interior support point B where

$$\sigma_{max} = \frac{145.5}{94.4} \times 12 = 18.5 \text{ ksi}$$

The W 16×58 is more than adequate.

For the main member design just completed, it is necessary to prescribe the details which will ensure that the conditions presumed in the design will be realized. For example, the beam in question may rest on a masonry or concrete wall whose load-carrying capacity requires that there be between the beam and the wall a bearing plate to spread out the reaction resultants over a greater area. There also is the possibility that the web of the beam, in and of itself, is insufficient in thickness at these support locations, and doubler plates or stiffeners are required. These possibilities are illustrated in fig. 7.11.

For masonry walls, it usually is assumed that the allowable bearing stress is 250 lb/in². *For a 3000 lb/in² concrete*, an allowable bearing stress of 1750 lb/in² can be assumed. For both of these cases, for reasonably proportioned plates, it is common practice to assume that the pressure from the wall on the plate will be distributed uniformly over the bearing surface.

In fig. 7.11, $B \times C \times t$ are the dimensions of the bearing plate. Normally, C is limited to a maximum value of 12 in. Therefore, with C given, B can be determined from the allowable pressure and the support reaction. The thickness of the bearing

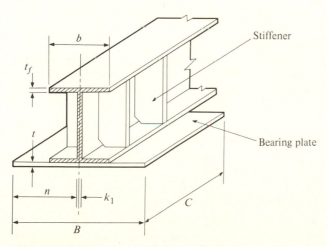

Figure 7.11 Bearing-plate and stiffener arrangements.

plate is determined from a consideration of the bending capacity of the bearing plate about either (i) the web toe of the fillet of the rolled section, or (ii) the overhanging portion of the plate beyond the width of the flange. In both cases, as noted above, it is assumed that a uniform upward pressure of F_p exists on the plate.

Assuming that there are no stiffeners (or that any that might be present will not contribute to the bending capacity of the flange and bearing plate), the bending moment acting on the bearing plate or the flange will be

(i): $$\text{Moment} = F_p \times n \times \tfrac{1}{2}n$$

(ii): $$\text{Moment} = F_p \times \tfrac{1}{2}(B - b)^2$$

where b is the width of the flange of the rolled shape. Since $M = S \times \sigma_{\text{all}}$ and

$$S = \frac{I}{c} = \frac{(t + t_f)^2}{6} \quad \text{or} \quad \frac{t^2}{6} \tag{7.14}$$

the thickness of the bearing plate can be determined from the relationships

(i): $$t + t_f = \sqrt{\frac{3F_p n^2}{\sigma_{\text{all}}}} \tag{7.15}$$

(ii): $$t = \sqrt{\frac{3F_p[(B - b)/2]^2}{\sigma_{\text{all}}}} \tag{7.16}$$

where b is the flange width, and t_f is the flange thickness of the rolled shape. For purposes of design, it usually is assumed that the allowable normal stress—due to bending of the bearing plate—equals

$$\sigma_{\text{all}} = 0.75\sigma_y$$

Assume that the beam in question (example 7.13) is supported on masonry walls having allowable bearing stresses of 250 lb/in². Further, presume that the bearing plates (if required) are made of A-36 steel [that is, $\sigma_y = 36$ ksi, or $\sigma_{\text{all}} = (0.75) \times (36) = 27$ ksi]. Still further, presume that the bearing length along the beam at the three support points is constrained to be 10 in ($C = 10$ in). Table 7.3 summarizes the design solutions of the bearing plates.

Concentrated loads or reactions which act over relatively short lengths of wide-flange members can produce normal stresses in the webs of the members—perpendicular to the flanges—of such magnitude that yielding in that direction may occur. Crumpling of the web and even collapse of the member may follow shortly thereafter. It is necessary, therefore, to ensure that these stresses, increased by whatever safety factor is desired, do not exceed the yield value. The analytical determination of the magnitude of the normal stress in the web of a wide-flange member in the vicinity of a concentrated load is a complicated problem in the theory of elasticity. It has been observed, however, that a reasonably good approximation can be obtained by presuming a uniform stress distribution along a section defined by lines extending from the ends of the loading area into the web at an angle of 45°. The most critically stressed plane would be at the toe of the fillet where the web joins the flange. An average stress at that section—under working loads—equal to $0.75\sigma_y$ is considered to be adequate for most design situations.

Table 7.3 Base plate designs

	Support location A	Support location B	Support location C
Reaction	14.91 kips	72.02 kips	33.06 kips
C	10 in	10 in	10 in
B_{min}	$\dfrac{14.91}{(10)(0.250)} = 6$ in	$\dfrac{72.02}{(10)(0.250)} = 29$ in	$\dfrac{33.06}{(10)(0.250)} = 13.2$ in
Selected B	$8\frac{1}{2}$ in (width of flange)	30 in	14 in
n	$\frac{1}{2}(8\frac{1}{2}) - 0.813 - 3.44$ in	14.19 in	6.19 in
$\dfrac{B - b}{2}$	—	10.75 in	2.75 in
$(t + t_f)_i$	0.48 in	2.32 in	1.00 in
$(t)_{ii}$	—	1.76 in	0.45 in

Normal stresses in the web due to the application of concentrated forces can be reduced in several different ways. For example, the load could be spread over a longer length. Equally possible, the thickness of the web could be increased in the critical areas. A third possibility would be to add stiffeners to the member under the loads. These stiffeners should bear against the flanges on the side of load application and be attached to the web along its depth so that the load will be gradually transferred into the member. Bearing stiffeners of this type were illustrated in fig. 7.11.

For the W 16×58 member of example 7.13, the web thickness is $\frac{7}{16}$ in. The distance from the outside of the flange to the toe of the fillet in the web is $1\frac{3}{8}$ in. The vertical normal stresses in the web at each of the support locations are therefore

$$\sigma_A = \frac{14.91}{(\frac{7}{16})(10 + 1.375)} = 3.00 \text{ ksi}$$

$$\sigma_B = \frac{72.02}{(\frac{7}{16})(10 + 2.75)} = 12.91 \text{ ksi}$$

$$\sigma_C = \frac{33.06}{(\frac{7}{16})(10 + 1.375)} = 6.64 \text{ ksi}$$

Since all these stresses are less than $0.75\sigma_y = (0.75)(36) = 27$ ksi, no stiffeners are required.

Example 7.14 To illustrate the design procedure for wide-flange beams subjected to combined bending and torsion, consider a 24-ft-long simply supported beam subjected to a uniformly distributed load of 1.2 kips/ft that is applied 6 inches eccentric to the plane of the web. (This loading could represent a beam supporting a brick wall that is placed out of the plane of the web because of architectural requirements.) The beam and loading essentially are those shown in fig. 4.71. It is assumed that the boundary conditions are as follows.†

$$\text{At } z = 0 \quad u = u'' = 0 \quad \text{and} \quad \beta = \beta'' = 0$$

$$\text{At } z = l \quad u = u'' = 0 \quad \text{and} \quad \beta = \beta'' = 0$$

The material properties and the factor of safety to be presumed are

$$E = 29,000 \text{ ksi}$$

$$G = 11,154 \text{ ksi}$$

$$\sigma_y = 65 \text{ ksi}$$

$$FS = 2.00$$

Longitudinal normal stresses result from both primary bending σ_b and warping σ_ω:

$$(\sigma_b)_{\max} = \frac{M_{\max}}{S_x} = \frac{1}{S_x}\left[\frac{qz}{2}(l - z)\right] \qquad \text{[eq. (2.15)]} \qquad (7.17)$$

$$(\sigma_\omega)_{\max} = -\frac{E \, db}{4}\frac{d^2\beta}{dz^2} \qquad \text{[eq. (2.45)]} \qquad (7.18)$$

where

$$\frac{d^2\beta}{dz^2} = \frac{qe}{G\kappa_T}\left(\frac{\cosh \lambda l - 1}{\sinh \lambda l}\sinh \lambda z - \cosh \lambda z + 1\right)$$

$$\lambda = \sqrt{\frac{G\kappa_T}{EI_\omega}}$$

Maximum normal stresses occur at $z = l/2$. Therefore,

$$(\sigma_b)_{\max} = \frac{ql^2}{8S_x} \qquad (7.19)$$

$$(\sigma_\omega)_{\max} = -\frac{E \, db}{4G\kappa_T}qe\left(\frac{\cosh \lambda l/2 - \sinh \lambda l/2}{\sinh \lambda l} + 1\right) \qquad (7.20)$$

St. Venant shearing stresses are maximum on the surface, at the midpoint of the flange:

$$(\tau_{SV})_{\max} = Gt\frac{d\beta}{dz} \qquad (7.21)$$

---

† It is to be noted that in chaps. 2 and 4 the angle of rotation about the longitudinal axis of the member was defined as θ—as is customary in many strength of materials texts. In chap. 6 and in this chapter, however, the same angle has been defined as β—as is most often the case in stability texts.

where

$$\frac{d\beta}{dz} = \frac{qe}{G\kappa_T}\left(\frac{\cosh \lambda l - 1}{\lambda \sinh \lambda l}\cosh \lambda z - \frac{1}{\lambda}\sinh \lambda z - \frac{l}{2} + z\right)$$

Maximum shearing stresses occur at $z = 0$ and $z = l$, where

$$(\tau_{sv})_{\max} = \frac{qet}{\kappa_T}\left[\frac{\cosh \lambda l - 1}{\lambda \sinh \lambda l} - \frac{l}{2}\right] \tag{7.22}$$

The warping shearing stress in the flange (τ_ω) is defined by the equation

$$(\tau_\omega)_{\max} = -\frac{E\,db^2}{16}\frac{d^3\beta}{dz^3} \qquad [\text{eq. (2.47)}] \tag{7.23}$$

where

$$\frac{d^3\beta}{dz^3} = \frac{qe}{G\kappa_T}\left[\frac{\lambda(\cosh \lambda l - 1)}{\sinh \lambda l}\cosh \lambda z - \lambda \sinh \lambda z\right]$$

Again, the maximum values occur at $z = 0$ and $z = l$.

$$(\tau_\omega)_{\max} = -\frac{E\,db^2}{16G\kappa_T}\frac{\lambda(\cosh \lambda l - 1)}{\sinh \lambda l}qe \tag{7.24}$$

For $e = 6$ in, $q = 1.2$ kips/ft and an assumed cross section of W 27×94, the solution would be as follows:

$$d = 26.91 \text{ in} \qquad \lambda = \sqrt{\frac{G\kappa_T}{EI_\omega}} = 0.008562\dagger$$

$$b = 9.990 \text{ in} \qquad \lambda L = 2.4659$$

$$t = 0.747 \text{ in}$$

$$S_x = 243 \text{ in}^3 \qquad \frac{\lambda L}{2} = 1.2329$$

$$\kappa_T = 4.06 \text{ in}^4$$

$$I_\omega = 21,300 \text{ in}^6$$

$$\cosh 2.4659 = 5.930 \qquad \cosh 1.2329 = 1.861$$

$$\sinh 2.4659 = 5.845 \qquad \sinh 1.2329 = 1.570$$

$$(\sigma_b)_{\max} = \pm\left(\frac{1.2}{12}\right)\left[\frac{(288)^2}{(8)(243)}\right] = \pm 4.27 \text{ ksi}$$

$$(\sigma_\omega)_{\max} = \pm 27.11 \text{ ksi}$$

---

† Because of the smallness of this quantity for most wide-flange shapes, it is not uncommon to find listed in various tables the reciprocal of λ, rather than λ itself.

$$a = \frac{1}{\lambda} = \sqrt{\frac{EI_\omega}{G\kappa_T}}$$

or

$$(\sigma_z)_{max} = 4.27 + 27.11 = 31.38 \text{ ksi}$$

$$(\tau_{SV})_{max} = -5.02 \text{ ksi}$$

$$(\tau_\omega)_{max} = -0.47 \text{ ksi}$$

$$(\tau)_{max} = -5.49 \text{ ksi}$$

Since the allowable normal stress is $\sigma_y/\text{FS} = 32.5$ ksi, the W 27×94 is adequate.

7.2.4 Beam Columns

Steel beam columns are designed in a fashion similar to steel beams. When the member is constrained to deflect only in the plane of primary bending—due to bending about the weak axis of the cross section, or bending about the strong axis but with adequate lateral bracing—in-plane elastic or inelastic bending strength usually governs the design. However, when primary bending is about the strong axis of the cross section and there is insufficient lateral support, the mode of failure is more often than not that of lateral-torsional buckling. In this latter case, depending upon the particular cross-sectional properties and the unsupported length, the member may buckle in either the elastic or the inelastic range.

As indicated in chaps. 2, 4, and 6, there are available a number of different analytical methods for handling the beam-column design problem. Some are much more complicated than others, but these give more accurate results. They, in turn, require that there be known considerably more about the loading, the end or lateral restraints, the properties of the cross section, etc. As a multipurpose design tool—one which is sufficiently comprehensive to allow inclusion of the most significant design parameters, and still is sufficiently simple for routine calculational purposes—the interaction equation (or curve) approach is most often used. For the in-plane case, the governing interaction equations are (4.143) and (6.149). For elastic lateral-torsional buckling, eq. (6.122) governs the design. Modification using eq. (6.125) is required for the inelastic case.

Example 7.15 A W 8×31 member 30 ft long is pin-connected at its ends. In the weak direction of bending, the member is continuously restrained along its length against lateral deformations. At one end of the member, acting in the plane of the web, there is applied an end-moment M_0. In addition, equal and opposite compressive axial forces are applied through the centroid of the section at the ends. Assuming $E = 29 \times 10^3$ ksi, $\sigma_y = 30$ ksi, and a desired factor of safety of FS $= 1.00$, determine the value of M_0 for

Case 1: $P = 50$ kips   and Case 2: $P = 150$ kips

The properties of the cross section in question are

$$A = 9.12 \text{ in}^2 \qquad E = 29 \times 10^3 \text{ ksi}$$

$$S_x = 27.4 \text{ in}^3 \qquad \sigma_y = 30 \text{ ksi}$$

$$I_x = 110 \text{ in}^4 \qquad L = 30 \text{ ft} = 360 \text{ in}$$

$$r_x = 3.47 \text{ in}$$

Therefore,

$$P_{e_x} = \frac{\pi^2 E I_x}{L^2} = 242.9 \text{ kips}$$

$$P_y = A\sigma_y = 273.6 \text{ kips}$$

$$M_y = S_x \sigma_y = 822 \text{ in} \cdot \text{kips}$$

The governing interaction equation is (4.140), or

$$\frac{P}{P_y} + \frac{M_0}{M_y}(\zeta) = 1.0 \qquad\qquad (7.25)$$

where

$$\zeta = \frac{1}{\sin 2u} \qquad 2u = \pi\sqrt{\frac{P}{P_{e_x}}}$$

(ζ also can be obtained from fig. 4.52, for $\beta = 0$.)

Case 1 for $\qquad\qquad P = 50 \text{ kips}$:

$$\frac{P}{P_{e_x}} = \frac{50}{242.9} = 0.206$$

$$\frac{P}{P_y} = \frac{50}{273.6} = 0.183$$

$$\zeta = 1.000 \quad \left(\text{since } \frac{P}{P_{e_x}} < 0.250\right)$$

Substituting these values into the interaction equation yields

$$\frac{P}{P_y} + \frac{M_0}{M_y}(\zeta) = 1.0$$

$$0.183 + \frac{M_0}{M_y} = 1.000$$

or

$$\frac{M_0}{M_y} = 0.817$$

The allowable end moment, then, is

$$M_0 = (0.817)(822) = 672 \text{ in} \cdot \text{kips}$$

Case 2 for $\qquad P = 150 \text{ kips}$:

$$\frac{P}{P_{ex}} = \frac{150}{242.9} = 0.618$$

$$\frac{P}{P_y} = \frac{150}{273.6} = 0.548$$

$$2u = \pi \sqrt{\frac{P}{P_{ex}}} = 2.47$$

$$\zeta = \frac{1}{\sin 2u} = \frac{1.000}{0.6222} = 1.6072$$

Therefore

$$\frac{P}{P_y} + \frac{M_0}{M_y}(\zeta) = 1.000$$

$$0.548 + \frac{M_0}{M_y}(1.6072) = 1.000$$

$$\frac{M_0}{M_y} = 0.281$$

or

$$M_0 = (0.281)(822) = 231 \text{ in} \cdot \text{kips}$$

Example 7.16 For the beam column shown in fig. 7.12, select an 8-in wide-flange section such that at lateral-torsional buckling the member can sustain a loading of $P = 100$ kips and $M_0 = 550$ in·kips. (Consider only the elastic buckling case.)

The elastic lateral-torsional buckling load for a beam column subjected to equal end moments is governed by eq. (6.122)

$$(M_0)_{cr}^2 = r_0^2(P_{ey} - P)(P_{ez} - P) \tag{7.26}$$

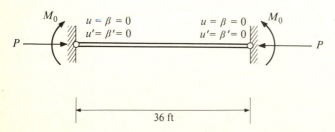

36 ft

Figure 7.12

where $r_0^2 = $ (the polar radius of gyration)$^2 = \dfrac{I_x + I_y}{A}$

$$P_{e_y} = \text{Euler buckling load about the weak axis} = \frac{\pi^2 E I_y}{(K_y L)^2} \qquad (7.27)$$

$$P_{e_z} = \text{Elastic torsional buckling load} = \frac{1}{r_0^2}\left[\frac{\pi^2}{(K_z L)^2} E I_\omega + G \kappa_T \right] \qquad (7.28)$$

For the prescribed boundary conditions, $K_y = 0.7$ and $K_z = 0.7$. It will be assumed that $E = 30 \times 10^3$ ksi and $G = 11.54 \times 10^3$ ksi. Table 7.4 summarizes the solutions for three 8-in wide-flange members. Of the three sections, the W 8 × 35 is the obvious design choice.

Table 7.4 Summary of trial solutions

	W 8 × 31	W 8 × 35	W 8 × 40
I_x	110 in^4	126	146
I_y	37.0 in^4	42.5	49.0
r_x	3.47 in	3.50	3.53
r_y	2.01 in	2.03	2.04
A	9.12 in^2	10.3	11.8
κ_T	0.534 in^4	0.768	1.12
I_ω	529 in^6	618	725
r_0^2	16.118 in^2	16.359	16.525
P_{e_y}	119.8 kips	137.6	158.7
P_{e_z}	488.6 kips	664.1	924.2
(For $P = 100$ kips) $(M_0)_{cr}$	352.2 in · kips	589.0	894.1

Example 7.17 A W 24 × 76 member is subjected to single-curvature-type bending moments of 200 ft · kips about the major axis of bending at each end. Assume that both ends of the member are simply supported in all three reference directions. What is the magnitude of the maximum axial compression that simultaneously can be applied without danger of lateral-torsional buckling? The column has an unsupported length of 24 ft.

The geometrical properties for the W 24 × 76 and the material properties to be assumed are

$$A = 22.4 \text{ in}^2 \qquad I_\omega = 11{,}100 \text{ in}^6$$

$$S_x = 176 \text{ in}^3 \qquad \kappa_T = 2.70 \text{ in}^4$$

$$I_x = 2096 \text{ in}^4 \qquad r_0^2 = \frac{I_x + I_y}{A} = 97.0 \text{ in}^2$$

$$I_y = 76.5 \text{ in}^4 \qquad E = 29 \times 10^3 \text{ ksi}$$

$$r_x = 9.68 \text{ in} \qquad G = 11.154 \times 10^3 \text{ ksi}$$

$$r_y = 1.92 \text{ in}$$

The limiting axial-load buckling values are then

$$P_{ex} = \frac{\pi^2 E I_x}{L^2} = 7233 \text{ kips}$$

$$P_{ey} = \frac{\pi^2 E I_y}{L^2} = 264 \text{ kips}$$

$$P_{ez} = \frac{1}{r_0^2} \left(\frac{\pi^2}{L^2} E I_\omega + G\kappa_T \right) = 705 \text{ kips}$$

Therefore

$$(M_0)_{cr}^2 = r_0^2 (P_{ey} - P)(P_{ez} - P)$$

or

$$(200 \times 12)^2 = (97.0)(264 - P)(705 - P)$$

The two roots of this quadratic equation are

$$P_{cr} = 156 \text{ kips} \qquad \text{and} \qquad P_{cr} = 813 \text{ kips}$$

Therefore, the beam column will first buckle in the lateral-torsional mode at the lower value, $P_{cr} = 156$ kips.

Example 7.18 Determine the ultimate-strength end bending moments M_0 for the beam column shown in fig. 7.13. It is to be assumed that bending is about the strong axis of the cross section, and deformations occur only in the plane of that loading. The ultimate-strength interaction curves of fig. 6.35 are to be used to obtain the solution.

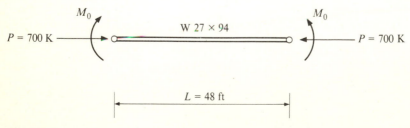

Figure 7.13

The geometrical and material properties for the shape in question are given as follows:

$$r_x = 10.9 \text{ in} \qquad \sigma_y = 50 \text{ ksi}$$

$$A = 27.7 \text{ in}^2 \qquad E = 30 \times 10^3 \text{ ksi}$$

$$S_x = 243 \text{ in}^3 \qquad \sigma_r^* = 0.3\sigma_y$$

The strong axis slenderness ratio for the member is

$$\frac{L}{r_x} = \frac{48 \times 12}{10.9} = 52.8$$

But fig. 6.35 was computed on the basis of an assumed yield stress value of 33 ksi. Therefore, the modified slenderness ratio for use in fig. 6.35 is

$$\left(\frac{L}{r_x}\right)_{50} = \left(\frac{L}{r_x}\right)\sqrt{\frac{50}{33}} = 65 \tag{7.29}$$

The yield value under pure axial thrust is

$$P_y = A\sigma_y = (27.7)(50) = 1385 \text{ kips}$$

or

$$\frac{P}{P_y} = \frac{700}{1385} = 0.505$$

The appropriate cross-curve (for $P/P_y = 0.505$) is shown in fig. 7.14, with the solution for $(L/r_x)_{50} = 65$ indicated. Therefore $M_0/M_y = 0.28$, or $M_0 = 0.28 S_x \sigma_y = (0.28)(243)(50) = 3400 \text{ in} \cdot \text{kips}$.

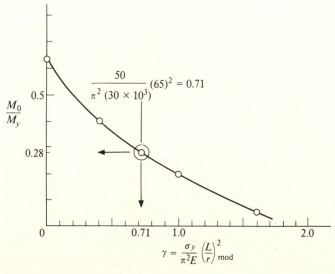

Figure 7.14 Ultimate-strength solution.

Had the bending moment been applied only at one end of the member, fig. 6.36 would have been used, and the solution would have been

$$\frac{M_0}{M_y} = 0.54 \qquad \text{or} \qquad M_0 = 6560 \text{ in} \cdot \text{kips}$$

It is to be noted that without figs. 6.35 and 6.36, it would have been necessary to determine by trial and error, using numerical methods for integration, the maximum strength values—a most time-consuming and tedious process.

Example 7.19 A W 18×64 beam column of 400-in unsupported length is subjected to an applied bending moment of M_0 at one end of the member and $\frac{1}{2}M_0$ at the other end—these primary moments being so directed as to produce single curvature deformations. The axial thrust acting on the member is 100 kips. If $\sigma_r^* = 0.5\sigma_y$ and $\sigma_y = 36$ ksi, determine the maximum allowable value of M_0 using interaction equation (6.149). $E = 30,000$ ksi, $G = 11,540$ ksi.

For the W 18×64 member:

$$A = 18.9 \text{ in}^2 \qquad r_x = 7.46 \text{ in}$$

$$I_x = 1050 \text{ in}^4 \qquad r_y = 2.00 \text{ in}$$

$$S_x = 118 \text{ in}^3 \qquad \kappa_T = 2.41 \text{ in}^4$$

$$I_y = 75.8 \text{ in}^4 \qquad I_\omega = 5600 \text{ in}^6$$

$$S_y = 17.4 \text{ in}^3$$

Since $L/r_y = 200$, the member will most likely buckle in the elastic range of behavior. (The reader should verify this assumption using the final result that will be obtained.)

Interaction equation (6.149) is

$$\frac{P}{P_u} + \frac{M_0}{M_u(1/C_m)} \frac{1}{1 - P/P_{ex}} = 1.0 \qquad (7.30)$$

where C_m is a coefficient introduced to take into account the loading and support conditions of a beam column when it is part of a structure.

In this equation

$$P = 100 \text{ kips}$$

$$P_u = P_{ey} = \frac{\pi^2 E I_y}{L^2} = 140 \text{ kips}$$

$$P_{ex} = \frac{\pi^2 E I_x}{L^2} = 1943 \text{ kips}$$

$$M_u = (C_1)\frac{\pi}{L}\sqrt{EI_y \, G\kappa_T + \frac{\pi^2}{L^2} EI_y \, EI_\omega} = (2314 \text{ in} \cdot \text{kips})(C_1)$$

where C_1 = moment modification factor for lateral buckling cases other than uniform moment.

From table 6.6,

$$C_1 = 1.75 - 1.05(0.5) + 0.3(0.5)^2 = 1.30$$

Substituting these values into the interaction equation yields

$$M_0 = 815.2 \text{ in·kips}$$

It here should be noted that if the member had been laterally supported along its entire length, the same interaction equation could have been used. P_u and $C_1 M_u$, however, would have had to be modified. For the in-plane bending case,

$$P_u = P_{ex} = \frac{\pi^2 E I_x}{L^2}$$

and

$$C_1 M_u = \text{maximum in-plane bending moment} = M_y \text{ or } M_p$$

Because of the generality of application, the interaction equation demonstrated in this example is the one most often used in present-day engineering practice.

7.2.5 Design of the Components of a Rigid Frame

The pin-base, rigid portal frame shown in fig. 7.15 is subjected to two different kinds of loading. Along the uppermost rafters it is subjected to a uniformly distributed, vertical load of q lb/ft. At location D there is applied a concentrated horizontal force of Q acting either to the right or to the left.

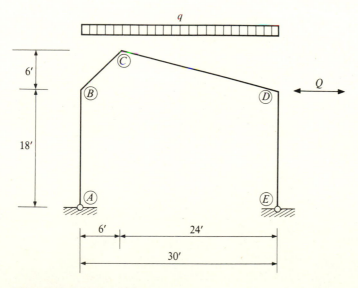

Figure 7.15 Typical rigid frame.

The structure in question is one time indeterminate, the same as was the two-span continuous beam case in example 7.13. Since it is the purpose of this chapter to develop procedures for the selection of individual members and not for analyzing indeterminate structures, it will be assumed that methods (or tables) are available for the determination of the magnitudes of the individual reactions. (As noted, chap. 8 will deal exclusively with methods for solving such problems.)

The loading and material properties for the structure in question are as follows:

$$\sigma_y = 60 \text{ ksi} \qquad\qquad q = 1500 \text{ lb/ft}$$

$$E = 29 \times 10^3 \text{ ksi} \qquad Q = 10{,}000 \text{ lb}$$

$$G = 11.2 \times 10^3 \text{ ksi}$$

Buckling or initial yielding are specified as the design criteria; that is, the loads already have been increased by the desired margin of safety.

Since all four of the straight members of the frame are beam columns, the interaction equation (6.149) will be used to determine the adequacy of a given design:

$$\frac{P}{P_u} + \frac{C_m M}{M_u(1 - P/P_{ex})} \leq 1.0 \qquad\qquad (7.31)$$

where P and M are due to the applied loads, and P_u and M_u are, respectively, the maximum axial force and the maximum bending moment that can be sustained by the selected member. $1 - P/P_{ex}$ is the moment amplification factor due to axial thrust times the deflection in the plane of bending. C_m is a factor to take into account other possible adjustments of the allowable load-carrying capacity when the individual members are considered to be integral parts of structures. It is to be understood that P_u and M_u are dependent upon both the end conditions (effective length factors) and the lateral support conditions along the member in question.

Two distinct and separate loading cases must be examined—Q acting to the left, and Q acting to the right. The member selected must satisfy both conditions.

As noted earlier, indeterminate structural design is basically an iterative process: Relative member sizes (or at least relative bending stiffnesses) must be assumed before any solutions can be formed. For example, assume that all of the members in the frame are of the same cross-sectional shape and size and that Q acts to the right. Using the methods of chap. 8, the horizontal and vertical components of force as well as end bending moments at the connection will be those indicated in fig. 7.16.

Further, assume that the cross section chosen for examination is the W 10×45, whose properties are as follows:

$$I_x = 249 \text{ in}^4 \qquad A = 13.2 \text{ in}^2 \qquad r_0^2 = 22.9 \text{ in}^2$$

$$I_y = 53.2 \text{ in}^4 \qquad S_x = 49.1 \text{ in}^3 \qquad E = 29{,}000 \text{ ksi}$$

$$I_\omega = 1200 \text{ in}^6 \qquad r_x = 4.33 \text{ in} \qquad G = 11{,}154 \text{ ksi}$$

$$\kappa_T = 1.5 \text{ in}^4 \qquad r_y = 2.0 \text{ in} \qquad \sigma_y = 60 \text{ ksi}$$

Column AB will be checked for adequacy.

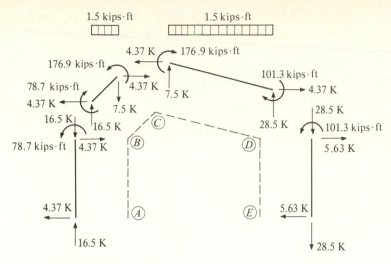

Figure 7.16 Components of loading on each of the members of the frame.

First, assume that the frame is supported laterally by the walls and roof. Deformations of the various members then will be constrained to occur only in the plane of the frame. For this case

$$P_u = P_{ex} \quad \text{and} \quad M_u = M_y = S_x \sigma_y$$

To obtain the buckling load about the x-x axis of the cross section if the member in question (column AB) is subjected to axial force alone, there must first be estimated the effective length factor K_x. Since the top of the column (point B) will deflect horizontally under applied loads and the base is defined as pin-connected, the effective length factor will be between 1.8 and 2.2—and closer to the 1.8 value. Selecting $K_x = 2.0$,

$$P_u = \frac{\pi^2 E I_x}{(K_x L)^2} = \frac{\pi^2 (29{,}000)(249)}{(2 \times 18 \times 12)^2} = 382 \text{ kips}$$

Since the effective slenderness ratio is

$$\frac{2 \times 18 \times 12}{4.33} = 100$$

and the limiting slenderness value for inelastic action is

$$\sqrt{\frac{2\pi^2 E}{\sigma_y}} = 98$$

the elastic buckling load is the maximum allowable for the member in question. The maximum permissible bending moment M_u is given by

$$M_u = S_x \sigma_y = 49.1 \times 60 = 2946 \text{ in} \cdot \text{kips}$$

As noted earlier, C_m depends upon a number of factors—primarily the loading and support conditions of the beam column. In present-day practice† it is assumed that a given member can be described by one of the following:

1. A beam column that is subjected to end moments, and for which joint translation is possible:

$$C_m = 0.85 \qquad (7.32)$$

2. A beam column that is subjected to end moments, and for which joint translation is prevented:

$$C_m = 0.6 - 0.4 \frac{M_1}{M_2} \geq 0.4 \qquad (7.33)$$

3. A beam column subjected to transverse loading:

$$C_m = 1 + \psi \left(\frac{\sigma_{appl}}{\sigma_{all, x}} \right) \qquad (7.34)$$

where

$$\psi = \frac{\pi^2 \delta_0 E I_x}{M_0 L^2} - 1 \qquad (7.35)$$

δ_0 = maximum deflection due to transverse loading
M_0 = maximum bending moment between the supports due to transverse loading

For the member in question, joint B can deflect horizontally. Therefore

$$C_m = 0.85$$

The interaction equation, then, becomes

$$\frac{16.5}{382} + \frac{(0.85)(78.7)(12)}{(2946)(1 - 16.5/382)} = 0.04 + 0.29 = 0.33 < 1.0$$

The member is more than adequate to sustain the imposed loading. (Before selecting another, lighter cross section for examination, however, it should first be determined whether or not a similar reserve in strength is available for the other members of the frame. The assumption was made in the frame analysis that all members have the same unit bending stiffnesses, EI_x.)

Had the frame been supported laterally at locations A, B, D, and E, and continuously along members BC and CD (because of the roofing), beam column AB must be reexamined based upon different allowable values. In the interaction equation, the following values would remain the same:

$$C_m = 0.85 \qquad P = 16.5 \text{ kips} \qquad P_{ex} = 382 \text{ kips} \qquad M = 78.7 \text{ ft·kips}$$

All other values must be reconsidered.

---

† See, for example, the American Institute of Steel Construction, Commentary on Specification, contained in the Appendix to the Specification.

For P_u, besides P_{ex}, there also must be determined P_{ey} and P_{ez}. Since it is specified that joint B is supported laterally,

$$K_y = K_z = 1.0$$

Thus,

$$P_{ey} = \frac{\pi^2 EI_y}{(K_y L)^2} = 326 \text{ kips}$$

$$P_{ez} = \frac{1}{r_0^2} \left[\frac{\pi^2}{(K_z L)^2} EI_\omega + G\kappa_T \right] = 1052 \text{ kips}$$

Weak-axis buckling, therefore, governs and

$$P_u = 326 \text{ kips}$$

(It is to be noted that $K_y L/r_y = 108$. The member buckles in the elastic range.)

For M_u it is necessary to determine the lateral buckling moment for a member subjected to an end bending moment at one end only.

$$M_{cr} = C_1 \frac{\pi}{L} \sqrt{EI_y G\kappa_T + \frac{\pi^2}{L^2} EI_y EI_\omega}$$

$$= 1.75 \times 2804 = 4907 \text{ in} \cdot \text{kips}$$

This value is greater than the initial yield bending moment. Therefore

$$M_u = M_y = 2946 \text{ in} \cdot \text{kips}$$

The interaction equation is, therefore,

$$\frac{16.5}{326} + \frac{(0.85)(78.7)(12)}{2946(1 - 16.5/382)} = 0.05 + 0.29 = 0.34 < 1.0$$

The W 10×45 section is still considerably more than adequate—even if the column is not laterally supported along its length.

The moment magnification term $1 - P/P_{ex}$ is the same for both cases. P_u, however, is different. It may equal P_{ex}, P_{ey}, or P_{ez}, depending upon the particular cross section, the end restraints, and the lateral support. The buckling term in the denominator of the moment magnification term is always the buckling load in the plane of the applied moments.

7.2.6 Steel Connections

Connections in steel structures are treated as if they are separate design components—whether the main members are directly attached one to the other, or whether they are attached to some intermediate (relatively short) element. In all cases the purposes of the connections are to transmit the forces and moments to the adjoining member(s), and to provide the desired degree of stiffness at the points of junction.

As noted in chap. 1, slender beams and columns are considered to be one-dimensional structural components. The various methods of analysis and design that have been discussed in this text are for those types of members. Rigorously speaking, "connections" consist mostly of two-dimensional components (mainly plates) subjected to complicated combinations of bending, axial, and shearing stress. While the same general principles of elasticity and plasticity hold for these types of elements, and—at least conceptually—there are available mathematical methods for solving problems of this sort, once they have been adequately defined, purely theoretical solutions to connections problems are always difficult, and frequently nearly impossible, to obtain. For these and other reasons, the theoretical basis for design criteria of connections has not been fully developed. Many of the present-day guides were formulated on the basis of experimental evidence, experience, and the use of very simple, conservative analytical models.

Design criteria currently are based upon either working stresses or ultimate strength—depending upon the governing specification involved, the characteristic of the loading, the type of connection, etc. In this section, connection design will be based upon the American Institute of Steel Construction Design Specification.

Steel connections generally are classified into three categories: fully rigid connections, simple connections, and semirigid connections. These are referred to by the AISC as type 1, type 2, and type 3 connections, respectively. They are characterized as follows:

AISC type 1 connection. Rigid framing, where full continuity is provided at the connections so that the original angles between the intersecting members are held virtually constant, that is, with rotational restraint on the order of 90 percent or more of that necessary to prevent any angle change. Such connections are used for both working-stress and ultimate-strength design methods.

AISC type 2 connection. Simple framing, where it is assumed that, insofar as gravity loading is concerned, the ends of beams and girders are connected for shear only, and are free to rotate under gravity loading. The framing may be considered simple if the original angle between intersecting pieces may change up to 80 percent of the amount it would theoretically change if frictionless hinged connections could be used.

AISC type 3 connection. Semirigid framing assumes that the connections of beams and girders possess a dependable and known moment capacity intermediate between the rigidity of type 1 and the flexibility of type 2. Semirigid connections are not used in ultimate-strength design and only rarely in working-stress design. This is primarily because of the difficulty involved in evaluating the degree of restraint. Types 1 and 2 connections are the ones most frequently used in steel construction.

Simple connections All of the following connections permit enough end rotation so that the connected beams may be assumed to be simply supported, and suitable for type 2 construction.

Lap or butt connections (*see fig. 7.17*) When two members to be joined are essentially in the same plane, and the loading is tension, these are the simplest types of connec-

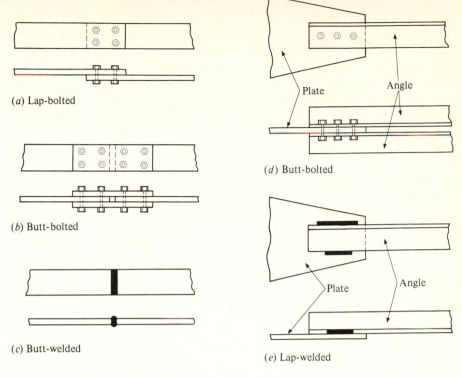

(a) Lap-bolted

(b) Butt-bolted

(c) Butt-welded

(d) Butt-bolted

(e) Lap-welded

Figure 7.17 Lap and butt connections.

tions to use. They can be arranged so that they will or will not require additional pieces of metal to complete the connection. Two basic types of configurations are observed: butt joints and lap joints.

Framing angles (*see fig. 7.18*) Structural members framing perpendicular to one another, usually require the use of angles to effect the connection. Bolts or welds may be used as the fasteners. Once the members are positioned, it is customary to temporarily hold them in place with drift pins, bolts, or tack welds until the permanent fastening can be accomplished.

   The number of high-strength bolts needed as fasteners for this type of connection is determined based on direct shear, neglecting any eccentricity of loading. For welding or bolting with common bolts, numbers, lengths, and sizes are determined considering the eccentric loadings. The fasteners are designed in accordance with procedures which will be outlined later in this section.

Beam seats (*see fig. 7.19*) Beam seats provide a ledge or shelf on which the beam rests while the permanent connection is made. The clip angle at the top is assumed to carry no load, and only provides lateral support. Flexible beam seats are the simplest

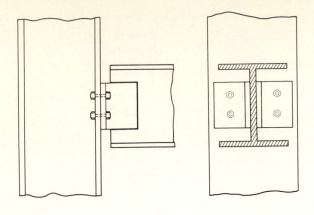

(*a*) Beam-to-column connection

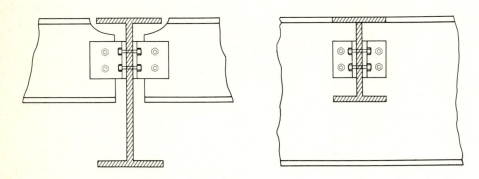

(*b*) Beam-to-beam connection

Figure 7.18 Framing-angle connections.

and most desirable. Because the thickness of the seat angle provides the primary resistance against bending, the outstanding leg must be stiffened when the load becomes too large.

As with the case of framing-angle connections, the seat connections are intended to transfer only the vertical component of the reaction. No significant restraining moment should be applied to the beam.

The thickness of the seat angle is determined by the flexural stress acting on the critical section(s) of the angle. These are indicated in fig. 7.19*b*. As a practical matter, rarely will the beam be left unattached to the seat—and design procedures presume a critical section $\frac{3}{8}$ inch from the face of the angle.

The bending moment on the critical section of the angle and on the connection to the column flange is determined by taking the beam-end reaction times its distance from the section in question. The beam reaction is assumed to occur at the centroid of the bearing stress distribution (see fig. 7.20). The required bearing length N

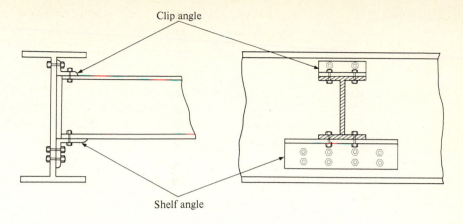

(*a*) Flexible beam seats (beam-to-girder connection)

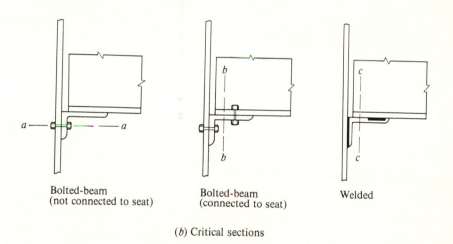

Bolted-beam
(not connected to seat)

Bolted-beam
(connected to seat)

Welded

(*b*) Critical sections

Figure 7.19 Flexible-seat connections.

(AISC—1.10.10.1), measured from the end of the beam, may be computed from

$$N = \frac{Q}{(0.75 \times \sigma_y)t_w} - k \geq k$$

where t_w = thickness of the web of the beam

k = distance from the outer face of the flange to the web toe of the fillet

Generally, $N \geq 3$ inches, for a standard 4-in seat width.

$$e_f = a + \tfrac{1}{2}N$$

or

$$e = e_f - t - \tfrac{3}{8}$$

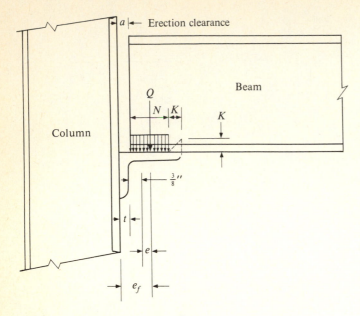

Figure 7.20 AISC support assumptions.

At the critical section, the bending stress would be

$$\frac{M}{S} = \frac{6 \times Q \times e}{(bt^2)} \leq \sigma_{\text{all}}$$

where the allowable stress most commonly used is $0.75\sigma_y$. Based upon this, the required thickness of the seat angle is

$$t = \sqrt{\frac{8 \times Q \times e}{b \times \sigma_y}}$$

When reactions are heavier than that capable of being supported by unstiffened seats, stiffeners must be added. Bolted angle stiffeners and welded T-shape stiffeners are shown in fig. 7.21. The assumed bearing stresses on stiffened seats are shown in fig. 7.22. As before, the seat width is controlled by the bearing length N. The thickness of the stiffener is determined based upon the maximum allowable normal stress due to the combined action of compression and bending $Q/A + (Q \times e)/S$.

Direct web connections (*see fig. 7.23*) Another common type of simple connection in steel construction is the direct web connection. This is often seen in cases where the beam web is directly secured to the column flange. An example of one such spandrel connection is shown in fig. 7.23. The erection angle is used only to facilitate construction. It is assumed to carry no load. Again, the fasteners are designed only to transmit shear.

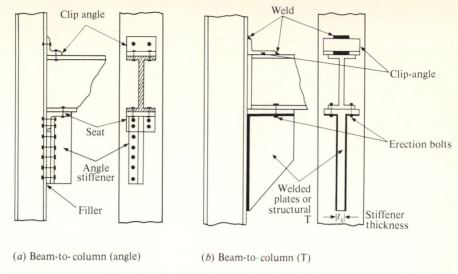

Clip angle

Weld

Clip-angle

Seat

Angle stiffener

Filler

Erection bolts

Welded plates or structural T

Stiffener thickness

t_s

(*a*) Beam-to-column (angle)

(*b*) Beam-to-column (T)

Figure 7.21 Stiffened beam seats.

Bracketed connections (*see fig. 7.24*) When the two members to be connected do not intersect, a bracketed type of connection may be required. One common type of such connection is the T seat with a triangular stiffener.

Fully rigid connections Fully rigid connections are designed to effect complete transfer of moment, shear, and thrust with little or no relative deformation within the joint.

Rigid framing may be used to its greatest advantage in both *plastic design* and in working-stress design, when the members to be connected are *compact*. In this section only the design procedures for working stress situations will be described.

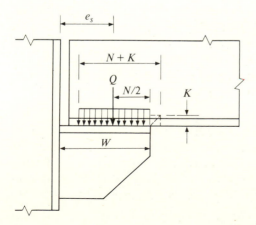

e_s

$N + K$

Q

$N/2$

K

W

Figure 7.22 Assumed bearing stress on stiffened seats.

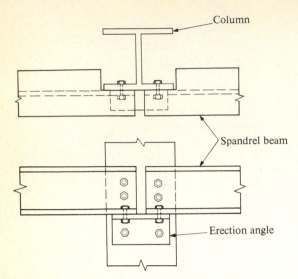

Column

Spandrel beam

Erection angle

Figure 7.23 Spandrel connection.

Beam-to-column connections For continuous beam-to-column connections, it is common practice to design the various elements so that the flanges resist the full bending moment, and the web resists the shear force. Typical connections of this type are shown in fig. 7.25*a* to *e*.

The most common of the connections in question is shown in fig. 7.25*a*. The flanges of the two members are attached by using a direct flange-to-flange groove weld. (Normally this would be the full thickness of the smaller connected element.) The web of the beam is attached to the column by a fillet weld. The sizes of the various welds are determined from both shear and tensile stress considerations. For this type of connection, local buckling of the web as well as crippling near the compression flange of the beam should be checked.

When the forces to be transmitted are large and it is undesirable to change the

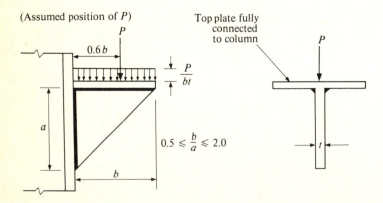

(Assumed position of P)

P

$0.6b$

$\dfrac{P}{bt}$

a

b

$0.5 \leqslant \dfrac{b}{a} \leqslant 2.0$

Top plate fully connected to column

P

t

Figure 7.24 Triangular bracket plate.

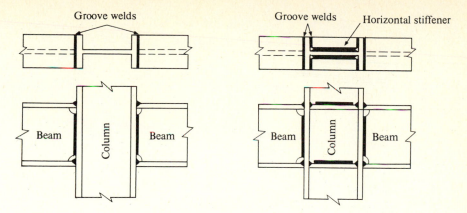

Figure 7.25a Direct flange-to-flange connections.

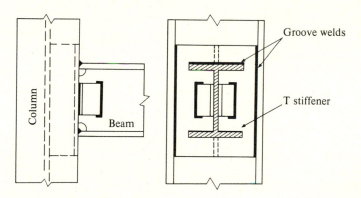

Figure 7.25b T-section stiffener connection.

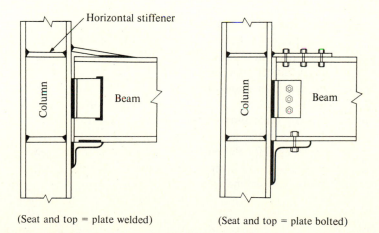

(Seat and top = plate welded)   (Seat and top = plate bolted)

Figure 7.25c Top-plate attachment.

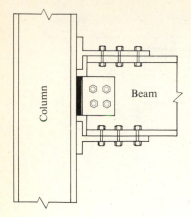

Figure 7.25d Split-T top and bottom plates.

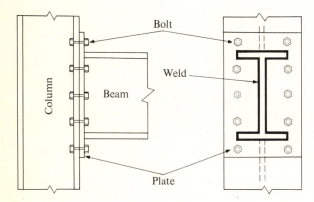

Figure 7.25e Welded end plates.

depth of the beam section, stiffeners may be required. These are shown in the second of the two examples of fig. 7.25a.

To facilitate beam-to-column connections when the beam is perpendicular to the plane of the web of the column, a separate attachment plate often is placed between the column flanges along its outside edge. For beams with large end moments, T stiffeners of the type shown in fig. 7.25b may be required. These types of attachment also are useful in four-way beam connection systems.†

Another common type of connection is shown in fig. 7.25c. Here top plates are used to transmit the tensile forces in the beam flange, and seat angles or brackets are used at the bottom. Split T's for the same purpose are illustrated in fig. 7.25d.

Connections via end plates, illustrated in fig. 7.25e, also are used. It is to be recognized, however, that the bolts used for these types of connections are subjected to combined shear and normal stresses, and both of these conditions must be taken into account in selecting the particular fastener size and spacing.

† See Omer W. Blodgett and James F. Lincoln, *Design of Welded Structures*. Arc Welding Foundation, Cleveland, 1966.

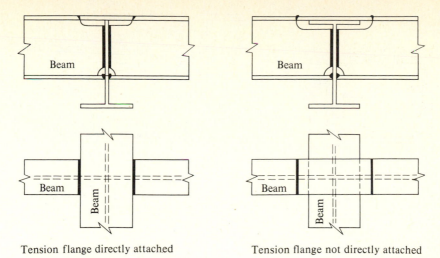

Tension flange directly attached   Tension flange not directly attached

Figure 7.26 Intersecting beam connections.

Beam-to-beam connections Rigid beam-to-beam connections presume full transfer of moments and shearing forces. Examples of these types of connections are shown in fig. 7.26. In the first case, the top flanges are rigidly attached one to the other using groove welds. In the second, a tension plate is employed to transmit the flange force. When tension flanges are rigidly attached, there is created in the element a biaxial state of stress, which increases the possibility of brittle fracture. Hence, these types of connections are less desirable for dynamic loading.

Rigid-frame knees When two members are joined and those members have their webs lying in the same plane, the junction is called a *knee joint*. Typical knee joints are shown in fig. 7.27.

Knee connections transfer the end moment, end shear, and end thrust from the beam to the column, and vice versa. Moreover, the knee must deform in a manner consistent with the overall analysis of the structure in question. In most cases, it can be assumed that the span length of a given member extends from center to center of adjacent knees, and that the moment of inertia varies in accordance with the properties of the cross section taken at right angles to a line extending from the center of the knee at one end to that at the other end. Bending moments and shears are determined using appropriate methods of structural analysis, taking into account the variable moment of inertia in the knees.

In ultimate-strength design (so-called *plastic design*), knees must be capable of large inelastic rotations—to satisfy rotation capacity requirements of plastic hinges.

Beam splices There are a number of reasons why a rolled beam or plate girders may be spliced: the full length may not be available from the rolling mill or the supplier; the fabricator may find it more economical to use a splice, rather than transport the entire length in one piece; the designer may desire a change in section to fit the strength requirement along the length; etc.

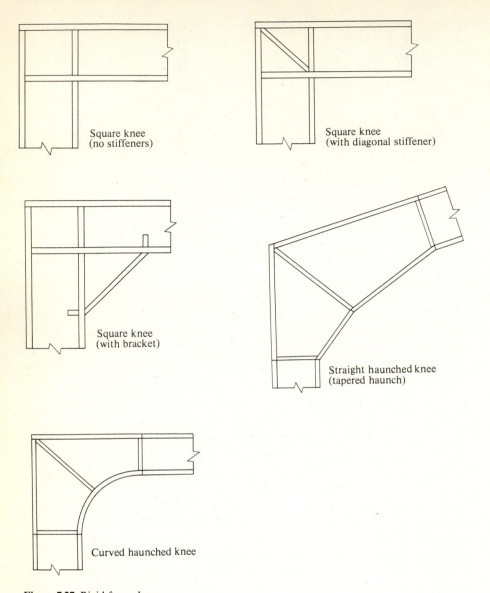

Square knee
(no stiffeners)

Square knee
(with diagonal stiffener)

Square knee
(with bracket)

Straight haunched knee
(tapered haunch)

Curved haunched knee

Figure 7.27 Rigid frame knees.

Splices must be proportioned so that they develop the desired moment and shear capacity on each side of the joint—as well as at the joint, itself. While a number of different types of fasteners have been used to accomplish this, most splices are made now by welding. Typical field-splice examples are shown in fig. 7.28.

Column bases The design of column base plates involves two main considerations. First, the compressive forces in the columns are transmitted to the supporting mediums via these plates. Their size and proportions, therefore, must be such that the

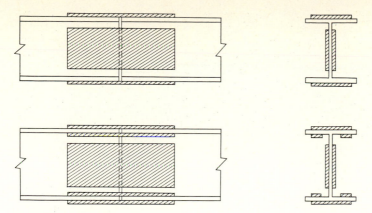

Figure 7.28 Beam splices.

resulting bearing stresses in the medium are within permissible limits. (These allow-able values are given in most design codes and specifications.) In addition, the connections themselves must be sufficient to ensure the desired constraint. For more refined frame analyses it may be necessary to evaluate the moment-rotation charac-teristics of the entire anchorage system—including the base plate, anchor bolts, and concrete footing—to determine its bending stiffness and strength. Fortunately, such an analysis is seldom required.

Structural fasteners An integral part of any connection is, of course, the structural fastener. As noted earlier, metal members may be fastened together with rivets, bolts, or welding. For many years rivets were the accepted standard means of connecting members. In recent years, however, high-strength bolts and welding have become the norm. For completeness, then, a brief discussion of the design considerations for these types of fasteners will be given here.

High-strength bolts In general, high-strength bolts used in structural connections are sized and spaced to prevent failure in tension, shear, or combined tension and shear. The actual stress in any given bolt depends upon the type of connection being considered and the load carried. Types of connections which produce pure tension, pure shear, and combined tension and shear are shown in fig. 7.29.

If the fasteners are determined to be in a state of pure shear, the following is an outline of design considerations for the connection:

1. An adequate net cross-sectional area must be provided to carry the tensile load.
2. An adequate gross area must be available to carry the compressive load.
3. An adequate number of bolts must be selected so that the allowable nominal shearing stress in the connectors is not exceeded.
4. An adequate number of bolts must be used or there must be selected a plate of adequate thickness so that the allowable bearing stress is not exceeded—for bearing-type joints.

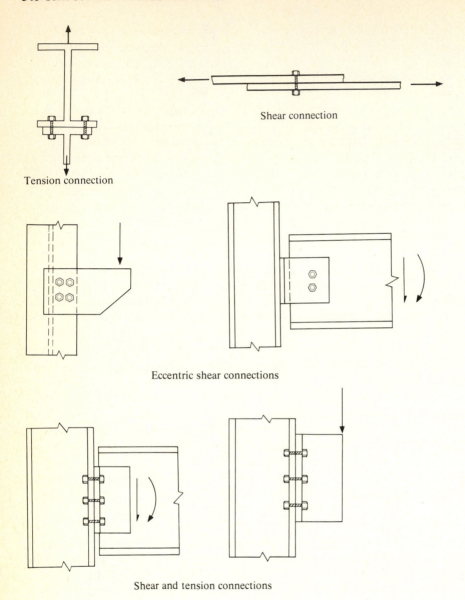

Figure 7.29 Bolted connections.

5. An adequate edge distance must be provided so that the connectors cannot "shear out."† (When not more than two connectors are used in a line parallel to the direction of stress, special provisions may be required by Design Codes.)
6. Maximum edge distances must not be exceeded, so that "dishing"‡ will not occur.

---

† "Shear out" is failure in the plate between the connector hole and the end of the plate by shearing in the direction of the load or by transverse tension.

‡ "Dishing" is buckling of the plate between the connector and the edge of the plate.

7. There must be a reasonable spacing between connectors, measured center-to-center, so that tearing of the plates cannot occur, and so that wrenches easily can be applied for bolt tightening. (The current AISC Specification—section 1.16.4—states that the maximum spacing is $2\frac{2}{3}$ times the bolt diameter, but preferably 3 bolt diameters.)
8. Long lines of connectors should be avoided; use compact joints if at all feasible. (Compact joints result when connector spacing both transverse and parallel to the direction of stress is approximately five times the diameter of the bolts, and the length of the joint does not exceed about five pitches in the direction of stress.)
9. Long bolt "grips" should be avoided to minimize bending of the fasteners—with resulting nonuniform stress. Furthermore, the minimum pretensioning force causes greater elongation and a reduction in the clamping force, when long grips are used. (The grip, or total thickness of pieces fastened by the bolt, should not exceed five diameters.)

In a large number of commonly used connections, both shear and tension occur simultaneously and must be considered in the design. For example, if it is determined that both a large moment and a large shearing force are to be transmitted, then the interaction between the two will limit the effective amount of nominal shearing stress and nominal tensile stress that can be applied to a given bolt.

Welds The versatility and ease of welding make these types of connections desirable for a wide range of usages.

Welded connections often are made up of groups of simple welds, each initially selected to resist a given component of force. When the whole is examined, a particular location will in all likelihood, therefore, be subjected to more than one type of stress. Examples of various types of welded connections and loadings are shown in fig. 7.30.

In general, welds are subjected to the same basic types of stresses and loadings that were indicated in the preceding discussion on bolted connections. The actual stress distribution in welds, however, is very complex, and an exact analysis is exceedingly complicated. Fortunately, experiments have shown that for reasonably proportioned joints, adequate results can be obtained assuming that the plates being fastened are rigid and that only the welds deform. This allows the consideration of nominal stresses for design purposes—the same as was the case for bolts.

If a welded joint is subjected to pure shear, pure compression, or pure tension, it can be assumed that stresses are uniform over the entire length of the welds. If a weld is subjected to pure bending or pure torsion, it can be assumed that the stresses vary linearly depending upon the distance from the center of rotation. If two or more of these simple cases occur simultaneously, the stress at a given point along the weld can be assumed to be the vector sum of the elementary cases.

For welded connections, the following procedure would be used to ensure adequacy of the joint:

1. Establish the effective throat dimension t_e of the various welds, and determine the cross section of the weld group.
2. Using a reference coordinate system, determine the centroid of the weld group.

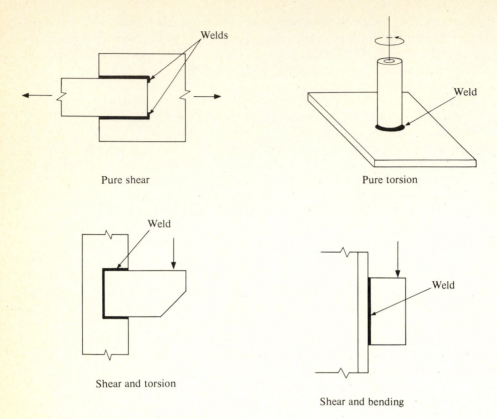

Figure 7.30 Types of loading on welds.

3. Determine the forces and moments that must be resisted—referenced to that centroid.
4. At critical points along the welds, compute the individual stresses resulting from direct shear, torsion, and moment.
5. Combine these individual (component) stresses vectorially.
6. The nominal combined stresses at these critical points should not exceed the allowable nominal stress given in the controlling specification.

7.2.7 References

Because steel connections and fasteners have been extensively researched, and there exists a considerable body of experience having to do with these types of structural elements, the reader is referred to the following published works for more detailed information:

1. American Institute of Steel Construction, *Manual of Steel Construction*, 7th ed., New York, 1973.
2. Charles G. Salmon and Bruce G. Johnston, *Steel Structures Design and Behavior*, pp. 84–226, 689–776, Intext, New York, 1971.
3. Omer W. Blodgett, *Design of Welded Structures*, Lincoln Arc Welding Foundation, Cleveland, 1966.

4. Charles G. Salmon, Analysis of Triangular Bracket-Type Plates, *J. Eng. Mech. Div., Am. Soc. Civ. Eng.,* vol. 88, no. EM 6, pp. 41–87, December 1962.
5. Charles G. Salmon, D. R. Buettner, and T. C. O'Sheriden, Laboratory Investigation of Unstiffened Triangular Bracket Plates, *J. Struct. Div., Am. Soc. Civ. Eng.,* vol. 90, no. ST 2, pp. 257–278, April 1964.
6. Lynn S. Beedle et al., *Structural Steel Design,* pp. 550–555, Ronald, New York, 1964.
7. Welding Research Council and the American Society of Civil Engineers, Commentary on Plastic Design in Steel, *ASCE Man. Eng. Practice,* no. 41, 1961.
8. R. T. Douty and W. McGuire, High Strength Bolted Moment Connections, *J. Struct. Div., Am. Soc. Civ. Eng.,* vol. 91, no. ST 2, pp. 101–128, April 1965.
9. American Welding Society, *Welding Handbook,* 7th ed., vol. 1, *Fundamentals of Welding,* New York, 1976.

7.3 DESIGN OF REINFORCED CONCRETE MEMBERS

Unlike metal structures, poured-in-place concrete structures—either plain or reinforced—are quite literally manufactured on the site of the work. In many cases construction is with local materials. It also should be recognized that the relative fluidity of fresh concrete makes possible the creation of members of almost any desired shape. Since concrete has good fire-resistant properties, concrete-covered reinforcing rods are both rustproof and fireproof. But possibly the most important design advantage of reinforced concrete is the fact that it capitalizes on the best properties of two readily available, economical materials: the high compressive strength of concrete, and the high tensile strength (and ductility) of steel reinforcing bars.

On the other side of the ledger, freshly mixed concrete is in a plastic state that requires temporary containment in its ultimate desired form. The cost of these forms may represent a significant portion of the cost of the structure in question. But more than that, concrete takes time to solidify and develop its useful strength. The forms may have to stay in place for possibly up to two weeks while this occurs. It is to be recognized, too, that the quality of the raw materials used in concrete construction varies considerably. Because of the relative ease with which the ingredients can be mixed and placed, construction and inspection are not always performed by experienced individuals, and this personal element further contributes to a variation in behavior.

In spite of all this—and because of it—since the beginning of the twentieth century reinforced concrete has become a useful and important structural material.

Design of structural members in reinforced concrete has been discussed several times in the preceding portion of this text. Up to this point, however, only a part of the total design problem has been considered: the selection of a permissible cross-sectional geometry. In each case, the prime concern was for equilibrium between the externally imposed loads or moments (at the most highly stressed cross section) and the internal moment capacity of the selected member at that location.

Procedures for sizing members were illustrated in chaps. 2 and 4. In chap. 2, both singly and doubly reinforced beams were considered, as were T beams and composite members. Short columns also were discussed. In chap. 4, the beam-column problem was discussed, and design examples were given.

Design of reinforced concrete members involves much more than the selection of a cross-sectional geometry based upon the behavior at a particular location. Variation in maximum-strength requirements along the length can and should be taken into account. True, if a given member is subjected to a constant bending moment along its entire length, then the requirements for design are the same at all cross sections. But for most real cases, variations do occur, and these should be taken into account in the design.

Design codes for reinforced concrete Design codes are for the primary purpose of ensuring safe and serviceable structures. Their provisions are based upon structural theory, experimentation, and experience. If the material in question has properties that are relatively constant and well-defined, and if the range of possible application is well understood, then a significant part of the Code can be directly derived from strength of materials and structural theory. On the other hand, when material properties vary considerably, and when usage is constantly changing, experimentation and empirically derived formulas are relied upon much more heavily. Also, increasing the factor of safety becomes necessary. Codes relating to the usage of reinforced concrete fall largely in this latter category.

The Code developed by the American Concrete Institute, which previously was referred to in this text, is one of the more complete specifications available for concrete design. A companion Commentary document, which discusses the general thinking behind each of the provisions of the Code, as well as listing the basic research sources, also is available.

As noted in previous sections of this text, no real Code provisions, as such, will be reproduced here. Instead, typical code-type requirements will be developed or introduced as the need arises. These will be used in the design examples.

7.3.1 Cross-Section Details

While the general concept of cross-sectional selection was discussed in the earlier chapters, and several actual design examples were there carried out, consideration of particular material placement problems within the overall dimensions should be further examined before proceeding. Among those "other design considerations" are spacing of the reinforcing bars, maximum and minimum percentages of steel, choice of bar sizes, protection of the steel by concrete cover, and the effective moment of inertia of the cross section.

Bar Spacing Reinforcing bars can be placed too close together. Equally true, they can be too far apart. If they are too close, they tend to prevent the coarser parts of the aggregate from easily passing through or even from passing through at all. The result is a segregation of the concrete mix into parts having characteristics markedly different from that assumed. (It should be noted that the tendency toward separation is natural and ever present. The wetter the mix, the more the heavier parts tend to migrate downward. For this reason, excessive tamping, rodding, or vibration should be avoided.) Bars placed too far apart effectively leave large parts of the concrete unreinforced.

Any code provision relating to spacing of reinforcing bars should take into account the overall cross-sectional dimensions of the member in question, the size and number of reinforcing bars to be used, and the size of the largest aggregate. In this chapter, the following design requirements will be presumed:

The clear distance between parallel bars, both horizontally and vertically, shall not be less than the nominal bar size, or 1 in, whichever is the larger. Bars placed in layers shall be placed directly over one another. (C.1)

Shrinkage and temperature steel Concrete shrinks as it hardens. Therefore, concrete members contract as they age. Members also contract due to decreases in temperature. If the member in question is rigidly attached either at its ends or along its length by unyielding supports, then these contractions can result in cracks.

In typical one-dimensional structural members the steel reinforcement designed to resist bending is usually adequate to prevent cracks due to contraction. In two-dimensional members such as slabs and walls, however, such is not always the case, and additional steel may be required to prevent cracks at right angles to the supported sides.

To prevent (or control) cracks due to temperature or shrinkage, it is common practice to specify a certain minimum percentage of steel that must be contained within the element cross section in question.

The minimum positive flexural steel ratio shall not be less than that given by

$$p_{min} = \frac{A_s}{bd} = \frac{200}{\sigma_y} \tag{C.2}$$

When no flexural steel is required due to applied loads, sufficient reinforcement shall be provided so that

$$\frac{A_s}{bd} \geq 0.0018 \tag{C.3}$$

In no case shall reinforcing bars be placed farther apart than five times the slab thickness nor more than 18 in.

Bond All reinforced concrete design is based upon the assumption that the two materials act together; that there is perfect adhesion between the concrete and the steel reinforcement; that the concrete and the steel are bonded together.

Bond is due to both adhesion and sliding resistance. It is to be noted, however, that sliding resistance does not come into play until adhesive resistance is overcome. Bars rolled with lugs and projections—so-called *deformed bars*—have a considerably larger resistance to sliding than do *plain bars*. (Today, deformed bars are used almost exclusively.)

The selection of the proper bar size(s) is an important design consideration. Should several smaller bars—requiring relatively greater labor and material costs—be used to provide the required steel area or should there be selected fewer larger bars having the same total cross-sectional area? The allowable bond strength of the materials governs this decision.

The bond stress acting on a given set of bars in a reinforced concrete beam is given by the equation

$$u = \frac{vb}{\Sigma_0} \tag{7.36}$$

where v = shearing unit stress at the section in question
b = width of the beam
Σ_0 = sum of the perimeters of all of the longitudinal tensile reinforcing bars at the section

Allowable bond stresses are given in the various design codes, and depend upon the type of bar and its location in the beam.

Anchorage Reinforcing rods must extend into the concrete a sufficient distance to prevent their pulling out. In terms of the stress in the steel rod and the allowable bond stress between the concrete and the steel, this distance should be at least

$$L_d = \frac{\sigma_s D}{4u_{\text{all}}} \tag{7.37}$$

where L_d = length of embedment
σ_s = stress in the steel
D = diameter of bar in question
u_{all} = allowable unit bond stress

When there is insufficient length available for a straight section of reinforcement (for anchorage), hooks may be used. Hooks are complete semicircular (or 90°) turns of the reinforcing rods with extending straight sections beyond the turn.

Concrete cover The steel bars in reinforced concrete members need to be covered for a variety of reasons. Potentially, if exposed, these bars can corrode due to moisture or other corrosive agents. Steel also must be insulated from the damaging effects of fire.

The minimum amount of cover is dependent upon the particular application in question and the bar sizes involved. The following minimum cover provisions will be specified:

$$
\begin{array}{lll}
& \textit{Cast-in-place concrete} & \\
(a) & \text{Exposed to the earth} & 3 \text{ in} \\
& \text{Exposed to the weather} & 1\frac{1}{2} \text{ in} \\
(b) & \text{No. 5 bars, or smaller} & 1\frac{1}{2} \text{ in} \\
& \text{No. 6 bars, or larger} & 2 \text{ in}
\end{array} \tag{C.4}
$$

(Corresponding provisions would have to be given for other types of concrete.)

7.3.2 Basic Design Concepts

Bending moments, shears, or thrust vary along the length of most real structural members. For economical designs, then, cross-sectional properties should vary along the length. In reinforced concrete members, this can be accomplished either by changing the depth or width of the member or by changing the amount of steel that is contained along the length. Sometimes both of these possibilities are used. However, since forming costs are major costs in concrete construction, changing the overall dimensions along the length should be carefully considered from a cost-effectiveness point of view. More often, varying the steel ratio is the method used for changing properties.

Changing the number or size of bars from section to section along the length, however, introduces other considerations. For example, bending analysis assumes that the tensile forces in the steel bars are "developed" at the points of maximum stress. This, in turn, requires that there be provided additional *anchorage lengths* of the steel bars beyond these points. (This was discussed in the preceding section.) For a given situation, the design question arises as to whether a smaller number of larger bars—requiring longer development length—should be used, or there should be specified a larger number of smaller bars—with shorter lengths for anchorage. Ease of placement of materials and relative labor costs are not inconsequential considerations when making these decisions.

In chap. 2 it was shown that shear is the consequence of variation in bending moment along the member. Capacity to resist shear, then, is critical to the design of structural members. In reinforced concrete design, this becomes a significant factor, for as was earlier shown, shear has associated with it *diagonal tension* stresses, and concrete is exceedingly weak in tension. Therefore, in many cases it is necessary to provide additional secondary steel reinforcement rods to ensure against premature diagonal cracking. (Sloping bars placed at right angles to the direction of potential cracks would be one method of reinforcing to prevent diagonal tension failure. Generally, it is not the most economical method. The usual procedure is to add no. 3 and no. 4 bars, bent into the general shape of a U in the transverse direction at those locations along the beam where diagonal tensile stresses require their use. These bent reinforcing bars are called *stirrups*. Their design and usage will be discussed more in detail in later sections of this chapter.) The need for additional secondary steel reinforcement to prevent cracking due to diagonal tension also arises when three-dimensional frame members are subjected to twisting moments.

For members containing compressive as well as tensile steel—such as columns and doubly reinforced beams—the longitudinal bars must be placed and restrained in such ways that they are prevented from buckling or otherwise moving from their presumed design location. (Lateral ties or spiral, hooping-type reinforcement are used for this purpose.) In addition, a sufficient number of longitudinal bars must be called for so that the member has no apparent weak directions. (These ideas were illustrated briefly in fig. 2.39 and in example 4.16.)

7.3.3 Design of Compression Members

For compression member design, it will be required that the following steel reinforcement provisions be met:

The minimum and maximum percentages of longitudinal steel for compression members shall be 1 and 8 percent of the gross cross-sectional area, respectively. The minimum number of bars in a tied column shall be four, while the minimum number for a column with a circular cross-sectional arrangement of bars shall be six. (C.5)

Lateral ties, at least no. 3 in size, shall be provided to constrain bars no. 10 and smaller. No. 4 ties must be used for larger longitudinal bars. (Ties are to be spaced according to the provisions of example 4.16.) Ties are to be arranged so that every longitudinal bar is constrained within a 90° bend in the tie. (C.6)

Spirals no smaller in size than no. 3 shall be used to constrain longitudinal reinforcing bars when such bars are arranged in a circular pattern. The clear spacing between spirals shall be not less than 1 inch or $1\frac{1}{2}$ times the maximum size of the aggregate, nor no greater than 3 in. The actual spiral size and spacing shall be selected so that the spiral reinforcement ratio is not less than that given by

$$p_s = 0.45\left(\frac{A_g}{A_c} - 1\right)\frac{\sigma'_c}{\sigma_y} \tag{7.38}$$

where p_s = *the ratio of the volume of the spiral reinforcement to the core volume*
  A_c = *area of the concrete core, with the core diameter taken from the outside of the spiral steel*
  A_g = *gross area of the column cross section*
  σ'_c = *ultimate compressive strength of the concrete*
  σ_s = *allowable compressive unit stress of the longitudinal reinforcing steel*

(The volume of the spiral, per inch of column length, is equal to $p_s A_c$.)

Example 7.20 In example 2.14, the cross-sectional dimensions and reinforcing required to carry a specified loading were given for a round, spiral column. The cross section was shown in fig. 2.40. A no. 4 ($\frac{1}{2}$-in diameter) spiral was chosen for the lateral reinforcement. It is the purpose of this example to determine for the previously defined design (using the above-listed design provision C.7) the pitch of the spiral reinforcement.

Figure 7.31 is a cross-sectional view and a partial longitudinal view of the member in question. The core and gross areas for the dimensions given in fig. 2.40 are, respectively,

$$A_c = \frac{\pi}{4}d_c^2 = 283.53 \text{ in}^2$$

$$A_g = \frac{\pi}{4}d^2 = 380.13 \text{ in}^2$$

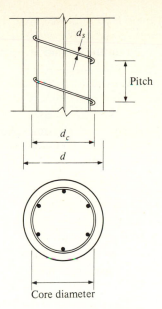

Figure 7.31 Circular concrete column.

Therefore, from (C.7),

$$p_s = (0.45)\left(\frac{380.13}{283.53} - 1\right)\left(\frac{3.5}{60}\right)$$

$$= 0.0089$$

Since a 1-in length of the core contains 283.53 in^3, there is required per inch of length of column a volume of spiral steel equal to

$$0.0089 A_c = (0.0089)(283.53) = 2.523 \text{ in}^3$$

A complete turn of the spiral ($18\frac{1}{2}$-in-diameter circle), however, gives a volume of 11.624 in^3. Therefore, the pitch required by the equation in design provision C.7 is

$$\text{Pitch} = \frac{11.624}{2.523} = 4.607 \text{ in}$$

This is greater than the 3-in maximum specified. The smaller value, therefore, governs.

Since the calculated pitch is considerably greater than the limiting value, a smaller diameter spiral—a no. 3—should be examined for adequacy. This exercise is left to the reader.

7.3.4 Design of Beams

Design of reinforced concrete beams usually is accomplished in three distinct and separate phases: (1) determination of the cross-sectional dimensions and reinforcement, (2) determination of the additional lengths of reinforcement required for anchorage, and (3) determination of the size and location of web reinforcement. The first of these phases was to some extent discussed in chaps. 2 and 4. Therefore, in the examples that will follow, primary attention will be focused on the second and third phases.

In reinforced concrete design, it must be understood that weight of the member itself is significant and must be taken into account. The general procedure is to first assume a probable cross-sectional size (based on prior experience, anticipated number of reinforcing bars, etc.) and compute its weight. Then, when the dimensions of the actual beam have been established, the two must be checked to ensure that adequate provision has been made. Because it is primarily dependent upon the number and placement of reinforcing bars, usual procedure is to first assume the width of a beam and then to compute its depth.

Example 7.21 The beam in question is 45 ft long. It is supported on two simple supports located 10 ft from each end as shown in fig. 7.32. An applied, uniformly distributed vertical live load of w kips/ft is given. Shown are shear and bending-moment diagrams.

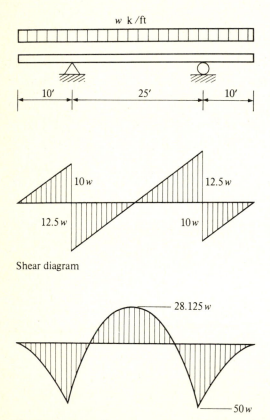

Shear diagram

Bending-moment diagram

Figure 7.32 Simple beam with overhangs.

The following values are specified:

$$w = 3.0 \text{ kips/ft}$$

$$\sigma'_c = 4000 \text{ lb/in}^2$$

$$\sigma_y = 60,000 \text{ lb/in}^2$$

For the given material properties, the minimum and maximum steel ratios for singly reinforced cross sections can be determined.

From (C.2)

$$p_{min} = \frac{200}{60,000} = 0.003 \tag{7.39}$$

From eq. (2.76a)

$$p_{max} = (0.75)\frac{(0.85)k_1\sigma'_c}{\sigma_y}\frac{87,000}{87,000 + \sigma_y} = 0.0214 \tag{7.40}$$

The maximum moment capacity for a rectangular cross-section beam can be calculated from eq. (2.79), which is here repeated as eq. (7.41).

$$M_{max} = bd^2\left[p\sigma_y\left(1 - 0.59p\frac{\sigma_y}{\sigma'_c}\right)\right] \tag{7.41}$$

where p = actual steel ratio = A_s/bd.

As an initial estimate, it will be assumed that $b = 12$ in, $d = 18$ in, and there will be an additional 2 in of concrete required for cover. (The overall depth of the beam is 20 in.) The total dead load of the beam is

$$w = 3.00 + (1)(1)(\tfrac{20}{12})(0.150) = 3.25 \text{ kips/ft}$$

Using the definition contained in example 2.16, the load factor to be used in the design is

$$LF = 1.7 + 1.4\frac{DL}{LL} = 1.82$$

The cross-sectional dimensions at the points of maximum moment would be determined as follows:

Maximum positive moment

$$M = (28.125w)(LF) = (28.125)(3.25)(1.82) = 167.1 \text{ kip} \cdot \text{ft}$$

For the given material properties and bending moment, the steel ratio p can be determined from eq. (7.41).

$$(167.1)(1000)(12) = (12)(18^2)\left\{p(60,000)\left[1 - (0.59)p\frac{60,000}{4000}\right]\right\}$$

or

$$p = 0.0095$$

Therefore,

$$A_s = (0.0095)(12)(18) = 2.05 \text{ in}^2$$

This steel area can be supplied by two no. 7 bars and one no. 9 bar, giving a total area of steel of 2.20 in². This amount of positive steel reinforcement, however, is required only at the point of maximum positive bending moment. Less is needed elsewhere.

Steel reinforcing bars should be arranged in the cross section symmetrically with respect to the vertical centroidal axis. If this is not done, the member will twist when loaded. With this and other minimum requirements relating to continuing steel in mind, it will be assumed that the no. 9 bar will be discontinued at some point along the center segment of the beam.

The bending moment capacity of a reinforced concrete beam ($b = 12$ in, $d = 18$ in) containing two no. 7 bars ($A_s = 1.20$ in², $p = (1.20)/(18)(12) = 0.0056 > 0.003$) is again computed from eq. (7.41).

$$M = \frac{(12)(18^2)}{(12)}\left\{(0.0056)(60)\left[1 - (0.59)(0.0056)\frac{(60)}{(4)}\right]\right\}$$

$$= 103.47 \text{ kip} \cdot \text{ft}$$

The points on the beam where this moment occurs are obtained from statical considerations:

$$M = -\frac{wz^2}{2} + 22.5w(z - 10)$$

where $w = (3.25)(\text{LF}) = 5.915$ kips/ft. Solving this quadratic equation yields

$$z = \begin{cases} 17.9 \text{ ft} \\ 27.1 \text{ ft} \end{cases}$$

The no. 9 bar is therefore not required 17.9 ft from either end of the beam.

Maximum negative moments The bending moment diagram shown in fig. 7.32 assumes single point supports; that is, the reaction forces are presumed concentrated loads. In reality, they are spread out over finite distances, and this will tend to reduce the value of the maximum negative moment. A number of approximate methods are available to correct the value shown in fig. 7.32. One of the more convenient ones is to reduce the peak value by $(V)(aL)/3$, where V is the computed shear at the center-line of the support and aL is the support width. For this example, the full (un-rounded) peak value determined from a consideration of center-to-center dimensions will be used. The design moment, then, is

$$M = (50)(3.25)(1.82) = 295.75 \text{ kip} \cdot \text{ft}$$

For a singly reinforced, rectangular concrete beam of $b = 12$ in and $d = 18$ in, the maximum bending moment capacity is [from eq. (7.41)]

$$(295.75)(12) = (12)(18^2)\{p(60)[1 - (0.59)(p)(15)]\}$$

giving

$$p = 0.0182 \quad \text{and} \quad A_s = (0.0182)(18)(12) = 3.92 \text{ in}^2$$

Four no. 9 bars will be used—having an area of $A_s = 4.00$ in². Two of these bars will be cut off where they are no longer needed.

The capacity of a 12×18 in beam with two no. 9 bars $[p = 2.0/(12)(18) = 0.0093]$ is

$$M_{max} = (12)(18^2)\{(0.0093)(60)[1 - (0.59)(0.0093)(15)]\}$$

$$= 165.91 \text{ kip} \cdot \text{ft}$$

In the cantilever, this value of moment is located at $z = 7.49$ ft from the end. In the center span, it occurs at $z = 11.90$ ft from the end of the beam.

The resulting required steel layout is shown in fig. 7.33. Note that the steel is placed symmetrical about the vertical axis of the beam at *all* locations.

It should be recognized that it has been presumed in fig. 7.33 that separate straight bars in the top and bottom of the beam are to be used. While such an arrangement is possible and permissible, and perhaps even simpler, many engineers prefer not to cut the bars when they are no longer needed but to bend them up (or down) to the opposite face of the adjoining segments. These bent bars are better anchored, and—as will be shown later—the inclined portions of the continuous bars provide reinforcement in the web against diagonal tensile stresses. Also, if the same size bars are required at the top and the bottom, continuous reinforcement along the beam can be specified—thus reducing the labor costs and the possibility of error of placing that is inherent in the use of a multiplicity of short straight bars. (Bent bars usually are bent at an angle of 45°, but other angles are used when necessary.)

Bending of bars is feasible only when concrete beams are reinforced by groups of bars. If the bars in a given segment of the beam are placed (vertically) in two rows, then a bar from the outer row cannot be bent until the bar in the inner row in the same vertical plane is cut or bent out of the way.

In the previous section on anchorage, it was pointed out that each bar must have an embedment length beyond the point of maximum moment sufficient to develop the required stress in the reinforcing bar [see eq. (7.37)]. Because loadings or material properties vary considerably, a conservative design assumption usually is made, and lengths of embedment for all bars are selected so that the full yield strength (σ_y) of the bars can be realized using as u_{all} the maximum safe bond stress that can be developed.

$$L_d = \frac{\sigma_y(D)}{4(u_b)} = \frac{\sigma_y(A_s)}{\phi(u_b)} \qquad (7.42)$$

where ϕ is the perimeter of the bar.

Because of placement, vibration of wet mix, etc., concrete properties vary. Therefore, u_b varies from point-to-point within the member. For this reason most codes define the embedment length empirically. For example, the ACI Code defines the length for deformed bars as the product of a basic development length (which primarily depends upon the bar size) and modification factors which take into account other factors, such as the location of the bar in the beam.

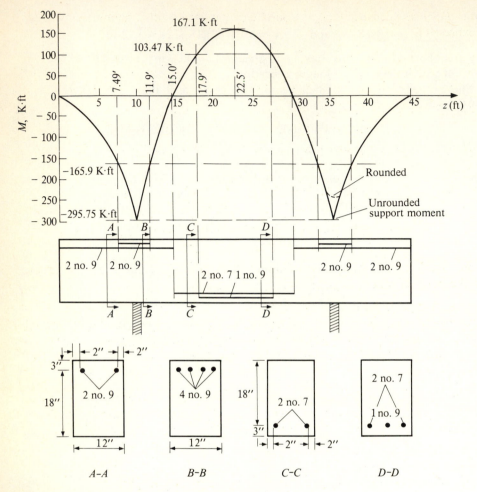

Figure 7.33 Theoretical cutoff points for reinforcing bars.

For no. 11 bars and smaller:

For bars in tension, the basic development length is given by

$$\frac{0.04A_s\sigma_y}{\sqrt{\sigma_c'}} \tag{7.43}$$

but not less than $0.0004D\sigma_y$, or 12 in, where D is the diameter of the bar. Several of the modification factors referred to are:

For bars in the top of the beam = 1.4

For $\sigma_y > 60{,}000$ lb/in^2 $\qquad = 2 - \dfrac{60{,}000}{\sigma_y}$

For bars in compression, the basic embedment length is computed from the expression

$$\frac{0.02D\sigma_y}{\sqrt{\sigma_c'}} \tag{7.44}$$

but not less than $0.0003D\sigma_y$, or 8 in.

Thus, for a no. 8 top bar in tension, with $\sigma_y = 60$ ksi and $\sigma_c' = 4$ ksi,

$$L_d = (1.4)\left(0.04\frac{A_s\sigma_y}{\sqrt{\sigma_c'}}\right)$$

$$= \frac{(1.4)(0.04)(0.79)(60,000)}{(\sqrt{4000})} = 41.97 \text{ in}$$

The bar, therefore, should extend a minimum distance of 42 inches in both directions from the point of maximum moment.

Codes also specify that every reinforcing bar shall be extended by a distance at least equal to 12 bar diameters or the effective depth of the member, whichever is greater, beyond the point where they are no longer needed to resist stress.

If the desired stress in the reinforcing bar cannot be developed by bond alone, a hook at the end of the bar may be used. Several ACI standard hooks for main and secondary reinforcement are shown in fig. 7.34.

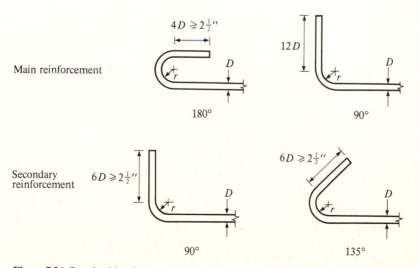

Figure 7.34 Standard hooks.

For these hooks, r is measured to the inside of the bar, and minimum values are as follows:

Bar size	Minimum diameter of bend
No. 3 to no. 8	6 D
No. 9 to no. 11	8 D
No. 14 and 18	10 D

The capacity of standard hooks in tension can be obtained from the relationship

$$\sigma_{hook} = \xi \sqrt{\sigma'_c} \tag{7.45}$$

where σ_{hook} is the effective stress developed in the bar, just outside the hook, and ξ is a factor given in table 7.5.

Table 7.5 Values of ξ

Bar size	$\sigma_y = 60$ ksi		$\sigma_y = 40$ ksi
	Top bars	Other	All bars
No. 3 to no. 5	540	540	360
No. 6	450	540	360
No. 7 to no. 9	360	540	360
No. 10	360	480	360
No. 11	360	420	360

Source: Table 12.8.1 of the American Concrete Institute Building Code, ACI 318-11, pp. 44, 1974.

The effective length of embedment of the hook is determined from eq. (7.46) with the stress in the steel given by eq. (7.45):

$$L_e = 0.04 \frac{A_s \sigma_s}{\sqrt{\sigma'_c}} = 0.04 \frac{A_s \xi \sqrt{\sigma'_c}}{\sqrt{\sigma'_c}} = 0.04 A_s \xi \tag{7.46}$$

Therefore

$$L_d = L_e + (\text{addition embedment lengths beyond hook}) \tag{7.47}$$

It should be noted that hooks are not effective in adding to the compressive strength of bars.

When both positive and negative steel are required in a given beam (that is, for continuous beams with both positive and negative bending moments), the ACI Code stipulates still further requirements concerning the distribution of longitudinal steel:

At least 1/4 of the positive moment steel must be continued uninterrupted along the beam, and at least 6 inches into the support. For simple beams at least 1/3 of the positive steel must be handled in this fashion. When such steel is a part of the primary load carrying system, it shall be anchored into the support to develop its full yield strengths value.

At simple supports and at points of inflection positive steel shall be limited in diameter such that

$$L_d \leq \frac{M_T}{V_u} + L_a \tag{7.48}$$

where M_T = the flexural strength of the beam, if all reinforcement at the section is stressed to σ_y
     V_u = the maximum shear of the section, and
     L_a = the sum of the embedment length beyond the center of the support and any hook effective lengths—but limited to the effective depth of the member or 12 D, whichever is greater.

This requirement forces the designer to use more smaller bars and thereby reduce the intensity of bond stress at the location.

At least 1/3 of the total negative moment steel required at the support must be extended beyond the point of inflection, a distance of not less than 1/6 of the clear span, or the effective beam depth, or 12 D, whichever is the largest.

These requirements are summarized for a general case in fig. 7.35. (The discontinuity in the moment diagrams shown at the point of inflection is to emphasize the fact that different loading conditions may be responsible for the design positive and negative moments.)

Finally, bars which are cut off in the tension zone of the member, because of the stress concentrations produced, tend to cause premature local cracking and a reduction in shear capacity and ductility. For this reason, no bar may be terminated in a tension zone unless one of the following conditions is met:

1. *The shear at the cutoff point does not exceed 2/3 of that permitted, including the shear strength of the furnished web reinforcement.*
2. *Shear reinforcement in excess of that required is provided along each terminated bar over a distance from the point of termination to 3/4 of the effective depth of the member.*
3. *The continuing bars provide twice the area required for flexure at the point, and the shear does not exceed 3/4 of that permitted.*

For example 7.21, the use of no. 7 bars at the bottom of the beam at the points of inflection is first checked. The section moment capacity with two no. 7 bars was computed to be 103.97 kip·ft. The factored shear at the inflection point (determined from the shear diagram) is 44.36 kips. Thus, from

$$\frac{M_T}{V_u} + L_a$$

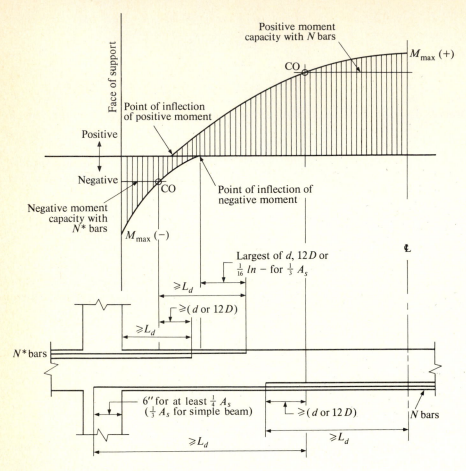

Figure 7.35 ACI code requirements for reinforcing bars.

(where L_a in this case is the effective member depth), L_d can be no greater than

$$\frac{(103.47)(12)}{(44.36)} + 18 = 46 \text{ in}$$

For a no. 7 bottom bar, the actual required development length is

$$L_d = \frac{0.04 A_s \sigma_y}{\sqrt{\sigma_c'}} = \frac{(0.04)(0.6)(60,000)}{\sqrt{4000}} = 22.76 \text{ in.}$$

Since this is less than the maximum allowed, the no. 7 bar is satisfactory.

For the beam in question, no. 7 and no. 9 bars were chosen for the bottom reinforcement, and no. 9 bars for the top. For these, the required anchorage considerations and lengths are

No. 7 bars $D = 0.875$ in, $12 \, D = 10.5$ in
                               Use $d = 18$ in

No. 9 bars $\quad\quad\quad D = 1.125$ in, $12\,D = 13.54$ in
$$\text{Use } d = 18 \text{ in}$$

The development lengths are:
No. 7 (bottom) bars

$$L_d = \frac{0.04 A_s \sigma_y}{(\sqrt{\sigma_c'})} = \frac{(0.04)(0.60)(60{,}000)}{(\sqrt{4000})} = 22.6 \text{ in}$$

No. 9 (bottom) bars

$$L_d = \frac{(0.04)(1.00)(60{,}000)}{(\sqrt{4000})} = 37.95 \text{ in}$$

No. 9 (top) bars

$$L_d = (1.4)(37.95) = 53.13 \text{ in}$$

For the top no. 9 bars extending to the end of the beam, sufficient anchorage length is not available. Anchorage in this case will be provided by a 180° standard hook plus whatever extra extension is necessary. The bar stress developed in the hook is given by

$$\sigma_{\text{hook}} = \xi \sqrt{\sigma_c'}$$

where ξ (from table 7.5) is taken as 360. Using eq. (7.46)

$$L_e = 0.04 A_s \xi = (0.04)(1.00)(360) = 14.4 \text{ in}$$

Since an anchorage of 18 inches is required, the hook must have an extension of at least 3.60 inches.

On the top of the beam, two no. 9 bars (that is, one-half of the negative steel) are to be extended beyond the inflection points. On the bottom, two no. 7 bars will be extended to the supports.

The final arrangement of primary longitudinal reinforcement for the beam is shown in fig. 7.36. Here, the theoretically required steel is shown solid, and the required extensions are shown open.

Shear and diagonal tension reinforcement It was shown in chap. 2 for a homogeneous rectangular cross-section beam subjected to bending and shear that the transverse shearing stress at a given cross section varies in the form of a symmetrical parabola—from zero at the top and bottom faces of the member to a maximum value at the neutral axis. A reinforced concrete beam, however, is nonhomogeneous, and the shearing stress diagram differs considerably from that just described. Above the neutral axis—where the concrete is in compression—the behavior is similar. Below that axis—where for bending analysis purposes it is presupposed that concrete is incapable of resisting tensile forces—the shearing stress remains essentially constant at the maximum value.

It also was shown in chap. 2 that normal stresses and transverse shearing stresses at a given point in a member can be resolved into principal (normal) stresses at the

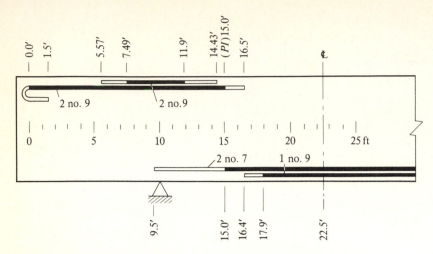

Figure 7.36 Final arrangement of primary longitudinal reinforcing bars.

same point. At the neutral axis—where, by definition, the normal stress in the longitudinal direction is zero—the maximum, diagonal, tensile stress acts in a direction inclined 45° to the longitudinal direction, and has a magnitude equal to the transverse shearing stress. For a given beam there is the distinct possibility that these diagonal tension stresses can equal or exceed the tensile capacity of the concrete. When this occurs, the member cracks at that point, in a direction perpendicular to the direction of maximum tensile stress.

Two kinds of cracks are observed in tests of reinforced concrete beams. The first of these start at the outside edge of the beam on the tension side, in regions of large bending moment and small shear. They are due to the elongation of the longitudinal reinforcing bars. The bending design equations previously derived *presume* that such cracks will occur. The second kind of cracks develop due to the combined action of longitudinal tension and transverse shear. They are due to "excessive diagonal tension" and start at or near the neutral axis of the cross section, in regions of high transverse shear. Normal bending analysis *requires* that such cracks be prevented. Therefore, additional reinforcing bars may have to be provided to ensure against this type of behavior.

To be most effective, web steel to prevent failure by diagonal tension should be perpendicular to the potential crack. To realize such a solution in practice, however, noting that the direction varies considerably along the beam, would be excessively expensive. Therefore, real members only contain approximations to this ideal situation.

The most commonly used methods of arranging web reinforcement to prevent failure due to excessive diagonal tension are (see fig. 7.37):

1. Stirrups or ties placed perpendicular to the longitudinal steel
2. Stirrups or ties places at 45° or more to the longitudinal steel
3. Longitudinal bars bent up at 30° or more
4. Combinations of bent bars and stirrups

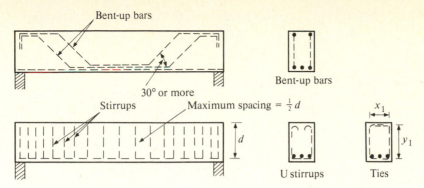

Figure 7.37 Typical patterns of web reinforcement.

Web reinforcement is spaced in such a way that every potential crack line is crossed by at least one line of steel, and in every case shall be adequately anchored at its ending by bending around a longitudinal bar, a hook, or extending the bar.

The basic design concept governing the sizing of web reinforcement to resist diagonal tension is that it should be adequate, in and of itself, to resist the vertical component of the excessive diagonal tension. (While codes do not permit concrete to be used to support bending tension, they *do* permit it to be used to assume some of the diagonal tension caused by shear.)

It must be admitted at the outset that the behavior of reinforced concrete beams subjected to shearing forces is not, at this stage, totally understood. Many tests have been carried out, but there is not an acceptable best method available for determining the actual shearing stresses that will be present under a given set of circumstances, or even of predicting the shearing resistance that can safely be developed.

Prior to cracking, the entire cross section is effective in carrying shear. After cracks have formed, however, the manner in which the section responds changes considerably. It is common practice in design, therefore, to deal with an average shearing stress at a given section, rather than attempt to determine the actual one present at a given time, and to compensate for the error so introduced by modifying the allowable shearing stress by empirically determined factors based upon the extensive test referred to above. Thus, for simplicity, the nominal shearing stress at a given section is computed from the equation

$$v_u = \frac{V_u}{b_w d} \tag{7.49}$$

where V_u = the ultimate transverse shearing force resulting from the factored load
b_w = the width of the section in shear
d = the effective depth of the member measured from the compression face to the centroid of the tension steel

It is accepted practice that whenever the transverse shearing stress v_u exceeds one-half of the allowable shearing strength of the concrete, web reinforcement should be provided. When shearing reinforcement is required, the area provided should not be less than

$$A_V \geq 50 \frac{b_w s}{\sigma_y} \tag{7.50}$$

where A_V = the effective cross-sectional area of the web reinforcement
σ_y = the yield stress of the web reinforcement (not to exceed 60,000 lb/in²)
s = the spacing of the web reinforcement, inches
b_w = the width of the web of the beam, inches

According to the ACI Code, the shearing stress v_c that results in diagonal cracking due to bending and transverse shear can be determined from the empirical relationship

$$v_c = 1.9\sqrt{\sigma_c'} + 2500\rho_w \frac{V_u d}{M_u} \leq 3.5\sqrt{\sigma_c'} \tag{7.51}$$

M_u is the factored bending moment occurring at the point where V_u is taken, and ρ_w is equal to $A_s/b_w d$, with A_s the longitudinal tensile reinforcement. It is further required that $V_u d/M_u \leq 1.0$.

It also is permissible, if desired, to determine v_c from the less accurate—but conservative—relationship

$$v_c = 2\sqrt{\sigma_c'} \tag{7.52}$$

For members subjected to axial compression as well as bending, the shearing stress carried by the concrete should be no greater than

$$v_c = 2\left(1 + \frac{N_u}{2000 A_g}\right)\sqrt{\sigma_c'} \tag{7.53}$$

where N_u is the axial force acting on the section in question (positive when in compression), with N_u/A_g expressed in lb/in².

For cross sections subjected to torsional shearing stresses V_{tu} in excess of $1.5\sqrt{\sigma_c'}$, the maximum shearing stress in the concrete is given by

$$v_c = \frac{2\sqrt{\sigma_c'}}{\sqrt{1 + \left(\dfrac{v_{tu}}{1.2v_u}\right)^2}} \tag{7.54}$$

where

$$v_{tu} = 3T_u/x^2 y \tag{7.55}$$

T_u is the factored torsional moment, and x and y are the shorter and longer dimensions of the rectangular section.

Closed stirrups or ties are required whenever $v_{tu} > 1.5\sqrt{\sigma_c'}$. The minimum torsional reinforcement must be such that the total area of shear (web) reinforcement will be at least equal to

$$A_V + 2A_T = 50 \frac{b_w s}{\sigma_y} \tag{7.56}$$

where A_T is the area of *one leg* of the closed stirrup. When the torsional stress v_{tu} is greater than $1.5\sqrt{\sigma_c'}$, the torsional reinforcement must be designed to sustain the torsional shearing stress in excess of that carried by the concrete, which is

$$v_{tc} = \frac{2.4\sqrt{\sigma_c'}}{\sqrt{1 + (1.2v_u/v_{tu})^2}} \qquad (7.57)$$

In no case shall the torsional stress v_{tu} be greater than

$$(v_{tu})_{\max} = \frac{12.0\sqrt{\sigma_c'}}{\sqrt{1 + (1.2v_u/v_{tu})^2}} \qquad (7.58)$$

In summary:

Bending shear: *If $v_u < \frac{1}{2}v_c$, no web reinforcement is required.*
   If $\frac{1}{2}v_c \leq v_u \leq v_c$, minimum web steel required, which is given by

$$A_V = 50\frac{b_w s}{\sigma_y}$$

   If $v_u > v_c$, web reinforcement must be provided for all shear in excess of v_c.

Bending shear plus torsional shear: *If $v_{tu} < 1.5\sqrt{\sigma_c'}$, torsion may be neglected.*
   If $v_{tu} \geq \sqrt{\sigma_c'}$, v_c is given by eq. (7.54), and the torsional shear reinforcement must carry the torsional shearing stresses in excess of V_{tc} which was defined in eq. (7.57).

Construction of the nominal shearing stress diagram (that is, $v_u = V_u/b_w d$) is helpful when determining the regions of required web reinforcement. On this diagram would then be superimposed the shearing stress capacity of the concrete. This is shown in fig. 7.38 for one-half of the beam considered in example 7.21. Since the beam is not subjected to torsion, v_c can be calculated conservatively from eq. (7.52).

$$v_c = 2\sqrt{\sigma_c'} = 2\sqrt{4000} = 126.49 \text{ lb/in}^2$$

$\frac{1}{2}v_c$, then, equals 63.24 lb/in^2.

In the outermost portions of the overhang and in the very center of the center span (where $v_u \leq 63.24$), no web reinforcement is required. In the region between the left-hand support and 5.40 ft to the left of it, and between that same support and 7.90 ft to the right of it (where $v_u \geq 126.49$ lb/in^2), web reinforcement is needed to support the excessive shearing stresses shown shaded in the figure.[†] Between these two extremes (where $63.24 \leq v_u < 126.49$), minimum web reinforcement is needed. (Had eq. (7.51) been used to define v_c, there would have resulted a smaller shaded area, with less web reinforcement. The simpler relationship was used to illustrate the procedure.)

---

† It is common engineering practice to take into account the confining effects of concentrated reactions by assuming that the maximum nominal shearing stress v_u is that associated with a distance d from the centerline of the reaction.

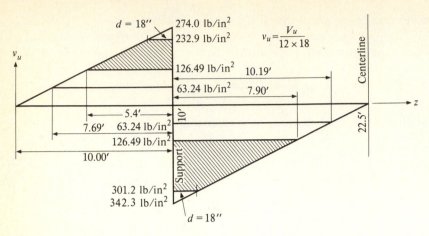

Figure 7.38 Regions of required web reinforcement.

As indicated earlier, diagonal tension cracks occur at 45° to the longitudinal axis of the member. They start at the neutral axis and progress downward to the tension reinforcement. If it is assumed that these cracks develop at intervals of s along the member in the region in question, then the spacing required for a given web reinforcement bar size, or the bar size required for a given spacing of web reinforcement, can be determined from equilibrium considerations. Consider, for example, the simply supported beam shown in fig. 7.39. In this figure it is presumed that vertical stirrups provide the web reinforcement, and these are spaced at s intervals in the region in question. Two cracks are indicated. Along the cracked surface no forces can act. At the neutral axis, along A-B, the total shearing force is vsb. Perpendicular to B-C, the diagonal compressive force C acts. The stirrup will act in tension T. Since

$$T = A_V \sigma_y = vsb = \frac{V}{bd} sb$$

the cross-sectional area of the vertical stirrup must be, at least,

$$A_V = \frac{vbs}{\sigma_y} \tag{7.59}$$

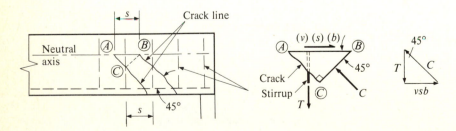

Figure 7.39 Forces in vertical stirrups.

Had A_V been given, the stirrups spacing would be

$$s = \frac{A_V \sigma_y}{vb} \tag{7.60}$$

In both equations, v is the shear to be carried by the vertical steel, or

$$v = v_u - v_c \tag{7.61}$$

For an inclined bar a similar procedure would be followed. Figure 7.40 is a simple beam with two cracks shown and with the web reinforcement inclined to the longitudinal axis by the angle α. Again, the presumed crack spacing is s, as is the web reinforcement. Equilibrium of the segment A-B-C of the beam requires that

$$T \cos(\alpha - 45) = \frac{vsb}{\sqrt{2}}$$

or

$$T(\cos \alpha \cos 45° + \sin \alpha \sin 45°) = \frac{vsb}{\sqrt{2}}$$

But

$$T = A_V \sigma_y = \frac{vsb}{\sin \alpha + \cos \alpha}$$

Therefore

$$A_V = \frac{(v_u - v_c)sb}{\sigma_y(\sin \alpha + \cos \alpha)} \tag{7.62}$$

In terms of a given cross-sectional area of web steel, the spacing is

$$s = \frac{A_V \sigma_y(\sin \alpha + \cos \alpha)}{(v_u - v_c)b} \tag{7.63}$$

When both vertical and inclined web reinforcement are used, the required area is taken as the sum of the two parts (acting separately).

As noted earlier, the maximum spacing s for web reinforcement to be effective must be such that each potential 45° crack is intercepted by at least one set of web

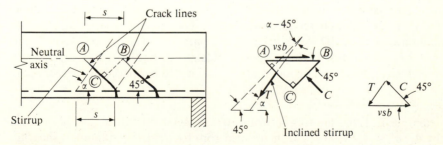

Figure 7.40 Forces in inclined stirrups.

reinforcement bars. For vertical U stirrups, this means that the stirrups can be no further apart than $\frac{1}{2}d$. For deep beams this value may be quite large, and an additional upper limit of 24 in is normally stipulated. When the shearing stress is very large ($v_u - v_c > 4\sqrt{\sigma_c'}$), it is further required that the above-defined maximum spacing be reduced by a factor of one-half, and in no case shall $v_u - v_c$ exceed $8\sqrt{\sigma_c'}$. For inclined web reinforcement, it is normally considered that only three-quarters of the inclined portion of the bar is effective as shear reinforcement. Therefore, so that each potential 45° crack line will be crossed by at least one bar, it is necessary that the bars be no farther apart (in the longitudinal direction) than $\frac{3}{8}d$.

Web reinforcement to sustain torsional shearing stresses consists of both closed stirrups (or ties) placed perpendicular to the longitudinal axis of the member at the point in question, and additional longitudinal bars. It is emphasized that this longitudinal reinforcement is in addition to that required for other purposes—such as bending resistance, for example. (Providing the area and spacing are adequate, a single bar can serve both purposes, with some of its area contributing to each.)

An equilibrium analysis, similar to the one carried out above for bending shear reinforcement, yields the required web steel area for closed stirrups.

$$A_T = \frac{(v_{tu} - v_{tc})s \sum x^2 y}{3\alpha_1 x_1 y_1 \sigma_y} \tag{7.64}$$

A_T is the cross-sectional area of *one leg* of the stirrup, x_1 and y_1 are the minimum and maximum, *center* to *center*, dimensions of the closed stirrups as shown in fig. 7.37, and $\alpha_1 = 0.66 + 0.33 y_1/x_1$, but not more than 1.5. The maximum spacing of the closed stirrups cannot exceed $\frac{1}{4}(x_1 + y_1)$ or 12 in, whichever is the smaller.

The required cross-sectional area of additional longitudinal steel is computed from the expressions

$$A_l = 2A_t \frac{x_1 + y_1}{s}$$

or $\tag{7.65}$

$$A_l = \left(\frac{400sx}{\sigma_y} \frac{u_{tu}}{u_{tu} + u_u} - 2A_t\right)\frac{x_1 + y_1}{s}$$

whichever is the greater. In the second of these equations, x is the shorter cross-sectional dimension, and the area of steel so defined is the minimum area required for equilibrium. The first equation ensures that there is as much additional longitudinal steel as there is vertical web reinforcement steel. The minimum size bar used for longitudinal steel is no. 3 (D = 3/8 in), and the bars are to be spaced around the perimeter of the cross section of the beam with at least one bar located in each corner to brace the closed stirrups. These longitudinal bars are to be spaced no further apart than 12 in.

For example 7.21, the limiting shearing stress associated with $s_{max} = \frac{1}{2}d$ or 24 in is given by $4\sqrt{\sigma_c'} = 4\sqrt{4000} = 252.98$ lb/in^2. From fig. 7.38, the maximum shearing stress acting on the beam occurs just to the inside of the supports and equals

$(v_u - v_c)_{\max} = 301.2 - 126.49 = 174.71 \text{ lb/in}^2$. This is considerably less than the maximum allowed, which has associated with it $s = \frac{1}{2}d$.

Equation (7.60) also defines a maximum allowable spacing—depending upon the size of this stirrup. For a no. 3 bar single U stirrup grade 40 steel the maximum spacing is

$$s = \frac{A_V \sigma_y}{vb} = \frac{(2)(0.11)(40,000)}{(50)(12)} = 14.66 \text{ in}$$

Since a larger bar stirrup would have associated with it a larger value of s, the no. 3 with a maximum spacing of $s = d/2 = 9$ in as determined above will be used.

The range of spacings required for the no. 3 stirrups is calculated using the values of the maximum shearing stresses in the various regions:

In the center span, at the support, where $(v_u - v_c)_{\max} = 174.71 \text{ lb/in}^2$

$$s_{\text{reqd}} = \frac{A_V \sigma_y}{b(u_u - u_c)} = \frac{(0.22)(40,000)}{(12)(174.71)} = 4.2 \text{ in}$$

In the cantilever, at the support, where $(v_u - v_c)_{\max} = 106.4 \text{ lb/in}^2$

$$s_{\text{reqd}} = \frac{(0.22)(40,000)}{(12)(106.4)} = 6.9 \text{ in}$$

To facilitate establishment of the final stirrup arrangement, it is convenient to determine, beforehand, the points along the member where, according to eq. (7.60), various spacings can commence. For example, for $s = 9$ in

$$v_u - v_c = v = \frac{(0.22)(40,000)}{(9)(12)} = 81.48 \text{ lb/in}^2$$

Using the nominal shearing stress diagram of fig. 7.38, this level is found by considering similar triangles, and occurs at 28.92 inches from the support in the cantilever section, and at 58.86 inches from the support in the center section of the beam. Similarly, other spacings could be considered. Using this information, the web shear reinforcement is located as follows, with the first stirrup at $\frac{1}{2}s$ from the support:

In the cantilever portion of the beam (all dimensions in inches)

		Total distance
One space at 3 in	= 3	3
Three spaces at 6 in	= 18	21
Two spaces at 7 in	= 14	35
Seven spaces at 9 in	= 63	98

In the center portion of the beam (between the supports, in inches)

		Total distance
One space at 2 in	= 2	2
Ten spaces at 4 in	= 40	42
Three spaces at 6 in	= 18	60
Seven spaces at 9 in	= 63	123

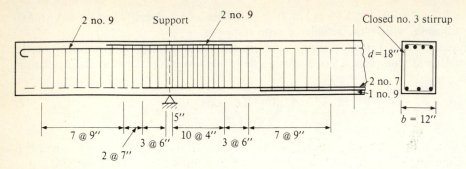

Figure 7.41 Spacing of web reinforcement.

The final arrangement of the web steel is shown in fig. 7.41. For ease of manufacture, closed stirrups will be used and two of the no. 9 bars on the top and the two no. 7 bars on the bottom will be extended for the full length of the beam, as indicated.

Example 7.22 This example will highlight a number of the problems associated with reinforced concrete design for combined bending and torsion. The member in question is part of a rigid frame as indicated in fig. 7.42. The applied loading is a single concentrated vertical force of 120 kips. This force is applied at the center point along the beam; however, it is eccentric to the centroid of the cross section by 4 in. Both bending and torsional moments are induced in the member, as are shearing forces.

For the purposes of this example, the beams and columns adjoining the ends of the member in question are sufficiently rigid so that it can be presumed fixed at its ends.

The yield stress of both the longitudinal and web reinforcement steel is given as $\sigma_y = 60,000$ lb/in^2. The concrete has a strength of $\sigma_c' = 4000$ lb/in^2.

If it is presumed that the overall cross-sectional dimensions of the beam are 16 in × 24 in, the weight of the member per foot of length will be 400 lb. The bending moment, torsional moment, and transverse shear diagrams are shown in fig. 7.43 for these combined loadings.

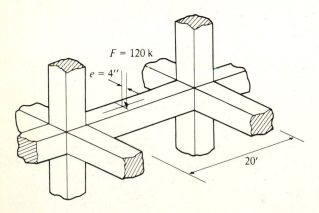

Figure 7.42

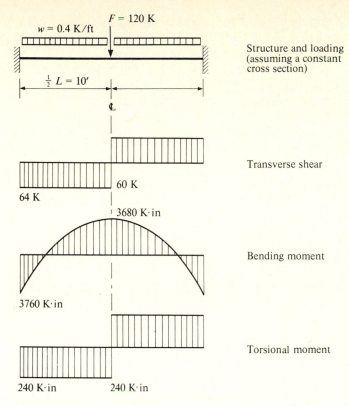

$F = 120$ K

$w = 0.4$ K/ft

$\frac{1}{2} L = 10'$

Structure and loading (assuming a constant cross section)

Transverse shear

60 K

64 K

3680 K·in

Bending moment

3760 K·in

Torsional moment

240 K·in   240 K·in

Figure 7.43 Shear, bending-moment, and twisting-moment diagrams.

The load factor to be used is

$$\text{LF} = 1.7 + 1.4\,\frac{\text{DL}}{\text{LL}} = 1.7 + 1.4\,\frac{(0.4)(20)}{120} = 1.79$$

The shears, bending moments and torsional moments at the factored loads, therefore, are

At the ends of the beam:

$$V_u = (1.79)(64) = 114.56 \text{ kips}$$

$$M_u = (1.79)(3760) = 6730.4 \text{ kip·in}$$

$$T_u = (1.79)(240) = 429.6 \text{ kip·in}$$

At the centerline of the beam:

$$V_u = (1.79)(60) = 107.4 \text{ kips}$$

$$M_u = (1.79)(3680) = 6587.2 \text{ kip·in}$$

$$T_u = (1.79)(240) = 429.6 \text{ kip·in}$$

It will be assumed that the 16-in × 24-in member contains one layer of top and one layer of bottom longitudinal steel—not necessarily extending over the entire length of the member. For initial dimensioning purposes, it further will be presumed that these are no. 11 bars (D = 1.41 in), and the closed stirrups (or ties) will be no. 4's (D = 0.5 in). The general dimensions of the cross section are as shown in fig. 7.44. The effective depth of the member is

$$d = 24.0 - 2.0 - \frac{1.41}{2} = 21.3 \text{ in}$$

It follows that

$$A_g = (b)(d) = (16)(21.3) = 348.0 \text{ in}^2$$

and (from C.2)

$$p_{\min} = \frac{A_s}{bd} = \frac{200}{60,000} = 0.003$$

From eq. (2.79) or (7.41),

$$M_{\max} = bd^2 \left[pd^2 \left(1 - 0.59p \frac{\sigma_y}{\sigma_c'} \right) \right]$$

For the two critical moment values,

$$M = 6730.4 \text{ kip·in} \qquad p = 0.0185 \qquad \text{and} \qquad A_s = 6.30 \text{ in}^2$$
$$M = 6587.2 \text{ kip·in} \qquad p = 0.0179 \qquad \text{and} \qquad A_s = 6.10 \text{ in}^2$$

Four no. 11 bars provide $A_s = 6.24 \text{ in}^2$. It is interesting to note that if it is assumed that the concrete in compression has an effective depth equal to the distance from the outside surface of the beam to the compression steel, then

$$M = 18.5 A_s \sigma_y = 6730.4 \text{ kip · in}$$

or

$$A_s = 6.06 \text{ in}^2$$

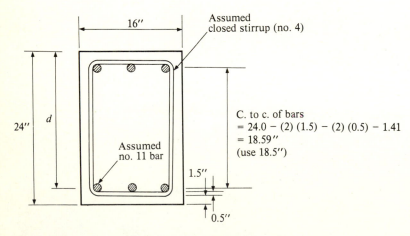

Figure 7.44 Assumed cross section and reinforcing bars.

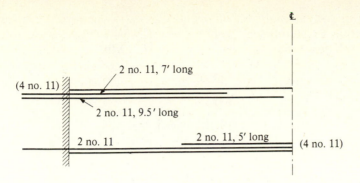

Figure 7.45 Tentative longitudinal reinforcing-bar plan.

At the top of the beam, the anchorage length for no. 11 bars is determined from eq. (7.43) to be

$$(L_d)_{top} = (1.4)(0.04)\frac{A_s\sigma_y}{\sqrt{\sigma'_c}} = 82.87 \text{ in}$$

At the bottom of the beam,

$$(L_d)_{bot} = 59.19 \text{ in}$$

The longitudinal reinforcing bar plan to carry the bending moment is shown in fig. 7.45. Since the bars extend over so much of the length, it is desirable to require their use over the entire span—at both the top and the bottom. (It is, of course, to be understood that the bars must be properly anchored into the adjoining columns and beams at the ends of the member.)

Consider now the additional problems introduced by the torsional moments. From eq. (7.54), the maximum shearing stress that the concrete can carry is

$$v_c = \frac{2\sqrt{\sigma'_c}}{\sqrt{1 + \left(\dfrac{v_{tu}}{1.2v_u}\right)^2}}$$

where

$$v_{tu} = \frac{3T_u}{x^2 y} = \frac{(3)(429.6)}{(16)^2(24)} = 209.76 \text{ lb/in}^2$$

(From eq. (7.53), $(v_{tu})_{max} = 350.1 \text{ lb/in}^2$.) The nominal transverse shearing stresses at the ends and centerline of the beam are, respectively,

$$(v_u)_{end} = \frac{114.56}{(16)(21.3)} = 336.2 \text{ lb/in}^2$$

$$(v_u)_{centerline} = \frac{107.4}{(16)(21.3)} = 315.1 \text{ lb/in}^2$$

Therefore

$$(v_c)_{end} = \frac{2\sqrt{4000}}{\sqrt{1 + \left[\dfrac{209.76}{(1.2)(336.2)}\right]^2}} = 112 \text{ lb/in}^2$$

$$(v_c)_{centerline} = \frac{2\sqrt{4000}}{\sqrt{1 + \left[\dfrac{209.76}{(1.2)(315.1)}\right]^2}} = 111 \text{ lb/in}^2$$

At the ends of the beam, the shearing stress to be carried by the web reinforcement is $v_u - v_c = 336 - 112 = 224 \text{ lb/in}^2$. At the centerline, $v_u - v_c = 204 \text{ lb/in}^3$. A linear variation exists between these two extremes. However, since their values are so close, the maximum will be used to establish a uniformly spaced web reinforcement throughout the entire length of the member. Since the shear to be carried is less than $4\sqrt{\sigma_c'} = 253 \text{ lb/in}^2$, the maximum spacing of the web steel will be $d/2$.

For closed stirrups:

$$A_V = \frac{(v_u - v_c)bs}{\sigma_y}$$

where A_V is the cross-sectional area of *both* legs of the stirrup. Substituting into this expression the values for b, σ_y, and $v_u - v_c$, the required areas are:

$$\begin{aligned}
\text{For } s = 2 \text{ in} \qquad & A_V = 0.119 \text{ in}^2 \\
= 3 \text{ in} \qquad & = 0.179 \text{ in}^2 \\
= 4 \text{ in} \qquad & = 0.239 \text{ in}^2 \\
= 5 \text{ in} \qquad & = 0.299 \text{ in}^2 \\
= 6 \text{ in} \qquad & = 0.359 \text{ in}^2
\end{aligned}$$

Assuming that the stirrups are no. 5's and the longitudinal bars are no. 11's, the centerline dimensions (x_1 and y_1) of the web reinforcement will be those shown in fig. 7.46. From the determined values above:

$$v_{tu} = 209.76 \text{ lb/in}^2 \qquad \text{and} \qquad v_u = 336.2 \text{ lb/in}^2$$

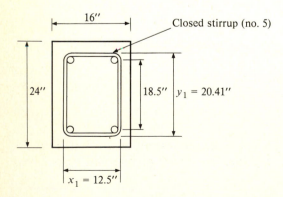

Figure 7.46 Stirrup geometry.

Substituting these into eq. (7.57) yields

$$v_{tc} = \frac{2.4\sqrt{\sigma_c'}}{\sqrt{1 + \left(1.2 \times \dfrac{v_u}{v_{tu}}\right)^2}} = 70.02 \text{ lb/in}^2$$

Therefore,

$$v_{tu} - v_{tc} = 209.76 - 70.02 = 139.74 \text{ lb/in}^2$$

If these values are now substituted into eq. (7.64), the required cross-sectional area of the web steel is determined.

$$\alpha_1 = (0.66) + (0.33)\left(\frac{20.41}{12.50}\right) = 1.20$$

$$A_T = \frac{(v_{tu} - v_{tc})(s)(16)^2(24)}{(3)(1.20)(20.41)(12.50)(60,000)} = 0.016s$$

(Note that in *this* expression A_T is the cross-sectional area of *one* leg of the stirrup.) For various values of s,

$$
\begin{array}{ll}
s = 2 \text{ in} & A_T = 0.032 \text{ in}^2 \\
 = 3 \text{ in} & = 0.048 \text{ in}^2 \\
 = 4 \text{ in} & = 0.064 \text{ in}^2 \\
 = 5 \text{ in} & = 0.080 \text{ in}^2 \\
 = 6 \text{ in} & = 0.096 \text{ in}^2
\end{array}
$$

The total cross-sectional area of *one* leg of the stirrup, therefore, will be, with $A = A_T + \frac{1}{2}A_V$,

$$
\begin{array}{ll}
s = 2 \text{ in} & A = 0.092 \text{ in}^2 \\
 = 3 \text{ in} & = 0.138 \text{ in}^2 \\
 = 4 \text{ in} & = 0.184 \text{ in}^2 \\
 = 5 \text{ in} & = 0.229 \text{ in}^2 \\
 = 6 \text{ in} & = 0.276 \text{ in}^2
\end{array}
$$

No. 5 stirrups $(A = 0.31 \text{ in}^2)$ placed on 6-in spacing over the entire length of the member will be more than adequate.

The anchorage length for such no. 5 bars is

$$L_d = 1.7\left(\frac{0.04A_s\sigma_y}{\sqrt{\sigma_c'}}\right) = (1.7)(11.76) = 20 \text{ in}$$

Therefore, the sides of the stirrup should be overlapped for the full depth of the main longitudinal reinforcement, with the overlap being placed on alternate sides along the member.

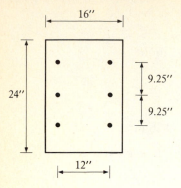

Figure 7.47 Additional longitudinal steel required because of torsion.

The additional area of longitudinal steel required because of torsion is determined from eq. (7.65). Since A_T for a 6-in spacing of stirrups is 0.096 in^2, A_l will be the larger of

$$A_l = 2A_T \frac{x_1 + y_1}{s} = 1.04 \text{ in}^2$$

or

$$A_l = \left(\frac{400sx}{\sigma_y} \frac{v_{tu}}{v_{tu} + v_u} - 2A_T\right)\frac{x_1 + y_1}{s} = 0.293 \text{ in}^2$$

Therefore $A_l = 1.04$ in^2. This area should be more or less evenly spaced around the cage, with no open distance greater than 12 in. Assuming 6 parts as shown in fig. 7.47, the additional steel (in each part) would be $1.04/6 = 0.173$ in^2. This steel can be combined with that required for bending. If the two center bars at the top and at the bottom are selected as no. 11 bars, there is remaining to be provided at the corners at the top

$$6.06 - (2)(1.56) = 2.94 \text{ in}^2$$

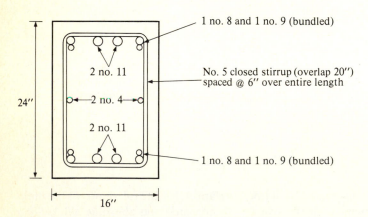

Figure 7.48 Steel reinforcement design.

Therefore, each corner must contain at least $2.94/2 = 1.47$ in^2 of longitudinal (bending) steel, and an additional 0.173 in^2 of steel for torsion. The total needed in each corner, then, is 1.64 in^2—which can be supplied by "bundling" one no. 8 bar and one no. 9 bar ($A_s = 1.79$ in^2). Figure 7.48 is a cross section which meets these requirements.

7.3.5 Beam Columns

The behavior of beam columns was discussed in chaps. 4 and 6. In chap. 4 (sec. 4.5.1), for members constrained to deform in the plane of the applied moments, interaction relationships were developed based upon first yielding. These were of the general form [eq. (4.140)].

$$\frac{P}{P_y} + \frac{M}{M_y}(\zeta) = 1.0$$

where $P_y = A\sigma_y$, $M_y = S\sigma_y$, and ζ is the so-called moment magnification factor, which takes into account the $P - \Delta$ effect for given conditions of loading or supports.

In chap. 6 [eq. (6.149)], this interaction equation was extended to include the possibility of other types of failure. There,

$$\frac{P}{P_u} + \frac{M}{M_u}(\delta) = 1.0 \tag{7.66}$$

where P_u is the axial-force carrying capacity of the member when axial force alone exists, M_u is the bending-moment carrying capacity of the member when moment alone exists, and δ is the moment magnification factor given by

$$\delta = \frac{C_m}{1 - P/P_{cr}} \tag{7.67}$$

In this expression P_{cr} is the critical (flexural) buckling load of the member in the plane of the applied moments, and C_m is a modification factor which accounts for moment gradients, different loading conditions, support conditions, etc. If the member is pinned at its ends, $P_{cr} = P_e$. For other values

$$P_{cr} = \frac{\pi^2 EI}{(KL)^2} \tag{7.68}$$

where K is the effective length factor (see table 6.1), L is the unsupported length of the member in the plane of bending, and EI is the member's flexural stiffness.

Both the American Institute of Steel Construction and the American Concrete Institute have suggested for members braced at their ends against sidesway, and for no transverse loading along the length, that

$$C_m = 0.6 + 0.4\frac{M_1}{M_2} \geq 0.4 \tag{7.69}$$

In this expression M_1 is the smaller applied end moment, and M_2 is the larger. The ratio M_1/M_2 is to be taken as positive if the moments tend to produce single curvature bending in the member, and negative if double curvature is induced.

When dealing with reinforced concrete members, the quantity EI takes on special significance. This is due to the fact that the members are nonhomogeneous, and the material behaves in a nonlinear fashion, creeps with time and develops tension cracks. For these reasons, codes provide *empirically determined* expressions for EI. For example, the American Concrete Institute suggests that the following equation be used:

$$EI = \frac{\dfrac{E_c I_g}{5} + E_s I_s}{1 + \beta_d} \tag{7.70}$$

where E_c = modulus of elasticity of concrete in lb/in^2 calculated from
$\quad\quad E_c = 57{,}000\sqrt{\sigma_c'}$ for normal weight concrete
$\quad I_g$ = moment of inertia of the gross section
$\quad E_s$ = modulus of elasticity of steel (29×10^6 lb/in^2)
$\quad I_s$ = moment of inertia of the longitudinal reinforcement about the centroidal axis of the cross section
$\quad \beta_d$ = the ratio of the maximum dead-load design moment to the maximum total-load design moment. (This quantity is always positive, and represents the influence of creep.)

From the above, it should be evident that the sizing of a reinforced concrete beam column is a relatively complex task. For example, to compute the flexural stiffness, it is first necessary that the column size be known. The same is true when determining the effective length factor K.

To aid the designer, codes (or commentaries to codes) usually spell out various limiting values, and sometimes even approximate value which can be used to start the iterative process. For example, the ACI Code (318-71) states that for a frame braced against sidesway, the slenderness effects may be neglected when KL/r is less than $34 - 12\,M_1/M_2$. When sidesway is allowed, slenderness can be neglected when KL/r is less than 22.

Example 7.23 The 20-ft-long reinforced concrete beam column to be designed is an interior column of a multistory building. At both ends of the member there are attached (in a continuous fashion) another column and two horizontal beams. All members are in the same plane.

The structure in question is indeterminate. By methods that will be discussed in chap. 8, it is found that the loading on the column is equivalent to an eccentrically applied, axial compressive thrust. The eccentricity at both ends is 4 in, and the thrust is 140 kips (dead load = 80 kips, and live load = 60 kips). $\sigma_c' = 3{,}500$ lb/in^2 and $\sigma_y = 60{,}000$ lb/in^2.

It is given that sidesway is prevented. Moreover, based upon other calculations, it is known that I/L for each of the four connecting beams is 6.5 in^3.

A requirement of the design is that the percentage of steel p_s be as close as practical to 0.015. It also is required that the load factor be determined from the special relationship LF = 1.5DL + 1.8LL.

The design values for the axial thrust and the end-moments are

$$P_u = (1.5)(80) + (1.8)(60) = 228 \text{ kips}$$

$$M_u = (228)(4) = 912 \text{ kip·in}$$

E_c for the concrete is given by $E_c = 57,000\sqrt{\sigma_c'} = 3.37 \times 10^6 \text{ lb/in}^2$, and $E_s = 29 \times 10^6 \text{ lb/in}^2$.

The process is begun by assuming cross-sectional dimensions for the column; for example, 14 in × 16 in ($A_g = 224 \text{ in}^2$). For the desired percentage of steel, $A_s = (224)(0.015) = 3.36 \text{ in}^2$. The quantities I_g and I_s in eq. (7.70) are therefore

$$I_g = \left(\frac{1}{12}\right)(14)(16)^3 = 4779 \text{ in}^4$$

and

$$I_s \simeq (2)\left(\frac{3.36}{2}\right)\left(\frac{16}{2} - 2\right)^2 = 121 \text{ in}^4$$

If maximum creep is presumed, $\beta_d = 1.0$, and $(EI)_{col}$ becomes

$$(EI)_{col} = \left[\frac{(3.37 \times 10^6)(4779)}{5} + (29 \times 10^6)(121)\right]\left(\frac{1}{2}\right) = 3365 \times 10^6 \text{ lb·in}^2$$

Therefore

$$\left(\frac{EI}{L}\right)_{col} = \frac{3365 \times 10^6}{(20)(12)} = 14.02 \times 10^6 \text{ lb·in}$$

and

$$\left(\frac{EI}{L}\right)_{beam} = (3.37 \times 10^6)(6.5) = 21.91 \times 10^6 \text{ lb·in}$$

To determine the effective length of the member, the alignment chart shown in fig. 6.11 will be used.

$$G = \frac{\sum (EI/L)_{col}}{\sum (EI/L)_{beam}} = \frac{(2)(14.02 \times 10^6)}{(2)(21.91 \times 10^6)} = 0.64$$

This value holds for both the top and bottom of the column. The effective length factor (sidesway prevented) is $K = 0.72$.

For a rectangular column, the radius of gyration r can be approximated by $r \simeq 0.3d$. (For a round column, $r \simeq 0.25d$.) Therefore, $r \simeq (0.3)(16) = 4.8$ in, and

$$\frac{KL}{r} = \frac{(0.72)(20)(12)}{4.80} = 36.0$$

This value is greater than $34 - 12M_1/M_2 = 34 - (12)(1) = 22$. The member, then, is "long," and the influence of slenderness must be taken into account.

The critical flexural buckling load (in the plane of the frame) is

$$P_{cr} = \frac{\pi^2(EI)_{col}}{(KL)^2} = \frac{\pi^2(3365 \times 10^6)}{[(0.72)(240)]^2} = 1.112 \times 10^6 \text{ lb}$$

From eq. (7.69)

$$C_m = 0.6 + 0.4(1.0) = 1.0$$

Therefore, from eq. (7.67), the moment magnification term is

$$\delta = \frac{C_m}{1 - P/P_{cr}} = \frac{1.0}{1 - 228/1112} = 1.258$$

The cross section must, therefore, be able to sustain at its centerline section a moment of

$$(M_{\mathbb{Q}})_U = (912)(1.258) = 1147 \text{ kip} \cdot \text{in}$$

The beam column interaction charts in chap. 4 (fig. 4.61), for $\gamma = 0.75$, $e/h\dagger \simeq 5.03/16 \simeq 0.314$, and $p\mu = (0.015)(20.2) = 0.30$, give

$$P_U = 435.1 \text{ kips}$$

$$(M_{\mathbb{Q}})_U = 2189.5 \text{ kip} \cdot \text{in}$$

These values are considerably in excess of the required capacity. A smaller section is called for.

Repeating the process for a 13-in × 13-in cross-section member with $p_s = 0.015$ yields:

$$A_s = 2.54 \text{ in}^2$$

$$I_g = 2380 \text{ in}^4$$

$$I_s \simeq 59.54 \text{ in}^4$$

$$(EI)_{col} = 1665 \times 10^6 \text{ lb} \cdot \text{in}^2$$

$$\left(\frac{EI}{L}\right)_{col} = 6.94 \times 10^6 \text{ lb} \cdot \text{in}$$

$$\left(\frac{EI}{L}\right)_{beam} = 21.91 \times 10^6 \text{ lb} \cdot \text{in}$$

$$G = 0.317$$

$$K = 0.635$$

$$\frac{KL}{r} = 39.08$$

$$P_{cr} = 0.708 \times 10^6 \text{ lb}$$

$$\delta = 1.475$$

† $e_{\mathbb{Q}}$ at the centerline section is approximately equal to e at the end of the member times δ.

Required $(M_{\xi})_U = 1345$ kip·in.

$$\gamma = 0.69 \qquad \frac{e}{h} = 0.454 \qquad \mu p = 0.30$$

Delivered $P_U = 251.4$ kips. Delivered $(M_{\xi})_U = 1483$ kip·in.

This cross section also delivers more than is required. These two trials, however, demonstrate the procedure and give an indication of the rate of convergence to the desired value.

7.3.6 References

1. Charles S. Whitney, Plastic Theory of Reinforced Concrete Design, *Trans. Am. Soc. Civ. Eng.*, vol. 107, pp. 251–326, 1942.
2. ACI-ASCE Committee 327, Ultimate Strength, *Proc. Am. Concr. Inst.*, vol. 52, no. 7, pp. 504–524, 1956.
3. American Concrete Institute, *Building Code Requirements for Reinforced Concrete* (ACI 318-71), 1971.
4. American Concrete Institute, Commentary on Building Code Requirements for Reinforced Concrete (ACI 318-71), 1971.
5. The American Association of State Highway and Transportation Officials, Standard Specifications for Highway Bridges, 11th ed., 1973.
6. Portland Cement Association, Notes on ACI 318-71 Building Code Requirements with Design Applications, PCA Publication no. EBO7OD, 1972.
7. American Concrete Institute, *ACI Manual of Standard Practice for Detailing Reinforced Concrete Structures*, ACI 315-74, 1974.
8. American Concrete Institute, *ACI Manuals of Concrete Practice*, pts, 1, 2, and 3, 1976, 1977.
9. Concrete Reinforcing Steel Institute, *CRSI Handbook*, 1975.
10. G. Winter et al., *Design of Concrete Structures*, 8th ed., McGraw-Hill, New York, 1972.
11. Phil M. Ferguson, *Reinforced Concrete Fundamentals*, 3d ed., Wiley, New York, 1972.
12. C. K. Wang and C. G. Salmon, Reinforced Concrete Design, 2d ed., Intext, New York, 1973.
13. R. P. Johnson, *Composite Structures of Steel and Concrete*, Wiley, New York, 1976.
14. H. S. Iyenger, State of the Art Report on Composite or Mixed Steel-Concrete Construction for Buildings, *ASCE*, 1977.

7.4 COMPOSITE CONSTRUCTION

When two materials act integrally together in a given structural element, it is said that a *composite member* exists. In the strictest sense, then, plain concrete and steel rods acting together in what has heretofore been defined and referred to as reinforced concrete is an example of *composite construction*. In civil engineering practice, however, the term has been more restrictively used. Today, the term most often is applied to those combinations of elements which, in and of themselves, are major load-carrying units. For example, a steel wide-flange member encased in concrete—where the two are so interconnected that they act essentially as one—is a composite member. Bare steel beams to which have been attached concrete slabs at their top flanges are composite members. Precast—or prestressed—concrete beams with concrete slabs poured on their tops are another example, as are timber beams reinforced along their lengths by attached steel bars or plates.

Composite construction is advantageous in obtaining added strength. The size of steel beams supporting floor slabs can be reduced markedly if the two are constrained to act as a single monolithic unit. Composite construction also can be used effectively to obtain added bending stiffness. Floor deflections can be reduced, and lateral bracing can be significantly modified.

Another advantage of steel beam–concrete slab composite construction is that the steel beam often can be used to support the forms for the concrete, and thus reduce the construction cost of shoring. If shores are used to support the steel beam while the concrete cures, the steel beams normally will not be significantly stressed until the shores are removed. (It should be noted that if the steel beams are not shored, special attention must be given to ensure that the beams are adequately braced in the lateral direction to carry the dead load of the concrete.)

It must be noted that when composite beams frame into steel girders and columns, care should be taken to avoid cracking over the girders due to the deflection of the composite beam. If possible, the steel framing should be made continuous. Otherwise heavy bars should be used in the slab to carry the tension across these members.

Elastic analysis of composite beams was discussed in sec. 2.9.2 (see example 2.20). Many design aids and tables are available from steel manufacturing companies to facilitate this approach. Some codes also permit ultimate-strength design for composite construction—much in the same fashion as ultimate-strength design of reinforced concrete beams. These methods will not be considered here.

As noted above, for members to behave in a composite fashion, it is necessary that there be provided positive physical connections between the elements at various points along the member (see fig. 2.53). When concrete slabs and steel beams are used together, it is common practice to periodically weld on the top flanges of the beams some kind of device to act as a mechanical means of ensuring interaction. Certain of these devices also hold the two elements together vertically, which is most desirable in bridge design where vibrations are a problem. "Lugs," or "studs," spirals, bent rods, and short pieces of channels, angles, or T's are all suitable connectors for these purposes.

When precast-concrete beams are combined with a poured-in-place slab, ties or stirrups extending from the beam into the slab can serve as connecting devices.

Using the elastic theory discussed in chap. 2, the spacing of shear connectors varies inversely with the transverse shear V. Numerous tests of interconnected steel beams and concrete slabs, however, have indicated that a uniform spacing throughout the length of the member is just as satisfactory. (It should be noted that this conclusion also can be supported using the methods of simple plastic theory.) For such cases, the number of connectors is determined from a consideration of the maximum possible horizontal shear that exists at the interface between the two materials. That number is then spaced evenly along the entire length of the beam. The maximum shear force is usually taken as one-half of the ultimate concrete compressive force or the steel tensile stress resultant, whichever is the least. Special spacing provisions must, of course, be made for unusual moment or shear conditions; such as, for example, where there are several concentrated lateral loads.

It is to be emphasized that the shear connectors should be designed to resist *all* of the longitudinal shear. Bond between the steel I beam and the slab should not be relied upon for any resistance.

In steel beam–concrete slab composite construction, the slab normally is designed as a one-way reinforced system, and it is assumed continuous over the intermediate supporting steel beams. Design of the slabs is most easily accomplished by considering a 1-ft-wide strip. (Minimum steel would, of course, be used in the perpendicular direction.)

Example 7.24 A composite steel beam–concrete slab system 30 ft in length is to be designed to support a bearing wall of DL + LL = 5.0 kips/ft (acting along the centerline), and a uniformly distributed live load on the slab of LL = 150 lb/ft², as shown in fig. 7.49. (It is assumed that the bearing wall load acts on the slab as a line load.) The beams are simply supported at their ends.

Shored construction is to be presumed, and the steel beams are to be standard, rolled wide-flange shapes. (No coverplates are to be added.) It is given that $\sigma'_c = 3000$ lb/in², σ_y (of reinforcing rods) = 40,000 lb/in², and $n = 9$. The yield strength of the steel beam is $\sigma_y = 36,000$ lb/in². Considering just the center beam and slab, the loads are

DL (on slab): $(\frac{4}{12})(150) = 50$ lb/ft², $50 \times 8 = 400$ lb/ft of beam
DL (of beam) estimated at 100 lb/ft of beam
LL (on slab) = $(150)(8) = 1200$ lb/ft of beam

The total uniformly distributed load along the member is

$$w = (400 + 100 + 1200) = 1700 \text{ lb/ft} = 1.7 \text{ kips/ft}$$

The concentrated load (due to the bearing wall) is

$$Q = (5.0)(8) = 40 \text{ kips}$$

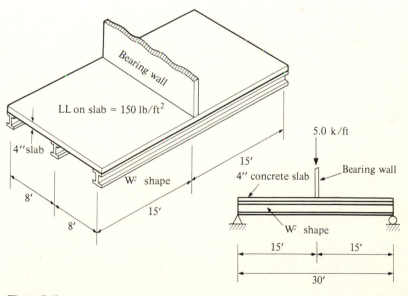

Figure 7.49

Using the recommended safety factors for composite construction of the American Institute of Steel Construction, the allowable stresses are

$$\sigma_c = 0.45\sigma_c' = (0.45)(3) = 1.35 \text{ ksi}$$

$$\sigma_s = 0.66\sigma_y = (0.66)(36) = 24.0 \text{ ksi}$$

The effective widths of the flange (that is, the concrete slab)—see fig. 2.53—should be taken as the lesser of the following:

$$b_{\text{eff}} = \tfrac{1}{4} \text{ span} = \frac{(30)(12)}{4} = 90 \text{ in}$$

$$b_{\text{eff}} = \text{beam spacing} = 96 \text{ in or}$$

$$b_{\text{eff}} = (\text{width of steel flange}) + (2)(8)(\text{slab thickness})$$

$$= (\text{assume } 12 \text{ in}) + (2)(8)(4) = 76 \text{ in}$$

The latter value governs.

The maximum applied bending moment on the composite beam (which occurs directly under the bearing wall) is

$$M = \frac{wL^2}{8} + \frac{QL}{4} = 491.25 \text{ kip·ft} = 5895 \text{ kip·in}$$

The required transformed section modulus (see example 2.20) is therefore

$$S_{\text{top}} = \frac{M}{n\sigma_c} = \frac{5895}{(9)(1.35)} = 485 \text{ in}^3$$

$$S_{\text{bot}} = \frac{M}{\sigma_s} = \frac{5895}{24} = 245 \text{ in}^3$$

Assuming as a trial section a W 24 × 94

Section	I_0	A	y_0 from bottom	Ay_0	d to NA	Ad^2
Beam	2690	27.7	12.15	336.50	−7.76	1670
Slab	45	33.7	26.29	886.07	6.38	1372
Σ	2735	61.4		1222.63		3042

$$\bar{y} = \frac{1222.63}{61.4} = 19.91 \text{ in} \qquad I_T = 2735 + 3042 = 5777 \text{ in}^4$$

$$S_{\text{top}} = \frac{5777}{8.38} = 690 \text{ in}^3 \qquad \text{and} \qquad S_{\text{bot}} = \frac{5777}{19.91} = 290 \text{ in}^3$$

The design is more than adequate. In fact, a W 24 × 84 will be sufficient.

Shear-connector design Equation (2.100) is the governing relationship for shear flow in bending members. For a composite beam, at the interface between the concrete slab and the steel beam

$$\text{Shear flow} = vb = \frac{VQ_T}{I_T}$$

where Q_T is the first moment of the transformed concrete area about the neutral axis of the composite beam, and I_T is the moment of inertia of the entire transformed cross section. If q is the load capacity of a given connector, equilibrium requires that

$$q = \int_m^n vb \, dx = \int_m^n \frac{VQ_T}{I_T} \, dx = \frac{Q_T}{I_T} \int_m^n V \, dx$$

If the shear diagram varies linearly between the two locations $x = m$ and $x = n$,

$$q = \frac{Q_T}{I_T} \frac{V_m + V_n}{2} s \tag{7.71}$$

or

$$s = \frac{2qI_T}{Q_T(V_m + V_n)} \tag{7.72}$$

where s is the spacing of the connectors along the beam at $x = (m + n)/2$.

The strength of individual hooked or headed (stud) shear connectors is given in table 7.6. These values correspond to q, with a load factor of 1.0, and are from design values published by the Nelson Stud Welding Division of Gregory Industries of Lorain, Ohio.

Table 7.6 Ultimate strength of hooked or headed (stud) shear connectors, in pounds

Connector size (dimensions in inches)	Concrete strength σ_c', lb/in²		
	3,000	3,500	4,000
$\frac{1}{2}$ diameter × 2 long	5,100	5,500	5,900
$\frac{5}{8}$ diameter × $2\frac{1}{2}$ long	8,000	8,600	9,200
$\frac{3}{4}$ diameter × 3 long	11,500	12,500	13,300
$\frac{7}{8}$ diameter × $3\frac{1}{2}$ long	15,600	16,800	18,000

Equation (7.72) presumes elastic behavior of the connectors. The resulting spacing diagram will vary along the member, depending upon the variation in V. It was noted earlier in this section, however, at ultimate load all shear connections participate equally—independent of spacing. Therefore, if N is the total number of connectors to be used in, say, one-half a beam, then

$$Nq = \frac{Q_T}{I_T} \times \left(\text{area under the shear diagram from 0 to } \frac{L}{2} \right) \tag{7.73}$$

The spacing will be uniform.

two 3″ × ¾ steel shear connection, 32 per ½ beam

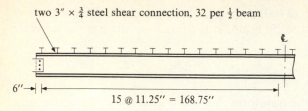

6″ ⟶| |←
15 @ 11.25″ = 168.75″

Figure 7.50 Distribution of stud connectors.

The maximum possible shear force that can be developed over any given length of beam will be the lesser of

$$F_c = 0.85\sigma'_c A_c$$

or

$$F_s = \sigma_y A_s$$

Against these should be applied the desired factor of safety. The American Institute of Steel Construction has suggested that $\frac{1}{2}F$ would be appropriate for design. For the composite beam in question (W 24 × 84 with $b_{\text{eff}} = 73.02$ in)

$$F_c = (0.85)(3000)(73.02)(4) = 744.8 \text{ kips}$$

$$F_s = (36)(24.7) = 889.2 \text{ kips}$$

Therefore $F = 744.8$ kips. Assuming $\frac{3}{4}$-in-diameter studs 3 in long, $q = 11.5$ kips. Therefore,

$$Nq = \tfrac{1}{2}F$$

or

$$N = \left(\frac{1}{2}\right)(7.448)\left(\frac{1}{11.5}\right) = 32.4$$

Since studs normally are fastened in pairs, use 16 rows of 2 connectors each, spaced at $11\frac{1}{4}$ in over each half of the beam. The first row should be started at 6 in from the end.

In normal situations, the slabs now would be designed to carry the loads in the direction perpendicular to the steel beams. A 1-ft-wide strip would be taken out, and the reinforcement would be selected as if the member were a continuous beam. It should be recognized, however, that the continuous beam strips in question are "indeterminate" and the methods of chap. 8 are needed to determine the reactions and the bending-moment diagrams. Approximate (conservative) coefficients are available in several of the codes to facilitate the handling of this problem. However, since the design process is essentially the same as that which has been demonstrated in chap. 2, the slab-design problem will not be considered further here.

CHAPTER

EIGHT

ANALYSIS AND DESIGN OF SIMPLE INDETERMINATE STRUCTURES

It was noted in chap. 3 that "statically determinate structures are those for which the equations of statics are sufficient to establish within the structure the distribution of forces and moments." Indeterminate structures, then, by definition, are structures for which the laws of statics (that is, the equations of static equilibrium) are *not* sufficient to fully define the reactions and stresses in terms of known quantities such as the applied load(s) and the length(s) of the various interconnected members that make up the structure in question.

Structures may be indeterminate externally or internally, or a combination of the two. They may be subjected to any of the various types or kinds of loading discussed in the earlier chapters of this volume.

Indeterminate structures in the elastic range respond to applied loads in the normal fashion, that is, with deformations increasing proportionally to the loads. However, under special conditions of loading or geometrical properties of the structure, they may buckle. Buckling can involve only one member or a group of members, or the entire structure may be involved. Depending on the degree of redundancy of the structure and the particular member involved, buckling of an individual member (or even a limited group of members) may not necessarily limit the capacity of the structure to sustain increased loading beyond that particular buckling load. If the structure is capable of redistributing the increases to other members, then it responds to the increases as if it were a new structure of reduced redundancy. (The buckled member would be replaced by the force or moment that existed in that member at the time of buckling.) It is to be understood, however, that overall frame buckling represents a sudden, dramatic shift in the manner in which the loads are sustained— and this type of buckling is usually considered to be a sufficient design limitation.

Throughout this chapter it generally will be assumed that the elastic limit of the material has not been exceeded; that Hooke's law is valid; that superposition holds; that deflections and deformations are relatively small; and that forces and deformations increase gradually. Unless otherwise specified, it also will be assumed that deflections are sufficiently small to allow a first-order formulation of the problem (that is, the equilibrium equations will be defined in the undeformed state). For certain types of problems, however, the influence of the reduced bending stiffness due to axial thrust (a second-order consideration) will be taken into account.

It should be noted that several indeterminate structures were solved in chap. 6, although it was not there indicated (see, for example, figs. 6.5, 6.7, and 6.9). The important point to be made here is that those solutions were based upon a formulation of the problem that allowed consideration of deformational as well as statical conditions. It should be evident that a direct extension of those methods that were there used to more "realistic" (that is, complicated) structures would be formidable. Theoretically, such methods are capable of handling these types of problems. Practically, however, the work that would be required to obtain solutions would be prohibitive for most cases.

Many practical civil engineering type structures are subjected to loads that result in relatively low levels of axial forces in the various members. Therefore, when solving indeterminate structural problems, it is not uncommon to discount (at least during a "first pass") the influence on bending stiffness of axial thrust. For continuous bridge-type beam structures, this assumption is obvious. For low buildings, it has been demonstrated to be adequate for a large number of cases. Depending upon the design control condition, the effect may or may not be serious when multistory frames are considered. Most rigid-frame structures that are made up of reinforced concrete beams and columns have relatively short and stocky members. Again, for such cases, the influence of axial thrust on the bending stiffness has been shown to be relatively small.

In chaps. 2 and 3, the concept of a *design envelope* to provide assistance in selecting member sizes was developed. There, with the resultant forces, moments, and shears known, the problem was to choose the optimum proportions of the cross section in question. For indeterminate structures it will be demonstrated that the ratios of the bending stiffness of the various members comprising the structure may have a profound effect upon the resultants in the individual members. These ratios, then, provide an added dimension to the optimum design process.

It is the primary purpose of this chapter to develop several of the more common methods for analyzing (and designing) indeterminate systems. Only relatively simple structures will be considered—since it is "procedure" that will be a major concern. When faced with larger and more complex structural configurations, a computer should be used to facilitate the effective and efficient handling of the many variables. Chapter 9 deals with this subject.

8.1 INTRODUCTION

From the standpoint of material economy, it is more advantageous to use an indeterminate structure than it is to use a determinate one. Consider, for example, the two beams shown in fig. 8.1. Both beams are of length L and are subjected to a uniformly

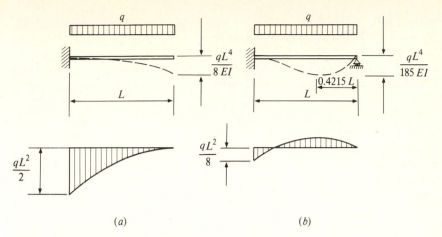

Figure 8.1 Bending moments and deflections for two beams.

distributed, lateral load of q lb/ft. The first member, (a), is statically determinate. The propped cantilever beam shown in (b) is indeterminate to the first degree. (That is, there is one more unknown component of end reaction than there are equations of static equilibrium.)

The bending moment and deflection diagrams for both these cases can be determined using the equations given in table 4.1, taking into account the appropriate boundary conditions. For the determinate (a): The maximum bending moment occurs at the fixed support and equals $qL^2/2$; and the maximum deflection occurs at the free end and equals $qL^4/8EI$. For the indeterminate (b): The maximum bending moment occurs at the fixed support and equals $qL^2/8$; and the maximum deflection occurs within the span as shown in the figure and equals $qL^4/185EI$. If bending normal stresses control the design, (b) is more efficient than (a), by a factor of 4.0. If maximum deflection is the critical design condition, the indeterminate structure is superior to the determinate one by a factor of approximately 13. Clearly, then, by either measure, the indeterminate structure is the more efficient of the two in supporting the given load. Additional savings would be realized by making the structure even more redundant, for example, by fixing the right-hand end of the beam against rotation.

8.1.1 Stiffness and Flexibility Methods of Analysis

In general, there are two basic methods used to solve indeterminate structural problems: the *force* (or *flexibility*) method, and the *displacement* (or *stiffness*) method. In the force method, the redundant forces (or stress resultants) are taken as the unknown quantities to be determined. In the stiffness method, the unknowns are translations, or rotations—usually assumed to occur at one or more joints of the structure. (The unknown forces or moments used in the first method are frequently referred to as *static redundants*. The unknown displacements of the second method are *kinematic redundants*.)

In both methods, problem formulation relies heavily upon the principles of superposition, and sets of simultaneous algebraic equations which must be solved result. The size (order) of the set that results from the application of the force method corresponds to the number of static redundants in the problem. The set resulting from the displacement method has a size equal to the kinematic redundants. (It is to be understood that since the principle of superposition holds only when the behavior is elastic and deformations are small, both these methods are valid only for linear systems.)

In the abstract sense, the question of which method should be used to solve a particular problem is relatively easy to answer: The method used should be the one which involves the least number of unknowns. There are, of course, exceptions to this general rule; for example, when only one or two particular quantities are desired, or when the stability of a given member or group of members is in question. (It should be understood, however, that such cases are exceptions.)

Consider the four-member, indeterminate structures shown in fig. 8.2. Each of the members has its outer end fixed. At A, they are all rigidly joined. A concentrated load Q is applied as shown in (a). To solve the problem by the force method, it is first necessary to determine the number of static redundants. As shown in (b), there are 9 of these. Thus, for this frame, the force method would involve the solution of a 9×9 set of equations. On the other hand, as indicated in (c), the entire deformed shape of the structures can be described in terms of the single angle θ_A. For the displacement method, then, only one equation need be solved. The choice of method in this case should be obvious. Other cases would be handled in a similar way, as will be demonstrated in this chapter.

The basic concepts of stiffness and flexibility can be illustrated by considering the behavior of the axially loaded spring shown in fig. 8.3. In this case it is assumed that the spring in question responds to loading in a linearly elastic fashion. For an applied load of P, there results a total change in length of the spring of ΔL. Denoting the spring constant as c, where the units are, for example, pounds per inch per inch:

$$k = \frac{P}{\Delta L} = cL \qquad \text{lb/in} \qquad (8.1)$$

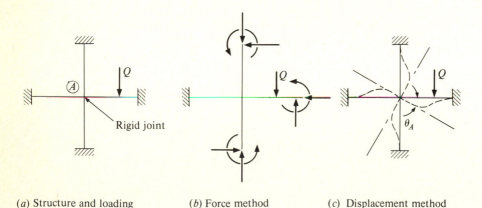

(a) Structure and loading   (b) Force method   (c) Displacement method

Figure 8.2 Number of redundants as a function of the chosen method of solution.

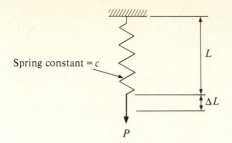

Figure 8.3

where k is defined as the stiffness of the member (the amount of force required to produce one unit of displacement in the total spring). Flexibility (usually denoted by the symbol u) is defined as the deformation that corresponds to a unit value of the applied load P; that is,

$$u = \frac{\Delta L}{P} \quad \text{in/lb} \tag{8.2}$$

It should be evident, at least for this case, that stiffness and flexibility are the inverse of each other.

$$k = \frac{1}{u} = u^{-1} \quad \text{and} \quad u = \frac{1}{k} = k^{-1} \tag{8.3}$$

These concepts can be extended readily to more complex configurations. For example, if a uniform cross-section, axially loaded, tension member of the type defined in fig. 8.4a were the member in question, the corresponding values would be

$$u = \frac{\Delta L}{P} = \frac{L}{AE} \quad k = \frac{P}{\Delta} = \frac{AE}{L} \quad \text{and} \quad k = u^{-1}$$

In fig. 8.4b is shown a solid, circular cross-section shaft of length L subjected to an applied end torque of T. Using eq. (4.32), the flexibility of this shaft would be

$$u = \frac{\beta}{T} = \frac{L}{G\kappa_T}$$

and the stiffness would be

$$k = u^{-1} = \frac{G\kappa_T}{L}$$

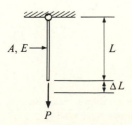

Figure 8.4a Stiffness and flexibility of an axially loaded member.

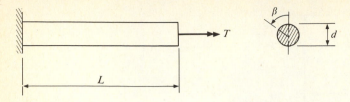

Figure 8.4b

In general, stiffness is defined as the stress resultant which produces a unit of deformation. Similarly, flexibility—by definition—is the magnitude of the deformation caused by a unit value of the stress resultant. Stiffness and flexibility, then, are the inverse one of the other.

In each of the systems examined, there obviously was but one degree of deformational freedom. In fig. 8.4a, for example, the tension member could deform only in the direction of the applied load P. Similarly, in fig. 8.4b, rotation occurred only about the longitudinal axis. It is to be understood that these are limiting, special cases. More general possibilities must be defined.

Consider, for example, the cantilever beam shown in fig. 8.5. Moreover, presume that two separate and distinct loading conditions exist. The first (case A) is a vertical force of F_1 applied at the free end of the member. The second (case B) is an applied end-bending moment of magnitude F_2. Both of these loadings produce at the free end of the beam (and for that matter, everywhere along the beam) vertical displacements and bending-type rotations. Defining the vertical deflection at the free end as *direction 1* and the rotation at that same point as *direction 2*, and noting, for example, that flexibility has been defined as displacement per unit of applied load, four separate flexibility quantities must be defined for the beams and loadings in question. For ease of solution and to facilitate discussion a double subscript notation will be used to identify these various quantities. u_{ij} will stand for the flexibility at the ith point (or

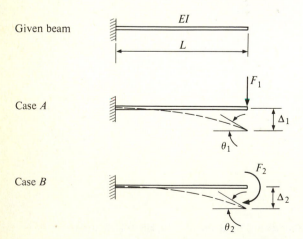

Figure 8.5 Cantilever beam subjected to two end loadings.

in the ith direction or both) resulting from the application of a unit load or moment at the jth point (or in the jth direction or both).

The four flexibilities in question are indicated in fig. 8.6, where (again, for emphasis)

$$\text{Flexibility} = u_{ij} \begin{cases} \text{position or direction of displacement} \\ \text{position or direction of unit load or moment} \end{cases}$$

The values of these flexibilities are computed readily using the unit load method of sec. 5.5.

$$u_{ij} = \int_0^L \frac{m_j m_i}{EI} \, dz \tag{8.4}$$

m_j is the moment resulting from a unit load (or moment) in the jth direction, and m_i is the moment caused by a unit load (or moment) in the ith direction or at the ith position. For this example, m_1 and m_2 are shown in fig. 8.7. (Moment values are plotted on the compression side of the member.) Therefore

$$u_{11} = \int_0^L \frac{m_1^2}{EI} \, dz = \frac{L^3}{3EI}$$

$$u_{21} = \int_0^L \frac{m_2 m_1}{EI} \, dz = \frac{L^2}{2EI}$$

$$u_{22} = \int_0^L \frac{m_2^2}{EI} \, dz = \frac{L}{EI} \tag{8.5}$$

$$u_{12} = \int_0^L \frac{m_1 m_2}{EI} \, dz = u_{21} = \frac{L^2}{2EI}$$

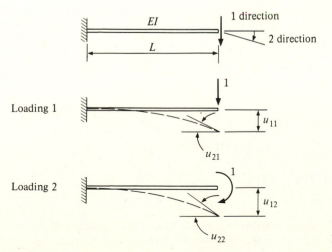

Figure 8.6 Notation for flexibility terms.

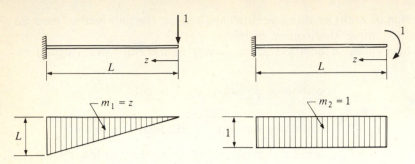

Figure 8.7 Cantilever beam subjected to unit loadings.

If the concentrated load F_1 is applied as shown in fig. 8.5, Δ_1 and θ_1 can be computed directly from these flexibility values.

$$\Delta_1 = F_1 u_{11}$$
$$\theta_1 = F_1 u_{21} \tag{8.6}$$

If the moment F_2 is applied,

$$\Delta_2 = F_2 u_{12}$$
$$\theta_2 = F_2 u_{22} \tag{8.7}$$

If both are applied simultaneously, Δ is the total vertical displacement of the free end, and θ is the total bending rotation at that point; that is,

$$\Delta = F_1 u_{11} + F_2 u_{12}$$
$$\theta = F_1 u_{21} + F_2 u_{22}$$

In matrix form this would be written as†

$$\begin{bmatrix} u_{11} & u_{12} \\ u_{21} & u_{22} \end{bmatrix} \begin{bmatrix} F_1 \\ F_2 \end{bmatrix} = \begin{bmatrix} \Delta \\ \theta \end{bmatrix} \tag{8.8}$$

$$\underset{\substack{\text{Flexibility} \\ \text{matrix}}}{} \qquad \underset{\substack{\text{Force} \\ \text{vector}}}{} \quad \underset{\substack{\text{Displacement} \\ \text{vector}}}{}$$

or

$$[U][F] = [D] \tag{8.9}$$

in which

$$[U] = \begin{bmatrix} \dfrac{L^3}{3EI} & \dfrac{L^2}{2EI} \\[2ex] \dfrac{L^2}{2EI} & \dfrac{L}{EI} \end{bmatrix} = \text{flexibility matrix}$$

† For a discussion of matrix formulation or matrix representation, see R. L. Ketter and S. P. Prawel, *Modern Methods of Engineering Computation*, chap. 4, McGraw-Hill, New York, 1969.

The flexibility matrix $[U]$ will always be square and symmetrical and will have as many rows as the number of degrees of freedom of the system being described. This formulation is nothing more than a direct application of the method of superposition discussed in chap. 4.

The stiffness of the beam at its free end can be derived in a similar fashion, noting that—by definition—stiffness is the stress resultant necessary to produce a unit of deformation in the direction in question. It also can be obtained by inverting the flexibility matrix.†

$$\text{Stiffness matrix} = [K] = [U]^{-1} = \begin{bmatrix} k_{11} & k_{12} \\ \\ k_{21} & k_{22} \end{bmatrix} = \begin{bmatrix} \dfrac{12EI}{L^3} & -\dfrac{6EI}{L^2} \\ \\ -\dfrac{6EI}{L^2} & \dfrac{4EI}{L} \end{bmatrix} \qquad (8.10)$$

By definition, each element k_{ij} of the stiffness matrix represents the stress resultant induced in the ith direction (or the ith position) by a unit of deformation in the jth direction (or the jth position) *with all other deformations prevented*. This is shown diagrammatically in fig. 8.8a. It is to be recognized that a direct solution for the various k_{ij}'s requires the solution of two indeterminate structural problems.

It is to be noted in fig. 8.8a that the force and moment vectors indicated—which, in effect, are the various terms of the stiffness matrix—are shown in assumed directions consistent with those of fig. 8.7. This is in spite of the fact that one could reasonably conclude in advance, from elementary equilibrium considerations, that certain of the vectors would actually act in the opposite sense. It is extremely important when solving indeterminate structural problems to define and methodically hold to a consistent sign system. For the cases shown, it has been assumed that forces and moments acting on the ends of the member are plus when they tend to produce clockwise rotations of the member. Similarly, clockwise rotations at the ends or relative end deflections that cause overall clockwise rotations of the entire member are presumed positive.

For the first of the loading conditions shown in fig. 8.8a, the moment diagrams

---

† For a discussion of matrix inversion see R. L. Ketter and S. P. Prawel, *Modern Methods of Engineering Computation*, chap. 6, McGraw-Hill, New York, 1969.

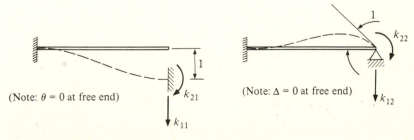

(Note: $\theta = 0$ at free end)

(Note: $\Delta = 0$ at free end)

Figure 8.8a Unit deformations for determination of stiffness terms.

Moment due to k_{11}

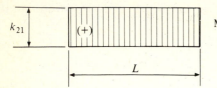

Moment due to k_{21}

Figure 8.8b

would be those shown in fig. 8.8b. Using the moment-area method of sec. 4.3.2 to define the deformations of the right-hand end of the member:

$$\theta = 0 = \frac{1}{2}\frac{k_{11}L}{EI}L + \frac{k_{21}}{EI}L$$

and

$$\Delta = 1 = \frac{1}{2}\frac{k_{11}L}{EI}L\frac{2L}{3} + \frac{k_{21}}{EI}L\frac{L}{2}$$

Solving these simultaneously yields

$$k_{11} = \frac{12EI}{L^3} \tag{8.11}$$

and

$$k_{21} = -\frac{6EI}{L^2} \tag{8.12}$$

If the same procedure were used for the second of the two indicated loading conditions, the following would be obtained:

$$\theta = 1 = \frac{1}{2}\frac{k_{12}L}{EI}L + \frac{k_{22}}{EI}L$$

$$\Delta = 0 = \frac{1}{2}\frac{k_{12}L}{EI}L\frac{2L}{3} + \frac{k_{22}}{EI}L\frac{L}{2}$$

from which it can be determined that

$$k_{12} = -\frac{6EI}{L^2} \tag{8.13}$$

$$k_{22} = \frac{4EI}{L} \tag{8.14}$$

[These are the same values that are given in eq. (8.10), which were determined by inverting the earlier derived flexibility matrix.]

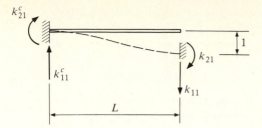

Figure 8.8c

The above discussion was concerned with stiffness at the free end of the cantilever. Using the same definitions, stiffness at the fixed end due to deformations or loads at the free end could be developed. For example, a unit deformation in the 1 direction at the free end produces, in addition to the force and moment k_{11} and k_{21} at the free end, k_{11}^c and k_{21}^c at the fixed end (or with respect to the location of imposed displacement, the far end) as shown in fig. 8.8c. Similarly, a unit rotation at the free end produces a vertical reaction k_{12}^c and bending moment k_{22}^c at the left end, which is in addition to k_{12} and k_{22} at the right-hand unit-displacement end. In both of these cases, the superscript c has been used to indicate the *carry-over effect*.

The *carry-over stiffness matrix* is defined as

$$[K^c] = \begin{bmatrix} k_{11}^c & k_{12}^c \\ k_{21}^c & k_{22}^c \end{bmatrix} \tag{8.15}$$

The values of the individual elements in this matrix can be determined from a consideration of static equilibrium.

$$[K^c] = \begin{bmatrix} k_{11} & k_{12} \\ -k_{21} - k_{11}L & -k_{22} - k_{12}L \end{bmatrix} = \begin{bmatrix} \dfrac{12EI}{L^3} & -\dfrac{6EI}{L^2} \\ -\dfrac{6EI}{L^2} & \dfrac{2EI}{L} \end{bmatrix} \tag{8.16}$$

The ratio k_{ij}^c/k_{ij} is referred to as the carry-over factor c_{ij}. For examples:

$$c_{22} = \frac{k_{22}^c}{k_{22}} = \frac{2EI/L}{4EI/L} = +\frac{1}{2}$$

$$c_{21} = \frac{k_{21}^c}{k_{21}} = \frac{-6EI/L^2}{-6EI/L^2} = +1$$

The plus signs denote that the carry-over moment and shear at the left-hand end of the member will act in a direction and in a sense consistent with the chosen clockwise positive sign system. This is not to say that they necessarily will act upward or clockwise—since ratios have been taken—but that the two ends will act consistently, as per the assumed directions. For example, if the vertical reaction on the right is upward, then the carry-over effect to the left-hand end will result in a downward vertical reaction of equal magnitude. Similarly, a clockwise moment on the right will result in a clockwise moment on the left equal to one-half that value.

Example 8.1 As an example of a three-degree-of-freedom system, consider the portal-type two-dimensional determinate structure shown in fig. 8.9a. The vertical members are each of length αL and have unit bending stiffnesses of βEI. The structure is rigidly attached at point A and is free at D. The three deformational directions in question are those indicated as 1, 2, and 3 at point D:

Direction 1, vertical (assumed positive upward)
Direction 2, horizontal (assumed positive to the right)
Direction 3, rotation (assumed positive counterclockwise)

The flexibilities u_{ij} are defined as shown in fig. 8.9b. For $j = 1$, a unit force is applied in the vertical direction. For $j = 2$, the unit force is horizontal. A unit moment corresponds to $j = 3$. As before, each flexibility term can be determined using the basic unit load equation (8.4).

$$u_{ij} = \sum_k \int^{L_k} \frac{m_i m_j}{EI_k} \, ds \qquad i = j = 1, 2, 3 \tag{8.17}$$

In this expression k refers to any one of the three members which make up the structure. The moments m_i and m_j used to compute u_{ij} are those shown in fig. 8.9c. The resulting flexibilities are

$$u_{11} = \int \frac{m_1^2}{EI} \, ds = \frac{L^3}{EI}\left(\frac{1}{3} + \frac{\alpha}{\beta}\right)$$

$$u_{12} = \int \frac{m_1 m_2}{EI} \, ds = \frac{L^3}{2EI}\left(\alpha + \frac{\alpha^2}{\beta}\right)$$

$$u_{13} = \int \frac{m_1 m_3}{EI} \, ds = \frac{L^2}{EI}\left(\frac{1}{2} + \frac{\alpha}{\beta}\right)$$

$$u_{21} = \int \frac{m_1 m_2}{EI} \, ds = \frac{L^3}{2EI}\left(\alpha + \frac{\alpha^2}{\beta}\right)$$

$$u_{22} = \int \frac{m_2^2}{EI} \, ds = \frac{L^3}{EI}\left(\alpha^2 + \frac{2\alpha^3}{3\beta}\right)$$

$$u_{23} = \int \frac{m_2 m_3}{EI} \, ds = \frac{L^2}{EI}\left(\alpha + \frac{\alpha^2}{\beta}\right)$$

$$u_{31} = \int \frac{m_3 m_1}{EI} \, ds = \frac{L^2}{EI}\left(\frac{1}{2} + \frac{\alpha}{\beta}\right)$$

$$u_{32} = \int \frac{m_3 m_2}{EI} \, ds = \frac{L^2}{EI}\left(\alpha + \frac{\alpha^2}{\beta}\right)$$

$$u_{33} = \int \frac{m_3^2}{EI} \, ds = \frac{L}{EI}\left(1 + \frac{2\alpha}{\beta}\right)$$

It should be evident that

$$u_{ij} = u_{ji}$$

The flexibility matrix is then

$$[u_{ij}] = \begin{bmatrix} \dfrac{L^3}{EI}\left(\dfrac{1}{3}+\dfrac{\alpha}{\beta}\right) & \dfrac{L^3}{2EI}\left(\alpha+\dfrac{\alpha^2}{\beta}\right) & \dfrac{L^2}{EI}\left(\dfrac{1}{2}+\dfrac{\alpha}{\beta}\right) \\[3ex] \dfrac{L^3}{2EI}\left(\alpha+\dfrac{\alpha^2}{\beta}\right) & \dfrac{L^3}{EI}\left(\alpha^2+\dfrac{2\alpha^3}{3\beta}\right) & \dfrac{L^2}{EI}\left(\alpha+\dfrac{\alpha^2}{\beta}\right) \\[3ex] \dfrac{L^2}{EI}\left(\dfrac{1}{2}+\dfrac{\alpha}{\beta}\right) & \dfrac{L^2}{EI}\left(\alpha+\dfrac{\alpha^2}{\beta}\right) & \dfrac{L}{EI}\left(1+\dfrac{2\alpha}{\beta}\right) \end{bmatrix} \qquad (8.18)$$

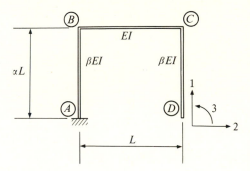

Figure 8.9a

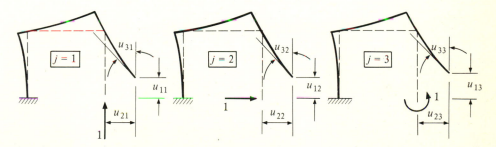

Figure 8.9b Loading and deformations for determining flexibilities.

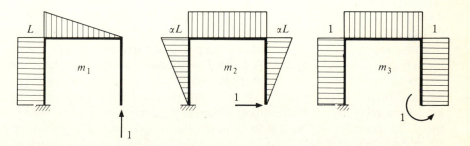

Figure 8.9c Moment diagrams associated with unit loadings.

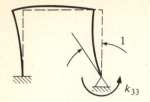

Note: Hinge introduced at free
end to ensure that no other
displacements occur.

k_{33}

Figure 8.9d Physical meaning of k_{33}.

As noted above, the stiffness matrix can be obtained directly by inverting the flexibility matrix. For the special case where $\alpha = \frac{1}{2}$ and $\beta = 1$, the resulting stiffness matrix is

$$[K] = [U^{-1}] = \begin{bmatrix} k_{11} & k_{12} & k_{13} \\ \\ k_{21} & k_{22} & k_{23} \\ \\ k_{31} & k_{32} & k_{33} \end{bmatrix} = \begin{bmatrix} \dfrac{3EI}{L^3} & 0 & -\dfrac{1.5EI}{L^2} \\ \\ 0 & \dfrac{19.2EI}{L^3} & -\dfrac{7.2EI}{L^2} \\ \\ -\dfrac{1.5EI}{L^2} & -\dfrac{7.2EI}{L^2} & \dfrac{3.45EI}{L} \end{bmatrix} \quad (8.19)$$

Both matrices (8.18) and (8.19) are symmetrical; that is,

$$u_{ij} = u_{ji} \quad \text{and} \quad k_{ij} = k_{ji}$$

It is extremely important to understand the physical meaning and limitations imposed on the stiffness terms k_{ij}. k_{33}, for example, is the magnitude of the force developed in direction 3 due to a unit deformation in direction 3, *with all other displacements at that free end prevented?* This case is illustrated in fig. 8.9d.

In the preceding examples, all the deformations in question occurred at a single given point on the structure. This need not be the case. Consider, for example, the following conceptual example:

Example 8.2 The structure in question is a simply supported beam. Vertical deflections at three different locations along the member are assumed as the unknowns. For flexibility calculations, the presumed structure, loading, and deformations are those shown in fig. 8.10a. For stiffness calculations, the situation would be that shown in fig. 8.10b. (Realistic directions for the various "forces" k_{ij} have been selected. Had a more constant pattern been chosen, signs would have been accounted for in resulting subsequent calculations. Even for the structures illustrated, however, it is to be noted that the directions assumed for k_{ij}, when $i \neq j$, are consistent. This has been done to allow a separate check of the numerical calculations from the known relationships $k_{ij} = k_{ji}$.)

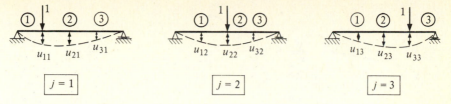

Figure 8.10a Loadings and deformations for determining flexibilities.

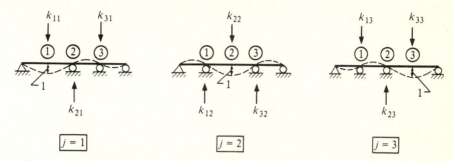

Figure 8.10b Loadings and deformations for determining stiffnesses.

8.1.2 Stiffness in the Presence of Axial Thrust

It was demonstrated in sec. 4.5 that the influence of compressive axial force is to reduce a member's overall effective bending resistance, and thereby to cause greater deformations. Tensile forces, on the other hand, reduce deformations. For both cases it is possible to determine the interrelationship among the loading parameters, the unit bending stiffnesses, and the resulting deformations. For a constant value of P then, assuming that buckling does not occur, one can define stiffness and flexibility including the influence of axial thrust.

Because of the general applicability of the stiffness method to a variety of different problem situations, since for this method it is assumed for each segment that all but one of the deformation parameters are zero, and that that particular one is unity, only stiffness calculations will here be discussed. It is presumed that the uniform cross-section member in question is of length L, has a unit bending stiffness of EI, a cross-sectional area of A, and in addition to imposed end moments or end shears is subjected to a constant axial compressive force P.

Two separate cases must be examined as shown in fig. 8.11: (*a*) the case of end-bending rotation with no lateral displacement of the ends of the member, and (*b*) the case where end rotations are held at zero and the two ends are relatively displaced. In the term k_{ij}, the subscript i refers to the resulting or imposed loading and the j to deformation parameter.

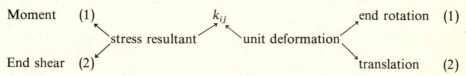

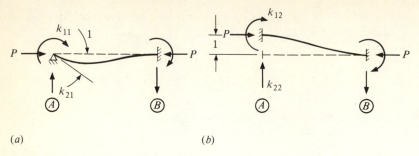

(a) (b)

Figure 8.11 Loadings and deformations for determining stiffnesses.

In case (a): the induced moment due to the imposed unit rotation at end A of the member AB would be denoted as k_{11}, since the prescribed deformation is a rotation (that is, $j = 1$), and the loading in question is a moment (that is, $i = 1$). Correspondingly, the end shear for case (a) would be k_{21}. For case (b): the imposed deformation is a translation and, therefore, $j = 2$.

In terms of the notation used in chap. 4, the structure and loading shown as case (a) in fig. 8.11 would be that illustrated in fig. 8.12. It is to be recognized that

$$k_{11} = \frac{M_A}{\theta_A} \tag{8.20}$$

and

$$k_{11}^c = \frac{M_B}{\theta_A} \tag{8.21}$$

The carry-over effect is defined by the relationship

$$c_{AB} = \frac{M_B}{M_A} = \frac{k_{11}^c}{k_{11}} \tag{8.22}$$

Since for the structure and loading in question there are no applied or induced lateral loads along the member, the governing differential equation (from table 4.6) is

$$\frac{d^4y}{dz^4} + k^2 \frac{d^2y}{dz^2} = 0$$

where

$$k^2 = \frac{P}{EI}$$

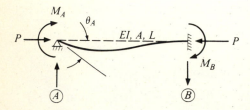

Figure 8.12 General structure and loading for determining stiffnesses, including the effect of axial thrust.

Assuming a coordinate system that has its origin at the left-hand end of the member, the boundary conditions to be satisfied are

$$\text{At } z = 0 \qquad y = 0 \text{ and } y'' = -\frac{M_A}{EI}$$

$$\text{At } z = L \qquad y = 0 \text{ and } y' = 0$$

The general solution of the differential equation is

$$y = C_1 \sin kz + C_2 \cos kz + C_3 z + C_4$$

When this solution is substituted into the boundary conditions, the values of the constants become

$$C_1 = \frac{M_A}{k^2 EI} \left[\frac{1 - \cos \lambda - \lambda \sin \lambda}{\sin \lambda - \lambda \cos \lambda} \right]$$

$$C_2 = \frac{M_A}{k^2 EI}$$

$$C_3 = \frac{M_A}{k^2 EI} \left[\frac{k(1 - \cos \lambda)}{\sin \lambda - \lambda \cos \lambda} \right]$$

$$C_4 = -\frac{M_A}{k^2 EI}$$

The equation of the elastic curve is then

$$y = \frac{M_A}{k^2 EI} \left[\frac{1 - \cos \lambda - \lambda \sin \lambda}{\sin \lambda - \lambda \cos \lambda} \right] \sin kz + \frac{M_A}{k^2 EI} \cos kz$$

$$+ \frac{M_A}{k^2 EI} \left[\frac{k(1 - \cos \lambda)}{\sin \lambda - \lambda \cos \lambda} \right] z - \frac{M_A}{k^2 EI} \qquad (8.23)$$

In these equations

$$\lambda = kL = \pi \sqrt{\frac{P}{P_e}}$$

where

$$P_e = \frac{\pi^2 EI}{L^2}$$

The end rotation θ_A can be determined by differentiating with respect to z eq. (8.23) and then setting $z = 0$.

$$\theta_A = \frac{M_A}{k^2 EI} \left[\frac{k(2 - 2 \cos kL - kL \sin kL)}{\sin kL - kL \cos kL} \right] \qquad (8.24)$$

The stiffness k_{11} is therefore

$$k_{11} = \frac{M_A}{\theta_A} = \frac{EI}{L} \left[\frac{kL(\sin kL - kL \cos kL)}{2(1 - \cos kL) - kL \sin kL} \right]$$

or

$$k_{11} = \frac{4EI}{L} \left[\frac{kL(\sin kL - kL \cos kL)}{8(1 - \cos kL) - 4 kL \sin kL} \right]$$

or

$$k_{11} = \frac{4EI}{L} \left[\frac{3\psi(u)}{4\psi^2(u) - \phi^2(u)} \right] \tag{8.25}$$

where $\psi(u)$ and $\phi(u)$ are defined in table 4.7. Since $4EI/L$ is the value of the stiffness when axial thrust is neglected, the particular influence of thrust on this stiffness term can be determined from the expression

$$k_{11} = \bar{k}_{11} K_{11}(u)$$

where

$$\bar{k}_{11} = \frac{4EI}{L} \tag{8.26}$$

and

$$K_{11}(u) = \frac{3\psi(u)}{4\psi^2(u) - \phi^2(u)}$$

$K_{11}(u)$ is shown as a function of the ratio P/P_e in fig. 8.13. (It should be recognized that the critical buckling load for the member in question—a member fixed at one end and pin-connected at the other end—is equal to $P_{cr} = 2.046 P_e$, where P_e is the Euler buckling load for a member pin connected at both ends.)

The induced bending moment at the far end of the member due to the applied moment M_A can be determined from the relationship

$$M_B = EI \frac{d^2 y}{dz^2} \Big|_{z=L} = M_A \frac{kL - \sin kL}{kL \cos kL - \sin kL}$$

The carry-over factor is then

$$c_{AB} = \frac{M_B}{M_A} = \frac{1}{2} \frac{\phi(u)}{\psi(u)} \tag{8.27}$$

Considering the entire member as a free body diagram, equilibrium of moments about the right-hand end gives

$$k_{11} + k_{21}(L) + k_{11}^c = 0$$

from which the end-shear stiffness term k_{21} can be defined.

$$k_{21} = \bar{k}_{21} K_{21}(u)$$

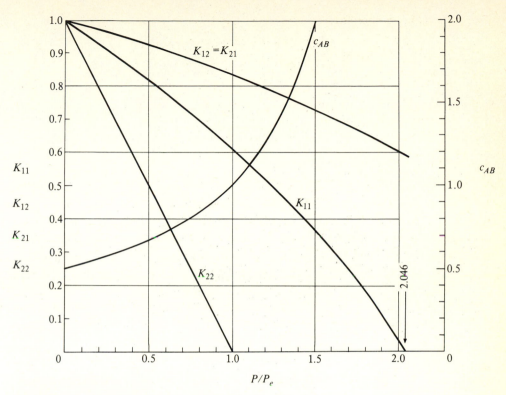

Figure 8.13 Beam-bending stiffnesses as a function of axial thrust.

where

$$\bar{k}_{21} = -\frac{6EI}{L^2}$$ (8.28)

and

$$K_{21}(u) = \frac{1}{2\psi(u) - \phi(u)}$$

This relationship also is shown in fig. 8.13.

For the second loading condition shown in fig. 8.11 (that is, for an imposed unit end displacement), solutions can be obtained in exactly the same fashion. The resulting stiffness terms would be

$$k_{22} = \bar{k}_{22}\, K_{22}(u)$$

where

$$\bar{k}_{22} = \frac{12EI}{L^3}$$ (8.29)

$$K_{22}(u) = \frac{1}{2\psi(u) - \phi(u)} - \frac{(2u)^2}{12}$$

and

$$k_{12} = \bar{k}_{12} \, K_{12}(u)$$

where

$$\bar{k}_{12} = -\frac{6EI}{L^2}$$

$$K_{12}(u) = \frac{1}{2\psi(u) - \phi(u)}$$

(8.30)

These relationships also are shown in fig. 8.13.

8.1.3 Analysis of Indeterminate Systems

It should be evident that both the flexibility method and the stiffness method are concerned with the interrelationships that exist between forces and deformations in a given structure. The basic difference in the two approaches, when solving indeterminate structural problems, is that in the flexibility (or force) formulation the object is to determine a set of carefully selected unknown forces or moments (usually referred to as *redundants*); whereas, in the stiffness method, particular, preselected deformations are presumed to be the unknowns. The flexibility formulation, then, generally assumes the form

$$[U][F] = [D]$$

(8.31)

where $[U]$ is called the flexibility matrix, $[F]$ is a column vector whose elements are the selected redundant forces or moments and $[D]$ is a displacement vector that ensures compatability. The stiffness formulation requires that

$$[K][D] = [F]$$

(8.32)

where $[K]$ is defined as the stiffness matrix, and $[D]$ is a column vector of the unknown deformations.

In the flexibility method, equilibrium is presumed to be satisfied at the outset, and the basic problem is the forcing of compatibility. This method is analogous to a complementary energy formulation. The stiffness method, on the other hand, presumes that all the deformations are compatible, and it is equilibrium which must be forced. This latter method, then, is analogous to the minimum potential principle.

While each of these methods could be used to solve any given structural problem, there are usually distinct advantages to be gained by selecting one as opposed to the other. If for the same problem one formulation involves fewer unknowns, then it is usually the best method to use for the solution of that problem.

8.2 FLEXIBILITY METHOD: THE METHOD OF CONSISTENT DEFORMATIONS

An indeterminate structure, made determinate by the introduction of unknown forces or moments for the redundants, will have an infinite number of possible values or combinations of values of those redundants that will satisfy the statical equations of equilibrium. There will be one—and only one—set, however, that will yield deformations that are geometrically compatible with the originally defined problem.

The procedure that normally is used when solving problems by this method requires that the given indeterminate structure first be made determinate by the removal of a sufficient number of unknowns. (The choice of one particular set of unknowns as opposed to another will depend on a variety of factors, not the least of which is the personal preference of the analyst. It is to be noted, however, that in a number of cases a judicious selection can materially reduce the amount of work required to obtain an answer.) Moreover, for complex structures, the accuracy of the solution is influenced by the selection.

The determinate structure that results from the removal of the redundants is often referred to as the *Base Structure*. For that structure, deformations in the directions of each of the chosen redundants can be determined using the methods developed in chaps. 4 and 5. Assigning reference numbers to these directions, a flexibility matrix can be established. Since each of the coefficients u_{ij} represents a displacement in the ith direction due to the application of a unit force in the jth direction, superposition can be used to define the composite effect of the various applied and redundant forces or moments, and these expressions can be written in such a fashion that they force the originally specified geometry. There results a set of simultaneous algebraic equations equal in number to the number of redundants, which contain as the unknowns the magnitude of the redundants.

Consider for illustration of the basic concept the propped cantilever beam problem shown in fig. 8.14. The member is fixed at end A and is simply supported at end B. Two lateral loads P_1 and P_2 are presumed acting. As shown in the second of the sketches, four components of unknown end reactions act on the structure: H_A, V_A, V_B, and M_A. There are, however, only three component equations of static equilibrium available for the solution of two-dimensional structural problems. The beam is, then, indeterminate to the first degree. Selecting V_B as the redundant, a solution could be obtained considering the loading diagrams of fig. 8.15. Δ_B is the deflection of the determinate (base) structure at location B in the direction of V_B. Δ'_B is the corresponding deflection due to the application of V_B. Obviously, to ensure the originally specified condition that there be no vertical movement of the end support

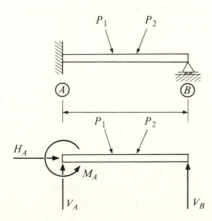

Figure 8.14 Propped cantilever beam subjected to concentrated lateral loads.

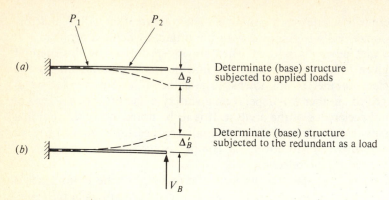

(a) Determinate (base) structure subjected to applied loads

(b) Determinate (base) structure subjected to the redundant as a load

Figure 8.15 Deformations due to determinate and redundant loadings.

at B, V_B must be selected such that Δ_B (which is downward and has a unique value dependent on the known quantities P_1, P_2, EI, and L) is exactly counterbalanced by Δ'_B (which is upward and whose magnitude is dependent on the magnitude of V_B). That is,

$$\Delta_B - \Delta'_B = 0$$

Using the unit load method of Chapter 5.

$$\Delta_B = \int \frac{M_{\text{det}} m_B \, ds}{EI}$$

$$\Delta'_B = V_B \int \frac{m_B^2 \, ds}{EI}$$

or

$$V_B = \frac{\int M_{\text{det}} m_B \, ds}{\int m_B^2 \, ds} \tag{8.33}$$

In these equations M_{det} has been used to designate the moment along the determinate member, which is a function of z, due to the applied forces P_1 and P_2. m_B is the corresponding moment due to the application of a unit force at B in the assumed direction of V_B. It is to be recognized that if the resulting value for V_B as determined from eq. (8.33) is positive, then V_B has been assumed in the correct direction. If the result is negative, the magnitude of V_B is correct, but the assumed direction is in error.

Once the value of V_B has been obtained, the values of M_A, V_A, and H_A can be determined for the originally specified problem, that is, for the indeterminate structures shown in fig. 8.15. Knowing these reaction components, bending moments, thrusts, and shears at all points along the member can be determined.

Example 8.3 As a numerical example of the process just described, consider the structure and loading shown in fig. 8.16a. Again, the member is a propped cantilever beam which is one time redundant. However, this time a uniformly distributed vertical load of q lb/ft is applied over the entire length. Since no

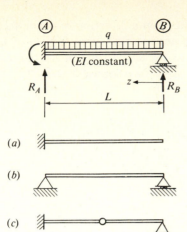

(a)

(b)

(c)

Figure 8.16a Propped cantilever beam subjected to uniformly distributed lateral loading (selection of possible redundants).

horizontal loads are specified, it is reasonable to presume that the horizontal reaction at end A is zero. The remaining components of reactions are then M_A, R_A, and R_B. Any of these, or for that matter any internal stress resultant, could be selected as the redundant. Several of these possibilities will be considered in this example, thus allowing a comparison of the relative merits of each.

Case (a): In this case the base structure is a cantilever beam. The governing loadings, moment diagrams and deflections are those shown in fig. 8.16b. In both cases the moments have been plotted on the compression side of the member. When multiplying their values together, then, independent of whether one is called positive and the other negative, those on the same side will have like signs, and those on the opposite, unlike. Since it is the product of moments which is important when using the unit load method, it is only this sameness or difference that is of significance. For example,

$$\Delta_B = \int_0^L \frac{Mm}{EI}\, dz = \int_0^L \frac{qz^2}{2EI}(-z)\, dz = -\frac{qL^4}{8EI}$$

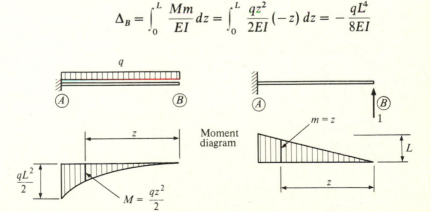

Figure 8.16b Moment diagrams due to determinate and redundant loadings—vertical reaction at B selected as redundant.

(The negative sign indicates that end B deflects in a direction opposite to that assumed shown in the second sketch by the unit vertical force applied at B.)

The vertical deflection of point B due to the application of the redundant force R_B can be determined by multiplying by R_B the response of the structure to a unit load at that point. That is,

$$\Delta_B' = R_B(u_{11}) = R_B \int_0^L \frac{m^2}{EI} dz = R_B \frac{L^3}{3EI}$$

Therefore, for zero deflection to exist at point B in the indeterminate structure,

$$\Delta_B + \Delta_B' = 0$$

or

$$\frac{-qL^4}{8EI} + R_B \frac{L^3}{3EI} = 0$$

This gives for the redundant reaction

$$R_B = +\tfrac{3}{8}qL \tag{8.34}$$

(Since the value is positive, R_B acts in the direction of the assumed unit load in the second sketch of fig. 8.16b.)

Case (b): In this case the base structure is presumed to be a simple beam; that is, the selected redundant is the end moment M_A. The deformation in question— that is, the one designated as the unknown force in the formulation—is a rotation at end A. The necessary condition for solution is

$$\theta_A + \theta_A' = 0$$

In terms of the applied and unit load moments

$$\int \frac{Mm}{EI} dz + M_A \int \frac{m^2}{EI} dz = 0$$

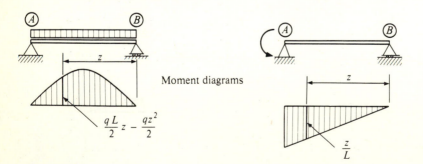

Moment diagrams

Figure 8.16c Moment diagrams due to determinate and redundant loadings—moment at A selected as redundant.

or

$$\int_0^L \frac{\left(\frac{qL}{2}z - \frac{qz^2}{2}\right)\left(-\frac{z}{L}\right)dz}{EI} + M_A \int_0^L \frac{\left(-\frac{z}{L}\right)\left(-\frac{z}{L}\right)dz}{EI} = 0$$

This yields the control equation

$$-\frac{qL^3}{24EI} + M_A \frac{L}{3EI} = 0$$

The redundant end moment is therefore

$$M_A = +\frac{qL^2}{8} \tag{8.35}$$

Again, the plus sign signifies that the counterclockwise direction initially chosen for the unit moment will be the direction of the redundant moment M_A. (The reader should verify that this value of M_A will yield the same value for R_B as was derived in the preceding case.)

Case (c): Next, consider the same problem but presume that the structure is made determinate by the introduction of a pin (a hinge) at the midspan of the beam as shown in fig. 8.16d. Deformation consistent with the original geometry requires that

$$\theta_C + M_C u_{33} = 0$$

or

$$\int_C^A \frac{M_{AC}m_{AC}}{EI}dz + \int_B^C \frac{M_{BC}m_{BC}}{EI}dz + M_C\left[\int_C^A \frac{m_{AC}^2}{EI}dz + \int_C^B \frac{m_{BC}^2}{EI}dz\right] = 0$$

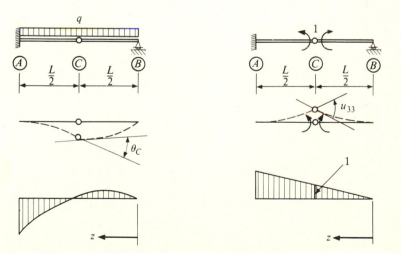

Figure 8.16d Deflections and moment diagrams due to determinate and redundant loadings—moment at C selected as redundant.

Carrying out these integrations and solving for M_C yields

$$M_C = +\tfrac{1}{16}qL^2 \tag{8.36}$$

(Again, the reader should verify that this value of M_C will yield the same values for M_A and R_B that were obtained in the preceding cases.) It should be evident that, because of the increased number of computations required, the choice of M_C as the redundant was not a wise one.

Example 8.4 As a second example, consider the fixed-base rectangular portal frame shown in fig. 8.17a. The frame is three times statically indeterminate. Selecting as the redundants the reactions at point D, that is, M_D, V_D and H_D; the loading, moment diagram and conceptual deflection diagram for the determinate system subjected to the applied load Q are those shown in fig. 8.17b.

Denoting the 1, 2, and 3 directions as the vertical, the horizontal and the rotation at D, the base structure deformations can be determined from the relationship

$$\Delta_i = \sum \int \frac{Mm_i}{EI}\,dz \qquad i = 1, 2, 3 \tag{8.37}$$

where the summation is used to indicate that all the members in the structure

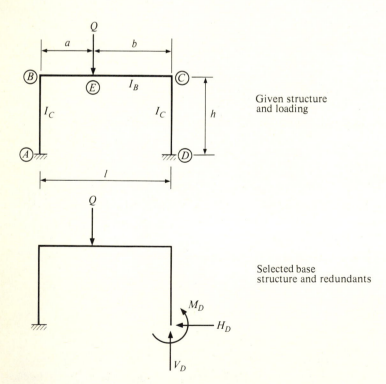

Given structure and loading

Selected base structure and redundants

Figure 8.17a

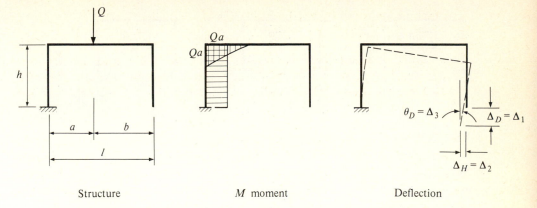

Figure 8.17b Moment diagrams and deflections due to determinate loading.

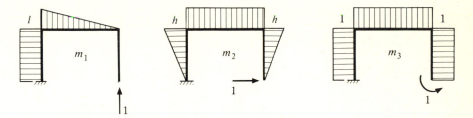

Figure 8.17c Moment diagrams due to unit redundant loadings.

must be taken into account. The m_i values are moments due to unit load in the 1, 2, and 3 directions. They are shown in fig. 8.17c.

The deformations due to the redundants can be determined using the flexibility coefficients corresponding to unit load considerations;

$$u_{ij} = \sum \int \frac{m_i m_j}{EI} dz \qquad (8.38)$$

Figure 8.17d indicates the subscript and component correspondence when using these equations. Since deformations at D must be equal to zero, the following three equations can be written:

$$V_D u_{11} + H_D u_{12} + M_D u_{13} + \Delta_1 = 0$$
$$V_D u_{21} + H_D u_{22} + M_D u_{23} + \Delta_2 = 0$$
$$V_D u_{31} + H_D u_{32} + M_D u_{33} + \Delta_3 = 0$$

or

$$\begin{bmatrix} u_{11} & u_{12} & u_{13} \\ u_{21} & u_{22} & u_{23} \\ u_{31} & u_{32} & u_{33} \end{bmatrix} \begin{bmatrix} V_D \\ H_D \\ M_D \end{bmatrix} = \begin{bmatrix} -\Delta_1 \\ -\Delta_2 \\ -\Delta_3 \end{bmatrix} \qquad (8.39)$$

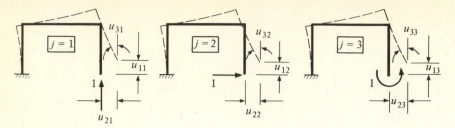

Figure 8.17d Deflections due to unit redundant loadings.

Taking into account the appropriate signs for the various integrations and solving simultaneously the equations listed as (8.39) yields the following values for the redundants:

$$V_D = Q\left\{1 - \frac{b}{l}\left[1 + \frac{(a/l)(b/l - a/l)}{6\gamma + 1}\right]\right\}$$

$$H_D = Q\frac{3ab}{2lh(\gamma + 2)} \tag{8.40}$$

$$M_D = \frac{Qab}{l}\left[\frac{1}{2(\gamma + 2)} + \frac{b/l - a/l}{2(6\gamma + 1)}\right]$$

where

$$\gamma = \frac{h}{l}\frac{I_B}{I_C} = \frac{(I_B/l)}{(I_C/h)} \tag{8.41}$$

γ is a relative measure of the bending stiffness of the beam to the column.

The influence of axial thrust was not considered in the example just carried out, even though it is obvious that the columns (and for that matter, even the beam) are subjected to compressive forces. For most real, single-story, civil engineering type portal frames, the ratio of the magnitude of the axial thrust to the yield strength in the various members is quite small, usually less than 10 percent. The ratios of thrusts to the Euler buckling loads also are small [note eq. (4.145)]. Unless the loading is unusual, it is common practice to discount the influence of thrust on bending stiffness when solving indeterminate structural problems of this type. An example of where this definitely would not be an appropriate assumption is the case where a heavy crane is attached directly to the columns of the frame.

In example 8.4 it was assumed that the structure was subjected to one and only one type of loading. If this were a real case, there undoubtedly would be other loadings to consider. Having solved this one case, however, the work required to solve another would be less; many of the terms (especially the flexibility matrix entries) could be reused directly. (This is because the flexibility matrix is a property of the given base structure and not of the applied loading.)

To illustrate how the previously derived values could be incorporated into other situations, consider the analysis of the frame shown in fig. 8.4, but this time subjected simultaneously to the following three different conditions of loading:

1. A varying lateral load (fig. 8.18*a*)
2. A horizontal misalignment of the foundation to the right at *D* (fig. 8.18*b*)
3. A temperature increase of $+\Delta_T$ (fig. 8.18*c*)

In case (*a*) the lateral loading produces deformations at *D* in all the component directions (1), (2), and (3). (Note that those deformations indicated in the sketch are in the assumed positive directions.) For case (*b*), it is specified that the foundation for the frame has been misconstructed by a horizontal amount equal to ΔH_b. (The frame equally well could have been fabricated short by the amount indicated, and the effect would be the same.) For this case, it is presumed that the vertical deflection and the rotational restraint are correct; that is, there will be no vertical deflection nor rotation at *D*. In case (*c*), the assumption is made that the entire structure, not just one or two of the members, is subjected to the total increased temperature. More than that, it is presumed that the thermal coefficients of expansion for both of the columns are equal. There will then result in the base structure a horizontal displacement at *D* equal to ΔH_c, but ΔV_c and $\Delta \theta_c$ will equal zero. (Had the columns been of different lengths, or had their thermal coefficients been different, there also would have resulted a component of deformation in the vertical direction.)

For the resulting deformations at *D* to be consistent with those specified in the original statement of the indeterminate problem, the following equation would have to be satisfied:

$$\begin{bmatrix} u_{11} & u_{12} & u_{13} \\ u_{21} & u_{22} & u_{23} \\ u_{31} & u_{32} & u_{33} \end{bmatrix} \begin{bmatrix} V_D \\ H_D \\ M_D \end{bmatrix} + \begin{bmatrix} \Delta V_a \\ \Delta H_a + \Delta H_b + \Delta H_c \\ \Delta \theta_a \end{bmatrix} = \begin{bmatrix} 0 \\ 0 \\ 0 \end{bmatrix} \qquad (8.42)$$

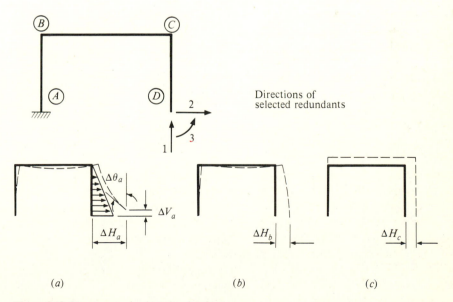

(*a*)   (*b*)   (*c*)

Figure 8.18 Structure and selected redundants.

In general matrix form:

$$[U][F] + [\Delta_Q + \Delta_T + \Delta_Y] = [0] \tag{8.43}$$

where $[U]$ and $[F]$ represent, as before, the flexibility and redundant matrices, and $[\Delta_Q + \Delta_T + \Delta_Y]$ is a column vector of components of deformations at the points and in the directions of the various redundants. Δ_Q represents the components due to the applied external loading; Δ_T represents the deformations of the base structure due to temperature changes; and Δ_Y, support settlements.

Example 8.5 Truss structures also can be indeterminate, and the method of consistent deformations can be used to determine the value(s) of the redundant(s). Consider the structure shown in fig. 8.19a in which all members are presumed to be pin-connected at their ends (and are below their buckling loads). Externally, the structure is indeterminate to the first degree. Internally there are two redundants. It is given that all members have a cross-sectional area equal to A, and the same material is used throughout (that is, all members have the same E). Assuming as the redundants the vertical reaction at C and the forces in members BC and CF, the base structure and applied loading are those shown in fig. 8.19b. Also indicated on this diagram are the resulting forces in each of the members due to this loading. (Tension has been presumed positive, and compression negative.)

For the assumed base structure and loading shown in fig. 8.19b, there will be a vertical deflection at point C, and there also will result relative movements between points B and C and between points C and F. These are the deformations of the base structure that must be determined and subsequently eliminated in the consistent deformation formulation:

$$\text{Between } B \text{ and } C \qquad \Delta_{BC} = 0$$
$$\text{Between } C \text{ and } F \qquad \Delta_{CF} = 0$$
$$\text{At support } C \qquad \Delta_{CV} = 0$$

To determine these deformations, and also to establish the flexibility matrix for the base structure, the loadings shown in fig. 8.19c would be considered.

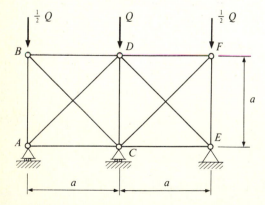

Figure 8.19a

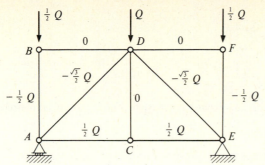

Figure 8.19b Forces in members due to determinate loading.

The desired deflections would be computed from the general relationship derived earlier as eq. (5.67); that is,

$$\Delta = \sum \frac{P}{AE} pl \qquad (8.44)$$

In this equation Δ is the deflection (or the relative displacement between any two points in the truss) in the direction of the assumed unit forces, P is the bar force due to the applied loading, and p is the bar force due to the unit loads.

For the base structure,

$$\Delta_{BC} = \Delta_1 = \sum \frac{P}{AE} p_a l$$

$$\Delta_{CF} = \Delta_2 = \sum \frac{P}{AE} p_b l \qquad (8.45)$$

$$\Delta_{CV} = \Delta_3 = \sum \frac{P}{AE} p_c l$$

Correspondingly, the flexibilities would be determined from the relationship

$$u_{ij} = \sum \frac{p_i p_j}{AE} l \qquad i, j = a, b, c \qquad (8.46)$$

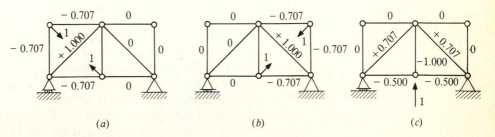

Figure 8.19c Forces in members due to unit redundant loadings.

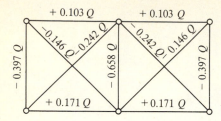

Figure 8.19d Resulting forces in members of structures.

The equations of consistent deformations are then

$$\begin{bmatrix} u_{11} & u_{12} & u_{13} \\ u_{21} & u_{22} & u_{23} \\ u_{31} & u_{32} & u_{33} \end{bmatrix} \begin{bmatrix} F_{AC} \\ F_{CF} \\ R_C \end{bmatrix} + \begin{bmatrix} \Delta_1 \\ \Delta_2 \\ \Delta_3 \end{bmatrix} = \begin{bmatrix} 0 \\ 0 \\ 0 \end{bmatrix} \qquad (8.47)$$

where F_{BC} and F_{CP} are, respectively, the redundant bar forces in members BC and CF, and R_C is the redundant vertical reaction at C. Solving these equations simultaneously yields

$$F_{BC} = -0.146Q$$

$$F_{CF} = -0.146Q$$

$$R_C = +0.864Q$$

(The signs indicate that members BC and CF are in compression, and the reaction R_C is in the direction of the assumed unit force at C.) Knowing these values, the remaining bar forces may be determined from static equilibrium equations. These forces are shown on the structure in fig. 8.19d.

8.2.1 Three-Moment Equation

In the preceding examples, the computational work required to obtain solutions was not discussed. It should be evident, however, that as the degree of indeterminacy increases, if all the elements of the flexibility matrix have values, the amount of work required to obtain a solution increases rapidly. A choice of redundants that limits the number of entries in the flexibility matrix is therefore to be desired.

For continuous-beam-type problems, the computational work can be markedly reduced if it is assumed that the redundants are the bending moments over the interior supports. The deformational terms for such an assumption are the end rotations of the resulting determinate simple beams. It is to be recognized, however, that it is the total change in angle of rotation between the ends of the two adjacent beams that must be computed, not just the rotation of one of the members.

Consider two adjacent spans in the interior of a multispan continuous beam (see fig. 8.20). It is presumed that the left-hand span in question is subjected to a uniformly distributed load of w_L, and the right-hand span to a load of w_R. The left-hand span is presumed to have a unit bending stiffness of EI_L and a length of L_L; the right-hand span is EI_R and L_R.

Considering location 0: the total change in angle between the right-hand end of

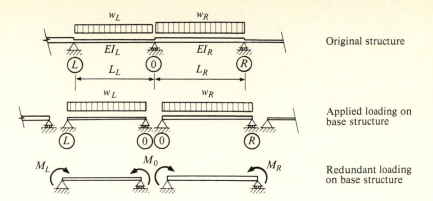

Figure 8.20 Continuous beam subjected to uniformly distributed loadings.

the left-hand simple beam and the left-hand end of the right-hand simple beam, due to the distributed loads w_L and w_R, is

$$\theta_0 = \frac{w_L L_L^3}{24EI_L} + \frac{w_R L_R^3}{24EI_R} \tag{8.48}$$

The corresponding rotation (at point 0) due to the redundant moments is

$$\theta_0' = M_j u_{ij} = M_L \frac{L_L}{6EI_L} + M_0 \left(\frac{L_L}{3EI_L} + \frac{L_R}{3EI_R} \right) + M_R \frac{L_R}{6EI_R}$$

For consistency of deformations $\theta_0 + \theta_0' = 0$, or

$$M_L \frac{L_L}{EI_L} + 2M_0 \left(\frac{L_L}{EI_L} + \frac{L_R}{EI_R} \right) + M_R \frac{L_R}{EI_R} + \frac{1}{4} \left(\frac{w_L L_L^3}{EI_L} + \frac{w_R L_R^3}{EI_R} \right) = 0 \tag{8.49}$$

(This formulation does not require that the actual value of the slope of the continuous beam be zero at the support. Rather, it requires that the slope "just to the left" of the support be equal to that "just to the right"; or, in other words, that the change in slope when going from the left-hand member to the right-hand member—at the support—be zero.)

Had the distributed lateral loads been concentrated ones (as shown in fig. 8.21), the value of θ_0 would be

$$\theta_0 = \frac{Q_L L_L^2}{EI_L} (\alpha_L - \alpha_L^3) + \frac{Q_R L_R^2}{EI_L} (\alpha_R - \alpha_R^3) \tag{8.50}$$

Had support settlements occurred (see fig. 8.22),

$$\theta_0'' = \frac{y_R - y_0}{L_R} - \frac{y_0 - y_L}{L_L} \tag{8.51}$$

The basic three-moment equation, then, including all of these possibilities is that given as eq. (8.52).

$$M_L \frac{L_L}{EI_L} + 2M_0 \left(\frac{L_L}{EI_L} + \frac{L_R}{EI_R} \right) + M_R \frac{L_R}{EI_R} + 6E(\theta_{0,cL} + \theta_{0,cR}) + 6E(\theta_0'') = 0 \tag{8.52}$$

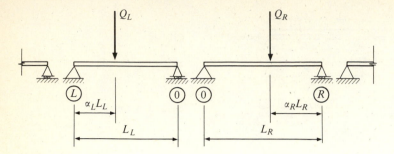

Figure 8.21 Beams subjected to concentrated lateral loads.

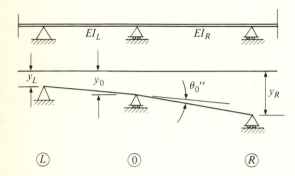

Figure 8.22 Support settlement.

where $\theta_{0,\,CL}$ and $\theta_{0,\,CR}$ are end rotations of simple beams due to the applied loading. Their values for various loadings are given in table 4.4. Including the effect of axial thrust, the equation would be

$$M_L \frac{L_L}{EI_L} \phi_L + 2M_0 \left(\frac{L_L}{EI_L} \psi_L + \frac{L_R}{EI_R} \psi_R \right) + M_R \left(\frac{L_R}{EI_R} \phi_R \right)$$
$$+ 6E(\theta_{0,CL} + \theta_{0,CR}) + 6E(\theta_0'') = 0 \qquad (8.52a)$$

where ϕ and ψ are as defined in chap. 4, and $\theta_{0,\,CL}$ and $\theta_{0,\,CR}$ are the appropriate end slopes of the simple beam column.

Example 8.6 The five-span (four times redundant) continuous-beam structure shown in Fig. 8.23 is subjected to a number of different loading conditions. One particular one is that indicated in the figure. Using the three-moment method, determine the moments at each of the interior supports. (It is assumed that E is constant throughout the structure.) Applying eq. (8.52), in turn, to each of the two-span sets yields:

For spans A to C: $2M_B\left(\dfrac{40}{2} + \dfrac{30}{1}\right) + M_C\dfrac{30}{1} + \dfrac{1}{4}\dfrac{(1)(40)^3}{2} = 0$

or $\qquad\qquad 100M_B + 30M_C + 8000 = 0$

For spans B to D: $30M_B + 100M_C + 20M_D + 15{,}000 = 0$

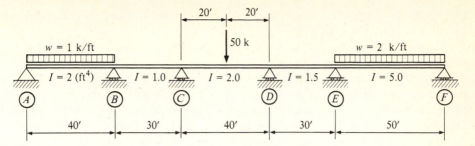

Figure 8.23

For spans C to E: $20M_C + 80M_D + 20M_E + 15{,}000 = 0$

For spans D to F: $20M_D + 60M_E + 12{,}500 = 0$

Therefore, in matrix form, the governing set of equations is

$$
\begin{bmatrix}
100 & 30 & 0 & 0 \\
30 & 100 & 20 & 0 \\
0 & 20 & 80 & 20 \\
0 & 0 & 20 & 60
\end{bmatrix}
\begin{bmatrix}
M_B \\ M_C \\ M_D \\ M_E
\end{bmatrix}
=
\begin{bmatrix}
-8000 \\ -15{,}000 \\ -15{,}000 \\ -12{,}500
\end{bmatrix}
\tag{8.53}
$$

The solution is

$$
\begin{aligned}
M_B &= -46.17 \ \text{ft} \cdot \text{kips} \\
M_C &= -112.76 \ \text{ft} \cdot \text{kips} \\
M_D &= -116.98 \ \text{ft} \cdot \text{kips} \\
M_E &= -169.34 \ \text{ft} \cdot \text{kips}
\end{aligned}
\tag{8.54}
$$

Had other conditions of loading been specified for this continuous beam structure, the flexibility matrix would be identical to that derived—so long as the redundants selected were the moments at the interior support points. All that would change would be the column vector of determinate rotations.

8.2.2 The Elastic-Center Method

The flexibility method can be used to solve any type or kind of indeterminate structural problem. The method ultimately reduces to a set of linear algebraic equations that must be solved for the unknown redundant forces. As illustrated in the preceding section, a judicious choice of redundants can materially reduce the amount of work required to obtain a solution.

For structures that form a single closed loop with the supporting medium, not only will the selection of the redundants be important, but the location of where these unknown forces or moments are presumed to be acting will materially alter the amount of work required to solve any given problem. (Structural types that fall into this category are arches, single-story, single-bay rigid frames, rings, nonprismatic beams, etc.)

Consider the fixed-end arch shown in fig. 8.24. The structure is three times redundant, and it will be assumed that the three left-hand end components of reaction are those redundants: V_A, H_A, and M_A. The shape of the centerline of the arch is presumed known in terms of the coordinate axes x and y, whose origin also is presumed at the left-hand end of the member. In developing the flexibility matrix for this choice of redundants, deflections in the vertical direction at A (that is, in the V_A direction) will be noted by the subscript 1. Those in the horizontal (or H_A) direction will be noted by 2. Rotations (corresponding to the M_A direction) will be given by subscript 3.

For consistent deformations:

$$\begin{bmatrix} u_{11} & u_{12} & u_{13} \\ u_{21} & u_{22} & u_{23} \\ u_{31} & u_{32} & u_{33} \end{bmatrix} \begin{bmatrix} V_A \\ H_A \\ M_A \end{bmatrix} + \begin{bmatrix} \Delta'_1 \\ \Delta'_2 \\ \Delta'_3 \end{bmatrix} = \begin{bmatrix} 0 \\ 0 \\ 0 \end{bmatrix}$$

or

$$\begin{bmatrix} u_{11} & u_{12} & u_{13} \\ u_{21} & u_{22} & u_{23} \\ u_{31} & u_{32} & u_{33} \end{bmatrix} \begin{bmatrix} V_A \\ H_A \\ M_A \end{bmatrix} = \begin{bmatrix} -\Delta'_1 \\ -\Delta'_2 \\ -\Delta'_3 \end{bmatrix} \qquad (8.55)$$

where Δ'_1, Δ'_2, and Δ'_3 are the components of deformation of the base structure in the directions 1, 2, and 3, respectively, due to the applied external loadings Q_1 and Q_2.

Eq. (8.55) can be solved in a number of different ways, but for the purpose of this particular development, expansion by the method of determinants will be used. (Only the solution for V_A will be illustrated.)

$$V_A = \frac{\begin{vmatrix} -\Delta'_1 & u_{12} & u_{13} \\ -\Delta'_2 & u_{22} & u_{23} \\ -\Delta'_3 & u_{32} & u_{33} \end{vmatrix}}{\begin{vmatrix} u_{11} & u_{12} & u_{13} \\ u_{21} & u_{22} & u_{23} \\ u_{31} & u_{32} & u_{33} \end{vmatrix}} = \frac{-\Delta'_1 \begin{vmatrix} u_{22} & u_{23} \\ u_{32} & u_{33} \end{vmatrix} + \Delta'_2 \begin{vmatrix} u_{12} & u_{13} \\ u_{32} & u_{33} \end{vmatrix} - \Delta'_3 \begin{vmatrix} u_{12} & u_{13} \\ u_{22} & u_{23} \end{vmatrix}}{u_{11} \begin{vmatrix} u_{22} & u_{23} \\ u_{32} & u_{33} \end{vmatrix} - u_{21} \begin{vmatrix} u_{12} & u_{13} \\ u_{32} & u_{33} \end{vmatrix} + u_{31} \begin{vmatrix} u_{12} & u_{13} \\ u_{22} & u_{23} \end{vmatrix}}$$

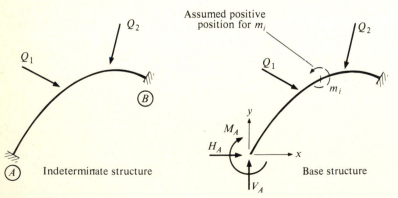

Figure 8.24 Fixed-end arch.

or

$$V_A = - \left[\frac{\Delta_1'(u_{22}u_{33} - u_{32}u_{23}) - \Delta_2'(u_{12}u_{33} - u_{32}u_{13}) + \Delta_3'(u_{12}u_{23} - u_{22}u_{13})}{u_{11}(u_{22}u_{33} - u_{32}u_{23}) - u_{21}(u_{12}u_{33} - u_{32}u_{13}) + u_{31}(u_{12}u_{23} - u_{22}u_{13})} \right]$$

(8.56)

where

$$\Delta_1' = \int \frac{Mm_1}{EI} ds = \int \frac{M}{EI}(x)\, ds$$

$$\Delta_2' = \int \frac{Mm_2}{EI} ds = \int \frac{M}{EI}(-y)\, ds$$

$$\Delta_3' = \int \frac{Mm_3}{EI} ds = \int \frac{M}{EI} ds$$

$$u_{11} = \int \frac{m_1^2}{EI} ds = \int \frac{x^2}{EI} ds$$

$$u_{22} = \int \frac{m_2^2}{EI} ds = \int \frac{y^2}{EI} ds$$

$$u_{33} = \int \frac{m_3^2}{EI} ds = \int \frac{1}{EI} ds$$

$$u_{13} = u_{31} = \int \frac{m_1 m_3}{EI} ds = \int \frac{x}{EI} ds$$

$$u_{23} = u_{32} = \int \frac{m_2 m_3}{EI} ds = \int \frac{-y}{EI} ds$$

$$u_{12} = u_{21} = \int \frac{m_1 m_2}{EI} ds = \int \frac{-xy}{EI} ds$$

It is clear from the complexities of solving eq. (8.56) that a more usable expression is desired.

Two things should be remembered. First, the redundants were rather arbitrarily chosen. Secondly, without questioning whether or not a better selection could be made, the coordinate system was selected through the left-hand end of the arch. (Directions of the coordinates also were defined without any real concern for whether or not another choice would have been more advantageous.) If the origin of the coordinate system had been selected such that

$$\int \frac{x}{EI} ds = \int \frac{y}{EI} ds = 0$$

(8.57)

then four of the terms in the flexibility matrix would equal zero; that is,

$$u_{31} = u_{13} = u_{23} = u_{32} = 0$$

and eq. (8.56) would have reduced to

$$\bar{V}_A = - \left[\frac{\Delta_1' - \Delta_2'(u_{12}/u_{22})}{u_{11} - (u_{12}^2/u_{22})} \right] \tag{8.58}$$

If, in addition, the chosen directions of the coordinate system are the principal directions, then

$$\int \frac{xy}{EI} ds = 0$$

and $u_{21} = u_{12} = 0$. This would yield for the value of the vertical redundant force

$$\bar{V}_A = - \frac{\Delta_1'}{u_{11}} \tag{8.59}$$

Making the same assumptions concerning the location and directions of the coordinate axes, the corresponding equations for $\bar{H}_A$ and $\bar{M}_A$ would be

$$\bar{H}_A = - \frac{\Delta_2'}{u_{22}} \tag{8.60}$$

and

$$\bar{M}_A = - \frac{\Delta_3'}{u_{33}} \tag{8.61}$$

(Bars over the redundant forces have been introduced to emphasize that these are presumed to act at the centroid—that is, the *elastic center*—of the ds/EI system and not at the originally chosen location of the redundants.) It is to be recognized—from eqs. (8.59), (8.60), and (8.61)—that a force applied to the base structure through the elastic center of the system will cause that structure to deflect *only* in the direction of the applied force or moment.

For a fixed-end symmetrical arch, the situation would be that shown in fig. 8.25. The given indeterminate structure is cut at A to yield the base structure shown in the second sketch. (The three chosen redundants at the cut section are V_A, H_A, and M_A.)

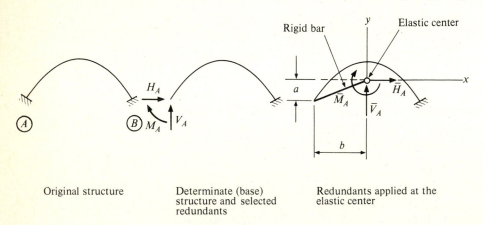

Original structure   Determinate (base)   Redundants applied at the
                            structure and selected   elastic center
                            redundants

Figure 8.25 Assumed determinate structure and elastic center of fixed-end arch.

If it is presumed that the unknowns are applied through the elastic center of the system, as shown in the third sketch, then these two sets of unknown forces are related by the expressions

$$H_A = \bar{H}_A$$
$$V_A = \bar{V}_A \tag{8.62}$$

and

$$M_A = \bar{M}_A + \bar{H}_A a - \bar{V}_A b$$

since the connecting bar is—by definition—rigid.

It should be recognized that the procedure used to develop the expressions given above is not dependent upon where the structure is cut to make it determinate. For example, the arch shown in fig. 8.25 could equally well have been cut at the centerline section, and the redundants could have been selected as the vertical shear, axial thrust, and bending moments at that section (fig. 8.26).

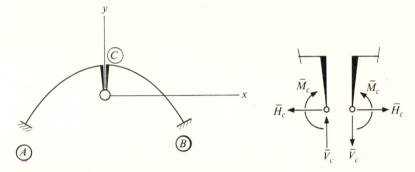

Figure 8.26 Redundant forces at the elastic center.

Example 8.7 A 10-ft-high, 30-ft-span fixed-base portal frame of uniform cross section is subjected to a concentrated horizontal force at the beam-column junction of 10 kips. Determine the bending-moment diagram throughout the structure.

The elastic center (that is, the center of gravity) of the structural system is located at that particular point where

$$\Sigma \int y \frac{ds}{EI} = \Sigma \int x \frac{ds}{EI} = 0$$

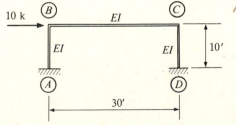

Figure 8.27

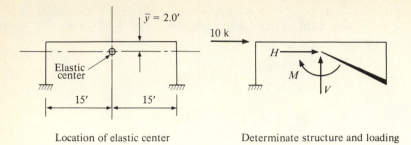

| Location of elastic center | Determinate structure and loading |

Figure 8.28 Elastic center and redundant forces on rigid portal frame.

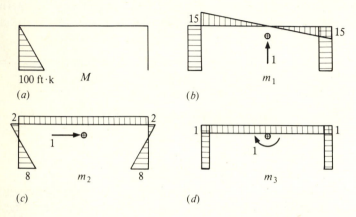

(a) (b)

(c) (d)

Figure 8.29 Moment diagrams due to unit redundant forces applied through elastic center.

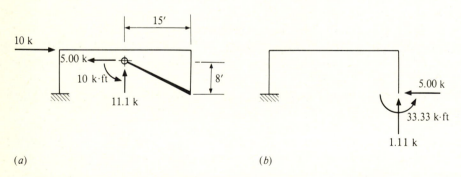

(a) (b)

Figure 8.30 Redundant forces at elastic center and at the right-hand support.

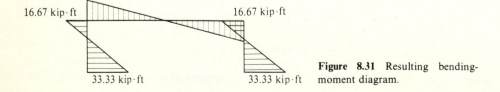

Figure 8.31 Resulting bending-moment diagram.

Considering the location of the x axis: Moments about the upper horizontal beam could be taken of the actual and the assumed concentrated cases. These must be equal. Therefore,

$$\frac{10}{EI}(5) + \frac{10}{EI}(5) = \left(\frac{10}{EI} + \frac{10}{EI} + \frac{30}{EI}\right)\bar{y}$$

or

$$\bar{y} = \frac{100}{50} = 2 \text{ ft}$$

For the y axis: It should be noted that the structure is symmetrical. Therefore, the elastic center must lie on the line of symmetry. Moment diagrams (plotted on the compression sides of the members) for the applied loading (M), and for each of the unit load cases (m), are shown in fig. 8.29a to d.

For the redundant V: $V = -\dfrac{\Delta_1}{u_{11}} = -\dfrac{-7500/EI}{6750/EI} = +1.111 \text{ kips}$

For H: $H = -\dfrac{\Delta_2}{u_{22}} = -\left[\dfrac{2333.33/EI}{466.67/EI}\right] = -5.00 \text{ kips}$

For M: $M = -\dfrac{500/EI}{50/EI} = -10 \text{ ft} \cdot \text{kips}$

These are shown in fig. 8.30a. Their equivalent values (at the right-hand support at D) are given in sketch (b). The moment diagram for the entire structure is shown in fig. 8.31.

In the above examples, the specified end attachments were fixed supports. Since such supports imply an infinite amount of bending stiffness, ds/EI at those locations reduces to zero. Had the supports been pin connections, however, the effective bending stiffness at those locations would have been zero, and the ds/EI values would have been infinity. Therefore, independent of the particular size or shape of the structure in question, when pin supports are specified, the elastic center will be located along an axis that passed through those supports. This is illustrated in fig. 8.32.

Another general conclusion concerning the elastic-center method that can be drawn: The unknown redundants are presumed to act at the elastic center (or, as it is sometimes called, the *neutral point*) of the structure in question. Using simple static equilibrium equations, these unknowns are then related to (that is, written in terms of) end reactions or internal stress resultants. At these cut sections, however, in the assumed base structure, there could have been introduced quite arbitrarily a set of additional forces or moments, which forces and moments could have altered the originally derived determinate bending-moment diagram (the M diagram) so that for all purposes it has the same shape and magnitude of another choice of redundants. The influence of these added terms, however, would have cancelled out in the subsequent transfer operation. Therefore, it can be concluded that the particular choice

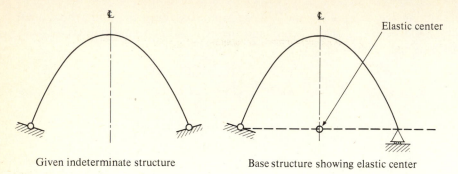

Given indeterminate structure   Base structure showing elastic center

Figure 8.32 Elastic center for a pin-based arch.

of determinate bending moment system used in the elastic-center method is not important to the solution process. Moreover, should a given structure be subjected simultaneously to a variety of different loading situations, quite different determinate systems could be selected for each, with the combination still being handled within the general formulation and operation.

The unit load method used in each of the examples above automatically takes care of the question of signs and assumed directions for the unknowns at the elastic center. (A resulting positive value for the unknown signifies that the assumed force or moment acts in the direction of the selected unit force or moment.) When other methods are used to calculate the various displacement terms, a physical picture of what occurs will usually be required—but in most single-cell cases this presents no real difficulty. In general, the redundants will *always* act in a direction which opposes the displacement of the elastic center.

8.2.3 The Column Analogy

The general elastic-center method involves the determination of a number of integrated quantities:

$$\int \frac{ds}{EI} \qquad \int x^2 \frac{ds}{EI} \qquad \int Mx \frac{ds}{EI}$$

$$\int x \frac{ds}{EI} \qquad \int y^2 \frac{ds}{EI} \qquad \int My \frac{ds}{EI}$$

$$\int y \frac{ds}{EI} \qquad \int xy \frac{ds}{EI} \qquad \int M \frac{ds}{EI}$$

To reduce the amount of work involved in obtaining a solution, the concept of a center of gravity (or elastic center) of the integrated ds/EI system was introduced. This idea can be given physical meaning if there is presumed a fictitious strip whose centerline has the same form as the centerline of the structure in question, and whose width is $1/EI$ at all locations (fig. 8.33). $\int ds/EI$, then, is the total area of this fictitious strip. In like fashion (see the second sketch of fig. 8.33), $\int M \, ds/EI$ is the volume of

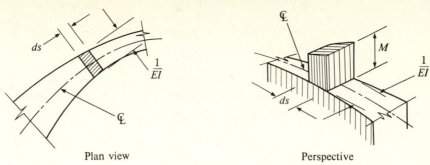

Plan view   Perspective

Figure 8.33 Cross section and loading on analogous column.

the M function (the determinate moment on the base structure due to the applied loading) acting on the strip. If an xy coordinate system were introduced and if those coordinates had their origin at the centroid of the fictitious strip system, the analogous relationships listed in table 8.1 could be defined.

Table 8.1 Relationship between column analogy and real systems

Real system	Analogous system	Usual notation for columns
$\int \dfrac{ds}{EI}$	Area of fictitious strip	A
$\int M \dfrac{ds}{EI}$	Total load on strip	P
$\int Mx \dfrac{ds}{EI}$	Moment of load on strip about the y axis	M_{yy}
$\int My \dfrac{ds}{EI}$	Moment of load on strip about the x axis	M_{xx}
$\int x^2 \dfrac{ds}{EI}$	Moment of inertia of strip about the y axis	I_{yy}
$\int y^2 \dfrac{ds}{EI}$	Moment of inertia of strip about the x axis	I_{xx}
$\int xy \dfrac{ds}{EI}$	Product of inertia of strip with reference to the xy axes	I_{xy}

$$X_1 = -\frac{\Delta_1' - \Delta_2'(u_{12}/u_{22})}{u_{11} - (u_{12}^2/u_{22})} = -\frac{M_{yy} - M_{xx}(I_{xy}/I_{xx})}{I_{yy} - (I_{xy}^2/I_{xx})}$$

$$X_2 = -\frac{\Delta_2' - \Delta_1'(u_{12}/u_{11})}{u_{22} - (u_{12}^2/u_{11})} = -\frac{M_{xx} - M_{yy}(I_{xy}/I_{yy})}{I_{xx} - (I_{xy}^2/I_{yy})}$$

$$X_3 = -\frac{\Delta_3'}{u_{33}} = -\frac{P}{A}$$

In most structural problems, insofar as the behaviors of the various members are concerned, it is not the values of the reactions, per se, that are of greatest importance. Rather, it is the consequences of the loads as they manifest themselves in terms of bending moments, shears, and thrusts along the various members that are of major significance.

The moment of any section in an indeterminate structure is made up of two parts: (a) that part which is due to the applied loading (on the chosen base, determinate system) and (b), that part which is due to the redundants. If M_{det} is defined as the determinate moment; X_1, X_2, and X_3 are the redundants (as listed in table 8.1); and x and y are the coordinate distances from the elastic center to the point on the structure in question; then the moment at that point is

$$M = M_{det} \pm [X_3 \pm X_1(x) \pm X_2(y)] \qquad (8.63)$$

(The simultaneous plus and minus signs are used to emphasize that the various terms can either add to or subtract from each other, depending on both the direction of the redundants and the location of the point in question. It is to be remembered that the moment due to the redundants will *always* reduce or be subtracted from the value of the determinate moment M_{det} at the location where M_{det} is a maximum.)

Substituting into eq. (8.63) the values listed in table 8.1 yields

$$X_3 \pm X_1(x) \pm X_2(y) = \frac{P}{A} \pm \frac{M_{yy}I_{xx} - I_{xy}M_{xx}}{I_{yy}I_{xx} - I_{xy}^2} x \pm \frac{M_{xx}I_{yy} - I_{xy}M_{yy}}{I_{xx}I_{yy} - I_{xy}^2} y$$

Separating terms and collecting M_{xx}'s and M_{yy}'s,

$$X_3 \pm X_1(x) \pm X_2(y) = \frac{P}{A} \pm M_{xx}\frac{I_{yy}(y) - I_{xy}(x)}{I_{xx}I_{yy} - I_{xy}^2} \pm M_{yy}\frac{I_{xx}(x) - I_{xy}(y)}{I_{xx}I_{yy} - I_{xy}^2} \qquad (8.64)$$

Using the methods in secs. 2.5 and 2.8, consider now the stress behavior of a short column subjected to an axial thrust P which acts along a centroidal axis, and two bending moments M_x and M_y which act about each of the coordinate axes. For plane sections before loading to remain plane after the loads are applied,

$$\sigma = ax + by + c$$

where σ is the stress at any point in the cross section. For equilibrium

$$\int \sigma \, dA = P$$

$$\int \sigma y \, dA = M_x$$

$$\int \sigma x \, dA = M_y$$

Solving these and substituting the results back into the original plane-section expression for stress yields

$$\sigma = \frac{P}{A} + M_x\frac{I_y(y) - I_{xy}(x)}{I_x I_y - I_{xy}^2} + M_y\frac{I_x(x) - I_{xy}(y)}{I_x I_y - I_{xy}^2} \qquad (8.65)$$

where P = the applied axial thrust
M_x = the applied moment about the x axis
M_y = the applied moment about the y axis
A = the area of the cross section
I_x = the moment of inertia about the x axis
I_y = the moment of inertia about the y axis
I_{xy} = the product of inertia of the area with respect to both the x and y axes

Equation (8.65) is to be compared with eq. (8.64).

As noted in the development of the elastic-center method, work can be materially reduced if a coordinate system can be found for which I_{xy} is zero. Such axes are referred to as the principal axes. Their directions can be determined analytically, or by use of the *Mohr's circle* graphical method.

If as indicated in fig. 8.34 a given dA element is referenced to both xy and uv coordinate systems, which systems have the same origin but are rotated through an angle β with respect to each other, then the interrelationships between the various geometrical properties associated with the two systems can be defined as follows:

$$I_u = \int v^2 \, dA = \frac{I_x + I_y}{2} + \frac{I_x - I_y}{2} \cos 2\beta - I_{xy} \sin 2\beta$$

$$I_v = \int u^2 \, dA = \frac{I_x + I_y}{2} + \frac{I_x - I_y}{2} \cos 2\beta + I_{xy} \sin 2\beta$$

$$I_{uv} = \int uv \, dA = \frac{I_x - I_y}{2} \sin 2\beta + I_{xy} \cos 2\beta$$

If now the product of inertia in the uv system is set equal to zero; that is, $I_{uv} = 0$; then

$$\tan 2\beta = -\frac{2I_{xy}}{I_x - I_y} \tag{8.66}$$

and the resulting u and v directions correspond to the principal axes of the system.

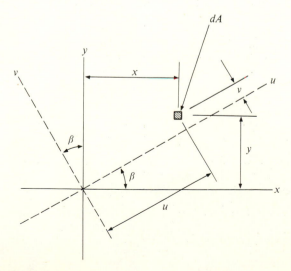

Figure 8.34

Example 8.8 Solve the problem defined in fig. 8.27 using the column analogy method. It will be assumed that the structure is made determinate by removing all three of the components of support at location D. The determinate moment diagram—due to the applied, external loading—is the diagram shown in sketch (a) of fig. 8.29. The analogous column and its loading is given in fig. 8.35. (When solving indeterminate structural problems by the column analogy method, it is common practice to assume as compressive column loads bending moments on the base structure which produce compression on the outside of the single cellular structures. The bending moments M in sketch (a) of fig. 8.29 are therefore upward forces on the column shown in fig. 8.35.)

Because of symmetry, the centroid will lie along the indicated centerline axis. The distance $\bar{y}$ will be

$$50\bar{y} = (2)(10)(5)$$

or

$$\bar{y} = 2.00 \text{ ft}$$

About the x and y axes, the various geometrical properties are

$$A = \frac{50}{EI}$$

$$I_x = 2\left[\left(\frac{1}{12}\right)\left(\frac{1}{EI}\right)(10)^3 + \left(\frac{1}{EI}\right)(10)(3)^2\right] = \frac{466.67}{EI}$$

$$I_y = \left(\frac{1}{12}\right)\left(\frac{1}{EI}\right)(30)^3 + (2)\left(\frac{1}{EI}\right)(10)(15)^2 = \frac{6750}{EI}$$

$$I_{xy} = 0\dagger$$

† The coordinate axes shown are therefore principal axes.

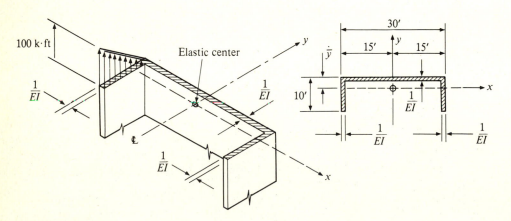

Figure 8.35

The loading terms are

$$P = \left(\frac{1}{2}\right)(-100)(10)\left(\frac{1}{EI}\right) = -\frac{500}{EI}$$

$$M_x = \left(-\frac{500}{EI}\right)(-4.667) = +\frac{2333.5}{EI}$$

$$M_y = \left(-\frac{500}{EI}\right)(-15) = +\frac{7500}{EI}$$

The bending moment at any location in the structure is

$$M = M_{det} \pm \left[\frac{P}{A} + \frac{M_x(y)}{I_x} + \frac{M_y(x)}{I_y}\right] \tag{8.67}$$

Substituting in the values derived above,

$$M = M_{det} \pm \left[\frac{-\dfrac{500}{EI}}{\dfrac{50}{EI}} + \frac{\dfrac{2333.5}{EI}(y)}{\dfrac{466.67}{EI}} + \frac{\dfrac{7500}{EI}(x)}{\dfrac{6750}{EI}}\right]$$

or

$$M = M_{det} \pm [-10.00 + (5.00)(y) + (1.11)(x)]$$

It should be noted that while the signs of the various terms have been rigorously assured within the bracketed term, both a plus and a minus have been indicated outside. Since it is known that the influence of the redundants is to reduce the magnitude of the maximum determinate (base structure) moment, the particular sign to be used would be determined from a consideration of the moment values at that point. At location A:

$$M_A = -100 \text{ ft·kips} \pm [-10.00 + (5.00)(-8.00) + (1.11)(-15.00)]$$
$$= -100 \text{ ft·kips} \pm [-66.66 \text{ ft·kips}]$$

The minus sign therefore will be used, and the governing equation becomes

$$M = M_{det} + [10.00 - (5.00)(y) - (1.11)(x)] \tag{8.68}$$

The equation yields the same moment diagram as that shown in fig. 8.31.

Example 8.9 The fixed-base structure shown in fig. 8.36 is subjected to a concentrated load Q. By the column analogy method, determine the moments at locations A, B, and C, and at the point of load application. The distances from the members to the centroid of the analogous column [see sketch (c) of fig. 8.36] are

$$\bar{x} = \bar{y} = \frac{(L/EI)(L/2)}{2L/EI} = \frac{L}{4}$$

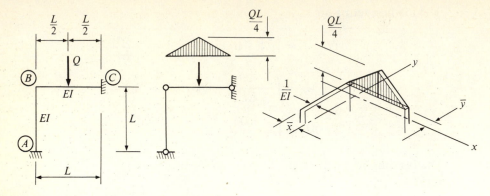

Given structure and loading   Assumed determinate system   Analogous column

(a)   (b)   (c)

Figure 8.36

The geometrical properties of the column about the x and y axes are

$$A = 2L\frac{1}{EI} = \frac{2L}{EI}$$

$$I_{xx} = \frac{L}{EI}\left(\frac{L}{4}\right)^2 + \frac{1}{12}\frac{1}{EI}L^3 + \frac{L}{EI}\left(\frac{L}{4}\right)^2 = \frac{5}{24}\frac{L^3}{EI}$$

$$I_{yy} = \frac{5}{24}\frac{L^3}{EI}$$

$$I_{xy} = \frac{L}{EI}\frac{L}{4}\frac{L}{4} + \frac{L}{EI}\left(-\frac{L}{4}\right)\left(-\frac{L}{4}\right) = \frac{1}{8}\frac{L^3}{EI}$$

The loading parameters are

$$P = \frac{1}{2}L\frac{QL}{4}\frac{1}{EI} = +\frac{1}{8}\frac{QL^2}{EI}$$

$$M_{xx} = \frac{1}{8}\frac{QL^2}{EI}\frac{L}{4} = +\frac{1}{32}\frac{QL^3}{EI}$$

$$M_{yy} = \frac{1}{8}\frac{QL^2}{EI}\frac{L}{4} = +\frac{1}{32}\frac{QL^3}{EI}$$

The moment at any location in the structure [using eq. (8.64)] is then

$$M = M_{\text{det}} - \left[\frac{\dfrac{1}{8}\dfrac{QL^2}{EI}}{\dfrac{2L}{EI}} + \frac{1}{32}\frac{QL^3}{EI}\frac{\dfrac{5}{24}\dfrac{L^3}{EI}x - \dfrac{1}{8}\dfrac{L^3}{EI}y}{\left(\dfrac{5}{24}\dfrac{L^3}{EI}\right)^2 - \left(\dfrac{1}{8}\dfrac{L^3}{EI}\right)^2} + \frac{1}{32}\frac{QL^3}{EI}\frac{\dfrac{5}{24}\dfrac{L^3}{EI}y - \dfrac{1}{8}\dfrac{L^3}{EI}x}{\left(\dfrac{5}{24}\dfrac{L^3}{EI}\right)^2 - \left(\dfrac{1}{8}\dfrac{L^3}{EI}\right)^2}\right]$$

Table 8.2 Analogous column loads for support movements

Support movement	Loading on analogous column

Collecting terms and simplifying yields

$$M = M_{det} - \frac{QL}{32}\left[2 + 3\left(\frac{x}{L} + \frac{y}{L}\right)\right]$$

(8.69)

Therefore:

$$M_A = 0 - \frac{QL}{32}\left[2 + 3\left(-\frac{1}{4} - \frac{3}{4}\right)\right] = +\frac{1}{32}QL$$

$$M_B = 0 - \frac{QL}{32}\left[2 + 3\left(-\frac{1}{4} + \frac{1}{4}\right)\right] = -\frac{1}{16}QL$$

$$M_C = 0 - \frac{QL}{32}\left[2 + 3\left(\frac{3}{4} + \frac{1}{4}\right)\right] = -\frac{5}{32}QL$$

and

$$M_Q = \frac{QL}{4} - \frac{QL}{32}\left[2 + 3\left(\frac{1}{4} + \frac{1}{4}\right)\right] = +\frac{9}{64}QL$$

The equivalent loading Mds/EI on the column in the column analogy method corresponds to a change in curvature (that is, $\Delta\phi$) in the actual structure.[†] Therefore, the influence of support settlement or misalignment can be handled directly in the column as a point loading. These (with appropriate signs indicated) are shown in table 8.2.

8.2.4 Influence Lines for Indeterminate Structures

In chap. 3, an influence line was defined as a graphic or pictorial representation of the magnitude of a particular stress resultant at one point in a structure, when a unit load is placed at various other locations in that structure. In chap. 5, it further was shown that the deformation at i due to a unit load at j is equal to the deformation at j due to a unit load at i; that is,

$$\delta_{ij} = \delta_{ji}$$

Therefore—independent of whether a given structure is determinate or indeterminate—the deflected shape of the member caused by a unit load at the point in question is the influence line for the deflection at the same point in the member. This leads directly to the Muller-Breslau principle:

If a stress resultant is considered to act through some small deformation in the direction of the resultant and thereby cause that structure to deform, then the deflected shape, to some scale, will be the influence line for that particular stress resultant.

Consider for illustration the two-span continuous beam shown in fig. 8.37a. This structure is one-time indeterminate, and the vertical reaction at A is assumed to be the unknown redundant. It is the influence line for that reaction that is desired.

---

† This concept was developed in chap. 4 in the section, Moment-Area Method.

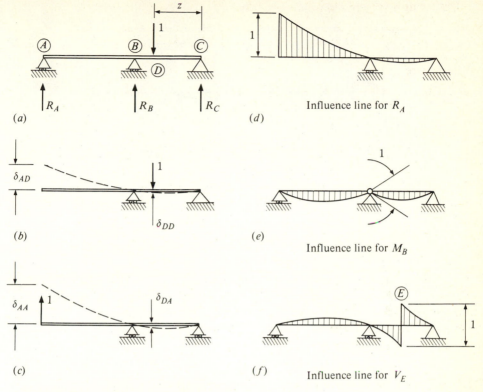

Figure 8.37 Influence lines for a two-span beam.

If, as indicated in sketch (*b*), the determinate structure is subjected to a unit load at location *D*, the deflection at *A* will be δ_{AD}. At *D*, the deflection is δ_{DD}. If now, as shown in sketch (*c*), a unit load is applied at *A*, the deflections at *A* and *D* will be δ_{AA} and δ_{DA}, respectively. For the deflection at *A* to equal zero, as was specified in the original statement of the problem

$$R_A \delta_{AA} - \delta_{AD} = 0$$

Therefore,

$$R_A = \frac{\delta_{AD}}{\delta_{AA}} = \frac{\delta_{DA}}{\delta_{AA}} \tag{8.70}$$

If δ_{AA} were selected equal to 1.0, the deflected shape of the resulting deformed structure would be the influence line for the reaction (see fig. 8.37*d*). Had the moment at *B* been selected as the redundant, the imposed deformation would be a unit rotation at *B*. This is shown in sketch (*e*) of fig. 8.37. For vertical shear at location *E* in the right-hand span, the imposed displacement and the resulting influence line is shown in sketch (*f*).

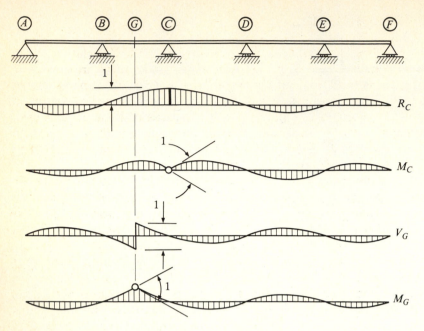

Figure 8.38 Qualitative influence lines for a five-span beam.

While the actual values of the influence line ordinates are needed in many cases, often it is only the shape of the curve that is of immediate concern. For example, fig. 8.38 is a qualitative indication of the shapes of a number of different influence lines for a five-span continuous beam. In like fashion, the configuration shown by the dashed lines in fig. 8.39 is the qualitative influence line for bending moment in location A of the three-span four-story rigid frame.

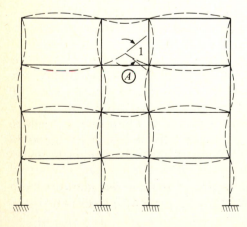

Figure 8.39 Qualitative influence lines for a multi-story, multispan rigid frame.

8.3 STIFFNESS METHODS

As noted earlier in this chapter, the unknown quantities in a stiffness solution are the displacements. In general, in a continuous structural system there are an infinite number of such terms, each of which can be identified and described analytically or experimentally. It is possible, however, to describe the entire deflected configuration of a given structure subjected to a defined loading in terms of a limited number of these. For example, in fig. 8.2 the prescription of the single term θ_A is sufficient to allow the entire deflection configuration of the nine-times redundant structure to be deduced.

The number of independent deformational terms required to fully describe the deflected shape of a structure is known as the *degree of freedom* of the system. In fig. 8.2 there is one degree of freedom, since knowledge of θ_A is sufficient to allow the definition of all other points in the structure. For the continuous beam shown in fig. 8.23 there would be four degrees of freedom: θ_B, θ_C, θ_D, and θ_E.

A structure is composed of both individual members and joints. Joints exist between members and at points of support. For each of the joints there is the possibility, in a two-dimensional situation, of having up to three degrees of freedom, that is, three components of deformation. In three-dimensional cases, six components can occur. (It is to be understood, however, that in a given situation, because of known geometrical constraints, the number of degrees of freedom associated with a given joint may be less than the maximum. In fig. 8.2, for example, joint A was constrained against movement in the horizontal and vertical directions. There was, therefore, only one deformational parameter possible at that joint.) Additional degrees of freedom can be introduced; for example, at points of change in cross section—even though these occur within a given member. Here again, three additional degrees of freedom would be introduced if the structure is two-dimensional.

When solving indeterminate structures by the flexibility method, it was first necessary to determine the degree of indeterminacy. (This was done by progressively removing components of reaction or internal stress resultants until the resulting structure was determinate.) When using the stiffness method, it is first necessary to determine the number of independent degrees of freedom in the given system. While this may appear to be a relatively simple matter (and, if large numbers of unknowns present no difficulty in obtaining a solution, this will obviously be the case), computational work can be reduced if particular deformational interrelationships are recognized from the outset.

In structure (*a*) of fig. 8.40 the column is pin-connected at A. It is therefore position-fixed but can have a rotation θ_A. At point B, if it is assumed that the shortening of the column can be neglected, there also is only one deformation parameter, θ_B. Therefore the structure has two degrees of freedom, θ_A and θ_B.

The two-bay structure shown in fig. 8.40*b* is typical of a structure that allows sidesway. Again, if axial shortening is neglected, there are five independent degrees of freedom: θ_A, θ_B, θ_C, θ_E, and Δ. (Note that the top of each of the columns moves horizontally through the same distance Δ.)

The frame shown in fig. 8.41 has seven degrees of freedom: five joint rotations (θ_A, θ_B, θ_C, θ_D, and θ_E) and two column translations (Δ_{AB} and Δ_{DE}). In this case it has

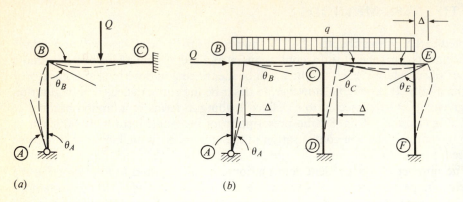

(a)   (b)

Figure 8.40 Deformations of typical indeterminate rigid frames.

been assumed that *B* and *D* move horizontally, independently of each other. However, making those assumptions, the horizontal and vertical displacement of *C* cannot also be assumed to be independent. Rather, it must be related to Δ_{AB} and Δ_{DE} and the geometry of the frame.

There are a number of different stiffness methods available for solving indeterminate structural problems: the direct stiffness method, the method of slope deflection, the moment distribution method, etc. In all these, the first major step in the solution is the determination of the basic relationships between individual member forces and member displacements. In these, it is assumed that the deformational components (at the points of connection to other members) are the unknown, independent variables, and the end shears, thrusts, and moments at these points are the dependent terms. Static equilibrium at the joints can then be forced. There results a set of linear algebraic expressions where the end deformational parameters are the unknowns.

Chapter 9 on matrix methods will deal almost exclusively with the direct stiffness method of solution. Therefore, in this section only the slope-deflection method will be extensively developed.

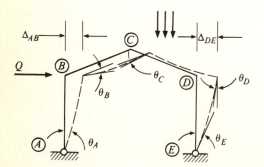

Figure 8.41 Pin-based, single-span gable frame.

8.3.1 The Slope-Deflection Method

The slope-deflection method is intended for use in structures where connections are rigid and where the load-carrying capacity is realized primarily through the bending stiffnesses of the various members. While it is possible to include the influence of axial thrust on the bending stiffness, it is assumed that axial shortening does not significantly change the lengths of the various members in the structure. Rather, end slopes and relative end rotations govern deformational variables.

Assume that a member AB of length L is subjected to lateral loads, end moments, end rotations, and relative support displacements as shown in fig. 8.42. The applied moment at the left-hand end is M_{AB}; on the right, it is M_{BA}. Correspondingly, the slopes are θ_A and θ_B. The relative displacement (perpendicular to the original position of the beam) of one end with respect to the other is assumed to be Δ. (The indicated clockwise moments, rotations and relative displacements are in the chosen positive directions.)

By definition, the system in question is elastic and deformations are small. Therefore, superposition of individual effects can be presumed. Considering only the influence of θ_A, θ_B and Δ on M_{AB} and M_{BA} (that is, neglecting the effects of the applied lateral load), it can be shown that for prismatic members

$$M_{AB} = \frac{2EI}{L}\left(2\theta_A + \theta_B - 3\frac{\Delta}{L}\right)$$

$$M_{BA} = \frac{2EI}{L}\left(\theta_A + 2\theta_B - 3\frac{\Delta}{L}\right)$$

(8.71)

For example, presuming that the member is both positionally and rotationally fixed at B and is subjected to a unit rotation at A, the resulting moment at A, from eq. (8.14), would be

$$M_{AB} = \frac{4EI}{L}\theta_A$$

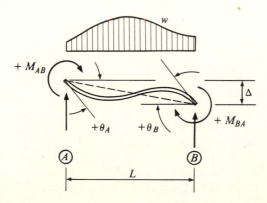

Figure 8.42 General deformation of a beam subjected to lateral loading and end-bending moments.

Table 8.3 Fixed-end moments

M^F_{AB}		M^F_{BA}

$$-\frac{QL}{8}$$

$$+\frac{QL}{8}$$

$$-Q\frac{ab^2}{L^2}$$

$$+Q\frac{a^2b}{L^2}$$

$$+M\frac{b}{L^2}(2a-b)$$

$$+M\frac{a}{L^2}(2b-a)$$

$$-\frac{qL^2}{12}$$

$$+\frac{qL^2}{12}$$

$$-\frac{qa^2}{12L^2}(3a^2-8aL+6L^2)$$

$$+\frac{qa^3}{12L^2}(4L-3a)$$

$$-\frac{q}{12L^2}[d^3(4L-3d)-b^3(4L-3b)]$$

$$+\frac{q}{12L^2}[a^3(4L-3a)-c^3(4L-3c)]$$

$$-\frac{q_0L^2}{20}$$

$$+\frac{q_0L^2}{30}$$

$$-\frac{5q_0L^2}{96}$$

$$+\frac{5q_0L^2}{96}$$

$$-\frac{q_0a^2}{60L^2}(3a^2-10aL+10L^2)$$

$$+\frac{q_0c^3}{60L^2}(5L-3a)$$

Had A been fixed and a rotation of θ_B induced at B—again presuming A and B do not deflect

$$M_{AB} = \frac{2EI}{L} \theta_B$$

If both A and B are prevented from rotating but translate laterally with respect to each other an amount Δ [see eq. (8.12)],

$$M_{AB} = -\frac{6EI}{L^2} \Delta$$

The applied lateral loads will cause an increase or a decrease in these values, depending upon the particular applied loading in question and upon whether it is the left-hand or the right-hand end of the member that is being considered. In general, downward loads will induce counterclockwise fixed-end moments at the left-hand support (opposite to that assumed as positive in the above derivation), and clockwise fixed-end moments at the right. Noting this problem (and presuming that the appropriate signs will be used in any given case), the influence of lateral loads can be directly added to that expression defined. That is

$$M_{AB} = \frac{2EI}{L} \left(2\theta_A + \theta_B - 3\frac{\Delta}{L} \right) + M_{AB}^F$$

$$M_{BA} = \frac{2EI}{L} \left(\theta_A + 2\theta_B - 3\frac{\Delta}{L} \right) + M_{BA}^F$$

(8.72)

where M_{AB}^F is the fixed-end moment at A due to the applied lateral loads, and M_{BA}^F is the fixed-end moment at B due to the applied lateral loads. These expressions could be written in more general form, in terms of rotational and translational stiffness, carryover factors, and fixed-end moments. Such general expressions would hold for members of varying cross section as well as for the special prismatic cases indicated above.

$$M_{AB} = k_{AA}\theta_A + c_{AB}k_{BB}\theta_B - k_\Delta \frac{\Delta}{L} + M_{AB}^F$$

$$M_{BA} = c_{BA}k_{AA}\theta_A + k_{BB}\theta_B - k_\Delta \frac{\Delta}{L} + M_{BA}^F$$

(8.73)

where k_{AA} and k_{BB} = the rotational stiffnesses at A and B, respectively

c_{AB} and c_{BA} = the carryover factors from A to B and B to A, respectively

k_Δ = the translational stiffness of the member

M_{AB}^F and M_{BA}^F = the fixed-end moments at A and B

Fixed-end moments for a number of prismatic beam cases are listed in table 8.3.

Example 8.10 The two-span uniform cross section continuous beam shown in fig. 8.43 is subjected to both a distributed load q and a concentrated load Q. The beam has two geometric degrees of freedom, θ_B and θ_C. (It is also two times

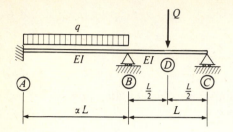

Figure 8.43

redundant.) Using the slope-deflection equations (8.72) (noting that Δ is zero),

$$M_{AB} = \frac{2EI}{\alpha L}(2\theta_A + \theta_B) - \frac{q(\alpha L)^2}{12}$$

$$M_{BA} = \frac{2EI}{\alpha L}(\theta_A + 2\theta_B) + \frac{q(\alpha L)^2}{12}$$

$$M_{BC} = \frac{2EI}{L}(2\theta_B + \theta_C) - \frac{QL}{8}$$

$$M_{CB} = \frac{2EI}{L}(\theta_B + 2\theta_C) + \frac{QL}{8}$$

(8.74)

(Equations (8.74), while accurate and geometrically compatible for any given individual member—or for that matter, for all members joining at a particular point in the structure—say nothing about static equilibrium within the system. This must be forced.)

Moment M_{BA}, which acts on the right-hand end of member AB, has been assumed to act in a clockwise positive fashion. This member, in turn, subjects joint B to a counterclockwise moment of equal magnitude. M_{BC}, likewise, causes a counterclockwise moment to be applied to joint B. However, for moment equilibrium of the joint (as an isolated entity), $M = 0$. Therefore,

$$M_{BA} + M_{BC} = 0 \qquad (8.75)$$

Similarly, at joint C,

$$M_{CB} = 0 \qquad (8.76)$$

Substituting expressions (8.74) into eqs. (8.75) and (8.76) yields

$$\left(\frac{4EI}{\alpha L} + \frac{4EI}{L}\right)\theta_B + \frac{2EI}{L}\theta_C = \frac{QL}{8} - \frac{q(\alpha L)^2}{12}$$

and

$$\frac{2EI}{L}\theta_B + \frac{4EI}{L}\theta_C = -\frac{QL}{8}$$

In matrix form, this would be

$$\begin{bmatrix} \dfrac{4EI}{L}\dfrac{1+\alpha}{\alpha} & \dfrac{2EI}{L} \\[3mm] \dfrac{2EI}{L} & \dfrac{4EI}{L} \end{bmatrix} \begin{bmatrix} \theta_B \\[3mm] \theta_C \end{bmatrix} = \begin{bmatrix} +\dfrac{QL}{8} - \dfrac{q(\alpha L)^2}{12} \\[3mm] -\dfrac{QL}{8} \end{bmatrix} \tag{8.77}$$

For $\alpha = 1.0$, $Q = 1$ kip, $q = 1$ kip/ft, and $L = 10$ ft,

$$\theta_B = -\frac{9.23}{EI}$$

and

$$\theta_C = +\frac{1.49}{EI}$$

When substituted back into expressions (8.74), there are obtained for the end moments

$$M_{AB} = -10.179 \text{ ft} \cdot \text{kips}$$

$$M_{BA} = +4.644 \text{ ft} \cdot \text{kips}$$

$$M_{BC} = -4.644 \text{ ft} \cdot \text{kips}$$

$$M_{CB} = 0$$

Using these values, reactions can be determined and bending-moment diagrams plotted.

Example 8.11 For the structure defined in fig. 8.43, it is presumed that lateral loads are zero, but that support B settles by an amount Δ. What will be the bending moments at A, B, and C?

Since there are no applied loads there will be no fixed-end moments. Equations (8.72) therefore, will be

$$M_{AB} = \frac{2EI}{\alpha L}\left(\theta_B - 3\frac{\Delta}{\alpha L}\right)$$

$$M_{BA} = \frac{2EI}{\alpha L}\left(2\theta_B - 3\frac{\Delta}{\alpha L}\right)$$

$$M_{BC} = \frac{2EI}{L}\left[2\theta_B + \theta_C - 3\left(-\frac{\Delta}{L}\right)\right] \tag{8.78}$$

$$M_{CB} = \frac{2EI}{L}\left[\theta_B + 2\theta_C - 3\left(-\frac{\Delta}{L}\right)\right]$$

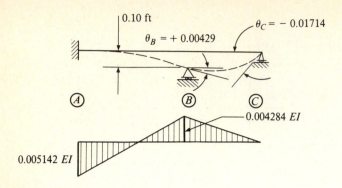

Figure 8.44 Support settlement of continuous beam.

For equilibrium at joints B and C:

$$M_{BA} + M_{BC} = 0$$

$$M_{CB} = 0$$

which yields

$$\begin{bmatrix} \dfrac{4EI}{L}\dfrac{1+\alpha}{\alpha} & \dfrac{2EI}{L} \\[2ex] \dfrac{2EI}{L} & \dfrac{4EI}{L} \end{bmatrix} \begin{bmatrix} \theta_B \\[2ex] \theta_C \end{bmatrix} = \begin{bmatrix} \dfrac{-6EI}{L^2}\Delta\dfrac{\alpha^2-1}{\alpha} \\[2ex] \dfrac{-6EI}{L^2}\Delta \end{bmatrix} \qquad (8.79)$$

For $\Delta = 0.1$ ft, $L = 10$ ft, and $\alpha = 1.0$,

$$\theta_B = +0.00429$$

$$\theta_C = -0.01714$$

for which the bending moments equal

$$M_{AB} = -0.005142EI$$

$$M_{BA} = -0.004284EI$$

$$M_{BC} = +0.004284EI$$

$$M_{CB} = 0$$

The elastic curve and the moment diagram associated with this condition of loading are shown in fig. 8.44.

Example 8.12 The two-bay single-story rigid frame shown in fig. 8.45 is to be analyzed for bending. The structure is six times statically indeterminate. There are, however, only four kinematic redundants (degrees of freedom). These are the

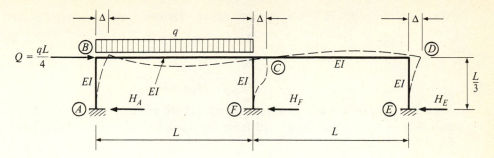

Figure 8.45

joint rotations θ_B, θ_C, and θ_D and a horizontal displacement parameter Δ. (For small deformations the horizontal displacements of joints B, C, and D can be assumed to be the same.)

Since there are five members, ten slope-deflection equations are required to describe the end-moments in terms of the unknown end deformations. These are

$$M_{AB} = \frac{2EI}{L/3}\left[(0) + \theta_B - 3\frac{\Delta}{L/3}\right] - 0$$

$$M_{BA} = \frac{2EI}{L/3}\left[(0) + 2\theta_B - 3\frac{\Delta}{L/3}\right] + 0$$

$$M_{BC} = \frac{2EI}{L}\left[2\theta_B + \theta_C - 3(0)\right] - \frac{qL^2}{12}$$

$$M_{CB} = \frac{2EI}{L}\left[\theta_B + 2\theta_C - 3(0)\right] + \frac{qL^2}{12}$$

$$M_{CF} = \frac{2EI}{L/3}\left[2\theta_C + (0) - 3\frac{\Delta}{L/3}\right] - 0$$

$$M_{FC} = \frac{2EI}{L/3}\left[(0) + \theta_C - 3\frac{\Delta}{L/3}\right] + 0$$

$$M_{CD} = \frac{2EI}{L}\left[2\theta_C + \theta_D - 3(0)\right] - 0$$

$$M_{DC} = \frac{2EI}{L}\left[2\theta_D + \theta_C - 3(0)\right] + 0$$

$$M_{DE} = \frac{2EI}{L/3}\left[2\theta_D + (0) - 3\frac{\Delta}{L/3}\right] - 0$$

$$M_{ED} = \frac{2EI}{L/3}\left[\theta_D + (0) - 3\frac{\Delta}{L/3}\right] + 0$$

At each of the joints B, C, and D, moment-equilibrium equations can be written:

$$M_{BA} + M_{BC} = 0$$

$$M_{CB} + M_{CF} + M_{CD} = 0 \tag{8.80}$$

$$M_{DC} + M_{DE} = 0$$

A fourth independent equation is required. (There are four unknown deformation parameters.) This is provided from horizontal equilibrium considerations for the frame as a whole.

$$H_A + H_F + H_E = \frac{qL}{4} \tag{8.81}$$

H_A, H_F, and H_E, however, can be expressed in terms of the end bending moments by considering each column individually as a free body (see fig. 8.46).

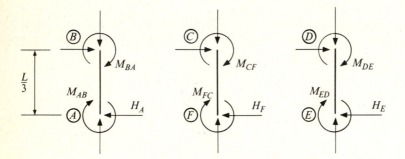

Figure 8.46 Shear equilibrium in the columns.

For example, if moments are summed about point B in column AB, the horizontal shear at the base of the column becomes

$$H_A = -\frac{M_{AB} + M_{BA}}{L/3}$$

Similarly,

$$H_F = -\frac{M_{CF} + M_{FC}}{L/3}$$

and

$$H_E = -\frac{M_{DE} + M_{ED}}{L/3}$$

Substituting these into eq. (8.81) yields the desired fourth, independent equilibrium equation:

$$M_{AB} + M_{BA} + M_{CF} + M_{FC} + M_{DE} + M_{ED} = -\frac{qL^2}{12} \tag{8.82}$$

Substituting into these four equations the slope-deflection expressions listed above yields the following set:

$$\begin{bmatrix} 16 & 2 & 0 & -54 \\ 2 & 20 & 2 & -54 \\ 0 & 2 & 16 & -54 \\ 18 & 18 & 18 & -324 \end{bmatrix} \begin{bmatrix} \theta_B \\ \theta_C \\ \theta_D \\ \Delta/L \end{bmatrix} = \frac{qL^3}{EI} \begin{bmatrix} 1/12 \\ -1/12 \\ 0 \\ -1/12 \end{bmatrix} \tag{8.83}$$

From these the unknown deformations can be determined. Subsequent substitution back into the slope-deflection expressions will give the desired end-moment values.

Example 8.13 The two-story frame shown in fig. 8.47 is six times indeterminate. There also are six degrees of freedom for the system. (The horizontal loads represent an approximation of the effects of the wind. In normal practice, such loads are presumed to act as if they were concentrated at the joints.)

The structure has a total of six degrees of freedom: θ_B, θ_C, θ_D, θ_E, Δ_{BA}, and Δ_{CB}. (Note that $\Delta_{BA} = \Delta_{EF}$ and $\Delta_{CB} = \Delta_{DE}$.) Assuming that the twelve slope-deflection equations have been written, the following four joint-equilibrium equations must be satisfied:

$$M_{BA} + M_{BE} + M_{BC} = 0$$
$$M_{CB} + M_{CD} = 0$$
$$M_{DC} + M_{DE} = 0 \tag{8.84}$$
$$M_{ED} + M_{EB} + M_{EF} = 0$$

In addition, two shear equilibrium equations are necessary. Considering a free-body diagram of the entire structure and loading above the horizontal beam *BE*, horizontal force equilibrium requires that

$$H_{BC} + H_{ED} = \frac{qL}{2} \tag{8.85}$$

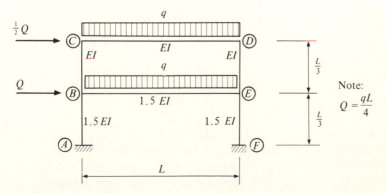

Figure 8.47

But H_{BC} and H_{ED} can be related to the moments at the ends of members BC and DE—as was done in example 8.12. Therefore equation (8.85) can be written as

$$M_{CB} + M_{BC} + M_{DE} + M_{ED} = -\frac{qL^2}{24} \tag{8.86}$$

If in the same fashion the entire structure now is considered, horizontal force equilibrium requires that

$$H_{AB} + H_{FE} = \tfrac{3}{8}qL$$

or

$$M_{BA} + M_{AB} + M_{FE} + M_{EF} = -\frac{qL^2}{8} \tag{8.87}$$

The algebraic set corresponding to eqs. (8.84), (8.86), and (8.87) that must be solved is

$$
\begin{bmatrix}
36 & 6 & 0 & 3 & -81 & -54 \\
6 & 16 & 2 & 0 & 0 & -54 \\
0 & 2 & 16 & 6 & 0 & -54 \\
3 & 0 & 6 & 36 & -81 & -54 \\
-81 & 0 & 0 & -81 & -972 & 0 \\
-54 & -54 & -54 & -54 & 0 & -628
\end{bmatrix}
\begin{bmatrix}
\theta_B \\
\theta_C \\
\theta_D \\
\theta_E \\
\Delta_{AB}/L \\
\Delta_{BC}/L
\end{bmatrix}
= \frac{1}{12}\frac{qL^3}{EI}
\begin{bmatrix}
+1 \\
+1 \\
-1 \\
-1 \\
+\tfrac{9}{2} \\
+\tfrac{3}{2}
\end{bmatrix}
\tag{8.88}
$$

A diagonal line has been shown on the stiffness matrix to emphasize that the system is symmetrical. (In order to write the set in this form, it was necessary to reverse the order of eqs. (8.86) and (8.87), and further to multiply each of them by a constant. This also could have been done for eq. (8.83)—multiply the third equation by 3.00—and the resulting set would have been symmetrical. Equations (8.77) are symmetrical in the form shown. This also is true for eq. (8.79). In general, these types of structural problems, when solved by the slope-deflection method, have symmetrical stiffness matrices.)

As before, once the deformation parameters have been obtained, values for the end moments can be determined by back substitution into the particular slope-deflection equations.

In each of the above examples, a particular set of usable, independent deformation parameters was more or less obvious from the outset. For certain cases, however (usually those where one or more of the members is sloping), this will not be true. Consider the frame shown in fig. 8.48. The geometry and loading is defined in sketch (a). From sketch (b) it would appear that there are three degrees of translational

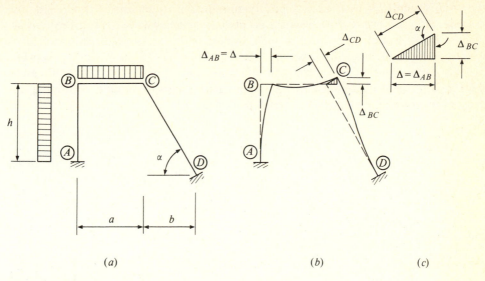

Figure 8.48 Deformation of a rigid frame.

freedom. Recalling, however, that it was assumed at the outset that member lengths do not change and, therefore, point B is constrained to move in a horizontal direction and point C must move perpendicular to member CD. The shaded triangular area of sketch (*b*), then, provides the information necessary to interrelate the various translational quantities. This triangle is shown enlarged in sketch (*c*). From it,

$$\Delta_{AB} = \Delta$$

$$\Delta_{BC} = \frac{\Delta}{\tan \alpha} \tag{8.89}$$

$$\Delta_{CD} = \frac{\Delta}{\sin \alpha}$$

The structure then has only one independent degree of translational freedom.

Sketch (*c*) of fig. 8.48 was relatively easy to deduce for the particular structure given. Had both legs of the frame been sloping, however, this would not necessarily have been the case. But before looking at that more general situation, consider again the problem defined in fig. 8.48.

When considering the question of the interrelationship between the various end-translational parameters, it frequently is helpful to consider the structure as if it were reduced to a "linkage system" by the introduction of hypothetical pin connections at each of the points of possible relative joint translation. For the case in question, this would mean that pins would be introduced at joints A, B, C, and D. Member AB, because it is positionally fixed at A, is then constrained to rotate about point A. Similarly, member CD must rotate about D. For small deformations, this means that point B of member AB (and therefore point B of member BC) must move horizontally—which is perpendicular to member AB. Point C of member CD (and therefore C of member BC) must move perpendicular to member CD. Considering member BC: rotation will be about a point—an instantaneous center of rotation—which allows B and C to move in their respective, constrained directions. This is shown in fig. 8.49, where member BC rotates about IC. The relationship between ρ_{AB}, ρ_{IC}, and ρ_{CD}, that is, between the rotations at A, IC, and D, respectively, can be defined directly from a consideration of the geometry of the frame. Assuming it is desired to express all rotations in terms of that at A:

$$\rho_{AB} = \rho_{AB}$$

$$\rho_{IC} = \rho_{AB} \frac{h}{a \tan \alpha} = \rho_{AB} \frac{h/a}{\tan \alpha} \tag{8.90}$$

$$\rho_{CD} = \rho_{IC} \frac{a/\cos \alpha}{b/\cos \alpha} = \rho_{AB} \frac{h/b}{\tan \alpha}$$

These could be expressed in terms of the end displacements noted in fig. 8.48b:

$$\Delta_{AB} = h\rho_{AB} = \Delta$$

$$\Delta_{BC} = a\rho_{IC} = \rho_{AB} \frac{(h/a)a}{\tan \alpha} = \frac{\Delta}{\tan \alpha}$$

$$\Delta_{CD} = \frac{b}{\cos \alpha} \rho_{CD} = \rho_{AB} \frac{(h/a)(a/b)b}{\tan \alpha \cos \alpha} = \frac{\Delta}{\sin \alpha}$$

which are the same equations given as (8.89).

It is to be noted that the original slope-deflection equations [eq. (8.72) and (8.73)] contain not the end-displacement term Δ_i, per se, but rather the term

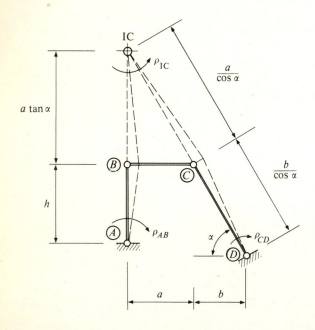

Figure 8.49 Relative rotations using the method of instantaneous centers.

$(\Delta/L)_i$, which is the rotation term ρ_i shown in fig. 8.49. The basic equations, then, could be written—possibly more usably—in the following forms:

$$M_{AB} = \frac{2EI}{L}(2\theta_A + \theta_B - 3\rho_{AB}) + M_{AB}^F$$

$$M_{BA} = \frac{2EI}{L}(\theta_A + 2\theta_B - 3\rho_{AB}) + M_{BA}^F$$

(8.91)

and

$$M_{AB} = k_{AA}\theta_A + c_{AB}k_{BB}\theta_B - k_\Delta\rho_{AB} + M_{BA}^F$$

$$M_{BA} = c_{BA}k_{AA}\theta_A + k_{BB}\theta_B - k_\Delta\rho_{AB} + M_{BA}^F$$

(8.92)

Consider now the more general three-member frame structure shown in fig. 8.50. It is assumed that the geometry of the frame is given. Therefore, the various indicated geometrical parameters are known. If ρ_{AB} is assumed as the unknown end-displacement variable, then

$$\rho_{AB} = \rho_{AB}$$

$$\rho_{IC} = \rho_{AB}\frac{a}{b} = \rho_{BC}$$

and

$$\rho_{CD} = \rho_{IC}\frac{c}{d} = \rho_{AB}\frac{a}{b}\frac{c}{d}$$

If the actual value of the perpendicular distance Δ_{AB} were desired, this could be found by multiplying ρ_{AB} by the distance from the point of rotation of member AB to point B. If the horizontal component were desired, it would be the angle of rotation times the vertical projection from A to B. In like fashion, Δ_{BC} would be equal to the angle of

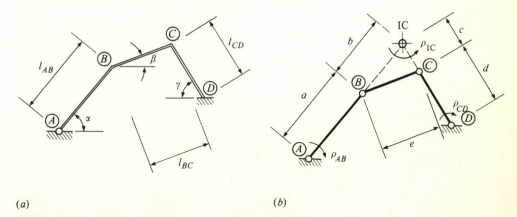

(*a*)　　　　　　　　　　　　　　　　　　　(*b*)

Figure 8.50 Example using the method of instantaneous centers.

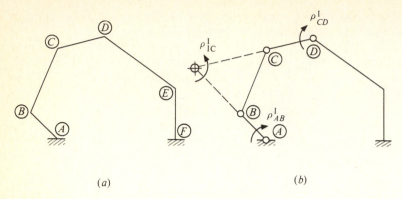

(a) (b)

Figure 8.51 Example using the method of instantaneous centers.

rotation ρ_{IC} times the length of member BC. In general, for linkage of the type described above, displacements are equal to the angle of rotation times the component of length of the member in a direction perpendicular to the desired displacement.

The question of the number of degrees of translational freedom required to solve more complex structural problems (using the slope-deflection method) also can be examined using the linkage-reduction concept described above. Consider the five-bar rigid-frame system shown in fig. 8.51a. Assuming that one of the desired independent translational parameters is ρ_{AB}, a possible three-bar linkage configuration that could be completely described would be that shown in sketch (b) of the figure. This presumes, however, that D, E, and F do not translate—a highly questionable assumption. The system therefore must have more than one translational degree of freedom. Assume now that B is position-fixed, and that a hypothetical pin is inserted at E. A new set of end-deformational relationships can be determined—in terms of this second, independent deformational parameter ρ_{BC}^{II}. The bars affected would be BC, CD, and DE. Again, it would have to be presumed that E remained fixed—or it would be impossible to define an instantaneous center for member CD. As a third independent case, assume B and C are position-fixed, with hypothetical pins in the structure at C, D, E, and F. From this, the relative movements of CD, DE, and EF can be described. Using these three independent cases, all possibilities can be defined. (It must be understood that consistency of signs must be maintained: clockwise rotations are positive.) To solve the structure shown in fig. 8.51a by the slope-deflection method, there then would be four joint rotation unknowns $(\theta_B, \theta_C, \theta_D, \text{ and } \theta_E)$ and three joint translation unknowns $(\rho_{AB}^{I}, \rho_{BC}^{II}, \text{ and } \rho_{CD}^{III})$. (Since the structure has only three redundants, the stiffness method of solution with seven degrees of freedom is obviously not the most desirable method.)

8.3.2 Slope-Deflection Equations Including the Influence of Axial Thrust

In developing the slope-deflection equations in the preceding section, it was assumed that axial forces either did not exist or were sufficiently small that they could be neglected. While this is a reasonable assumption for many real cases, there are a number of structures and loading where such is not a valid assumption, and the influence of thrust on bending stiffness must be taken into account.

Equations were developed in sec. 8.1.2 for the various bending stiffness coefficients and carry-over factors. These were expressed as the product of the stiffness with no axial thrust present times an axial load correction factor K_{ij}. These factors were shown to be functions of the ratio P/P_e, where P is the axial thrust in the member in question, and P_e is the Euler buckling load, presuming that both of the ends of the member are pinned. These modified stiffness and carry-over factors must be taken into account in the slope-deflection equations. For a member ij, the expressions for the end moments would be

$$\left. \begin{aligned} M_{ij} &= k_{ii}\theta_i + c_{ij}k_{jj}\theta_j - (k_{ii} + c_{ij}k_{ii})\rho_{ij} + M_{ij}^F \\ M_{ji} &= c_{ji}k_{jj}\theta_i + k_{jj}\theta_j - (k_{jj} + c_{ji}k_{jj})\rho_{ij} + M_{ji}^F \end{aligned} \right| \tag{8.93}$$

$k_{ii} + c_{ij}k_{ii}$ and $k_{jj} + c_{ji}k_{jj}$ represent the translational stiffnesses of the member. From the earlier derivations in chapter 4, assuming a prismatic member,

$$k_{ii} = k_{jj} = \frac{4EI}{L} \frac{3\psi(u)}{4\psi^2(u) - \phi^2(u)}$$

and

$$c_{ij} = c_{ji} = \frac{1}{2} \frac{\phi(u)}{\psi(u)}$$

where

$$\phi(u) = \frac{3}{u}\left(\frac{1}{\sin 2u} - \frac{1}{2u} \right)$$

$$\psi(u) = \frac{3}{2u}\left(\frac{1}{2u} - \frac{1}{\tan 2u} \right)$$

and

$$2u = \pi\sqrt{\frac{P}{P_e}}$$

It is to be understood that the fixed-end moment terms M_{ij}^F and M_{ji}^F in eq. (8.93) are functions of both the lateral and the axial loads. For example, it can be shown (using the equations listed in table 4.7) that for a fixed-end beam of length L the fixed-end moments for a uniformly distributed load of q over the entire span is

$$M_{ij}^F = -\frac{1}{4}qL^2 \frac{\chi(u)}{2\psi(u) + \phi(u)}$$

$$M_{ji}^F = +\frac{1}{4}qL^2 \frac{\chi(u)}{2\psi(u) + \phi(u)} \tag{8.93a}$$

For a concentrated load of Q at the midspan of a beam

$$M_{ij}^F = -\frac{1}{8} QL \frac{3\lambda(u)}{2\psi(u) + \phi(u)}$$

$$M_{ji}^F = +\frac{1}{8} QL \frac{3\lambda(u)}{2\psi(u) + \phi(u)} \tag{8.94}$$

$\chi(u)$, $\psi(u)$, $\phi(u)$, and $\lambda(u)$ values are given in table 4.8, and in the Appendix to this volume.

For rigid-frame structures, axial forces in the various members are not known at the outset. Solutions therefore are iterative in nature. The general approach usually used is to first estimate the magnitude of the axial force in each member. The modified bending-stiffness coefficients are then calculated using these estimates and the slope-deflection equations written. Using the same methods that were developed in the preceding examples, the translations, rotations, and end moments are then determined. From these the corresponding axial forces in each of the members can be ascertained. In most cases these computed axial thrusts differ from the initially

Table 8.4 Fixed-end moments, including influence of axial thrust

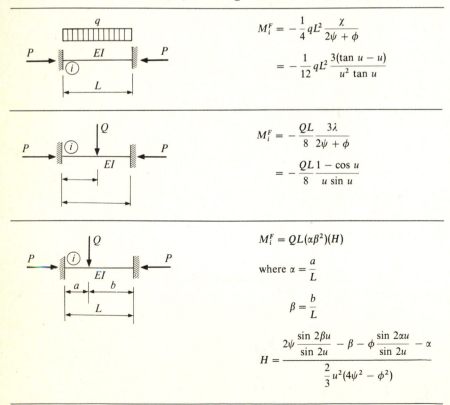

$$M_i^F = -\frac{1}{4} qL^2 \frac{\chi}{2\psi + \phi}$$

$$= -\frac{1}{12} qL^2 \frac{3(\tan u - u)}{u^2 \tan u}$$

$$M_i^F = -\frac{QL}{8} \frac{3\lambda}{2\psi + \phi}$$

$$= -\frac{QL}{8} \frac{1 - \cos u}{u \sin u}$$

$$M_i^F = QL(\alpha\beta^2)(H)$$

$$\text{where } \alpha = \frac{a}{L}$$

$$\beta = \frac{b}{L}$$

$$H = \frac{2\psi \dfrac{\sin 2\beta u}{\sin 2u} - \beta - \phi \dfrac{\sin 2\alpha u}{\sin 2u} - \alpha}{\dfrac{2}{3} u^2 (4\psi^2 - \phi^2)}$$

assumed values, and the process must be repeated. This cyclic process would continue until the axial loads used to define the stiffness are (within the desired limits) the same as those produced by the analysis.

The slope-deflection equations given above as eq. (8.93) also could be written with the end slopes as the dependent variables (that is, more or less in the flexibility form).

$$\theta_i = u_{ii} M_{ij} - u_{ij} M_{ji} + \rho_{ij} + \theta_{i0}$$
$$\theta_j = u_{jj} M_{ji} - u_{ji} M_{ij} + \rho_{ij} + \theta_{j0}$$

(8.95)

where

$$u_{ii} = u_{jj} = \frac{L}{3EI} \psi(u) \qquad u_{ij} = u_{ji} = \frac{L}{6EI} \phi(u)$$

(8.96)

θ_{i0} and θ_{j0} are the end slopes of a *simple beam* of length L subjected to an axial thrust P and the applied lateral loads Q. Clockwise rotations at the ends are positive. Counterclockwise rotations are negative.

Tables 8.4 and 8.5 give alternative equations for fixed-end moments and simple-beam end rotations, including the influence of compressive axial thrust, which are

Table 8.5 Simple-beam end rotations, including influence of axial thrust

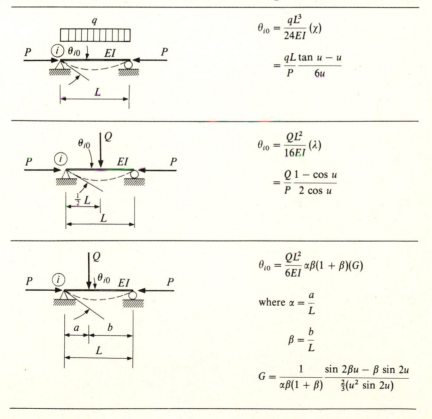

$$\theta_{i0} = \frac{qL^3}{24EI}(\chi)$$

$$= \frac{qL}{P} \frac{\tan u - u}{6u}$$

$$\theta_{i0} = \frac{QL^2}{16EI}(\lambda)$$

$$= \frac{Q}{P} \frac{1 - \cos u}{2 \cos u}$$

$$\theta_{i0} = \frac{QL^2}{6EI} \alpha\beta(1 + \beta)(G)$$

where $\alpha = \dfrac{a}{L}$

$$\beta = \frac{b}{L}$$

$$G = \frac{1}{\alpha\beta(1 + \beta)} \frac{\sin 2\beta u - \beta \sin 2u}{\frac{2}{3}(u^2 \sin 2u)}$$

suitable for use in the slope-deflection equations. In each case, the first listed equation separates out as a constant the value when $P = 0$.

For illustration of the general solutions procedure consider the following example:

The structure is the pin-based single-span rigid portal frame shown in fig. 8.52. A uniformly distributed vertical load is applied along the beam. A concentrated horizontal force of βqL acts at B. The solution will be obtained using the slope-deflection equations (8.95).

Assuming that axial thrust in BC is small, the simple-beam end slopes θ_{BO} and θ_{CO} are

$$\theta_{BO} = +\frac{qL^3}{24EI_2}$$

$$\theta_{CO} = -\frac{qL^3}{24EI_2}$$

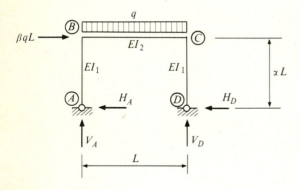

Figure 8.52 Single-span, pin-based portal frame subjected to generalized loading.

The general slope-deflection equations are then

$$\theta_{AB} = \left(\frac{\alpha L}{3EI_1}\psi_{AB}\right)M_{AB} - \left(\frac{\alpha L}{6EI_1}\phi_{AB}\right)M_{BA} + \frac{\Delta}{\alpha L} + (0)$$

$$\theta_{BA} = \left(\frac{\alpha L}{3EI_1}\psi_{AB}\right)M_{BA} - \left(\frac{\alpha L}{6EI_1}\phi_{AB}\right)M_{AB} + \frac{\Delta}{\alpha L} - (0)$$

$$\theta_{BC} = \left(\frac{L}{3EI_2}\psi_{BC}\right)M_{BC} - \left(\frac{L}{6EI_2}\phi_{BC}\right)M_{CB} + (0) + \frac{qL^3}{24EI_2}$$

$$\theta_{CB} = \left(\frac{L}{3EI_2}\psi_{BC}\right)M_{CB} - \left(\frac{L}{6EI_2}\phi_{BC}\right)M_{BC} + (0) - \frac{qL^3}{24EI_2} \qquad (8.97)$$

$$\theta_{CD} = \left(\frac{\alpha L}{3EI_1}\psi_{CD}\right)M_{CD} - \left(\frac{\alpha L}{6EI_1}\phi_{CD}\right)M_{DC} + \frac{\Delta}{\alpha L} + (0)$$

$$\theta_{DC} = \left(\frac{\alpha L}{3EI_1}\psi_{CD}\right)M_{DC} - \left(\frac{\alpha L}{6EI_1}\phi_{CD}\right)M_{CD} + \frac{\Delta}{\alpha L} - (0)$$

In these equations Δ is the horizontal translation of joint B. Since the bases of columns AB and CD are pinned

$$M_{AB} = M_{DC} = 0$$

It also has been assumed that the thrust in member BC is small. Therefore

$$\phi_{BC} = \psi_{BC} = 1.000$$

The structure is defined as rigid; that is, there will be no relative rotations of members at their points of connection. Therefore,

$$\theta_{AB} = \theta_A$$

$$\theta_{BA} = \theta_{BC} = \theta_B$$

$$\theta_{CB} = \theta_{CD} = \theta_C$$

$$\theta_{DC} = \theta_D$$

Substituting the appropriate slope-deflection equations [from (8.97)] into the second and third of the compatibility statements yields

$$\left(\frac{\alpha L}{3EI_1}\phi_{AB}\right)M_{BA} + \frac{\Delta}{\alpha L} = -\left(\frac{L}{3EI_2}\right)M_{BA} + \left(\frac{L}{6EI_2}\right)M_{CD} + \frac{qL^3}{24EI_2} \qquad (8.98)$$

and

$$-\left(\frac{L}{3EI_2}\right)M_{CD} + \left(\frac{L}{6EI_2}\right)M_{BA} - \frac{qL^3}{24EI_2} = \left(\frac{\alpha L}{3EI_1}\psi_{CD}\right)M_{CD} + \frac{\Delta}{\alpha L} \qquad (8.99)$$

At the joints, there also must be moment equilibrium.

$$M_{ji} + M_{jk} = 0$$

or

$$M_{ji} = -M_{jk} \qquad (8.100)$$

This requires for the frame in question that

$$M_{BC} = -M_{BA} \quad \text{and} \quad M_{CB} = -M_{CD}$$

A third independent equation can be determined from a consideration of overall horizontal force equilibrium.

$$H_A + H_D = \beta qL \qquad (8.101)$$

H_A and H_D can be expressed in terms of the end moments as was done for the previous cases; however, now the influence of axial thrust and Δ must be taken into account. Consider member AB (see fig. 8.53). Summing moments about point B,

$$H_A(\alpha L) + P_{AB}(\Delta) + M_{BA} = 0$$

or

$$H_A = -\frac{M_{BA} + P_{AB}(\Delta)}{\alpha L}$$

A similar expression could be obtained for H_D. Substituting these back into eq. (8.101) yields

$$M_{BA} + M_{CD} = \alpha \beta qL^2 - P_{AB}(\Delta) - P_{CD}(\Delta) \qquad (8.102)$$

The three unknowns M_{BA}, M_{CD}, and Δ are then defined by simultaneous solution of the set

$$
\begin{bmatrix}
\psi_{AB} + \dfrac{1}{\alpha}\dfrac{I_1}{I_2} & -\dfrac{1}{2\alpha} & \dfrac{3EI_1}{\alpha L} \\[2ex]
-\dfrac{1}{2\alpha} & \psi_{CD} + \dfrac{1}{\alpha}\dfrac{I_1}{I_2} & \dfrac{3EI_1}{\alpha L} \\[2ex]
1 & 1 & \dfrac{\pi EI_1}{\alpha L}\left[\left(\dfrac{P}{P_e}\right)_{AB} + \left(\dfrac{P}{P_e}\right)_{CD}\right]
\end{bmatrix}
\begin{bmatrix}
M_{BA} \\[2ex]
M_{CD} \\[2ex]
\dfrac{\Delta}{\alpha L}
\end{bmatrix}
=
\begin{bmatrix}
\dfrac{qL^2}{8\alpha}\dfrac{I_1}{I_2} \\[2ex]
-\dfrac{qL^2}{8\alpha}\dfrac{I_1}{I_2} \\[2ex]
\alpha \beta qL^2
\end{bmatrix}
$$

$$(8.103)$$

While the coefficient matrix has not been shown in a symmetrical form, this could be accomplished by multiplying eq. (8.102) by $3EI_1/\alpha L$ or by considering as the third variable $(3EI_1/\alpha L)(\Delta/\alpha L)$.

If P_{AB} and P_{CD} are assumed to be zero, ψ_{AB} and ψ_{CD} can be taken equal to unity

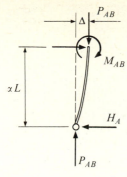

Figure 8.53 Deformation of the column.

and eq. (8.103) can be readily solved. If values of P_{AB} and P_{CD} other than zero are taken into account, ψ_{AB} and ψ_{CD} will be unknowns and an iterative process for their determination must be used. To facilitate the selection of "reasonable" approximations it is necessary that relative values of α and β be known. For example, if β is large, P_{AB} may be in tension. This would require that there be used different values of ψ from those tabulated, as noted in chap. 4.†

To illustrate how solutions are obtained, presume that β is so small that both columns are in compression. Since the applied horizontal force is to the right, P_{CD} will always be greater than P_{AB}. For example, assume that

$$_1\!\left(\frac{P}{P_e}\right)_{AB} = \frac{1}{4} \qquad \text{and} \qquad _1\!\left(\frac{P}{P_e}\right)_{CD} = \frac{1}{2} \tag{8.104}$$

This gives

$$_1(2u)_{AB} = \pi \sqrt{_1\!\left(\frac{P}{P_e}\right)_{AB}} = \frac{\pi}{2}$$

and

$$_1(2u)_{CD} = \frac{\pi}{1.414}$$

In turn,

$$_1\psi_{AB} = 1.210 \qquad \text{and} \qquad _1\psi_{CD} = 1.635 \tag{8.104a}$$

† For members in compression:

$$\psi = \frac{3}{2u}\left(\frac{1}{2u} - \frac{1}{\tan 2u}\right)$$

For members in tension:

$$\psi^+ = \frac{3}{2u}\left(\frac{1}{\tanh 2u} - \frac{1}{2u}\right)$$

(The presubscript 1 is used to indicate that this is the first assumption of values for P_{AB} and P_{CD}. Subsequent assumptions would be labeled 2, 3, etc.) The set corresponding to the values given in eq. (8.104) is

$$
\begin{bmatrix}
1.210 + \dfrac{1}{\alpha}\dfrac{I_1}{I_2} & -\dfrac{1}{2\alpha} & 1 \\[2ex]
-\dfrac{1}{2\alpha} & 1.635 + \dfrac{1}{\alpha}\dfrac{I_1}{I_2} & 1 \\[2ex]
1 & 1 & \pi^2\left(\dfrac{1}{4}+\dfrac{1}{2}\right)
\end{bmatrix}
\begin{bmatrix}
M_{BA} \\[2ex]
M_{CD} \\[2ex]
\dfrac{\Delta E I_1}{(\alpha L)^2}
\end{bmatrix}
=
\begin{bmatrix}
\dfrac{qL^2}{8\alpha}\dfrac{I_1}{I_2} \\[2ex]
-\dfrac{qL^2}{8\alpha}\dfrac{I_1}{I_2} \\[2ex]
-\alpha\beta qL^2
\end{bmatrix}
\qquad (8.105)
$$

For given values of α, β, I_1/I_2, etc., values can be determined for $_1 M_{BA}$, $_1 M_{CD}$, and $_1\Delta$. From these, using static equilibrium considerations, new values for P_{AB} and P_{CD} can be computed. These values would be used as a second approximation—$_2(P/P_e)_{AB}$ and $_2(P/P_e)_{CD}$—and new values of $_3(P/P_e)_{AB}$ and $_3(P/P_e)_{CD}$ would be determined. The process would be continued until the epsilon quantities are sufficiently small:

$$\varepsilon_{AB} = \left|_{k+1}P_{AB} - _kP_{AB}\right|$$

and

$$\varepsilon_{CD} = \left|_{k+1}P_{CD} - _kP_{CD}\right|$$

It can be shown that this process is convergent to the correct solution—providing structural instability does not occur.

For $L = 15$ ft, $\alpha = 1.0$, $E = 30 \times 10^6$ lb/in^2, $I_1 = 110$ in^4, $\beta = 0.02$, $q = 5$ kips/ft, and $I_1/I_2 = 0.5$, and assuming that $_1P_{AB} = _1P_{CD} = 0$, the solution is summarized in table 8.6. A summary of the load-deflection relationship q vs. Δ is shown in fig. 8.54. Also indicated is the solution which neglects the influence of axial thrust on bending stiffness.

Table 8.6 Iterations of axial force for example problem

Cycle	\multicolumn{6}{c}{Assumed values}					\multicolumn{2}{c}{Calculated}	
	P_{AB}, kips	P_{CD}, kips	$\left(\dfrac{P}{P_e}\right)_{AB}$	$\left(\dfrac{P}{P_e}\right)_{CD}$	M_{BA}, in·kips	M_{CD}, in·kips	Δ, in	$\left(\dfrac{P}{P_e}\right)_{AB}$	$\left(\dfrac{P}{P_e}\right)_{CD}$
1	0	0	0	0	347.14	−617.14	0.55227	0.035812	0.038797
2	36.000	39.000	0.035812	0.038797	315.01	−635.27	0.67003	0.035535	0.039074
3	35.721	39.279	0.035535	0.039074	314.98	−635.26	0.67040	0.035534	0.039074
4	35.720	39.279	0.035534	0.039074	314.98	−635.26	0.67040	0.035534	0.039074

Slope-deflection equations also can be used to determine the buckling loads of rigidly connected frames. In solving these types of problems it is important to note whether or not the structure can side sway, that is, have values for Δ. Consider the structure shown in fig. 8.55 (which, except for loading, is the same as that shown in fig. 8.52).

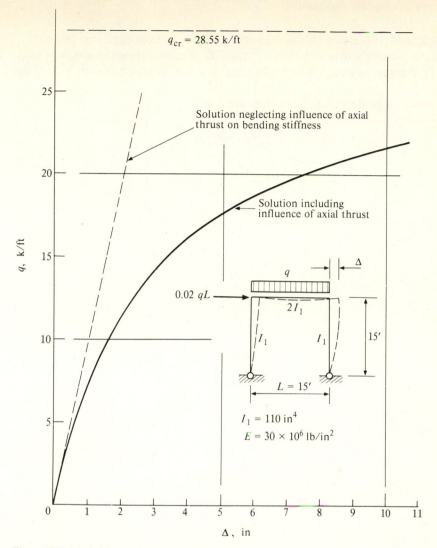

Figure 8.54 Load-deformation relationship.

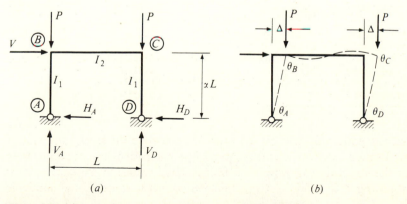

Figure 8.55 (a) Structure and loading. (b) Deformed structure.

Example 8.14 It is assumed that the vertical load is two concentrated forces of P, each acting directly over columns AB and CD. A concentrated, horizontal force of V is presumed acting at joint B. Deformations are indicated in sketch (b) of the figure.

The six slope-deflection equations are

$$\theta_{AB} = \frac{\alpha L}{EI_1}\left(-M_{BA}\frac{\phi_{AB}}{6}\right) + \frac{\Delta}{\alpha L}$$

$$\theta_{BA} = \frac{\alpha L}{EI_1}\left(M_{BA}\frac{\psi_{AB}}{3}\right) + \frac{\Delta}{\alpha L}$$

$$\theta_{BC} = \frac{L}{EI_2}\left(M_{BC}\frac{\psi_{BC}}{3} - M_{CB}\frac{\phi_{BC}}{6}\right)$$

$$\theta_{CB} = \frac{L}{EI_2}\left(M_{CB}\frac{\psi_{BC}}{3} - M_{BC}\frac{\phi_{BC}}{6}\right) \qquad (8.106)$$

$$\theta_{CD} = \frac{\alpha L}{EI_1}\left(M_{CD}\frac{\psi_{CD}}{3}\right) + \frac{\Delta}{\alpha L}$$

$$\theta_{DC} = \frac{\alpha L}{EI_1}\left(-M_{CD}\frac{\phi_{CD}}{6}\right) + \frac{\Delta}{\alpha L}$$

At joint B, $\theta_{BA} = \theta_{BC}$, or

$$\frac{\alpha L}{EI_1}\left(M_{BA}\frac{\psi_{AB}}{3}\right) + \frac{\Delta}{\alpha L} = \frac{L}{EI_2}\left(M_{BC}\frac{\psi_{BC}}{3} - M_{CB}\frac{\phi_{BC}}{6}\right)$$

But from equilibrium at that joint, $M_{BC} = -M_{BA}$. Therefore,

$$M_{BA}\left(\psi_{AB} + \frac{1}{\alpha}\frac{I_1}{I_2}\psi_{BC}\right) + M_{CB}\left(-\frac{1}{\alpha}\frac{I_1}{I_2}\frac{1}{2}\phi_{BC}\right) + \frac{3EI_1}{(\alpha L)^2}\Delta = 0 \quad (8.107)$$

Similarly, at joint C,

$$M_{BA}\left(-\frac{1}{\alpha}\frac{I_1}{I_2}\frac{1}{2}\phi_{BC}\right) + M_{CD}\left(\psi_{CD} + \frac{1}{\alpha}\frac{I_1}{I_2}\psi_{BC}\right) + \frac{3EI_1}{(\alpha L)^2}\Delta = 0 \quad (8.108)$$

For horizontal shear equilibrium,

$$H_A + H_D = V$$

But (considering each of the columns separately as free body diagrams, and summing moments about the uppermost points B and C),

$$H_A = -\frac{M_{BA} + P_{AB}(\Delta)}{\alpha L}$$

and

$$H_D = -\frac{M_{CD} + P_{CD}(\Delta)}{\alpha L}$$

Substituting these values into the horizontal force equilibrium equation yields the third necessary condition

$$M_{BA} + M_{CD} + (P_{AB} + P_{CD})\Delta + V(\alpha L) = 0$$

which can be written in the more usable form

$$M_{BA} + M_{CD} + \frac{\pi^2}{3}\left[\left(\frac{P}{P_e}\right)_{AB} + \left(\frac{P}{P_e}\right)_{CD}\right]\frac{3EI_1}{(\alpha L)^2}\Delta = -V(\alpha L) \qquad (8.109)$$

In matrix form, the set to be solved is then

$$
\begin{bmatrix}
\psi_{AB} + \xi\psi_{BC} & -\frac{1}{2}\xi\phi_{BC} & 1 \\
-\frac{1}{2}\xi\phi_{BC} & \psi_{CD} + \xi\psi_{BC} & 1 \\
1 & 1 & \zeta
\end{bmatrix}
\begin{bmatrix}
M_{BA} \\
M_{CD} \\
\frac{3EI_1}{(\alpha L)^2}\Delta
\end{bmatrix}
=
\begin{bmatrix}
0 \\
0 \\
-V(\alpha L)
\end{bmatrix}
\qquad (8.110)
$$

where

$$\xi = \frac{1}{\alpha}\frac{I_1}{I_2} \qquad \text{and} \qquad \zeta = \frac{\pi^2}{3}\left[\left(\frac{P}{P_e}\right)_{AB} + \left(\frac{P}{P_e}\right)_{CD}\right]$$

If it can be assumed that the thrust in member BC is small

$$\psi_{BC} = \phi_{BC} = 1.00$$

and eqs. (8.110) become

$$
\begin{bmatrix}
\psi_{AB} + \xi & -\frac{1}{2}\xi & 1 \\
-\frac{1}{2}\xi & \psi_{CD} + \xi & 1 \\
1 & 1 & \zeta
\end{bmatrix}
\begin{bmatrix}
M_{BA} \\
M_{CD} \\
\frac{3EI_1}{(\alpha L)^2}\Delta
\end{bmatrix}
=
\begin{bmatrix}
0 \\
0 \\
-V(\alpha L)
\end{bmatrix}
\qquad (8.111)
$$

For stability considerations, $V = 0$, and the set to be solved is homogeneous. From chap. 6, such an algebraic set can have values for the unknowns if (and only if) the determinant of the coefficients is equal to zero; that is,

$$
\begin{vmatrix}
\psi + \xi & -\frac{1}{2}\xi & 1 \\
-\frac{1}{2}\xi & \psi + \xi & 1 \\
1 & 1 & \zeta
\end{vmatrix}
= 0
\qquad (8.112)
$$

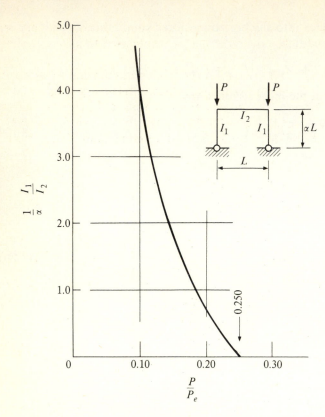

Figure 8.56 Buckling as a function of relative bending stiffnesses.

(In this expression the subscripts have been dropped from the ψ terms. Since buckling corresponds to first movement and since $V = 0$, $P_{AB} = P_{CD} = P$, and $\psi_{AB} = \psi_{CD} = \psi$.) Expanding the determinant, and solving for ξ,

$$\xi = \frac{6}{(2u)^2} - 2\psi \tag{8.113}$$

where

$$2u = \pi \sqrt{\frac{P}{P_e}} = \pi \sqrt{\left(\frac{P}{P_e}\right)_{AB}} = \pi \sqrt{\left(\frac{P}{P_e}\right)_{CD}}$$

This is shown graphically in fig. 8.56.

It is to be recognized that the coefficient matrix indicated in eq. (8.103) is essentially the same as that of eq. (8.111) (with the appropriate modification of the third variable—from $\Delta/\alpha L$ to $3EI_1 \Delta/(\alpha L)^2$). That is, the coefficient matrix is independent of the loading condition. Since stability is governed by that matrix, it was not necessary to reformulate the problem to determine the conditions for instability.

It also is to be observed (from fig. 8.54) that the maximum elastic load-carrying capacity for this structure (which is subjected to loadings which produce primary bending moments as well as axial loads in the various members) is the buckling load for the structure. This is the same phenomena that was demonstrated in fig. 4.47 for individual members subjected to axial thrust and bending moments.

Figure 8.57 is a graphic summary of several different buckling solutions. All the structures shown are of the general, single-span, single-story type. The ordinate values are the ratio of the relative stiffnesses, that is, $I_1/H \div I_2/L$. Three different abcissa values have been indicated: $2u$, which is equal to $L\sqrt{P/EI_1}$; P/P_e, where P_e is the Euler buckling load of the columns; and K, which is the

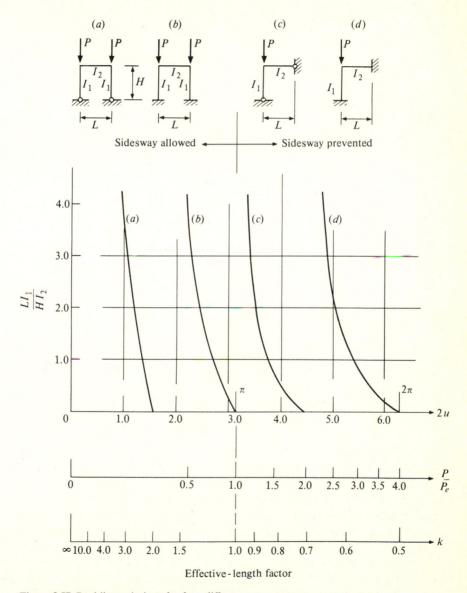

Figure 8.57 Buckling solutions for four different structures.

effective length factor for the columns [see eqs. (6.19) and (6.20)]. For those cases where sidesway is prevented: $\pi \leq 2u \leq 2\pi$; $1.0 \leq P/P_e \leq 4.0$; and $0.5 \leq K \leq 1.0$. Where sidesway is allowed: $0 \leq 2u \leq \pi$; $0 \leq P/P_e \leq 1.0$; and $1.0 \leq K \leq \infty$.

Example 8.15 The two-story rigid-frame structure shown in fig. 8.58 is to be analyzed using the slope-deflection equations (8.95). The unknowns selected are M_{21}, M_{23}, M_{32}, M_{45}, M_{54}, and M_{56}, and ρ_{12} and ρ_{23}. It is to be recognized (from static equilibrium considerations at the joints) that

$$M_{25} = -M_{21} - M_{23}$$
$$M_{52} = -M_{54} - M_{56}$$
$$M_{34} = -M_{32}$$
$$M_{43} = -M_{45}$$

$$(8.114)$$

At joint 2: $\qquad\qquad\qquad \theta_{21} = \theta_{25}$

$$M_{21}\left(\frac{1}{3} + \frac{\psi_{21}}{\xi}\right) + M_{23}\left(\frac{1}{3}\right) + M_{54}\left(-\frac{1}{6}\right) + M_{56}\left(-\frac{1}{6}\right) + \rho_{12}\left(\frac{EI_2}{L}\right) = \frac{1}{24}qL^2$$

$$\theta_{23} = \theta_{25}$$

$$M_{21}\left(\frac{1}{3}\right) + M_{23}\left(\frac{1}{3} + \frac{\psi_{23}}{\xi}\right) + M_{32}\left(-\frac{1}{6}\frac{\phi_{23}}{\xi}\right) + M_{54}\left(-\frac{1}{6}\right) + M_{56}\left(-\frac{1}{6}\right)$$

$$+ \rho_{23}\left(\frac{EI_2}{L}\right) = \frac{1}{24}qL^2$$

At joint 3: $\qquad\qquad\qquad \theta_{32} = \theta_{34}$

$$M_{23}\left(-\frac{1}{6}\frac{\phi_{23}}{\xi}\right) + M_{32}\left(\frac{1}{3} + \frac{\psi_{32}}{\xi}\right) + M_{45}\left(-\frac{1}{6}\right) + \rho_{23}\left(\frac{EI_2}{L}\right) = \frac{1}{24}qL^2$$

At joint 4: $\qquad\qquad\qquad \theta_{43} = \theta_{45}$

$$M_{32}\left(-\frac{1}{6}\right) + M_{45}\left(\frac{1}{3} + \frac{\psi_{45}}{\xi}\right) + M_{54}\left(-\frac{1}{6}\frac{\phi_{45}}{\xi}\right) + \rho_{23}\left(\frac{EI_2}{L}\right) = -\frac{1}{24}qL^2$$

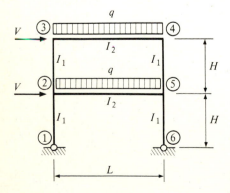

Figure 8.58

At joint 5: $\qquad\qquad\qquad\theta_{54} = \theta_{52}$

$$M_{21}\left(-\frac{1}{6}\right) + M_{23}\left(-\frac{1}{6}\right) + M_{45}\left(-\frac{1}{6}\frac{\phi_{45}}{\xi}\right)$$

$$+ M_{54}\left(\frac{1}{3} + \frac{\psi_{45}}{\xi}\right) + M_{56}\left(\frac{1}{3}\right) + \rho_{23}\left(\frac{EI_2}{L}\right) = -\frac{1}{24}qL^2$$

$$\theta_{56} = \theta_{52}$$

$$M_{21}\left(-\frac{1}{6}\right) + M_{23}\left(-\frac{1}{6}\right) + M_{54}\left(\frac{1}{3}\right) + M_{56}\left(\frac{1}{3} + \frac{\psi_{56}}{\xi}\right) + \rho_{12}\left(\frac{EI_2}{L}\right) = -\frac{1}{24}qL^2$$

For shear equilibrium of the frame above joints 2 and 5:

$$V = V_{32} + V_{45}$$

$$M_{23}(1) + M_{32}(1) + M_{45}(1) + M_{54}(1) + \rho_{23}(P_{23}H + P_{45}H) = -HV$$

For shear equilibrium of the total frame:

$$2V = V_{12} + V_{56}$$

$$M_{21}(1) + M_{56}(1) + \rho_1(P_{12}H + P_{56}H) = -2HV$$

where

$$\xi = \frac{L\,I_1}{H\,I_2}$$

In matrix form the set is

$$
\begin{bmatrix}
A & \frac{1}{3} & 0 & 0 & -\frac{1}{6} & -\frac{1}{6} & 1 & 0 \\
\frac{1}{3} & B & -E & 0 & -\frac{1}{6} & -\frac{1}{6} & 0 & 1 \\
0 & -E & B & -\frac{1}{6} & 0 & 0 & 0 & 1 \\
0 & 0 & -\frac{1}{6} & C & -F & 0 & 0 & 1 \\
-\frac{1}{6} & -\frac{1}{6} & 0 & -F & C & \frac{1}{3} & 0 & 1 \\
-\frac{1}{6} & -\frac{1}{6} & 0 & 0 & \frac{1}{3} & D & 1 & 0 \\
1 & 0 & 0 & 0 & 0 & 1 & Q_1 & 0 \\
0 & 1 & 1 & 1 & 1 & 0 & 0 & Q_2
\end{bmatrix}
\begin{bmatrix}
M_{21} \\
M_{23} \\
M_{32} \\
M_{45} \\
M_{54} \\
M_{56} \\
\rho_{12}\dfrac{EI_2}{L} \\
\rho_{23}\dfrac{EI_2}{L}
\end{bmatrix}
=
\begin{bmatrix}
\frac{1}{24}qL^2 \\
\frac{1}{24}qL^2 \\
\frac{1}{24}qL^2 \\
-\frac{1}{24}qL^2 \\
-\frac{1}{24}qL^2 \\
-\frac{1}{24}qL^2 \\
-2VH \\
-VH
\end{bmatrix}
$$

$$(8.115)$$

where $A = \frac{1}{3} + \frac{\psi_{12}}{\xi}$ $\quad E = \frac{1}{6}\frac{\phi_{23}}{\xi}$

$\quad\quad\quad B = \frac{1}{3} + \frac{\psi_{23}}{\xi}$ $\quad F = \frac{1}{6}\frac{\phi_{45}}{\xi}$

$\quad\quad\quad C = \frac{1}{3} + \frac{\psi_{45}}{\xi}$ $\quad Q_1 = \pi^2\xi\left[\left(\frac{P}{P_e}\right)_{12} + \left(\frac{P}{P_e}\right)_{56}\right]$

$\quad\quad\quad D = \frac{1}{3} + \frac{\psi_{56}}{\xi}$ $\quad Q_2 = \pi^2\xi\left[\left(\frac{P}{P_e}\right)_{23} + \left(\frac{P}{P_e}\right)_{45}\right]$

As was true for the case given in table 8.4, the solution will be iterative in nature. The various column thrusts can be obtained from free body diagrams of the top beam and the structure as a whole. Buckling loads are determined from a consideration of the determinant of the coefficient matrix. (Buckling occurs when the determinant is equal to zero. Of course, the lowest possible eigenvalue governs.)

Rigidly connected trusses can be analyzed for buckling using the heretofore developed slope-deflection equations. For *triangular subset trusses* (as opposed to Vierendeel trusses, which are rectangular in form), making the same assumptions as listed before, there will be no relative displacements of the ends of the member— other than those due to axial shortening, which are determined in the usual manner defined in chap. 5. It therefore can be assumed in stability analyses that ρ will be zero. Consider the truss of example 8.16 shown in fig. 8.59.

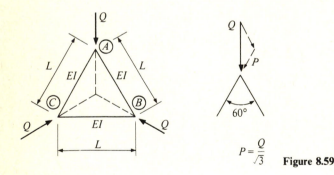

Figure 8.59

$$P = \frac{Q}{\sqrt{3}}$$

Example 8.16 The rigidly connected truss is composed of three bars AB, BC, and CA. Loads Q are applied in a symmetrical fashion such that the compressive thrust in each member is equal to $Q/\sqrt{3}$.

For equilibrium at the joints

$$M_{AB} + M_{AC} = 0$$

$$M_{BA} + M_{BC} = 0$$

$$M_{CB} + M_{CA} = 0$$

Since all axial forces are equal, the slope-deflection equations can be written in the form

$$M_{AB} = \frac{6EI}{L}\gamma^2(2\psi\theta_A + \phi\theta_B)$$

$$M_{BA} = \frac{6EI}{L}\gamma^2(2\psi\theta_B + \phi\theta_A)$$

$$M_{BC} = \frac{6EI}{L}\gamma^2(2\psi\theta_B + \phi\theta_C)$$

$$M_{CB} = \frac{6EI}{L}\gamma^2(2\psi\theta_C + \phi\theta_B)$$

$$M_{CA} = \frac{6EI}{L}\gamma^2(2\psi\theta_C + \phi\theta_A)$$

$$M_{AC} = \frac{6EI}{L}\gamma^2(2\psi\theta_A + \phi\theta_C)$$

where

$$\gamma = \frac{1}{4\psi^2 - \phi^2}$$

Substituting these into the equilibrium equations yields

$$\begin{bmatrix} 4\psi & \phi & \phi \\ \phi & 4\psi & \phi \\ \phi & \phi & 4\psi \end{bmatrix} \begin{bmatrix} \theta_A \\ \theta_B \\ \theta_C \end{bmatrix} = \begin{bmatrix} 0 \\ 0 \\ 0 \end{bmatrix} \tag{8.116}$$

Buckling will occur when the determinant of the coefficient matrix equals zero.

$$\begin{vmatrix} 4\psi & \phi & \phi \\ \phi & 4\psi & \phi \\ \phi & \phi & 4\psi \end{vmatrix} = 0$$

When expanded, there is obtained the eigenvalue equation

$$32\psi^3 - 6\psi\phi^2 + \phi^3 = 0 \tag{8.117}$$

This equation can be factored into the form

$$(4\psi - \phi)^2(2\psi + \phi) = 0 \tag{8.118}$$

whose solutions are

$$4\psi - \phi = 0$$
$$4\psi - \phi = 0 \tag{8.119}$$
$$2\psi + \phi = 0$$

The first two of these have a solution at

$$2u = 3.86 = \pi \sqrt{\frac{P}{P_e}} \tag{8.120a}$$

For the third of equations (8.119),

$$2u = 2\pi = \pi \sqrt{\frac{P}{P_e}} \tag{8.120b}$$

Equation (8.120a) controls and

$$\frac{P_{\text{crit}}}{EI} = \left(\frac{3.86}{L}\right)^2$$

or

$$P_{\text{crit}} = \frac{14.90 EI}{L^2}$$

Therefore,

$$Q_{\text{crit}} = \frac{(14.90)(\sqrt{3})EI}{L^2} \tag{8.121}$$

It is interesting to note that solutions (8.120a) occur with $\theta_A = +\theta_B$ and $\theta_A = -\theta_B$. Solution (8.120b) corresponds to $\theta_A = \theta_B = \theta_C = 0$.

8.4 DYNAMIC RESPONSE OF SIMPLE INDETERMINATE STRUCTURES

As noted in sec. 4.7, a lumped-mass model is frequently used to analyze the dynamic response of structural systems. This is particularly true for indeterminate structures, which by their very nature are more complex than determinate ones. In these types of analyses, masses are presumed to be concentrated at certain specific preselected points within the structure. The displacements at these points are the dependent variables. The number of them to be found constitutes the number of degrees of freedom of the system in question. It is to be noted, however, that while the mass of the structure is assumed localized, the elastic properties of the various members remain distributed.

The quality of the solution obtained from such a lumped-mass model depends for the most part on the assumed number of degrees of freedom. For example, a three-story single-span indeterminate frame would be modeled in several ways—as indicated in fig. 8.60. If each of the floors is presumed to be very rigid in comparison to the columns, the column floor joints will undergo very little rotation as the structure displaces laterally. This is shown in model (a). As a second approximation, floor flexibilities could be presumed with the masses concentrated, as indicated in model (b). Model (c) presumes a three-degree-of-freedom system where the weight of the structure is relatively small, but the structure retains its lateral stiffness.

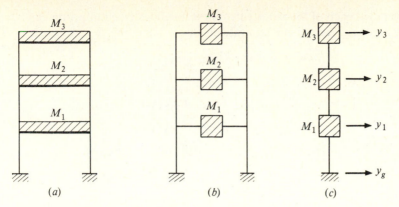

Figure 8.60 Multistory single-span frame.

The examples shown in fig. 8.60 are natural models. The introduction of additional translational degrees of freedom would probably not yield a corresponding increase in the accuracy of solution obtained. Other cases are not so clear-cut, and the search for a good model can be a significant problem in itself.

Once the model to be used has been chosen, the equations of motion for the "discretized" structure can be formulated using the concept of stiffness as discussed earlier in this chapter. This is accomplished by first investigating the response of the model to a series of static loads, as shown in fig. 8.61. Here y_g represents any lateral support or foundation motion. To define the stiffness coefficients for the structure, a series of unit lateral displacements is introduced, in turn, in each degree-of-freedom direction. The relative forces for the four displacement states are the desired stiffness coefficients, and these can be obtained using any of the methods heretofore described. Herein lies the principal difference between a determinate and an indeterminate analysis: An indeterminate structure requires an indeterminate analysis to ascertain the stiffness coefficients. Other than that, indeterminate problems are treated in exactly the same way as any other multidegree-of-freedom system.

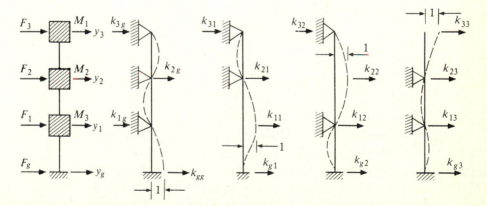

Figure 8.61 Stiffnesses and assumed unit displacements of the multistory structure.

Having determined each of the required stiffness coefficients, equilibrium can be defined by superimposing each of the separate effects.

$$F_g = k_{gg} y_g + k_{g1} y_1 + k_{g2} y_2 + k_{g3} y_3$$

$$F_1 = k_{1g} y_g + k_{11} y_1 + k_{12} y_2 + k_{13} y_3 \qquad (8.122)$$

$$F_2 = k_{2g} y_g + k_{21} y_1 + k_{22} y_2 + k_{23} y_3$$

$$F_3 = k_{3g} y_g + k_{31} y_1 + k_{32} y_2 + k_{33} y_3$$

In chap. 4 it was shown that equilibrium for a problem of this type is defined by an equation of the form [eq. (4.158)]

$$M\ddot{y} + C\dot{y} + Ky = f(t) \qquad (8.123)$$

where M, C, and K represent, respectively, the mass, the damping, and the stiffness of the system. For the above described multidegree-of-freedom system, eqs. (8.122) take the place of the single Ky term in eq. (8.123).

As indicated earlier, the term y_g represents a lateral motion of the supporting medium of the structure. This could be the motion of another structure to which the one being studied is attached, or it could be a ground motion, as in an earthquake. In any event, however, it would most likely be zero or a known quantity. Equations (8.122), then, can be considered in two parts:

$$(a) \quad F_g = k_{gg} y_g + k_{g1} y_1 + k_{g2} y_2 + k_{g3} y_3$$

and

$$(b) \quad F_1 = k_{1g} y_g + k_{11} y_1 + k_{12} y_2 + k_{13} y_3$$

$$F_2 = k_{2g} y_g + k_{21} y_1 + k_{22} y_2 + k_{23} y_3$$

$$F_3 = k_{3g} y_g + k_{31} y_1 + k_{32} y_2 + k_{33} y_3$$

The second set would be used to calculate y_1, y_2, and y_3, and the equivalent static forces F_1, F_2, and F_3. Using case (a) would give the corresponding base shear.

For the structure under consideration, the equations of motion are therefore

$$M_1 \ddot{y}_1 + C_1 \dot{y}_1 + k_{11} y_1 + k_{12} y_2 + k_{13} y_3 = f_1(t) - k_{1g} y_g(t)$$

$$M_2 \ddot{y}_2 + C_2 \dot{y}_2 + k_{21} y_1 + k_{22} y_2 + k_{23} y_3 = f_2(t) - k_{2g} y_g(t) \qquad (8.124)$$

$$M_3 \ddot{y}_3 + C_3 \dot{y}_3 + k_{31} y_1 + k_{32} y_2 + k_{33} y_3 = f_3(t) - k_{3g} y_g(t)$$

For given initial conditions, dynamic forces $f_i(t)$, and time-dependent support displacements $y_g(t)$, these equations can be solved for the responses y_1, y_2, and y_3 (as a function of time).

The natural frequencies of the structure are associated with the undamped free vibration case. For this

$$M_1 \ddot{y}_1 + k_{11} y_1 + k_{12} y_2 + k_{13} y_3 = 0$$

$$M_2 \ddot{y}_2 + k_{21} y_1 + k_{22} y_2 + k_{23} y_3 = 0 \qquad (8.125)$$

$$M_3 \ddot{y}_3 + k_{31} y_1 + k_{32} y_2 + k_{33} y_3 = 0$$

For these types of problems [see eq. (4.191)]

$$y_i = A_i \sin \omega t$$

where A_i is the amplitude of y_i and ω is the natural frequency of vibration. The equations of motion then become

$$
\begin{aligned}
(k_{11} - M_1\omega^2)A_1 &+ k_{12}A_2 &+ k_{13}A_3 &= 0 \\
k_{21}A_1 &+ (k_{22} - M_2\omega^2)A_2 &+ k_{23}A_3 &= 0 \quad (8.126) \\
k_{31}A_1 &+ k_{32}A_2 &+ (k_{33} - M_3\omega^2)A_3 &= 0
\end{aligned}
$$

For solutions other than $y_i \equiv 0$, the determinant of the coefficient matrix must equal zero.

$$
\begin{vmatrix}
k_{11} - M_1\omega^2 & k_{12} & k_{13} \\
k_{21} & k_{22} - M_2\omega^2 & k_{23} \\
k_{31} & k_{32} & k_{33} - M_3\omega^2
\end{vmatrix} = 0
$$

This can be solved for the critical frequencies.

Example 8.17 For the structure shown in fig. 8.52: assuming $L = 15$ ft, $\alpha = 1.0$, $q = 10$ kips/ft (this includes the weight of the roof, the snow load, etc.), $I_1 = 110$ in^4 (a W 8×31 beam), and $E = 30 \times 10^6$ lb/in^2 (structural steel), it is desired to determine the seismic base shear for which the structure must be designed, if eq. (1.6) governs. The frame is presumed located in a zone 3 seismic risk area (see fig. 1.5) and is not considered to be an essential building.

From eq. (1.6), $V = ZIKCSW$

where Z = zone factor = 0.75
I = occupancy factor = 1.0
K = ductility factor = 0.67
W = total weight = applied vertical loads + member weights
$\quad = (10)(15) + (0.031)(15)(3) = 151.4$ kips
C = seismic coefficient = $1/15\sqrt{T}$
S = site-structure resonance coefficient = 1.5

and

T = natural period of the structure

The structure in question can be modeled as a one-degree-of-freedom system with the mass concentrated at the level of the beam. For such a system, the natural frequency of vibration is given by eq. (4.165).

$$f = \frac{1}{T} = \frac{1}{2\pi}\sqrt{\frac{K}{M}}$$

where M is the mass of the structure and the applied loads and K is the lateral stiffness. (K is the force required to cause one unit of lateral deformation at the top of the frame.) It was previously determined that for a load of $q = 10$ kips/ft, a lateral force of 3.0 kips caused a lateral deflection of $\Delta = 1.7075$ in. The stiffness K, then, is

$$K = \frac{3.00}{1.7075} = 1.757 \text{ kips/in}$$

The natural frequency is

$$f = \frac{1}{2\pi} \sqrt{\frac{(1.757)(32.2)(12)}{151.4}} = 0.337 \text{ Hz}$$

Therefore,

$$T = \frac{1}{f} = \frac{1}{0.337} = 2.967 \text{ s}$$

$$C = \frac{1}{15\sqrt{2.967}} = 0.0387$$

The base shear (that is, the equivalent lateral load to be applied) is

$$V = ZIKCSW = (0.75)(1.0)(0.67)(0.0387)(1.5)(151.4) = 4.42 \text{ kips}$$

The frame must be designed to withstand this magnitude of force applied at the level of the beam.

8.5 DESIGN OF INDETERMINATE STRUCTURES

The general procedures for design developed in chap. 7 can be applied directly to indeterminate structures. As noted earlier in this chapter, however, one further complication is present for these types of systems: a solution cannot be obtained unless (and until) an assumption is made concerning the relative sizes of the various members. That assumption can be refined by further trials—until the desired degree of accuracy is obtained.

In most cases, for structures that carry their loads primarily by bending, the influence of axial thrust on bending stiffness is neglected during the first (or early) trials. This variable is introduced only when relative proportions of members have been comparatively well established and ranges of variables ascertained.

There is no unique best way to approach the design problem. Depending on the complexity of the structure, the experience of the designer, the availability of materials (and for that matter, even the availability of particular member sizes), and the relative cost of material to labor, various legitimate assumptions (simplifications) can be made. In general, the more complicated the structure and loading, the more likely will be the choice of particular ratios of member sizes from the outset. For simpler systems, however, the various terms can be kept more or less in general form until the near-final stages. The following example is presented not because it is a typical problem often encountered, nor to provide a model for subsequent emulation, but rather because its simplicity allows the variables to be kept " open " throughout most of the process. A variety of types of situations can be examined for this structure.

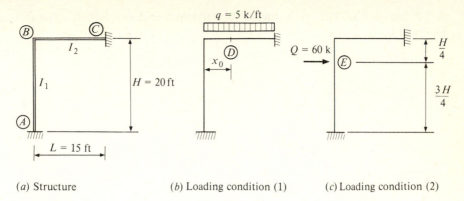

(a) Structure   (b) Loading condition (1)   (c) Loading condition (2)

Figure 8.62

Example 8.18 The structure and loading conditions are shown in fig. 8.62. The column has a moment of inertia in the direction of primary bending (that is, about an axis perpendicular to the plane of the frame) of I_1. The moment of inertia of the beam is I_2. It is assumed that sufficient lateral bracing is provided so that there will be no possibility of buckling out of the plane of the frame. Local buckling of the various elements of the resulting cross section also will be discounted.

Two independent loading conditions are prescribed, as shown in sketches (b) and (c). Their combined effect also must be taken into account—as a third loading possibility. It is to be assumed that the members are rectangular in cross section, the modulus of elasticity is 4 kips/in², the allowable combined stress cannot exceed 2 kips/in², and the material has a density of 120 lb/ft³.

To simplify the equations, a parameter $\gamma = I_2/I_1$ will be introduced. It will be assumed that positive bending-moment values correspond to compressive stresses on the outer faces of the structure. (For the beam, this would be the upper surface. For the column, moments that cause compression on the left-hand side would be plus.)

Loading condition 1: Using any of the methods for solution of indeterminate structures developed in this chapter, and assuming that the influence of axial thrust on the bending stiffness of the members can be neglected, the following equations are obtained:

$$M_A = \left[+ \frac{1}{8(3 + 4\gamma)} \right] qL^2 \tag{8.127a}$$

$$M_B = \left[- \frac{1}{4(3 + 4\gamma)} \right] qL^2 \tag{8.127b}$$

$$M_C = \left[- \frac{1}{4} \frac{1 + 2\gamma}{3 + 4\gamma} \right] qL^2 \tag{8.127c}$$

At some point D [see sketch (b)], located x_0 from B along member BC, the bending moment will have a maximum positive value.

$$x_0 = \frac{3L}{2}\left[\frac{1+\gamma}{3+4\gamma}\right]$$

$$M_D = M_{max} = \left[\frac{9\gamma^2 + 10\gamma + 3}{8(3+4\gamma)^2}\right]qL^2 \tag{8.127d}$$

Loading condition 2:

$$M_A = \left[-\frac{1}{32}\frac{15+8\gamma}{3+4\gamma}\right]QL \tag{8.128a}$$

$$M_B = \left[-\frac{3}{4}\frac{\gamma}{3+4\gamma}\right]QL \tag{8.128b}$$

$$M_C = \left[+\frac{3}{8}\frac{\gamma}{3+4\gamma}\right]QL \tag{8.128c}$$

$$M_E = \left[+\frac{1}{128}\frac{81+48\gamma}{3+4\gamma}\right]QL \tag{8.128d}$$

Loading condition 3:

$$M_A = \left[+\frac{1}{8(3+4\gamma)}\right]qL^2 + \left[-\frac{1}{32}\frac{15+8\gamma}{3+4\gamma}\right]QL \tag{8.129a}$$

$$M_B = \left[-\frac{1}{4(3+4\gamma)}\right]qL^2 + \left[-\frac{3}{4}\frac{\gamma}{3+4\gamma}\right]QL \tag{8.129b}$$

$$M_C = \left[-\frac{1}{4}\frac{1+2\gamma}{3+4\gamma}\right]qL^2 + \left[+\frac{3}{8}\frac{\gamma}{3+4\gamma}\right]QL \tag{8.129c}$$

$$M_D = \left[+\frac{9\gamma^2 + 10\gamma + 3}{8(3+4\gamma)^2}\right]qL^2 + \left[-\frac{3\gamma(7\gamma+3)}{16(3+4\gamma)^2}\right]QL \tag{8.129d}$$

$$M_E = \left[-\frac{5}{32}\frac{1}{3+4\gamma}\right]qL^2 + \left[+\frac{1}{128}\frac{81+48\gamma}{3+4\gamma}\right]QL \tag{8.129e}$$

(While it is true that location D—the point of maximum positive moment in beam BC—for loading condition (1) will be different from that for loading condition (3), it is considered in this example that the difference in moment values will be relatively small, whichever location is considered. Therefore, it has been assumed that the location associated with loading (1) will be examined, and the two cases added at that point for loading (3).)

The axial forces in the two members also can be defined in terms of the quantities listed above:

Loading condition 1:

$$P_{AB} = \frac{3}{2}\left[\frac{1+\gamma}{3+4\gamma}\right]qL$$

$$P_{BC} = \frac{9}{4}\left[\frac{1}{8(3+4\gamma)}\right]qL$$

Loading condition 2:

$$P_{AB} = \frac{9}{8}\left[\frac{\gamma}{3+4\gamma}\right]Q$$

$$P_{BC} = \frac{27}{128}\left[\frac{\gamma}{3+4\gamma}\right]Q$$

Loading condition (3) would be the sum of the above two situations.

Before design can proceed, it is necessary that there be assumed a value for the unit weight of the beam. (This weight will be additive to the 5 kips/ft q loading.) The determination of an approximate unit weight can be accomplished in a variety of fashions, not the least desirable of which is to review previous similar designs. However, assuming no experience, a consideration of limiting cases also can be used to arrive at "suitable" preliminary values.

For the beam: Both ends of the member fixed:

$$M_B = M_C = -\frac{1}{12}qL^2$$

$$M_D = +\frac{1}{24}qL^2$$

End C fixed, end B pinned:

$$M_C = -\frac{1}{8}qL^2$$

$$M_D = +\frac{9}{128}qL^2$$

For the column: Both ends of the member fixed:

$$M_A = -\frac{3}{64}QH$$

$$M_B = -\frac{9}{64}QH$$

$$M_E = +\frac{9}{128}QH$$

End A fixed, end B pinned:

$$M_A = -\frac{15}{128} QH$$

$$M_E = +\frac{81}{512} QH$$

The design bending moment values therefore should be contained within the following bounds:

For the beam:

$$-0.04167 \leq \frac{M_B}{qL^2} \leq 0$$

$$-0.125 \leq \frac{M_C}{qL^2} \leq -0.0833$$

$$+0.04167 \leq \frac{M_D}{qL^2} \leq +0.0703$$

For the column:

$$-0.1172 \leq \frac{M_A}{QH} \leq -0.04688$$

$$-0.1406 \leq \frac{M_B}{QH} \leq 0$$

$$+0.07031 \leq \frac{M_E}{QH} \leq +0.1582$$

Assuming maximum bending-moment values but neglecting any axial forces that may exist:

$$M_{\text{beam}} = 0.125qL^2 = 1688 \text{ in} \cdot \text{kips}$$

$$M_{\text{col}} = 0.158QH = 2275 \text{ in} \cdot \text{kips}$$

If the beam is rectangular in cross-sectional form and its width is equal to one-half its depth, then

$$M_{\text{beam}} = S\sigma_{\text{all}} = \tfrac{1}{12}(d_2)^3(2.00) = 1688 \text{ in} \cdot \text{kips}$$

or

$$d_2 \doteq 22 \text{ in}$$

The weight of such a member would be

$$W = \frac{A(12)}{1728}(\text{density}) = (22)(11)(12)(120) = 202 \text{ lb/ft}$$

Therefore, it will be assumed that the beam weighs 200 lb/ft, and the q loading to which the structure will be subjected is

$$q = 5 \text{ kips/ft} + 0.200 \text{ kip/ft} = 5.2 \text{ kips/ft}$$

(Should this value be significantly different from that ultimately determined for the design, it will be necessary to ascertain the relative importance of the difference, and possibly even to carry out an entirely new set of calculations using the more refined value for the weight of the member.)

Substituting $q = 5.2$ kips/ft, $Q = 60$ kips, $L = 15$ ft, and $H = 20$ ft into eqs. (8.121), (8.128), and (8.129), values can be determined for M_A, M_B, M_C, M_D, M_E, $P_{\text{col}} = P_{AB}$, and $P_{\text{beam}} = P_{BC}$, for each of the conditions of loading specified—providing there is first assumed a value for γ.

Using the subscript 2 to denote the beam, and 1, the column, the ratio of the moments of inertia of the two members is

$$\gamma = \frac{I_2}{I_1} = \frac{\frac{1}{12}b_2 d_2^3}{\frac{1}{12}b_1 d_1^3} = \frac{b_2}{b_1}\left(\frac{d_2}{d_1}\right)^3 \tag{8.130}$$

Maximum normal stresses in each of the members depend on the axial force and the maximum bending moment—for each of the three specified conditions of loading. The particular loading that causes maximum normal stress in the beam may not be the critical one for the column. Moreover, as γ changes, the location of the control section also may change, as may the critical condition of loading. In short, to arrive at the correct design solution, all possibilities must be examined for each selected value of γ.

Both axial force and bending moments are present in the members in question. The prescribed design control condition is the maximum normal stress ($\sigma_{\text{max}} \leq \sigma_{\text{all}}$). Therefore,

$$\sigma_{\text{all}} = \frac{P_{\text{col}}}{b_1 d_1} + \frac{M_{\text{col}}}{\frac{1}{6}b_1 d_1^2} \tag{8.131}$$

and

$$\sigma_{\text{all}} = \frac{P_{\text{beam}}}{b_2 d_2} + \frac{M_{\text{beam}}}{\frac{1}{6}b_2 d_2^2} \tag{8.132}$$

Equations (8.130), (8.131), and (8.132) constitute the three necessary conditions for design. But there are four unknown cross-sectional parameters: $d_1, d_2, b_1,$ and b_2. A unique solution, therefore, cannot be found—unless further conditions (such at least total weight, least cost, etc.) are specified.

If it is assumed that d_2 is known, then from eq. (8.131)

$$b_2 = \frac{1}{\sigma_{\text{all}}}\left(\frac{P_{\text{beam}}}{d_2} + \frac{6M_{\text{beam}}}{d_2^2}\right) \tag{8.133}$$

Operating similarly on eq. (8.132)

$$b_1 = \frac{1}{\sigma_{\text{all}}}\left(\frac{P_{\text{col}}}{d_1} + \frac{6M_{\text{col}}}{d_1^2}\right) \tag{8.134}$$

Dividing eq. (8.133) by (8.134) and noting the relationship given in (8.130), there can be written the following quadratic equation in the unknown d_1:

$$d_1^2 + d_1 \frac{6M_{col}}{P_{col}} = \frac{1}{\gamma} \frac{P_{beam}d_2^2 + 6M_{beam}d_2}{P_{col}} \tag{8.135}$$

This can be solved by "completing-the-square," or any other such method. Back substitution of that value into eq. (8.130) yields

$$b_1 = b_2 \left[\frac{1}{\gamma} \left(\frac{d_2}{d_1} \right)^3 \right] \tag{8.136}$$

A summary of the resulting possible design solutions for the structure in question is shown in fig. 8.63.

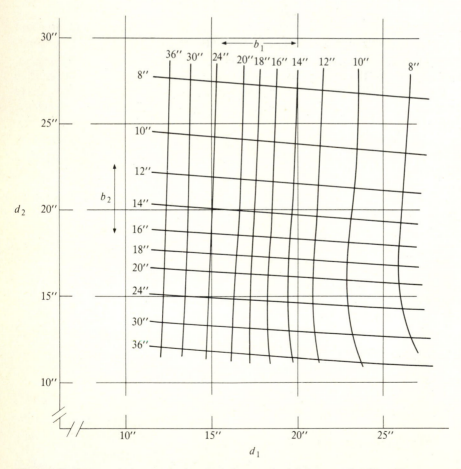

Figure 8.63 Design solutions neglecting the influence of axial thrust on bending stiffnesses.

Solutions which take into account the influence of axial thrust on bending stiffnesses are not so easily obtained. As noted earlier, not only does such a solution require an assumption of the ratio of the moments of inertia γ but also the coefficients of the unknown end moments in the slope-deflection equations presume a knowledge of the axial forces in the members, which, in turn, requires that the values of the end moments be known. Solution, then, will be trial and error in nature.

The stiffness matrix for the structure will be the same for each of the conditions of loading. All that will vary will be the column vector of displacements. (For a formulation of the eq. (8.95) type, these would be the simple-beam end slopes listed in table 8.5.)

The slope-deflection equations for the structure in question are

$$\theta_{AB} = \frac{H}{EI_1}\left(M_{AB}\frac{\psi_{AB}}{3} - M_{BA}\frac{\phi_{AB}}{6}\right) + \theta_{AB,0}$$

$$\theta_{BA} = \frac{H}{EI_1}\left(M_{BA}\frac{\psi_{AB}}{3} - M_{AB}\frac{\phi_{AB}}{6}\right) - \theta_{BA,0}$$

$$\theta_{BC} = \frac{L}{EI_2}\left(M_{BC}\frac{\psi_{BC}}{3} - M_{CB}\frac{\phi_{BC}}{6}\right) + \theta_{BC,0}$$

$$\theta_{CB} = \frac{L}{EI_2}\left(M_{CB}\frac{\psi_{BC}}{3} - M_{BC}\frac{\phi_{BC}}{6}\right) - \theta_{CB,0}$$

At the column base, $\theta_A = \theta_{AB} = 0$. Therefore,

$$M_{AB}\frac{\psi_{AB}}{3} - M_{BA}\frac{\phi_{AB}}{6} = 0 \tag{8.137}$$

At joint B, $\theta_{BA} = \theta_{BC} = \theta_B$. (Also, for equilibrium of moments at that joint, $M_{BC} = -M_{BA}$.)

$$M_{AB}\left[-\frac{\xi\phi_{AB}}{6}\right] + M_{BA}\left[\frac{\xi\psi_{AB}}{3} + \frac{\psi_{BC}}{3}\right] + M_{CB}\left[\frac{\phi_{BC}}{6}\right] = \theta_{BA,0} \tag{8.138}$$

where $\xi = I_2 H/I_1 L$.

At joint C, $\theta_{CB} = \theta_C = 0$. Therefore,

$$M_{BA}\frac{\phi_{AC}}{6} + M_{CB}\frac{\psi_{BC}}{3} = \theta_{CB,0} \tag{8.139}$$

In matrix form, the governing equations (8.137), (8.138), and (8.139) are

$$\begin{bmatrix} 2\xi\psi_{AB} & -\xi\phi_{AB} & 0 \\ -\xi\phi_{AB} & 2(\xi\psi_{AB} + \psi_{BC}) & \phi_{BC} \\ 0 & \phi_{BC} & 2\psi_{BC} \end{bmatrix} \begin{bmatrix} M_{AB} \\ M_{BA} \\ M_{CB} \end{bmatrix} = \begin{bmatrix} \text{Column} \\ \text{vector of} \\ \text{displacements} \end{bmatrix} \tag{8.140}$$

where

$$
\begin{bmatrix} \text{Column} \\ \text{vector of} \\ \text{displacements} \end{bmatrix}
$$

$$
= \underbrace{\begin{bmatrix} 0 \\ \dfrac{qL^2}{4}(\chi_{BC}) \\ \dfrac{qL^2}{4}(\chi_{BC}) \end{bmatrix}}_{\substack{\text{Loading} \\ \text{condition} \\ (1)}} \text{ or } \underbrace{\begin{bmatrix} -\dfrac{5}{128}\xi QH(G_1)_{AB} \\ +\dfrac{7}{128}\xi QH(G_2)_{AB} \\ 0 \end{bmatrix}}_{\substack{\text{Loading} \\ \text{condition} \\ (2)}} \text{ or } \underbrace{\begin{bmatrix} -\dfrac{5}{128}\xi QH(G_1)_{AB} \\ +\dfrac{7}{128}\xi QH(G_2)_{AB} + \dfrac{qL^2}{4}(\chi_{BC}) \\ \dfrac{qL^2}{4}(\chi_{BC}) \end{bmatrix}}_{\substack{\text{Loading} \\ \text{condition} \\ (3)}}
$$

$$(8.141)$$

The axial thrusts in the members are

$$
P_{AB} = \frac{qL^2}{2} - \frac{M_{BC} + M_{CB}}{L} \qquad = -\frac{M_{BC} + M_{CB}}{L} \qquad = \frac{qL^2}{2} - \frac{M_{BC} + M_{CB}}{L}
$$

and

$$
P_{BC} = \underbrace{-\frac{M_{AB} + M_{BA}}{H}}_{\substack{\text{Loading} \\ \text{condition} \\ (1)}} \qquad = \underbrace{-\frac{M_{AB} + M_{BA} + \frac{3}{4}QH}{H}}_{\substack{\text{Loading} \\ \text{condition} \\ (2)}} \qquad = \underbrace{-\frac{M_{AB} + M_{BA} + \frac{3}{4}QH}{H}}_{\substack{\text{Loading} \\ \text{condition} \\ (3)}}
$$

In these,

$$
\xi = \frac{I_2 H}{I_1 L} = \gamma \frac{H}{L} \qquad u = \frac{L}{2}\sqrt{\frac{P}{EI}} = \frac{\pi}{2}\sqrt{\frac{P}{P_e}}
$$

$$
\psi = \frac{3}{2u}\left(\frac{1}{2u} - \frac{1}{\tan 2u}\right) \qquad G_1 = \frac{96}{15}\frac{\sin(u/2) - \frac{1}{4}\sin 2u}{u^2 \sin(2u)}
$$

$$
\phi = \frac{3}{u}\left(\frac{1}{\sin 2u} - \frac{1}{2u}\right) \qquad G_2 = \frac{96}{21}\frac{\sin \frac{3}{2}u - \frac{3}{4}\sin 2u}{u^2 \sin 2u}
$$

$$
\chi = \frac{3(\tan u - u)}{u^3}
$$

Based on the earlier calculations summarized in fig. 8.63, it is known that the maximum bending moment in the beam occurs under loading condition (3) at joint B. The axial thrust in the beam also is greatest for the combined loading. For the column, either loading (2) or (3) will govern, depending upon the value of γ. Moreover, the point of maximum stress will be either at joint B or at load point E.

For the beam:

$$\sigma_{all} = \frac{P_{beam}}{b_2 d_2} + \frac{M_{B,\,beam}}{\frac{1}{6}b_2 d_2^2} \qquad \text{(loading condition 3 governs)} \qquad (8.142)$$

For the column:

$$\sigma_{all} = \frac{P_{col}}{b_1 d_1} + \frac{M_{B,\,col}}{\frac{1}{6}b_1 d_1^2} \qquad (8.143a)$$

or

$$\sigma_{all} = \frac{P_{col}}{b_1 d_1} + \frac{M_E}{\frac{1}{6}b_1 d_1^2} \frac{1}{1 - (P/P_e)_{col}} \qquad (8.143b)$$

where

$$M_E = \frac{3}{16} QH + \frac{1}{4}(M_{AB} - 3M_{BA})\dagger \qquad \begin{array}{l}\text{(both loading conditions 2} \\ \text{and 3 must be considered)}\end{array} \qquad (8.144)$$

If instead of preselecting γ and d_2 and then solving for d_1, b_1, and b_2 using eqs. (8.133), (8.135), and (8.136) as was done in the preceding case, values for both d_1 and d_2 are chosen, a somewhat easier computational solution can be realized. From eq. (8.142), considering loading condition (3):

$$b_2 = \frac{1}{\sigma_{all}}\left(\frac{P_{beam}}{d_2} + \frac{6M_{B,\,beam}}{d_2^2}\right) \qquad (8.145)$$

From eqs. (8.143a) and (8.144b), considering both loading conditions (2) and (3):

$$b_1 = \frac{1}{\sigma_{all}}\left(\frac{P_{col}}{d_1} + \frac{6M_{B,\,col}}{d_1^2}\right) \qquad (8.146a)$$

or

$$b_1 = \frac{1}{\sigma_{all}}\left(\frac{P_{col}}{d_1} + \frac{6M_E}{d_1^2}\frac{1}{1 - (P/P_e)_{col}}\right) \qquad (8.146b)$$

The largest computed value for b_1 controls the design. Finally,

$$\gamma = \frac{b_2}{b_1}\left(\frac{d_2}{d_1}\right)^3 = \frac{I_2}{I_1} \qquad (8.147)$$

This value of the ratio of the moments of inertia is used in the simultaneous equation formulation (8.141).

The solution procedure would be to first assume a value for γ, say 1.0. Initially, axial load values in the members also would be assumed. Then, for preselected values

---

† M_E is the primary bending moment at location E, which neglects the influence of the axial thrust times the deflection at that point. The $1/(1 - P/P_e)$ term is a conservative modifying factor to account for this $P - \Delta$ effect (see chap. 4).

of d_1 and d_2, say, for example, $d_1 = d_2 = 20$ in, compute b_1 and b_2 using eqs. (8.145) and (8.146). These cross-sectional dimensions will have associated with them values for γ and P which probably are different from those assumed. Using the new values, the process will be repeated and, again, values for b_1, b_2, and γ would be computed. The process would be repeated until the desired degree of accuracy is obtained. It is to be remembered, however, that this particular solution corresponds to only one possible set of values—that for the preselected d_1 and d_2. The same process would have to be repeated for other values. Figure 8.64 is a summary of a number of possible design solutions using this procedure. (These curves should be compared against those shown in fig. 8.63.)

For the solutions given in fig. 8.64, fig. 8.65 is a contour plot of w_2, the unit weight of member 2. It is to be recalled that an assumption concerning that unit weight was made in the initial stages of the solution; that is, $w_{\text{assumed}} = 200$ lb/ft.

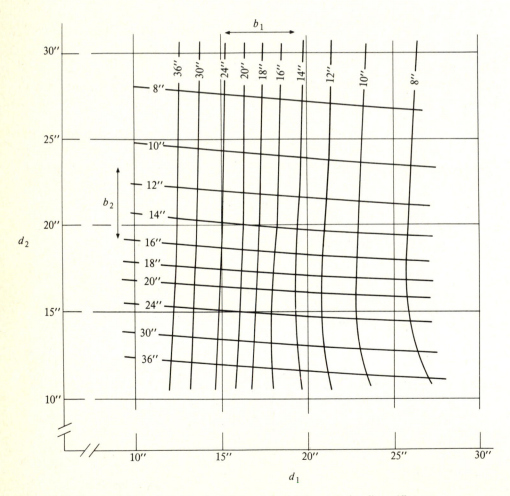

Figure 8.64 Design solutions including the influence of axial thrust on bending stiffnesses.

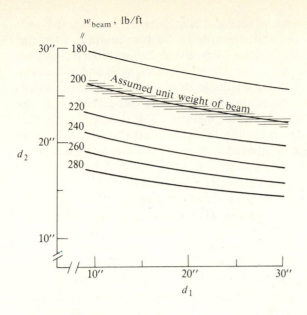

Figure 8.65 Unit weight of beam for various designs.

Should the design chosen have a value significantly different from that 200 lb/ft value, a completely new set of calculations would have to be carried out using a revised w_2.

Figure 8.66 is a contour plot of the total weight of the structure for the designs defined in fig. 8.64. Since for many materials cost is more or less directly related to weight, cost advantages of one particular "main frame" design as opposed to another can be judged from this diagram. It is, of course, to be understood that

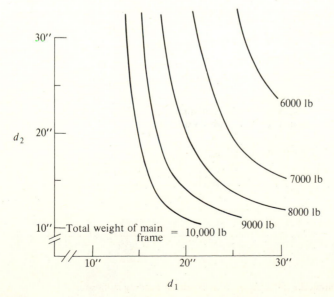

Figure 8.66 Total weight on main frame.

certain materials or fabrication or construction techniques have associated with them practical, minimum cross-sectional dimensions. Such minimums were not presumed in the above calculations. They could, however, be indicated directly on figs. 8.63 and 8.64. Similarly, at the other end of the spectrum, as d_1 and d_2 become large, the costs associated with providing adequate lateral bracing (so that lateral-torsional buckling will not occur) become abnormally high. Practical limits for this situation also could be indicated directly on the figures.

It must be understood that design example 8.18 just concluded was carried out to a degree of detail and generality seldom (if ever) done in a real design situation. More often than not, based on either past experience or construction or fabrication considerations, an initial assumption of the unit weights and ratios of the stiffnesses of the various members is made. A preliminary (relatively crude) approximation is carried out to ascertain the relative importance of axial thrust on the loading stiffnesses of the various members, and a preliminary indeterminate analysis is carried out. A better approximation of the values of the axial thrusts would then be obtained using the moment values of that analysis. With these new P's and the moment values, actual member sizes would be selected, and a new ratio of the moments of inertia determined. The process would be repeated, always taking into account the actual member sizes available for use in the structure until the control conditions are met and the desired accuracy achieved. (In general, prefabricated shapes come in specific, discrete sizes. A continuum cannot be presumed. Therefore, more often than not actual members are selected for examination, even in the preliminary design stages.) Once a particular set of members is known which satisfies the imposed loading condition and the design specification for the material in question, a decision would be made as to whether or not other possible combinations of members would yield a more desirable solution. Usually at least one other choice is examined—if for no reason other than to provide the designer with a greater degree of confidence in the analysis and numerical computations carried out for that first design. For more complicated indeterminate structures or loadings, it is not unusual to consider several different possible design solutions.

8.6 PROBLEMS

8.1 The structure shown consists of two springs of equal unit stiffness γ. Determine the flexibility and stiffness matrices with respect to the indicated coordinate axes x and y.

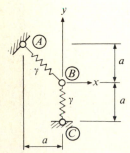

Figure P8.1

8.2 If the x and y axes were rotated counterclockwise 30°, what would be the corresponding matrices?

8.3 A tension member AB of uniform cross-sectional area A, length l, and modulus of elasticity E is inclined at an angle of θ to the x axis. What will be the generalized stiffness matrix for the indicated x and y axes?

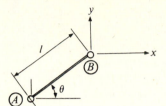

Figure P8.3

8.4 All members in the truss are of the same cross section and material. What will be the forces in each of the members due to the applied load Q?

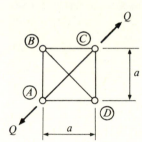

Figure P8.4

8.5 For the truss shown in prob. 8.4, member CD is subjected to a change in temperature of ΔT. What will be the resulting forces in each of the members?

8.6 All vertical and horizontal members of the two-story truss have cross-sectional areas of 4 in². All diagonals are 2 in². If the same material is used throughout, determine the forces in each of the members.

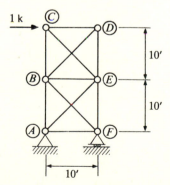

Figure P8.6

8.7 For a continuous uniform cross-section beam of three equal length spans, determine the influence lines for
(a) Moments at the interior support points
(b) Each of the vertical reactions

8.8 A uniform cross section beam of length $2L$ is supported at its midpoint by an elastic spring of stiffness γ. Determine the influence lines for

    (a) The moment at B
    (b) The moment at D
    (c) The vertical shear at D

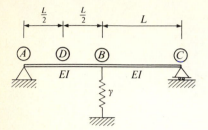

Figure P8.8

8.9 Given that the three-dimensional structure and loading shown are symmetrical, assuming that the redundant is the bending moment at section C, M_C, and using Castigliano's method for determining deformations, determine the values of M_C. Construct the bending and twisting moment diagrams for each of the members.

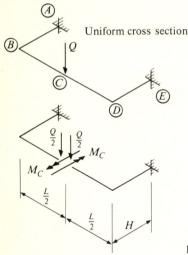

Figure P8.9

8.10 Points A and D of the portal frame shown are connected by a cable of cross-sectional area A, whose modulus of elasticity is E. Determine the force in the cable due to the applied loads Q and q. What will be the bending moments at C? What will be the bending moment at the centerline of member BC?

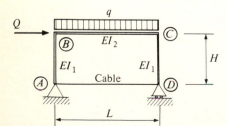

Figure P8.10

8.11 The pinned-base gable frame shown is subjected to a uniformly distributed vertical load of q. What will be the value of the bending moment at point D?

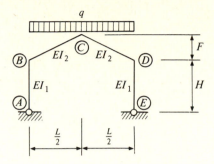

Figure P8.11

8.12 If the distributed vertical load had been applied only over member BC, what would be the bending moment value at D?

8.13 For the structure and loading shown, determine the horizontal deflection of point D.

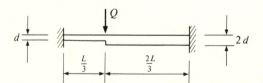

$$EI = 18 \times 10^6 \text{ k·in}$$

Figure P8.13

8.14 Using the column analogy method, determine the fixed-end moments for the variable depth beam of constant width shown.

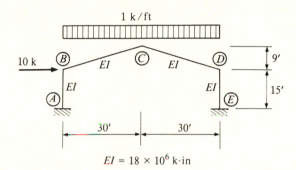

Figure P8.14

8.15 For the beam shown, determine the carry-over factors c_{AB} and c_{BA}.

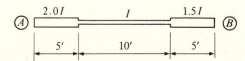

Figure P8.15

8.16 The rigidly connected triangular structure shown is subjected to an externally applied moment (in the plane of the frame) at point B. Determine the end moments in each of the members of the frame.

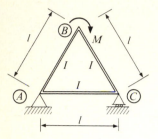

Figure P8.16

8.17 For a fixed-end parabolic arch whose centerline is defined by the equation

$$y = \frac{4h}{l^2} z^2$$

(origin at the crown of the arch), it is assumed that the cross section varies as

$$I = I_C \sec \frac{dy}{dz}$$

Determine the influence lines for the bending moment, shear, and thrust at the crown.

8.18 If for the structure defined in prob. 8.17 the ends were fixed, what would be the influence lines?

8.19 Linearly tapered rectangular cross-section members of constant thickness are frequently used as component members in continuous beams. Solution of the indeterminate structural problem can be carried out using modified slope-deflection equations which take into account such variation in depth.

(a) Derive these equations, assuming that the basic form of the slope deflection equations is to be the same.

(b) What will be the fixed-end moments for these tapered members for a uniformly distributed lateral load, and a concentrated load at a generalized point along the members?

8.20 The structure shown is a viaduct bent. Neglecting the influence of axial thrust on bending stiffnesses, determine the bending moments in the members. If it is assumed that bending moments are zero at the midpoints of BC, CD, DE, and BE, how does the resulting moment diagram compare with the exact solution derived in the first part of this problem?

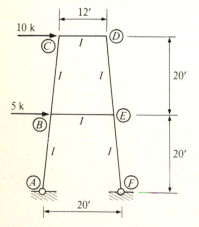

Figure P8.20

8.21 A semicircular ring of radius R and constant cross section is pin-connected as indicated. If the ring is subjected to a uniformly distributed radial force of q, what will be the horizontal reactions at the pin connections? Plot the bending-moment diagram.

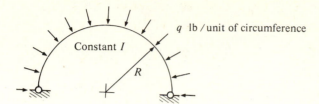

Figure P8.21

8.22 If the ends were fixed, what would be the values?

8.23 A section of concrete culvert 1 ft long is subjected to the soil pressure shown. Assuming uniform thickness of the concrete, determine the bending moments in the culvert.

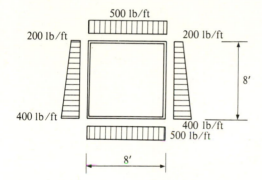

Figure P8.23

8.24 The axis of the arch shown is defined by the equation

$$y = H \sin \frac{\pi z}{L}$$

Assuming the cross section varies as

$$I = I_c \sec \frac{dy}{dz}$$

determine the elastic center and compute the horizontal and vertical reactions and bending moment at A for a vertical load of Q applied at point C.

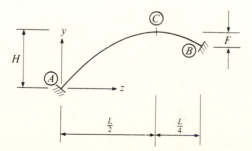

Figure P8.24

8.25 For the structure shown, determine the buckling load as a function of the relative bending stiffnesses LI_1/HI_2. Plot a curve of these values similar to that given as fig. 8.57.

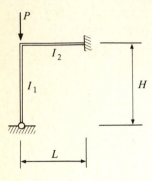

Figure P8.25

MATRIX ANALYSIS OF STRUCTURES

.

Since the early 1950s, a considerable amount of attention has been given to the use of digital computers and matrix formulations in the analysis of structural problems. The methods developed are comparatively new and—in a practical sense—can be used only in conjunction with computers. Generally, they are not intended for nor are they well suited to hand computations.

While the use of these concepts in structural analysis is relatively new, it will be recognized upon close examination that much of the basic theory involved follows directly from older procedures. In most cases, the methods can best be described as extensions or modifications of the basic flexibility and stiffness approaches that were discussed in chap. 8. The primary virtue of the methods lies in the fact that these modifications are accomplished in such ways that maximum utilization of the inherent organizational properties of matrix formulation is realized. The result is a completely generalized approach, the basic elements of which can be used for a very wide range of problem types with little or no alteration.

In general, the solution of a particular structural problem is formulated as a series of matrix operations. Since these operations can be stored in very general forms within a computer, the formulation can be made to be essentially automatic. The analyst need only supply the basic problem data such as member geometry, loading conditions, etc. Quite sophisticated man-computer interactive programs have been developed using this approach.

In the application of these methods, the actual structure is first replaced by an equivalent, idealized structure composed of discrete parts referred to as *elements*. These elements are connected to one another at their points of junction—referred to as *node points*. It is then presumed that the response of the real structure can be adequately described in terms of the response, at these node points, of the idealized structure.

There are several different types of elements that are used in the matrix formulation of structural problems. The particular one that is selected for a given situation depends upon several factors. The geometry of the element is an obvious inclusion for all the cases. A predetermined set of load-deformational response properties also must be available and capable of being expressed as an element stiffness matrix or an element flexibility matrix. These matrices describe the relationships between the nodal forces and the nodal displacements of the element. (They could be defined using the previously developed procedures.)

As pointed out in chap. 8, there are two basic methods commonly used: the displacement (or stiffness) method, and the force (or flexibility) method. The displacement method considers equilibrium of forces or moments. Such a formulation yields nodal displacements. The force method is formulated from a consideration of geometrical compatibility, and yields forces at the node points. Because the stiffness method is based on equilibrium, it is comparatively easy to understand and deal with. This is reflected in the greater popularity of this method by practicing engineers. For this reason, the emphasis in this chapter will be upon the direct stiffness method of solution.

Two basic problems exist in the development of a workable matrix solution procedure. Fortunately, they can be considered more or less as separate one from the other. The first of these problems is the development of a very general sequence of matrix operations that provides the overall framework for solution. This framework, ideally, should be independent of any particular structural configuration. The second problem, and by far the more difficult of the two, is the development of various elements and their corresponding stiffness matrices for use within the general framework referred to above as the first problem.

The basic solution procedure will be demonstrated using as examples very simple structural configurations. This is done so that the details of the process will not be obscured by the complexity of the structure. The important factor to note is the generality of the process, and how the method can be used regardless of the kind of *element* selected. After the basic process has been evolved, attention will be focused on the development of various elements and their application to different types of structural problems.

Stiffness properties of structural elements can be formulated in a number of different ways. Almost any of the methods described in chaps. 4 and 5 can be used for this purpose. Solution of the differential equation is the method used in most of the main body of this chapter. In the appendix, however, stiffnesses will be developed by first presuming a deformed shape (a *displacement function*) within the element. Then, using energy methods, usually minimizing the total potential energy, the nodal displacements will be related to the nodal forces, yielding the elemental stiffness (or flexibility) matrix. (Obviously, if the presumed displacement function corresponds to the exact function, the matrix solution results will be the same as those that would have been obtained using any of the other "classical" methods of solution. If the function is an approximation, then the results of the matrix solutions also will be approximations.)

In general, if the system of elements, nodes, and kinematic redundants is chosen in such a way that they are capable of providing for full continuity in deformation, slope, etc., an exact solution can be obtained. This is often the case for structures which consist of one-dimensional members. If, however, more nodes are required than are provided to achieve full continuity in every direction (as would be the case for most elements having more than one dimension; for example, a plate element), the solution will be approximate. It is to be recognized that the quality of the approximation depends upon how well the presumed system corresponds to the actual one. Only those problems that fall into the first category of elements defined above (that is, one-dimensional members) will be considered in this chapter. Problems of the type in the second category are considered beyond the scope of this text.

9.1 THE MATRIX DISPLACEMENT OR STIFFNESS METHOD

In the application of the matrix displacement method, the real structure is replaced (that is, modeled) by a set of elements that are connected to one another at their node points. The load-deformation characteristics of these elements are predetermined and described by the elemental stiffness matrix $[K]$. Just as before, any element of the matrix, say k_{ij}, is defined as the stress resultant (force) in the ith direction due to a unit of displacement in the jth direction—with all other displacements maintained at zero. Thus, the matrix equation that describes the equilibrium of the element shown in fig. 9.1 is given by

$$[K][D] = [F]$$

or

$$\begin{bmatrix} K_{11} & K_{12} & K_{13} & K_{14} \\ K_{21} & K_{22} & K_{23} & K_{24} \\ K_{31} & K_{32} & K_{33} & K_{34} \\ K_{41} & K_{42} & K_{43} & K_{44} \end{bmatrix} \begin{bmatrix} D_1 \\ D_2 \\ D_3 \\ D_4 \end{bmatrix} = \begin{bmatrix} F_1 \\ F_2 \\ F_3 \\ F_4 \end{bmatrix} \tag{9.1}$$

where the 1, 2, 3, and 4 directions are as shown in the figure.

The individual entries of the matrix $[K]$ are determined by displacing the element a unit amount in the jth direction, while holding it "in place" in each of the other numbered directions, and then calculating the forces necessary to maintain the resulting deformed shape. The particular force in the ith direction is K_{ij}. Knowing the stiffness matrix, equilibrium of the element in the 3 direction, for example, is given by

$$K_{31}D_1 + K_{32}D_2 + K_{33}D_3 + K_{34}D_4 = F_3 \tag{9.2}$$

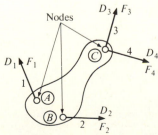

Element 1

Figure 9.1 Rigid element subjected to external loading.

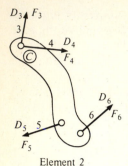

Element 2

Figure 9.2 Rigid element subjected to external loading.

If a second element is now defined, for example, the one shown in fig. 9.2; which has prescribed nodal deformations at node C—which is defined as the same point as C in the first element—the two can be joined at that point of connection. The ability to react to force at that point is the sum of the contributions of each of the elements.

For element 2 in fig. 9.2, equilibrium is given by the set

$$
\begin{bmatrix}
K_{33} & K_{34} & K_{35} & K_{36} \\
K_{43} & K_{44} & K_{45} & K_{46} \\
K_{53} & K_{54} & K_{55} & K_{56} \\
K_{63} & K_{64} & K_{65} & K_{66}
\end{bmatrix}
\begin{bmatrix}
D_3 \\
D_4 \\
D_5 \\
D_6
\end{bmatrix}
=
\begin{bmatrix}
F_3 \\
F_4 \\
F_5 \\
F_6
\end{bmatrix}
\tag{9.3}
$$

In the 3 direction,

$$
K_{33} D_3 + K_{34} D_4 + K_{35} D_5 + K_{36} D_6 = {}_2 F_3 \tag{9.4}
$$

where $_2 F_3$ represents the force in the 3 direction resulting from the deformation of element 2.

If elements 1 and 2 are joined at node point C, the force in the 3 direction resulting from the deformation of both elements is

$$
_T F_3 = {}_1 F_3 + {}_2 F_3 \tag{9.5}
$$

where the presubscript T indicates the total, and 1 and 2 indicate the effects of elements 1 and 2, respectively. Thus

$$
K_{31} D_1 + K_{32} D_2 + {}_1 K_{33} D_3 + {}_1 K_{34} D_4
$$
$$
+ {}_2 K_{33} D_3 + {}_2 K_{34} D_4 + K_{35} D_5 + K_{36} D_6 = {}_T F_3 \tag{9.6}
$$

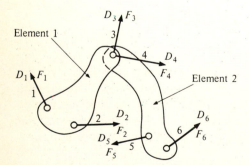

Figure 9.3 Linkage system subjected to external loading.

The elements joined together are shown in fig. 9.3. The complete equilibrium statement of this composite in each of the six prescribed directions is then given in matrix form in eq. (9.7).

$$
\begin{bmatrix}
K_{11} & K_{12} & K_{13} & K_{14} & 0 & 0 \\
K_{21} & K_{22} & K_{23} & K_{24} & 0 & 0 \\
K_{31} & K_{32} & {}_1K_{33} + {}_2K_{33} & {}_1K_{34} + {}_2K_{34} & K_{35} & K_{36} \\
K_{41} & K_{42} & {}_1K_{43} + {}_2K_{43} & {}_1K_{44} + {}_2K_{44} & K_{45} & K_{46} \\
0 & 0 & K_{53} & K_{54} & K_{55} & K_{56} \\
0 & 0 & K_{63} & K_{64} & K_{65} & K_{66}
\end{bmatrix}
\begin{bmatrix}
D_1 \\ D_2 \\ D_3 \\ D_4 \\ D_5 \\ D_6
\end{bmatrix}
$$

matrix (9.1) — matrix (9.3)

$$
=
\begin{bmatrix}
F_1 \\
F_2 \\
{}_1F_3 + {}_2F_3 \\
{}_1F_4 + {}_2F_4 \\
F_5 \\
F_6
\end{bmatrix}
\qquad (9.7)
$$

As indicated on the matrix itself, the resulting set is nothing more than the superimposed effects of each, written in an "expanded form." The expanded form contains zeros as indicated in the following:

Element 1:
$$
\begin{bmatrix}
K_{11} & K_{12} & K_{13} & K_{14} & 0 & 0 \\
K_{21} & K_{22} & K_{23} & K_{24} & 0 & 0 \\
K_{31} & K_{32} & {}_1K_{33} & {}_1K_{34} & 0 & 0 \\
K_{41} & K_{42} & {}_1K_{43} & {}_1K_{44} & 0 & 0 \\
0 & 0 & 0 & 0 & 0 & 0 \\
0 & 0 & 0 & 0 & 0 & 0
\end{bmatrix}
\begin{bmatrix}
D_1 \\ D_2 \\ D_3 \\ D_4 \\ D_5 \\ D_6
\end{bmatrix}
=
\begin{bmatrix}
F_1 \\ F_2 \\ {}_1F_3 \\ {}_1F_4 \\ F_5 \\ F_6
\end{bmatrix}
$$

Element 2:
$$
\begin{bmatrix}
0 & 0 & 0 & 0 & 0 & 0 \\
0 & 0 & 0 & 0 & 0 & 0 \\
0 & 0 & {}_2K_{33} & {}_2K_{34} & K_{35} & K_{36} \\
0 & 0 & {}_2K_{43} & {}_2K_{44} & K_{45} & K_{46} \\
0 & 0 & K_{53} & K_{54} & K_{55} & K_{56} \\
0 & 0 & K_{63} & K_{64} & K_{65} & K_{66}
\end{bmatrix}
\begin{bmatrix}
D_1 \\ D_2 \\ D_3 \\ D_4 \\ D_5 \\ D_6
\end{bmatrix}
=
\begin{bmatrix}
F_1 \\ F_2 \\ {}_2F_3 \\ {}_2F_4 \\ F_5 \\ F_6
\end{bmatrix}
$$

The reason for doing this is that the matrices in the (9.1) and (9.3) forms are not compatible—that is, they are not immediately subject to superposition. The addition of the zeros, corrects this situation. Moreover, the addition of the zeros accurately reflects the influence of the various forces and deformations on the elements. For example, D_5 and D_6 have no influence on ${}_1F_1, \ldots, {}_1F_4$ and, therefore, the zeros reflect their contributions to those forces.

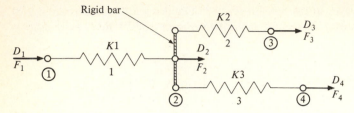

Figure 9.4 Three-spring system subjected to external loading.

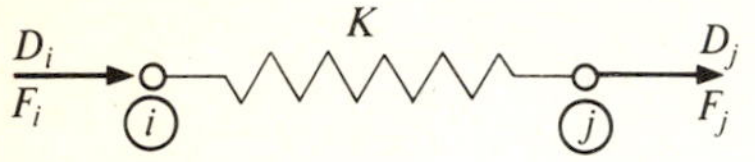

$$\frac{D_i}{F_i} \quad \overset{K}{\wedge\!\wedge\!\wedge\!\wedge} \quad \frac{D_j}{F_j}$$

Figure 9.5 Single-spring system.

Providing that the directions at a common node point of the force and displacement are the same, other elements can be joined to these and their stiffnesses combined at that common node. By properly adding together stiffness properties of various elements, the overall equilibrium of an entire structure can be developed and the behavior of the total system defined.

To illustrate the process, consider the simple three-spring rigid-bar structure shown in fig. 9.4. It is specified that the displacement D_2 (at node point 2) is the same for each of the three indicated node points on that rigid bar (that is, the bar does not rotate). The structure, then, has four degrees of freedom: D_1, D_2, D_3, and D_4. The K's denote the spring constants for each of the three springs as indicated.

The simple spring element shown in fig. 9.5 will be used to model each of the springs in the system. The element forces and displacements are assumed to be positive in the directions shown. Proceeding as before, there is introduced a unit of displacement in the D_i and D_j directions, holding the other constant, as illustrated in fig. 9.6. The elemental equilibrium is thus described by the matrix equation

$$\begin{bmatrix} K & -K \\ -K & K \end{bmatrix} \begin{bmatrix} D_i \\ D_j \end{bmatrix} = \begin{bmatrix} F_i \\ F_j \end{bmatrix} \tag{9.8}$$

or

$$K \begin{bmatrix} 1 & -1 \\ -1 & 1 \end{bmatrix} \begin{bmatrix} D_i \\ D_j \end{bmatrix} = \begin{bmatrix} F_i \\ F_j \end{bmatrix}$$

The 2×2 coefficient matrix $[K]$ is the elemental stiffness matrix for the spring.

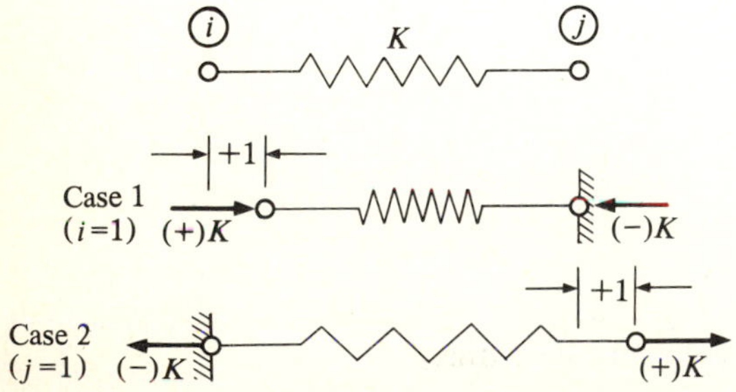

Figure 9.6

For the structure shown in fig. 9.4, this element stiffness formulation would be used three times. Substituting the node numbers for the i's and j's, there is obtained:

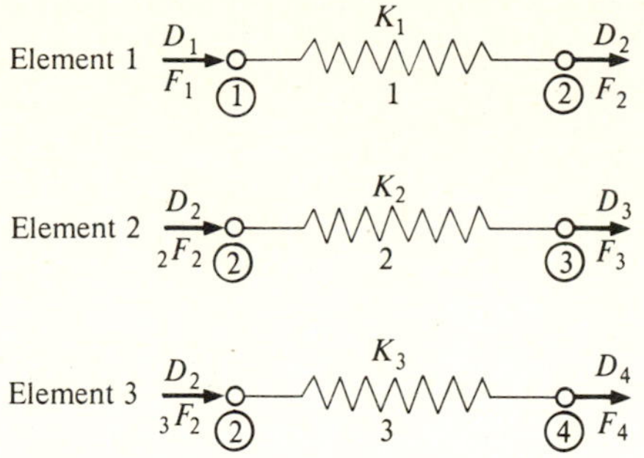

Element 1
$$\begin{bmatrix} K_1 & -K_1 \\ -K_1 & K_1 \end{bmatrix}\begin{bmatrix} D_1 \\ D_2 \end{bmatrix} = \begin{bmatrix} F_1 \\ F_2 \end{bmatrix}$$

Element 2
$$\begin{bmatrix} K_2 & -K_2 \\ -K_2 & K_2 \end{bmatrix}\begin{bmatrix} D_2 \\ D_3 \end{bmatrix} = \begin{bmatrix} {}_2F_2 \\ F_3 \end{bmatrix}$$

Element 3
$$\begin{bmatrix} K_3 & -K_3 \\ -K_3 & K_3 \end{bmatrix}\begin{bmatrix} D_2 \\ D_4 \end{bmatrix} = \begin{bmatrix} {}_3F_2 \\ F_4 \end{bmatrix}$$

Figure 9.7

where the forces ${}_2F_2$ and ${}_3F_2$ are the components of the force F_2 acting within elements 2 and 3, respectively.

Since the structure in question has four degrees of freedom, the master (or assembled) stiffness matrix for the entire structure will be of size 4×4. This assembly is accomplished by adding together the stiffness contributions of each element. For example, at node 2, the assembly of the elemental stiffness properties into the master stiffness matrix must reflect the fact that

$$F_2 = {}_1F_2 + {}_2F_2 + {}_3F_2 = -K_1 D_1 + K_1 D_2 + K_2 D_2 - K_2 D_3 + K_3 D_2 - K_3 D_4 \quad (9.9)$$

As indicated, the size of the master stiffness matrix is 4×4. Therefore, prior to assembly, each elemental stiffness formulation must be expanded to the same size as that master matrix. This is demonstrated in the following:

Element 1:
$$\begin{bmatrix} K_1 & -K_1 & 0 & 0 \\ -K_1 & K_1 & 0 & 0 \\ 0 & 0 & 0 & 0 \\ 0 & 0 & 0 & 0 \end{bmatrix}\begin{bmatrix} D_1 \\ D_2 \\ D_3 \\ D_4 \end{bmatrix} = \begin{bmatrix} F_1 \\ F_2 \\ 0 \\ 0 \end{bmatrix}_1$$

Element 2:
$$\begin{bmatrix} 0 & 0 & 0 & 0 \\ 0 & K_2 & -K_2 & 0 \\ 0 & -K_2 & K_2 & 0 \\ 0 & 0 & 0 & 0 \end{bmatrix}\begin{bmatrix} D_1 \\ D_2 \\ D_3 \\ D_4 \end{bmatrix} = \begin{bmatrix} 0 \\ {}_2F_2 \\ F_3 \\ 0 \end{bmatrix}_2$$

Element 3:
$$\begin{bmatrix} 0 & 0 & 0 & 0 \\ 0 & K_3 & 0 & -K_3 \\ 0 & 0 & 0 & 0 \\ 0 & -K_3 & 0 & K_3 \end{bmatrix}\begin{bmatrix} D_1 \\ D_2 \\ D_3 \\ D_4 \end{bmatrix} = \begin{bmatrix} 0 \\ {}_3F_2 \\ 0 \\ F_4 \end{bmatrix}_3$$

In each of these, the F's referred to are acting on that particular element. (The subscript after the force vector is intended to reflect this.)

Adding these three individual elemental matrices yields the master formulation for the problem.

<table>
<tr><td rowspan="2">Master
formulation:</td><td></td></tr></table>

$$
\begin{bmatrix}
K_1 & -K_1 & 0 & 0 \\
-K_1 & (K_1 + K_2 + K_3) & -K_2 & -K_3 \\
0 & -K_2 & K_2 & 0 \\
0 & -K_3 & 0 & K_3
\end{bmatrix}
\begin{bmatrix}
D_1 \\ D_2 \\ D_3 \\ D_4
\end{bmatrix}
=
\begin{bmatrix}
F_1 \\ F_2 \\ F_3 \\ F_4
\end{bmatrix}
\qquad (9.10)
$$

The F's with no presubscripts refer to the total loads acting on the structure in question at the specified node points. The D's are common to both the total structure and to the individual elements.

Particular facts concerning the master matrix should be noted. For example, for each row in the matrix,

$$
\sum_{j=1}^{n} K_{ij} = 0 \qquad i = 1, \ldots, n
$$

This reflects equilibrium. The matrix is symmetrical, as required by Maxwell's law. Thirdly, while not immediately apparent, the matrix $[K]$ is *singular*—that is, the value of the determinant is zero. Thus, in the form shown, and for the conditions prescribed, no unique solution to the problem exists. (It is to be realized that the system, as defined, will move as a rigid body if any external force is applied. This master matrix singularity merely reflects this overall condition of geometric instability.) The problem, then, has not been completely prescribed, inasmuch as no *deformational constraint(s)* or boundary conditions have been stipulated.

To complete the statement of the problem, assume that it is specified that nodes 3 and 4 undergo no displacements. Thus, the terms D_3 and D_4 are known and are zero. Correspondingly, the forces F_3 and F_4 are reactions. Further, assume that a particular load is applied at F_1 and has a magnitude of $+1$ kip. The problem then becomes that shown in fig. 9.8. The unknown displacements are D_1 and D_2 and the unknown forces (reactions) are $R_3 = F_3$ and $R_4 = F_4$. If these are introduced into the master formulation, there is obtained

$$
\begin{bmatrix}
K_1 & -K_1 & 0 & 0 \\
-K_1 & (K_1 + K_2 + K_3) & -K_2 & -K_3 \\
0 & -K_2 & K_2 & 0 \\
0 & -K_3 & 0 & K_3
\end{bmatrix}
\begin{bmatrix}
D_1 \\ D_2 \\ D_3 \\ D_4
\end{bmatrix}
=
\begin{bmatrix}
1 \\ 0 \\ R_3 \\ R_4
\end{bmatrix}
\qquad (9.11)
$$

Since there are only two unknown displacements (D_1 and D_2), the system described by eq. (9.11) is overdeterminate. Two of the equations must be omitted. This can be accomplished by dividing the set into two parts. The column vector $[D]$ can be partitioned into D_U and D_B, where D_U contains all the unknown displacements, and D_B holds the unknown boundary conditions. Similarly, the force vector

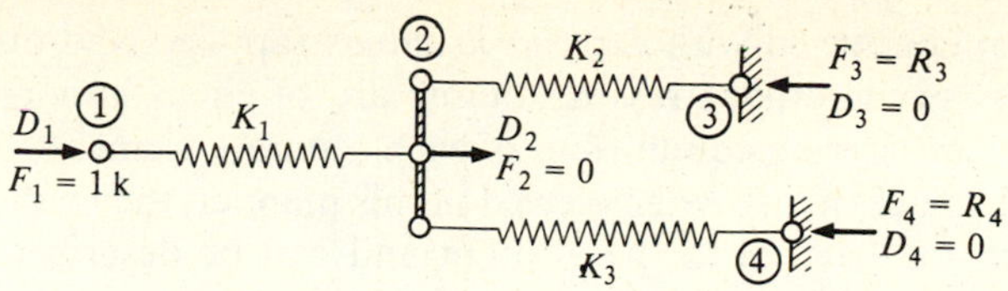

Figure 9.8 Three-spring system with known constraints.

can be partitioned into F_A and F_R, where F_A has all the elements of the applied loads and F_R the reaction components. $[K]$ would similarly be partitioned into four parts. The equations then become

$$\begin{bmatrix} [K_A] & \vdots & [K_B] \\ \cdots\cdots & \cdots & \cdots\cdots \\ [K_C] & \vdots & [K_D] \end{bmatrix} \begin{bmatrix} [D_U] \\ \cdots \\ [D_B] \end{bmatrix} = \begin{bmatrix} [F_A] \\ \cdots \\ [F_R] \end{bmatrix} \tag{9.12}$$

where

$$[K_A] = \begin{bmatrix} K_1 & -K_1 \\ -K_1 & K_1 + K_2 + K_3 \end{bmatrix} \qquad [K_B] = \begin{bmatrix} 0 & 0 \\ -K_2 & -K_3 \end{bmatrix}$$

$$[K_C] = \begin{bmatrix} 0 & -K_2 \\ 0 & -K_3 \end{bmatrix} \qquad\qquad [K_D] = \begin{bmatrix} K_2 & 0 \\ 0 & K_3 \end{bmatrix}$$

and

$$[D_U] = \begin{bmatrix} D_1 \\ D_2 \end{bmatrix} \qquad\qquad [D_B] = \begin{bmatrix} 0 \\ 0 \end{bmatrix}$$

$$[F_A] = \begin{bmatrix} 1 \\ 0 \end{bmatrix} \qquad\qquad [F_R] = \begin{bmatrix} R_3 \\ R_4 \end{bmatrix}$$

Multiplying the partitioned matrices (9.12),

$$[K_A][D_U] + [K_B][D_B] = [F_A]$$

$$[K_C][D_U] + [K_D][D_B] = [F_R]$$

Since in this case $[D_B]$ is a null matrix (that is, all entries are zero),

$$[K_A][D_U] = [F_A]$$

and

$$[K_C][D_U] = [F_R]$$

Thus the set

$$\begin{bmatrix} K & -K_1 \\ -K_1 & K_1 + K_2 + K_3 \end{bmatrix} \begin{bmatrix} D_1 \\ D_2 \end{bmatrix} = \begin{bmatrix} 1 \\ 0 \end{bmatrix} \tag{9.13}$$

which defines the unknown displacements. Substituting these values into eq. (9.14),

$$\begin{bmatrix} 0 & -K_2 \\ 0 & -K_3 \end{bmatrix} \begin{bmatrix} D_1 \\ D_2 \end{bmatrix} = \begin{bmatrix} R_3 \\ R_4 \end{bmatrix} \tag{9.14}$$

yields the reactions at the constrained nodes 3 and 4.

So far this example has been concerned with *external loadings* (applied and/or reactive forces). For design purposes, internal stress resultants are of equal importance. These can be obtained by considering equilibrium of each separate element—now knowing the various node displacements. When viewed in this manner, the loads that are internal to the structure are external to the element and can be described as the elemental nodal forces necessary to maintain the deformed shape of the element, if it were to be removed from the loaded structure. Thus for the element in question (fig. 9.5), the forces are defined by the relationships

$$\begin{bmatrix} K & -K \\ -K & K \end{bmatrix} \begin{bmatrix} D_i \\ D_j \end{bmatrix} = \begin{bmatrix} F_i \\ F_j \end{bmatrix}$$

Assuming that the displacements are known,

$$F_i = KD_i - KD_j = K(D_i - D_j)$$

and

$$F_j = -F_i$$

For the problem under consideration,

$$D_1 = \frac{K_1 + K_2 + K_3}{K_1(K_2 + K_3)} \quad \text{and} \quad D_2 = \frac{1}{K_2 + K_3}$$

The reactions are then

$$R_3 = -K_2 D_2 = -\frac{K_2}{K_2 + K_3}$$

and

$$R_4 = -K_3 D_2 = -\frac{K_3}{K_2 + K_3}$$

The internal stress resultants for member 2-3 are given by the equations

$$\begin{bmatrix} K_2 & -K_2 \\ -K_2 & K_2 \end{bmatrix} \begin{bmatrix} \dfrac{1}{K_2 + K_3} \\ 0 \end{bmatrix} = \begin{bmatrix} {}_2F_2 \\ F_3 \end{bmatrix}$$

This yields

$$_2F_2 = \frac{K_2}{K_2 + K_3} \quad \text{and} \quad F_3 = -\frac{K_2}{K_2 + K_3} = R_3$$

This example illustrates most of the basic steps that make up displacement-type matrix analysis. These steps are well defined, as will be evident from the summary listing which follows. Assuming that the elements are selected and located within the structure:

1. The stiffness matrix for each element used in the structure is written in terms of *its* own deformation parameters.

2. The master stiffness matrix is then assembled by adding the stiffnesses of each element that contribute to a common node. (This is accomplished by expanding the elemental matrix to the size of the master matrix by introducing column and row zeros to correspond to displacements that are not involved in that element.†)

3. The deformation matrix (vector) is then partitioned into two parts, $[D_U]$ and $[D_B]$. ($[D_U]$ contains, as elements, all the to-be-found displacements of the structure. $[D_B]$ consists of the known boundary conditions.) More often than not, this will require that rows within the displacement matrix be rearranged so as to collect all of the known displacements into one vector. (If any row within the $[D]$ matrix is changed in position, the corresponding columns in the $[K]$ matrix also must be changed in exactly the same manner. The same would be true for the F's.)

4. The load matrix is then partitioned into two parts $[F_A]$ and $[F_R]$. (The matrix $[F_A]$ represents the applied forces—and is known. $[F_R]$ is the reaction vector, and, in general, is unknown. While it may be necessary to rearrange the ordering of the terms to facilitate partitioning—just as was the case described above for the D's in step 3—this presents no difficulty. Those points where constraints are defined, also correspond to the location of reactions. Thus, only one rearrangement is necessary, with $[D]$ and $[F]$ requiring the same change.) To maintain the validity of the entire set, the rows of $[K]$ must be changed to correspond to any changes in $[F]$. Thus, if in the D matrix the elements D_i and D_j are interchanged, the elements F_i and F_j also must be interchanged, as must be both the columns *and* the rows i and j of $[K]$.

5. At this point in the solution process, the master stiffness matrix $[K]$ can be expressed in the 2×2 form

$$\begin{bmatrix} [K_A] & [K_B] \\ [K_C] & [K_D] \end{bmatrix}$$

with the master formulation being

$$\begin{bmatrix} [K_A] & [K_B] \\ [K_C] & [K_D] \end{bmatrix} \begin{bmatrix} [D_U] \\ [D_B] \end{bmatrix} = \begin{bmatrix} [F_A] \\ [F_R] \end{bmatrix} \tag{9.15}$$

The following then can be written:

$$[K_A][D_U] + [K_B][D_B] = [F_A] \quad \text{and} \quad [K_C][D_U] + [K_D][D_B] = [F_R]$$

or

$$[K_A][D_U] = [F_A] - [K_B][D_B]$$

and

$$[D_U] = [K_A^{-1}]\big[[F_A] - [K_B][D_B]\big] \tag{9.16}$$

---

† It should here be noted that in practice, various shortcuts can be employed in the execution of this assembly. The most elementary form of assembly is the one described.

where the D_U's are the unknown displacements. Knowing $[D_U]$ and $[D_B]$, the reactions can be determined from the relationships

$$[F_R] = [K_C][D_U] + [K_D][D_B] \tag{9.17}$$

6. Internal stress resultants are determined by substituting the now-known displacements back into the element equilibrium equations. The element displacement vector is multiplied by the element stiffness matrix, yielding a force vector—the internal stress resultant.

While these steps more or less describe the process that was used in the preceding example, in practice, for more general structures and loadings, one additional step almost always is needed. Whenever the displacement directions for one element do not correspond to those for another to which it joins, or if the directions for a given element do not correspond to the reference directions chosen for the entire structure, the local element directions must be rotated (transformed) to correspond to those for the entire system. Only if all are referenced to a common coordinate system can the elements be added to one another.

There should, therefore, be inserted into the list of steps, directly after step 1:

1a. Translate the elemental stiffness matrix from its *local* coordinate system to the *reference* system of directions for the entire structure. (To accomplish this, a *transformation matrix*† appropriate to the elements employed is used.)

While the steps that are listed above will automatically lead to the required solution, often various shortcuts can be employed to advantage. For example, based on the preceding case, it is noted that the introduction of the two conditions $D_3 = D_4 = 0$—to facilitate the master formulation—in effect amounts to adding two additional equations to the system. Two equations therefore must be dropped. This can be accomplished by noting the subscript of the known displacement and then dropping each equation (that is, each row) having that value of i from the matrix, and each column having the same value of j. The elements which remain after these eliminations will correspond to the elements of a nonsingular set that described the unknown displacements.

$$[K_A][D_U] = [F_A] \tag{9.18}$$

When using the process of striking the rows and columns, just described, there will result the elimination of the elements of $[K_C]$. These usually will become available, however, during the process of computation of the internal stress resultants. This process also eliminates the need for rearranging rows and columns. (Of course, the remaining elements should be rewritten into a smaller set.)

If any known displacement is other than zero, the same procedure would be followed—with $[F]$ appropriately modified.

$$F_i = \text{load}_i - D_l K_{i,l} \qquad i = 1, \ldots, n \tag{9.19}$$

† The development of appropriate transformation matrices will be considered later in this chapter.

where l defines the position in the displacement vector of the known finite displacement and load$_i$ is the original ith element of the load vector.

For the example just concluded, the master formulation was

$$\begin{bmatrix} K_1 & -K_1 & 0 & 0 \\ -K_1 & K_1 + K_2 + K_3 & -K_2 & -K_3 \\ 0 & -K_2 & K_2 & 0 \\ 0 & -K_3 & 0 & K_3 \end{bmatrix} \begin{bmatrix} D_1 \\ D_2 \\ D_3 \\ D_4 \end{bmatrix} = \begin{bmatrix} F_1 \\ F_2 \\ F_3 \\ F_4 \end{bmatrix}$$

The known displacements (the specified boundary conditions) were given as $D_3 = D_4 = 0$. The loading was defined as $F_1 = 1$ kip, and $F_2 = 0$. Following the shortcut described above, rows 3 and 4 and columns 3 and 4 would be eliminated.

$$\begin{bmatrix} K_1 & -K_1 & 0 & 0 \\ -K_1 & K_1 + K_2 + K_3 & -K_2 & -K_3 \\ \hline 0 & -K_2 & K_2 & 0 \\ 0 & -K_3 & 0 & K_3 \end{bmatrix} \begin{bmatrix} D_1 \\ D_2 \\ D_3 \\ D_4 \end{bmatrix} = \begin{bmatrix} 1 \\ 0 \\ F_3 \\ F_4 \end{bmatrix}$$

This gives

$$\begin{bmatrix} K_1 & -K_1 \\ -K_1 & K_1 + K_2 + K_3 \end{bmatrix} \begin{bmatrix} D_1 \\ D_2 \end{bmatrix} = \begin{bmatrix} 1 \\ 0 \end{bmatrix}$$

which is the same as eq. (9.13).

Suppose that D_1 and D_4 had been specified as equaling zero, with an applied load of 1 kip for F_3. The rows and columns numbered 1 and 4 would be eliminated leaving

$$\begin{bmatrix} K & -K_1 & 0 & 0 \\ -K_1 & K_1 + K_2 + K_3 & -K_2 & -K_3 \\ 0 & -K_2 & K_2 & 0 \\ 0 & -K_3 & 0 & K_3 \end{bmatrix} \begin{bmatrix} D_1 \\ D_2 \\ D_3 \\ D_4 \end{bmatrix} = \begin{bmatrix} F_1 \\ 0 \\ 1 \\ F_4 \end{bmatrix}$$

or

$$\begin{bmatrix} K_1 + K_2 + K_3 & -K_2 \\ -K_2 & K_2 \end{bmatrix} \begin{bmatrix} D_2 \\ D_3 \end{bmatrix} = \begin{bmatrix} 0 \\ 1 \end{bmatrix}$$

These would be solved for D_2 and D_3

Consider now the particular case where $D_1 = 0.02$ in and $D_3 = 0.01$ in. From the above,

$$\begin{bmatrix} K_1 & -K_1 & 0 & 0 \\ -K_1 & K_1 + K_2 + K_3 & -K_2 & -K_3 \\ 0 & -K_2 & K_2 & 0 \\ 0 & -K_3 & 0 & K_3 \end{bmatrix} \begin{bmatrix} 0.02 \\ D_2 \\ 0.01 \\ D_4 \end{bmatrix} = \begin{bmatrix} F_1 \\ F_2 \\ F_3 \\ F_4 \end{bmatrix}$$

or

$$\begin{bmatrix} 0 & -K_1 & 0 & 0 \\ 0 & K_1 + K_2 + K_3 & 0 & -K_3 \\ 0 & K_2 & 0 & 0 \\ 0 & -K_3 & 0 & K_3 \end{bmatrix} \begin{bmatrix} 0.02 \\ D_2 \\ 0.01 \\ D_4 \end{bmatrix} = \begin{bmatrix} F_1 - 0.02K_1 \\ F_2 + 0.02K_1 + 0.01K_2 \\ F_3 - 0.01K_2 \\ F_4 \end{bmatrix}$$

Striking the appropriate rows and columns,

$$\begin{bmatrix} K_1 + K_2 + K_3 & -K_3 \\ -K_3 & K_3 \end{bmatrix} \begin{bmatrix} D_2 \\ D_4 \end{bmatrix} = \begin{bmatrix} F_2 + 0.02K_1 + 0.01K_2 \\ F_4 \end{bmatrix}$$

If both applied loads F_2 and F_4 are zero, then

$$\begin{bmatrix} K_1 + K_2 + K_3 & -K_3 \\ -K_3 & K_3 \end{bmatrix} \begin{bmatrix} D_2 \\ D_4 \end{bmatrix} = \begin{bmatrix} 0.02K_1 + 0.01K_2 \\ 0 \end{bmatrix}$$

which can be solved for D_2 and D_4.

In this section, an attempt has been made to develop an insight and thereby a degree of confidence in the overall methodology employed in the matrix displacement method. It should be reemphasized that, while the particular example discussed was elementary, the procedures developed were general ones and can be used in future cases, regardless of how complex the geometry of the structure becomes. All that changes from one application to another is the element itself. Therefore, the major part of the discussion which follows is devoted to the development of various elements that are commonly used, and to their application in the analysis of particular types of structures.

Before leaving this section, it should be noted that at no time did the question of redundancy arise. This is characteristic of the stiffness method of solution. Independent of whether the structure in question is determinate or indeterminate, the solution procedure would be the same.

9.2 AXIAL-FORCE MEMBERS

A typical axial-force element is shown in fig. 9.9. The element is L unit long, has a constant cross-sectional area of A, and a Young's modulus of elasticity of E. $\bar{X}$ represents the local coordinate system. As shown, the element has two degrees of freedom in the local system. These are the displacements at the nodes i and j, $D_{\bar{X}i}$ and $D_{\bar{X}j}$, respectively. The forces and displacements indicated in the figure are in the assumed positive sense. Recognizing that the spring constant K in the previous example can here be represented by the quantity AE/L, equilibrium is defined by the equation

$$\begin{bmatrix} \dfrac{AE}{L} & -\dfrac{AE}{L} \\ -\dfrac{AE}{L} & \dfrac{AE}{L} \end{bmatrix} \begin{bmatrix} D_{\bar{X}i} \\ D_{\bar{X}j} \end{bmatrix} = \begin{bmatrix} F_{\bar{X}i} \\ F_{\bar{X}j} \end{bmatrix} \tag{9.20}$$

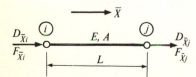

Figure 9.9 Axial-force member.

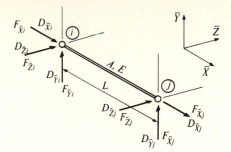

Figure 9.10

While the formulation given in eq. (9.20) accurately describes the behavior of the element in its own coordinate system, practicality dictates that the element be described by another, more general set of coordinates. This new set of reference coordinates will be denoted by the directions X, Y, and Z. The force and displacement vectors in eq. (9.20), therefore, must be transformed into these coordinates. The more general case, however, can be solved just as readily. Therefore, consider the element defined in the $\bar{X}$, $\bar{Y}$, and $\bar{Z}$ system shown in fig. 9.10. (It is to be understood that this reformulation in no way changes the characteristics of the element.) For equilibrium, the conditions prescribed in eq. (9.21) must be satisfied:

$$
\begin{bmatrix}
\dfrac{AE}{L} & 0 & 0 & -\dfrac{AE}{L} & 0 & 0 \\[2mm]
0 & 0 & 0 & 0 & 0 & 0 \\[1mm]
0 & 0 & 0 & 0 & 0 & 0 \\[1mm]
-\dfrac{AE}{L} & 0 & 0 & \dfrac{AE}{L} & 0 & 0 \\[2mm]
0 & 0 & 0 & 0 & 0 & 0 \\[1mm]
0 & 0 & 0 & 0 & 0 & 0
\end{bmatrix}
\begin{bmatrix}
D_{\bar{X}i} \\[1mm]
D_{\bar{Y}i} \\[1mm]
D_{\bar{Z}i} \\[1mm]
D_{\bar{X}j} \\[1mm]
D_{\bar{Y}j} \\[1mm]
D_{\bar{Z}j}
\end{bmatrix}
=
\begin{bmatrix}
F_{\bar{X}i} \\[1mm]
F_{\bar{Y}i} \\[1mm]
F_{\bar{Z}i} \\[1mm]
F_{\bar{X}j} \\[1mm]
F_{\bar{Y}j} \\[1mm]
F_{\bar{Z}j}
\end{bmatrix}
\qquad (9.21)
$$

Figure 9.11 shows the relationship of the local system of coordinates, denoted by $\bar{X}$, $\bar{Y}$, and $\bar{Z}$, and the reference system, X, Y, and Z. The governing equations relating the two are

$$\bar{X} = X \cos \theta_{\bar{X}X} + Y \cos \theta_{\bar{X}Y} + Z \cos \theta_{\bar{X}Z}$$

$$\bar{Y} = X \cos \theta_{\bar{Y}X} + Y \cos \theta_{\bar{Y}Y} + Z \cos \theta_{\bar{Y}Z} \qquad (9.22)$$

$$\bar{Z} = X \cos \theta_{\bar{Z}X} + Y \cos \theta_{\bar{Z}Y} + Z \cos \theta_{\bar{Z}Z}$$

In matrix form

$$
\begin{bmatrix}
\bar{X} \\
\bar{Y} \\
\bar{Z}
\end{bmatrix}
=
\begin{bmatrix}
a_{\bar{X}X} & a_{\bar{X}Y} & a_{\bar{X}Z} \\
a_{\bar{Y}X} & a_{\bar{Y}Y} & a_{\bar{Y}Z} \\
a_{\bar{Z}X} & a_{\bar{Z}Y} & a_{\bar{Z}Z}
\end{bmatrix}
\begin{bmatrix}
X \\
Y \\
Z
\end{bmatrix}
\qquad (9.22a)
$$

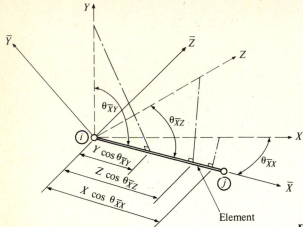

Figure 9.11 Rotation of coordinate axes.

where the a's are the cosines of the angles between the axes in question. For the element

$$
\begin{bmatrix}
\bar{X}_i \\
\bar{Y}_i \\
\bar{Z}_i \\
---- \\
\bar{X}_j \\
\bar{Y}_j \\
\bar{Z}_j
\end{bmatrix}
=
\begin{bmatrix}
a_{\bar{X}X} & a_{\bar{X}Y} & a_{\bar{X}Z} & & & \\
a_{\bar{Y}X} & a_{\bar{Y}Y} & a_{\bar{Y}Z} & & 0 & \\
a_{\bar{Z}X} & a_{\bar{Z}Y} & a_{\bar{Z}Z} & & & \\
-------- & & & -------- & & \\
 & & & a_{\bar{X}X} & a_{\bar{X}Y} & a_{\bar{X}Z} \\
 & 0 & & a_{\bar{Y}X} & a_{\bar{Y}Y} & a_{\bar{Y}Z} \\
 & & & a_{\bar{Z}X} & a_{\bar{Z}Y} & a_{\bar{Z}Z}
\end{bmatrix}
\begin{bmatrix}
X_i \\
Y_i \\
Z_i \\
--- \\
X_j \\
Y_j \\
Z_j
\end{bmatrix}
$$

or

$$[\bar{X}] = [T_R][X] \tag{9.23}$$

where $[T_R]$ is the transformation matrix.

It is to be noted that this transformation applies equally well to the displacements $[\bar{D}]$ and the forces $[\bar{F}]$.

$$[\bar{D}] = [T_R][D]$$

and

$$[\bar{F}] = [T_R][F]$$

In all of these, the barred quantities are in the local system, and the unbarred ones are in the reference system.

Substituting these relationships into the element local equilibrium formulation,

$$[\bar{K}][\bar{D}] = [\bar{F}]$$

yields

$$[\bar{K}][T_R][D] = [T_R][F]$$

Premultiplying both sides of this matrix equation by the inverse of the transformation matrix, that is, by $[T_R^{-1}]$, we obtain

$$[T_R^{-1}][\bar{K}][T_R][D] = [F] \tag{9.24}$$

For this and most cases,† the transformation matrix $[T_R]$ has the property of orthogonality. Therefore, the inverse is the same as the transpose $[T_R^T]$.

$$[T_R^T][\bar{K}][T_R][D] = [F] \tag{9.25}$$

or

$$[K][D] = [F] \tag{9.26}$$

where all the terms are in the reference system. Thus, the stiffness matrix described in the reference system is

$$[T_R^T][\bar{K}][T_T] = [K] \tag{9.27}$$

To form the master stiffness matrix for any given problem, it is necessary that there be provided both the stiffness matrix in the local coordinates $[\bar{K}]$ and the corresponding transformation matrix $[T_R]$. These multiplied together as indicated in eq. (9.27) yield the required $[K]$. (It should here be indicated that this process for forming $[K]$ is common to *all* of the elements that will be studied.)

As far as methodology is concerned, the two matrices $[T_R]$ and $[\bar{K}]$ are usually read into a computer as data, and the machine performs the actual matrix multiplication. On occasion, however, when $[\bar{K}]$ and $[T_R]$ are not complex or large, the multiplication may be done by hand, and the element described in terms of the reference, once and for all times. Consider, for example, the two-dimensional case shown in fig. 9.12. The transformation matrix $[_1 T_R]$ is given by

$$[_1 T_T] = \begin{bmatrix} \cos \theta_{\bar{X}X} & \cos \theta_{\bar{X}Y} \\ \cos \theta_{\bar{Y}X} & \cos \theta_{\bar{Y}Y} \end{bmatrix} = \begin{bmatrix} a_{\bar{X}X} & a_{\bar{X}Y} \\ a_{\bar{Y}X} & a_{\bar{Y}Y} \end{bmatrix} \tag{9.28}$$

---

† See H. C. Martin, *Introduction to Matrix Methods of Structural Analysis*, p. 224, McGraw-Hill, New York, 1966.

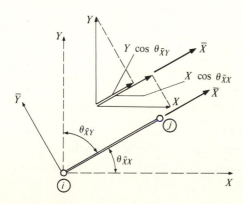

Figure 9.12 Rotation of coordinate axes.

and the transformation for the element by

$$[T_R] = \begin{bmatrix} a_{\bar{X}X} & a_{\bar{X}Y} & 0 & 0 \\ a_{\bar{Y}X} & a_{\bar{Y}Y} & 0 & 0 \\ \hline 0 & 0 & a_{\bar{X}X} & a_{\bar{X}Y} \\ 0 & 0 & a_{\bar{Y}X} & a_{\bar{Y}Y} \end{bmatrix} \tag{9.29}$$

Equilibrium of the element in the local system, for the indicated two-dimensional case, is described by the equations

$$\begin{bmatrix} \dfrac{AE}{L} & 0 & -\dfrac{AE}{L} & 0 \\ 0 & 0 & 0 & 0 \\ -\dfrac{AE}{L} & 0 & \dfrac{AE}{L} & 0 \\ 0 & 0 & 0 & 0 \end{bmatrix} \begin{bmatrix} D_{\bar{X}i} \\ D_{\bar{Y}i} \\ D_{\bar{X}j} \\ D_{\bar{Y}j} \end{bmatrix} = \begin{bmatrix} F_{\bar{X}i} \\ F_{\bar{Y}i} \\ F_{\bar{X}j} \\ F_{\bar{Y}j} \end{bmatrix} \tag{9.30}$$

$$\underbrace{\qquad\qquad\qquad\qquad\qquad\qquad}_{[\bar{K}]}$$

Therefore, in the reference system, the stiffness matrix $[K]$ is given by

$$[K] = \underbrace{\begin{bmatrix} a_{\bar{X}X} & a_{\bar{Y}X} & 0 & 0 \\ a_{\bar{X}Y} & a_{\bar{Y}Y} & 0 & 0 \\ 0 & 0 & a_{\bar{X}X} & a_{\bar{Y}X} \\ 0 & 0 & a_{\bar{X}Y} & a_{\bar{Y}Y} \end{bmatrix}}_{[T_R^T]} \underbrace{\begin{bmatrix} \dfrac{AE}{L} & 0 & -\dfrac{AE}{L} & 0 \\ 0 & 0 & 0 & 0 \\ -\dfrac{AE}{L} & 0 & \dfrac{AE}{L} & 0 \\ 0 & 0 & 0 & 0 \end{bmatrix}}_{[\bar{K}]} \underbrace{\begin{bmatrix} a_{\bar{X}X} & a_{\bar{X}Y} & 0 & 0 \\ a_{\bar{Y}X} & a_{\bar{Y}Y} & 0 & 0 \\ 0 & 0 & a_{\bar{X}X} & a_{\bar{X}Y} \\ 0 & 0 & a_{\bar{Y}X} & a_{\bar{Y}Y} \end{bmatrix}}_{[T_R]} \tag{9.31}$$

Multiplying,

$$[K] = \frac{AE}{L} \begin{bmatrix} [B] & [-B] \\ \hline [-B] & [B] \end{bmatrix} \tag{9.32}$$

where

$$[B] = \begin{bmatrix} a_{\bar{X}X}^2 & a_{\bar{X}X}a_{\bar{X}Y} \\ a_{\bar{X}X}a_{\bar{X}Y} & a_{\bar{X}Y}^2 \end{bmatrix}$$

Since $[K]$ in the reference system now has been defined, there is no further need to consider the local system. The only data necessary to form the matrix $[K]$ are A, E, and L for the member in question, and the angles $\theta_{\bar{X}X}$ and $\theta_{\bar{X}Y}$.

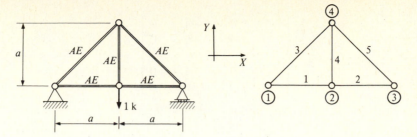

Figure 9.13

Example 9.1 As an example of the use of this process, consider the simple truss shown in fig. (9.13). It is desired to determine the bar forces due to the 1 kip load. Each member has the same values for A and E.

First, the system is replaced by the axial force elements 1 through 5 shown in the second sketch of the figure. The node points (that is, the joints) are numbered 1 to 4. Thus, element 5 is located between nodes 3 and 4 *in that order*. (This is required to maintain the general i–j notation used in defining the element and its matrix.) Next, using the general stiffness matrix for this element type in the reference coordinates X and Y, the stiffness matrix for each of the elements is written. This is conveniently done using table 9.1.

Table 9.1 Transformations, example 9.1

Member	Area	Length	Node i	Node j	$\theta_{\bar{X}X}$	$\theta_{\bar{X}Y}$	$a^2_{\bar{X}X}$	$a_{\bar{X}X}a_{\bar{X}Y}$	$a^2_{\bar{X}Y}$
1	A	a	1	2	0°	90°	1.0	0	0
2	A	a	2	3	0°	90°	1.0	0	0
3	A	$\sqrt{2}\,a$	1	4	45°	45°	0.5	0.5	0.5
4	A	a	2	4	90°	0°	0	0	1.0
5	A	$\sqrt{2}\,a$	3	4	135°	45°	0.5	-0.5	0.5

The element stiffness matrices are then

Element 1:

$$
\frac{AE}{a}
\begin{array}{cc}
\qquad \overbrace{}^{\text{Node 1}} \quad \overbrace{}^{\text{Node 2}} \\
\begin{array}{cc}
x \quad y & x \quad y
\end{array} \\
\left[
\begin{array}{cc:cc}
1 & 0 & -1 & 0 \\
0 & 0 & 0 & 0 \\
\hdashline
-1 & 0 & 1 & 0 \\
0 & 0 & 0 & 0
\end{array}
\right]
\begin{array}{l} {\scriptstyle \}1} \\ \\ {\scriptstyle \}2} \end{array}
\end{array}
\tag{9.33a}
$$

Element 2:
$$\frac{AE}{a} \begin{bmatrix} 1 & 0 & -1 & 0 \\ 0 & 0 & 0 & 0 \\ \hdashline -1 & 0 & 1 & 0 \\ 0 & 0 & 0 & 0 \end{bmatrix} \begin{matrix} 2 \\ \\ 3 \end{matrix}$$

(9.33b)

Element 3:
$$\frac{AE}{\sqrt{2}a} \begin{bmatrix} 0.5 & 0.5 & -0.5 & -0.5 \\ 0.5 & 0.5 & -0.5 & -0.5 \\ \hdashline -0.5 & -0.5 & 0.5 & 0.5 \\ -0.5 & -0.5 & 0.5 & 0.5 \end{bmatrix} \begin{matrix} 1 \\ \\ 4 \end{matrix}$$

(9.33c)

Element 4:
$$\frac{AE}{a} \begin{bmatrix} 0 & 0 & 0 & 0 \\ 0 & 1 & 0 & -1 \\ \hdashline 0 & 0 & 0 & 0 \\ 0 & -1 & 0 & 1 \end{bmatrix} \begin{matrix} 2 \\ \\ 4 \end{matrix}$$

(9.33d)

Element 5:
$$\frac{AE}{\sqrt{2}a} \begin{bmatrix} 0.5 & -0.5 & -0.5 & 0.5 \\ -0.5 & 0.5 & 0.5 & -0.5 \\ \hdashline -0.5 & 0.5 & 0.5 & -0.5 \\ 0.5 & -0.5 & -0.5 & 0.5 \end{bmatrix} \begin{matrix} 3 \\ \\ 4 \end{matrix}$$

(9.33e)

General notation:
$$\frac{AE}{L} \begin{bmatrix} [B] & [-B] \\ \hdashline [-B] & [B] \end{bmatrix} \begin{matrix} \text{Node } i \\ \\ \text{Node } j \end{matrix}$$

(9.33f)

The node numbers shown with each matrix are the nodes where the stiffnesses are to be added. For example, element 3 lies between nodes 1 and 4. Thus the 2 × 2 submatrix having the subscript 1 − 1 will lie in the first two rows and first two columns of the master matrix. The element matrices are written in this way to avoid the need to expand each matrix to the size of the master matrix. If element 3 were written in its expanded form, it would be as shown in matrix (9.34).

$$
\begin{bmatrix}
0.5 & 0.5 & & & & & -0.5 & -0.5 \\
 & & & 0 & & 0 & & \\
0.5 & 0.5 & & & & & -0.5 & -0.5 \\
\hline
 & 0 & & 0 & & 0 & & 0 \\
\hline
 & 0 & & 0 & & 0 & & 0 \\
\hline
-0.5 & -0.5 & & & & & 0.5 & 0.5 \\
 & & & 0 & & 0 & & \\
-0.5 & -0.5 & & & & & 0.5 & 0.5
\end{bmatrix}
\begin{matrix} x \\ \\ y \\ \\ x \\ \\ y \\ \\ x \\ \\ y \\ \\ x \\ \\ y \end{matrix}
\quad (9.34)
$$

(This is a 2 × 2 null matrix)

Superposition of these element matrices results in the following master equilibrium formulation

$$
\frac{AE}{a}
\begin{bmatrix}
1+\frac{\sqrt{2}}{4} & \frac{\sqrt{2}}{4} & -1 & 0 & 0 & 0 & -\frac{\sqrt{2}}{4} & -\frac{\sqrt{2}}{4} \\
\frac{\sqrt{2}}{4} & \frac{\sqrt{2}}{4} & 0 & 0 & 0 & 0 & -\frac{\sqrt{2}}{4} & -\frac{\sqrt{2}}{4} \\
-1 & 0 & 2 & 0 & -1 & 0 & 0 & 0 \\
0 & 0 & 0 & 1 & 0 & 0 & 0 & -1 \\
0 & 0 & -1 & 0 & 1+\frac{\sqrt{2}}{4} & -\frac{\sqrt{2}}{4} & -\frac{\sqrt{2}}{4} & \frac{\sqrt{2}}{4} \\
0 & 0 & 0 & 0 & -\frac{\sqrt{2}}{4} & \frac{\sqrt{2}}{4} & \frac{\sqrt{2}}{4} & -\frac{\sqrt{2}}{4} \\
-\frac{\sqrt{2}}{4} & -\frac{\sqrt{2}}{4} & 0 & 0 & -\frac{\sqrt{2}}{4} & \frac{\sqrt{2}}{4} & \frac{\sqrt{2}}{4}+\frac{\sqrt{2}}{4} & \frac{\sqrt{2}}{4}-\frac{\sqrt{2}}{4} \\
-\frac{\sqrt{2}}{4} & -\frac{\sqrt{2}}{4} & 0 & -1 & \frac{\sqrt{2}}{4} & -\frac{\sqrt{2}}{4} & \frac{\sqrt{2}}{4}-\frac{\sqrt{2}}{4} & H
\end{bmatrix}
\begin{bmatrix}
D_{X1} \\ D_{Y1} \\ D_{X2} \\ D_{Y2} \\ D_{X3} \\ D_{Y3} \\ D_{X4} \\ D_{Y4}
\end{bmatrix}
=
\begin{bmatrix}
F_{X1} \\ F_{Y1} \\ F_{X2} \\ F_{Y2} \\ F_{X3} \\ F_{Y3} \\ F_{Y4} \\ F_{Y4}
\end{bmatrix}
$$

where $H = 1 + \frac{\sqrt{2}}{4} + \frac{\sqrt{2}}{4}$

The boundary constraints for the problem are $D_{X1} = D_{Y1} = D_{Y3} = 0$. The loading is $F_{Y2} = -1$ kip and $F_{X2} = F_{X3} = F_{X4} = F_{Y4} = 0$. The reactions are F_{X1}, F_{Y1}, and F_{Y3}.

After striking the appropriate rows and columns and rearranging, the set to be solved for the unknown displacements is

$$\frac{AE}{a}\begin{bmatrix} 2 & 0 & -1 & 0 & 0 \\ 0 & 1 & 0 & 0 & -1 \\ -1 & 0 & 1+\dfrac{\sqrt{2}}{4} & -\dfrac{\sqrt{2}}{4} & \dfrac{\sqrt{2}}{4} \\ 0 & 0 & -\dfrac{\sqrt{2}}{4} & \dfrac{\sqrt{2}}{2} & 0 \\ 0 & -1 & \dfrac{\sqrt{2}}{4} & 0 & 1+\dfrac{\sqrt{2}}{2} \end{bmatrix}\begin{bmatrix} D_{X2} \\ D_{Y2} \\ D_{X3} \\ D_{X4} \\ D_{Y4} \end{bmatrix} = \begin{bmatrix} 0 \\ -1 \\ 0 \\ 0 \\ 0 \end{bmatrix}$$

If AE/a equals 1.0, the nodal displacements are as follows:

Node	D_X, in	D_Y, in
1	0	0
2	5.0000000×10^{-1}	-2.9142136
3	1.0000000	0
4	5.0000000×10^{-1}	-1.9142136

The individual bar forces and reactions are next computed. This can be done in one of two ways. Since the displacements are in the reference coordinate system, the form $[K][D] = [F]$ could be used for each element, computing the element loads in the reference system. These could then be converted to the local system using the equations $[\bar{F}] = [T_R][F]$. On the other hand, the displacements of the element nodes could be converted from the reference system to the local system using the equations $[\bar{D}] = [T_R][D]$. The local element forces would then be determined from $[\bar{F}] = [\bar{K}][\bar{D}]$. Both these methods involve the conversion of a vector from the reference to the local system. Both methods also involve a matrix multiplication to extract the forces. Since the size of the element stiffness matrix in the first case is two times as large as that in the second, the second approach is preferable.

Consider the determination of the forces in element 3:

$$[\bar{D}] = [T_R][D]$$

where

$$[T_R] = \begin{bmatrix} \dfrac{1}{\sqrt{2}} & \dfrac{1}{\sqrt{2}} & 0 & 0 \\ -\dfrac{1}{\sqrt{2}} & \dfrac{1}{\sqrt{2}} & 0 & 0 \\ 0 & 0 & \dfrac{1}{\sqrt{2}} & \dfrac{1}{\sqrt{2}} \\ 0 & 0 & -\dfrac{1}{\sqrt{2}} & \dfrac{1}{\sqrt{2}} \end{bmatrix}$$

Therefore,

$$
\begin{bmatrix} \bar{D}_{\bar{X}1} \\ \bar{D}_{\bar{Y}1} \\ \bar{D}_{\bar{X}4} \\ \bar{D}_{\bar{Y}4} \end{bmatrix} = \frac{1}{\sqrt{2}} \begin{bmatrix} 1 & 1 & 0 & 0 \\ -1 & 1 & 0 & 0 \\ 0 & 0 & 1 & 1 \\ 0 & 0 & -1 & 1 \end{bmatrix} \begin{bmatrix} D_{X1} \\ D_{Y1} \\ D_{X4} \\ D_{Y4} \end{bmatrix}
$$

But since $D_{X1} = D_{Y1} = 0$, and

$$
\begin{bmatrix} \bar{D}_{\bar{X}4} \\ \bar{D}_{\bar{Y}4} \end{bmatrix} = \frac{1}{\sqrt{2}} \begin{bmatrix} 1 & 1 \\ -1 & 1 \end{bmatrix} \begin{bmatrix} D_{X4} \\ D_{Y4} \end{bmatrix},
$$

Defining $\bar{D}_{\bar{Y}4} = 0$, $\bar{D}_{\bar{X}4} = \dfrac{1}{\sqrt{2}}(D_{X4} + D_{Y4}) = -1.00$. Then $[\bar{F}] = [\bar{K}][\bar{D}]$, or

$$
\begin{bmatrix} \bar{F}_{X1} \\ \bar{F}_{X4} \end{bmatrix} = \frac{AE}{\sqrt{2}\,a} \begin{bmatrix} 1 & -1 \\ -1 & 1 \end{bmatrix} \begin{bmatrix} 0 \\ \bar{D}_{X4} \end{bmatrix}
$$

which gives

$$
\bar{F}_{X4} = -\bar{F}_{X1} = \frac{AE}{\sqrt{2}\,a}\,\bar{D}_{X4} = \frac{1}{\sqrt{2}}(-1.00) = -0.70710678 \text{ kips}
$$

Using this procedure, the bar forces in all of the members are:

Element	Node	$\bar{F}_{xi}$, kips	$\bar{F}_{xj}$, kips
1	1–2	$-5.0000000 \times 10^{-1}$	5.0000000×10^{-1}
2	2–3	$-5.0000000 \times 10^{-1}$	5.0000000×10^{-1}
3	1–4	7.0710678×10^{-1}	$-7.0710678 \times 10^{-1}$
4	2–4	-1.0000000	1.0000000
5	3–4	7.0710678×10^{-1}	$-7.0710678 \times 10^{-1}$

Suppose now, for the same structure, that it is desired to determine the reactions, node displacements, and stress resultants in the various members due to a 1-kip load acting horizontally to the right at node 4. Since the truss geometry is unchanged, and since the constraints are the same, the only modification is the applied load vector. (The element F_{4X} becomes $+1.0$.) Had the loading been specified as a given node displacement, changes of the type illustrated in the spring example would be required. (That is, modifications in the load matrix would be made using factors consisting of the given displacements times the appropriate stiffness coefficients.)

Example 9.2 Assume that an additional member 6 is added to the truss assumed in the preceding example, as is shown in fig. 9.14. The length and cross-sectional area of this new member are $a/3$ and A, respectively.

To include this additional member—or for that matter, any additional members—all that is required is that the stiffness of the member (in the reference system of directions) be added to the existing master formulation. It is to be understood, however, that the master formulation would be expanded by two

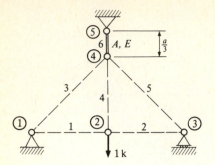

Figure 9.14

more rows, and two more columns. Since for the new member 6, $\theta_{XX} = 90°$ and $\theta_{XY} = 0°$, the element stiffness matrix is

$$[K_6] = \frac{AE}{a} \begin{bmatrix} \overset{4}{0} & 0 & \overset{5}{0} & 0 \\ 0 & 3 & 0 & -3 \\ \hdashline 0 & 0 & 0 & 0 \\ 0 & -3 & 0 & 3 \end{bmatrix} \begin{matrix} 4 \\ \\ \\ 5 \end{matrix}$$

The portion of the master formulation affected by this addition is

$$\frac{AE}{a} \begin{array}{c} 3 \\ \\ 4 \\ \\ 5 \end{array} \begin{bmatrix} \overset{3}{1+\frac{\sqrt{2}}{4}} & -\frac{\sqrt{2}}{4} & \overset{4}{-\frac{\sqrt{2}}{4}} & \frac{\sqrt{2}}{4} & \overset{5}{0} & 0 \\ -\frac{\sqrt{2}}{4} & \frac{\sqrt{2}}{4} & \frac{\sqrt{2}}{4} & -\frac{\sqrt{2}}{4} & 0 & 0 \\ \hdashline -\frac{\sqrt{2}}{4} & \frac{\sqrt{2}}{4} & \frac{\sqrt{2}}{2} & 0 & 0 & 0 \\ \frac{\sqrt{2}}{4} & -\frac{\sqrt{2}}{4} & 0 & \frac{\sqrt{2}}{2}+3 & 0 & -3 \\ \hdashline 0 & 0 & 0 & 0 & 0 & 0 \\ 0 & 0 & 0 & -3 & 0 & 3 \end{bmatrix} \begin{bmatrix} D_{X3} \\ D_{Y3} \\ D_{X4} \\ D_{Y4} \\ D_{X5} \\ D_{Y5} \end{bmatrix} = \begin{bmatrix} F_{X3} \\ F_{Y3} \\ F_{X4} \\ F_{Y4} \\ F_{X5} \\ F_{Y5} \end{bmatrix}$$

Since node 5 is constrained in both the X and Y directions, the pair of rows and columns corresponding to this node can be dropped. The set to be solved for the unknown displacements is, then,

$$\frac{AE}{a} \begin{bmatrix} 2 & 0 & -1 & 0 & 0 \\ 0 & 1 & 0 & 0 & -1 \\ -1 & 0 & 1+\frac{\sqrt{2}}{4} & -\frac{\sqrt{2}}{4} & \frac{\sqrt{2}}{4} \\ 0 & 0 & -\frac{\sqrt{2}}{4} & \frac{\sqrt{2}}{4} & 0 \\ 0 & -1 & \frac{\sqrt{2}}{4} & 0 & \frac{\sqrt{2}}{2}+4 \end{bmatrix} \begin{bmatrix} D_{X2} \\ D_{Y2} \\ D_{X3} \\ D_{X4} \\ D_{Y4} \end{bmatrix} = \begin{bmatrix} 0 \\ -1 \\ 0 \\ 0 \\ 0 \end{bmatrix}$$

Comparing this to the corresponding set in example 9.1, we note that the effect of adding member 6 to the truss is to increase the stiffness of the structure at node 4 in the Y direction by the addition of the number 3.

It should be noted that the truss shown in fig. 9.13 is statically determinate. The truss shown in fig. 9.14 is indeterminate to the first degree. The application of the matrix displacement method, however, was the same for both cases. Not only were there no additional steps required to solve the indeterminate problem, other than to introduce the added stiffness matrix, but also the question was not even raised as to whether or not indeterminacy existed.

Example 9.3 The crane problem shown in fig. 5.21 is to be solved by the matrix displacement method. For the various members, the geometry is indicated in the following table.

Table 9.2 Direction cosines and transformations, example 9.3

Member	Length, ft	AE, lb	$\theta_{\bar{X}X}$	$\theta_{\bar{X}Y}$	$a_{\bar{X}X}$	$a_{\bar{X}Y}$	$a_{\bar{X}X}^2$	$a_{\bar{X}X}a_{\bar{X}Y}$	$a_{\bar{X}Y}^2$
AC	50.00	15.00	53.13	36.87	0.600	0.800	0.360	0.480	0.640
BC	40.00	150.00	90.00	0	0	1.000	0	0	1.000
CD	63.24	15.00	18.43	71.57	0.949	0.316	0.901	0.300	0.100
BD	84.84	150.00	45.00	45.00	0.707	0.707	0.500	0.500	0.500

Since

$$[K] = \frac{AE}{L}\left[\begin{array}{c:c} [B] & [-B] \\ \hdashline [-B] & [B] \end{array}\right] \qquad \text{where } [B] = \begin{bmatrix} a_{\bar{X}X}^2 & a_{\bar{X}X}a_{\bar{X}Y} \\ a_{\bar{X}X}a_{\bar{X}Y} & a_{\bar{X}Y}^2 \end{bmatrix}$$

the $(AE/L)[B]$ matrices for the various members become:

Member AC:
$$\begin{bmatrix} 0.90 & 1.20 \\ 1.20 & 1.60 \end{bmatrix} \times 10^4$$

Member BC:
$$\begin{bmatrix} 0 & 0 \\ 0 & 31.25 \end{bmatrix} \times 10^4$$

Member CD:
$$\begin{bmatrix} 1.78 & 0.59 \\ 0.59 & 0.20 \end{bmatrix} \times 10^4$$

Member BD:
$$\begin{bmatrix} 7.37 & 7.37 \\ 7.37 & 7.37 \end{bmatrix} \times 10^4$$

The master formulation is then

$$
\begin{bmatrix}
0.90 & 1.20 & 0 & 0 & -0.90 & -1.20 & 0 & 0 \\
1.20 & 1.60 & 0 & 0 & -1.20 & -1.60 & 0 & 0 \\
0 & 0 & 7.37 & 7.37 & 0 & 0 & -7.37 & -7.37 \\
0 & 0 & 7.37 & 38.62 & 0 & -31.25 & -7.37 & -7.37 \\
-0.90 & -1.20 & 0 & 0 & 2.68 & 1.79 & -1.78 & -0.59 \\
-1.20 & -1.60 & 0 & -31.25 & 1.79 & 33.05 & -0.59 & -0.20 \\
0 & 0 & -7.37 & -7.37 & -1.78 & -0.59 & 9.15 & 7.96 \\
0 & 0 & -7.37 & -7.37 & -0.59 & 0.20 & 7.96 & 7.57
\end{bmatrix}
$$

$$
\times
\begin{bmatrix}
D_{AX} \\
D_{AY} \\
D_{BX} \\
D_{BY} \\
D_{CX} \\
D_{CY} \\
D_{DX} \\
D_{DY}
\end{bmatrix}
= Q
\begin{bmatrix}
R_{AX} \\
R_{AY} \\
R_{BX} \\
R_{BY} \\
0 \\
0 \\
0 \\
-1.0
\end{bmatrix}
\times 10^{-4}
$$

and the equations to be solved are

$$
\begin{bmatrix}
2.68 & 1.79 & -1.78 & -0.59 \\
1.79 & 33.05 & -0.59 & -0.20 \\
-1.78 & -0.59 & 9.15 & 7.96 \\
-0.59 & -0.20 & 7.96 & 7.57
\end{bmatrix}
\begin{bmatrix}
D_{CX} \\
D_{CY} \\
D_{DX} \\
D_{DY}
\end{bmatrix}
= Q
\begin{bmatrix}
0 \\
0 \\
0 \\
-1.0
\end{bmatrix}
\times 10^{-4}
$$

The vertical deflection at point D is

$$
D_{DY} = -0.4046 \text{ in}†
$$

---

† This value for D_{DY} (-0.4046 in) is slightly different from that obtained in chap. 5 for the same problem (-0.41418 in). This is due to the rounding off of terms. If four decimal places are carried throughout the entire matrix computational process, the following value would be obtained

$$
D_{DY} = -0.4141 \text{ in}
$$

Example 9.4 The structure in question is shown in fig. 8.19. All members have the same cross-sectional areas and are of the same material. Q is assumed to be 1 kip. The geometrical properties are given in table 9.3.

Table 9.3 Geometrical properties for example 9.4

Member	L	$\times \dfrac{AE}{a}$	$\theta_{\bar{X}X}$	$\theta_{\bar{X}Y}$	$a_{\bar{X}X}$	$a_{\bar{X}Y}$	$a_{\bar{X}X}^2$	$a_{\bar{X}X}a_{\bar{X}Y}$	$a_{\bar{X}Y}^2$
AC	a	1	0	90°	1.000	0	1.000	0	0
CE	a	1	0	90°	1.000	0	1.000	0	0
BD	a	1	0	90°	1.000	0	1.000	0	0
DF	a	1	0	90°	1.000	0	1.000	0	0
AB	a	1	90°	0	0	1.000	0	0	1.000
CD	a	1	90°	0	0	1.000	0	0	1.000
EF	a	1	90°	0	0	1.000	0	0	1.000
AD	1.414a	0.707	45°	45°	0.707	0.707	0.500	0.500	0.500
CF	1.414a	0.707	45°	45°	0.707	0.707	0.500	0.500	0.500
CB	1.414a	0.707	135°	45°	−0.707	0.707	0.500	−0.500	0.500
ED	1.414a	0.707	135°	45°	−0.707	0.707	0.500	−0.500	0.500

The $(AE/a)[B]$ matrices for the various members are:

$$
Members \begin{cases} AC \\ CE \\ BD \\ DF \end{cases} : \qquad \frac{AE}{a} \begin{bmatrix} 1.0 & 0 \\ 0 & 0 \end{bmatrix}
$$

$$
Members \begin{cases} AB \\ CD \\ EF \end{cases} : \qquad \frac{AE}{a} \begin{bmatrix} 0 & 0 \\ 0 & 1.0 \end{bmatrix}
$$

$$
Members \begin{cases} AD \\ CF \end{cases} : \qquad 0.707 \frac{AE}{a} \begin{bmatrix} 0.500 & 0.500 \\ 0.500 & 0.500 \end{bmatrix} = \frac{AE}{a} \begin{bmatrix} 0.354 & 0.354 \\ 0.354 & 0.354 \end{bmatrix}
$$

$$
Members \begin{cases} CB \\ ED \end{cases} : \qquad \frac{AE}{a} \begin{bmatrix} 0.354 & -0.354 \\ -0.354 & 0.354 \end{bmatrix}
$$

Using these for the master matrix formulation, and then striking all rows and columns associated with the reactions, the following 8 × 8 set is obtained:

$$
\begin{bmatrix}
2.708 & -1.0 & -0.354 & 0.354 & 0 & 0 & -0.354 & 0.354 \\
 & 1.354 & 0 & 0 & -0.354 & 0.354 & 0 & 0 \\
 & & 1.354 & -0.354 & -1.0 & 0 & 0 & 0 \\
 & & & 1.354 & 0 & 0 & 0 & 0 \\
 & & & & 2.708 & 0 & -1.0 & 0 \\
 & \text{symmetrical} & & & & 1.708 & 0 & 0 \\
 & & & & & & 1.354 & 0.354 \\
 & & & & & & & 1.354
\end{bmatrix}
$$

$$
\times
\begin{bmatrix}
D_{CX} \\
D_{EX} \\
D_{BX} \\
D_{BY} \\
D_{DX} \\
D_{DY} \\
D_{FX} \\
D_{FY}
\end{bmatrix}
=
\frac{a}{AE}
\begin{bmatrix}
0 \\
0 \\
0 \\
-0.500 \\
0 \\
-1.000 \\
0 \\
-0.500
\end{bmatrix}
$$

The solutions are

$$D_{CX} = 0.1717\,\frac{a}{AE} \qquad D_{DX} = 0.1717\,\frac{a}{AE}$$

$$D_{EX} = 0.3434\,\frac{a}{AE} \qquad D_{DY} = -0.6566\,\frac{a}{AE}$$

$$D_{BX} = 0.06805\,\frac{a}{AE} \qquad D_{FX} = 0.2753\,\frac{a}{AE}$$

$$D_{BY} = -0.3964\,\frac{a}{AE} \qquad D_{FY} = -0.3964\,\frac{a}{AE}$$

from which

$$R_{AX} = 0$$

$$R_{AY} = 0.5681 \text{ kips}$$

$$R_{CY} = 0.8638 \text{ kips}$$

$$R_{EY} = 0.5681 \text{ kips}$$

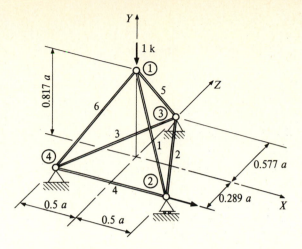

0.817 a

1 k

Y

Z

5

6

3

3

1

2

0.577 a

4

2

0.289 a

X

0.5 a

0.5 a

Figure 9.15

Example 9.5 Consider the three-dimensional truss structure shown in fig. 9.15. This tetrahedron consists of six individual members, all having the same length a and same cross-sectional area A. The members and nodes are numbered as indicated. Also shown are the dimensions of the structure and the presumed reference axis system. A single, vertical downward load of 1 kip is presumed acting at node 1.

For this type of structure, the general three-dimensional axial load element defined in fig. 9.10 is required. In its own coordinate system, the local element stiffness matrix is

$$[\bar{K}] = \frac{AE}{L}
\begin{bmatrix}
\begin{array}{ccc|ccc}
1 & 0 & 0 & -1 & 0 & 0 \\
0 & 0 & 0 & 0 & 0 & 0 \\
0 & 0 & 0 & 0 & 0 & 0 \\
\hline
-1 & 0 & 0 & 1 & 0 & 0 \\
0 & 0 & 0 & 0 & 0 & 0 \\
0 & 0 & 0 & 0 & 0 & 0
\end{array}
\end{bmatrix}
\begin{array}{l}
\bar{X} \\ \bar{Y} \\ \bar{Z} \\ \\ \bar{X} \\ \bar{Y} \\ \bar{Z}
\end{array}
$$

Node i $\bar{X}$ $\bar{Y}$ $\bar{Z}$ Node j $\bar{X}$ $\bar{Y}$ $\bar{Z}$ Node i Node j (9.35)

The transformation matrix is

$$[T_R] = \begin{bmatrix}
[\Lambda] & 0 \\
\hline
0 & [\Lambda]
\end{bmatrix}
\begin{array}{l} i \\ j \end{array}$$

(9.36)

where

$$[\Lambda] = \begin{bmatrix}
a_{\bar{X}X} & a_{\bar{X}Y} & a_{\bar{X}Z} \\
a_{\bar{Y}X} & a_{\bar{Y}Y} & a_{\bar{Y}Z} \\
a_{\bar{Z}X} & a_{\bar{Z}Y} & a_{\bar{Z}Z}
\end{bmatrix}$$

and

$$a_{\bar{X}X} = \cos \theta_{\bar{X}X} \cdots, \text{ etc.}$$

The element stiffness matrix, in the reference system, is formed using the expression

$$[K] = [T_R^T][\bar{K}][T_R]$$

Performing this matrix multiplication, the stiffness matrix $[K]$ becomes

$$[K] = \frac{AE}{L} \begin{bmatrix} i & j \\ [B] & \vdots & [-B] \\ \hline [-B] & \vdots & [B] \end{bmatrix} \begin{matrix} i \\ \\ j \end{matrix} \tag{9.37}$$

where

$$[B] = \begin{bmatrix} a_{\bar{X}X}^2 & a_{\bar{X}X}a_{\bar{X}Y} & a_{\bar{X}X}a_{\bar{X}Z} \\ a_{\bar{X}X}a_{\bar{X}Y} & a_{\bar{X}Y}^2 & a_{\bar{X}Y}a_{\bar{X}Z} \\ a_{\bar{X}X}a_{\bar{X}Z} & a_{\bar{X}Y}a_{\bar{X}Z} & a_{\bar{X}Z}^2 \end{bmatrix} \tag{9.38}$$

The data required for each element are the three direction cosines $a_{\bar{X}X}$, $a_{\bar{X}Y}$, and $a_{\bar{X}Z}$. For member 1, the angles in question are shown in fig. 9.16. (Note that the direction $\bar{X}$ is taken from node i to node j, with i having the smaller node number. Table 9.4 is developed, considering each of the elements in this same way.)

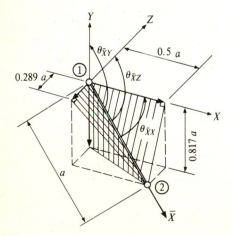

Figure 9.16 Component dimensions of member 1-2.

Table 9.4 Geometry and transformations for three-dimensional truss, example 9.5

Element	Length	Area	ith node	jth node	$a_{\bar{X}X}$	$a_{\bar{X}Y}$	$a_{\bar{X}X}$	$a^2_{\bar{X}X}$	$a_{\bar{X}X}a_{\bar{X}Y}$	$a_{\bar{X}X}a_{\bar{X}Z}$	$a^2_{\bar{X}Y}$	$a_{\bar{X}Y}a_{\bar{X}Z}$	$a^2_{\bar{X}Z}$
1	a	A	1	2	0.500	−0.817	−0.289	0.250	−0.409	−0.145	0.667	0.236	0.084
2	a	A	2	3	−0.500	0	0.866	0.250	0	−0.409	0	0	0.667
3	a	A	3	4	−0.500	0	−0.866	0.250	0	0.409	0	0	0.667
4	a	A	2	4	−1.000	0	0	1.000	0	0	0	0	0
5	a	A	1	3	0	−0.817	0.577	0	0	0	0.667	−0.471	0.333
6	a	A	1	4	−0.500	−0.817	−0.289	0.250	0.409	0.145	0.667	0.236	0.084

The " B" matrix for each is

$$[B_1] = \frac{AE}{a} \begin{bmatrix} 0.250 & -0.409 & -0.145 \\ & 0.667 & 0.236 \\ \text{symmetrical} & & 0.084 \end{bmatrix} \qquad [B_2] = \frac{AE}{a} \begin{bmatrix} 0.250 & 0 & -0.409 \\ & 0 & 0 \\ \text{symmetrical} & & 0.667 \end{bmatrix}$$

$$[B_3] = \frac{AE}{a} \begin{bmatrix} 0.250 & 0 & 0.409 \\ & 0 & 0 \\ \text{symmetrical} & & 0.667 \end{bmatrix} \qquad [B_4] = \frac{AE}{a} \begin{bmatrix} 1.000 & 0 & 0 \\ & 0 & 0 \\ \text{symmetrical} & & 0 \end{bmatrix}$$

$$[B_5] = \frac{AE}{a} \begin{bmatrix} 0. & 0 & 0 \\ & 0.667 & -0.471 \\ \text{symmetrical} & & 0.333 \end{bmatrix} \qquad [B_6] = \frac{AE}{a} \begin{bmatrix} 0.250 & 0.409 & 0.145 \\ & 0.667 & 0.236 \\ \text{symmetrical} & & 0.084 \end{bmatrix}$$

The assembled master matrix consists of the superimposed 3×3 $[B_i]$ matrices. The master formulation becomes

$$\begin{bmatrix} [B_1] + [B_5] + [B_6] & -[B_1] & -[B_5] & -[B_6] \\ -[B_1] & [B_1] + [B_2] + [B_4] & -[B_2] & -[B_4] \\ -[B_5] & -[B_2] & [B_2] + [B_3] + [B_5] & -[B_3] \\ -[B_6] & -[B_4] & -[B_3] & [B_3] + [B_4] + [B_6] \end{bmatrix} \begin{bmatrix} D_{1X} \\ D_{1Y} \\ D_{1Z} \\ D_{2X} \\ D_{2Y} \\ D_{2Z} \\ D_{3X} \\ D_{3Y} \\ D_{3Z} \\ D_{4X} \\ D_{4Y} \\ D_{4Z} \end{bmatrix} = \begin{bmatrix} F_{1X} \\ F_{1Y} \\ F_{1Z} \\ F_{2X} \\ F_{2Y} \\ F_{2Z} \\ F_{3X} \\ F_{3Y} \\ F_{3Z} \\ F_{4X} \\ F_{4Y} \\ F_{4Z} \end{bmatrix}$$

The constraints are

$$D_{2Y} = D_{3X} = D_{3Y} = D_{3Z} = D_{4X} = D_{4Y} = D_{4Z} = 0$$

Therefore, F_{2Y}, F_{3X}, F_{3Y}, F_{3Z}, F_{4X}, F_{4Y}, and F_{4Z} are reactive forces. The applied loads are

$$F_{1X} = F_{1Z} = F_{2Z} = 0$$

and

$$F_{1Y} = -1.0 \text{ kips} \qquad \text{and} \qquad F_{2X} = +1.0 \text{ kips}$$

The unknown displacements that are to be determined are D_{1X}, D_{1Y}, D_{1Z}, D_{2X}, and D_{2Z}. Introducing all of the above information into the master formulation (that is, striking appropriate rows and columns and condensing) the desired displacements are defined by the set

$$
\begin{array}{c}
\begin{array}{ccccc} X & Y & Z & X & Y \end{array}
\end{array}
$$

$$
\begin{array}{c} X \\ Y \\ Z \\ X \\ Y \end{array}
\left[
\begin{array}{ccc|cc}
(0.500) & 0 & 0 & (-0.250) & (0.145) \\
 & (2.000) & 0 & (0.409) & (-0.236) \\
 & & (0.401) & (0.145) & (-0.084) \\
\hline
\multicolumn{3}{c|}{\text{symmetrical}} & (1.500) & (-0.554) \\
 & & & & (0.084)
\end{array}
\right]
\left[
\begin{array}{c}
D_{1X} \\ D_{1Y} \\ D_{1Z} \\ \hline D_{2X} \\ D_{2Z}
\end{array}
\right]
=
\left[
\begin{array}{c}
0 \\ -a/AE \\ 0 \\ \hline a/AE \\ 0
\end{array}
\right]
$$

If a is defined as 10 in, E for all the members is equal to 30×10^6 lb/in^2, and each bar has a cross-sectional area of 2.00 in^2, the node displacements are

Node	D_X, in	D_Y, in	D_Z, in
1	7.0633978×10^{-5}	$-1.1273394 \times 10^{-4}$	4.0715362×10^{-5}
2	1.8930699×10^{-4}	0	$-8.3112509 \times 10^{-5}$
3	0	0	0
4	0	0	0

Suppose, now, that the force in bar 6 is desired. Knowing the displacements, the process would be as follows:

$$[\bar{D}] = [T_R][D] \qquad \text{or} \qquad \begin{bmatrix} \bar{D}_i \\ \bar{D}_j \end{bmatrix} = \begin{bmatrix} \Lambda & 0 \\ 0 & \Lambda \end{bmatrix} \begin{bmatrix} D_i \\ D_j \end{bmatrix}$$

Therefore,

$$[\bar{D}_i] = [\Lambda][D_i]$$

In particular

$$\bar{D}_{Xi} = a_{\bar{X}X} D_{Xi} + a_{\bar{X}Y} D_{Yi} + a_{\bar{X}Z} D_{Zi}$$

It also is known that

$$[\bar{F}] = [\bar{K}][\bar{D}] \qquad \text{or} \qquad \begin{bmatrix} \bar{F}_{Xi} \\ \bar{F}_{Xj} \end{bmatrix} = \frac{AE}{a} \begin{bmatrix} 1 & -1 \\ -1 & 1 \end{bmatrix} \begin{bmatrix} \bar{D}_{Xi} \\ \bar{D}_{Xj} \end{bmatrix}$$

For member 6,

$$i = 1 \qquad \text{and} \qquad j = 4$$

Therefore,

$$_6 \bar{F}_{X1} = \frac{AE}{a} (\bar{D}_{Xi} - \bar{D}_{Xj})$$

but since $\bar{D}_{Xj} = 0$

$$_6 \bar{F}_{X1} = \frac{AE}{a} \bar{D}_{X1} = \frac{AE}{a} (_6 a_{\bar{X}X} D_{X1} + _6 a_{\bar{X}Y} D_{Y1} + _6 a_{\bar{X}Z} D_{Z1})$$

or

$$_6 \bar{F}_{X1} = -\frac{AE}{a} (0.500 D_{X1} + 0.817 D_{Y1} + 0.289 D_{Z1})$$

(The presubscript 6 is used to indicate that the quantity in question refers to that associated with member 6.) Using this process, the components of the local bar forces for each of the members are determined to be:

Elements	Nodes	$\bar{F}_{Xi}$, lb	$\bar{F}_{Xj}$, lb
1	1-2	4.0796704×10^2	-4.0796704×10^2
2	2-3	-1.3617435×10^2	1.3617435×10^2
3	3-4	0	0
4	2-4	-1.1358419×10^3	1.1358419×10^3
5	1-3	4.0855322×10^2	-4.0855322×10^2
6	1-4	4.0855322×10^2	-4.0855322×10^2

9.3 BEAM BENDING

A beam-bending element of length L, located between nodes i and j and having no lateral loads along the member, is shown in fig. 9.17. The local coordinate directions are $\bar{X}$, $\bar{Y}$, and $\bar{Z}$. Node forces and displacements are $F_{\bar{X}i}$, $F_{\bar{Y}i}$, $F_{\bar{Z}i}$, $F_{\bar{X}j}$, $F_{\bar{Y}j}$, and $F_{\bar{Z}j}$, and $D_{\bar{X}i}$, $D_{\bar{Y}i}$, $D_{\bar{Z}i}$, $D_{\bar{X}j}$, $D_{\bar{Y}j}$, and $D_{\bar{Z}j}$, respectively. Bending moments about the $\bar{Z}$ axis are $M_{\bar{Z}i}$ and $M_{\bar{Z}j}$, and about the $\bar{Y}$ axis are $M_{\bar{Y}i}$ and $M_{\bar{Y}j}$. The twisting moments

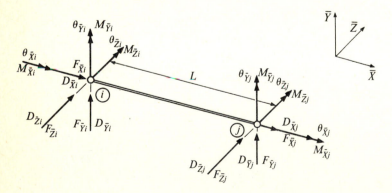

Figure 9.17 Beam element subject to end-bending and twisting moments.

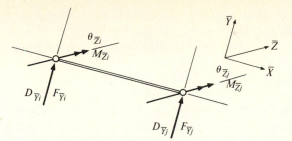

Figure 9.18 Beam element subjected to end-bending moments.

about the longitudinal axis $\bar{X}$ at the nodes are $M_{\bar{X}i}$ and $M_{\bar{X}j}$. Deformations at the nodes corresponding to these various components are $\theta_{\bar{Z}i}$, $\theta_{\bar{Z}j}$, $\theta_{\bar{Y}i}$, $\theta_{\bar{Y}j}$, $\theta_{\bar{X}i}$ and $\theta_{\bar{X}j}$. For this coordinate system, neglecting shear deformations (see sec. 4.4), the element has twelve degrees of freedom, each shown in its assumed positive direction in the figure. (It is to be recognized that this assumed sign convention, which is extremely convenient for use with computers, is *different* from that which was used in the slope-deflection method developed in chap. 8. There, clockwise moments, rotations, end shears, etc., acting on the member were assumed positive. Here, vectors—forces, moments, displacements and rotations—are positive when in the direction of the assumed plus coordinate direction.) The local displacement matrix (shown horizontally) is then

$$[\bar{D}] = \{D_{\bar{X}i} \quad D_{\bar{Y}i} \quad D_{\bar{Z}i} \quad \theta_{\bar{X}i} \quad \theta_{\bar{Y}i} \quad \theta_{\bar{Z}i} \quad D_{\bar{X}j} \quad D_{\bar{Y}j} \quad D_{\bar{Z}j} \quad \theta_{\bar{X}j} \quad \theta_{\bar{Y}j} \quad \theta_{\bar{Z}j}\} \qquad (9.38a)$$

In most instances, this full complement of displacements is not needed to describe a given problem. For example, when only end moments about one of the principal axes exist, and when shearing deformations are neglected, only three degrees of freedom at each node are required—as opposed to six in the general

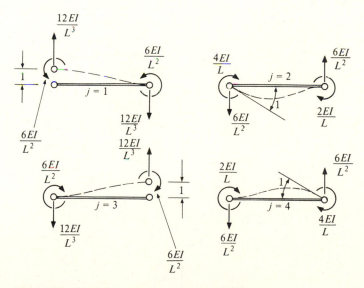

Figure 9.19 Unit deformations and resulting stiffnesses.

formulation. This represents a considerable reduction in required computer storage. For the beam-bending problem defined by the element shown in fig. 9.18, the stiffness matrices $[K]$ and $[\bar{K}]$ are 6×6 in size, as is the transformation matrix. The deformations are $D_{\bar{X}}$, $D_{\bar{Y}}$, and $\theta_{\bar{Z}}$ at i and j.

As defined in chap. 8 and also in earlier parts of this chapter, each element of the local stiffness matrix $[\bar{K}]$ is given by the quantity $\bar{k}_{ij}$, where the i and j represent not the node locations indicated in fig. 9.18, but the direction and location of the resulting force (or moment) due to a particular imposed displacement (or rotation)—with all other deformations maintained at zero. (In this system i and j can take on six values depending on whether the direction in question is the $\bar{X}$, or $\bar{Y}$, or θ direction, and whether the location is node i or node j.) Thus, if no lateral loads exist between node points i and j, $[\bar{K}]$ is defined by the four displacement states shown in fig. 9.19. The end moments and end shears associated with the various imposed unit deformations are in equilibrium in the directions indicated in fig. 9.19. Their signs, however—when used in the stiffness matrix—would have to be determined from a separate examination as to whether or not they are in the plus direction of the appropriate coordinate axis. $[\bar{K}]$ for this element and loading is given by

$$
[\bar{K}] = \begin{bmatrix}
0 & 0 & 0 & 0 & 0 & 0 \\
0 & \dfrac{12EI}{L^3} & -\dfrac{6EI}{L^2} & 0 & -\dfrac{12EI}{L^3} & -\dfrac{6EI}{L^2} \\
0 & -\dfrac{6EI}{L^2} & \dfrac{4EI}{L} & 0 & \dfrac{6EI}{L^2} & \dfrac{2EI}{L} \\
0 & 0 & 0 & 0 & 0 & 0 \\
0 & -\dfrac{12EI}{L^3} & \dfrac{6EI}{L^2} & 0 & \dfrac{12EI}{L^3} & \dfrac{6EI}{L^2} \\
0 & -\dfrac{6EI}{L^2} & \dfrac{2EI}{L} & 0 & \dfrac{6EI}{L^2} & \dfrac{4EI}{L}
\end{bmatrix}
\begin{matrix} X \\ Y \\ \theta \\ X \\ Y \\ \theta \end{matrix}
\tag{9.39}
$$

with column headings X, Y, θ (node i) and X, Y, θ (node j).

or

$$
[\bar{K}] = \begin{bmatrix}
[\bar{B}_{11}] & [\bar{B}_{12}] \\
[\bar{B}_{21}] & [\bar{B}_{12}]
\end{bmatrix}
\tag{9.39a}
$$

where the $[\bar{B}_{ij}]$'s are the appropriate partitioned submatrices of eq. (9.39).

The zeros in matrix (9.39) provide the degrees of freedom required for transformation from local to reference directions. Equilibrium in the local system is given by

$$
[\bar{K}][\bar{D}] = [\bar{F}]
$$

or

$$
\begin{bmatrix}
[\bar{B}_{11}] & [\bar{B}_{12}] \\
[\bar{B}_{21}] & [\bar{B}_{22}]
\end{bmatrix}
\begin{bmatrix}
\bar{D}_1 \\
\bar{D}_2
\end{bmatrix}
=
\begin{bmatrix}
\bar{F}_1 \\
\bar{F}_2
\end{bmatrix}
\tag{9.40}
$$

where

$$[\bar{D}_1] = \begin{bmatrix} D_{\bar{X}i} \\ D_{\bar{Y}i} \\ \theta_{\bar{Z}i} \end{bmatrix} \quad [\bar{D}_2] = \begin{bmatrix} D_{\bar{X}j} \\ D_{\bar{Y}j} \\ \theta_{\bar{Z}j} \end{bmatrix} \quad [\bar{F}_1] = \begin{bmatrix} F_{\bar{X}i} \\ F_{\bar{Y}i} \\ M_{\bar{Z}i} \end{bmatrix} \quad \text{and} \quad [\bar{F}_2] = \begin{bmatrix} F_{\bar{X}j} \\ F_{\bar{Y}j} \\ M_{\bar{Z}j} \end{bmatrix}$$

The transformation matrix for this element also is 6×6 in size.

$$[\bar{D}] = [T_R][D]$$

or

$$\begin{bmatrix} \bar{D}_1 \\ \hline \bar{D}^2 \end{bmatrix} = \begin{bmatrix} [\Lambda] & 0 \\ \hline 0 & [\Lambda] \end{bmatrix} \begin{bmatrix} D_1 \\ \hline D_2 \end{bmatrix}$$

where

$$[\Lambda] = \begin{bmatrix} a_{\bar{X}X} & a_{\bar{X}Y} & 0 \\ a_{\bar{Y}X} & a_{\bar{Y}Y} & 0 \\ 0 & 0 & 1 \end{bmatrix}$$

The a_{ij}'s are as previously defined.

The element stiffness matrix in the reference system is given by:

$$[K] = \begin{bmatrix} [B_{11}] & [B_{12}] \\ \hline [B_{21}] & [B_{22}] \end{bmatrix} = [T_R^T][\bar{K}][T_R]$$

$$= \begin{bmatrix} [\Lambda^T] & 0 \\ \hline 0 & [\Lambda^T] \end{bmatrix} \begin{bmatrix} [\bar{B}_{11}] & [\bar{B}_{12}] \\ \hline [\bar{B}_{21}] & [\bar{B}_{22}] \end{bmatrix} \begin{bmatrix} [\Lambda] & 0 \\ \hline 0 & [\Lambda] \end{bmatrix}$$

or

$$[K] = \begin{bmatrix} [\Lambda^T \ \bar{B}_{11} \ \Lambda] & [\Lambda^T \ \bar{B}_{12} \ \Lambda] \\ \hline [\Lambda^T \ \bar{B}_{21} \ \Lambda] & [\Lambda \ \bar{B}_{22} \ \Lambda] \end{bmatrix} \tag{9.41}$$

Example 9.6 Consider the beam shown in fig. 9.20. It is assumed that the member is prismatic, of length l, and is subjected to a concentrated load Q acting at a point a from the left-hand support.

Because the master formulations developed thus far have not considered the possibility of a lateral load acting between the nodes, the beam is modeled by the two elements and three nodes shown in the right-hand sketch.

If the nodes are numbered as indicated, the local and reference coordinates coincide. For this case, then, coordinate transformation is unnecessary. Further, since no axial forces (or displacements) are present, the stiffness matrix can be reduced to a 4×4 size.

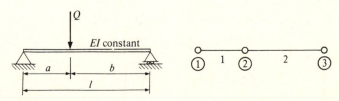

Figure 9.20

$$[K] = [\bar{K}] = \begin{bmatrix} \dfrac{12EI}{l^3} & -\dfrac{6EI}{l^2} & -\dfrac{12EI}{l^3} & -\dfrac{6EI}{l^2} \\ -\dfrac{6EI}{l^2} & \dfrac{4EI}{l} & \dfrac{6EI}{l^2} & \dfrac{2EI}{l} \\ -\dfrac{12EI}{l^3} & \dfrac{6EI}{l^2} & \dfrac{12EI}{l^3} & \dfrac{6EI}{l^2} \\ -\dfrac{6EI}{l^2} & \dfrac{2EI}{l} & \dfrac{6EI}{l^2} & \dfrac{4EI}{l} \end{bmatrix} \begin{matrix} i \\ \\ j \end{matrix} = \begin{bmatrix} [B_{11}] & [B_{12}] \\ [B_{21}] & [B_{22}] \end{bmatrix} \begin{matrix} i \\ j \end{matrix}$$

Moreover,

For element 1:
$$[_1K] = \begin{bmatrix} [_1B_{11}] & [_1B_{12}] \\ [_1B_{21}] & [_1B_{22}] \end{bmatrix}$$

For element 2:
$$[_2K] = \begin{bmatrix} [_2B_{11}] & [_2B_{12}] \\ [_2B_{21}] & [_2B_{22}] \end{bmatrix}$$

$$[K] = \begin{bmatrix} [_1B_{11}] & [_1B_{12}] & 0 \\ [_1B_{21}] & [_1B_{22}] + [_2B_{11}] & [_2B_{12}] \\ 0 & [_2B_{21}] & [_2B_{22}] \end{bmatrix} \begin{matrix} 1 \\ 2 \\ 3 \end{matrix}$$

or

$$[K] = \begin{bmatrix} \dfrac{12EI}{a^3} & -\dfrac{6EI}{a^2} & -\dfrac{12EI}{a^3} & -\dfrac{6EI}{a^2} & 0 & 0 \\ -\dfrac{6EI}{a^2} & \dfrac{4EI}{a} & \dfrac{6EI}{a^2} & \dfrac{2EI}{a} & 0 & 0 \\ -\dfrac{12EI}{a^3} & \dfrac{6EI}{a^2} & \dfrac{12EI}{a^3} + \dfrac{12EI}{b^3} & \dfrac{6EI}{a^2} - \dfrac{6EI}{b^2} & -\dfrac{12EI}{b^3} & -\dfrac{6EI}{b^2} \\ -\dfrac{6EI}{a^2} & \dfrac{2EI}{a} & \dfrac{6EI}{a^2} - \dfrac{6EI}{b^2} & \dfrac{4EI}{a} + \dfrac{4EI}{b} & \dfrac{6EI}{b^2} & \dfrac{2EI}{b} \\ 0 & 0 & -\dfrac{12EI}{b^3} & \dfrac{6EI}{b^2} & \dfrac{12EI}{b^3} & \dfrac{6EI}{b^2} \\ 0 & 0 & -\dfrac{6EI}{b^2} & \dfrac{2EI}{b} & \dfrac{6EI}{b^2} & \dfrac{4EI}{b} \end{bmatrix} \begin{matrix} 1 \\ \\ 2 \\ \\ 3 \\ \\ \end{matrix}$$

The displacement boundary conditions for this problem are

$$D_{Y1} = D_{Y3} = 0$$

Introducing these boundary conditions and the loading $F_{Y2} = -Q$ yields the set†

$$
\begin{bmatrix}
\dfrac{4EI}{a} & \dfrac{6EI}{a^2} & \dfrac{2EI}{a} & 0 \\[2mm]
\dfrac{6EI}{a^2} & 12EI\left(\dfrac{1}{a^3} + \dfrac{1}{b^3}\right) & 6EI\left(\dfrac{1}{a^2} + \dfrac{1}{b^2}\right) & -\dfrac{6EI}{b^2} \\[2mm]
\dfrac{2EI}{a} & 6EI\left(\dfrac{1}{a^2} - \dfrac{1}{b^2}\right) & 4EI\left(\dfrac{1}{a} + \dfrac{1}{b}\right) & \dfrac{2EI}{b} \\[2mm]
0 & -\dfrac{6EI}{b^2} & \dfrac{2EI}{b} & \dfrac{4EI}{b}
\end{bmatrix}
\begin{bmatrix}
\theta_1 \\[2mm] D_{Y2} \\[2mm] \theta_2 \\[2mm] \theta_3
\end{bmatrix}
=
\begin{bmatrix}
0 \\[2mm] -Q \\[2mm] 0 \\[2mm] 0
\end{bmatrix}
\qquad (9.42)
$$

This can be solved for the unknown displacements.

If $a = b = \frac{1}{2}$ and $l = 1.0$, the resulting displacements are

$$\theta_1 = -\theta_3 = \frac{3}{48}\frac{Q}{EI} \qquad \text{and} \qquad \theta_2 = 0$$

and

$$D_{Y2} = -\frac{1}{48}\frac{Q}{EI}$$

To compute the reactions and the internal stress resultants, the displacement values would be substituted back into each of the equilibrium equations. For example, $_1M_2$ (that is, M_2 in element 1) would be determined as follows:

$$
\begin{bmatrix}
[_1B_{11}] & [_1B_{12}] \\
[_1B_{21}] & [_1B_{22}]
\end{bmatrix}
\begin{bmatrix}
D_1 \\ D_2
\end{bmatrix}
=
\begin{bmatrix}
_1F_1 \\ _1F_2
\end{bmatrix}
$$

or

$$[_1B_{21}][D_1] + [_1B_{22}][D_2] = [_1F_2]$$

Therefore

$$
\begin{bmatrix}
-\dfrac{12EI}{a^3} & \dfrac{6EI}{a^2} \\[2mm]
-\dfrac{6EI}{a^2} & \dfrac{2EI}{a}
\end{bmatrix}
\begin{bmatrix}
D_{Y1} \\[2mm] \theta_1
\end{bmatrix}
+
\begin{bmatrix}
\dfrac{12EI}{a^3} & \dfrac{6EI}{a^2} \\[2mm]
\dfrac{6EI}{a^2} & \dfrac{4EI}{a}
\end{bmatrix}
\begin{bmatrix}
D_{Y2} \\[2mm] \theta_2
\end{bmatrix}
=
\begin{bmatrix}
_1F_{Y2} \\[2mm] _1M_2
\end{bmatrix}
$$

---

† This reduction can be accomplished either by rearranging rows and columns and partitioning or by striking rows and columns associated with the given boundary conditions.

Since $D_{Y1} = 0$,

$$_1M_2 = \frac{2EI}{a}\theta_1 + \frac{6EI}{a^2}D_{Y2} + \frac{4EI}{a}\theta_2$$

or

$$_1M_2 = -\frac{1}{4}Q$$

Example 9.7 A square, uniform cross section rigid frame (shown in fig. 9.21) is subjected to diagonally opposite forces Q. It is desired to determine the deformations of the structure as a function of the applied load.

The arrangement of nodes and elements to be used is given in fig. 9.22. Also indicated is the presumed reference coordinate system. This particular system is chosen because of the ease with which the displacement boundary conditions can be imposed, and because of the symmetry of the structure and loading about the selected x axis. (Owing to this symmetry, the half-frame, from nodes 1-2-3, carries one-half the applied load Q.)

The stiffness matrix associated with the element loading shown in fig. 9.18 and given in eq. (9.39) presumes no axial thrust in the element—only end-bending moments and end shears. The presence of axial compressive forces, however, can markedly influence bending stiffnesses (see fig. 8.13). But this is not the only problem, as will be demonstrated in the following.

Since the local coordinates for each of the elements do not agree with the reference directions, there first will be developed the transformation matrices. The tabular arrangement used for the previous problems also will be here employed.

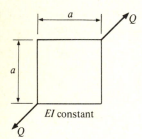

Figure 9.21

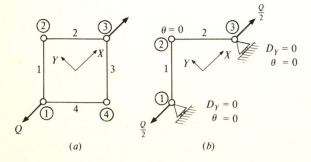

(a)   (b)

Figure 9.22 Structure and loading assumed for analysis.

Table 9.5 Elements of transformation matrix

Member	Node i	Node j	$\theta_{\bar{x}X}$	$\theta_{\bar{x}Y}$	$\theta_{\bar{y}X}$	$\theta_{\bar{y}Y}$	$a_{\bar{x}X}$	$a_{\bar{x}Y}$	$a_{\bar{y}X}$	$a_{\bar{y}Y}$
1	1	2	45°	45°	135°	45°	0.707	0.707	−0.707	0.707
2	2	3	45°	135°	45°	45°	0.707	−0.707	0.707	0.707

The transformation submatrices $[_l\Lambda]$ are

Element 1:
$$[_1\Lambda] = \begin{bmatrix} 0.707 & 0.707 & 0 \\ -0.707 & 0.707 & 0 \\ 0 & 0 & 1 \end{bmatrix}$$

Element 2:
$$[_2\Lambda] = \begin{bmatrix} 0.707 & -0.707 & 0 \\ 0.707 & 0.707 & 0 \\ 0 & 0 & 1 \end{bmatrix}$$

Each of the element stiffness matrices are given by

$$[_l K] = [_l T_R^T][_l \bar{K}][_l T_R] = \begin{bmatrix} [_l\Lambda^T] & 0 \\ \hline 0 & [_l\Lambda^T] \end{bmatrix} \begin{bmatrix} [_l\bar{B}_{11}] & [_l\bar{B}_{12}] \\ \hline [_l\bar{B}_{21}] & [_l\bar{B}_{22}] \end{bmatrix} \begin{bmatrix} [_l\Lambda] & 0 \\ \hline 0 & [_l\Lambda] \end{bmatrix}$$

$$= \begin{bmatrix} [_lB_{11}] & [_lB_{12}] \\ \hline [_lB_{21}] & [_lB_{22}] \end{bmatrix} \begin{matrix} i \\ \\ j \end{matrix}$$

The master matrix is assembled by locating the submatrices $[_l B_{ij}]$ for each element in its proper node row and node column. For this problem

$$[K] = \begin{bmatrix} [_1B_{11}] & [_1B_{12}] & 0 \\ \hline [_1B_{21}] & [_1B_{22}] + [_2B_{11}] & [_2B_{12}] \\ \hline 0 & [_2B_{21}] & [_2B_{22}] \end{bmatrix} \begin{matrix} 1 \\ \\ 2 \\ \\ 3 \end{matrix}$$

In the master equilibrium formulation,

$$[K] \begin{bmatrix} [D_1] \\ [D_2] \\ [D_3] \end{bmatrix} = \begin{bmatrix} [F_1] \\ [F_2] \\ [F_3] \end{bmatrix}$$

where

$$[D_1] = \begin{bmatrix} D_{X1} \\ D_{Y1} \\ \theta_1 \end{bmatrix} \qquad [D_2] = \begin{bmatrix} D_{X2} \\ D_{Y2} \\ \theta_2 \end{bmatrix} \qquad \text{and} \qquad [D_3] = \begin{bmatrix} D_{X3} \\ D_{Y3} \\ \theta_3 \end{bmatrix}$$

and

$$[F_1] = \begin{bmatrix} -\frac{1}{2}Q \\ F_{Y1} \\ M_1 \end{bmatrix} \qquad [F_2] = \begin{bmatrix} 0 \\ 0 \\ 0 \end{bmatrix} \qquad \text{and} \qquad [F_3] = \begin{bmatrix} \frac{1}{2}Q \\ F_{Y3} \\ M_3 \end{bmatrix}$$

As indicated in fig. 9.22b, the displacement boundary conditions are

$$D_{Y1} = \theta_1 = \theta_2 = D_{Y3} = \theta_3 = 0$$

Adding to these the conditions for symmetry,

$$D_{X1} = -D_{X3} \qquad \text{and} \qquad D_{X2} = 0$$

the equilibrium formulation is

$$\frac{EI}{a^3} \begin{bmatrix} 6 & 6 \\ 12 & 12 \end{bmatrix} \begin{bmatrix} D_{X1} \\ D_{Y2} \end{bmatrix} = \begin{bmatrix} -\frac{1}{2}Q \\ 0 \end{bmatrix}$$

It is to be recognized, however, that the coefficient matrix is *singular;* that is, its determinant equals zero. Therefore, for the problem as defined, no unique solution exists.

This singularity can be easily explained by considering more carefully the manner in which the load is transmitted from one member to the other member in the frame. For example, end shear in member 2-3 will manifest itself as an axial thrust in member 1-2. Both elements 1-2 and 2-3 must, therefore, possess both axial and bending stiffnesses—thus the reason for the geometric instability indicated by the matrix singularity.

In the previous section, in a general reference system, it was determined that the stiffness matrix for an axial load system [eq. (9.32)] is

$$[K]_{\text{axial}} = \frac{AE}{L} \begin{bmatrix} [B] & \vdots & [-B] \\ \hdashline [-B] & \vdots & [B] \end{bmatrix} \qquad \text{where } [B] = \begin{bmatrix} a_{\bar{X}X}^2 & a_{\bar{X}X}a_{\bar{X}Y} & 0 \\ a_{\bar{X}X}a_{\bar{X}Y} & a_{\bar{X}Y}^2 & 0 \\ 0 & 0 & 0 \end{bmatrix}$$

The a's are the same direction cosines that were used for the bending transformations. Including these axial stiffnesses values in the master formulation, the equilibrium equations become

$$\begin{bmatrix} \dfrac{6EI}{a^2} + \dfrac{AE}{2a} & \dfrac{6EI}{a^3} - \dfrac{AE}{2a} \\[2ex] \dfrac{12EI}{a^3} - \dfrac{AE}{a} & \dfrac{12EI}{a^3} + \dfrac{AE}{a} \end{bmatrix} \begin{bmatrix} D_{X1} \\[2ex] D_{Y2} \end{bmatrix} = \begin{bmatrix} -\frac{1}{2}Q \\[2ex] 0 \end{bmatrix} \qquad (9.43)$$

which can be solved, and yields the deflections

$$D_{X1} = -\frac{Qa^3}{48EI}\left(1 + \frac{12I}{a^2 A}\right)$$

$$D_{Y2} = -\frac{Qa^3}{48EI}\left(1 - \frac{12I}{a^2 A}\right)$$

The effect of axial deformations now can be removed from the solution by presuming A equals infinity. For this case

$$D_{X1} = D_{Y2} = -\frac{Qa^2}{48EI}$$

which is the same solution that would be obtained by any of the methods discussed in chaps. 4, 5, and 8.

A solution obtained in the above example—where it is assumed that axial load has associated with it an axial stiffness, but that the bending and shear stiffnesses are independent of thrust—is at best an approximation to the true situation. For relatively small values of P the approximation may be adequate; however, as P increases, the error becomes increasingly larger.

Figure 9.23 Beam element subjected to end-bending moments and axial thrust.

Beam bending including the influence of axial thrust The two-dimensional element in question is shown in fig. 9.23. The local stiffness matrix $[\bar{K}]$, when a net *compressive axial force* exists in the member, is given by

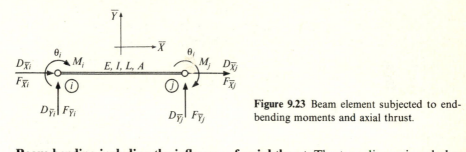

$$(9.44)$$

where the α terms are the axial-load modifying factors (derived in chap. 8),† and are given by

$$\alpha_1 = \frac{3\psi}{4\psi^2 - \phi^2} \qquad\qquad \alpha_2 = \frac{3\phi}{4\psi^2 - \phi^2}$$

$$\alpha_3 = \frac{1}{2\psi - \phi} \qquad\qquad \alpha_4 = \frac{1}{2\psi - \phi} - \frac{(2u)^2}{12} \qquad\qquad (9.45)$$

$$\psi = \frac{3}{2u}\left(\frac{1}{2u} - \frac{1}{\tan 2u}\right) \qquad\qquad \phi = \frac{3}{u}\left(\frac{1}{\sin 2u} - \frac{1}{2u}\right)$$

and

$$u = \frac{L}{2}\sqrt{\frac{P}{EI}} = \frac{\pi}{2}\sqrt{\frac{P}{P_e}}$$

where P_e is the Euler buckling load.

If the net axial force on the member is a *tensile force*, the local stiffness matrix is

$$[\bar{K}] = \begin{bmatrix}
\dfrac{AE}{L} & 0 & 0 & -\dfrac{AE}{L} & 0 & 0 \\[2ex]
0 & \dfrac{12EI}{L^3}\beta_4 & -\dfrac{6EI}{L^2}\beta_3 & 0 & -\dfrac{12EI}{L^3}\beta_4 & -\dfrac{6EI}{L^2}\beta_3 \\[2ex]
0 & -\dfrac{6EI}{L^2}\beta_3 & \dfrac{4EI}{L}\beta_1 & 0 & \dfrac{6EI}{L^2}\beta_3 & \dfrac{2EI}{L}\beta_2 \\[2ex]
-\dfrac{AE}{L} & 0 & 0 & \dfrac{AE}{L} & 0 & 0 \\[2ex]
0 & -\dfrac{12EI}{L^3}\beta_4 & \dfrac{6EI}{L^2}\beta_3 & 0 & \dfrac{12EI}{L^3}\beta_4 & \dfrac{6EI}{L^2}\beta_3 \\[2ex]
0 & -\dfrac{6EI}{L^2}\beta_3 & \dfrac{2EI}{L}\beta_2 & 0 & \dfrac{6EI}{L^2}\beta_2 & \dfrac{4EI}{L}\beta_1
\end{bmatrix}
\begin{matrix}
D_{\bar{X}i} \\[2ex] D_{\bar{Y}i} \\[2ex] \theta_i \\[2ex] D_{\bar{X}j} \\[2ex] D_{\bar{Y}j} \\[2ex] \theta_j
\end{matrix}$$

$$\begin{matrix} D_{\bar{X}i} & D_{\bar{Y}i} & \theta_i & D_{\bar{X}j} & D_{\bar{Y}j} & \theta_i \end{matrix}$$

$$(9.46)$$

---

† These bending and shear stiffness terms are from eq. (8.26), (8.27), (8.28), (8.29), and (8.30).

where

$$\beta_1 = \frac{3\psi^+}{4(\psi^+)^2 - (\phi^+)^2} \qquad \beta_2 = \frac{3\phi^+}{4(\psi^+)^2 - (\phi^+)^2}$$

$$\beta_3 = \frac{1}{2\psi^+ - \phi^+} \qquad \beta_4 = \frac{1}{2\psi^+ - \phi^+} + \frac{2u^2}{12} \qquad (9.47)$$

$$\psi^+ = \frac{3}{2u}\left(\frac{1}{\tanh 2u} - \frac{1}{2u}\right) \qquad \phi^+ = \frac{3}{u}\left(\frac{1}{2u} - \frac{1}{\sinh 2u}\right)$$

and

$$u = \frac{L}{2}\sqrt{\frac{P}{EI}} = \frac{\pi}{2}\sqrt{\frac{P}{P_e}}$$

It is to be recognized that with the exception of the β's as opposed to the α's, the element stiffness matrix (9.46) is identical to (9.45). Moreover, with the exception of ψ^+ for ψ and ϕ^+ for ϕ, eq. (9.47) are the same as (9.45). In defining u, the numerical values of the Euler buckling load is used in both cases as a reference quantity— independent of whether P is in tension or in compression. (Values for ψ, ϕ, ψ^+, and ϕ^+ are tabulated in the Appendix for various values of P/P_e.)

Polynomial approximations to the above trigonometric functions $\alpha_1, \ldots, \alpha_4$ and $\beta_1, \ldots, \beta_4$ can be developed. Using a least-squares type of curve-fitting approximation, assuming consideration of six points and a range of interest of $0 \leq |P/P_e| \leq 1.20$, the second-order polynomials summarized in table 9.6 are obtained.

Table 9.6 Second-order polynomial approximations to stiffness axial-force multipliers

Compression	Tension
$\alpha_1 = 1.0000 - 0.3194\dfrac{P}{P_e} - 0.0650\left(\dfrac{P}{P_e}\right)^2$	$\beta_1 = 1.0000 + 0.3184\dfrac{P}{P_e} - 0.0253\left(\dfrac{P}{P_e}\right)^2$
$\alpha_2 = 1.0000 + 0.1488\dfrac{P}{P_e} + 0.0872\left(\dfrac{P}{P_e}\right)^2$	$\beta_2 = 1.0000 - 0.1583\dfrac{P}{P_e} + 0.0328\left(\dfrac{P}{P_e}\right)^2$
$\alpha_3 = 1.0000 - 0.1638\dfrac{P}{P_e} - 0.0138\left(\dfrac{P}{P_e}\right)^2$	$\beta_3 = 1.0000 + 0.1640\dfrac{P}{P_e} - 0.0101\left(\dfrac{P}{P_e}\right)^2$
$\alpha_4 = 1.0000 - 0.9853\dfrac{P}{P_e} - 0.0148\left(\dfrac{P}{P_e}\right)^2$	$\beta_4 = 1.0000 + 0.9853\dfrac{P}{P_e} - 0.0102\left(\dfrac{P}{P_e}\right)^2$

Example 9.8 As an example of the solution procedure when the influence of axial forces on bending stiffness is included, consider the one-story, two-bay rigid frame shown in fig. 9.24. The beams have cross-sectional areas of A_b and moments of inertia of $2I$. Correspondingly, the column areas are A_c, and their

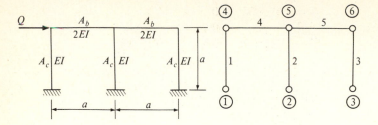

Figure 9.24

moments of inertia are I. The same material is used throughout (that is, E is the same for all members). Shown in the second sketch of the figure are the assumed numbers of the various elements and nodes.

The element stiffness matrices correspond to either eq. (9.44) or eq. (9.46), depending on whether the axial force is in compression or in tension. This presents somewhat of a problem, however, since the magnitude and direction of the various deformations must be known before there can be made any real assessment of the axial forces in the various members. Thus, in effect, the solution process cannot be started until a solution is known. To resolve this dilemma, an iterative approach of the type used in chap. 8 will be employed.

For this example problem it will be assumed—as an initial set of starting values—that the axial forces in elements 1, 2, and 3 are zero, and that in elements 4 and 5 the axial forces are compressive and equal $-\frac{1}{2}Q$ and $-\frac{1}{4}Q$, respectively. Using these values the structure can be analyzed, and new values for the various axial loads can be obtained.

The geometry associated with the various elements in the structure in question is indicated in table 9.7.

Table 9.7 Geometrical properties for example 9.8

Member	Node i	Node j	Length	Unit bending rigidity	Area	$\theta_{\bar{X}X}$	$\theta_{\bar{X}Y}$	$\theta_{\bar{Y}X}$	$\theta_{\bar{Y}Y}$	$a_{\bar{X}X}$	$a_{\bar{X}Y}$	$a_{\bar{Y}X}$	$a_{\bar{Y}Y}$
1	1	4	a	EI	A_c	90°	0°	180°	90°	0	1.000	−1.000	0
2	2	5	a	EI	A_c	90°	0°	180°	90°	0	1.000	−1.000	0
3	3	6	a	EI	A_c	90°	0°	180°	90°	0	1.000	−1.000	0
4	4	5	a	$2EI$	A_b	0°	90°	90°	0°	1.000	0	0	1.000
5	5	6	a	$2EI$	A_b	0°	90°	90°	0°	1.000	0	0	1.000

The transformation matrices are, therefore,

$$\textit{For the columns:} \quad [\Lambda_c] = \begin{bmatrix} 0 & 1.000 & 0 \\ -1.000 & 0 & 0 \\ 0 & 0 & 1.000 \end{bmatrix}$$

For the beams:
$$[\Lambda_b] = \begin{bmatrix} 1.000 & 0 & 0 \\ 0 & 1.000 & 0 \\ 0 & 0 & 1.000 \end{bmatrix} = [I]$$

Next, the presumed axial forces and the modification factors α or β are determined.

Table 9.8 Axial-force multipliers for first cycle, example 9.8

Member	Cycle	Axial load	P_e	$\dfrac{P}{P_e}$	ψ's	ϕ's	α's	β's
1	1	0	$\dfrac{\pi^2 EI}{a^2}$	0	1.0000	1.0000	1.0000	1.0000
2	1	0	$\dfrac{\pi^2 EI}{a^2}$	0	1.0000	1.0000	1.0000	1.0000
3	1	0	$\dfrac{\pi^2 EI}{a^2}$	0	1.0000	1.0000	1.0000	1.0000
4	1	$-\dfrac{1}{2}Q$	$\dfrac{2\pi^2 EI}{a^2}$	$-\dfrac{1}{4}\dfrac{Q}{P_e}$	$_4\psi^1$	$_4\phi^1$	$_4\alpha^1$	†
5	1	$-\dfrac{1}{4}Q$	$\dfrac{2\pi^2 EI}{a^2}$	$-\dfrac{1}{8}\dfrac{Q}{P_e}$	$_5\psi^1$	$_5\phi^1$	$_5\alpha^1$	†

† Since the assumed thrusts are in compression, there will be no values for the β's for cycle 1.

The notation used here is

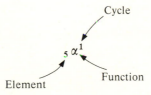

Having the first set of axial-load modification factors, and the transformation matrices, the master stiffness matrix can be constructed. The submatrix notation that will be used is

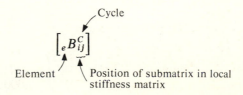

The master formulation is then

$$
\begin{bmatrix}
[_1B^1_{11}] & 0 & 0 & [_1B^1_{12}] & 0 & 0 \\
0 & [_2B^1_{11}] & 0 & 0 & [_2B^1_{12}] & 0 \\
0 & 0 & [_3B^1_{11}] & 0 & 0 & [_3B^1_{12}] \\
[_1B^1_{21}] & 0 & 0 & [_1B^1_{22}]+[_4B^1_{11}] & [_4B^1_{12}] & 0 \\
0 & [_2B^1_{21}] & 0 & [_4B^1_{21}] & [_2B^1_{22}]+[_4B^1_{22}]+[_5B^1_{11}] & [_5B^1_{12}] \\
0 & 0 & [_3B^1_{21}] & 0 & [_5B^1_{21}] & [_3B^1_{22}]+[_5B^1_{22}]
\end{bmatrix}
\begin{bmatrix}
D^1_{X1}\\D^1_{Y1}\\\theta^1_1\\ \hline D^1_{X2}\\D^1_{Y2}\\\theta^1_2\\ \hline D^1_{X3}\\D^1_{Y3}\\\theta^1_3\\ \hline D^1_{X4}\\D^1_{Y4}\\\theta^1_4\\ \hline D^1_{X5}\\D^1_{Y5}\\\theta^1_5\\ \hline D^1_{X6}\\D^1_{Y6}\\\theta^1_6
\end{bmatrix}
=
\begin{bmatrix}
F_{X1}\\F_{Y1}\\M_1\\ \hline F_{X2}\\F_{Y2}\\\theta_2\\ \hline F_{X3}\\F_{Y3}\\\theta_3\\ \hline F_{X4}\\F_{Y4}\\\theta_4\\ \hline F_{X5}\\F_{Y5}\\\theta_5\\ \hline F_{X6}\\F_{Y6}\\\theta_6
\end{bmatrix}
$$

(columns: 1 $D_{X1}\;D_{Y1}\;\theta_1$, 2 $D_{X2}\;D_{Y2}\;\theta_2$, 3 $D_{X3}\;D_{Y3}\;\theta_3$, 4 $D_{X4}\;D_{Y4}\;\theta_4$, 5 $D_{X5}\;D_{Y5}\;\theta_5$, 6 $D_{X6}\;D_{Y6}\;\theta_6$)

All displacements at nodes 1, 2, and 3 are defined as equaling zero. The set to be solved is, therefore

$$
\begin{bmatrix}
[_1B^1_{22}]+[_4B^1_{11}] & [_4B^1_{12}] & 0 \\
[_4B^1_{22}] & [_2B^1_{22}]+[_4B^1_{22}]+[_5B^1_{11}] & [_5B^1_{12}] \\
0 & [_5B^1_{21}] & [_3B^1_{22}]+[_5B^1_{22}]
\end{bmatrix}
\begin{bmatrix}
D^1_{X4}\\D^1_{Y4}\\\theta^1_4\\ \hline D^1_{X5}\\D^1_{Y5}\\\theta^1_5\\ \hline D^1_{X6}\\D^1_{Y6}\\\theta^1_6
\end{bmatrix}
=
\begin{bmatrix}
Q\\0\\0\\ \hline 0\\0\\0\\ \hline 0\\0\\0
\end{bmatrix}
$$

(columns: $D_{X4}\;D_{Y4}\;\theta_4$, $D_{X5}\;D_{Y5}\;\theta_5$, $D_{X6}\;D_{Y6}\;\theta_6$)

Once the D's and θ's are known, the internal stress resultants for each element can be determined.

$$
\begin{array}{cc}
i & j
\end{array}
$$
$$
\begin{bmatrix}
[_eB^1_{11}] & \vdots & [_eB^1_{12}] \\
\hdotsfor{3} \\
[_eB^1_{21}] & \vdots & [_eB^1_{22}]
\end{bmatrix}
\begin{bmatrix}
[D^1_i] \\
\hdotsfor{1} \\
[D^1_j]
\end{bmatrix}
=
\begin{bmatrix}
[_eF^1_i] \\
\hdotsfor{1} \\
[_eF^1_j]
\end{bmatrix}
$$

where

$$
[D^1_i] = \begin{bmatrix} D^1_{Xi} \\ D^1_{Yi} \\ \theta^1_i \end{bmatrix}
\quad
[D^1_j] = \begin{bmatrix} D^1_{Xj} \\ D^1_{Yj} \\ \theta^1_j \end{bmatrix}
\quad
[_eF^1_i] = \begin{bmatrix} _eF^1_{Xi} \\ _eF^1_{Yi} \\ _eM^1_i \end{bmatrix}
\quad \text{and} \quad
[_eF^1_j] = \begin{bmatrix} _eF^1_{Xj} \\ _eF^1_{Yj} \\ _eM^1_j \end{bmatrix}
$$

The stress resultants at node 4, in element 4, for example, are given by

$$
[_4F^1_4] = [_4B^1_{11}][D^1_4] + [_4B^1_{12}][D^1_5]
$$

$$
\text{and} \quad [_4F^1_4] = \begin{bmatrix} _4F^1_{X4} \\ _4F^1_{Y4} \\ _4M^1_4 \end{bmatrix}
$$

Following this computation there would be made a comparison between these values and the presumed axial forces. (For element 4, comparison would be between the assumed $-\frac{1}{2}Q$ and the calculated $_4F^1_{X4}$.) If the difference is too large, a new set of computations must be carried out, using as the assumed axial thrust values those computed at the end of cycle 1. This process would be repeated until the desired accuracy is achieved.

For $Q = 400$ kips, $a = 15$ ft, $E = 30 \times 10^3$ kips, $I = 225$ in^4, $A_b = 10.6$ in^2, and $A_c = 17.0$ in^2, table 9.9 summarizes the assumed and computed axial thrust values for several cycles of iteration. In these calculations the stiffness values α and β are computed using the exact trigonometric functions listed in eqs. (9.45) and (9.47).

Table 9.9 Assumed and computed axial-thrust values for several cycles of iteration, example 9.8

	Assumed					Computed				
Cycle	P_1, kips	P_2, kips	P_3, kips	P_4, kips	P_5, kips	P_1, kips	P_2, kips	P_3, kips	P_4, kips	P_5, kips
1	0	0	0	−200	−100	+93.7	−1.57	−92.2	−275	−123
2	+93.7	−1.57	−92.2	−275	−123	+96.9	−2.04	−94.8	−277	−120
3	+96.9	−2.04	−94.8	−277	−120	+97.0	−2.06	−94.9	−277	−120
4	+97.0	−2.06	−94.9	−277	−120	+97.0	−2.06	−94.9	−277	−120

Note: +, tension; −, compression.

9.4 MIXED ELEMENTS

In many instances, it is necessary to model a structure using more than one kind of element. Consider, for example, the king-post truss shown in fig. 9.25. Elements 1 and 2 are beam elements. Elements 3, 4, and 5 are truss—or axial-load—elements.

It is to be recognized that at nodes 1, 2, and 3, three degrees of freedom are required for each connecting element. The beam elements satisfy this condition. The axial-load elements, however, do not—they involve no reference to a rotational deformation θ. The stiffness matrix for the bars, therefore, must be modified to include this deformation. This is easily accomplished by adding two rows and two columns of zeros to the local stiffness matrix. For the bar elements, then,

$$
[\bar{K}] =
\begin{array}{c}
 \\

\end{array}
\overset{\displaystyle i \qquad\qquad\qquad j}{
\left[
\begin{array}{ccc:ccc}
\dfrac{AE}{L} & 0 & 0 & -\dfrac{AE}{L} & 0 & 0 \\[2mm]
0 & 0 & 0 & 0 & 0 & 0 \\[1mm]
0 & 0 & 0 & 0 & 0 & 0 \\ \hdashline
-\dfrac{AE}{L} & 0 & 0 & \dfrac{AE}{L} & 0 & 0 \\[2mm]
0 & 0 & 0 & 0 & 0 & 0 \\[1mm]
0 & 0 & 0 & 0 & 0 & 0
\end{array}
\right]}
\begin{array}{c}
\\[2mm] i \\[6mm] \\[2mm] j \\[6mm]
\end{array}
\qquad (9.48)
$$

The general transformation matrix for the truss elements is the same as that used for the beam elements. Thus, the bar stiffness in the reference system is

$$
[K] =
\overset{\displaystyle i \qquad\quad j}{
\left[
\begin{array}{c:c}
[B_{11}] & [B_{12}] \\ \hdashline
[B_{21}] & [B_{22}]
\end{array}
\right]}
\begin{array}{c}
i \\[4mm] j
\end{array}
$$

where

$$
[B_{11}] = [B_{22}] = [-B_{12}] = [B_{21}]
$$

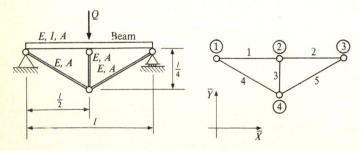

Figure 9.25 King-post truss.

and

$$[B_{11}] = \frac{AE}{L} \begin{bmatrix} a_{\bar{X}X}^2 & a_{\bar{X}X} a_{\bar{X}Y} & 0 \\ a_{\bar{X}X} a_{\bar{X}Y} & a_{\bar{X}Y}^2 & 0 \\ 0 & 0 & 0 \end{bmatrix}$$

The beam-element stiffness matrices are the same as those previously defined. The master formulation for this example is

$$\begin{bmatrix} [_1B_{11}] + [_4B_{11}] & [_1B_{12}] & 0 & [_4B_{12}] \\ [_1B_{21}] & [_1B_{22}] + [_2B_{11}] + [_3B_{11}] & [_2B_{12}] & [_3B_{12}] \\ 0 & [_2B_{21}] & [_2B_{22}] + [_5B_{11}] & [_5B_{12}] \\ [_4B_{21}] & [_3B_{21}] & [_5B_{21}] & [_3B_{22}] + [_4B_{22}] + [_5B_{22}] \end{bmatrix} \begin{bmatrix} [D_1] \\ [D_2] \\ [D_3] \\ [D_4] \end{bmatrix} = \begin{bmatrix} [F_1] \\ [F_2] \\ [F_3] \\ [F_4] \end{bmatrix}$$

where the $[D]$'s and the $[F]$'s, including the boundary conditions and the reactions, are

$$[D_1] = \begin{bmatrix} 0 \\ 0 \\ \theta_1 \end{bmatrix} \quad [D_2] = \begin{bmatrix} D_{X2} \\ D_{Y2} \\ \theta_2 \end{bmatrix} \quad [D_3] = \begin{bmatrix} D_{X3} \\ 0 \\ \theta_3 \end{bmatrix} \quad [D_4] = \begin{bmatrix} D_{X4} \\ D_{Y4} \\ \theta_4 \end{bmatrix}$$

$$[F_1] = \begin{bmatrix} R_{X1} \\ R_{Y1} \\ 0 \end{bmatrix} \quad [F_2] = \begin{bmatrix} 0 \\ -Q \\ 0 \end{bmatrix} \quad [F_3] = \begin{bmatrix} 0 \\ R_{Y3} \\ 0 \end{bmatrix} \quad [F_4] = \begin{bmatrix} 0 \\ 0 \\ 0 \end{bmatrix}$$

Striking the rows and columns associated with D_{X1}, D_{Y1}, and D_{Y3}, the remaining equations can be solved for the unknown components of displacement at the various nodes.

9.5 DISTRIBUTED LOADING

In the development of both the axial load and beam elements, it was stated at the outset that all loads were assumed to be applied to the structure at the node points. Neglecting the weights of the members in question, this assumption is a good one for truss-type members. For beams, such is not the case. However, where a single, concentrated lateral load was applied to the member in question, this presents no real difficulty, as was demonstrated in example 9.6. All that is required for solution is the introduction of an additional node at the point of load application. When distributed loads are involved, however, a new approach must be devised.

One method that is frequently used to take into account this effect is to substitute a statically equivalent set of concentrated loads or moments for the distributed load, and proceed as before. It is to be recognized that the use of such a procedure leads to approximate results, and the quality of those approximations depends on the number and placement of the assumed concentrated loads. Consider, for example, the simple beam shown in fig. 9.26. The loading in question is a uniformly distributed lateral load q, and the span length is L. The actual bending-moment diagram is the parabola shown in sketch a of the figure. If it is assumed that the entire beam is to be one

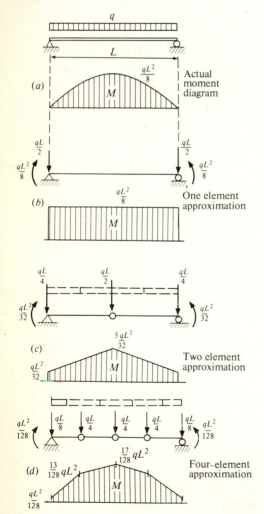

Figure 9.26 Approximations for distributed loads.

element (that is, no node points along the length), the concentrated loads or moments, which must be applied at the node points at the ends of the element will be those shown in sketch *b*. (The basic object is to determine a set of loads that are statically equivalent to the qL distributed load. However, concentrated end moments also are required to ensure that the resulting moment diagram—for the approximation—everywhere along the member is greater than or equal to the true moment diagram; thus, the need for the end moments $\frac{1}{8}qL^2$.) If the beam is sub-divided into two parts (that is, composed of two elements), the equivalent loading and moment diagram would be that shown in sketch *c*. (Note that a decision has been made in this case—that concentrated moments are to be introduced only at the ends of the member, not necessarily at all of the element end points.) To circumscribe the actual bending moment diagram (*a*), the matching of the actual and the approxi-mate moment values will occur only at the quarter-points of the beam. The approxi-mate moment diagram will be greater at all other points. In sketch *d* there is shown the equivalent loading system for a four-element assumption. Again, the resulting moment diagram everywhere circumscribes the actual moment diagram.

While the above-described method is a simple and straightforward way to repre-sent a distributed load—to almost any degree of accuracy desired—it has one very serious drawback. This is the increased computer storage required for the extra nodes. For many larger problems, it is the limited availability of computer storage capacity that determines the size of problem that can be handled. Obviously, when this is the case, this procedure for handling distributed loads leaves much to be desired.

A second method for taking into account distributed loads presumes that the final deformed shape and stress resultants associated with a given loading can be defined as the sum of two independent solutions. The first of these consists of the beam and distributed loading, along with added end moments and shears of sufficient magnitude *to cause zero relative deflection and slope* at the ends of the loaded member. This is shown as case I in fig. 9.27. (For this beam and loading, the required shear is $\frac{1}{2}ql$, and the end moments are $\frac{1}{12}ql^2$.)

Case II consists of all of the applied loading *other than* the distributed loading, plus a set of shears and end moments equal and opposite in sense to those assumed in case I. The sum of cases I and II corresponds to the actual solution in question. For this example, the equal end-rotation angles θ are

$$\theta = \frac{ql^3}{24EI}$$

They are in the direction shown.

In general, the required deformations are given by the matrix equations

$$[K][D] = [F_A] + [F_F]$$

where $[F_A]$ = column matrix of applied loads—other than the distributed load
$[F_F] = (-1)$ times the fixed-end moments and shear matrix

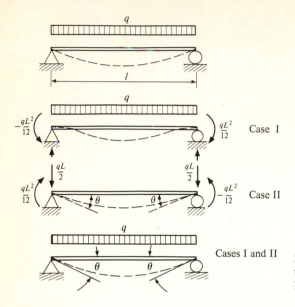

Figure 9.27 Alternative method for taking into account distributed loads.

The vector $[F_A] + [F_F]$ corresponds to the case II loading just described. The actual stress resultants are the sums of those from cases I and II. (Note the similarity between this process and the slope-deflection method developed in chap. 8.)

Example 9.9 Consider the two-span beam shown in fig. 9.28. Case I and case II loadings are shown on the element sketches to the right.

Since the deformations and stress resultants for case I are known, only case II will be considered here.

$$[K][D] = [F_A] + [F_F]$$

where†

$$[F_A] = \{R_{Y1} \quad 0 \quad -Q \quad 0 \quad R_{Y3} \quad 0 \quad R_{Y4} \quad 0\}$$

and

$$[F_F] = \left\{0 \quad 0 \quad 0 \quad 0 \quad -\frac{ql}{2} \quad \frac{ql^2}{12} \quad -\frac{ql}{2} \quad -\frac{ql^2}{12}\right\}$$

The boundary conditions are $D_{Y1} = D_{Y3} = D_{Y4} = 0$.

---

† To conserve space, these vectors have been shown horizontally, as opposed to their normal vertical form.

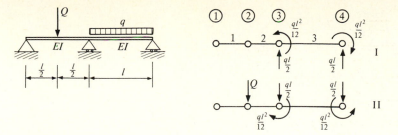

Figure 9.28

The set of equations to be solved is, therefore,

$$\frac{EI}{l}\begin{bmatrix} 8 & \dfrac{24}{l} & 4 & 0 & 0 \\[2mm] \dfrac{24}{l} & \dfrac{192}{l^2} & 0 & -\dfrac{24}{l} & 0 \\[2mm] 4 & 0 & 16 & 4 & 0 \\[2mm] 0 & -\dfrac{24}{l} & 4 & 12 & 2 \\[2mm] 0 & 0 & 0 & 2 & 4 \end{bmatrix}\begin{bmatrix} \theta_1 \\[2mm] D_{Y2} \\[2mm] \theta_2 \\[2mm] \theta_3 \\[2mm] \theta_4 \end{bmatrix} = \begin{bmatrix} 0 \\[2mm] -Q \\[2mm] 0 \\[2mm] 0 \\[2mm] 0 \end{bmatrix} + \begin{bmatrix} 0 \\[2mm] 0 \\[2mm] 0 \\[2mm] \dfrac{1}{12}ql^2 \\[2mm] -\dfrac{1}{12}ql^2 \end{bmatrix}$$

For $l = 10$ ft, $Q = 1$ kip and $q = 200$ lb/ft, and assuming that the beam is made of reinforced concrete ($E = 3 \times 10^6$ lb/in^2)[†] 12 in wide and 18 in deep ($I \cong 18^3$ in^4), the response matrix is

$$[D] = \frac{l}{EI}\{0.260 \quad -0.716 \quad -0.091 \quad 0.104 \quad -0.469\}$$

and $l/EI = 8.23 \times 10^{-6}$/kip·ft.

From that part of the master matrix not used in the deformation solutions, the case II reactions would be determined.

$$\frac{EI}{l}\begin{bmatrix} -2.4 & -0.96 & -2.4 & 0 & 0 \\ 0 & -0.96 & 2.4 & 1.8 & -0.6 \\ 0 & 0 & 0 & 0.6 & 0.6 \end{bmatrix}\frac{l}{EI}\begin{bmatrix} 0.260 \\ -0.716 \\ -0.091 \\ 0.104 \\ -0.469 \end{bmatrix} = \begin{bmatrix} F_{Y1} \\ F_{Y3} \\ F_{Y4} \end{bmatrix}$$

[†] The value used for E in this case is a rough approximation. For actual values see the appropriate ACI standard.

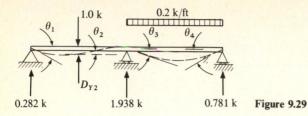

0.282 k   1.938 k   0.781 k   **Figure 9.29**

This yields

$$[F_R] = \begin{bmatrix} 0.282 \\ 0.938 \\ -0.219 \end{bmatrix}$$

The final reactions are the sums of those from cases I and II (See table 9.10).

Table 9.10 Summary of solutions, example 9.9

Reaction	F_{Y1}, kips	M_1, ft·kips	F_{Y3}, kips	M_3, ft·kips	F_{Y4}, kips	M_4, ft·kips
Case I	0	0	1.000	-1.667	1.000	1.667
Case II	0.282	0	0.938	1.667	-0.219	-1.667
Actual = sum	0.282	0	1.938	0	0.781	0

The results are shown in fig. 9.29.

Example 9.10 Consider the spring-supported structure and the corresponding model shown in fig. 9.30. There is applied to the beam a distributed vertical load of q over the entire $2l$ span. A concentrated load Q acts at the center of the beam, over the spring support. The spring constant for the elastic support at the center is k.

The case I loading is shown in fig. 9.31. The indicated stress resultants are those required to remove the node deformations resulting from the distributed load. Figure 9.32 illustrates the loading corresponding to case II. This loading consists of the concentrated load Q and a set of end shears and end moments

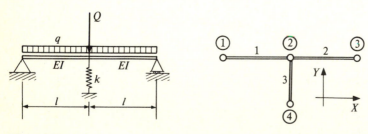

Figure 9.30

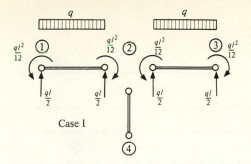

Case I

Figure 9.31 Case I loading for distributed load.

equal but opposite in direction to those of case I. For elements 1 and 2 the stiffness matrix is

$$
\begin{array}{cccc}
Y_i & \theta_i & Y_j & \theta_j
\end{array}
$$

$$
\begin{bmatrix}
\dfrac{12EI}{l^3} & -\dfrac{6EI}{l^2} & -\dfrac{12EI}{l^3} & -\dfrac{6EI}{l^2} \\[2ex]
-\dfrac{6EI}{l^2} & \dfrac{4EI}{l} & \dfrac{6EI}{l^2} & \dfrac{2EI}{l} \\[2ex]
-\dfrac{12EI}{l^3} & \dfrac{6EI}{l^2} & \dfrac{12EI}{l^3} & \dfrac{6EI}{l^2} \\[2ex]
-\dfrac{6EI}{l^2} & \dfrac{2EI}{l} & \dfrac{6EI}{l^2} & \dfrac{4EI}{l}
\end{bmatrix}
\begin{array}{c}
Y_i \\[2ex] \theta_i \\[2ex] Y_j \\[2ex] \theta_j
\end{array}
$$

For element 3,

$$
\begin{array}{cccc}
X_i & Y_i & X_j & Y_j
\end{array}
$$

$$
\begin{bmatrix}
0 & 0 & 0 & 0 \\
0 & k & 0 & -k \\
0 & 0 & 0 & 0 \\
0 & -k & 0 & k
\end{bmatrix}
\begin{array}{c}
X_i \\ Y_i \\ X_j \\ Y_j
\end{array}
$$

The master matrix for case II is then given by

$$
[K][D] = [F_A] + [F_F]
$$

The case II master formulation is shown in eq. (9.49).

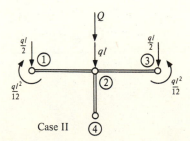

Case II

Figure 9.32 Case II loading for distributed load.

Matrix equation (9.49) — a symmetric 12×12 stiffness system with column headers $X_1,\,Y_1,\,\theta_1,\;X_2,\,Y_2,\,\theta_2,\;X_3,\,Y_3,\,\theta_3,\;X_4,\,Y_4,\,\theta_4$ and corresponding row labels:

$$
\begin{bmatrix}
0 & 0 & 0 & 0 & 0 & 0 & 0 & 0 & 0 & 0 & 0 & 0 \\[4pt]
0 & \dfrac{12EI}{l^3} & -\dfrac{6EI}{l^2} & 0 & -\dfrac{12EI}{l^3} & -\dfrac{6EI}{l^2} & 0 & 0 & 0 & 0 & 0 & 0 \\[6pt]
0 & -\dfrac{6EI}{l^2} & \dfrac{4EI}{l} & 0 & \dfrac{6EI}{l^2} & \dfrac{2EI}{l} & 0 & 0 & 0 & 0 & 0 & 0 \\[6pt]
0 & 0 & 0 & 0 & 0 & 0 & 0 & 0 & 0 & 0 & 0 & 0 \\[4pt]
0 & -\dfrac{12EI}{l^3} & \dfrac{6EI}{l^2} & 0 & \dfrac{24EI}{l^3} & 0 & 0 & -\dfrac{12EI}{l^3} & -\dfrac{6EI}{l^2} & 0 & 0 & 0 \\[6pt]
0 & -\dfrac{6EI}{l^2} & \dfrac{2EI}{l} & 0 & 0 & \dfrac{8EI}{l} & 0 & \dfrac{6EI}{l^2} & \dfrac{2EI}{l} & 0 & 0 & 0 \\[6pt]
0 & 0 & 0 & 0 & 0 & 0 & 0 & 0 & 0 & 0 & 0 & 0 \\[4pt]
0 & 0 & 0 & 0 & -\dfrac{12EI}{l^3} & \dfrac{6EI}{l^2} & 0 & \dfrac{12EI}{l^3}+k & \dfrac{6EI}{l^2} & 0 & -k & 0 \\[6pt]
0 & 0 & 0 & 0 & -\dfrac{6EI}{l^2} & \dfrac{2EI}{l} & 0 & \dfrac{6EI}{l^2} & \dfrac{4EI}{l} & 0 & 0 & 0 \\[6pt]
0 & 0 & 0 & 0 & 0 & 0 & 0 & 0 & 0 & 0 & 0 & 0 \\[4pt]
0 & 0 & 0 & 0 & 0 & 0 & 0 & -k & 0 & 0 & k & 0 \\[4pt]
0 & 0 & 0 & 0 & 0 & 0 & 0 & 0 & 0 & 0 & 0 & 0
\end{bmatrix}
\begin{bmatrix}
D_{X1} \\ D_{Y1} \\ \theta_1 \\ D_{X2} \\ D_{Y2} \\ \theta_2 \\ D_{X3} \\ D_{Y3} \\ \theta_3 \\ D_{X4} \\ D_{Y4} \\ \theta_4
\end{bmatrix}
=
\begin{bmatrix}
R_{X1} \\ R_{Y1} \\ 0 \\ 0 \\ -Q \\ 0 \\ R_{X3} \\ R_{Y3} \\ 0 \\ R_{X4} \\ R_{Y4} \\ 0
\end{bmatrix}
+
\begin{bmatrix}
0 \\ -\dfrac{ql}{2} \\ \dfrac{ql^2}{12} \\ 0 \\ -ql \\ 0 \\ 0 \\ -\dfrac{ql}{2} \\ \dfrac{ql^2}{12} \\ 0 \\ 0 \\ 0
\end{bmatrix}
\tag{9.49}
$$

The displacement boundary conditions are

$$D_{X1} = D_{Y1} = D_{X3} = D_{Y3} = D_{X4} = D_{Y4} = \theta_4 = 0$$

Also, since no axial shortening occurs in members 1 and 2, $D_{X2} = 0$. From symmetry, $\theta_2 = 0$ and $\theta_3 = -\theta_1$. The required displacements are, therefore, given by

$$
\begin{bmatrix}
\dfrac{4EI}{l} & \dfrac{6EI}{l^2} \\[2ex]
\dfrac{12EI}{l^3} & \dfrac{24EI}{l^3} + k
\end{bmatrix}
\begin{bmatrix}
\theta_1 \\[2ex]
D_{Y2}
\end{bmatrix}
=
\begin{bmatrix}
0 \\[2ex]
-Q
\end{bmatrix}
+
\begin{bmatrix}
\dfrac{ql^2}{12} \\[2ex]
-ql
\end{bmatrix}
$$

which leads to

$$\theta_1 = \frac{3(4Q + 5ql)l^2}{8(6EI + kl^3)} + \frac{ql^3}{48EI}$$

and

$$D_{Y2} = -\frac{(4Q + 5ql)l^3}{4(6EI + kl^3)}$$

The case II stress resultants for element 1 are given by

$$[K][D] = [_1F]$$

where

$$
[D] =
\begin{bmatrix}
0 \\
\theta_1 \\
D_{Y2} \\
0
\end{bmatrix}
\qquad \text{and} \qquad
[_1F] =
\begin{bmatrix}
F_{Y1} \\
M_1 \\
F_{Y2} \\
M_2
\end{bmatrix}
$$

or

$$F_{Y1} = -\frac{6EI}{l^2}\theta_1 - \frac{12EI}{l^3}D_{Y2}$$

$$F_{Y2} = \frac{6EI}{l^3}\theta_1 + \frac{12EI}{l^3}D_{Y2} = -F_{Y1}$$

$$M_1 = \frac{4EI}{l}\theta_1 + \frac{6EI}{l^2}D_{Y2}$$

$$M_2 = \frac{2EI}{l}\theta_1 + \frac{6EI}{l^2}D_{Y2}$$

yielding

$$F_{Y1} = \frac{6EI(4Q + 5ql)}{8(6EI + kl^3)} - \frac{ql}{8} = -F_{Y2}$$

$$M_1 = \frac{ql^2}{12}$$

and

$$M_2 = \frac{3EI(4Q + 5ql)}{4(6EI + kl^3)} + \frac{ql^3}{24}$$

The final stress resultants for element 1 are the sums of those from cases I and II.

$$F_{Y1} = \frac{6EI(4Q + 5ql)}{8(6EI + kl^3)} - \frac{ql}{8} + \frac{ql}{2} = \frac{3}{8}ql + \frac{6EI(4Q + 5ql)}{8(6EI + kl^3)}$$

$$M_1 = \frac{ql^2}{12} - \frac{ql^2}{12} = 0$$

$$F_{Y2} = -\frac{6EI(4Q + 5ql)}{8(6EI + kl^3)} + \frac{ql}{8} + \frac{ql}{2} = \frac{5}{8}ql - \frac{6EI(4Q + 5ql)}{8(6EI + kl^3)}$$

$$M_2 = \frac{ql^2}{24} + \frac{3EI(4Q + 5ql)}{4(6EI + kl^3)} + \frac{ql^2}{12} = \frac{ql^2}{8} + \frac{3EI(4Q + 5ql)}{4(6EI + kl^3)}$$

The final displacements are those given as θ_1 and D_{Y2} for case II.

9.6 TEMPERATURE PROBLEMS AND CONSTRUCTION MISALIGNMENTS

Problems involving temperature or thermal loading can be treated in a manner very much like that just described for distributed loading. If the case I node forces are those needed to eliminate the thermal deformations, the case II loads and the actual deformations are given by

$$[K][D] = [F_A] + [F_F] + [F_T] \tag{9.50}$$

where $[F_A]$ and $[F_F]$ are as previously defined, and $[F_T]$ is the thermal load vector, the elements of which consist of forces equal and opposite to the temperature stress resultants used in case I. As with the distributed loads, the final stress resultants are given by the sum of cases I and II.

Example 9.11 Consider the member shown in fig. 9.33. The beam is subjected to a concentrated lateral load Q at the midspan. In addition, there is a temperature gradient ΔT over the depth of the beam.

$$\Delta T = T_t - T_b = T_{\text{top of beam}} - T_{\text{bottom of beam}}$$

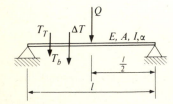

Figure 9.33

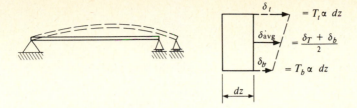

Figure 9.34 Deflection and strain distribution for temperature gradient.

It is assumed that the temperature varies linearly, through the depth and α is the coefficient of thermal expansion for the material in question.

A temperature loading of this type leads to both axial and bending deformations. If the member is unconstrained, no stresses will result. If resistance is present, however, stresses will be induced.

For this beam, there will be a tendency both to bend and to elongate. This is illustrated in fig. 9.34. To remove these strains there will have to be applied node loadings that result in strains, equal and opposite to the following:

$$\varepsilon_t = \frac{T_t \alpha \, dz}{dz} = T_t \alpha$$

$$\varepsilon_b = \frac{T_b \alpha \, dz}{dz} = T_b \alpha$$

If the material responds in a linear-elastic fashion, the corresponding stresses are

$$\sigma_t = E T_t \alpha$$

$$\sigma_b = E T_b \alpha$$

The average value is

$$\sigma_{\text{avg}} = \frac{E\alpha}{2} (T_t + T_b)$$

The stress resultants required to maintain zero thermal deformations at the ends are the integrated effects of these stresses.

$$P_T = \int_A \sigma_{\text{avg}} \, dA = \sigma_{\text{avg}} A = \frac{E\alpha A}{2} (T_t + T_b)$$

and

$$M_T = \int_A \sigma y \, dA = \frac{E\alpha I}{d} (T_t - T_b)$$

(9.51)

where d is the depth of the beam in the direction of the thermal gradient.

The case I and case II loadings are those shown in fig. 9.35.

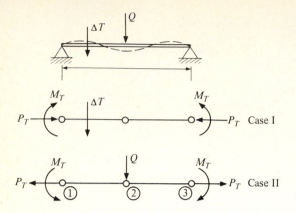

Figure 9.35 Superposition.

The matrix solution would be carried out in the same fashion as for the distributed load problem. For case II:

$$[K][D] = [F_A] + [F_T]$$

where

$$[F_A] = \{R_{1X} \quad R_{1Y} \quad 0 \quad 0 \quad -Q \quad 0 \quad R_{3X} \quad R_{3Y} \quad 0\}$$

and

$$[F_T] = \{-P_T \quad 0 \quad -M_T \quad 0 \quad 0 \quad 0 \quad P_T \quad 0 \quad M_T\}$$

Using the standard beam bending stiffness matrix, noting the boundary conditions

$$D_{X1} = D_{Y1} = D_{X3} = D_{Y3} = 0$$

and taking into account the effect of symmetry,

$$D_{X2} = \theta_2 = 0$$

the governing equations for determining the unknown deformations are

$$
\begin{bmatrix}
\dfrac{4EI}{l} & \dfrac{6EI}{l^2} \\[3mm]
\dfrac{6EI}{l^2} & \dfrac{24EI}{l^3}
\end{bmatrix}
\begin{bmatrix}
\theta_1 \\[3mm]
D_{Y2}
\end{bmatrix}
=
\begin{bmatrix}
-M_T \\[3mm]
-Q
\end{bmatrix}
$$

Solving this set yields the following:

$$\theta_1 = \frac{l}{4EI}\left[\frac{2}{5}Ql - \frac{8}{5}M_T\right]$$

and

$$D_{Y2} = -\frac{l^3}{30EI}\left[2Q - \frac{3M_T}{l}\right]$$

Table 9.11 Final stress resultants for element 1

	Case I	Case II	Sum (final solution)
F_{X1}	P_T	0	P_T
F_{Y1}	0	$\frac{1}{5}Q + \frac{6}{5}\frac{M_T}{l}$	$\frac{1}{5}Q + \frac{6}{5}\frac{M_T}{l}$
M_1	M_T	$-M_T$	0
F_{X2}	$-P_T$	0	$-P_T$
F_{Y2}	0	$-\frac{1}{5}Q - \frac{6}{5}\frac{M_T}{l}$	$-\frac{1}{5}Q - \frac{6}{5}\frac{M_T}{l}$
M_2	$-M_T$	$-\frac{1}{5}Ql - \frac{1}{5}M_T$	$-\frac{1}{5}Ql - \frac{6}{5}M_T$

Substituting these into the equilibrium formulation, taking into account the known boundary conditions, there is obtained

$$F_{X1} = F_{X2} = 0$$

$$F_{Y1} = -F_{Y2} = \frac{1}{5}Q + \frac{6}{5}\frac{M_T}{l}$$

$$M_1 = -M_T$$

$$M_2 = -\frac{1}{5}Ql - \frac{1}{5}M_T$$

The final stress resultants for element 1 are given in table 9.11.

Example 9.12 A rigid, simple portal-type frame is supposed to have the dimensions shown in the first sketch of fig. 9.36: span L, height L, and all members having a unit bending stiffness of EI. Assume that the top member is fabricated too long by an amount equal to Δ. The deformations and stress resultants that are caused by using this member in the frame are to be determined. Assuming nodes at the bases of each of the columns and at the points of junction of the columns and beams, the case I loading would be the axial force required to shorten the top member by an amount Δ.

The boundary conditions for this example problem are

$$D_{X1} = D_{Y1} = \theta_1 = D_{X4} = D_{Y4} = \theta_4 = D_{Y2} = D_{Y3} = 0$$

From symmetry, it also is known that

$$\theta_2 = -\theta_3 \qquad \text{and} \qquad D_{X2} = -D_{X3}$$

Using these conditions—and both the axial and bending stiffness matrices—leads to the following equilibrium formulation:

$$\begin{bmatrix} \dfrac{12EI}{L^3} + \dfrac{2AE}{L} & -\dfrac{6EI}{L^2} \\[3mm] -\dfrac{6EI}{L^2} & \dfrac{6EI}{L^2} \end{bmatrix} \begin{bmatrix} D_{X2} \\[3mm] \theta_2 \end{bmatrix} = \begin{bmatrix} 0 \\[3mm] 0 \end{bmatrix} + \begin{bmatrix} -\Delta\dfrac{AE}{L} \\[3mm] 0 \end{bmatrix}$$

The solution of this set is

$$D_{X2} = -\frac{\Delta A L^2}{6I + 2AL^2}$$

and

$$\theta_2 = -\frac{\Delta A L}{6I + 2AL^2}$$

The beam stress resultants are then given by

$$
\begin{bmatrix}
\dfrac{AE}{L} & 0 & 0 & -\dfrac{AE}{L} & 0 & 0 \\[2mm]
0 & \dfrac{12EI}{L^3} & -\dfrac{6EI}{L^2} & 0 & -\dfrac{12EI}{L^3} & -\dfrac{6EI}{L^2} \\[2mm]
0 & -\dfrac{6EI}{L^2} & \dfrac{4EI}{L} & 0 & \dfrac{6EI}{L^2} & \dfrac{2EI}{L} \\[2mm]
-\dfrac{AE}{L} & 0 & 0 & \dfrac{AE}{L} & 0 & 0 \\[2mm]
0 & -\dfrac{12EI}{L^3} & \dfrac{6EI}{L^2} & 0 & \dfrac{12EI}{L^3} & \dfrac{6EI}{L^2} \\[2mm]
0 & -\dfrac{6EI}{L^2} & \dfrac{2EI}{L} & 0 & \dfrac{6EI}{L^2} & \dfrac{4EI}{L}
\end{bmatrix}
\begin{bmatrix}
D_{X2} \\[2mm] D_{Y2} \\[2mm] \theta_2 \\[2mm] D_{X3} \\[2mm] D_{Y3} \\[2mm] \theta_3
\end{bmatrix}
$$

$$
=
\begin{bmatrix}
F_{X2} = \dfrac{2AE}{L} D_{X2} \\[3mm]
F_{Y2} = 0 \\[3mm]
M_2 = \dfrac{2EI}{L}\theta_2 \\[3mm]
F_{X3} = -\dfrac{2AE}{L} D_{X2} \\[3mm]
F_{Y3} = 0 \\[3mm]
M_3 = -\dfrac{2EI}{L}\theta_2
\end{bmatrix}
=
\begin{bmatrix}
-ALG\Delta \\[3mm]
0 \\[3mm]
-IG\Delta \\[3mm]
ALG\Delta \\[3mm]
0 \\[3mm]
IG\Delta
\end{bmatrix}
$$

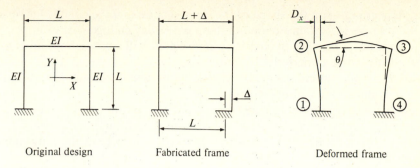

| Original design | Fabricated frame | Deformed frame |

Figure 9.36

where $G = \dfrac{2AE}{6I + 2AL^2}$

As before, the final stress resultants and deformations for the beam are the sum of those from cases 1 and 2 as listed in table 9.12.

Table 9.12 Case 1, case 2, and final beam-stress resultants

	Case 1	Case 2	Sum (final solution)
F_{X2}	$\dfrac{AE}{L}\Delta$	$-\dfrac{2A^2EL}{6I + 2AL^2}\Delta$	$\dfrac{AE}{L}\dfrac{3I - 2AL^2}{3I - AL^2}\Delta$
F_{Y2}	0	0	0
M_2	0	$-\dfrac{2AEI}{6I + 2AL^2}\Delta$	$-\dfrac{2AEI}{6I + 2AL^2}\Delta$
F_{X3}	$-\dfrac{AE}{L}\Delta$	$\dfrac{2A^2EL}{6I + 2AL^2}\Delta$	$-\dfrac{AE}{L}\dfrac{3I - 2AL^2}{3I - AL^2}\Delta$
F_{Y3}	0	0	0
M_3	0	$\dfrac{2AEI}{6I + 2AL^2}\Delta$	$\dfrac{2AEI}{6I + 2AL^2}\Delta$

9.7 ELEMENTS SUBJECTED TO TORSION

In sec. 2.7, the general problem of members subjected to torsion was discussed. It was shown that resistance of slender members to these types of imposed moments can be thought of as consisting of two parts: (*a*) that known as St. Venant or *uniform* torsion, and (*b*) that due to warping of the cross section (referred to as *warping torsion*).

$$T = T_{sv} + T_{\omega}$$

Moreover, depending upon the shape of the cross section and the conditions of end support, the relative importance of one of these parts may change. For example, for small deformations, a member of circular cross section will not warp when twisted within the elastic range of material behavior. On the other hand, if a thin-walled, open cross section such as an I shape is twisted, warping may account for the major part of the member's torsional resistance.

Stiffness formulations of the general type considered in this chapter require that both St. Venant's and warping behavior be taken into account.

The total torque acting on a member is a vector that acts in the longitudinal direction of the member. In fig. 9.17, for example, $M_{\bar{X}i}$ and $M_{\bar{X}j}$ are the applied torques, and $\theta_{\bar{X}i}$ and $\theta_{\bar{X}j}$ are the resulting displacement parameters. The St. Venant torque is a vector that acts in the same direction as these end-torsional moments. This is not true for the warping torque T_ω. If the member in question is of the wide-flange type, the warping torque can more easily be represented in terms of the flange moment M_F (see fig. 2.30), and this flange moment can be represented as a vector that acts in the direction of the y axis (see fig. 9.17). Its displacement parameter is $du_F/a\bar{X}$. While the element does not directly involve this term, the proportional quantity $d\theta_X/d\bar{X}$ can be defined.† This flange rotation acts normal to the $\bar{X}$ axis, and in the direction of the $\bar{Y}$ axis—as does the flange moment.

This formulation leads to a value of the total rotation $\theta_{\bar{X}}$, and its first derivative $\theta'_{\bar{X}}$ at each node. The stress resultants obtained are the total torque $M_{\bar{X}}$, and the flange moment M_ω.‡ From these, the St. Venant torque can be found:

$$T_{SV} = G\kappa_T \theta'_X$$

Since T and T_{SV} are known, T_ω, the warping torque, can be calculated.

If the quantity θ'_X is taken as positive when θ_X is increasing, the torsion element in question is that shown in fig. 9.37. Defining the quantity θ'_X and θ_ω, the element equilibrium formulation is given by

$$
\begin{array}{cccc}
\theta_{\bar{X}i} & \theta_{\omega i} & \theta_{\bar{X}j} & \theta_{\omega j}
\end{array}
$$
$$
\begin{bmatrix}
[B_{11}] & [B_{12}] \\
[B_{21}] & [B_{22}]
\end{bmatrix}
\begin{bmatrix}
\theta_{\bar{X}i} \\
\theta_{\omega i} \\
\theta_{\bar{X}j} \\
\theta_{\omega j}
\end{bmatrix}
=
\begin{bmatrix}
M_{\bar{X}i} \\
M_{\omega i} \\
M_{\bar{X}j} \\
M_{\omega j}
\end{bmatrix}
$$

where each of the $[B_{ij}]$'s is a 2×2 matrix.

The differential equation governing the deformation of members subjected to end torques [from eq. (2.44)] is

$$\theta''''_{\bar{X}} - \lambda^2 \theta''_{\bar{X}} = 0$$

---

† For details, see J. L. Krahula, Analysis of Bent and Twisted Bars Using the Finite Element Method, *AIAA J.*, vol. 5, no. 6, p. 1194, June, 1967.

‡ For wide-flange type shapes, the quantity M_ω is given by $M_F \times d$, where d is the depth of the section. The use of M_F leads to some simplification in the stiffness matrix.

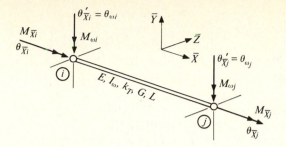

Figure 9.37 Element subjected to torsion.

where

$$\lambda^2 = \frac{G\kappa_T}{EI_\omega}$$

The general solution to the equation is

$$\theta_{\bar{X}} = A \sinh \lambda\bar{X} + B \cosh \lambda\bar{X} + C\bar{X} + D$$

This equation and the proper displacement boundary conditions allow the definition of the torsional stiffness matrix. For example, for the first column of the matrix, the conditions to be imposed are

$$\theta_{\bar{X}i} = +1.0$$

$$\theta_{\bar{X}j} = 0$$

$$\theta_{\omega i} = \theta'_{\bar{X}i} = 0$$

$$\theta_{\omega j} = \theta'_{\bar{X}j} = 0$$

and the entries are the corresponding stress resultants. This displaced state is illustrated in fig. 9.38. The corresponding stress resultants are

$$M_{\bar{X}j} = -M_{\bar{X}j} = -G\kappa_T \sqrt{\frac{G\kappa_T}{EI_\omega}} \frac{\sinh \lambda L}{2(\cosh \lambda L - 1) - \lambda L \sinh \lambda L}$$

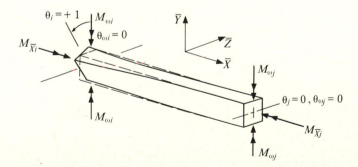

Figure 9.38 Unit displacement and boundary conditions for determining the first column of the stiffness matrix.

and

$$M_{\omega i} = M_{\omega j} = -G\kappa_T \frac{\cosh \lambda L - 1}{2(\cosh \lambda L - 1) - \lambda L \sinh \lambda L}$$

Column 2 of the stiffness matrix would be computed from the boundary conditions

$$\theta_{\bar{x}i} = 0$$
$$\theta_{\bar{x}j} = 0$$
$$\theta_{\omega i} = \theta'_{\bar{x}i} = +1.0$$
$$\theta_{\omega j} = \theta'_{\bar{x}j} = 0$$

This is illustrated in fig. 9.39. The stiffness terms are

$$M_{\bar{x}i} = -M_{\bar{x}j} = -G\kappa_T \frac{\cosh \lambda L - 1}{2(\cosh \lambda L - 1) - \lambda L \sinh \lambda L}$$

$$M_{\omega i} = \frac{G\kappa_T}{\lambda} \frac{\sinh \lambda L - \lambda L \cosh \lambda L}{2(\cosh \lambda L - 1) - \lambda L \sinh \lambda L}$$

and

$$M_{\omega j} = \frac{G\kappa_T}{\lambda} \frac{\lambda L - \sinh \lambda L}{2(\cosh \lambda L - 1) - \lambda L \sinh \lambda L}$$

Repeating this operation for units of displacement at the jth end of the element and collecting the resulting terms, the following torsional stiffness matrix is obtained:

$$[\bar{K}] = H \begin{bmatrix} (-\lambda \sinh \lambda L) & & & (\text{symmetrical}) \\ -(\cosh \lambda L - 1) & \frac{1}{\lambda}(\sinh \lambda L - \lambda L \cosh \lambda L) & & \\ \hline (\lambda \sinh \lambda L) & (\cosh \lambda L - 1) & (-\lambda \sinh \lambda L) & \\ -(\cosh \lambda L - 1) & \frac{1}{\lambda}(\lambda L - \sinh \lambda L) & (\cosh \lambda L - 1) & \frac{1}{\lambda}(\sinh \lambda L - \lambda L \cosh \lambda L) \end{bmatrix}$$

$$(9.52)$$

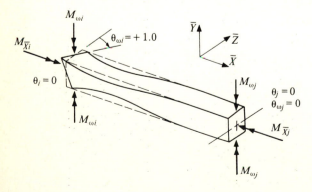

Figure 9.39 Unit displacement and boundary conditions for determining second column of the stiffness matrix.

where

$$H = \frac{G\kappa_T}{2(\cosh \lambda L - 1) - \lambda L \sinh \lambda L}$$

In shorthand notation, this could be written as

$$[\bar{K}] = \left[\begin{array}{cc|cc} -T_1 & -T_2 & T_1 & -T_2 \\ -T_2 & T_3 & T_2 & T_4 \\ \hline T_1 & T_2 & -T_1 & T_2 \\ -T_2 & T_4 & T_2 & T_3 \end{array}\right] = \left[\begin{array}{c|c} [B_{11}] & [B_{12}] \\ \hline [B_{21}] & [B_{22}] \end{array}\right] \quad (9.53)$$

where T_1 through T_4 are as defined by comparing the two matrices.

Example 9.13 Consider the fixed-end beam shown in fig. 9.40. The member is $2L$ in long and is subjected to a concentrated torque of $+T$ applied at its midpoint. As was done in the beam-bending problem, the member will be divided into two parts. The nodes and elements are numbered as indicated in the sketch.

The master equilibrium formulation for this problem is

$$\begin{array}{ccccc} \theta_{\bar{X}1} \; \theta_{\omega 1} & & \theta_{\bar{X}2} \; \theta_{\omega 2} & & \theta_{\bar{X}3} \; \theta_{\omega 3} \end{array}$$
$$\left[\begin{array}{c|c|c} [_1B_{11}] & [_1B_{12}] & 0 \\ \hline [_1B_{21}] & \begin{array}{c} [_1B_{22}] \\ + [_2B_{11}] \end{array} & [_2B_{12}] \\ \hline 0 & [_2B_{21}] & [_2B_{22}] \end{array}\right] \left\{\begin{array}{c} \theta_{\bar{X}1} \\ \theta_{\omega 1} \\ \hline \theta_{\bar{X}2} \\ \theta_{\omega 2} \\ \hline \theta_{\bar{X}3} \\ \theta_{\omega 3} \end{array}\right\} = \left\{\begin{array}{c} M_{\bar{X}1} \\ M_{\omega 1} \\ \hline M_{\bar{X}2} \\ M_{\omega 2} \\ \hline M_{\bar{X}3} \\ M_{\omega 3} \end{array}\right\}$$

Since both ends of the member are fixed, the displacement boundary conditions are

$$\theta_{\bar{X}1} = \theta_{\bar{X}3} = \theta_{\omega 1} = \theta_{\omega 3} = 0$$

From symmetry

$$\theta_{\omega 2} = 0$$

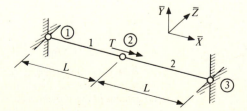

Figure 9.40

Striking all of these rows and columns, the problem reduces to

$$-\left[\frac{EI_\omega\sqrt{G\kappa_T/EI_\omega}}{2(\cosh \lambda L - 1) - \lambda \sinh \lambda L}\right]2\lambda^2(\sinh \lambda L)\theta_{\bar{x}2} = -2T_1\theta_{\bar{x}2} = T$$

or

$$\theta_{X2} = -\frac{T}{2T_1}$$

The corresponding stress resultants are computed as before from the displacements and the element stiffness matrices. Those for element 1, for example, are given by

$$\begin{bmatrix} -T_1 & -T_2 & T_1 & -T_2 \\ -T_2 & T_3 & T_2 & T_4 \\ T_1 & T_2 & -T_1 & T_2 \\ -T_2 & T_4 & T_2 & T_3 \end{bmatrix}\begin{bmatrix} 0 \\ 0 \\ -T/2T_1 \\ 0 \end{bmatrix} = \begin{bmatrix} M_{\bar{X}1} \\ M_{\omega1} \\ M_{\bar{X}2} \\ M_{\omega2} \end{bmatrix}$$

or

$$M_{\bar{X}1} = -\frac{T}{2T_1}(T_1) = -\frac{T}{2} \quad \text{and} \quad M_{\text{sv}} = 0$$

$$M_{\omega1} = -\frac{T}{2T_1}(T_2) = -\frac{T_2}{2T_1}(T)$$

and

$$M_{\bar{X}2} = \frac{T}{2T_1}(T_1) = \frac{T}{2} \quad M_{\text{sv}} = 0$$

$$M_{\omega2} = -\frac{T}{2T_1}(T_2) = -\frac{T_2}{2T_1}(T)$$

These stress resultants are shown in fig. 9.41.

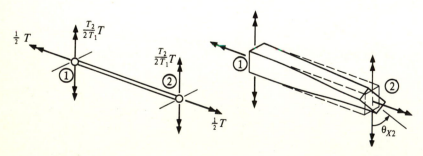

Figure 9.41 Resulting reactions.

9.8 PRISMATIC ELEMENTS SUBJECTED TO GENERALIZED END LOADINGS

The stiffness matrix in local coordinates for a prismatic member subjected to a generalized state of end stress resultants now can be defined. Using the notation shown in fig. 9.42, the equilibrium formulation is

$$[\bar{K}][\bar{D}] = [\bar{F}]$$

or

$$\begin{bmatrix} [\bar{B}_{11}] & \vdots & [\bar{B}_{12}] \\ \cdots\cdots & + & \cdots\cdots \\ [\bar{B}_{21}] & \vdots & [\bar{B}_{22}] \end{bmatrix} \begin{bmatrix} [\bar{D}_i] \\ \cdots \\ [\bar{D}_j] \end{bmatrix} = \begin{bmatrix} [\bar{F}_i] \\ \cdots \\ [\bar{F}_j] \end{bmatrix} \tag{9.54}$$

where

$$[\bar{D}_i] = \begin{bmatrix} D_{Xi} \\ D_{Yi} \\ D_{Zi} \\ \theta_{Xi} \\ \theta_{\omega i} \\ \theta_{Yi} \\ \theta_{Zi} \end{bmatrix} \quad [\bar{D}_j] = \begin{bmatrix} D_{Xj} \\ D_{Yj} \\ D_{Zj} \\ \theta_{Xj} \\ \theta_{\omega j} \\ \theta_{Yj} \\ \theta_{Zj} \end{bmatrix} \quad [\bar{F}_i] = \begin{bmatrix} F_{Xi} \\ F_{Yi} \\ F_{Zi} \\ M_{Xi} \\ M_{\omega i} \\ M_{Yi} \\ M_{Zi} \end{bmatrix} \quad [\bar{F}_j] = \begin{bmatrix} F_{Xj} \\ F_{Yj} \\ F_{Zj} \\ M_{Xj} \\ M_{\omega j} \\ M_{Yj} \\ M_{Zj} \end{bmatrix}$$

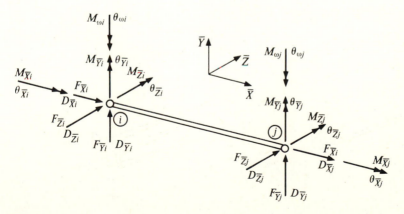

Figure 9.42 Element subjected to generalized end stress resultants.

and

$$[\bar{B}_{11}] = \begin{bmatrix} \dfrac{AE}{L} & 0 & 0 & 0 & 0 & 0 & 0 \\[2mm] 0 & \dfrac{12EI_{\bar{z}}}{L^3} & 0 & 0 & 0 & 0 & -\dfrac{6EI_{\bar{z}}}{L^2} \\[2mm] 0 & 0 & \dfrac{12EI_{\bar{Y}}}{L^3} & 0 & 0 & \dfrac{6EI_{\bar{Y}}}{L^2} & 0 \\[2mm] 0 & 0 & 0 & -T_1 & -T_2 & 0 & 0 \\[2mm] 0 & 0 & 0 & -T_2 & T_3 & 0 & 0 \\[2mm] 0 & 0 & \dfrac{6EI_{\bar{Y}}}{L^2} & 0 & 0 & \dfrac{4EI_{\bar{Y}}}{L} & 0 \\[2mm] 0 & -\dfrac{6EI_{\bar{z}}}{L^2} & 0 & 0 & 0 & 0 & \dfrac{4EI_{\bar{z}}}{L} \end{bmatrix}$$

(9.55)

$$[\bar{B}_{12}] = [\bar{B}_{21}] = \begin{bmatrix} -\dfrac{AE}{L} & 0 & 0 & 0 & 0 & 0 & 0 \\[2mm] 0 & -\dfrac{12EI_{z}}{L^3} & 0 & 0 & 0 & 0 & \dfrac{6EI_{z}}{L^2} \\[2mm] 0 & 0 & -\dfrac{12EI_{Y}}{L^3} & 0 & 0 & -\dfrac{6EI_{Y}}{L^2} & 0 \\[2mm] 0 & 0 & 0 & T_1 & -T_2 & 0 & 0 \\[2mm] 0 & 0 & 0 & T_2 & T_4 & 0 & 0 \\[2mm] 0 & 0 & \dfrac{6EI_{Y}}{L^2} & 0 & 0 & \dfrac{2EI_{Y}}{L} & 0 \\[2mm] 0 & \dfrac{6EI_{z}}{L^2} & 0 & 0 & 0 & 0 & \dfrac{2EI_{z}}{L} \end{bmatrix}$$

(9.56)

$$[\bar{B}_{22}] = \begin{bmatrix} \dfrac{AE}{L} & 0 & 0 & 0 & 0 & 0 & 0 \\[2ex] 0 & \dfrac{12EI_Z}{L^3} & 0 & 0 & 0 & 0 & \dfrac{6EI_Z}{L^2} \\[2ex] 0 & 0 & \dfrac{12EI_Y}{L^3} & 0 & 0 & -\dfrac{6EI_Y}{L^2} & 0 \\[2ex] 0 & 0 & 0 & -T_1 & T_2 & 0 & 0 \\[2ex] 0 & 0 & 0 & T_2 & T_3 & 0 & 0 \\[2ex] 0 & 0 & -\dfrac{6EI_Y}{L^2} & 0 & 0 & \dfrac{4EI_Y}{L} & 0 \\[2ex] 0 & \dfrac{6EI_Z}{L^2} & 0 & 0 & 0 & 0 & \dfrac{4EI_Z}{L} \end{bmatrix} \qquad (9.57)$$

Example 9.14 To illustrate the use of this generalized situation, consider the balcony frame shown in fig. 9.43. The frame consists of three steel W 10×49† wide-flange shapes welded to one another to form right-angle corners. It is assumed that the connections are so fabricated that the flanges of the beams can be considered to be continuous across the connection. The reference coordinate system is shown in the figure. The element model-numbering system is also indicated. Because of symmetry, only one-half the frame is considered.

† Properties for this section are listed in the AISC Steel Construction Manual: $A = 14.4 \text{ in}^2$, $I_Y = 93.0 \text{ in}^4$, $I_Z = 272.9 \text{ in}^4$, $\kappa_T = 1.38 \text{ in}^4$, and $I_\omega = 2070 \text{ in}^6$.

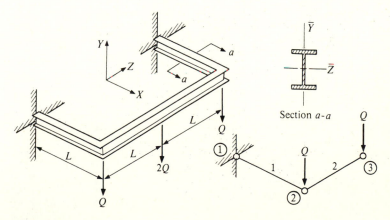

Figure 9.43

The first step in the solution is the transformation of element 2 from its local to the reference coordinate system. (No transformation is required for element 1, because its local system and the reference system already agree.) The required transformation is accomplished by simply rotating the element 90° in the XZ plane. Any displacements or stress resultants in the Y direction are not altered. The transformation matrix for this situation is

$$[T_R] = \begin{bmatrix} [\Lambda] & \vdots & 0 \\ \cdots & \cdots & \cdots \\ 0 & \vdots & [\Lambda] \end{bmatrix}$$

where

$$[\Lambda] = \begin{bmatrix} a_{\bar{X}X} & 0 & a_{\bar{X}Z} & 0 & 0 & 0 & 0 \\ 0 & 1 & 0 & 0 & 0 & 0 & 0 \\ a_{\bar{Z}X} & 0 & a_{\bar{Z}Z} & 0 & 0 & 0 & 0 \\ 0 & 0 & 0 & a_{\bar{X}X} & 0 & 0 & a_{\bar{X}Z} \\ 0 & 0 & 0 & 0 & 1 & 0 & 0 \\ 0 & 0 & 0 & 0 & 0 & 1 & 0 \\ 0 & 0 & 0 & a_{\bar{Z}X} & 0 & 0 & a_{\bar{Z}Z} \end{bmatrix}$$

As before, the element stiffness matrix in the reference system is given by

$$[K] = [T_R^T][\bar{K}][T_R]$$

or

$$[K] = \begin{bmatrix} [\Lambda^T] & 0 \\ 0 & [\Lambda^T] \end{bmatrix} \begin{bmatrix} [\bar{B}_{11}] & [\bar{B}_{12}] \\ [\bar{B}_{21}] & [\bar{B}_{22}] \end{bmatrix} \begin{bmatrix} [\Lambda] & 0 \\ 0 & [\Lambda] \end{bmatrix} = \begin{bmatrix} [\Lambda^T\bar{B}_{11}\Lambda] & \vdots & [\Lambda^T\bar{B}_{12}\Lambda] \\ \cdots & \cdots & \cdots \\ [\Lambda^T\bar{B}_{21}\Lambda] & \vdots & [\Lambda^T\bar{B}_{22}\Lambda] \end{bmatrix} \quad (9.58)$$

where the $[\bar{B}_{ij}]$'s are defined by eqs. (9.55), (9.56), and (9.57). For element 2, $[\Lambda]$ is

$$[{}_2\Lambda] = \begin{bmatrix} 0 & 0 & 1 & 0 & 0 & 0 & 0 \\ 0 & 1 & 0 & 0 & 0 & 0 & 0 \\ -1 & 0 & 0 & 0 & 0 & 0 & 0 \\ 0 & 0 & 0 & 0 & 0 & 0 & 1 \\ 0 & 0 & 0 & 0 & 1 & 0 & 0 \\ 0 & 0 & 0 & 0 & 0 & 1 & 0 \\ 0 & 0 & 0 & -1 & 0 & 0 & 0 \end{bmatrix}$$

Using this matrix as prescribed in eq. (9.58) yields the master stiffness matrix

$$[K] = \begin{matrix} & \begin{matrix} 1 & \quad\quad 2 & \quad\quad 3 \end{matrix} \\ \begin{bmatrix} [{}_1B_{11}] & \vdots & [{}_1B_{12}] & \vdots & 0 \\ \cdots & & \cdots & & \cdots \\ [{}_1B_{21}] & \vdots & \begin{matrix}[{}_1B_{22}]\\ + [{}_2B_{11}]\end{matrix} & \vdots & [{}_2B_{12}] \\ \cdots & & \cdots & & \cdots \\ 0 & \vdots & [{}_2B_{21}] & \vdots & [{}_2B_{22}] \end{bmatrix} & \begin{matrix} 1 \\ \\ 2 \\ \\ 3 \end{matrix} \end{matrix}$$

In expanding form, it is given as eq. (9.59).

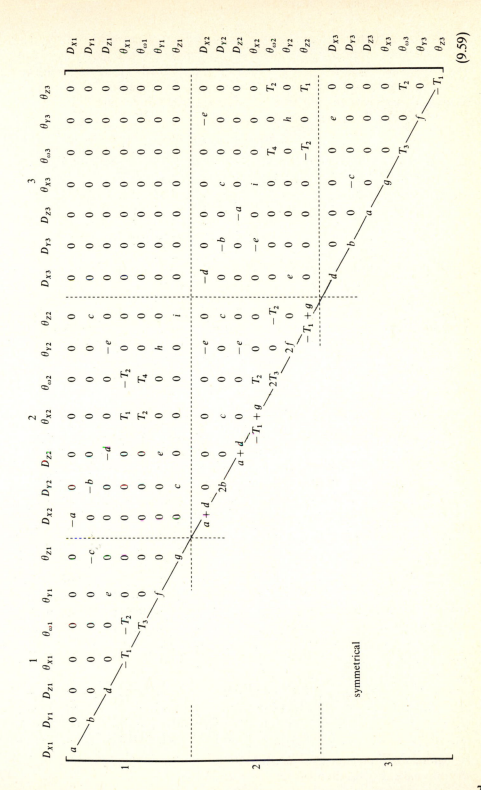

$$(9.59)$$

$$a = \frac{AE}{L} \qquad b = \frac{12EI_Z}{L^3} \qquad c = \frac{6EI_Z}{L^2} \qquad d = \frac{12EI_Y}{L^3}$$

$$e = \frac{6EI_Y}{L^2} \qquad f = \frac{4EI_Y}{L} \qquad g = \frac{4EI_Z}{L} \qquad h = \frac{2EI_Y}{L}$$

$$i = \frac{2EI_Z}{L} \qquad \lambda = \sqrt{\frac{G\kappa_T}{EI_\omega}}$$

$$T_1 = G\kappa_T \frac{\lambda \sinh \lambda L}{2(\cosh \lambda L - 1) - \lambda L \sinh \lambda L}$$

$$T_2 = G\kappa_T \frac{\cosh \lambda L - 1}{2(\cosh \lambda L - 1) - \lambda L \sinh \lambda L}$$

$$T_3 = \frac{G\kappa_T}{\lambda} \frac{\sinh \lambda L - \lambda L \cosh \lambda L}{2(\cosh \lambda L - 1) - \lambda L \sinh \lambda L}$$

$$T_4 = \frac{G\kappa_T}{\lambda} \frac{\lambda L - \sinh \lambda L}{2(\cosh \lambda L - 1) - \lambda L \sinh \lambda L}$$

The boundary conditions for this problem are

$$D_{X1} = D_{Y1} = D_{Z1} = \theta_{X1} = \theta_{\omega 1} = \theta_{Y1} = \theta_{Z1} = D_{Z3} = \theta_{X3} = \theta_{\omega 3} = \theta_{Y3} = 0$$

Striking the rows and columns associated with these parameters yields the following simultaneous set:

$$
\begin{bmatrix}
a+d & 0 & 0 & 0 & 0 & -e & 0 & -d & 0 & 0 \\
 & 2b & 0 & c & 0 & 0 & c & 0 & -b & 0 \\
 & & a+d & 0 & 0 & -e & 0 & 0 & 0 & 0 \\
 & & & -T_1+g & T_2 & 0 & 0 & 0 & -c & 0 \\
 & & & & 2T_3 & 0 & -T_2 & 0 & 0 & T_2 \\
 & & & & & 2f & 0 & e & 0 & 0 \\
 & & & & & & -T_1+g & 0 & 0 & T_1 \\
 & \text{(symmetrical)} & & & & & & d & 0 & 0 \\
 & & & & & & & & b & 0 \\
 & & & & & & & & & -T_1
\end{bmatrix}
\begin{bmatrix}
D_{X2} \\ D_{Y2} \\ D_{Z2} \\ \theta_{X2} \\ \theta_{\omega 2} \\ \theta_{Y2} \\ \theta_{Z2} \\ D_{X3} \\ D_{Y3} \\ \theta_{Z3}
\end{bmatrix}
=
\begin{bmatrix}
0 \\ -Q \\ 0 \\ 0 \\ 0 \\ 0 \\ 0 \\ 0 \\ -Q \\ 0
\end{bmatrix}
$$

The displacements are

$$
[D] =
\begin{bmatrix}
D_{X2} \\ D_{Y2} \\ D_{Z2} \\ \theta_{X2} \\ \theta_{\omega 2} \\ \theta_{Y2} \\ \theta_{Z2} \\ D_{X3} \\ D_{Y3} \\ \theta_{Z3}
\end{bmatrix}
=
\begin{bmatrix}
0 \\
-4.7490 \text{ in} \\
0 \\
-1.9711 \times 10^{-2} \text{ rad} \\
-9.3322 \times 10^{-5} \text{ rad} \\
0 \\
3.9575 \times 10^{-2} \text{ rad} \\
0 \\
-7.1167 \text{ in} \\
3.4434 \times 10^{-2} \text{ rad}
\end{bmatrix}
$$

As before, the individual element stiffness matrices would be used to determine the internal stress resultants acting on each element.

9.9 NONPRISMATIC AND CURVED ELEMENTS

It has been assumed in the work thus far presented in this chapter that elements are prismatic. While such an assumption would allow the handling of a very large number of real cases, there are exceptions: variable cross section bridge girders, welded tapered members for use in rigid frames, tapered concrete members, circular or spiral beams, parabolic roof beams, circular truss dome members, etc. All these cases result from the desire to use more efficiently materials or to present an esthetically more pleasing design.

The matrix method of structural analysis is not restricted in its use to prismatic elements. All that the method, per se, requires is that there be defined and readily available stiffness parameters associated with the ends of the element in question. What happens in between the node locations in the actual structure can be handled as a separate problem, once the node deformations and end stress resultants are determined.

When considering tapered elements, two procedures are available. The simpler one of these presumes that the tapered member in question is modeled by a series of prismatic elements, as illustrated in fig. 9.44. This is, at best, an approximation, the quality of which depends upon the number of segments assumed in the model. While the procedure is simple and straightforward, and can give solutions to almost any desired degree of accuracy, it leads to much larger stiffness formulations and much greater required computer storage capacity. For this reason, it is often better to use the actual tapered member as the element and either derive its exact stiffness matrix, using the procedures heretofore defined, or develop tables of stiffness coefficients for various types or kinds of tapers and ratios of end geometries, and then call upon these as required. (See, for example, *Handbook of Frame Constants* published by the Portland Cement Association.)

Consider the linearly tapered, rectangular cross-section member shown in fig. 9.45. By definition—the taper in the plane of bending is linear. Therefore, at any point x from node i

$$d_x = d_i - (d_i - d_j)\frac{x}{L} \tag{9.60}$$

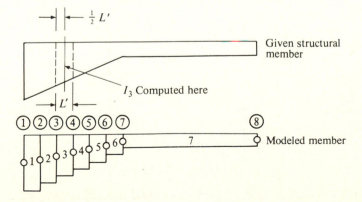

Figure 9.44 Approximate structural model for tapered member.

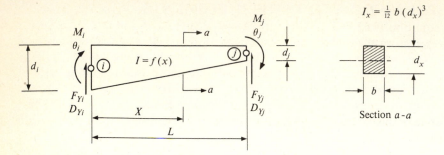

Figure 9.45 Linearly tapered rectangular cross-section element.

and

$$I_x = \frac{1}{12} bd_x^3 = \frac{1}{12} b \left[d_i - (d_i - d_j) \frac{x}{L} \right]^3 \tag{9.61}$$

For reasonable degrees of taper, it can be assumed that the second-order beam-bending equation holds; that is,

$$EI_x \frac{d^2 y}{dx^2} = -M_x$$

Since

$$M_x = F_{Yj}(L - x) - M_j$$

The basic differential equation to be solved is

$$\frac{d^2 y}{dx^2} = -\frac{1}{E} \left\{ \frac{F_{Yj}(L - x) - M_j}{\frac{1}{12} b [d_i - (d_i - d_j)(x/L)]^3} \right\} \tag{9.62}$$

The boundary conditions to be used depend on the particular column of the stiffness matrix desired.

The stiffness formulation for a general element of this type can be written in the form

$$
\begin{array}{cccc}
D_{Yi} & \theta_i & D_{Yj} & \theta_j
\end{array}
$$

$$
\left[
\begin{array}{c|c}
[B_{11}] & [B_{12}] \\
\hline
[B_{21}] & [B_{22}]
\end{array}
\right]
\left\{
\begin{array}{c}
D_{Yi} \\
\theta_i \\
\hline
D_{Yj} \\
\theta_j
\end{array}
\right\}
=
\left\{
\begin{array}{c}
F_{Yi} \\
M_i \\
\hline
F_{Yj} \\
M_j
\end{array}
\right\}
\tag{9.63}
$$

If it is assumed that $D_{Yi} = \theta_i = 0$ [that is, the boundary conditions for use with differential equation (9.62) will be (1) at $x = 0$, $y = 0$, and (2) at $x = 0$, $y' = 0$], there will be obtained from the differential equation and boundary condition solution

$$\begin{bmatrix} D_{Yj} \\ \theta_j \end{bmatrix} = \begin{bmatrix} a_{11} & a_{12} & F_{Yj} \\ a_{21} & a_{22} & M_j \end{bmatrix}$$

where $[A]$ is one of the flexibility submatrices for the element. But since for the case in question

$$[B_{22}] \begin{bmatrix} D_{Yj} \\ \theta_j \end{bmatrix} = \begin{bmatrix} F_{Yj} \\ M_j \end{bmatrix}$$

the desired submatrix $[B_{22}]$ will be given by

$$[B_{22}] = [A^{-1}] = \begin{bmatrix} a_{11} & a_{12} \\ a_{21} & a_{22} \end{bmatrix}^{-1} \tag{9.64}$$

The remaining elements of the stiffness matrix can be found in a similar fashion—or by using static equilibrium equations.

Unfortunately the general solution of eq. (9.62) for the more general conditions of end loading leads to an expression involving Bessel functions of both the first and second kind (J_n and Y_n).[†] Even in their alternate approximate trigonometric forms, these are anything but easy to handle. Approximate methods for obtaining the a_{ij}'s are therefore called for, and several of these methods are described in the appendix to this chapter.

The stiffness matrix for a curved element can be developed in the same manner. (It is to be understood that as with tapered elements, the many types of curves that are possible preclude the development of a general, overall curved element. The example that will here be considered is for one particular type of curve.) Consider the circular (in the plane of the end moments) element of uniform cross section shown in fig. 9.46. Assuming the end loadings and deformations shown, the general stiffness formulation for this element is

$$\begin{bmatrix} [B_{11}] & \vdots & [B_{12}] \\ \hdashline [B_{21}] & \vdots & [B_{22}] \end{bmatrix} \begin{bmatrix} D_{Ti} \\ D_{Ri} \\ \theta_i \\ \hline D_{Tj} \\ D_{Rj} \\ \theta_j \end{bmatrix} = \begin{bmatrix} F_{Ti} \\ F_{Ri} \\ M_i \\ \hline F_{Tj} \\ F_{Rj} \\ M_j \end{bmatrix} \tag{9.65}$$

† See Charles M. Fogel and R. L. Ketter, Elastic Strength of Tapered Columns, *Proc. Am. Soc. Civ. Eng.*, vol. 88, ST. 5, October 1962.

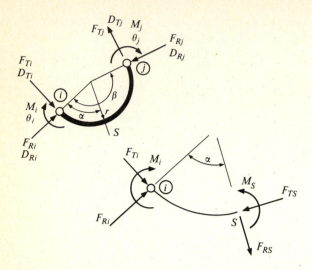

Figure 9.46 Curved element.

If the element is fixed at node j, then the submatrix $[B_{11}]$ can be determined from the relationships

$$[B_{11}] \begin{bmatrix} D_{Ti} \\ D_{Ri} \\ \theta_i \end{bmatrix} = \begin{bmatrix} F_{Ti} \\ F_{Ri} \\ M_i \end{bmatrix} \tag{9.66}$$

The deformations D_{Ti}, D_{Ri}, and θ_i can be determined by any of the methods developed in chaps. 4 and 5. This time, Castigliano's method will be used. From eq. (5.55), the deformations at the i node are

$$D_{Ti} = \frac{\partial W}{\partial F_{Ti}} \qquad D_{Ri} = \frac{\partial W}{\partial F_{Ri}} \qquad \text{and} \qquad \theta_i = \frac{\partial W}{\partial M_i} \tag{9.67}$$

where W is the strain energy of the system. The moment, shear and axial force at a general point s along the member are

$$M_s = M_i + F_{Ti} r(1 - \cos \alpha) - F_{Ri} r \frac{(1 - \cos \alpha) \cos \alpha}{\sin \alpha}$$

$$F_{Ts} = F_{Ti} \cos \alpha + F_{Ri} \sin \alpha \tag{9.68}$$

$$F_{Rs} = F_{Ri} \cos \alpha - F_{Ti} \sin \alpha$$

Assuming that *only bending strain energy is of importance*,

$$W = \frac{1}{2} \int_0^{\beta r} \frac{M_s^2}{EI} ds = \frac{1}{2} \int_0^{\beta} \left[M_i + F_{Ti} r(1 - \cos \alpha) - F_{Ri} r \frac{(1 - \cos \alpha) \cos \alpha}{\sin \alpha} \right]^2 \frac{r}{EI} d\alpha \tag{9.69}$$

Therefore,

$$D_{Ti} = \frac{r}{EI} \int_0^\beta \left[M_i + F_{Ti} r(1 - \cos \alpha) - F_{Ri} r \frac{(1 - \cos \alpha) \cos \alpha}{\sin \alpha} \right] r(1 - \cos \alpha) \, d\alpha$$

$$D_{Ri} = \frac{r}{EI} \int_0^\beta \left[M_i + F_{Ti} r(1 - \cos \alpha) - F_{Ri} r \frac{(1 - \cos \alpha) \cos \alpha}{\sin \alpha} \right] r(1 - \cos \alpha) \frac{(\cos \alpha)}{(\sin \alpha)} \, d\alpha$$

$$\theta_i = \frac{r}{EI} \int_0^\beta \left[M_i + F_{Ti} r(1 - \cos \alpha) - F_{Ri} r \frac{(1 - \cos \alpha) \cos \alpha}{\sin \alpha} \right] 1 \, d\alpha$$

or

$$
\begin{bmatrix} D_{Ti} \\ \\ D_{Ri} \\ \\ \theta_i \end{bmatrix} = \frac{r^2}{EI}
\begin{bmatrix}
\frac{3\beta}{2} - 2 \sin \beta + \frac{\sin 2\beta}{4} & \cos \beta + \frac{\sin^2 \beta}{2} - 1 & \beta - \sin \beta \\ \\
\cos \beta + \frac{\sin^2 \beta}{2} - 1 & \frac{\beta}{2} - \frac{\sin 2\beta}{4} & \cos \beta - 1 \\ \\
\beta - \sin \beta & \cos \beta - 1 & \beta
\end{bmatrix}
\begin{bmatrix} F_{Ti} \\ \\ F_{Ri} \\ \\ M_i \end{bmatrix}
$$

The stiffness matrix is then

$$
[B_{11}] = \frac{r^2}{EI}
\begin{bmatrix}
\frac{3\beta}{2} - 2 \sin \beta + \frac{\sin 2\beta}{4} & \cos \beta + \frac{\sin^2 \beta}{2} - 1 & \beta - \sin \beta \\ \\
\cos \beta + \frac{\sin^2 \beta}{2} - 1 & \frac{\beta}{2} - \frac{\sin^2 \beta}{4} & \cos \beta - 1 \\ \\
\beta - \sin \beta & \cos \beta - 1 & \beta
\end{bmatrix}^{-1}
\tag{9.70}
$$

The remainder of the stiffness matrix can be obtained by applying the equations of statics.

9.10 THE INFLUENCE OF VERTICAL SHEAR ON BENDING STIFFNESS

The energy method that was used in the preceding section to define the various entries in the stiffness matrix for curved elements also can be used to determine stiffness expressions which include the influence of vertical shear. For example, if it is assumed that the element in question is prismatic and is subjected to end-bending moments and end shears, the total strain energy stress in the member is

$$W = \frac{1}{2EI} \int_0^L M^2 \, dx + \frac{\lambda}{2GA} \int_0^L V^2 \, dx$$

For element i-j:

$$M = M_i + F_{Yi} x$$

and

$$V = F_{Yi}$$

The deformations at the ith node of the element are therefore

$$D_{Yi} = \frac{\partial W}{\partial F_{Yi}} = \frac{1}{EI} \int_0^L M \frac{\partial M}{\partial F_{Yi}} dx + \frac{\lambda}{GA} \int_0^L V \frac{\partial V}{\partial F_{Yi}} dx$$

$$= \frac{1}{EI} \int_0^L (M_i + F_{Yi} x) x \, dx + \frac{\lambda}{GA} \int_0^L F_{Yi} \, dx$$

or

$$D_{Yi} = \frac{1}{EI} \left(\frac{M_i L^2}{2} + \frac{F_{Yi} L^3}{3} \right) + \frac{\lambda}{GA} F_{Yi} L$$

Similarly,

$$\theta_i = \frac{\partial W}{\partial M_i} = \frac{1}{EI} \left(M_i L + \frac{1}{2} F_{Yi} L^2 \right)$$

The set to be solved is then

$$\begin{bmatrix} \dfrac{L^3}{3EI} + \dfrac{\lambda L}{GA} & \dfrac{L^2}{2EI} \\ \\ \dfrac{L^2}{2EI} & \dfrac{L}{EI} \end{bmatrix} \begin{bmatrix} F_{Yi} \\ \\ M_i \end{bmatrix} = \begin{bmatrix} D_{Yi} \\ \\ \theta_i \end{bmatrix} \qquad (9.71)$$

Inverting this flexibility matrix, there is obtained the stiffness form

$$\begin{bmatrix} \dfrac{12EIGA}{L^3GA + 12\lambda LEI} & -\dfrac{6EIGA}{L^2GA + 12EI\lambda} \\ \\ -\dfrac{6EIGA}{L^2GA + 12EI\lambda} & \dfrac{4(GAL^2 + 3\lambda EI)(EI)^2}{GAEIL^3 + 12\lambda L(EI)^2} \end{bmatrix} \begin{bmatrix} D_{Yi} \\ \\ \theta_i \end{bmatrix} = \begin{bmatrix} F_{Yi} \\ \\ M_i \end{bmatrix} \qquad (9.72)$$

or

$$[B_{11}][D_i] = [F_i]$$

[Note that eq. (9.72) reduces to the earlier defined stiffness terms when $\lambda = 0$.]

To compute the remaining terms of the first column of the local stiffness matrix, the equations of static equilibrium are used:

$$\sum F_y = 0 \qquad K_{41} = -K_{11} = -\frac{12EIGA}{L^3GA + 12\lambda LEI}$$

and

$$\sum M_j = 0 \qquad K_{31} + K_{21} + K_{11}L = 0$$

or

$$K_{31} = - \frac{6EIGA}{L^2GA + 12EI\lambda}$$

where

K_{11} is the vertical end shear at node i for $D_{Yi} = 1.0$

K_{21} is the end moment at node i for $D_{Yi} = 1.0$

K_{31} is the vertical end shear at node j for $D_{Yi} = 1.0$

K_{41} is the end moment at node j for $D_{Yi} = 1.0$

The column 2 entries would be obtained by subjecting the element to a unit end rotation. The final element stiffness matrix is

$$[\bar{K}] = \frac{1}{1+S} \begin{bmatrix} \dfrac{12EI}{L^3} & -\dfrac{6EI}{L^2} & -\dfrac{12EI}{L^3} & -\dfrac{6EI}{L^2} \\[2mm] -\dfrac{6EI}{L^2} & \dfrac{4EI}{L}\left(1+\dfrac{S}{4}\right) & \dfrac{6EI}{L^2} & \dfrac{2EI}{L}\left(1-\dfrac{S}{2}\right) \\[2mm] -\dfrac{12EI}{L^3} & \dfrac{6EI}{L^2} & \dfrac{12EI}{L^3} & \dfrac{6EI}{L^2} \\[2mm] -\dfrac{6EI}{L^2} & \dfrac{2EI}{L}\left(1-\dfrac{S}{2}\right) & \left(\dfrac{6EI}{L^2}\right) & \dfrac{4EI}{L}\left(1+\dfrac{S}{2}\right) \end{bmatrix} \tag{9.73}$$

where

$$S = \frac{12EI\lambda}{GAL^2}$$

The quantity S is a measure of the contribution of the vertical shear to the overall deformation of the element. For a cross section of 12-in depth and 6-in width, approximate values of S as a function of L are given in the following table. Both rectangular and wide-flange shapes are listed.

L	Approximate values for S†	
	Rectangular cross section ($\lambda = 1.5$)	Wide-flange cross section ($\lambda = 1.0$)
10 ft	0.045	0.040
2 ft	1.10	1.00
1 ft	4.50	4.00

† For rectangular shapes: $S \doteqdot 4.5\left(\dfrac{d}{L}\right)^2$

For wide-flange shapes: $S \doteqdot 4.0\left(\dfrac{d}{L}\right)^2$

9.11 DYNAMIC RESPONSE USING MATRIX METHODS

When the response of a structure varies with time—such as, for example, when the imposed loading is dynamic—the effects of mass, inertia, and damping, as well as the time-dependent loading, must be taken into account in the equilibrium formulation. The resulting equation will be of the general form

$$F_I + F_D + F_S = F_A + F_T \tag{9.74}$$

where

F_I is the equivalent inertia-force vector

F_D is the equivalent damping-force vector

F_S is the equivalent stiffness-force vector

F_A is the steady-state applied-force vector

F_T is the time-varying applied-force vector

For a system having n coupled degrees of freedom, eq. (9.74) represents a set of n simultaneous equations:

$$M_{11}\ddot{D}_1 + M_{12}\ddot{D}_2 + \cdots + M_{1n}\ddot{D}_n + C_{11}\dot{D}_1 + C_{12}\dot{D}_2 + \cdots + C_{1n}\dot{D}_n$$
$$+ K_{11}D_1 + K_{12}D_2 + \cdots + K_{1n}D_n = F_{A1} + F_{T1}$$
$$M_{21}\ddot{D}_1 + M_{22}\ddot{D}_2 + \cdots + M_{2n}\ddot{D}_n + C_{21}\dot{D}_1 + C_{22}\dot{D}_2 + \cdots + D_{2n}\dot{D}_n$$
$$+ K_{21}D_1 + K_{22}D_2 + \cdots + K_{2n}D_n = F_{A2} + F_{T2} \tag{9.75}$$
$$\cdots\cdots\cdots\cdots\cdots\cdots\cdots\cdots\cdots\cdots\cdots\cdots\cdots\cdots\cdots\cdots\cdots\cdots$$
$$M_{n1}\ddot{D}_1 + M_{n2}\ddot{D}_2 + \cdots + M_{nn}\ddot{D}_n + C_{n1}\dot{D}_1 + C_{n2}\dot{D}_2 + \cdots + C_{nn}\dot{D}_n$$
$$+ K_{n1}D_1 + K_{n2}D_2 + \cdots + K_{nn}D_n = F_{An} + F_{Tn}$$

In matrix form these could be written as

$$[M][\ddot{D}] + [C][\dot{D}] + [K][D] = [F_A] + [F_T] \tag{9.76}$$

where

$$[M] = \begin{bmatrix} M_{11} & M_{12} & \cdots & M_{1n} \\ M_{21} & M_{22} & \cdots & M_{2n} \\ \cdots\cdots\cdots\cdots\cdots\cdots \\ M_{n1} & M_{n2} & \cdots & M_{nn} \end{bmatrix}$$

$$[C] = \begin{bmatrix} C_{11} & C_{12} & \cdots & C_{1n} \\ C_{21} & C_{22} & \cdots & C_{2n} \\ \cdots\cdots\cdots\cdots\cdots\cdots \\ C_{n1} & C_{n2} & \cdots & C_{nn} \end{bmatrix}$$

$$[K] = \begin{bmatrix} K_{11} & K_{12} & \cdots & K_{1n} \\ K_{21} & K_{22} & \cdots & K_{2n} \\ \cdots\cdots\cdots\cdots\cdots\cdots \\ K_{n1} & K_{n2} & \cdots & K_{nn} \end{bmatrix}$$

and

$$[\ddot{D}] = \begin{bmatrix} \ddot{D}_1 \\ \ddot{D}_2 \\ \vdots \\ \ddot{D}_n \end{bmatrix} \quad [\dot{D}] = \begin{bmatrix} \dot{D}_1 \\ \dot{D}_2 \\ \vdots \\ \dot{D}_n \end{bmatrix} \quad [D] = \begin{bmatrix} D_1 \\ D_2 \\ \vdots \\ D_n \end{bmatrix} \quad [F_A] = \begin{bmatrix} F_{A1} \\ F_{A2} \\ \vdots \\ F_{An} \end{bmatrix} \quad [F_T] = \begin{bmatrix} F_{T1} \\ F_{T2} \\ \vdots \\ F_{Tn} \end{bmatrix}$$

The matrices $[K]$, $[D]$, and $[F_A]$, as before, represent the stiffness matrix of the structure in question (in the directions of the node displacements), the node displacements, and the applied static forces. The vectors $[\ddot{D}]$ and $[\dot{D}]$ are, respectively, the node-point accelerations and velocities. The vector $[F_T]$ defines the externally applied dynamic forces.

The matrix $[M]$ is referred to as the mass matrix and defines the coupled inertial effects. As shown, each element M_{ij} represents the force acting in the ith direction due to a unit acceleration in the jth direction. (It is to be noted that the matrix $[M]$ will be symmetrical.)

The matrix $[C]$ represents the coupled viscous damping effects, and is called the damping matrix. Thus, an element C_{ij} represents the force in the ith direction that results from a unit velocity in the jth direction. For real structures the matrix $[C]$, as defined here, cannot directly represent all of the damping effects. This is because of the presence in structures of both solid damping and dry friction. Both of these phenomena are dependent to varying degrees on quantities other than the velocity of deformation. For this reason fictitious damping coefficients are often used in the matrix $[C]$ to approximate the actual damping that occurs in real structures. When data concerning such specific damping information is lacking it is not unusual to assume several different, reasonable values, and then to compare the resulting overall deformation solutions.

As noted in chaps. 4 and 8, there are two fundamentally different dynamic problems that arise. The first of these is the more general of the two and involves defining the deformation history for each degree of freedom, as caused by a particular dynamic loading. An example of this type of problem was given in sec. 4.7.2. The second type of problem involves the determination of the natural frequency of vibration of the structure. Both of these will be considered later in this section. At this stage, it is first necessary that the matrices $[M]$ and $[C]$ as given in eq. (9.76) be developed so that they have the same significance and characteristics as the previously developed stiffness matrix $[K]$.

The most elementary development for the mass matrix $[M]$ involves the assumption of lumped or concentrated masses—as illustrated in chap. 4. In such cases, there is no cross-coupling and the mass matrix is simply a diagonal matrix with each of the separate lumped masses appearing as one of the diagonal elements.

$$[M_L] = \begin{bmatrix} M_1 & 0 & 0 & \cdots & 0 \\ 0 & M_2 & 0 & \cdots & 0 \\ \multicolumn{5}{c}{\dotfill} \\ 0 & 0 & 0 & \cdots & M_n \end{bmatrix} \tag{9.77}$$

The subscript L has been used to denote this lumped mass assumption. Similarly, damping that is dependent upon specific node-point displacements or rotations—

with no coupling to other nodes, such as would be the case if, for example, joint friction were taken into account—can be represented by the lumped damping matrix

$$[C_L] = \begin{bmatrix} C_1 & 0 & 0 & \cdots & 0 \\ 0 & C_2 & 0 & \cdots & 0 \\ \cdots & \cdots & \cdots & \cdots & \cdots \\ 0 & 0 & 0 & \cdots & C_n \end{bmatrix} \tag{9.78}$$

In a more general dynamics matrix analysis, consideration must be given to the continuous nature of the structure. In actuality, masses are not lumped and the mass matrix will have off-diagonal elements that describe this structural continuity. The same holds for the damping matrix. Matrices that are derived considering the continuity of the structure are frequently referred to as *consistent-mass* or *consistent-damping* matrices—to differentiate them from the lumped system described above.

The displacement at any particular point within a given element can be related to the node-point displacements by equations of the general form

$$[\delta_{XYZ}] = [A][D]$$

where the matrix $[A]$ represents a displacement function and $[D]$ are the nodal displacements. The acceleration at that same point is therefore

$$[\ddot{\delta}_{XYZ}] = [A][\ddot{D}] \tag{9.79}$$

In turn, the equivalent components of inertial force per unit of volume, which must oppose the movement, are

$$[f_{I_{XYZ}}] = \rho[\ddot{\delta}_{XYZ}] = \rho[A][\ddot{D}] \tag{9.80}$$

where ρ is the mass of the material at point X, Y, Z.

The equivalent force matrix representing the effects of body forces is

$$F_p = \int_V [A^T][p]\, dV$$

p is the body force per unit of volume, and can have values in the X, Y, and Z directions. Since $f_{T_{XYZ}}$ represents just such a body force, the equivalent node-point forces corresponding to the inertial effects for a homogeneous material is

$$[F_I] = \int_V [A^T][f_{I_{XYZ}}]\, dV \tag{9.81}$$

or

$$[F_I] = \int_V [A^T]\rho[A]\, dV \times [\ddot{D}] \tag{9.82}$$

The consistent-mass matrix is therefore given by the relationship

$$[M] = \int_V [A^T]\rho[A]\, dV \tag{9.83}$$

The viscous damping matrix $[C]$ can be developed in a similar manner. If μ is the equivalent viscous damping force per unit of volume, then for all points within the element

$$[f_{Dxyz}] = \mu[A][\dot{D}] \tag{9.84}$$

represents the equivalent damping body force. The damping force vector is

$$[F_D] = \int_V [A^T]\mu[A]\,dV[\dot{D}] = [C][\dot{D}] \tag{9.85}$$

and the damping matrix $[C]$ is

$$[C] = \int_V [A^T]\mu[A]\,dV \tag{9.86}$$

For a homogeneous material, matrices $[M]$ and $[C]$ are developed directly from the displacement function $[A]$. For the axially loaded element shown in fig. 9.9 the displacement of any point is equal to

$$D_X = D_{Xi} + \frac{D_{Xj} - D_{Xi}}{L}X$$

This gives for $[A]$, in the local coordinate system,

$$[A] = \begin{bmatrix} 1 - \dfrac{X}{L} & \dfrac{X}{L} \end{bmatrix} \tag{9.87}$$

The mass matrix $[\bar{M}]$ is, then, given by

$$[\bar{M}] = \rho \int_V \begin{bmatrix} 1 - \dfrac{X}{L} \\[2mm] \dfrac{X}{L} \end{bmatrix} \begin{bmatrix} 1 - \dfrac{X}{L} & \dfrac{X}{L} \end{bmatrix} dV$$

or

$$[\bar{M}] = \rho(\text{area}) \int_V \begin{bmatrix} 1 - \dfrac{2X}{L} + \dfrac{X^2}{L^2} & \dfrac{X}{L} - \dfrac{X^2}{L^2} \\[3mm] \dfrac{X}{L} - \dfrac{X^2}{L^2} & \dfrac{X^2}{L^2} \end{bmatrix} dx$$

or

$$[\bar{M}] = \frac{\rho(\text{area})L}{6} \begin{bmatrix} 2 & 1 \\ 1 & 2 \end{bmatrix} \tag{9.88}$$

The corresponding damping matrix $[\bar{C}]$ is

$$[\bar{C}] = \frac{\mu(\text{area})L}{6} \begin{bmatrix} 2 & 1 \\ 1 & 2 \end{bmatrix} \tag{9.89}$$

Since both of these matrices represent a vector of node forces in the local coordinate system, the same transformation matrix that was used for the stiffness matrix [eq. (9.29)] holds. In the reference coordinate system

$$[M] = [T_R^T][\bar{M}][T_R]$$

where

$$[T_R] = \begin{bmatrix} [\Lambda] & 0 \\ \cdots & \cdots \\ 0 & [\Lambda] \end{bmatrix} \quad \text{and} \quad [\Lambda] = \begin{bmatrix} a_{\bar{X}X} & a_{\bar{X}Y} \\ a_{\bar{Y}X} & a_{\bar{Y}Y} \end{bmatrix}$$

The mass matrix $[M]$ in the reference system is then

$$[M] = \frac{\rho(\text{area})L}{6} \begin{bmatrix} 2[m] & [m] \\ \cdots & \cdots \\ [m] & 2[m] \end{bmatrix}$$

where

$$[m] = \begin{bmatrix} a_{\bar{X}X}^2 & a_{\bar{X}X}a_{\bar{X}Y} \\ a_{\bar{X}X}a_{\bar{X}Y} & a_{\bar{X}Y}^2 \end{bmatrix} \tag{9.90}$$

For a beam element such as that one shown in fig. 9.18, the mass matrix can be developed in precisely the same way. Assume that the displacement function between the nodes can be adequately represented by the cubic polynomial

$$Y = A_0 + A_1 X + A_2 X^2 + A_3 X^3 \tag{9.91}$$

Introducing the node displacements for boundary conditions in this equation, the general displacement function is

$$Y = D_{Yi} + \theta_i X + \left[\frac{3}{L^2}(D_{Yj} - D_{Yi}) - \frac{1}{L}(2\theta_i + \theta_j) \right] X^2$$
$$+ \left[\frac{2}{L^3}(D_{Yi} - D_{Yj}) + \frac{1}{L^2}(\theta_i + \theta_j) \right] X^3$$

$[A]$ then becomes

$$[A] = [A_1 \quad A_2 \quad A_3 \quad A_4]$$

where

$$A_1 = 1 - 3\left(\frac{X}{L}\right)^2 + 3\left(\frac{X}{L}\right)^3$$

$$A_2 = X\left[1 - 2\frac{X}{L} + \left(\frac{X}{L}\right)^2\right]$$

$$A_3 = X\left[3\frac{X}{L} - 2\left(\frac{X}{L}\right)^2\right]$$

$$A_4 = X\left[-\frac{X}{L} + \left(\frac{X}{L}\right)^2\right]$$

If the beam element is prismatic and the material is homogeneous, the mass matrix in the local system is that given by

$$[\bar{M}] = \rho(\text{area}) \int_0^L [A^T][A] \, dx$$

or

$$[\bar{M}] = \rho(\text{area}) \int_0^L [A_1 A_2 A_3 A_4]\{A_1 A_2 A_3 A_4\} \, dx$$

On integration this yields

$$[\bar{M}] = \frac{\rho(\text{area})L}{420} \begin{bmatrix} 156 & 22L & 54 & -13L \\ 22L & 4L^2 & 13L & -3L^2 \\ 54 & 13L & 156 & -22L \\ -13L & -3L^2 & -22L & 4L^2 \end{bmatrix} \tag{9.92}$$

In the reference system

$$[M] = [T_R^T][\bar{M}][T_R] \tag{9.93}$$

where

$$[T_R] = \begin{bmatrix} [\Lambda] & \vdots & 0 \\ \cdots & \cdots & \cdots \\ 0 & \vdots & [\Lambda] \end{bmatrix} \quad \text{and} \quad [\Lambda] = \begin{bmatrix} a_{\bar{x}X} & a_{\bar{x}Y} & 0 \\ a_{\bar{y}X} & a_{\bar{y}Y} & 0 \\ 0 & 0 & 1 \end{bmatrix}$$

The damping matrix $[\bar{C}]$ and $[C]$ for the beam element is

$$[\bar{C}] = \frac{\mu(\text{area})L}{420} \begin{bmatrix} 156 & 22L & 54 & -13L \\ 22L & 4L^2 & 13L & -3L^2 \\ 54 & 13L & 156 & -22L \\ -13L & -3L^2 & -22L & 4L^2 \end{bmatrix} \tag{9.94}$$

and

$$[C] = [T_R^T][\bar{C}][T_R] \tag{9.95}$$

where the transformation matrices are the same ones given above.

As it here has been defined, mass and damping coupling is considered to exist between node translations normal to the element and node rotations. Any interplay between these node deformations *is presumed not to exist*. If interaction of this later type is needed to accurately describe a particular problem, then a new mass matrix based on a different displacement function would have to be defined.

The master mass and damping matrices for the structure are the matrix sums of all the element matrices, expanded to the size of the master matrix. To this must be added any diagonal terms representing lumped parameters. (It is to be noted that the same shortcuts employed in assembling the master stiffness matrix for the solution of static loading problems can be used to put together the dynamic matrices. The same is true with regard to handling the boundary conditions.)

9.11.1 Problems Involving Undamped Free Vibrations

For the special case of undamped free vibration, the matrix equation of motion for the system [eq. (9.76)] becomes

$$[M][\ddot{D}] + [K][D] = 0 \tag{9.96}$$

where the matrices $[M]$, $[\ddot{D}]$, $[K]$, and $[D]$ are as previously defined.

It was shown in chap. 4 that in the absence of damping, a freely vibrating structure exhibits a natural frequency of motion. The form of the matrix equation describing this motion is

$$[D] = [D_0] \sin \omega t \tag{9.97}$$

where $[D_0]$ is a matrix of vibration amplitudes, and ω, as before, is the natural or circular frequency of the assemblage of elements. It then follows that

$$[\ddot{D}] = -\omega^2[D_0] \sin \omega t \tag{9.98}$$

Substituting this into eq. (9.96) yields

$$-[M][D_0]\omega^2 \sin \omega t + [K][D_0] \sin \omega t = 0$$

or

$$[K][D_0] - \omega^2[M][D_0] = 0 \tag{9.99}$$

which can be written in the form

$$[K - \omega^2 M][D_0] = [0] \tag{9.100}$$

If $[D_0]$ is to be anything other than a null matrix (that is, one where all of the elements are zero), the determinant of the coefficient matrix, written in terms of the parameter ω, must be equal to zero. That is,

$$|[K] - \omega^2[M]| = 0 \tag{9.101}$$

Equation (9.101) is referred to as the eigenvalue or characteristic value equation of the system. Its solution yields the desired natural frequencies.†

---

† Methods for solving these types of equations can be found in a number of references. See, for example, R. L. Ketter and S. P. Prawel, Jr., *Modern Methods of Engineering Computation*, McGraw-Hill, New York, 1969; or R. D. Cook, *Concepts and Applications of Finite Element Analysis*, Wiley, New York, 1974.

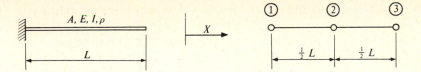

Figure 9.47

Example 9.15 Consider the fixed-ended cantilever member of uniform cross section and material properties shown in fig. 9.47. The natural frequencies of vibration in the axial direction are desired. It will be assumed that the two-element model shown in the second sketch is sufficient to describe the problem adequately. Because longitudinal (axial) vibrations are of concern, the element stiffness and mass matrices will be

$$[M] = \frac{\rho AL}{12} \begin{bmatrix} 2 & 1 \\ 1 & 2 \end{bmatrix} \quad \text{and} \quad [K] = \frac{2AE}{L} \begin{bmatrix} 1 & -1 \\ -1 & 1 \end{bmatrix}$$

where L is the total length of the member and A is the cross-sectional area.
The assembly of the element matrices in the form of eq. (9.100) is

$$\left\{ \frac{2AE}{L} \begin{bmatrix} 1 & -1 & 0 \\ -1 & 2 & -1 \\ 0 & -1 & 1 \end{bmatrix} - \frac{\omega^2 \rho AL}{12} \begin{bmatrix} 2 & 1 & 0 \\ 1 & 4 & 1 \\ 0 & 1 & 2 \end{bmatrix} \right\} \begin{bmatrix} D_{X1_0} \\ D_{X2_0} \\ D_{X3_0} \end{bmatrix} = \begin{bmatrix} 0 \\ 0 \\ 0 \end{bmatrix}$$

Striking the rows and columns (in both matrices) associated with the given boundary condition $D_{X1_0} = 0$, the determinant to be equated to zero is

$$\left| \frac{2AE}{L} \begin{bmatrix} 2 & -1 \\ -1 & 2 \end{bmatrix} - \frac{\omega^2 \rho AL}{12} \begin{bmatrix} 4 & 1 \\ 1 & 2 \end{bmatrix} \right| = 0$$

Combining these terms, and letting

$$v^2 = \frac{\omega^2 L^2 \rho}{24E}$$

there results the following determinant:

$$\left| \begin{matrix} 2(1 - 2v^2) & -(1 + v^2) \\ -(1 + v^2) & (1 - 2v^2) \end{matrix} \right| = 0$$

which can be written in the polynomial form

$$7v^4 - 10v^2 + 1 = 0$$

The roots of this equation are

$$v_1^2 = \frac{5 - 3\sqrt{2}}{7} \quad \text{and} \quad v_2^2 = \frac{5 + 3\sqrt{2}}{7}$$

The corresponding natural frequencies are

$$\omega_1 = 1.6114 \sqrt{\frac{E}{\rho L^2}} \quad \text{and} \quad \omega_2 = 5.6293 \sqrt{\frac{E}{\rho L^2}}$$

These values are, respectively, 2.6 percent and 19.5 percent different from corresponding values obtained from a rigorous, continuous solution. While better accuracy could have been achieved by using more bar elements to model the structure, the amount of computational effort required to extract the eigenvalues would increase significantly.

It is to be observed that whereas the real system has an infinite number of natural frequencies—because of the continuous nature of the structure—the matrix solution yields only as many natural frequencies as there are node-point degrees of freedom. Thus, the matrix solution in the case of dynamic problems does not give results that are "exact" in the same sense that they did for static loading cases. (The reader should consider carefully why this is so.) It is to be further noted that the higher modes of vibration are defined less accurately than are the lower ones. These characteristics of matrix solutions of dynamic problems occasionally lead to problems of interpretation of results for structural systems having a large number of degrees of freedom.

Example 9.16 Bending vibration problems would be handled in the same manner. Consider the cantilever beam shown in fig. 9.48. The member in question is of length L and density ρ and is rigidly attached at one end to a

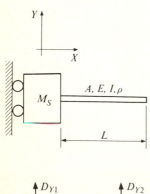

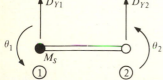

Figure 9.48

lumped mass M_S. Further, it is defined that this mass can move in the Y direction at node 1, but cannot rotate. For purpose of illustration, a one-element model representation of the system will be presumed. For this system:

$$[M_{\text{beam}}] = \frac{\rho A L}{420} \begin{bmatrix} 156 & & & \\ 22L & 4L^2 & \text{(symmetrical)} & \\ 54 & 13L & 156 & \\ -13L & -3L^2 & -22L & 4L^2 \end{bmatrix}$$

$$[M_{\text{lumped}}] = M_S \begin{bmatrix} 1 & & & \\ 0 & 0 & \text{(symmetrical)} & \\ 0 & 0 & 0 & \\ 0 & 0 & 0 & 0 \end{bmatrix}$$

and

$$[K] = \frac{EI}{L^3} \begin{bmatrix} 12 & & & \\ 6L & 4L^2 & \text{(symmetrical)} & \\ -12 & -6L & 12 & \\ 6L & 2L^2 & -6L & 4L^2 \end{bmatrix}$$

After eliminating the rows and columns corresponding to the given boundary condition of zero rotation at node 1, the set to be solved is

$$\left[\frac{EI}{L^3} \begin{bmatrix} 12 & \text{(symmetrical)} & \\ -12 & 12 & \\ 6L & -6L & 4L^2 \end{bmatrix} - \frac{\omega^2 \rho A L}{420} \begin{bmatrix} 156 + \dfrac{420 M_S}{\rho A L} & \text{(symmetrical)} & \\ 54 & 156 & \\ -13L & -22L & 4L^2 \end{bmatrix} \right] \begin{bmatrix} D_{Y1} \\ D_{Y2} \\ \theta_2 \end{bmatrix} = \begin{bmatrix} 0 \\ 0 \\ 0 \end{bmatrix}$$

For other than a trivial solution, the determinant of the coefficient matrix must equal zero. This leads to a cubic equation in ω^2. If it is assumed that M_S equals zero, the three natural frequencies that result are

$$\omega_1 = 0 \qquad \omega_2 = 5.606 \sqrt{\frac{EI}{\rho A L^4}} \qquad \text{and} \qquad \omega_3 = 43.870 \sqrt{\frac{EI}{\rho A L^4}}$$

Corresponding results for a lumped mass M_S of 1.0 and 3.0 times the mass of the beam are shown in the following table, along with an exact solution developed from closed-form equations. It is to be noted that in each case the first natural frequency is zero. (This reflects a rigid body mode and results from the original definition of freedom of motion in the Y direction.)

Mass ratio		Natural frequency of vibration, $$\omega = \eta \sqrt{\dfrac{EI}{\rho A L^4}}$$		
$\dfrac{M_S}{M_{\text{beam}}}$	Mode	One-element solution, η	Closed-form solution, η	Percent error
0	1	0	0	0
	2	5.606	5.593	0.2
	3	43.870	30.226	45.1
1.0	1	0	0	0
	2	4.229	4.219	0.2
	3	36.777	23.707	55.1
3.0	1	0	0	0
	2	3.835	3.822	0.3
	3	35.575	26.677	56.9

If four elements are used to model the beam, nine frequencies result—corresponding to the nine degrees of freedom in the model. These degrees of freedom are

$$[D] = \{D_{Y1} \quad D_{Y2} \quad \theta_2 \quad D_{Y3} \quad \theta_3 \quad D_{Y4} \quad \theta_4 \quad D_{Y5} \quad \theta_5\}$$

The results of the four-element formulation are given in table 9.13.

Table 9.13 Results of four-element formulation

Frequency no.	Mass ratio = 0			Mass ratio = 1.0			Mass ratio = 3.0		
	4-elem.	Exact	Percent error	4-elem.	Exact	Percent error	4-elem.	Exact	Percent error
1	0	0	0	0	0	0	0	0	0
2	5.594	5.593	0.0	4.219	4.219	0.0	3.821	3.822	0.0
3	30.290	30.226	0.2	23.731	23.707	0.1	22.701	22.677	0.1
4	75.436	74.639	1.1	63.934	63.457	0.8	62.813	62.331	0.8
5	140.680	138.791	1.4	124.344	122.722	1.3	123.254	121.551	1.4
6	248.448	222.683	11.6	229.768	201.725	13.9	228.710	200.506	14.1
7	392.238	326.314	20.2	368.131	300.442	22.5	366.996	299.209	22.7
8	603.554	449.684	34.2	582.223	418.894	39.0	581.326	417.647	39.2
9	954.671	592.793	61.0	953.153	557.081	71.1	953.087	555.822	71.5

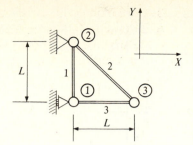

Figure 9.49

Example 9.17 Consider, next, the pin-connected truss shown in fig. 9.49. It is assumed that the cross-sectional areas are the same for all members, as are their material properties. The numerical designations of nodes and elements are those shown.

The stiffness and mass matrices for the various elements—in the reference system—are

$$[K] = \frac{AE}{L}\left[\begin{array}{c:c} [B] & [-B] \\ \hdashline [-B] & [B] \end{array}\right] \quad \text{and} \quad [M] = \frac{\rho A L}{3}\left[\begin{array}{c:c} [m] & \left[\dfrac{m}{2}\right] \\ \hdashline \left[\dfrac{m}{2}\right] & [m] \end{array}\right]$$

where

$$[B] = [m] = \begin{bmatrix} a_{\bar{X}X}^2 & a_{\bar{X}X}\,a_{\bar{X}Y} \\ a_{\bar{X}X}\,a_{\bar{X}Y} & a_{\bar{X}Y}^2 \end{bmatrix}$$

A tabular arrangement is convenient to define the matrix entries for the various elements.

Element	Length	$\theta_{\bar{X}X}$	$\theta_{\bar{X}Y}$	$a_{\bar{X}X}$	$a_{\bar{X}Y}$	$a_{\bar{X}X}^2$	$a_{\bar{X}X}\,a_{\bar{X}Y}$	$a_{\bar{X}Y}^2$
1	L	90°	0	0	1.000	0	0	1.000
2	$\sqrt{2}\,L$	45°	135°	0.707	−0.707	0.500	−0.500	0.500
3	L	0	90°	1.000	0	1.000	0	0

For element 1:
$$[_1B] = [_1m] = \begin{bmatrix} 0 & 0 \\ 0 & 1 \end{bmatrix}$$

For element 2:

$$[_2B] = \begin{bmatrix} \dfrac{\sqrt{2}}{4} & -\dfrac{\sqrt{2}}{4} \\[2ex] -\dfrac{\sqrt{2}}{4} & \dfrac{\sqrt{2}}{4} \end{bmatrix}$$

$$[_2m] = \begin{bmatrix} \dfrac{\sqrt{2}}{2} & -\dfrac{\sqrt{2}}{2} \\[2ex] -\dfrac{\sqrt{2}}{2} & \dfrac{\sqrt{2}}{2} \end{bmatrix}$$

For element 3:

$$[_3B] = [_3m] = \begin{bmatrix} 1 & 0 \\ 0 & 0 \end{bmatrix}$$

The assembled mass and stiffness matrices are

$$[M] = \tfrac{1}{3}\rho AL \begin{bmatrix} [_1m + {}_3m] & [\tfrac{1}{2}\,_1m] & [\tfrac{1}{2}\,_3m] \\ [\tfrac{1}{2}\,_1m] & [_1m + {}_2m] & [\tfrac{1}{2}\,_2m] \\ [\tfrac{1}{2}\,_3m] & [\tfrac{1}{2}\,_2m] & [_2m + {}_3m] \end{bmatrix}$$

and

$$[K] = \frac{AE}{L} \begin{bmatrix} [_1B + {}_3B] & [-{}_1B] & [-{}_3B] \\ [-{}_1B] & [_1B + {}_2B] & [-{}_2B] \\ [-{}_3B] & [-{}_2B] & [_2B + {}_3B] \end{bmatrix}$$

Introducing the displacement constraints at nodes 1 and 2, the formulation becomes

$$\left[\begin{bmatrix} 1 & 0 & 0 \\ 0 & 1+\dfrac{\sqrt{2}}{4} & -\dfrac{\sqrt{2}}{4} \\ 0 & -\dfrac{\sqrt{2}}{4} & \dfrac{\sqrt{2}}{4} \end{bmatrix} - \frac{\rho L^2 \omega^2}{3E} \begin{bmatrix} 1 & 0 & 0 \\ 0 & 1+\dfrac{\sqrt{2}}{2} & -\dfrac{\sqrt{2}}{2} \\ 0 & -\dfrac{\sqrt{2}}{2} & \dfrac{\sqrt{2}}{2} \end{bmatrix} \right] \begin{bmatrix} D_{Y1} \\ D_{X3} \\ D_{Y3} \end{bmatrix} = \begin{bmatrix} 0 \\ 0 \\ 0 \end{bmatrix}$$

The characteristic equation, then, is

$$\begin{vmatrix} 1-\lambda & 0 & 0 \\ 0 & 1+\dfrac{\sqrt{2}}{4}-\lambda\left(1+\dfrac{\sqrt{2}}{2}\right) & -\dfrac{\sqrt{2}}{4}+\lambda\dfrac{\sqrt{2}}{2} \\ 0 & -\dfrac{\sqrt{2}}{4}+\lambda\dfrac{\sqrt{2}}{2} & \dfrac{\sqrt{2}}{4}-\lambda\dfrac{\sqrt{2}}{2} \end{vmatrix} = 0$$

where

$$\lambda = \frac{\rho L^2 \omega^2}{3E}$$

The real roots of this equation are $\lambda = 1, 1, \frac{1}{2}$. For the particular case where $\lambda = 1.0$,

$$\omega = \sqrt{\frac{3E}{2\rho L^2}}$$

The mode shape for this natural frequency can be defined—at least relatively defined—by replacing ω in the above homogeneous set and solving for the displacements. The governing equations are

$$\begin{bmatrix} 0 & 0 & 0 \\ 0 & -\dfrac{AE}{L} & \dfrac{AE}{L} \\ 0 & \dfrac{AE}{L} & -\dfrac{AE}{L} \end{bmatrix} \begin{bmatrix} D_{Y1} \\ D_{X3} \\ D_{Y3} \end{bmatrix} = \begin{bmatrix} 0 \\ 0 \\ 0 \end{bmatrix}$$

Assuming D_{Y3} equal to one, the relative node displacements are

$$[D] = \{0 \quad -1 \quad +1\}$$

(It should be recognized that an infinite number of particular solutions could be found for these equations, but they would all have the same relative displacement ratios as those listed.) For the natural frequency corresponding to $\lambda = 1.0$ the deformational response of the node points would be

$$D_{Y1} = 0$$

$$D_{X3} = -\sin \omega t$$

$$D_{Y3} = \sin \omega t$$

9.11.2 Problems Involving Forced Vibrations

The more general dynamics problem involves the calculation of the time history of response to changing loads or deformations. This general topic was discussed in chap. 4.

The principle value of the matrix approach in solving these types of dynamic structural problems lies in the ease with which the equations of motion can be written—even for very complicated structures. The process is simply to implement eq. (9.76) using the appropriate mass, damping, and stiffness matrices. Introducing the boundary constraints yields a set of coupled differential equations that can be solved using, for example, the Runge-Kutta procedure.

To illustrate the general solution process, assume that the truss defined in example 9.17 is subjected to a time-dependent load Q that has no mass and acts vertically downward at node 3. Further, assume that that load is suddenly applied from the at rest position and is not removed. Still further, presume that it is only the maximum deformations that are of interest.

It is to be recalled that the effect of damping is to reduce the amplitude of vibration—with time. Since only the maximum values are desired, damping, therefore, need not be taken into account. The equations of motion, then, are

$$\frac{\rho AL}{3}\begin{bmatrix} 1 & 0 & 0 \\ 0 & 1+\dfrac{\sqrt{2}}{2} & -\dfrac{\sqrt{2}}{2} \\ 0 & -\dfrac{\sqrt{2}}{2} & \dfrac{\sqrt{2}}{2} \end{bmatrix}\begin{bmatrix} \ddot{D}_{Y1} \\ \ddot{D}_{X3} \\ \ddot{D}_{Y3} \end{bmatrix} + \frac{AE}{L}\begin{bmatrix} 1 & 0 & 0 \\ 0 & 1+\dfrac{\sqrt{2}}{4} & -\dfrac{\sqrt{2}}{4} \\ 0 & -\dfrac{\sqrt{2}}{4} & \dfrac{\sqrt{2}}{4} \end{bmatrix}\begin{bmatrix} D_{Y1} \\ D_{X3} \\ D_{Y3} \end{bmatrix} = \begin{bmatrix} 0 \\ 0 \\ -Q \end{bmatrix}$$

Inverting the mass matrix and multiplying yields

$$\begin{bmatrix} \ddot{D}_{Y1} \\ \ddot{D}_{X3} \\ \ddot{D}_{Y3} \end{bmatrix} + \begin{bmatrix} \dfrac{3E}{\rho L^2} & 0 & 0 \\ 0 & \dfrac{3E}{\rho L^2} & 0 \\ 0 & \dfrac{3E}{\rho L^2}\left(\dfrac{5\sqrt{2}}{4}+1\right) & \dfrac{3E}{\rho L^2}\dfrac{1+\sqrt{2}}{2} \end{bmatrix}\begin{bmatrix} D_{Y1} \\ D_{X3} \\ D_{Y3} \end{bmatrix} = \begin{bmatrix} 0 \\ -\dfrac{3Q}{\rho AL} \\ -\dfrac{2+\sqrt{2}}{\sqrt{2}}\dfrac{3Q}{\rho AL} \end{bmatrix}$$

This is a set of three second-order differential equations—of the initial-value type. The motion of nodes 1 and 3 are defined by these equations, plus the two sets of initial conditions

$$\begin{bmatrix} D_{Y1} \\ D_{X3} \\ D_{Y3} \end{bmatrix}_{t=0} = \begin{bmatrix} \dot{D}_{Y1} \\ \dot{D}_{X3} \\ \dot{D}_{Y3} \end{bmatrix}_{t=0} = 0$$

Consider the first of these differential equations,

$$\ddot{D}_{Y1} + \frac{3E}{\rho L^2}D_{Y1} = 0$$

with the boundary conditions

$$D_{Y1} = \dot{D}_{Y1} = 0 \qquad \text{at } t = 0$$

The solution is

$$D_{Y1} = C_1 \sin \omega t + C_2 \cos \omega t$$

which when differentiated gives

$$\dot{D}_{Y1} = C_1 \omega \cos \omega t - C_2 \omega \sin \omega t$$

Substituting these into the boundary conditions,

$$\begin{bmatrix} 0 & 1 \\ \omega & 0 \end{bmatrix}\begin{bmatrix} C_1 \\ C_2 \end{bmatrix} = \begin{bmatrix} 0 \\ 0 \end{bmatrix}$$

Since

$$\begin{bmatrix} 0 & 1 \\ \omega & 0 \end{bmatrix} \neq 0 \qquad \begin{bmatrix} C_1 \\ C_2 \end{bmatrix} = \begin{bmatrix} 0 \\ 0 \end{bmatrix}$$

and D_{Y1} must be zero. (Note that a static analysis of the truss indicates that there will be no force in element 1 for the loading defined.) The problem thus becomes

$$\begin{bmatrix} \ddot{D}_{X3} \\ \ddot{D}_{Y3} \end{bmatrix} + \frac{3E}{\rho L^2} \begin{bmatrix} 1 & 0 \\ \dfrac{5\sqrt{2}}{4} + 1 & \dfrac{1+\sqrt{2}}{2} \end{bmatrix} \begin{bmatrix} D_{X3} \\ D_{Y3} \end{bmatrix} = -\frac{3Q}{\rho AL} \begin{bmatrix} 1 \\ 2+\sqrt{2} \\ \sqrt{2} \end{bmatrix}$$

subject to

$$\begin{bmatrix} \dot{D}_{X3} \\ \dot{D}_{Y3} \end{bmatrix}_{t=0} = \begin{bmatrix} D_{X3} \\ D_{Y3} \end{bmatrix}_{t=0} = \begin{bmatrix} 0 \\ 0 \end{bmatrix}$$

Introducing the substitution

$$D_{X5} = D_{X3} \qquad \text{and} \qquad D_{Y5} = D_{Y3}$$

there will be obtained the following set of four first-order differential equations:

$$\begin{bmatrix} \dot{D}_{X3} \\ \dot{D}_{X5} \\ \dot{D}_{Y3} \\ \dot{D}_{Y5} \end{bmatrix} = -\frac{3E}{\rho L^2} \begin{bmatrix} 0 & -\dfrac{\rho L^2}{3E} & 0 & 0 \\ 1 & 0 & 0 & 0 \\ 0 & 0 & -\dfrac{\rho L^2}{3E} & 0 \\ \dfrac{5\sqrt{2}}{4} + 1 & 0 & \dfrac{1+\sqrt{2}}{4} & 0 \end{bmatrix} \begin{bmatrix} D_{X3} \\ D_{X5} \\ D_{Y3} \\ D_{Y5} \end{bmatrix} - \frac{3Q}{\rho AL} \begin{bmatrix} 0 \\ 1 \\ 0 \\ 2+\sqrt{2} \\ \sqrt{2} \end{bmatrix}$$

with the boundary conditions

$$\{D_{X3} \quad D_{X5} \quad D_{Y3} \quad D_{Y5}\}_{t=0} = 0$$

These can be solved using the fourth-order Runge-Kutta procedure described in chap. 4.

$$\Delta f_i = \tfrac{1}{6}[K_{1i} + 2K_{2i} + 2K_{3i} + K_{4i}] \qquad i = 1, 2, \dots, n$$

$$K_{1i} = hf_i[t_0, (X_{j0}, j = 1, 2, \dots, n)] \qquad i = 1, 2, \dots, n$$

$$K_{2i} = hf_i\left[t_0 + \frac{h}{2}, \left(X_{j0} + \frac{K_{1j}}{2}, j = 1, 2, \dots, n\right)\right] \qquad i = 1, 2, \dots, n$$

$$K_{3i} = hf_i\left[t_0 + \frac{h}{2}, \left(X_{j0} + \frac{K_{2j}}{2}, j = 1, 2, \dots, n\right)\right] \qquad i = 1, 2, \dots, n$$

$$K_{4i} = hf_i[t_0 + h, (X_{j0} + K_{3j}, j = 1, 2, \dots, n)] \qquad i = 1, 2, \dots, n$$

where n = number of first-order equations to be solved

$\quad h$ = step size in time (that is, the time increment in the numerical solution)

$\quad \Delta f_i$ = change in the function f_i over one step h

$\quad i$ = one line of the matrix equation listed above

$\quad X_{jo}$ = initial values at $t = t_0$ of the n dependent variables

If the applied load is a function of time rather than a constant as was the case just discussed, the solution for the undamped response would be accomplished in the same manner. All that need be changed is the load-vector column. Had initial displacements or velocities been specified, all these would modify would be the initial conditions. A forced displacement that is a function of time, however, represents an entirely different type of situation. Here, the forced displacements would have to be converted into equivalent node point forces, and these would have to be treated as if they were time-dependent loadings.

In the preceding example, the motion of node 3 was defined by the set

$$\frac{\rho A L}{3} \begin{bmatrix} 1 + \dfrac{\sqrt{2}}{2} & -\dfrac{\sqrt{2}}{2} \\ -\dfrac{\sqrt{2}}{2} & \dfrac{\sqrt{2}}{2} \end{bmatrix} \begin{bmatrix} \ddot{D}_{X3} \\ \ddot{D}_{Y3} \end{bmatrix} + \frac{AE}{L} \begin{bmatrix} 1 + \dfrac{\sqrt{2}}{4} & -\dfrac{\sqrt{2}}{4} \\ -\dfrac{\sqrt{2}}{4} & \dfrac{\sqrt{2}}{4} \end{bmatrix} \begin{bmatrix} D_{X3} \\ D_{Y3} \end{bmatrix} = \begin{bmatrix} 0 \\ -Q \end{bmatrix}$$

If the displacement D_{X3} were now prescribed as a function of time; that is,

$$D_{X3} = f_3(t)$$

and

$$\ddot{D}_{X3} = \ddot{f}_3(t)$$

then substitution of these into the set yields

$$\frac{\rho A L}{3} \frac{\sqrt{2}}{2} \ddot{D}_{Y3} + \frac{AE\sqrt{2}}{4L} D_{Y3} = -Q + \frac{\rho A L \sqrt{2}}{6} \ddot{f}_3(t) + \frac{AE\sqrt{2}}{4L} f_3(t)$$

which can be readily solved for D_{Y3}.

In all the above, the mass and damping effects that were referred to are those of the element itself. The mass is the mass of the structural member, and damping is that which occurs within the member. In many cases there needs to be introduced into the analysis other masses and damping forces. For example, the mass of a load Q—if it exists—must be taken into account. The frictional resistance to slippage of a connection is a type of damping force that also may be of significance and should be included. Since these forces, by and large, only depend upon location and direction, they should not be treated in the same way as the element mass and damping effects derived above. Actually, they are the same kind of thing that normally would be considered in a lumped mass model. For example, the effect of mass inertia of a concentrated load Q is given by

$$\frac{Q}{g} \ddot{D}_{Yi}$$

where g is the acceleration of gravity. In the preceding triangular truss problem the total mass effect, then, is

$$\frac{\rho AL}{3}\begin{bmatrix} 1+\dfrac{\sqrt{2}}{2} & -\dfrac{\sqrt{2}}{2} \\ -\dfrac{\sqrt{2}}{2} & \dfrac{\sqrt{2}}{2} \end{bmatrix}\begin{bmatrix} \ddot{D}_{X3} \\ \ddot{D}_{Y3} \end{bmatrix} + \frac{1}{g}\begin{bmatrix} 0 & 0 \\ 0 & +Q \end{bmatrix}\begin{bmatrix} \ddot{D}_{X3} \\ \ddot{D}_{Y3} \end{bmatrix}$$

or

$$\frac{\rho AL}{3}\begin{bmatrix} 1+\dfrac{\sqrt{2}}{2} & -\dfrac{\sqrt{2}}{2} \\ -\dfrac{\sqrt{2}}{2} & \dfrac{\sqrt{2}}{2}+\dfrac{3Q}{g\rho AL} \end{bmatrix}\begin{bmatrix} \ddot{D}_{X3} \\ \ddot{D}_{Y3} \end{bmatrix}$$

The variations of types and kinds of problems in structural dynamics are almost infinite in number: the kinds of damping that occur, partial rigidity of connections, the influence of axial thrust on bending stiffness, inelastic behavior, interaction of machine element and structure, etc. All these plus many other topics could be examined. That is not the purpose of this text. For more detailed discussion, the reader is referred to other more specialized works.

9.12 PROBLEMS

By matrix methods, solve the following previously defined problems:

9.1 Example 8.6.

9.2 Example 8.7.

9.3 Example 8.9.

9.4 Example 8.10—Assume $\alpha = 1.0$, $Q = 1$ kip, $q = 1$ kip/ft and $L = 10$ ft.

9.5 Example 8.11—Assume $\alpha = 1.0$, $\Delta = 0.1$ ft, $l = 10$ ft.

9.6 Example 8.12.

9.7 Example 8.13.

9.8 Example 8.15—Assume $L = 30$ ft, $H = 12$ ft, $V = 100$ kips, $q = 2$ kips/ft, $I_2 = 2I_1$.

9.9 Problem 8.4.

9.10 Problem 8.6.

9.11 Problem 8.9—Assume the cross section to be an 8-in-diameter, $\frac{1}{4}$-in-thick steel pipe.

9.12 Problem 8.11—Assume $L = 40$ ft, $H = 12$ ft, $F = 5$ ft, $q = 5$ kips/ft, $I_1 = I_2$.

9.13 Problem 8.13.

9.14 Problem 8.16.

9.15 Problem 8.20.

9.16 Determine the stiffness matrix for the element defined in problem 8.15.

9.17 Determine the stiffness matrix for the element defined in problem 8.24.

9.18 Example 6.1—Assume $\alpha L^3/EI = 0$, 5.0, 10.0, and 20.0.

9.19 Example 6.2.

9.20 Example 6.3—Assume $b/a = 0.25$, 0.50, 0.75, and 1.00.

9.21 Example 6.6—Assume $I_1/I_2 = 0.25$, 0.50, 0.75, and 1.00.

9.13 APPENDIX—ALTERNATIVE WAYS TO DERIVE STIFFNESS MATRICES

All the stiffness matrices derived in the first part of this chapter (secs. 9.1 to 9.8) were developed from closed-form solutions of appropriate differential equations subject to specified boundary conditions. While such a procedure leads to the required stiffness terms, for one reason or another, it may not be the best method for handling a given situation. For example, it was noted in sec. 9.9 that a rigorous solution of the differential equation associated with a constant-width, simply tapered, rectangular cross-section member leads to the handling of Bessel functions of both the first and second kind. Obviously, other variations in cross section would have other solutions. In fact, it must be understood that for some variations closed-form solutions may not even exist. For these, as well as other reasons, it is desirable to have alternative ways to define stiffness matrices. In this appendix, several of these will be discussed.

9.13.1 Methods Using Strain Energy

In chap. 5 it was shown that the strain energy stored in a one-dimensional member subjected to axial thrust, vertical shear, and bending and twisting moments is

$$W_i = \int_0^L \frac{P^2}{2AE} dx + \lambda \int_0^L \frac{V^2}{2AG} dx + \int_0^L \frac{M^2}{2EI} dx + \int_0^L \frac{T_{sv}^2}{2G\kappa_T} dx + \int_0^L \frac{T_\omega^2}{2EI_\omega} dx$$

(9.102)

This can be expressed in terms of the strains or displacements along the member:

$$W_i = \int_0^L \frac{AE}{2} \varepsilon^2 dx + \lambda \int_0^L \frac{AG}{2} \gamma^2 dx + \int_0^L \frac{EI}{2} \left(\frac{d^2 y}{dx^2}\right)^2 dx + \int_0^L \frac{G\kappa_T}{2} \left(\frac{d\theta_x}{dx}\right)^2 dx$$

$$+ \int_0^L \frac{EI_\omega}{2} \left(\frac{d^2\theta_x}{dx^2}\right) dx \quad (9.103)$$

However, when solving problems by the matrix method, it is necessary that there be available stiffness expressions which involve loads and displacements only at the node points. Equations of this type, therefore, must be developed, and these, in appropriate ways, must be related to either energy expressions (9.102) or (9.103).

Assume a general function of the form

$$W_i = \frac{1}{2} \sum_{k=1}^n F_k D_k$$

(9.104)

where F_k and D_k represent, respectively, a particular node force and a displacement defined by the column vectors

$$[D] = \begin{bmatrix} D_1 \\ D_2 \\ \vdots \\ D_n \end{bmatrix} = \{D_1 \quad D_2 \quad \cdots \quad D_n\} \quad \text{and} \quad [F] = \begin{bmatrix} F_1 \\ F_2 \\ \vdots \\ F_n \end{bmatrix} = \{F_1 \quad F_2 \quad \cdots \quad F_n\}$$

In expanded form, eq. (9.104) would be

$$W_i = \tfrac{1}{2}(F_1 D_1 + F_2 D_2 + F_3 D_3 + \cdots + F_n D_n) \tag{9.105}$$

or

$$W_i = \tfrac{1}{2}[F_1 \quad F_2 \quad F_3 \quad \cdots \quad F_n]\{D_1 \quad D_2 \quad D_3 \quad \cdots \quad D_n\} = \tfrac{1}{2}[F^T][D] \tag{9.106}$$

It is to be recalled, however, that the basic form of the element stiffness formulation is

$$[K][D] = [F]$$

The transpose of this equation would be equally valid; that is

$$[F^T] = [D^T][K^T]$$

or

$$[F^T] = [D_1 \quad D_2 \quad D_3 \quad \cdots \quad D_n]
\begin{bmatrix}
K_{11} & K_{21} & K_{31} & \cdots & K_{n1} \\
K_{12} & K_{22} & K_{32} & \cdots & K_{n2} \\
K_{13} & K_{23} & K_{33} & \cdots & K_{n3} \\
\multicolumn{5}{c}{\cdots\cdots\cdots\cdots\cdots} \\
K_{1n} & K_{2n} & K_{3n} & \cdots & K_{nn}
\end{bmatrix} \tag{9.107}$$

Substituting eq. (9.107) into (9.106) yields

$$W_i = \tfrac{1}{2}[F^T][D] = \tfrac{1}{2}[D^T][K^T][D] \tag{9.108}$$

or

$$W_i = \tfrac{1}{2}[D_1 \quad D_2 \quad \cdots \quad D_n]
\begin{bmatrix}
K_{11} & K_{21} & \cdots & K_{n1} \\
K_{12} & K_{22} & \cdots & K_{n2} \\
\multicolumn{4}{c}{\cdots\cdots\cdots\cdots} \\
K_{1n} & K_{2n} & \cdots & K_{nn}
\end{bmatrix}$$

Performing this matrix multiplication and collecting like terms, there is obtained

$$W_i = \sum_{i=1}^{n} \left(\tfrac{1}{2} D_i^2 K_{ii} + \sum_{j=i+1}^{n} D_i D_j K_{ij} \right) \tag{9.109}$$

If this expression is differentiated with respect to the node-point displacements, the values of the various entries in the stiffness matrix are obtained:

$$K_{ij} = K_{ji} = \frac{\partial^2 W_i}{\partial D_i \, \partial D_j} \tag{9.110}$$

The basic problem in the use of energy methods for determining stiffness expressions is the selection of appropriate displacement functions which can adequately describe the deformed shape of the element, and which can be substituted into eqs. (9.102) or (9.103).

It is to be noted that up to this point no restrictions have been placed on the type or kind of member under consideration. Equation (9.110) is for a general element having n degrees of freedom. Furthermore, no assumptions have been made, thus far, concerning the displacement function. Clearly, since this function can take on a variety of forms, it may or may not represent the exact solution to a given problem.

Herein lies the basic difference between the finite-element approach—as it recently has become known—and the method used in the main portions of this chapter—which is usually referred to as the matrix solution. If the selected displacement function does not contain the exact solution for the element in question, then the procedure—by definition—is of the finite-element type, and there will result an approximation to the exact solution. If on the other hand the presumed function contains the exact deformation solution, the resulting entries in the stiffness matrix will be the same as those obtained from closed-form equations.

To illustrate the general procedure, consider the development of the stiffness matrix for a simple beam-bending element (see fig. 9.18). Since no lateral loads are applied between the node points, the vertical shear along the member is constant. Moreover, since shear is proportional to the third derivative of the displacement, an assumed displacement function of the cubic polynomial form is appropriate.

$$Y(X) = P_3(X) = A_0 + A_1 X + A_2 X^2 \text{ and } A_3 X^3 \qquad (9.111)$$

(It is to be understood that this is only one of a number of assumptions that could have been made. For example, sine functions of the $n\pi X/L$ argument form could just as well have been assumed. From earlier work in chaps. 4 and 5, however, it is known that a polynomial will be the correct solution. Therefore the "obvious" choice was made, and it is to be expected that the resulting stiffness matrix will be "exact.") For this assumption the slopes and displacements at nodes i and j are given by

$$\begin{aligned}
D_{Yi} &= A_0 + A_1(0) + A_2(0)^2 + A_3(0)^3 = A_0 \\
\theta_i &= A_1 + 2A_2(0) + 3A_3(0)^2 = A_1 \\
D_{Yj} &= A_0 + A_1(L) + A_2(L)^2 + A_3(L)^3 \\
\theta_j &= A_1 + 2A_2(L) + 3A_3(L)^2
\end{aligned} \qquad (9.112)$$

Solving these equations for the A's in terms of the node deformations,

$$\begin{aligned}
A_0 &= D_{Yi} \\
A_1 &= \theta_i \\
A_2 &= \frac{3}{L^2}(D_{Yj} - D_{Yi}) - \frac{1}{L}(2\theta_i + \theta_j) \\
A_3 &= \frac{2}{L^3}(D_{Yi} - D_{Yj}) + \frac{1}{L^2}(\theta_i + \theta_j)
\end{aligned} \qquad (9.113)$$

The displacement function, expressed in terms of the end slopes and displacements, is therefore

$$\begin{aligned}
y(X) = D_{Yi} + \theta_i X &+ \left[\frac{3}{L^2}(D_{Yj} - D_{Yi}) - \frac{1}{L}(2\theta_i + \theta_j)\right]X^2 \\
&+ \left[\frac{2}{L^3}(D_{Yi} - D_{Yj}) + \frac{1}{L^2}(\theta_i + \theta_j)\right]X^3
\end{aligned} \qquad (9.114)$$

The second derivative of eq. (9.114) with respect to the distance along the member is

$$\frac{d^2 y(X)}{dX} = \frac{6}{L^2}(D_{Yj} - D_{Yi}) - \frac{2}{L}(2\theta_i + \theta_j) + \frac{12}{L^3}(D_{Yi} - D_{Yj})X + \frac{6}{L^2}(\theta_i + \theta_j)X \quad (9.115)$$

Assuming that the element is prismatic and that only bending strain energy is of concern, from eq. (9.103)

$$W_i = \frac{EI}{2} \int_0^L \left(\frac{d^2 y}{dx^2}\right)^2 dx$$

Substituting into this expression the values indicated in eqs. (9.115) and (9.112), there is obtained

$$W_i = \frac{EI}{2}(4A_2^2 L + 12A_2 A_3 L^2 + 12A_3^2 L^3) \quad (9.116)$$

Since from eq. (9.110)

$$K_{ij} = \frac{\partial^2 W_i}{\partial D_i \, \partial D_j}$$

$$K_{11} = \frac{\partial^2 W_i}{\partial D_{Yi}^2} = \frac{\partial}{\partial D_{Yi}}\left\{\frac{EI}{2}\left[8A_2 L \frac{\partial A_2}{\partial D_{Yi}} + 12L^2\left(A_2 \frac{\partial A_3}{\partial D_{Yi}} + A_3 \frac{\partial A_2}{\partial D_{Yi}}\right) + 24A_3 L^3 \frac{\partial A_3}{\partial D_{Yi}}\right]\right\}$$

or

$$K_{11} = \frac{\partial}{\partial D_{Yi}}\left[\frac{EI}{2}\left(-\frac{24}{L}A_2 + \frac{24}{L}A_2 - 36A_3 + 48A_3\right)\right]$$

or

$$K_{11} = 6EI \frac{\partial A_3}{\partial D_{Yi}} = 6EI \frac{2}{L^3} = \frac{12EI}{L^3} \quad (9.117)$$

Repeating this process for each column and row of the stiffness matrix, there is obtained

$$[K] = \begin{bmatrix} \dfrac{12EI}{L^3} & \dfrac{6EI}{L^2} & -\dfrac{12EI}{L^3} & \dfrac{6EI}{L^2} \\[2ex] \dfrac{6EI}{L^2} & \dfrac{4EI}{L} & -\dfrac{6EI}{L^2} & \dfrac{2EI}{L} \\[2ex] -\dfrac{12EI}{L^3} & -\dfrac{6EI}{L^2} & \dfrac{12EI}{L^3} & -\dfrac{6EI}{L^2} \\[2ex] \dfrac{6EI}{L^2} & \dfrac{2EI}{L} & -\dfrac{6EI}{L^2} & \dfrac{4EI}{L} \end{bmatrix} \quad (9.118)$$

It is to be recognized that while the absolute values of the various entries in this matrix are the same as those in eq. (9.39), the signs of several of the terms are different. This merely reflects the difference in the bending-sign convention (which was assumed in chaps. 4 and 5) and the special convention defined for matrix methods (fig. 9.17). For matrix (9.118), deflections and end shears upward at i and j are defined as positive, as are counterclockwise moments and rotations.

9.13.2 Determination of Stiffness Matrices Using Castigliano's Method

Castigliano's method for determining deflections [eq. (5.56)] also can be written in a form suitable for determining forces associated with particular assumed displacement functions:

$$F_j = \frac{\partial}{\partial D_j} W_i \tag{9.119}$$

where F_j = force in the direction and at the node defined by the subscript j
D_j = the corresponding node-point displacement
W_i = the strain energy stored in the element

Equation (9.119) is actually the jth equation of a set defined by $[K][D] = [F]$. The whole set is

$$[F] = \frac{\partial}{\partial \{D\}} W_i \tag{9.120}$$

That is, the force vector $[F]$ is equal to the first derivative with respect to the displacement vector of the strain energy.

For the axially loaded element shown in fig. 9.9:

$$[F] = \{F_{xi} \quad F_{xj}\}$$

$$[D] = \{D_{xi} \quad D_{xj}\}$$

The displacement function is

$$D_x = D_{xi} + \frac{D_{xj} - D_{xi}}{L}$$

or

$$D_x = [A][D]$$

where

$$[A] = \left[1 - \frac{X}{L} \quad \frac{X}{L}\right]$$

The strain-displacement relationship for axially loaded, prismatic members is

$$\varepsilon_x = \frac{d}{dx}(D_x) = \frac{d}{dx}[A][D]$$

Thus, the unit strain is

$$\varepsilon_x = \left[-\frac{1}{L} \quad \frac{1}{L}\right]\{D\} = \frac{1}{L}[-1 \quad 1]\{D\}$$

The total strain energy stored in the member in question is

$$W_i = \frac{1}{2} \int_V \varepsilon^T \sigma \, dV$$

Since $\sigma = E\varepsilon$, where E is Young's modulus of elasticity,

$$W_i = \frac{1}{2} \int_V \{D\}^T \frac{1}{L} \begin{bmatrix} -1 \\ 1 \end{bmatrix} \frac{E}{L} [-1 \quad 1] \{D\} \, dV$$

or

$$W_i = \frac{EA}{2L} \{D\}^T \begin{bmatrix} 1 & -1 \\ -1 & 1 \end{bmatrix} \{D\}$$

The force vector is then

$$[F] = \frac{\partial}{\partial \{D\}} W_i = \frac{EA}{L} \begin{bmatrix} 1 & -1 \\ -1 & 1 \end{bmatrix} \{D\}$$

and the element stiffness matrix is

$$[K] = \frac{EA}{L} \begin{bmatrix} 1 & -1 \\ -1 & 1 \end{bmatrix}$$

This is the same matrix that is given in eq. (9.20).

9.13.3 Determination of Stiffness Matrices Using the Principle of Virtual Displacements

Let $[\varepsilon]$ be a vector of strains that is related to a vector of node displacements $[D]$ by the matrix relationship

$$[\varepsilon] = [B][D] \tag{9.121}$$

Further, let $[\sigma]$ be a vector of stresses that is related to the strains $[\varepsilon]$ by the relationship

$$[\sigma] = [X][\varepsilon] \tag{9.122}$$

(For a homogeneous, isotropic material for which stresses increase linearly with strains, the relationships between the normal stresses and strains are $[X] = E$, Young's modulus of elasticity.)

If there is introduced into the system in question a set of virtual node point displacements $[D]$, there will be the following amount of virtual external work done:

$$W_e = [D^T][F] \tag{9.123}$$

where $[F]$ is a vector of node-point forces that correspond to the displacements $[D]$.

The virtual internal work done by this same disturbance is given by

$$W_i = \int_V [\varepsilon^T][\sigma] \, dV \tag{9.124}$$

where the integration, as indicated, is over the entire volume of the element. Since $[\varepsilon] = [B][D]$, $[\varepsilon^T] = [B^T][D^T]$ and

$$W_i = [D^T] \int_V [B^T][\sigma] \, dV \tag{9.125}$$

(It is to be noted that the displacements $[D]$ are not included within the integration—since they are node-point displacements.) In addition since $[\sigma] = [X][\varepsilon] = [X][B][D]$,

$$W_i = [D^T] \int_V [B^T][X][B] \, dV \times [D] \tag{9.126}$$

Equating the internal to external work done due to the virtual displacement,

$$[D^T][F] = [D^T] \int_V [B^T][X][B] \, dV \times [D]$$

or

$$[F] = \int_V [B^T][X][B] \, dV \times [D] \tag{9.127}$$

The stiffness matrix therefore is

$$[K] = \int_V [B^T][X][B] \, dV \tag{9.128}$$

To illustrate this method, consider the axially loaded element defined in fig. 9.9. The node-point displacement vector $[D]$ is given by

$$[D] = \{D_{Xi} \quad D_{Xj}\}$$

Since no loads are presumed to act between i and j, a general linear displacement function of the form

$$D_X = A_0 + A_1 X$$

will be selected. The boundary conditions that must be met are

$$\text{At } \bar{x} = 0 \qquad D_X = D_{Xi} = A_0 + A_1(0)$$

and

$$\text{At } \bar{x} = L \qquad D_X = D_{Xj} = A_0 + A_1(L)$$

This gives

$$A_0 = D_{Xi} \qquad \text{and} \qquad A_1 = \frac{D_{Xj} - D_{Xi}}{L}$$

This displacement function is then

$$D_X = D_{Xi} + \frac{D_{Yj} - D_{Xi}}{L}(X) \tag{9.129}$$

For elasticity theory,

$$\varepsilon_{xx} = \frac{\partial D_X}{\partial X} = \frac{1}{L}(D_{Xj} - D_{Xi})$$

or

$$\varepsilon_{xx} = \frac{1}{L}[-1 \quad 1]\begin{bmatrix} D_{Xi} \\ D_{Xj} \end{bmatrix}$$

Thus $[B]$, using the eq. (9.121) form, is

$$[B] = \frac{1}{L}[-1 \quad 1] \tag{9.130}$$

Since the material is assumed to be homogeneous and linearly elastic

$$\sigma = \varepsilon E$$

or

$$[X] = E$$

The stiffness matrix is therefore

$$[K] = \int_V [B^T][X][B] \, dV = A \int_0^L \frac{1}{L}\begin{bmatrix} -1 \\ 1 \end{bmatrix} E \times \frac{1}{L}[-1 \quad 1] \, dX = \frac{AE}{L}\begin{bmatrix} 1 & -1 \\ -1 & 1 \end{bmatrix}$$

As a second illustration of the use of the principle of virtual displacements in the establishment of stiffness matrices, consider the beam-bending element shown in fig. 9.18. (Again, it should be noted that the beam sign convention used in chaps. 4 and 5 are different from those defined for matrix analysis and therefore there will be differences in resulting signs of some of the terms.) Assume the cubic polynomial displacement function

$$y(X) = A_0 + A_1 X + A_2 X^2 + A_3 X^3 \tag{9.131}$$

In terms of the node-point displacements, this would be [see eq. (9.114)],

$$y(X) = D_{Yi} + \theta_i X + \left[\frac{3}{L^2}(D_{Yj} - D_{Yi}) - \frac{1}{L}(2\theta_i + \theta_j)\right]X^2$$

$$+ \left[\frac{2}{L^3}(D_{Yi} - D_{Yj}) + \frac{1}{L^2}(\theta_i + \theta_j)\right]X^3$$

For simple beam bending, the normal strains are given by

$$\varepsilon_{xx} = f(X, \xi)$$

$$\varepsilon_{yy} = 0$$

$$\varepsilon_{zz} = 0$$

where ξ is the perpendicular distance from the neutral axis of the cross section to the point in question. Moreover, it is known that

$$\sigma_{xx} = \frac{M_x}{I_x}\xi$$

$$\frac{\sigma_{xx}}{\varepsilon_{xx}} = E = [X]$$

and

$$EI_x\left(\frac{d^2 Y}{dX^2}\right) = M_x$$

The normal strains ε_{xx} can therefore be written as

$$\varepsilon_{xx} = \xi\left\{2\left[\frac{3}{L^2}(D_{Yj} - D_{Yi}) - \frac{1}{L}(2\theta_i + \theta_j)\right] + 6\left[\frac{2}{L^3}(D_{Yi} - D_{Yj}) + \frac{1}{L^2}(\theta_i + \theta_j)\right]X\right\}$$

or

$$\varepsilon_{xx} = \xi\left[\left(-\frac{6}{L^2} + \frac{12}{L^3}X\right) \left(-\frac{4}{L} + \frac{6}{L^2}X\right) \left(\frac{6}{L^2} - \frac{12}{L^3}X\right) \left(-\frac{2}{L} + \frac{6}{L^2}X\right)\right]$$
$$\times \{D_{Yi} \quad \theta_i \quad D_{Yj} \quad \theta_j\}$$

Thus

$$[B] = \xi\left[\left(-\frac{6}{L^2} + \frac{12}{L^3}X\right) \left(-\frac{4}{L} + \frac{6}{L^2}X\right) \left(\frac{6}{L^2} - \frac{12}{L^3}X\right) \left(-\frac{2}{L} + \frac{6}{L^2}X\right)\right]$$

Substituting this into eq. (9.128) yields

$$[K] = \int_V \xi \begin{bmatrix} -\dfrac{6}{L^2} + \dfrac{12}{L^3}X \\[2mm] -\dfrac{4}{L} + \dfrac{6}{L^2}X \\[2mm] \dfrac{6}{L^2} - \dfrac{12}{L^3}X \\[2mm] -\dfrac{2}{L} + \dfrac{6}{L^2}X \end{bmatrix}$$
$$\times E\xi\left[\left(-\frac{6}{L^2} + \frac{12}{L^3}X\right) \left(-\frac{4}{L} + \frac{6}{L^2}X\right) \left(\frac{6}{L^2} - \frac{12}{L^3}X\right) \left(-\frac{2}{L} + \frac{6}{L^2}X\right)\right] dV$$

If the cross section is rectangular in form, and has a width b and depth d, then

$$EI = E\int_{-1/2d}^{+1/2d} \xi^2 b \, d\xi$$

and

$$[K] = EI \int_0^L \begin{bmatrix} -\dfrac{6}{L^2} + \dfrac{12}{L^3}X \\[2mm] -\dfrac{4}{L} + \dfrac{6}{L^2}X \\[2mm] \dfrac{6}{L^2} - \dfrac{12}{L^3}X \\[2mm] -\dfrac{2}{L} + \dfrac{6}{L^2}X \end{bmatrix}$$

$$\times \left[\left(-\dfrac{6}{L^2} + \dfrac{12}{L^3}X \right) \quad \left(-\dfrac{4}{L} + \dfrac{6}{L^2}X \right) \quad \left(\dfrac{6}{L^2} - \dfrac{12}{L^3}X \right) \quad \left(-\dfrac{2}{L} + \dfrac{6}{L^2}X \right) \right] dX$$

Carrying out the indicated matrix multiplications and integration,

$$[K] = EI \begin{bmatrix} \dfrac{12}{L^3} & & \text{(symmetrical)} & \\[2mm] \dfrac{6}{L^2} & \dfrac{4}{L} & & \\[2mm] -\dfrac{12}{L^3} & -\dfrac{6}{L^2} & \dfrac{12}{L^3} & \\[2mm] \dfrac{6}{L^2} & \dfrac{2}{L} & -\dfrac{6}{L^2} & \dfrac{4}{L} \end{bmatrix} \tag{9.132}$$

which is the same as eq. (9.118), but different from eq. (9.39).

   In the most general case of loading, there must be included in the formulation other loading terms; for example, initial strains, initial stresses, body forces, surface forces, etc. Consider the element shown in fig. 9.50. At each node there are six displacement degrees of freedom—three translations and three rotations. The total

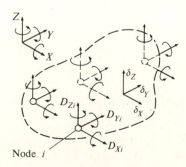

Node i

Figure 9.50 General three-dimensional element.

number of displacement parameters for the element is n. The force and displacement vectors are therefore

$$\text{Force} = [F] = \{F_1 \quad F_2 \quad F_3 \quad \cdots \quad F_n\}$$

$$\text{Displacements} = [D] = \{D_1 \quad D_2 \quad D_3 \quad \cdots \quad D_n\} \tag{9.133}$$

The displacements at any point (X, Y, Z) can be defined in terms of the node point displacements by the relationship

$$[\delta] = [A][D] \tag{9.134}$$

where, as before, the matrix $[A]$ represents the displacement function. The strain vector also will be presumed as previously defined;

$$[\varepsilon] = [B][D] \tag{9.135}$$

where $[B]$ is obtained from $[A]$ by employing the appropriate strain-displacement relationship.

The stress vector $[\sigma]$ is given by

$$[\sigma] = [X][\varepsilon - \varepsilon_I] + [\sigma_I] \tag{9.136}$$

where $[\sigma_I]$ represents the vector of initial stress, $[\varepsilon_I]$ represents the initial strains and $[X]$ is the matrix defining the stress-strain relationship.

The body force matrix $[p]$ (that is, the components of force that are proportional to the volume) is given by

$$[p] = \{p_X \quad p_Y \quad p_Z\} \tag{9.137}$$

The surface forces, which are forces per unit of area, are

$$[q] = \{q_X \quad q_Y \quad q_Z\} \tag{9.138}$$

If there is introduced into the system in question a virtual node point displacement of the form $[\Delta D]$, there will be a change in the displacement $[\delta]$ at (X, Y, Z) equal to $[\Delta\delta]$, where

$$[\Delta\delta] = [A][\Delta D] \tag{9.139}$$

Correspondingly, the strains will change.

$$[\Delta\varepsilon] = [B][\Delta D] \tag{9.140}$$

The external work done during the virtual disturbance is as follows:

    (a)   By the node point forces—$[\Delta D^T][F]$

    (b)   By the surface forces $-\int_A [\Delta\delta^T][q]\, dA$

The corresponding internal work done is

$$\int_V \left[[\Delta \varepsilon^T][\sigma] - [\Delta \delta^T][p] \right] dV$$

The minus sign signifies a loss in potential, and "A" and "V" represent the surface area and volume, respectively.

Since

$$[\Delta \varepsilon^T] = [\Delta D^T][B^T] \quad \text{and} \quad [\Delta \delta^T] = [\Delta D^T][A^T]$$

the change in strain energy associated with a virtual node-point displacement is

$$\Delta W_i = \int_V \left[[\Delta D^T][B^T][\sigma] - [\Delta D^T][A^T][p] \right] dV$$

or

$$\Delta W_i = \int_V [\Delta D^T][B^T]([X][\varepsilon - \varepsilon_I] + [\sigma_I]) \, dV - \int_V [\Delta D^T][A^T][p] \, dV \quad (9.141)$$

Equating the external and internal work:

$$[\Delta D^T][F] + \int_A [\Delta D^T][A^T][q] \, dA = \int_V [\Delta D^T][B^T]$$

$$\times \, ([X][\varepsilon - \varepsilon_I] + \sigma_I) \, dV - \int_V [\Delta D^T][A^T][p] \, dV$$

or

$$[F] = [K][D] - [F_{\varepsilon I}] + [F_{\sigma I}] - [F_p] - [F_q] \quad (9.142)$$

where

$$[F_{\varepsilon I}] = \text{equivalent force matrix representing the initial strain}$$

$$= \int_V [B^T][X][\varepsilon_I] \, dV$$

$$[F_{\sigma I}] = \text{equivalent force matrix representing the initial stress}$$

$$= \int_V [B^T][\sigma_I] \, dV$$

$$[F_p] = \text{equivalent force matrix representing the body forces}$$

$$= \int_V [A^T][p] \, dV$$

$$[F_q] = \text{equivalent force matrix representing the surface forces}$$

$$= \int_V [A^T][q] \, dA$$

The equivalent-force matrix for an externally applied, uniformly distributed load of magnitude q lb/ft on a beam-bending element (fig. 9.18) can be developed directly from these equations. Assuming a cubic polynomial displacement function

$$y(X) = A_0 + A_1 X + A_2 X^2 + A_3 X^3$$

and rearranging the terms so that they are expressed as multipliers of the node point displacements, there results

$$y(X) = \left(1 - \frac{3}{L^2} X^2 + \frac{2}{L^3} X^3\right) D_{Yi} + \left(X - \frac{2}{L} X^2 + \frac{1}{L^2} X^3\right) \theta_i$$
$$+ \left(\frac{3}{L^2} X^2 - \frac{2}{L^3} X^3\right) D_{Yj} + \left(-\frac{1}{L} X^2 + \frac{1}{L^2} X^3\right) \theta_j$$

This gives for matrix $[A]$

$$[A] = \left[\left(1 - \frac{3}{L^2} X^2 + \frac{2}{L^3} X^3\right) \quad \left(X - \frac{2}{L} X^2 + \frac{1}{L^2} X^3\right)\right.$$
$$\left.\left(\frac{3}{L^2} X^2 - \frac{2}{L^2} X^3\right) \quad \left(-\frac{1}{L} X^2 + \frac{1}{L^2} X^3\right)\right]$$

Thus for matrix $[F_q]$

$$[F_q] = q \int_0^L \begin{bmatrix} 1 - \dfrac{3}{L^2} X^2 + \dfrac{2}{L^3} X^3 \\[2mm] X - \dfrac{2}{L} X^2 + \dfrac{1}{L^2} X^3 \\[2mm] \dfrac{3}{L^2} X^2 - \dfrac{2}{L^3} X^3 \\[2mm] -\dfrac{1}{L} X^2 + \dfrac{1}{L^2} X^3 \end{bmatrix} dX$$

or

$$[F_q] = q \left| \frac{L}{2} \quad \frac{L^2}{12} \quad \frac{L}{2} \quad -\frac{L^2}{12} \right|$$

These are the same values that were obtained for the case I loading in example 9.9.

9.13.4 Determination of Stiffness Matrices Using Finite-Difference Methods

The finite-difference method for determining deformations that was discussed in sec. 4.3.5b also can be used to develop approximate stiffness matrices. Depending primarily upon the size of the interval of spacing chosen for numerical integration, almost any degree of accuracy can be achieved by this method.

When using the finite-difference method, it is necessary to go back to the original definition of the various stiffness terms—where K_{ij} was defined as the stress resultant or force acting in the ith direction due to a unit of displacement in the jth direction, with all other node displacements maintained at zero. Therefore there is imposed upon the element in question the jth displacement state, and the resulting boundary-value problem is solved numerically (via the finite-difference method) to determine the end stress resultants necessary to keep the element in that deformed position. The four separate cases shown in fig. 9.19 would have to be solved, and each solution would yield one column of the stiffness matrix. In effect, then N boundary-value problems would have to be considered to define an $N \times N$ matrix. It is, of course, to be understood that all these may not be required, if use is made of symmetry and the equations of statics.

APPENDIXES

Table of functions I

Compression

$$u = \frac{\pi}{2}\sqrt{\left(\frac{P}{P_e}\right)} \qquad \phi = \frac{3}{u}\left[\frac{1}{\sin 2u} - \frac{1}{2u}\right]$$

$$\psi = \frac{3}{2u}\left[\frac{1}{2u} - \frac{1}{\tan 2u}\right]$$

Tension

$$\phi^+ = \frac{3}{u}\left[\frac{1}{2u} - \frac{1}{\sinh 2u}\right]$$

$$\psi^+ = \frac{3}{2u}\left[\frac{1}{\tanh 2u} - \frac{1}{2u}\right]$$

$\dfrac{P}{P_e}$	Compression				Tension			
	ϕ	$\Delta\phi$	ψ	$\Delta\psi$	ϕ^+	$\Delta\phi^+$	ψ^+	$\Delta\psi^+$
0.00	1.0000		1.0000		1.0000		1.0000	
		0.0607		0.0345		−0.0547		−0.0314
0.05	1.0607		1.0345		0.9453		0.9686	
		0.0677		0.0381		−0.0495		−0.0287
0.10	1.1285		1.0727		0.8957		0.9398	
		0.0760		0.0424		−0.0450		−0.0264
0.15	1.2045		1.1150		0.8507		0.9134	
		0.0859		0.0474		−0.0411		−0.0244
0.20	1.2903		1.1624		0.8096		0.8891	
		0.0977		0.0534		−0.0377		−0.0225
0.25	1.3880		1.2159		0.7719		0.8665	
		0.1120		0.0607		−0.0346		−0.0209
0.30	1.5000		1.2766		0.7373		0.8456	
		0.1297		0.0697		−0.0319		−0.0195
0.35	1.6298		1.3463		0.7053		0.8261	
		0.1519		0.0809		−0.0295		−0.0182
0.40	1.7817		1.4272		0.6758		0.8078	
		0.1801		0.0951		−0.0274		−0.0171
0.45	1.9618		1.5223		0.6485		0.7907	
		0.2168		0.1136		−0.0254		−0.0160
0.50	2.1786		1.6359		0.6230		0.7747	
		0.2658		0.1383		−0.0237		−0.0151
0.55	2.4444		1.7742		0.5993		0.7596	
		0.3332		0.1721		−0.0221		−0.0143
0.60	2.7776		1.9463		0.5772		0.7453	
		0.4296		0.2205		−0.0207		−0.0135
0.65	3.2072		2.1668		0.5565		0.7319	
		0.5742		0.2929		−0.0194		−0.0128
0.70	3.7814		2.4597		0.5372		0.7191	
		0.8056		0.4088		−0.0182		−0.0121
0.75	4.5870		2.8685		0.5190		0.7070	
		1.2107		0.6115		−0.0171		−0.0115
0.80	5.7977		3.4801		0.5018		0.6955	
		2.0211		1.0169		−0.0161		−0.0110
0.85	7.8188		4.4970		0.4857		0.6845	
		4.0473		2.0302		−0.0152		−0.0104
0.90	11.8661		6.5272		0.4705		0.6741	
		12.1527		6.0831		−0.0144		−0.0100
0.95	24.0188		12.6103		0.4561		0.6641	
		$+\infty$		$+\infty$		−0.0136		−0.0095
1.00	$\pm\infty$		$\pm\infty$		0.4426		0.6545	
		$-\infty$		$-\infty$		−0.0129		−0.0091
1.05	−24.6274		−11.6985		0.4297		0.6454	
		12.1524		6.0836		−0.0122		−0.0087
1.10	−12.4750		−5.6149		0.4175		0.6367	
		4.0463		2.0309		−0.0116		−0.0084
1.15	−8.4287		−3.5840		0.4059		0.6283	
		2.0196		1.0178		−0.0110		−0.0081
1.20	−6.4091		−2.5663		0.3948		0.6202	
		1.2088		0.6126		−0.0105		−0.0077
1.25	−5.2003		−1.9536		0.3844		0.6125	
		0.8032		0.4101		−0.0100		−0.0075

Table I—*continued*

$\dfrac{P}{P_o}$	Compression				Tension			
	ϕ	$\Delta\phi$	ψ	$\Delta\psi$	ϕ^+	$\Delta\phi^+$	ψ^+	$\Delta\psi^+$
1.30	-4.3971		-1.5435		0.3744		0.6050	
		0.5713		0.2945		-0.0095		-0.0072
1.35	-3.8258		-1.2490		0.3648		0.5978	
		0.4262		0.2223		-0.0091		-0.0069
1.40	-3.3996		-1.0267		0.3557		0.5909	
		0.3294		0.1742		-0.0087		-0.0067
1.45	-3.0702		-0.8524		0.3470		0.5842	
		0.2615		0.1407		-0.0083		-0.0065
1.50	-2.8086		-0.7117		0.3387		0.5778	
		0.3868		0.2144		-0.0156		-0.0123
1.60	-2.4219		-0.4974		0.3232		0.5655	
		0.2693		0.1574		-0.0143		-0.0115
1.70	-2.1526		-0.3399		0.3089		0.5540	
		0.1950		0.1223		-0.0132		-0.0108
1.80	-1.9576		-0.2177		0.2957		0.5432	
		0.1445		0.0993		-0.0122		-0.0102
1.90	-1.8131		-0.1183		0.2835		0.5330	
		0.1081		0.0838		-0.0113		-0.0096
2.00	-1.7050		-0.0345		0.2722		0.5234	
		0.0806		0.0731		-0.0105		-0.0091
2.10	-1.6244		0.0385		0.2617		0.5144	
		0.0587		0.0657		-0.0098		-0.0086
2.20	-1.5657		0.1042		0.2519		0.5058	
		0.0404		0.0607		-0.0091		-0.0082
2.30	-1.5253		0.1650		0.2428		0.4976	
		0.0244		0.0577		-0.0085		-0.0078
2.40	-1.5010		0.2227		0.2343		0.4898	
		0.0096		0.0563		-0.0080		-0.0074
2.50	-1.4914		0.2790		0.2264		0.4824	
		-0.0048		0.0563		-0.0075		-0.0071
2.60	-1.4962		0.3353		0.2189		0.4754	
		-0.0196		0.0579		-0.0070		-0.0067
2.70	-1.5158		0.3931		0.2118		0.4686	
		-0.0356		0.0610		-0.0066		-0.0065
2.80	-1.5514		0.4541		0.2052		0.4622	
		-0.0539		0.0660		-0.0062		-0.0062
2.90	-1.6053		0.5202		0.1990		0.4560	
		-0.0758		0.0735		-0.0059		-0.0059
3.00	-1.6811		0.5937		0.1931		0.4500	
		-0.1033		0.0844		-0.0056		-0.0057
3.10	-1.7843		0.6781		0.1875		0.4443	
		-0.1396		0.1000		-0.0053		-0.0055
3.20	-1.9239		0.7781		0.1822		0.4388	
		-0.1900		0.1231		-0.0050		-0.0053
3.30	-2.1139		0.9011		0.1772		0.4336	
		-0.2642		0.1583		-0.0047		-0.0051
3.40	-2.3781		1.0595		0.1725		0.4285	
		-0.3817		0.2154		-0.0045		-0.0049
3.50	-2.7597		1.2749		0.1680		0.4236	
		-0.5857		0.3160		-0.0043		-0.0047
3.60	-3.3455		1.5909		0.1637		0.4189	
		-0.9922		0.5181		-0.0041		-0.0046
3.70	-4.3377		2.1089		0.1596		0.4143	
		-2.0065		1.0241		-0.0039		-0.0044
3.80	-6.3442		3.1331		0.1557		0.4099	
		-6.0602		3.0500		-0.0037		-0.0043
3.90	-12.4044		6.1831		0.1520		0.4056	
		$-\infty$		$+\infty$		-0.0036		-0.0041
4.00	$-\infty$		$+\infty$		0.1484		0.4015	

Table of functions II

Compression

$$\eta = \frac{12(2 \sec u - u^2 - 2)}{5u^4}$$

$$\lambda = \frac{2(1 - \cos u)}{u^2 \cos u}$$

$$\chi = \frac{3(\tan u - u)}{u^3}$$

Tension

$$\eta^+ = \frac{12(2 \operatorname{sech} u + u^2 - 2)}{5u^4}$$

$$\lambda^+ = \frac{2(\cosh u - 1)}{u^2 \cosh u}$$

$$\chi^+ = \frac{3(u - \tanh u)}{u^3}$$

$\dfrac{P}{P_e}$	Compression			Tension		
	η	λ	χ	η^+	λ^+	χ^+
0.00	1.0000	1.0000	1.0000	1.0000	1.0000	1.0000
0.05	1.0528	1.0541	1.0519	0.9522	0.9511	0.9530
0.10	1.1115	1.1143	1.1096	0.9088	0.9066	0.9103
0.15	1.1771	1.1815	1.1741	0.8691	0.8660	0.8712
0.20	1.2509	1.2572	1.2467	0.8328	0.8288	0.8355
0.25	1.3345	1.3430	1.3289	0.7993	0.7946	0.8025
0.30	1.4301	1.4411	1.4228	0.7685	0.7630	0.7722
0.35	1.5404	1.5543	1.5312	0.7399	0.7338	0.7440
0.40	1.6691	1.6864	1.6576	0.7134	0.7066	0.7179
0.45	1.8211	1.8425	1.8070	0.6886	0.6814	0.6935
0.50	2.0036	2.0299	1.9863	0.6656	0.6579	0.6708
0.55	2.2267	2.2591	2.2054	0.6440	0.6358	0.6496
0.60	2.5055	2.5455	2.4792	0.6238	0.6152	0.6296
0.65	2.8640	2.9139	2.8313	0.6048	0.5958	0.6109
0.70	3.3420	3.4051	3.3007	0.5869	0.5776	0.5933
0.75	4.0112	4.0929	3.9578	0.5701	0.5604	0.5766
0.80	5.0150	5.1247	4.9434	0.5542	0.5442	0.5609
0.85	6.6881	6.8445	6.5860	0.5391	0.5289	0.5461
0.90	10.0342	10.2844	9.8712	0.5249	0.5144	0.5320
0.95	20.0726	20.6045	19.7265	0.5113	0.5006	0.5186
1.00	$\pm \infty$	$\pm \infty$	$\pm \infty$	0.4985	0.4875	0.5059
1.05	− 20.0817	− 20.6780	− 19.6947	0.4862	0.4751	0.4939
1.10	− 10.0430	− 10.3576	− 9.8391	0.4746	0.4633	0.4824
1.15	− 6.6968	− 6.9177	− 6.5539	0.4635	0.4520	0.4714
1.20	− 5.0238	− 5.1978	-- 4.9113	0.4529	0.4412	0.4609
1.25	− 4.0199	− 4.1660	− 3.9256	0.4428	0.4310	0.4509
1.30	− 3.3508	− 3.4783	− 3.2685	0.4331	0.4211	0.4413
1.35	− 2.8727	− 2.9871	− 2.7991	0.4238	0.4117	0.4322
1.40	− 2.5142	− 2.6188	− 2.4470	0.4150	0.4027	0.4234
1.45	− 2.2354	− 2.3324	− 2.1732	0.4064	0.3941	0.4149
1.50	− 2.0124	− 2.1033	− 1.9541	0.3983	0.3858	0.4068
1.60	− 1.6778	− 1.7599	− 1.6253	0.3828	0.3702	0.3916
1.70	− 1.4389	− 1.5148	− 1.3905	0.3686	0.3558	0.3774
1.80	− 1.2597	− 1.3311	− 1.2143	0.3553	0.3424	0.3643
1.90	− 1.1203	− 1.1884	− 1.0772	0.3430	0.3300	0.3521
2.00	− 1.0089	− 1.0744	− 0.9674	0.3315	0.3184	0.3406
2.10	− 0.9177	− 0.9813	− 0.8776	0.3207	0.3076	0.3299
2.20	− 0.8417	− 0.9038	− 0.8027	0.3106	0.2974	0.3199
2.30	− 0.7774	− 0.8383	− 0.7393	0.3012	0.2879	0.3105

Table II—*continued*

$\dfrac{P}{P_e}$	Compression			Tension		
	η	λ	χ	η^+	λ^+	χ^+
2.40	− 0.7224	− 0.7824	− 0.6849	0.2922	0.2789	0.3016
2.50	− 0.6747	− 0.7340	− 0.6377	0.2838	0.2705	0.2932
2.60	− 0.6329	− 0.6918	− 0.5963	0.2759	0.2625	0.2853
2.70	− 0.5961	− 0.6547	− 0.5598	0.2684	0.2550	0.2778
2.80	− 0.5634	− 0.6218	− 0.5273	0.2613	0.2479	0.2707
2.90	− 0.5341	− 0.5925	− 0.4982	0.2545	0.2412	0.2640
3.00	− 0.5078	− 0.5662	− 0.4720	0.2481	0.2348	0.2576
3.10	− 0.4840	− 0.5426	− 0.4482	0.2420	0.2287	0.2515
3.20	− 0.4624	− 0.5212	− 0.4265	0.2362	0.2229	0.2457
3.30	− 0.4427	− 0.5018	− 0.4067	0.2307	0.2174	0.2402
3.40	− 0.4246	− 0.4842	− 0.3885	0.2254	0.2122	0.2349
3.50	− 0.4080	− 0.4680	− 0.3717	0.2203	0.2071	0.2298
3.60	− 0.3927	− 0.4533	− 0.3562	0.2155	0.2024	0.2250
3.70	− 0.3785	− 0.4397	− 0.3417	0.2109	0.1978	0.2204
3.80	− 0.3653	− 0.4273	− 0.3283	0.2065	0.1934	0.2159
3.90	− 0.3531	− 0.4158	− 0.3157	0.2022	0.1892	0.2117
4.00	− 0.3417	− 0.4053	− 0.3040	0.1981	0.1852	0.2076

Table of functions III

Compression

$$\alpha_1 = \left[\frac{3\psi}{4\psi^2 - \phi^2}\right]$$

$$\alpha_2 = \left[\frac{3\phi}{4\psi^2 - \phi^2}\right]$$

$$\alpha_3 = \left[\frac{1}{2\psi - \phi}\right]$$

$$\alpha_4 = \left[\frac{1}{(2\psi - \phi)} - \frac{(2u)^2}{12}\right]$$

Tension

$$\beta_1 = \left[\frac{3\psi^+}{4(\psi^+)^2 - (\phi^+)^2}\right]$$

$$\beta_2 = \left[\frac{3\phi^+}{4(\psi^+)^2 - (\phi^+)^2}\right]$$

$$\beta_3 = \left[\frac{1}{2\psi^+ - \phi^+}\right]$$

$$\beta_4 = \left[\frac{1}{(2\psi^+ - \phi^+)} + \frac{(2u)^2}{12}\right]$$

$\dfrac{P}{P_e}$	Compression				Tension			
	α_1	α_2	α_3	α_4	β_1	β_2	β_3	β_4
0.00	1.0000	1.0000	1.0000	1.0000	1.0000	1.0000	1.0000	1.0000
0.05	0.9834	1.0084	0.9917	0.9506	1.0163	0.9919	1.0082	1.0493
0.10	0.9667	1.0170	0.9834	0.9012	1.0325	0.9840	1.0163	1.0986
0.15	0.9497	1.0259	0.9751	0.8517	1.0484	0.9764	1.0244	1.1478
0.20	0.9324	1.0350	0.9666	0.8021	1.0642	0.9690	1.0324	1.1969
0.25	0.9149	1.0445	0.9581	0.7525	1.0797	0.9618	1.0404	1.2460
0.30	0.8972	1.0543	0.9496	0.7028	1.0951	0.9548	1.0483	1.2951
0.35	0.8792	1.0644	0.9409	0.6531	1.1103	0.9480	1.0562	1.3441
0.40	0.8610	1.0748	0.9323	0.6033	1.1253	0.9414	1.0640	1.3930
0.45	0.8424	1.0856	0.9235	0.5534	1.1402	0.9350	1.0718	1.4419
0.50	0.8236	1.0968	0.9147	0.5035	1.1549	0.9288	1.0795	1.4907
0.55	0.8045	1.1084	0.9058	0.4534	1.1694	0.9227	1.0872	1.5395
0.60	0.7851	1.1204	0.8968	0.4034	1.1838	0.9168	1.0948	1.5883
0.65	0.7653	1.1328	0.8878	0.3532	1.1980	0.9110	1.1023	1.6369
0.70	0.7452	1.1456	0.8787	0.3030	1.2121	0.9054	1.1099	1.6856
0.75	0.7248	1.1590	0.8695	0.2527	1.2260	0.9000	1.1173	1.7342
0.80	0.7040	1.1728	0.8603	0.2023	1.2398	0.8947	1.1248	1.7827
0.85	0.6828	1.1872	0.8509	0.1518	1.2535	0.8895	1.1322	1.8312
0.90	0.6612	1.2021	0.8415	0.1013	1.2670	0.8844	1.1395	1.8797
0.95	0.6393	1.2176	0.8320	0.0507	1.2804	0.8795	1.1468	1.9281
1.00	0.6173	1.2346	0.8231	0.0006	1.2937	0.8747	1.1540	1.9765
1.05	0.5940	1.2505	0.8128	−0.0508	1.3069	0.8700	1.1612	2.0248
1.10	0.5707	1.2679	0.8031	−0.1016	1.3199	0.8655	1.1684	2.0731
1.15	0.5469	1.2861	0.7933	−0.1526	1.3328	0.8610	1.1755	2.1214
1.20	0.5225	1.3050	0.7833	−0.2036	1.3456	0.8566	1.1826	2.1696
1.25	0.4977	1.3247	0.7733	−0.2547	1.3583	0.8524	1.1897	2.2177
1.30	0.4722	1.3453	0.7633	−0.3060	1.3709	0.8483	1.1967	2.2659
1.35	0.4462	1.3668	0.7531	−0.3573	1.3833	0.8442	1.2036	2.3140
1.40	0.4196	1.3893	0.7428	−0.4087	1.3957	0.8402	1.2106	2.3620
1.45	0.3922	1.4128	0.7324	−0.4602	1.4080	0.8364	1.2174	2.4100
1.50	0.3642	1.4373	0.7219	−0.5118	1.4201	0.8326	1.2243	2.4580
1.60	0.3060	1.4901	0.7007	−0.6153	1.4442	0.8253	1.2379	2.5538
1.70	0.2445	1.5481	0.6790	−0.7192	1.4678	0.8183	1.2513	2.6495
1.80	0.1793	1.6122	0.6569	−0.8235	1.4911	0.8116	1.2646	2.7451
1.90	0.1099	1.6833	0.6343	−0.9284	1.5141	0.8052	1.2778	2.8405
2.00	0.0357	1.7624	0.6113	−1.0337	1.5367	0.7991	1.2908	2.9358

Table III—*continued*

$\dfrac{P}{P_e}$	Compression				Tension			
	α_1	α_2	α_3	α_4	β_1	β_2	β_3	β_4
2.10	− 0.0439	1.8510	0.5877	− 1.1395	1.5590	0.7932	1.3037	3.0309
2.20	− 0.1299	1.9506	0.5636	− 1.2458	1.5810	0.7876	1.3165	3.1259
2.30	− 0.2232	2.0633	0.5390	− 1.3527	1.6027	0.7821	1.3292	3.2208
2.40	− 0.3251	2.1916	0.5138	− 1.4601	1.6241	0.7769	1.3417	3.3156
2.50	− 0.4375	2.3388	0.4880	− 1.5682	1.6452	0.7719	1.3541	3.4103
2.60	− 0.5622	2.5090	0.4615	− 1.6769	1.6660	0.7671	1.3664	3.5048
2.70	− 0.7023	2.7077	0.4344	− 1.7863	1.6866	0.7625	1.3786	3.5992
2.80	− 0.8612	2.9421	0.4066	− 1.8964	1.7070	0.7580	1.3907	3.6936
2.90	− 1.0441	3.2221	0.3780	− 2.0072	1.7271	0.7537	1.4026	3.7878
3.00	− 1.2580	3.5618	0.3486	− 2.1188	1.7470	0.7495	1.4145	3.8819
3.10	− 1.5130	3.9812	0.3184	− 2.2312	1.7666	0.7455	1.4262	3.9759
3.20	− 1.8243	4.5106	0.2874	− 2.3445	1.7860	0.7417	1.4379	4.0698
3.30	− 2.2157	5.1975	0.2554	− 2.4588	1.8052	0.7379	1.4495	4.1636
3.40	− 2.7271	6.1212	0.2224	− 2.5740	1.8242	0.7343	1.4609	4.2573
3.50	− 3.4297	7.4245	0.1883	− 2.6903	1.8430	0.7308	1.4723	4.3509
3.60	− 4.4667	9.3930	0.1532	− 2.8077	1.8616	0.7275	1.4836	4.4444
3.70	− 6.1713	12.6932	0.1169	− 2.9262	1.8800	0.7242	1.4948	4.5379
3.80	− 9.5435	19.3250	0.0793	− 3.0461	1.8983	0.7210	1.5059	4.6312
3.90	− 19.5834	39.2879	0.0404	− 3.1672	1.9163	0.7180	1.5169	4.7245
4.00	− 3345.7454	6691.4909	0.0000	− 3.2899	1.9342	0.7150	1.5278	4.8177

Areas and perimeters of round reinforcing bars

Bar designation	Diameter, inches		Number of bars					
			1	2	3	4	5	6
#3	$\frac{3}{8} = 0.375$	Area	0.11	0.22	0.33	0.44	0.55	0.66
		(Perimeter)	(1.178)	(2.36)	(3.53)	(4.71)	(5.89)	(7.07)
#4	$\frac{1}{2} = 0.500$	Area	0.20	0.40	0.60	0.80	1.00	1.20
		(Perimeter)	(1.571)	(3.14)	(4.71)	(6.28)	(7.85)	(9.42)
#5	$\frac{5}{8} = 0.625$	Area	0.31	0.62	0.93	1.24	1.55	1.86
		(Perimeter)	(1.963)	(3.93)	(5.89)	(7.85)	(9.82)	(11.78)
#6	$\frac{3}{4} = 0.750$	Area	0.44	0.88	1.32	1.75	2.21	2.65
		(Perimeter)	(2.356)	(4.71)	(7.07)	(9.42)	(11.78)	(14.14)
#7	$\frac{7}{8} = 0.875$	Area	0.60	1.20	1.80	2.41	3.01	3.61
		(Perimeter)	(2.749)	(5.50)	(8.25)	(11.00)	(13.74)	(16.49)
#8	$1 = 1.000$	Area	0.79	1.58	2.37	3.16	3.93	4.71
		(Perimeter)	(3.142)	(6.28)	(9.43)	(12.57)	(15.71)	(18.85)
#9	1.128	Area	1.00	2.00	3.00	4.00	5.00	6.00
		(Perimeter)	(3.544)	(7.09)	(10.63)	(14.18)	(17.72)	(21.26)
#10	1.270	Area	1.27	2.54	3.81	5.08	6.33	7.60
		(Perimeter)	(3.990)	(7.98)	(11.97)	(15.96)	(19.95)	(23.94)
#11	1.410	Area	1.56	3.12	4.68	6.24	7.81	9.37
		(Perimeter)	(4.430)	(8.86)	(13.29)	(17.72)	(22.15)	(26.58)

INDEX

INDEX